AF615396

Acoustical Imaging

Volume 12

Acoustical Imaging

Volume 1 Proceedings of the First International Symposium, December 1967, edited by A. F. Metherell, H. M. A. El-Sum, and Lewis Larmore

Volume 2 Proceedings of the Second International Symposium, March 1969, edited by A. F. Metherell and Lewis Larmore

Volume 3 Proceedings of the Third International Symposium, July 1970, edited by A. F. Metherell

Volume 4 Proceedings of the Fourth International Symposium, April 1972, edited by Glen Wade

Volume 5 Proceedings of the Fifth International Symposium, July 1973, edited by Philip S. Green

Volume 6 Proceedings of the Sixth International Symposium, February 1976, edited by Newell Booth

Volume 7 Proceedings of the Seventh International Symposium, August 1976, edited by Lawrence W. Kessler

Volume 8 Proceedings of the Eighth International Symposium, May 29–June 2, 1978, edited by A. F. Metherell

Volume 9 Proceedings of the Ninth International Symposium, December 3–6, 1979, edited by Keith Y. Wang

Volume 10 Proceedings of the Tenth International Symposium, October 12–16, 1980, edited by Pierre Alais and Alexander F. Metherell

Volume 11 Proceedings of the Eleventh International Symposium, May 4–7, 1981, edited by John P. Powers

Volume 12 Proceedings of the Twelfth International Symposium, July 19–22, 1982, edited by Eric A. Ash and C. R. Hill

Acoustical Imaging

Volume 12

Edited by

Eric A. Ash

University College
London, England

and

C. R. Hill

Institute of Cancer Research
Royal Marsden Hospital
Sutton, England

PLENUM PRESS · NEW YORK AND LONDON

The Library of Congress cataloged the first volume of this series as follows:

International Symposium on Acoustical Holography.

Acoustical holography; proceedings. v. 1-
New York, Plenum Press, 1967-

v. illus. (part col.), ports. 24 cm.

Editors: 1967- A. F. Metherell and L. Larmore (1967 with H. M. A. el-Sum)
Symposiums for 1967- held at the Douglas Advanced Research Laboratories, Huntington Beach, Calif.

1. Acoustic holography–Congresses–Collected works. I. Metherell. Alexander A., ed. II. Larmore, Lewis, ed. III. el-Sum, Hussein Mohammed Amin, ed. IV. Douglas Advanced Research Laboratories. v. Title.
QC244.5.I 5 69-12533

ISBN 0-306-41247-0

Proceedings of the Twelfth International Symposium on Acoustical Imaging, held July 19–22, 1982, in London, England

A Division of Plenum Publishing Corporation
233 Spring Street, New York, N.Y. 10013

Printed in the United States of America

PREFACE

The formation of images by ultrasound is a fascinating study, with well-established, yet rapidly growing, applications in medicine and with increasing relevance to a surprisingly disparate set of problems in the non-destructive examination of materials and components. The present volume is a record of the research presented at the Twelfth International Symposium on Acoustic Imaging, held in London during July 1982. Whilst, therefore, it offers primarily a snap-shot in time of a rapidly developing field, it is so organized that it will also serve as a high-speed entry into the literature for someone embarking, for the first time, on researches in this branch of applied science.

As in previous volumes, some of the work reported is concerned with topics which, whilst of critical importance to the performance of any imaging system, - e.g. transducers, signal processing - may not address themselves to image formation per se. A new departure is the inclusion of photo-acoustic imaging - a subject of rapidly growing importance for many of the same application areas relevant to acoustical imaging.

The editors, with the enthusiastic co-operation of the Publishers, have aimed to achieve the highest possible speed - a publication date within four months of the conference itself. In the pursuit of this endeavour, it is not impossible that the density of typos, and other minor errors which remain, is marginally higher than could have been achieved, given a more leisurely editing pace. The editors would like to express their sincere thanks to Dr David Sinclair and Mrs Lee Surry who have ensured nonetheless that a high standard of presentation has been achieved. Their gratitude also extends to all authors who, by careful attention to the "rules" of camera-ready production, and by their prompt delivery of the manuscrips, have played the key role in the creation of this volume and in allowing us rapidly to present it to the scientific community.

C. R. Hill
E. A. Ash

CONTENTS

ACOUSTICAL MICROSCOPY - 1

ACOUSTICAL MICROSCOPY - 2

NON-DESTRUCTIVE EVALUATION - 1

NON-DESTRUCTIVE EVALUATION - 2

SIGNAL PROCESSING - 1

SIGNAL PROCESSING - 2

TRANSDUCERS - 1

TRANSDUCERS - 2

MULTI ELEMENT ARRAYS

SCATTERING AND PROPAGATION

RECONSTRUCTION TOMOGRAPHY

IMAGING SYSTEMS - 1

CONSIDERATIONS OF CONTRAST IN THE HELIUM ACOUSTIC MICROSCOPE

S Christie and A F G Wyatt

Department of Physics
University of Exeter
Exeter UK

INTRODUCTION

Interest is now being shown in the ^{4}He acoustic microscope as it offers the potential for much improved resolution compared to the room temperature version using a normal liquid as the coupling medium.

The origins of contrast are well understood in the room temperature microscope to the extent that theoretical V(z) calculations, which depend on the elastic properties of the object material, have been shown to agree well with experimental results.

In this paper we intend to examine the origins of contrast in the ^{4}He microscope where we expect the situation to be significantly different due to the fact that the reflection coefficient at the ^{4}He-sample interface is very close to unity for all solids. We also consider the suggestion that the effects of attenuation in the sample can be detected (Heiserman[1,2]).

Figure 1. shows the basic acoustic microscope in the reflection mode. For room temperature microscopes the coupling medium is usually water which is sometimes at elevated temperatures in order to decrease its acoustic attenuation. Liquid metals[3], gases at high pressures[4] and cryogenic liquids[1] have also been used.

Liquid ^{4}He at 0.1K has a number of desirable properties that make it an attractive coupling medium. Firstly, there is no acoustic attenuation in the helium which is in contrast to the considerable attenuation shown by all other media. Secondly the low sound velocity makes the wavelength shorter for a given frequency and with the possibility of working at higher frequencies there is the

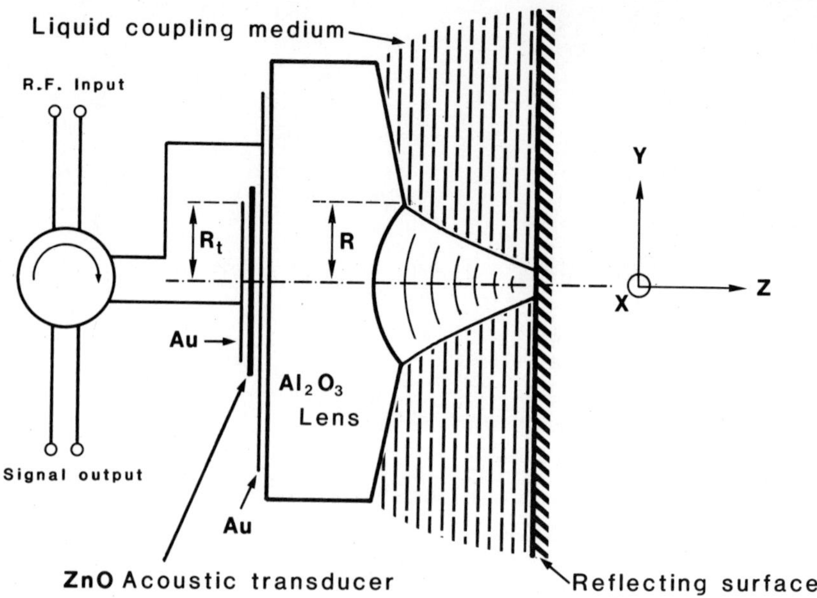

Fig. 1. Schematic diagram of the reflecting acoustic microscope.

potential for a significant increase in resolution. Thirdly, spherical aberration effects can be neglected since the velocity ratio between the sapphire lens and liquid ^{4}He is very small. Diffraction effects and lens figuring will limit the resolution of the system.

However, there are problems associated with using ^{4}He and to exemplify them we first consider the situation at normal incidence. The intensity reflection coefficient magnitude at the interface between materials 1 and 2 is given by:-

$$(Z_1 - Z_2)^2/(Z_1 + Z_2)^2 \quad (1)$$

where the acoustic impedance Z = ρc. For a general solid-liquid helium interface this is very close to unity. For the lens at the sapphire-^{4}He interface, only 0.3% of the incident wave is transmitted into the liquid ^{4}He, although this can be improved by quarter-wave matching layers as shown by Heiserman[1]. At the liquid ^{4}He-sample interface the reflection coefficient magnitude is approximately unity for all angles of incidence which indicates that changes in reflectivity will be small and difficult to detect.

THEORY AND RESULTS

The theory can be divided into two main sections. We first consider the complex reflection coefficient at a liquid ^{4}He-solid interface, taking into account the attenuation in the solid. Secondly, the relative output from the zinc-oxide transducer on the lens is calculated i.e., V(z), as a function of displacement z of the sample from the focal plane.

a. The Reflection Coefficient

For the calculation of the complex reflection coefficient, consider a plane wave incident on the interface which is ideally flat. The solid is assumed to be isotropic, then following Brekhovskikh[5] for the case of no attenuation and for the geometry as defined in figure 2, the reflection coefficient can be written:

$$R = |R|\ e^{i\phi} = \frac{Z_\ell \cos^2 2\beta_t + Z_t \sin^2 2\beta_t - Z_o}{Z_\ell \cos^2 2\beta_t + Z_t \sin^2 2\beta_t + Z_o} \tag{2}$$

where the impedances are defined as:-

$$Z_o = \frac{\rho_o c_o}{\cos\theta}, \quad Z_\ell = \frac{\rho c_\ell}{\cos\beta_\ell}, \quad Z_t = \frac{\rho c_t}{\cos\beta_t}$$

The incident and refracted angles, and acoustic velocities are related by Snell's law:-

$$\frac{\sin\theta}{c_o} = \frac{\sin\beta_\ell}{c_\ell} = \frac{\sin\beta_t}{c_t} \tag{3}$$

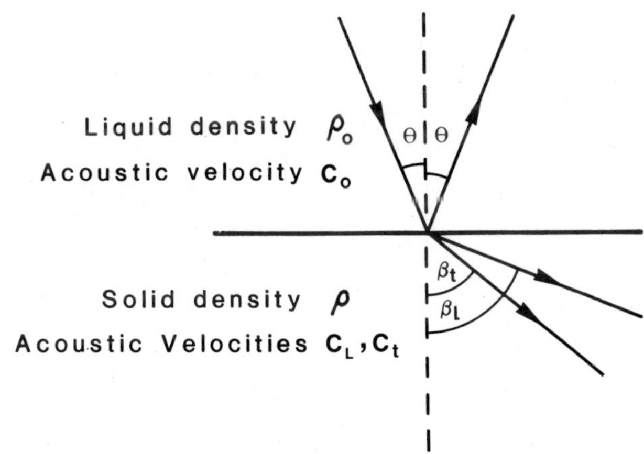

Fig. 2. Reflected and refracted acoustic waves at the liquid-sample (solid) interface.

It can be seen from this equation that, since θ is always real, then depending on the values of the velocities, β_ℓ and β_t can become complex. The transition from real to complex defines a critical angle and in the case of liquid ^{4}He there are two corresponding to the longitudinal and transverse waves in the solid, $\theta_{c\ell}$ and θ_{ct} respectively.

Now, attenuation in the solid may be represented by complex wavevectors in the solid (Merkolova[6]):

$$\bar{q}_{\ell,t} = q_{\ell,t} + i\,\alpha_{\ell,t} \tag{4}$$

where $\alpha_{\ell,t}$ is the attenuation coefficient for longitudinal or transverse waves in the solid. The angles $\beta_{\ell,t}$ are now complex for all angles of θ, and we write

$$\bar{\beta}_{\ell,t} = \beta'_{\ell,t} + i\beta''_{\ell,t} \tag{5}$$

Substitutions of equations (4) and (5) into equation (3) yield equations for β' and β'' (ℓ and t subscripts are implied in equations 6, 7 and 8).

$$\left(1 + \frac{\alpha^2}{q^2}\right) \frac{\sin\beta'}{\{1-(\alpha/q)^2\tan^2\beta'\}^{\frac{1}{2}}} = \frac{q_o}{q}\sin\theta \tag{6}$$

$$\tanh\beta'' = -\frac{\alpha}{q}\tan\beta' \tag{7}$$

The complex impedances $Z_{\ell,t}$ $(= Z' + iZ'')$ become:

$$Z = \rho c\{(\cos\beta'\cosh\beta'' + \frac{\alpha}{q}\sin\beta'\sinh\beta'') + i(\sin\beta'\sinh\beta'' - \frac{\alpha}{q}\cos\beta'\cosh\beta'')\} \{(1 + \alpha^2/q^2)(\cos^2\beta'\cosh^2\beta'' + \sin^2\beta'\sinh^2\beta'')\}^{-1} \tag{8}$$

Writing:-

$$\cos^2 2\bar{\beta}_t = m'_t + im''_t \tag{9}$$

and substituting equations (8) and (9) into (2) we obtain the following expressions for the modulus and phase of the reflection coefficient:

$$|R| = \left\{\frac{(M_1 - Z_o)^2 + M_2^2}{(M_1 + Z_o)^2 + M_2^2}\right\}^{\frac{1}{2}} \tag{10}$$

$$\phi = \tan^{-1}\left\{\frac{2M_2Z_o}{Z_o^2 - M_1^2 - M_2^2}\right\} \quad (11)$$

where

$$M_1 = Z'_\ell m'_t - Z''_\ell m''_t - Z'_t(m'_t - 1) + Z''_t m''_t$$

$$M_2 = - Z'_\ell m''_t - Z''_\ell m'_t + Z'_t m''_t + Z''_t(m'_t - 1) \quad (12)$$

The equations above summarize the effect of attenuation, in the sample, on the reflection coefficient. They have been evaluated on a DEC PDP 11/03 minicomputer. Figure 3 shows $|R|$ and ϕ for a ^{4}He-aluminium interface with no attenuation in either the coupling liquid or solid sample. It can be clearly seen, as anticipated, that the reflection modulus is very close to unity for angles less than the transverse critical angle and for greater angles is exactly unity with all the incident energy reflected back into the liquid. The two critical angles are both very small so that all changes in $|R|$ and ϕ occur for $\theta \lesssim 5^o$. The phase, ϕ, is zero up to the longitudinal critical angle, where there is a small change ($\simeq 6^o$). There is a

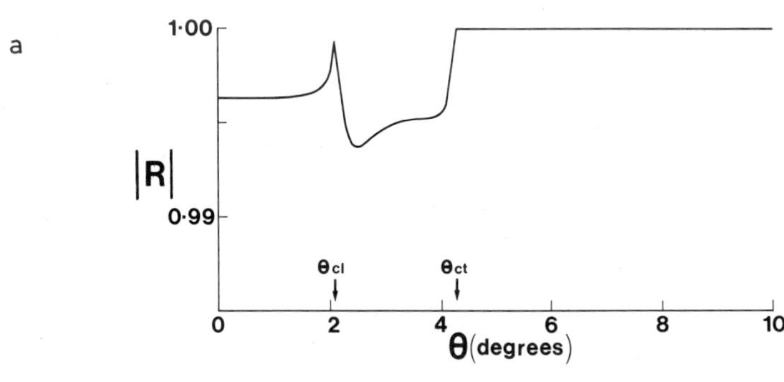

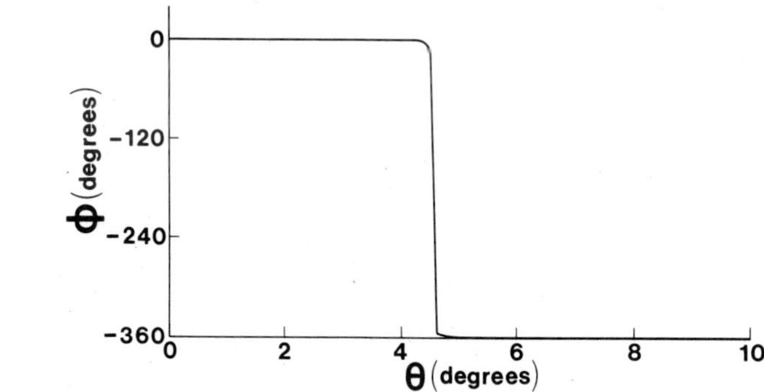

Fig. 3. (a) Modulus ($|R|$) and (b) phase (ϕ) of the reflection coefficient for a ^{4}He-aluminium interface with no attenuation in the aluminium, as a function of angle of incidence θ.

sharp transition at the Rayleigh critical angle (slightly greater than the transverse critical angle) and for larger angles the phase shift approaches -2π.

For a numerical example of an attenuating solid we consider aluminium. Besides having low acoustic impedances its attenuation is considerably different in the normal and superconducting states. This means in practice that the attenuation can be switched on by applying a magnetic field leaving all the other conditions unchanged.

The acoustic attenuation has been measured for aluminium up to 2 GHz by Rayne and Jones[7], they find: $\alpha^2/f=0.5$ dB cm^{-1} MHz^{-1} so for 1 GHz, $\alpha_\ell = 500$ dB cm^{-1}. The free electron model gives the ratio:

$$\frac{\alpha_t}{\alpha_\ell} = \frac{8}{\pi^2} \frac{c_\ell^2}{c_t^2} \qquad (13)$$

from which we calculate the transverse attenuation, $\alpha_t = 1686$ dB cm^{-1} at 1 GHz.

Figures 4 and 5 show the results of the calculation for the complex reflection coefficient for these attenuations.

It can be seen that the result of the above attenuation on the reflection coefficient is two fold. Firstly, in the modulus $|R|$ the details associated with the critical angles are smoothed out and $|R|$ is less than unity at all angles. Secondly, the incident angle at which the phase transition occurs is increased.

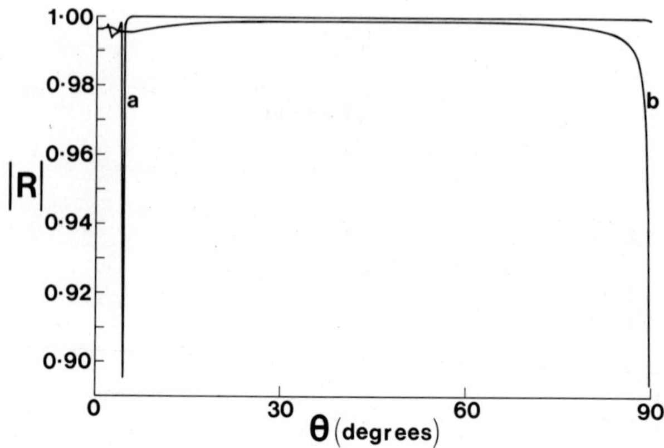

Fig. 4. Modulus $|R|$ of the reflection coefficient at a ^{4}He-aluminium interface with low attenuation; (a) $\alpha_\ell=3.2$ dB cm^{-1}, $\alpha_t=10.2$ dB cm^{-1} which is approximately the value of α which gives the deepest minimum at the Rayleigh angle, and (b) $\alpha_\ell=500$ dB cm^{-1}, $\alpha_t=1686$ dB cm^{-1} (which is a realistic attenuation for Aℓ in the normal state at 1 GHz).

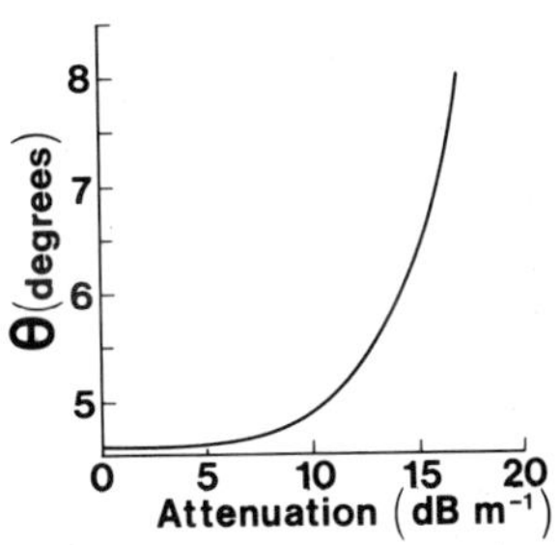

Fig. 5. The angle of incidence (θ) at which the discontinuity in phase occurs (see fig. 3.) as a function of attenuation (α_t).

With small values of attenuation, $|R|$ is very similar to the zero attenuation case shown in figure 3. As the attenuation increases a sharp dip in $|R|$ occurs at the Rayleigh angle. This dip reaches a minimum at particular values of α_ℓ and α_t, and it appears that changes in $|R|$ and ϕ are more sensitive to α_t than to α_ℓ. For yet higher values of α_ℓ and α_t the dip becomes less pronounced and $|R|$ shows a general decrease from unity over the whole angular range as shown in figure 4. The phase (ϕ) of the reflected wave is independent of attenuation at low values of attenuation but then increases rapidly with attenuation as shown in figure 5.

b. V(z)

We now consider the microscope response with liquid ^{4}He as the coupling medium and with the effects of attenuation in the sample. This problem has been considered by Atalar[8] and Wickramasinghe[9], for the room temperature version and we follow Atalar's analysis which uses the angular spectrum approach[10] with a paraxial approximation. This method consists of Fourier transforming the inhomogeneous beam in the lens into an angular spectrum of plane waves. These waves propagate to the sample, reflect and recombine in the lens. The calculation proceeds to evaluate variations in the signal that occur as the object is moved through the focal plane, i.e. V(z).

In practice there is a transducer on the lens which excites an acoustic wave which propagates along the c-axis of the sapphire lens to the concave spherical surface. The field at the back focal plane $u_1^+(x,y)$ can be determined as shown by Zemanek[11] and from this the field $u_2^+(x,y)$ at the front focal plane can be determined. Near this plane is the interface between the liquid and sample at which reflection takes place. Due to the decomposition into plane waves the complex reflectance function of the interface, equations 10 and 11, can be used. After reflection we have a reflected field travelling in the negative z-direction. By use of a Fourier transform we can propagate this field back through the lens to the back focal plane.

An assumption used in the theory is that of a paraxial approximation whereby only those parts of the acoustic fields which are nearly parallel to the symmetry axis are considered. This is reasonable as the field is much wider than the acoustic wavelength.

In the reflecting microscope the angular spectrum is symmetrically distributed about the normal to the interface since this is the axis of the actual acoustic beam.

The following expression can be obtained (Atalar[8]):

$$u_1^-(x,y) = -\exp\{i2k_o(z + f(1 + \bar{C}^2))\}$$

$$u_1^+(-x,-y)\ P_1(-x,-y)\ P_2(x,y)$$

$$\exp\{-ik_o z(x^2 + y^2)/f^2\}\ R(x/f,y/f) \qquad (14)$$

where the focal length $f = R_\ell\ (1 - \bar{C})^{-1}$ and $\bar{C}$ is the ratio of the longitudinal velocities in the liquid and lens. This expresses the reflected acoustic field at the lens back focal plane $u_1^+(x,y)$ in terms of the incident field at the same plane $u_1^-(-x,-y)$, the pupil functions of the lens P_1 and P_2 (for fields travelling towards the lens aperture and away from the lens aperture respectively), the reflectance function of the sample and the position of the object z. The first exponential is a constant phase factor and may be neglected. The second exponential only has an effect when the reflecting sample is not at the focal plane i.e. z = 0. It is effectively a phase curvature factor which alters the shape of the reflected wavefront.

When the reflected field is incident on the lens transducer it will be integrated over the field to yield the output voltage so that a maximum output is obtained when the wave fronts are parallel to the transducer. Any bending of the wavefronts, such as from the second exponential will reduce the output.

Using equation 14, and propagating the acoustic field to the transducer, the integrated field is:

$$V(z) = 2\pi\int_0^\infty r\ u_1^+(r)^2 P_1(r) P_2(r)\ R(r/f)\exp(-ik_o zr^2/f^2)\,dr \qquad (15)$$

where r is the radial distance across the lens aperture from the symmetry axis and the signal will be proportional to V(z).

The two pupil functions $P_1(r)$ and $P_2(r)$ are assumed to be equal by reciprocity and we take Atalar's[8] form of $u_1^+(r)^2 P_1(r) P_2(r)$ which is shown in figure 6.

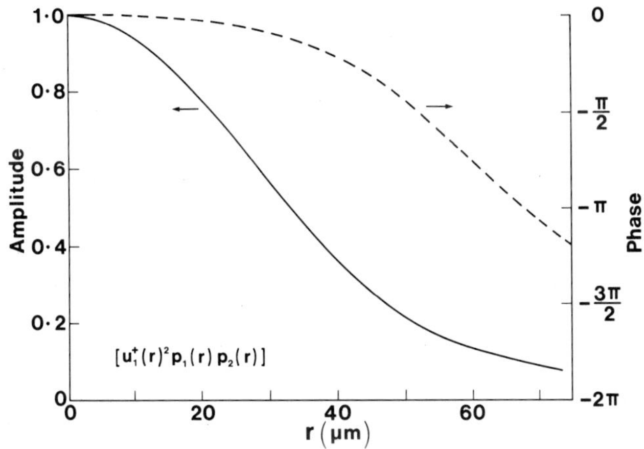

Fig. 6. Amplitude and phase of $u_1^+(r)^2\ P_1(r)P_2(r)$. After Atalar[6] *for a lens with transducer radius R_t = 105 μm, aperture radius R = 75 μm, radius of curvature = 104 μm.*

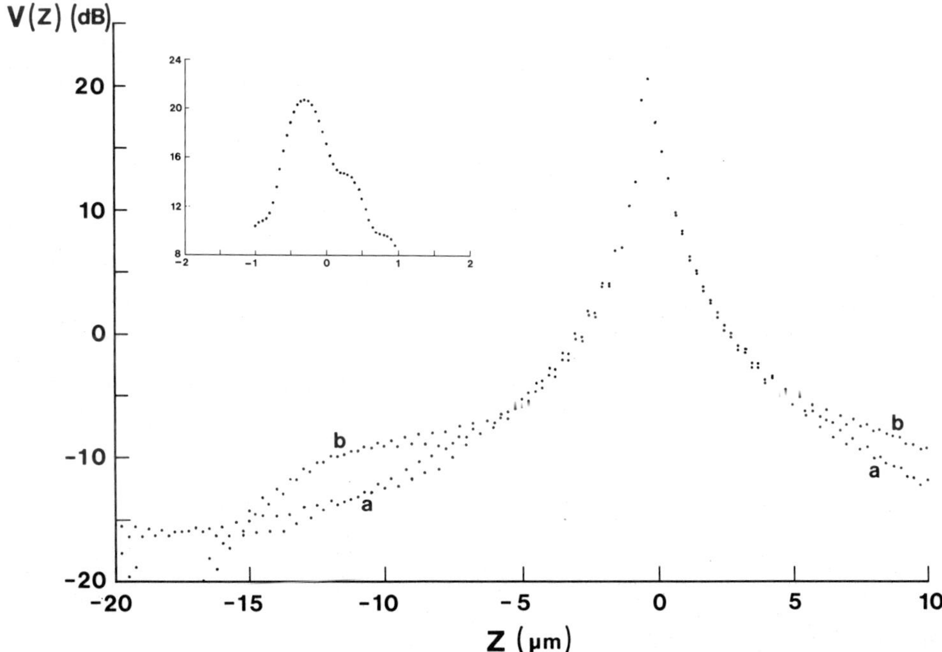

Fig. 7. V(z) curves for (a) zero attenuation and (b) for the realistic attenuation in aluminium (α_ℓ=500 dB cm^{-1}, α_t=1686 dB cm^{-1}). The inset shows detail of the peak. The double points arise from the oscillating nature of V(z).

be sensitive to variations in height across a specularly reflecting surface. For example a step from an evaporated layer of 0.12 μm thick will give a 3 dB change in signal.

So far we have considered flat, smooth surfaces in the preceeding analysis. If a surface is rough there will be variations in the diffuse to specular reflection ratio across the liquid-solid interface which will result in variations in the output signal. This could contribute to contrast when there are different materials within the scan length.

REFERENCES

1. J. Heiserman, D. Rugar and C.F. Quate, Cryogenic Acoustic Microscopy, J. Acoust. Soc. Am. 67: 1629
2. J. Heiserman, Cryogenic Acoustic Microscopy, in: "Scanned Image Microscopy", E.A. Ash, ed., Academic Press, London (1980)
3. J. Attal, Acoustic Microscopy: Imaging Microelectronic Circuits with Liquid Metals, in:"Scanned Image Microscopy", E.A. Ash, ed., Academic Press, London (1980)
4. C.R. Petts and H.K. Wickramasinghe, Acoustic Microscopy in gases, Elec. Lett., 16: 9 (1980)
5. L.M. Brekhovskikh, "Waves in Layered Media", Academic Press, London (1980)
6. V.M. Merkolova, Reflection of sound waves from the boundary between a liquid and a solid absorbing medium, Soviet Physics Acoustics, 15: 404 (1970)
7. J.A. Rayne and C.K. Jones, "Physical Acoustics", Vol. 7, pp 149-215, Academic Press, London (1970)
8. A. Atalar, An angular-spectrum approach to contrast in reflection acoustic microscope, J. Appl. Phys., 49: 5130 (1978)
9. H.K. Wickramasinghe, Contrast and imaging performance in the scanning acoustic microscope, J. Appl. Phys., 50: 554 (1979)
10. J.W. Goodman, "Introduction to Fourier Optics", McGraw-Hill, London (1968)
11. J. Zemanek, Beam Behaviour within the Nearfield of a Vibrating Piston, J. Acoust. Soc. Am., 49: 181 (1971).

ACOUSTIC MICROSCOPY AT TEMPERATURES LESS THAN 0.2°K

Daniel Rugar, John S. Foster and Joseph Heiserman

Edward L. Ginzton Laboratory
Stanford University
Stanford, California 94305

INTRODUCTION

The resolving power of the acoustic microscope is set primarily by the wavelength of the transmitted sound in the coupling fluid. The shortest wavelength that can be used is, in turn, determined by the acoustic loss in the fluid, which typically increases rapidly with frequency. To our knowledge, the shortest wavelength yet used for imaging in water is approximately 4000 Å at 3.8 GHz.[1] Since the attenuation at this frequency is extreme, approximately 14,000 dB/cm, the total path length through the fluid must be very short, typically less than 40 μm. To achieve shorter wavelengths and higher resolution, one can use a coupling fluid with lower acoustic attenuation and/or lower velocity than water. Among the fluids that potentially offer improved resolution performance are the cryogenic liquids[2] and high pressure inert gases.[3]

By far the best liquid to use for high resolution is liquid helium at very low temperatures. We have constructed a microscope which uses helium at temperatures less than 0.2°K and our initial results are presented in this paper. Imaging was performed at 980 MHz with a wavelength in the liquid of 2400 Å. Because of the unique properties of low temperature helium, the liquid path attenuation is less than 1 dB.

The performance of liquid helium as a medium for acoustic microscopy has been discussed previously.[2,4] Table 1 lists some relevant properties of helium at 1 GHz and provides a comparison to the properties of water. The velocity of sound in helium is quite slow, ranging from 183 m/s at 4.2°K (the normal boiling point) to 238 m/s at absolute zero. The acoustic attenuation in

Table 1. Acoustic Properties of Helium Compared to Water[11-13]

Liquid	Temp °K	Velocity m/s	1 GHz Attenuation dB/cm	Impedance 10^5 g/cm^2-s
Water	60°C	1550	950	1.5
Helium	4.2	183	19700	0.023
	1.95	227	6100	0.033
	0.2	238	8	0.035

helium is a strong function of temperature and is plotted in Fig. 1 for 1 GHz . At 4.2°K the liquid behaves classically and the attenuation is well accounted for by Navier-Stokes attenuation (viscous and heat conduction losses). As the temperature falls below the normal boiling point, the attenuation drops in accordance with classical predictions. At 2.17°K , the superfluid (lambda) phase transition occurs and is accompanied by a sharp peak in attenuation.

A local minimum in attenuation is found at 1.95°K . This minimum has been used previously for acoustic microscopy, though few images have been published. One previously unpublished image is shown in Fig. 2. The object is a grating of photoresist lines on silicon and has a spatial periodicity of 2500 Å . The image was taken at 840 MHz where the wavelength is 2700 Å . Despite the presence of the local minimum, the attenuation at 1.95°K is quite large, approximately 6100 dB/cm at 1 GHz . This high loss will make operation at frequencies much above 1 GHz difficult.

Continuing to lower temperatures, a broad peak in attenuation is found at 1.4°K , followed by a rapid decline. Below about 0.7°K , the attenuation falls approximately as T^4 , where T is temperature. At 0.1°K , the attenuation at 1 GHz is extremely small, less than 1 dB/cm . It is this nearly lossless propagation at low temperatures that makes helium such an attractive medium for acoustic microscopy.

The primary mechanism of acoustic attenuation for temperatures around 0.2°K is the scattering of the transmitted acoustic phonons by thermal phonons present in the liquid via three phonon processes.[5] We resort to a phonon description of the impressed sound wave because the usual hydrodynamic and thermodynamic description is no longer

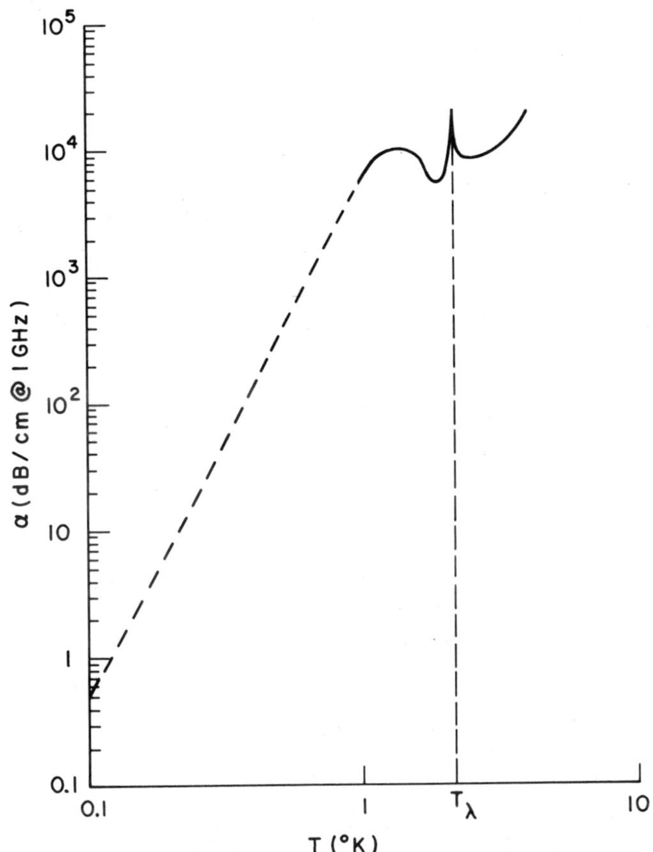

Fig. 1. Attenuation of sound in liquid helium at 1 GHz as a function of temperature. Data above 1°K (solid line) is from Imai and Rudnick.[12] *Low temperature data (dashed line) is extrapolated from the results of Abraham et al.*[11]

suitable; it is not possible to assign quantities such as temperature and pressure to various regions of the acoustic wave since the equilibration time of the thermal phonons is much longer than the period of the wave. The attenuation decreases with falling temperature since the number of thermal phonon scatterers decreases rapidly with the lowering of temperature. From a detailed analysis,[5] it is found that the (amplitude) attenuation behaves as

$$\alpha_{3pp} = A\ f\ T^4\ , \tag{1}$$

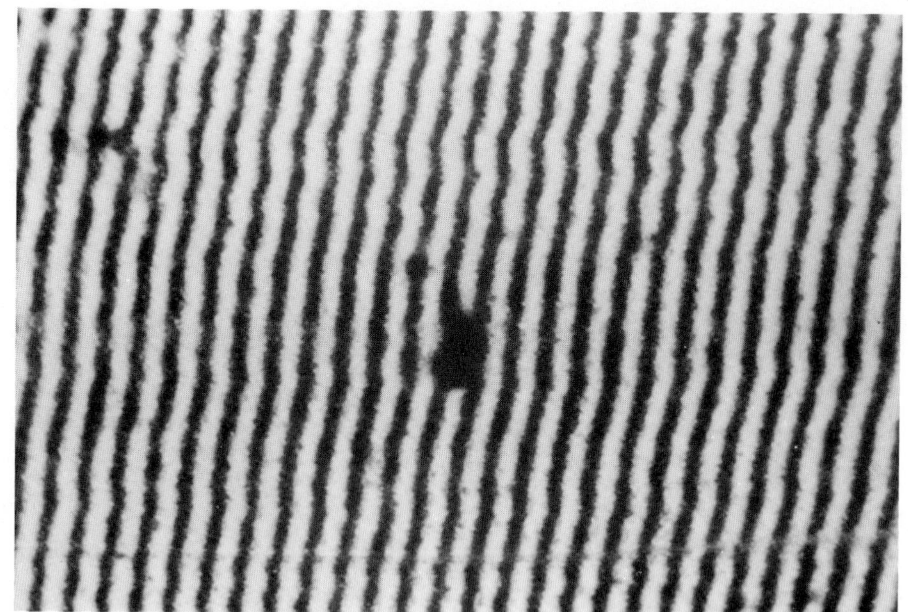

Fig. 2. Acoustic image of a grating with 2500 Å period taken at 840 MHz in 1.95°K helium.

where α_{3pp} is the attenuation due to the three phonon process, f is frequency and T is temperature. The constant A is given by

$$A = \frac{\pi^4 k_B^4}{30 \hbar^3} \frac{(1 + \mu)^2}{\rho c^6} \qquad (2)$$

where k_B is Boltzmann's constant, $\hbar$ is Planck's constant, ρ is the density of helium and c is the velocity of sound. The parameter μ is given by $\rho/c(\partial c/\partial \rho)$ and has the value 2.84.

Substituting numerical values into (2), we get

$$\alpha_{3pp} = 5 \times 10^3 \ f \ T^4 \ , \qquad (3)$$

where α_{3pp} is expressed in dB/cm , f in GHz and T in degrees Kelvin. Because the attenuation increases only linearly with frequency, rather than as the square, operation at frequencies considerably above 1 GHz should be possible. For example, the expected attenuation from three phonon thermal scattering at 10 GHz (λ = 238 Å) is only 5 dB/cm at 0.1°K .

CRYOGENIC APPARATUS

An apparatus suitable for acoustic microscopy at temperatures down to 0.05°K has been constructed. To achieve the low temperature of operation, the microscope is cooled using a ^3He-^4He dilution refrigerator.[6] The lens and mechanical scanning unit are contained in an experimental chamber which bolts to the mixing chamber of the refrigerator. The experimental chamber is filled with enough liquid helium, approximately 30 ml, to fully immerse the lens and mechanical scanner. The object to be examined by the microscope is mounted on the end of a two meter long focusing rod and inserted through an access tube which extends from a room temperature airlock to the experimental chamber. To reduce reflux heating due to superfluid film flow up the access tube, a solution of 3% ^3He in ^4He is used in the experimental chamber instead of pure ^4He .

The mechanical scanner is a critical component of the microscope and is shown schematically in Fig. 3. The acoustic lens is mounted on the end of a 10 cm long section of stainless steel semi-rigid coaxial cable (Uniform Tubes, UT-141SS). The coaxial cable serves both as the carrier of microwave signals to and from the acoustic lens, and as a flexible, spring-like support for the lens. When a picture is taken, the lens is translated in a two-dimensional raster pattern by means of orthogonally mounted drive coils. The velocity of the lens is sensed by two additional coils; only one pair of drive and sense coils is shown in Fig. 3. The velocity signals from the sense coils are electronically integrated to obtain information on the position of the lens. The sense coils are also used in a feedback arrangement to electronically damp the motion of the lens. With this simple scanning system, a raster scan can be executed with an accuracy better than 1000 Å .

A coarse adjustment of microscope focus is achieved using a room temperature differential micrometer connected to the sample holding rod. Focusing resolution of approximately 2000 Å is achieved in this way. The microscope can be focused with much finer accuracy by adjusting the voltage to a piezoelectric positioning element consisting of a 2.5 cm long PZT-5H tube. The piezoelectric positioner is located at the sample end of the focusing rod and has a focusing range of about 5000 Å .

DESIGN OF THE ACOUSTIC LENS

With the exception of the acoustic impedance matching layer, the acoustic lens used in our initial imaging experiment is of fairly traditional design. The lens element consists of a spherical depression polished in the end of a short sapphire rod. The radius of curvature is 80 μm and the opening half-angle is 15° (f/1.9 aperture). The sapphire rod is 2mm in length and a planar ZnO

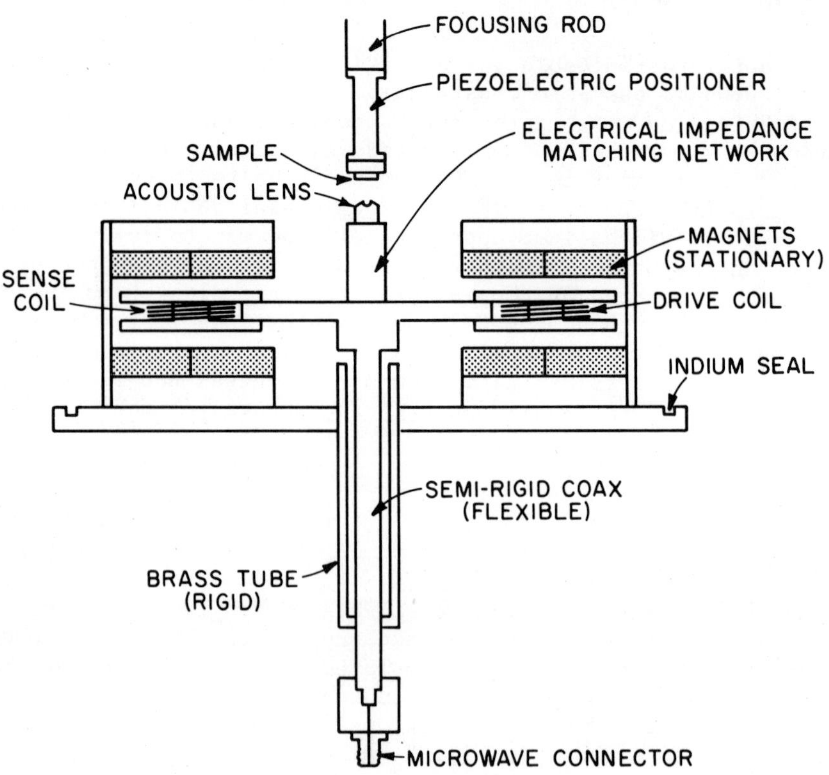

Fig. 3. Schematic diagram of the mechanical scanner used in conjunction with a dilution refrigerator.

transducer is used for acoustic wave generation.

Because of the large acoustic impedance mismatch between sapphire and helium, it is essential to fabricate a quarter-wave acoustic impedance matching layer (or anti-reflection coating) on the surface of the lens; without a matching layer, only 0.3% of the incident power would be transmitted across the interface. The ideal impedance of the matching layer would be the geometric mean of the sapphire impedance ($Z_s = 44.3 \times 10^5$ g/cm^2-s) and the helium impedance ($Z_{He} = .035 \times 10^5$ g/cm^2-s). Few solid materials possess such a low impedance, and those that do, such as plastics, are usually lossy or soft and, therefore, subject to damage. In lieu of finding a more suitable material, we use a quarter-wave layer of carbon to provide satisfactory impedance matching.

The carbon matching layer is fabricated by electron beam evaporation of a carbon target in a vacuum better than 10^{-6} torr.

The most suitable target material for the evaporation was found to be pyrolytic graphite,[7] as it is generally free of trapped gas and is resistant to breakage during the evaporation process. The sapphire lens is maintained at a temperature of 350^oC during deposition. Because the adhesion of carbon directly to sapphire is poor, interface layers consisting of 200 Å titanium followed by 1000 Å molybdenum are deposited. The carbon, which adheres well to the molybdenum, is deposited at a nominal rate of 20 Å/sec. The resulting carbon film is found to be smooth, well adhering and resistant to scratches. Examination by reflection electron diffraction indicates the film to be either amorphous or of very small grain size. The film may be thermally cycled without ill effect and does not appear to degrade with time.

Matching layer performance can be evaluated using a pulse echo technique.[2] A plot of plane wave power transmission into helium as a function of frequency is shown in Fig. 4. The solid dots are experimental data for helium at 4.2^oK. At the center frequency, the power transmission is approximately 5.5%. The solid line is the theoretical response for a matching layer with impedance 8.6×10^5 g/cm^2-s and helium at 4.2^oK. The dashed line is the theoretical transmission for a matching layer with the same impedance, but for helium near zero temperature. Transmission efficiency improves at the lower temperature since the impedance of helium increases by 50% from 4.2^oK to 0^oK. Maximum transmission efficiency, which occurs for a $\lambda/4$ layer thickness, is approximately 8% in the low temperature regime. The 3 dB bandwidth is about 26%.

It is interesting to note that the carbon matching layer developed for the helium microscope is ideal for the room temperature water microscope. The layers have been successfully used on water lenses at frequencies between 700 MHz and 3.8 GHz. Transmission efficiencies of 99% are typically observed for plane waves into water.

RESULTS

The frequency of operation for the initial imaging experiment was 980 MHz, with a corresponding helium wavelength of 2400 Å. The total insertion loss, which includes two-way transducer conversion loss, lens illumination loss, and acoustic impedance mismatch loss, was 53 dB. Despite this relatively low loss (room temperature microscopes typically operate with more than 90 dB of insertion loss), the signal-to-noise ratio of the imaging was less than 10 dB. This low signal-to-noise ratio is the result of the large nonlinear attenuation which is encountered in the liquid path for RF input power greater than approximately -30 dBm (1 microwatt). We will discuss this effect later.

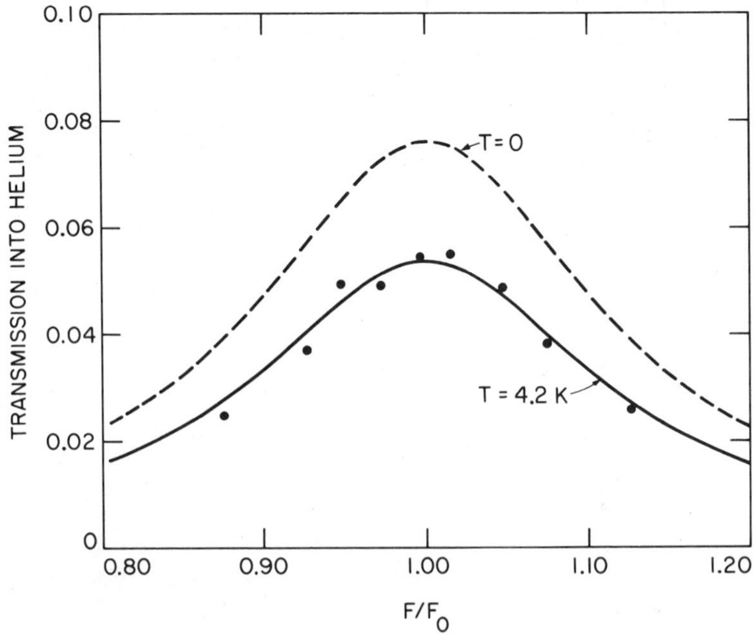

Fig. 4. Plane wave power transmission from sapphire into helium as a function of frequency when a carbon matching layer is used.

Two acoustic micrographs recorded between 50 and 80 mK are shown in Fig. 5. The sample is a 4 μm period grating consisting of 2 μm wide aluminum lines on a glass substrate. The grating is seen only faintly in Fig. 5(a) because the aluminum lines are thin (≈ 1000 Å) and because the depth of focus of our lens is rather large due to the small opening angle. Among the most prominent features in Fig. 5(a) are the structures seen with black outlines. These regions are believed to consist of a thin layer of frozen air which selectively condensed onto the aluminum grating lines. The grating lines can be seen with greater contrast in Fig. 5(b). The increased contrast is the result of operating the microscope in a highly nonlinear regime. The details of this contrast enhancing effect are not yet understood. The resolution of the imaging in these micrographs appears to be 4000 Å to 5000 Å, based on the apparent size of the smallest features. This is consistent with the expected resolution, taking into consideration the f/1.9 aperture of the lens.

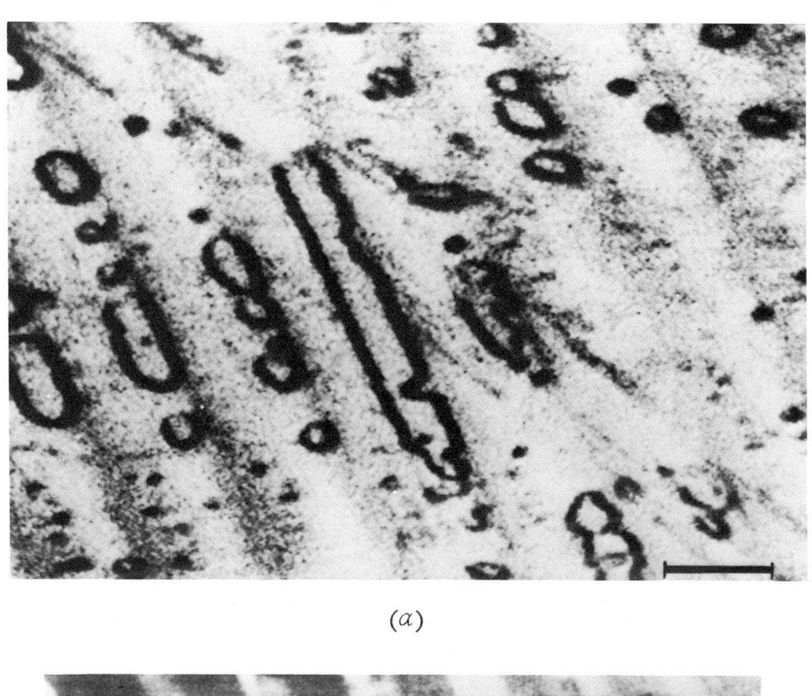

(*a*)

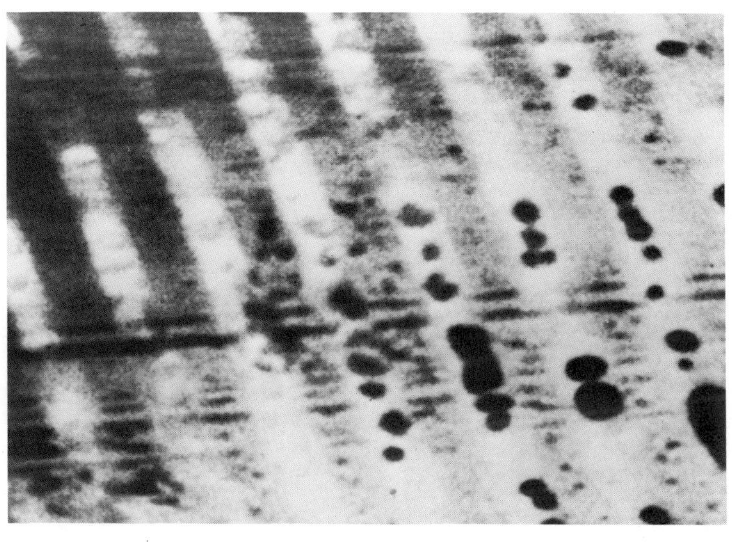

(*b*)

Fig. 5. Images taken in liquid helium at approximately 0.05^{o}K. The object is a 4 μm period grating consisting of aluminum lines on glass. Bar = 4 μm.

As mentioned above, we observed the onset of nonlinear excess attenuation in this initial imaging experiment. Because this effect potentially limits the achievable signal-to-noise ratio of the microscope, further discussion is merited. Nonlinear excess attenuation is the result of depletion of the transmitted fundamental wave by the generation of harmonics. For plane waves, nonlinear effects become important when the distance of propagation approaches a discontinuity length, which is given by[8]

$$L = 1/\beta Mk , \tag{4}$$

where $\beta = 1 + \mu$, M is the acoustic Mach number and $k = 2\pi/\lambda$. Because of the slow velocity of sound in helium, the acoustic Mach number tends to be higher than in other liquids, such as water.

For focused beams, nonlinear effects are significant when the confocal parameter of the focus is greater than or equal to a discontinuity length. Using the properties of Gaussian focused beams, the confocal length is related to the convergence angle of the beam according to[9]

$$b = \frac{2}{\pi} \frac{\lambda}{\theta^2} , \tag{5}$$

where θ is the convergence angle in radians. Thus, the Mach number at which nonlinear behavior begins is

$$\begin{aligned} M_o &= (\beta Lk)^{-1} = (\beta bk)^{-1} \\ &= \frac{\theta^2}{4\beta} . \end{aligned} \tag{6}$$

We see that the limiting Mach number at the focus is independent of frequency and depends quadratically on the convergence angle (i.e., f-number).

The intensity at the focus corresponding to this Mach number is

$$\begin{aligned} I_o &= \frac{1}{2} \rho c^3 M_o^{\ 2} \\ &= \frac{1}{32} \frac{\rho c^3}{\beta^2} \theta^4 \end{aligned} \tag{7}$$

To find the total power in the focal spot, we first find the area of the spot.

$$A = \pi \frac{w_o^2}{2} = \frac{\lambda^2}{2\pi\,\theta^2} , \tag{8}$$

where w_o is the $1/e$ radius of the focal spot. Thus, the power at the focus for the onset of nonlinear effects is

$$P_o = I_o A = \frac{1}{64} \frac{\rho c^3}{\pi \beta^2} \theta^2 \lambda^2 . \tag{9}$$

Substituting parameters for helium,

$$P_o = (6.6 \times 10^5 \frac{W}{m^2}) \theta^2 \lambda^2 . \tag{10}$$

Using the opening angle and wavelength appropriate for the 980 MHz lens, we find the power at the focus for the onset of nonlinear effects to be

$$P_o = 2.6 \times 10^{-9} \text{ W} = -56 \text{ dBm} . \tag{11}$$

This value is in good agreement with the experimentally estimated power at the focus for the onset of nonlinear effects.

The most significant consequence of the nonlinear excess attenuation is the limitation of signal-to-noise ratio in the received signal. The one-way insertion loss of a typical acoustic lens is about 27 dB . Thus, a −56 dBm signal at the focus corresponds to −83 dBm at the receiver port. With our current receiver electronics, the noise level is at −94 dBm , a value which is set by thermal noise, the noise figure of the amplifier and the bandwidth of the receiver. Therefore, we are left with a theoretical maximum signal-to-noise ratio of 11 dB .

Improving this marginally acceptable signal-to-noise ratio will be a high priority in the future. This is especially important in light of the fact that operation at higher frequencies will further reduce the received signal since the onset of excess attenuation will occur at lower power levels (see equation (10)). A significant improvement in signal-to-noise ratio should be obtainable by using cryogenically cooled pre-amplifiers.[10]

CONCLUSION

We have succeeded in constructing an acoustic microscope which uses liquid helium at temperatures in the 100 mK range as the acoustic coupling medium. The operating frequency was approximately 1 GHz and the observed imaging resolution was consistent with the

2400 Å wavelength and the f/1.9 aperture of the lens. Nonlinearity of the acoustic wave propagation was identified as a mechanism which limits the signal-to-noise ratio of the imaging. This limitation should not be critical, however, since means of improving noise performance of the receiver electronics are available.

The most significant consequence of this work is that a new regime of acoustic imaging has been opened. Since the acoustic attenuation in helium is so extremely small, acoustic liquid path loss will not limit achievable resolution, at least until the wavelength approaches a few hundred angstroms. Barring unforeseen technical difficulty, acoustic imaging with resolution significantly better than 1000 Å should soon be at hand.

ACKNOWLEDGEMENTS

This work was supported by the U.S. Office of Naval Research. D. Rugar is partially supported by the F. V. Hunt Postdoctoral Research Fellowship of the Acoustical Society of America. We thank C. F. Quate for his guidance in all phases of this work.

REFERENCES

1. B. Hadimioglu, private communication.
2. J. Heiserman, D. Rugar and C. F. Quate, Cryogenic acoustic microscopy, J. Acoust. Soc. Am. 67:1629 (1980).
3. C. R. Petts and H. K. Wickramasinghe, Acoustic microscopy in gases, Electron. Lett. 16:9 (1980).
4. J. Heiserman, Cryogenic acoustic microscopy: the search for ultrahigh resolution using cryogenic liquids, to be published _in_: "Proceedings of the 16th International Conference on Low Temperature Physics."
5. H. J. Maris, Phonon interactions in liquid helium, Rev. Mod. Physics 49:341 (1977).
6. Model DRI-420TL, SHE Inc., San Diego, CA.
7. Pfizer Inc., Easton, PA.
8. T. G. Muir, Nonlinear effects in acoustic imaging, _in_: "Proceedings of the Ninth International Symposium on Acoustical Imaging," K. Y. Wang, ed., Plenum Press, New York.
9. H. Kogelnik and T. Li, Laser beams and resonators, Appl. Optics 5:1550 (1966).
10. S. Weinreb, Low-noise cooled GASFET amplifiers, IEEE Trans. Microwave Theory Tech. 28:1041 (1980).
11. B. M. Abraham, Y. Eckstein, J. B. Ketterson, M. Kuchnir and J. Vignos, Sound propagation in liquid ^{4}He, Phys. Rev. 181:347 (1969).
12. J. S. Imai and I. Rudnick, Ultrasonic attenuation in liquid helium at 1 GHz, Phys. Rev. Lett. 22:694 (1969).

13. J. Wilks, "Liquid and Solid Helium," Oxford University Press, Oxford (1967).

NOTE ADDED IN PROOF

The operating frequency of the microscope was recently increased to 2.6 GHz (λ = 900 Å). Resolution of the order of a wavelength was observed.

PLANAR ACOUSTIC MICROSCOPE LENS

G.W. Farnell

Department of Electrical Engineering,
McGill University
Montreal, Canada H3A 2K6

C.K. Jen

Industrial Materials Research Institute,
National Research Council
Montreal, Canada, H4C 2K3

INTRODUCTION

The concept of a planar acoustic microscope lens to be used in the standard geometry of a scanning acoustic microscope (1) has been discussed previously (2,3). In this paper we wish to look a little more closely at the focussing mechanism of this lens and to consider some further extensions of the concept.

This planar lens has been called a Rayleigh-to-Compressional Conversion lens (RCC) lens because the underlying principle involves the generation of a Rayleigh wave by means of electrodes on the interface between a liquid and a solid and the energy in this Rayleigh wave then converts to a compressional wave in the liquid (4). The converted energy is in a beam of some one or two degrees width about the zenith angle ϕ_m corresponding to the angle where the Rayleigh wave phase matches to the compressional wave in the liquid. For a single linear impulse of stress on the interface the far-field displacement amplitude in the liquid has the angular dependence shown by the typical curve $M(\phi)$ of Fig. 1 drawn for kerosene on YZ $LiNbO_3$. The sharply peaked radiation at $\phi = \phi_m$ is this Rayleigh-to-compressional conversion.

In the prototype RCC lens a narrow ring of stress is produced on the interface by electrodes in the form of concentric circles plated on the piezoelectric substrate. Thus the energy from each region of the electrodes is directed mainly toward a common point on the axis, the geometric focus, producing a hollow converging conical beam in the liquid and a diffraction-limited focal spot at the geometrical focus whose location is determined by the electrode radii and the angle ϕ_m.

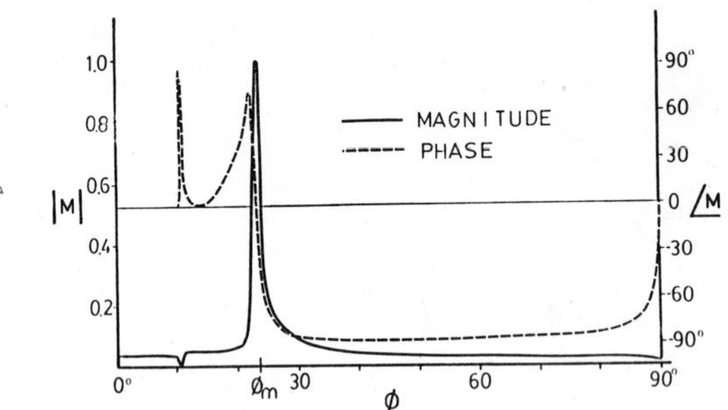

Fig. 1 Magnitude and phase of the function M(ϕ) which gives the far-field radiation pattern of a linear impulse.

If the concentric electrodes are fabricated as semicircles with separate terminals as indicated in Fig. 2a then they can be used in various combinations for transmitting and receiving (3). For example the combination A to D gives a standard transmission image of an object placed at the focus, A to B is a reflection image, while A to C is a dark-field image. This is one example of the geometric flexibility available with the RCC lens because of the simple fabrication technique, and, of course, if the coplanar semicircles are driven in parallel we have the prototype RCC lens.

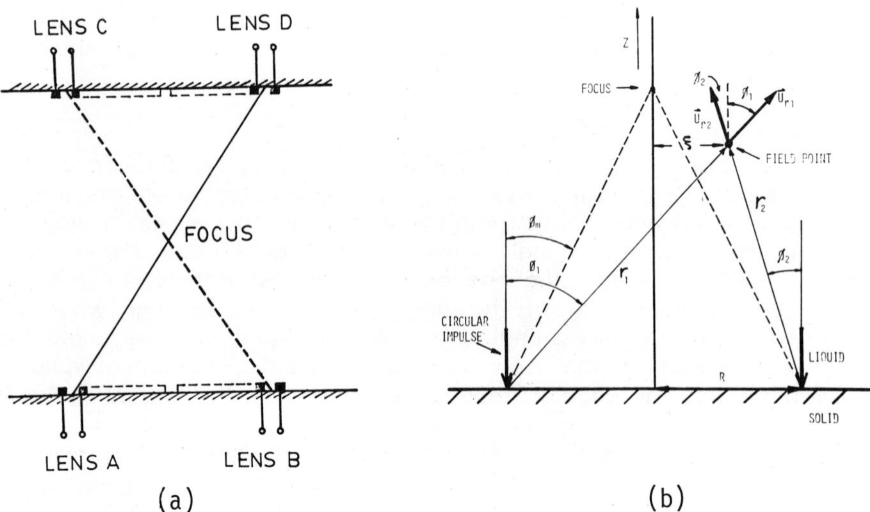

Fig. 2 (a) The RCC lens in a confocal geometry here with each lens split into two semicircles to which connections can be made separately.

(b) Geometry in a sagittal plane for calculating displacements. Source is a circular line impulse of stress of radius R on the interface.

RADIATION PATTERN

It is interesting to look a little more quantitatively at the radiation pattern in the liquid. If the solid is assumed to be isotropic and the excitation to be a circular impulse of stress varying sinusoidally in time, the displacement amplitudes in the liquid can be calculated and reduced to a relatively simple form. At a field point (ξ,z) in this axially symmetric case the displacement must lie in a sagittal plane such as shown in Fig. 2b. For field points more than a wavelenth off the axis, $k_c \xi > 2.5$, where k_c is the wave vector in the liquid, the displacement is composed of two terms:

$$\vec{u} = \vec{u}_{r1} + \vec{u}_{r2} \tag{1}$$

where $\vec{u}_{r1}$ and $\vec{u}_{r2}$ are radially outwards from the locations of the respective intersections of the circular impulse with the sagittal plane under consideration as illustrated in Fig. 2b. The lengths of these vectors are given by:

$$u_{r1} = \sqrt{k_c R/r_1 \xi} M(\phi_1) e^{jk_c r_1} \tag{2}$$

$$u_{r2} = \sqrt{k_c R/r_2 \xi} M(\phi_2) e^{jk_c r_2} \tag{3}$$

where $k_c = 2\pi/\lambda_c$ and λ_c is the wavelength of compressional waves in the liquid. Recalling the angular dependence of the function $M(\phi)$ as drawn in Fig. 1, it is seen that the displacement component u_{r1} will be large when the field point lies on the cone generator through the focus, i.e. $\phi_2 \cong \phi_m$ and the contribution of u_{r1} will be relatively small. This is the mathematical statement of the earlier remark that the wave propagates as a hollow cone. Figure 3 shows the component of the Poynting vector parallel to the axis, P_z, multiplied by the distance from the axis, ξ. The area under such a curve is proportional to the power crossing a plane perpendicular to the axis at the chosen value of z. In this case with no attenuation in the liquid the area under each curve is the same for different values of z though the amplitude decreases with z. The extremely thin conical shell which carries the energy from the assumed impulse to the focus should be noted.

Near the axis, $k_c\xi<2.5$, different approximations have to be used in the integration of the Bessel functions which led to Eqs. (2) and (3) with the result that the paraxial displacement has essentially but an axial component and this component is given by

$$u_z = -j\frac{k_c}{2} \sin 2\phi_o J_o(k_c \xi \sin\phi_o) M(\phi_o) e^{jk_c r_o} \tag{4}$$

where $r_o \cong r_1 \cong r_2$ and $\phi_o \cong \phi_1 \cong \phi_2$ in Fig. 2b.

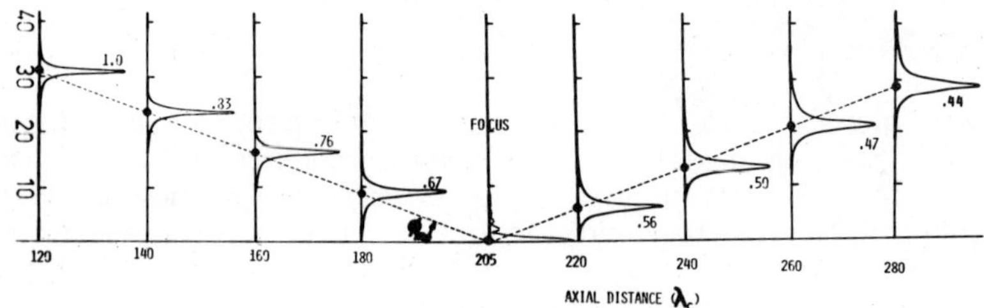

Fig. 3 Power flow in the liquid ξP_z as a function of radius ξ at various axial distances z where P_z is the axial component of the Poynting vector. Area under curves is power crossing a cross-section at z. Dotted curve is displacement u_z in the focal plane. Source is a circular impulse on interface between alcohol and $LiNbO_3$, attenuation in the liquid ignored.

The broader curve of Fig. 4 shows the measured amplitude along the axis, near the geometric focus for a lens of radius 94 λ_c using kerosene on $LiNbO_3$ at 20 MHz and it is seen to be of the form given by Eq. 4, $J_0 = 0$. The other curve gives the measured amplitude along a transverse scan through the axial maximum, i.e. the focal spot. Two points should be noted there with respect to Eq. 4, first, the axial profile near the focus depends primarily on $M(\phi)$ near ϕ_m and hence is monotonically varying as opposed to the axial dependence found with a full aperture. Second, the time delay to the focus is given by the length along the cone from the source ring to the focus. Note also that the effects of the decay or conversion length of the Rayleigh wave is built into the M function and affects the width and relative height of the peak.

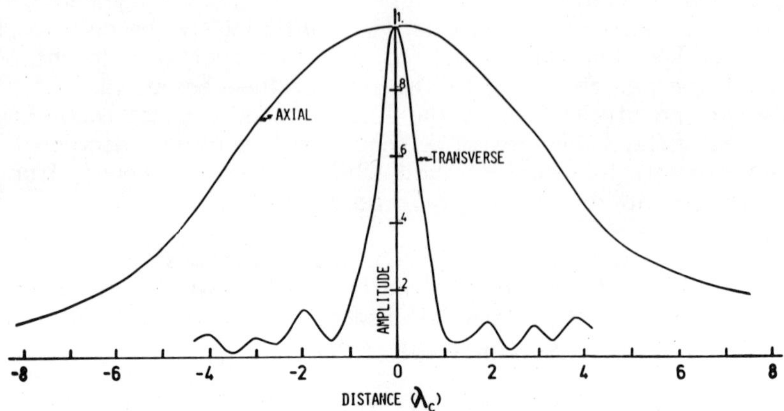

Fig. 4 Measured signal amplitude scattered from a point probe scanned axially and transversely from the geometric focus. Lens has two circular electrodes of radius 94 λ_c, kerosene on $LiNbO_3$.

The stress distribution produced by planar electrodes excited by an r-f voltage is not a spatial impulse but can be approximated crudely by a pair of oppositely phased impulses or more accurately by a distribution of impulses corresponding to the charge distribution transversely across the electrodes (5). For calculation purposes the particle displacements in the axially symmetric case can be evaluated by adding the contribution of each impulse as given by Eqs. (2), (3) and (4). In adding impulses the dependence of the phase of M on ϕ can be important because as seen in Fig. 1 this phase changes almost 180° across the narrow peak. The optimum impulse distribution to model the electrodes has not yet been determined.

For calculations in geometries without axial symmetry or with anisotropic substrates, the electrodes are assumed to be made up of straight-line segments whose contributions can be calculated separately using the velocities in the solid in the direction perpendicular to the segment and then the contributions of each separate segment can be added together.

Results measured in a microscope scanning geometry using the full-circle RCC lens in transmission and reflection and various combinations of semicircles in bright and dark-field configurations again in transmission and reception have been reported (3).

CONCENTRIC LENSES

One of the current interests is depth profiling of solid samples in the 50 MHz frequency range, and as a first approach consideration is being given to sets of concentric lenses. Concentric RCC lenses of radically different radii act independently of one another producing separate focii along the axis at spacings proportional to the different radii. The reflected signals from the separate beams can be distinguished one from another, in principle, either by exciting the rings separately and sequentially, by exciting simultaneously with different frequencies or by exciting in parallel and time gating the echoes. The latter is the simplest and some preliminary results are available. The initial lens consisted of three two-electrode concentric lenses of radii 26.2, 53.2 and 79.8 λ_c where λ_c = 38μm in the liquid, kerosene, at the operating frequency of 35 MHz and the lens was fabricated on PZT-5A from the mask of Fig. 5b. The traces of Fig. 5a are the reflected signals from a wire of 2.3 λ_c diameter placed at each focus in turn. In the first trace the reflected pulse arrives before the receiver becomes unsaturated from the feedthrough pulse which was not well compensated in this experiment. The transmitted pulse width was 0.5μs. Here the focii are quite independent and the time delay for each corresponds to a double transist of the slant height to the corresponding focus.

In passing it should be noted for experiments in the frequency range near 50 MHz PZT-5A has signal-to-noise advantages over Y-cut

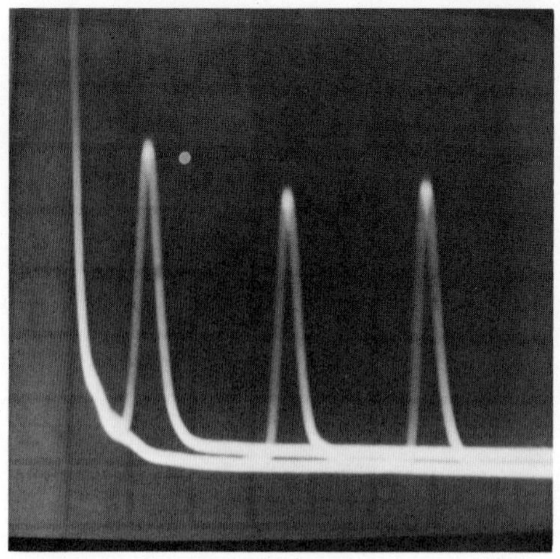

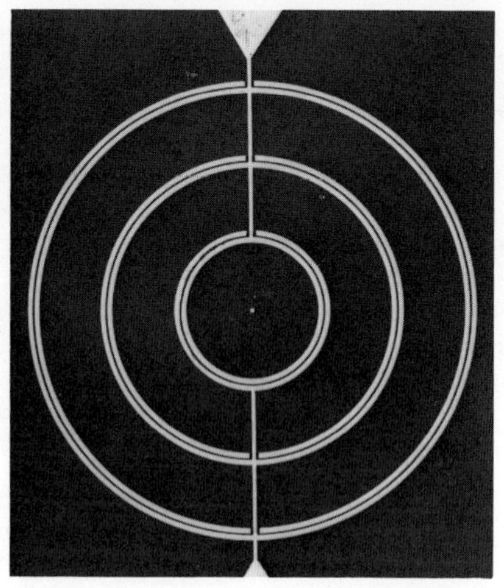

Fig. 5(a) Reflections from a wire transverse to the axis at three focal locations along the axis corresponding to the focii of the multiple lens in (b). Horizontal scale 1 μs per division.

(b) Photolithographic mask for three-ring lens. Fabricated on PZT with radii 26.6, 53.2 and 79.8 ϕ_c for operation at 35 MHz.

$LiNbO_3$ not only because of the higher electro-acoustic coupling but also because of lower excitation of bulk waves in the substrate which reflect back from its lower surface. With its low Rayleigh velocity, lenses on PZT have a greater cone angle in a given liquid than say with $LiNbO_3$ which must be taken into account if one is attempting to have the converging beam penetrate into a solid sample surface perpendicular to the axis.

The next question with multiple concentric lenses is the depth resolution of adjacent circles. For typical two-electrode geometries the 3dB depth of focus along the axis in the liquid is some $8\lambda_c$. Figure 6 shows the echo from a wire $1.9\lambda_c$ in diameter perpendicular to the axis and moved along this axis when the spacing between the gaps of two concentric two-elecrode lenses of a multiple set are separated by 8.8 λ_c as shown in the inset sketch and driven in parallel as common transmitter and receiver, that is a second nearby set of electrodes added to the inner ring of Fig. 5b. This 8.8 λ_c spacing thus gives about the minimum for time-of-flight resolution. Similar axial resolution applies for $LiNbO_3$ as the substrate.

Another consideration of importance in depth profiling in a solid is the character of the reflection from the surface itself. With a full aperture concave microscope lens, the amplitude of the reflected signal

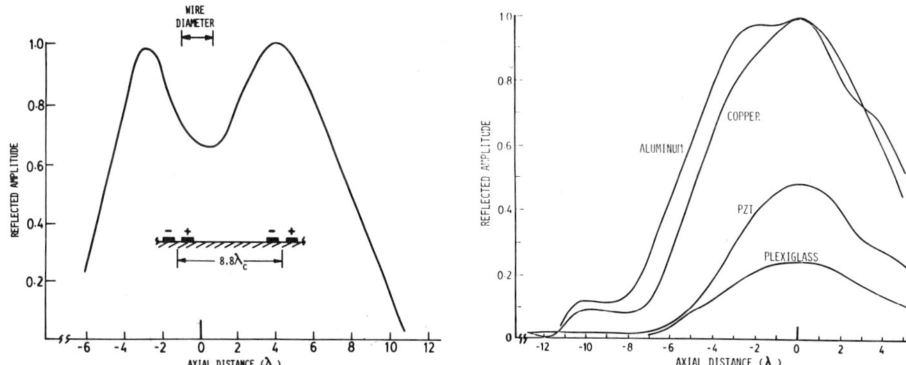

Fig. 6 (Left) Signal reflected from wire traversed axially through focal region of two adjacent lenses of a multiple set. Electrode spacing as shown, set added to inner ring of Fig. 5(b). Effective time gate was wider than change in delay with position.

Fig. 7 (Right) Reflection from surfaces moved perpendicular to lens axis (acoustic signature geometry) for materials having Rayleigh velocity greater than (aluminum and copper), equal to (PZT) and less than (plexiglass) that of PZT lens substrate.

from a flat surface varies as the distance between the lens and the surface is varied. These ripples come from the interference of rays incident on the surface at the Rayleigh excitation angle with the paraxial rays and the pattern is characteristic of the surface, the so-called acoustic signature (6). With the RCC lens where all the energy strikes the surface at about the same angle there is no evidence of acoustic signatures. Figure 6 shows the signal reflected from four flat surfaces as they are moved axially. The angle of incidence for the aluminum (and copper), PZT and plexiglass samples is respectively greater than, equal to, and less than the Rayleigh absorption angle. In these curves there is no evidence of the acoustic signature ripples.

It is the fact that most of the energy of the beam in the liquid strikes a solid sample surface at one angle of incidence only that accounts also for the low spherical aberration of the RCC lens for subsurface imaging.

Propagation loss in the liquid is not a very important factor at the frequencies under consideration here, however one must take into account the insertion loss introduced by the coupling of the electrical signal through the Rayleigh wave to the converging acoustic beam at the transmitting lens and the inverse process at the receiving lens. In the prototype geometry of Fig. 2a the electrodes are long and narrow and considered as an interdigital transducer there is but one finger pair (7)

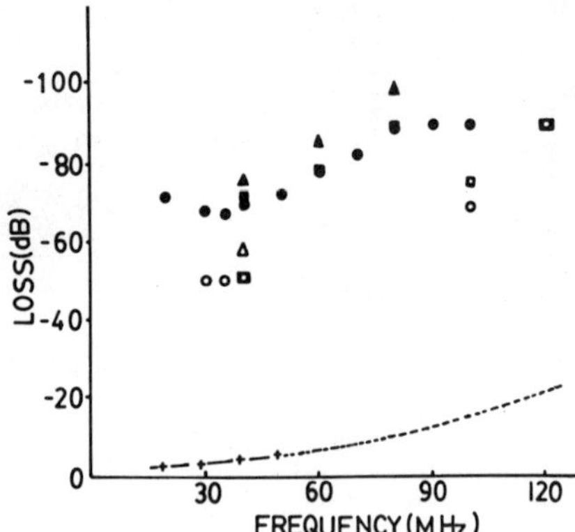

Fig. 8 Total loss for two lenses (radius 1.2 mm on $LiNbO_3$) in a confocal geometry. Solid points are untuned loss for different electrode widths. Open points are for single coil tuning of the lenses. Curve is propagation loss in the liquid.

thus it can be anticipated that the insertion loss in an untuned system will be high but the bandwidth will be large. Figure 8 gives the measured insertion loss in a 50-ohm system of a pair of confocal transducers on $LiNbO_3$. The solid points are the untuned loss for different combinations of gap and electrode width, and it is seen that the bandwidth is indeed very large. The open points indicate a decrease of about 20dB in the insertion loss when the transducers are coil tuned at specific frequencies. No attempt has been made to match the real part of the transducer impedance which is in part radiation resistance but largely ohmic resistance. The lower curve in Fig. 8 gives the lens-to-lens propagation loss in the liquid itself.

LATERAL BEAM MOVEMENT

The focal spot can be moved electronically along the axis by multiple concentric lenses as noted above. Large lateral displacements can be made by fabricating lenses side by side on the interface, but it is not obvious how electronic scanning of say a hundred spot diameters can be done with RCC lenses which are of diameter comparable to such scans. However, one attempt at very small angle scanning, dithering, has been made with a two-electrode lens the location of whose gap for one half of the lens is sketched in Fig. 9. The circles are concentric and the segments are spaced so that at the centre frequency adjacent segments have one wavelength difference in slant path length to a common focal point on the common axis. When the frequency is changed the phase from the different segments produces a step-wise approximation to the phase variation around the circle which would

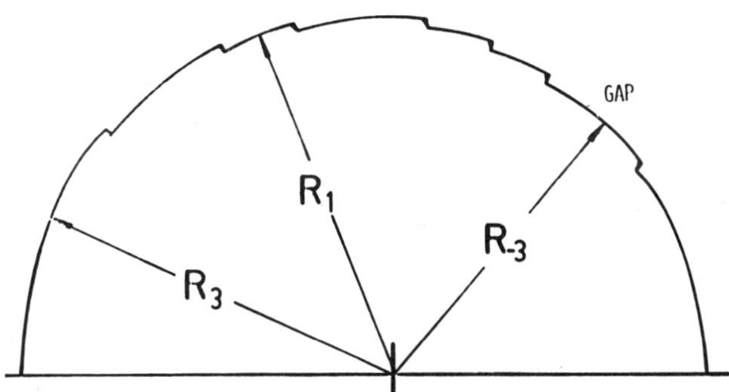

Fig. 9 The gap location of a two-electrode segmented lens for phase tilting. Other half of lens is mirror image.

simulate a tilting of the interface itself. The initial model of this lens on PZT produced a focal spot comparable to that of a simple circle near the centre frequency of 30 MHz and it could be moved transversely by some three spot sizes in the direction of the horizontal axis of Fig. 9 by varying the frequency ±4 MHz. One of the main deterents to further transverse motion is the very nature of the radiation itself in that individual segments do not "see" points very far off the axis because they lie outside the peak of Fig. 1.

In summary, development work is continuing on the RCC lens in an attempt to understand better its operation, its performance characteristics and its novel features.

This research is sponsored by the Natural Sciences and Engineering Research Council and by the National Research Council of Canada.

REFERENCES

1. Quate, C.F., Atalar, A. and Wickramasinghe, H.K.: "Acoustic Microscopy with Mechanical Scanning - A Review", Proc. IEEE, Vol. 67, pp. 1092-1114, 1979.

2. Farnell, G.W. and Jen, C.K.: "Planar Acoustic Microscopy Lens Using Rayleigh to Compressional Conversion", Elec. Lett., Vol. 16, pp. 541-543, 1980.

3. Farnell, G.W. and Jen, C.K.: "Experiments with the Planar Acoustic Microscope Lens", Proc. IEEE Ultrasonics Symposium, pp. 547-551, 1981.

4. Toda, K. and Murata, Y.: "Acoustic Focusing Device with an Interdigital Transducer", J. Acoust. Soc. Am., Vol. 61, pp. 1033-1036, 1977.

5. Ristic, V.M. and Hussein, A.: "Surface Charge and Field Distribution in a Finite SAW Transducer", IEEE Trans. MTT, Vol. MTT-27, pp. 897-901, 1979.

6. Weglein, R.D.: "A Model for Predicting Acoustic Material Signatures", Appl. Phys. Lett., Vol. 34, pp. 179-181, 1979.

7. Smith, W.R.: "Circuit Model Analysis and Design of Interdigital Transducers for Surface Acoustic Waves Devices", Physical Acoustics, Vol. XV, chapter 2, Academic Press 1981.

COVERSLIP INDUCED ARTIFACTS IN HIGH RESOLUTION SCANNING LASER ACOUSTIC MICROSCOPE IMAGES

R. K. Mueller and R. L. Rylander

Department of Electrical Engineering
University of Minnesota
Minneapolis, Minnesota 55455

A critical component of the scanning laser acoustic microscope (SLAM) is the optical reflector which covers the object and defines a plane where the sound field is to be observed. Ideally, this reflector should have no influence of its own on the resulting acoustic image but should be a perfectly flexible optically reflecting layer imbedded within the medium used to couple sound into and out of the object. In practice, however, it is most conveniently realized by a more or less conventional mirror; an aluminum or gold coating on the face of an optically transparent solid plate. This paper examines some of the interactions which occur at this interface and the artifacts which appear in the image as a result. In particular, surface waves in the reflector generated at discontinuities in the sound field lead to fringes aligned with edges of the object.

Acoustic microscopes have evolved into two basic forms: the scanning acoustic microscope (SAM) and the scanning laser acoustic microscope (SLAM). In the SAM, acoustic lenses are used to realize an acoustically defined spot which is scanned over an object to measure its local acoustic transmission properties. The resolution of this microscope is then limited by the acoustic wavelength. Higher resolution requires higher acoustic frequencies and/or coupling media with lower acoustic velocities. Attenuation of sound has been the major limitation in achieving resolution much beyond what is obtainable in optical microscopes.

In the SLAM, a focussed laser beam is used as a detector measuring local acoustic fields passing through an optical reflector by the phase modulation induced in the reflected light. The extent of the detector in this case is determined only by the laser spot size and is independent of the acoustic wavelength (for an interferometric type of detector[1]). This brings the possibility of acoustic images having the resolution of the optical microscope used to focus the laser beam while the acoustic wavelength may be much longer.

An instrument implementing this detector scheme has been constructed to demonstrate sub-acoustic wavelength resolution. Images made with this microscope instead of showing finer detail revealed sets of fringes correlated with the object boundaries which obscure fine object detail. Experiments verified that the fringes were acoustic in nature and not an artifact of the detection scheme. The fringes are due to surface waves generated at discontinuities in the acoustic field (object edges) as it passes through the essential optical reflector.

This paper examines some of the surface waves that are possible at various types of optically reflective coverslips. Photographs of experimentally observed fringes are included showing good agreement with the surface wave explanation.

The most convenient form of acousto-optic interface to use with the SLAM is a conventional mirror - an optically transparent solid with a reflective coating applied to one face. Another type of coverslip used consists of a thin aluminized mylar film stretched over a holder to keep the reflecting surface flat. The holder is then filled with water. It was hoped this structure would approximate the ideal of an optically reflective acoustically transparent layer immersed in the acoustic transmission medium, but films robust enough to allow objects to be mounted to them proved too thick to ignore their effects. The surface waves which resulted in both coverslip types imposed an acoustic limit to the resolution of object detail though the SLAM detector is capable of much finer measurements.

Figure 1 is an acoustic image of onion skin made using a cast epoxy coverslip and an acoustic frequency of 100 MHz. The cell walls are clearly defined showing the acoustic contrast available in biological specimens. Also apparent, however, is a granular texture within the cells. The relation of this spurious detail to object edges is more easily seen in Fig. 2, an acoustic image of a copper grid with bars 37 μm wide and square holes 90 μm across. The spurious detail now appears as straight fringes oriented parallel to the bar directions within the grid and fringes following the curvature of the outer edge of the target. Figure 3 is the video signal for a portion of a single scan line of the grid image. The locations of three bars are clearly shown as deep notches in the amplitude. The actual fringe depth is about 15 - 20% of the peak values within the open cell regions. Also present are fringes within the shadow (bar) regions which do not show up in the

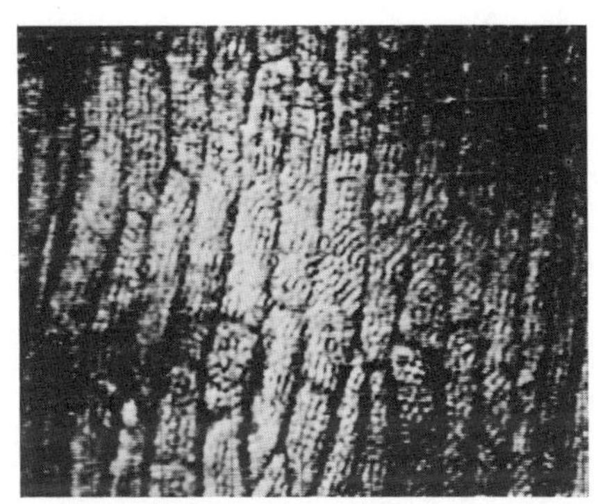

Fig. 1 Acoustic image of onion skin (cast epoxy coverslip, 100 MHz , 50x).

Fig. 2 Image of a copper grid, 37 μm wide bars, 90 μm wide holes.

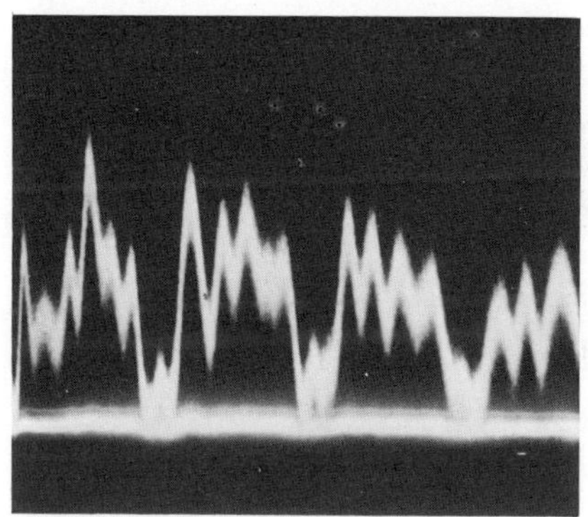

Fig. 3 Video signal from a portion of copper grid image.

Fig. 4 Bars of photoresist on a plexiglas coverslip, largest bars 30 μm x 150 μm .

brightness modulated two-dimensional displays due to the limited dynamic range. The presence of these fringe image artifacts can be explained very well by surface waves travelling at the reflective interface, generated by object edges.

The solid coverslips allow several types of surface waves to exist at their interface with a liquid. There are two true surface wave solutions (confined to the interface region), one real and one lossy, and a trapped wave solution where the water layer between the transducer generating the acoustic waves and the coverslip is considered to act as an acoustic waveguide. The coverslips are thick compared to a wavelength so that calculations of the various surface waves assume the solids to be semi-infinite. Which solution is appropriate for a particular coverslip material depends on the relationship of the solid acoustic parameters to those of the liquid coupling medium.

The real (undamped) surface wave solution will be examined first. Details of the calculation of surface waves at solid-liquid interfaces can be found in several references[2,3]. The equation by which the possible wavenumbers are determined always has one real root regardless of the relationship between the solid and liquid parameters. The solution corresponds to a wavelength shorter than any of the bulk waves in the solid or the liquid. No energy may then leave the boundary layer and the wave propagates without attenuation.

The parameters used for a plexiglas-water interface are:

water	density = $1.00\ g/cm^3$
	longitudinal sound velocity = 1.50×10^5 cm/s
	density = $1.18\ g/cm^3$
plexiglas	longitudinal sound velocity = 2.68×10^5 cm/s
	transverse sound velocity = 1.10×10^5 cm/s

The real root in this case corresponds to a wavelength of 9.52 μm, very close to the 10.36 μm wavelength for a Rayleigh wave on a free plexiglas surface. Calculation of the normal and tangential displacement amplitudes shows most of the wave energy to be on the plexiglas side of the boundary with nearly equal normal and tangential displacements.

Experimentally, this wave corresponds very well to fringes seen in acoustic images made using aluminized plexiglas cover-slips. In Fig. 4 a series of bars of photoresist are deposited directly on the aluminized face of a coverslip. Straight fringes of about 10 μm wavelength can be seen radiating from the edges of some of the larger bars with more circular fringes generated at smaller objects. A small air bubble outside the field of view is responsible for the longer wavelength fringes seen at the top of the image. Air bubbles, depending on their size, cast "soft" shadows since their edges are not in contact with the surface. The pattern is similar to a Fresnel zone pattern with the fringes becoming denser but less contrasty as one moves away from the air bubble where the scattered sound is incident with increasing obliquity.

A glass coverslip was also used with acoustic parameters:

$$\text{density} = 2.32\ \text{g/cm}^3$$

$$\text{longitudinal sound velocity} = 5.64 \times 10^5\ \text{cm/s}$$

$$\text{transverse sound velocity} = 3.28 \times 10^5\ \text{cm/s}$$

The real solution now corresponds to a wavelength of 14.96 μm, almost identical to the 15.00 μm wavelength for longitudinal waves in water and significantly shorter than the 30.13 μm wavelength for Rayleigh waves on a free glass surface. Evaluating normal and tangential displacement components in the glass and water now shows nearly all the energy to reside on the water side of the boundary with displacement parallel to the boundary more than 13 times the normal displacement, a result to be expected given the velocity so near that of an ordinary longitudinal wave in water. Since the SLAM responds to the normal component of surface displacement, this solution should have little influence on the image. The real (undamped) surface wave solutions then appear to be most appropriate for solid coverslip materials such as plexiglas with parameters similar to water.

A solution which fits well in the case of "stiff" coverslips (longitudinal and transverse velocities both greater than the water velocity) is the complex root which approaches the free-surface Rayleigh wave solution as the density of the liquid tends to zero. This "leaky" Rayleigh wave solution for a glass-water interface leads to a surface wave with wavelength less than 1% greater than the free-surface Rayleigh wavelength. Most of the energy exists in the glass with the portion in the water leaving the boundary so that the wave is attenuated by a factor 1/e in approximately ten wavelengths.

This wave is inhomogeneous in the solid, decaying as one moves away from the boundary, but not in the liquid. Because of the energy propagating into the depth of the liquid we must consider the effect of a water layer of finite thickness. Figure 5 is a plot of the allowed wavelengths as a function of the water depth for a symmetric glass-water-glass sandwich and a frequency of 100 MHz. As the water layer depth increases, so does the number of modes within this acoustic waveguide. Each new mode splits into two branches: one asymptotically approaching the water wavelength as the depth increases, and the other rapidly approaching the free-surface Rayleigh wavelength (of the solid) then gradually increasing to the shear wavelength upper limit where the real solutions end. The lowest order mode tends to the Rayleigh wavelength as the water depth goes to zero. The situation is that of Rayleigh waves on the surfaces of two identical solids in contact (but having a friction-free boundary), the waves matching perfectly with no interaction between the solids as if the surfaces were perfectly free.

Evaluating the displacement components for the two branches shows that the solutions which approach the water wavelength are similar to the real (undamped) surface wave solutions previously examined in their structure. The energy is localized primarily in the water and particle motion is mostly parallel to the interface surface. These waves should then have little influence on measurements made with a SLAM. The solutions with wavelengths near the Rayleigh wavelength have a structure very similar to the leaky Rayleigh wave solutions. Indeed, the main influence of a finite water depth seems to be a small change in the wavelength of the "leaky" solutions and eliminating their attenuation since no energy is now lost in the water.

Solutions were also calculated for an asymmetric quartz-water-glass structure since it is unlikely the transducer and coverslip would be made of the same material. The results were quite similar to the symmetric case above, the main difference occurring at the upper branches where the solutions are determined primarily by the solid properties. Here the solutions tend to an intermediate value between the quartz and glass Rayleigh wavelengths.

Figure 6 is an acoustic image of the edge of a piece of vinyl tape on a glass coverslip. Fringes of approximately 31 μm wavelength are clearly seen propagating away from the tape edge. This is very close to the 30.13 μm Rayleigh wavelength for glass. Adding glycerin to the water did not affect the contrast of the fringes as would have been the case if they were due only to guided wave solutions.

The surface waves in their various forms which exist at solid-liquid interfaces make the conventional mirror type of coverslips very much less than ideal for use in SLAMs. A structure more closely resembling the desired acoustically transparent optical reflector is a thin metallized film bounded by water on both sides. Coverslips of this type were made using aluminized mylar films of about 6 μm and 25 μm thickness.

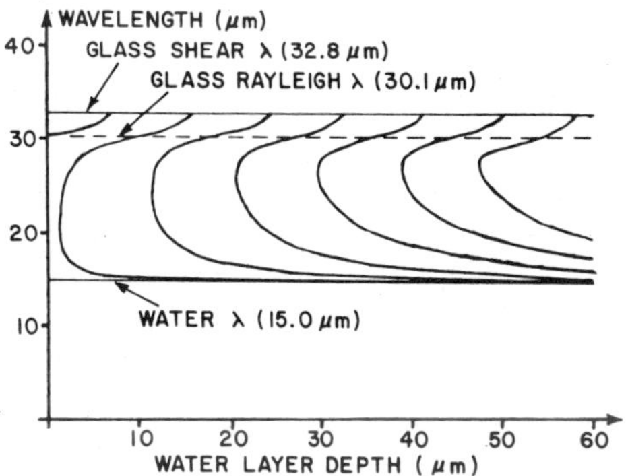

Fig. 5 Allowed wavelengths as a function of water depth in a glass - water - glass sandwich.

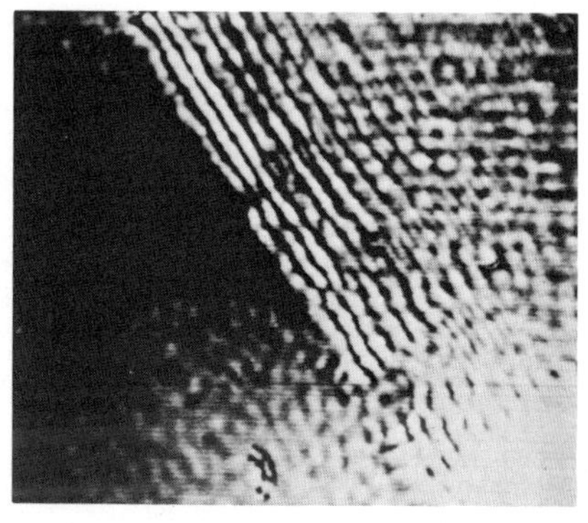

Fig. 6 Fringes radiating from edge of vinyl tape on a glass coverslip.

Fig. 7 Copper grid image using 25 μm thick mylar reflector.

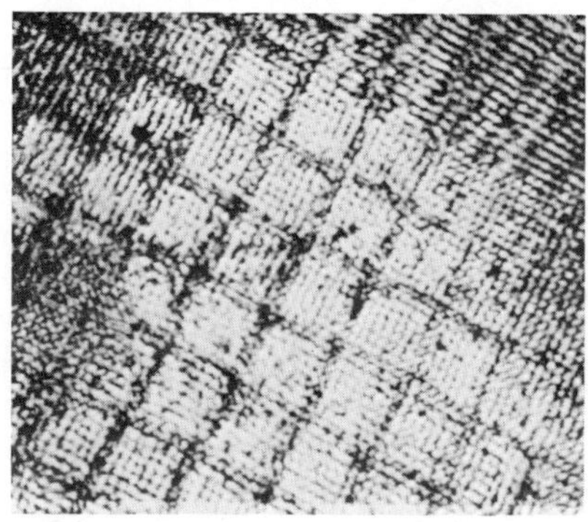

Fig. 8 Copper grid image using 6 μm thick mylar reflector.

Analyzing this structure is similar to the procedure used for the symmetric solid-liquid layer-solid sandwich but with the roles of the solid and liquid interchanged. The result is a pair of equations describing symmetric and antisymmetric boundary displacements. As in the liquid waveguide case, the number of possible modes increases as the plate (film) thickness increases but we shall be concerned with only the lowest order modes since the films should be as thin as practical. The forms of the symmetric and antisymmetric solutions also take on quite distinct characters as the film thickness tends to zero though both approach the Rayleigh wave solutions for thick plates.

The symmetric solution for thin films represents a longitudinal wave where most of the particle motion is along the direction of propagation (parallel to the film boundaries). The displacement normal to the film is just that due to the Poisson effect and is smaller than the longitudinal displacement by a factor of about $1/k_t d$, where k_t is the transverse wavenumber in the film and d is the film thickness. Since the SLAM is sensitive only to normal surface displacements, the symmetric solution will have little effect on images.

The antisymmetric solution, on the other hand, represents a flexural wave with large normal displacement of the film. The velocity of this flexural wave decreases from the velocity of a Rayleigh wave to zero as the transverse wave number-film thickness product $k_t d$ decreases. Even for the thinner of the two mylar films used experimentally (~ 6 μm), this product is about 3.5 making the free film velocity only a few percent slower than the Rayleigh wave velocity. The effect of water loading is a further reduction of the flexural wave velocity, again by only a few percent. Besides this lowest order flexural mode, other low order modes can exist in a film of this thickness (at the 100 MHz frequency used) further complicating the possible surface displacement structure.

Figures 7 and 8 are images of a copper grid made with aluminized mylar films 25 μm and 6 μm thick, respectively. The object is the same copper grid used in Fig. 2. Neither of the mylar films were really thin compared to a wavelength and the images show fringes due to flexural waves travelling slightly slower than the water velocity. The thinner of the two films does have slightly finer fringes, as expected since the water loading has more slowing influence here.

The consequence of the existence of Lamb waves in thin films and Rayleigh waves on the surfaces of solids is that an ideal coverslip is impossible to realize in practice. Because the entire image field is continuously insonified in the SLAM, normal surface deflections arise not only from the desired bulk waves transmitted by an object but also from discontinuity generated waves travelling along the solid-liquid interfaces. The waves moving along the acousto-optic interface interfere with those passing through the interface producing the ubiquitous fringes in high resolution SLAM images. The SAM does not suffer from such artifacts since the insonification occurs only at the point of measurement.

Fringes result not only from bulk solid or film coverslips but from any acousto-optic interface which represents a discontinuity in the acoustic transmission medium capable of supporting some form of surface wave. An object in contact with such an interface will inevitably couple some of the energy incident as a bulk wave into surface wave modes. If the object one is interested in imaging consists of subsurface flaws within a bulk solid, no "coverslip" will be required (assuming the solid's surface is optically reflective or can be coated with a reflector). Surface waves will still be generated at the solid-air interface in the form of genuine Rayleigh waves.

While the SLAM is capable of optically limited measurement of periodic displacements of a reflecting surface, the lack of an ideal coverslip makes images of the acoustic fields passing through objects dependent on the acoustic frequency for resolution. Fringes due to undesired surface waves obscure fine detail and the size of these fringes can be reduced only by increasing the acoustic frequency.

REFERENCES

1. R. K. Mueller and R. L. Rylander, "New demodulation scheme for laser-scanned acoustic imaging systems," J. Opt. Soc. Am., Vol. 69, p. 407, March 1979.

2. I. A. Victorov, Rayleigh and Lamb Waves, (Plenum, New York, 1967).

3. W. M. Ewing, W. S. Jardetzky, and F. Press, Elastic Waves in Layered Media, (McGraw-Hill, New York, 1957).

ULTRASONIC FOCUSSING IN ABSORPTIVE FLUIDS

M. Nikoonahad and E.A. Ash

Department of Electronic and Electrical Engineering
University College London
Torrington Place, London, WC1E 7JE

ABSTRACT

In acoustic microscopy we are concerned with the focussing of ultrasound through absorptive fluid couplers. The resolution of the microscope increases linearly with frequency; on the other hand, for most classical fluids the absorption coefficient is proportional to the square of frequency. This imposes an inherent limitation on the maximum operable frequency.

In this paper, we present a rigorous formulation for calculating the field distribution at different planes in an attenuating medium, with the prime objective of studying the effect of a finite attenuation coefficient on the focussing performance of a lens. This analysis is based on the techniques of Fourier transform and a solution for the elementary waves into which the field is expanded is obtained. The field distribution for typical lenses, in the vicinity of the focal plane, has been calculated, with different path losses in the propagating medium.

INTRODUCTION

The Fourier optics approach to the analysis of focussing and diffraction structures, leads to a complete formulation, and in addition provides a clear physical insight into the nature of the phenomena. It is, however, in its normal form, confined to loss-less systems. The reason for this restriction is simply stated: In Fourier optics one seeks to derive a field distribution in a plane BB', given a source distribution in a parallel plane AA', Figure 1.

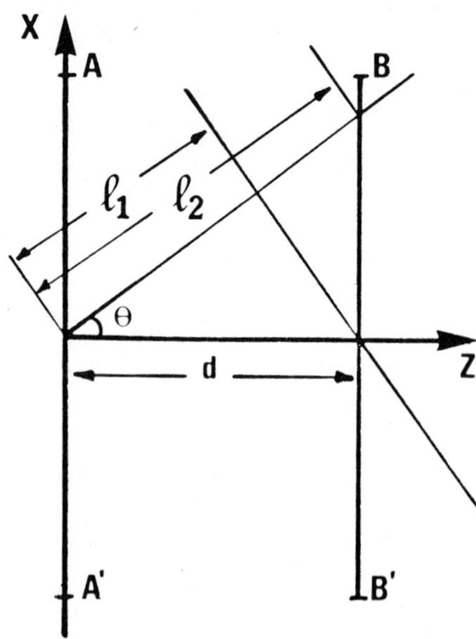

Fig.1. The lengths involved in the propagation of the angular spectrum from plane AA' to BB'.

The basic method is to expand the field in AA' into a set of plane waves, recombining them at BB' after including the phase change due to propagation. The expansion and subsequent recombination are effected by taking the appropriate Fourier transforms. The existence of these transforms implies that the amplitude of the individual constituent plane waves is constant on the planes such as AA' and BB'. It is clear that for plane waves in a lossy medium the planes of constant amplitude will be the equiphase planes normal to the propagation vector. The amplitude distribution on a plane such as AA' will vary exponentially with distance, increasing without limit in one direction.

In most cases this problem is not of any great importance. One can adopt the approach so widely used in analysing losses in guided wave structures - to solve the lossless case, and then include losses as a perturbation. Physical intuition would strongly suggest that, provided the losses are not too great, and provided that the angular spectrum of the waves is not too large, this approach will give results which are a good approximation to the truth. In propagating a typical plane wave one would simply include an attenuation factor $\exp(-\alpha\ell)$, where ℓ is the appropriate propagation length from plane AA' to BB'. However, even here there is an issue which is not readily resolved by purely intuitive arguments: By reference to Figure 1, should one choose $\ell_1 = d.\cos\theta$, or $\ell_2 = d/\cos\theta$? The former, which has been used by some authors[1], would suggest that higher spatial frequencies are relatively enhanced in propagating from AA' to BB'; the latter, which has been used by others[2], would suggest that they are preferentially attenuated. Clearly it is an issue which is of interest only if $\exp(+\alpha\ell)$ is large, and if the range of θ is also large. Such situations are found in ultrasonic imaging, and particularly in high resolution acoustic microscopy, where the attenuations involved may be over 30dB, and the angular range in excess of $\pm 50^\circ$. An even more dramatic instance arises in analysis of thermal wave imaging, where the real and imaginary components of the propagation vector are comparable[3].

The aim of the present paper is to present an analytical approach in which the losses are taken into account *ab initio*, and which can be used to compute focussing and diffraction fields in a manner which is inherently as simple as that of classical Fourier optics. At the same time, we will develop a picture of the basic waves which take the place of plane waves in the lossless case, and which provide a satisfactory understanding of the physical reality.

THEORETICAL FORMULATION

Basic Formulation

We will assume that the propagation phenomena are adequately conveyed by a scalar model, and that the ultrasonic field U is described by the Helmholtz equation

$$\nabla^2 U + k^2 U = 0 \qquad (1)$$

We assume further that the loss mechanism is such that it can be represented by a complex propagation constant,

$$k \equiv \beta - j\alpha \qquad (2)$$

We have for simplicity restricted the analysis and subsequent computations to the two-dimensional case with variations confined to the (x,z) plane. The extension to three dimensions is straightforward. For the two-dimensional case, we can then write equation 1 in the form,

$$\frac{\partial^2 U}{\partial x^2} + \frac{\partial^2 U}{\partial z^2} + k^2 U = 0 \qquad (3)$$

We define the Fourier transform of U(x,z) by

$$T(f,z) = \int_{-\infty}^{+\infty} U(x,z)\, \varepsilon^{+j2\pi fx} dx \qquad (4)$$

where f is the real transform variable. We can Fourier transform equation 3 directly,

$$\frac{\partial^2 T}{\partial z^2} + (k^2 - 4\pi^2 f^2)T = 0$$

which has the solution

$$T(f,z) = T(f,0)\ \varepsilon^{-jk_z z} \tag{5}$$

$$\left.\begin{aligned} \text{where } k_z^2 &= k^2 - k_x^2 \\ \text{and } k_x^2 &\equiv 4\pi^2 f^2 \end{aligned}\right\} \tag{6}$$

Superficially equation 5 looks like the standard formulation in terms of a plane wave spectrum. It is, however, important to appreciate that T(f,z) is not a homogenuous plane wave; we will see later that it is a truncated inhomogeneous plane wave, extending over a part of the (x,z) plane only.

We can now apply the inverse of equation 4 to the T(f,z) found in equation 5, with the final result that,

$$U(x,z) = \int_{-\infty}^{+\infty} T(f,z)\ \varepsilon^{-j2\pi fx}\ df \tag{7}$$

Equations 4-7 suffice to compute U(x,z) from U(x,0); it is the basis of the results presented in the next section.

Nature of the Elementary Waves

We have seen that the Fourier optics approach appears to remain rigorously correct in the loss present case - but that the elementary waves into which the field is expanded are the T(f,z), instead of the usual plane waves. In this section we will explore their nature, and indicate the ways in which they differ from plane waves. In the lossless case one can resolve the wave vector k into components,

$$\left.\begin{aligned} k_z &= k\cos\theta \\ k_x &= k\sin\theta \end{aligned}\right\} \tag{8}$$

We retain this formulation for the lossy case, but now regard equation 8 as a <u>definition</u> of the complex quantity θ,

$$\theta \equiv \theta' + j\theta'' \tag{9}$$

From equations 6,8,9, we can then write,

$$k_x \equiv 2\pi f = (\beta - j\alpha)\ \sin\ (\theta' + j\theta'') \tag{10}$$

We have defined f as a real variable, so that the imaginary part of the right hand side of equation 10 must be zero. Expanding the sine function, we find,

$$\tanh\ \theta'' = \frac{\alpha}{\beta}\ \tan\ \theta' \quad \text{for} \quad |\theta'| \leqslant \tan^{-1}\left(\beta/\alpha\right) \tag{11}$$

and $k_x = \beta \sin\theta'\ \cosh\theta'' + \alpha\ \cos\theta'\ \sinh\theta''$ (12)

Similarly, from equation 8 we find

$$k_z = \beta\cos\theta'\ \cosh\theta'' - \alpha\ \sin\theta'\ \sinh\theta''$$

$$- j(\alpha\ \cos\theta'\ \cosh\theta'' + \beta\ \sin\theta'\ \sinh\theta'') \tag{13}$$

From equation 7 we see that the elementary contribution to the spectrum is $dU(x,z)$

$$dU(x,z) = T(f,0)\ \varepsilon^{-jk_x x}\ \varepsilon^{-jk_z z} \tag{14}$$

The equi-phase front of this wave can be found by putting

$$k_x x + \text{Re}\ (k_z z) = 0 \tag{15}$$

It is clear therefore that the elementary waves retain planar wave fronts. Using equation 11, it is readily shown that this implies

$$-(\frac{x}{z}) = \frac{\beta^2 \cot\theta' - \alpha^2 \tan\theta'}{(\alpha^2 + \beta^2)} \equiv \cot\ \phi \tag{16}$$

where ϕ is the angle between the normal to the wave front and the z axis - i.e. the equivalent of θ in Figure 1. It is therefore illuminating to write equation 16 in the form,

$$\cot\phi = F.\cot\theta' \tag{17}$$

$$\text{where } F = \left(1 - (\frac{\alpha}{\beta})^2\ \tan^2\theta'\right)\left(1 + (\frac{\alpha}{\beta})^2\right)^{-1} \tag{18}$$

The factor F is then a measure of the extent by which the direction of propagation of the elementary waves deviate from the direction of the corresponding plane wave, when losses are neglected.

For $(\alpha/\beta)^2 << 1$ we can approximate equation 18

$$F \simeq 1 - (\frac{\alpha}{\beta})^2\ \sec^2\ \theta' \tag{19}$$

Although equation 19 assumes "low losses", it is worth noting that it is a good approximation to cases which one normally regards as a very high loss situation (-such as a loss of 1dB per wavelength). It is of particular interest to obtain an expression for the actual angular deviation $\delta = (\phi-\theta')$ arising from the presence of losses. Assuming this deviation is relatively small, we find from equations 18 and 19,

$$\delta \equiv \phi-\theta \simeq \left(\frac{\alpha}{\beta}\right)^2 \tan\phi \simeq \left(\frac{\alpha}{\beta}\right)^2 \tan\theta' \qquad (20)$$

$$\frac{\delta}{\phi} << 1$$

We have already seen that the wavefronts of the elementary wave remain plane. However, the amplitude of the field varies along the phase front. If we denote distance along the phase front by z', such that

$$z = z'.\sin\phi \qquad (21)$$

$$dU(z') = dU(0)\ \varepsilon^{-\mathrm{Re}(jk_z z'\sin\phi\ )}$$

From equation 13 we obtain

$$dU(z') = dU(0)\,\varepsilon^{-z'\sin\phi(\alpha\cos\theta'\ \cosh\theta'' + \beta\sin\theta'\ \sinh\theta'')} \qquad (22)$$

We see that the amplitude decays exponentially along the phase front. It is important to appreciate that the waves are defined only for $z > 0$, and hence $z' > 0$. There is therefore no problem with unlimited growth in the minus z' direction. The constant amplitude fronts of the waves are the planes z = constant. The waves can be regarded as physically wholly valid; they satisfy the wave equation; one could devise a source which would launch a single such wave. In some respects one can ascribe a greater degree of physical reality to these waves than to the plane waves used in the analysis of lossless systems, by the fact that the energy of a finite amplitude elementary wave evaluated along a phase front is also finite.

Figures 2(a) through (d) show some examples of the elementary waves, and indicate the changes as the relative loss, as measured by $(\alpha/\beta)^2$ is progressively increased. We see that the waves are decaying in the z-direction, and hence also along the phase fronts, but have a constant amplitude in the x-direction. The direction of propagation of the waves is perturbed by the loss factor.

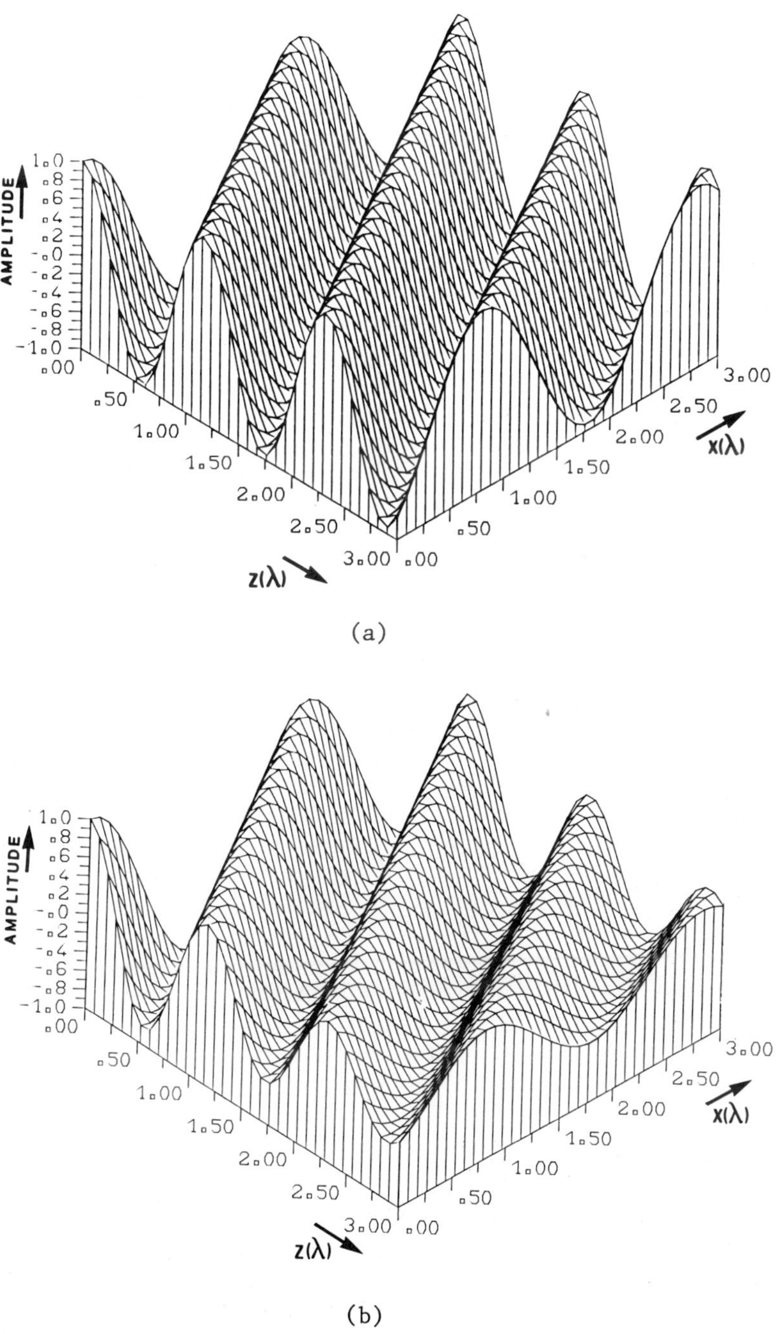

(a)

(b)

Fig. 2. One component of angular spectrum ($\phi = 30^{o}$) for different values of $(\alpha/\beta)^2$: (a) 0; (b) 0.002; (c) 0.02; (d) 0.2.

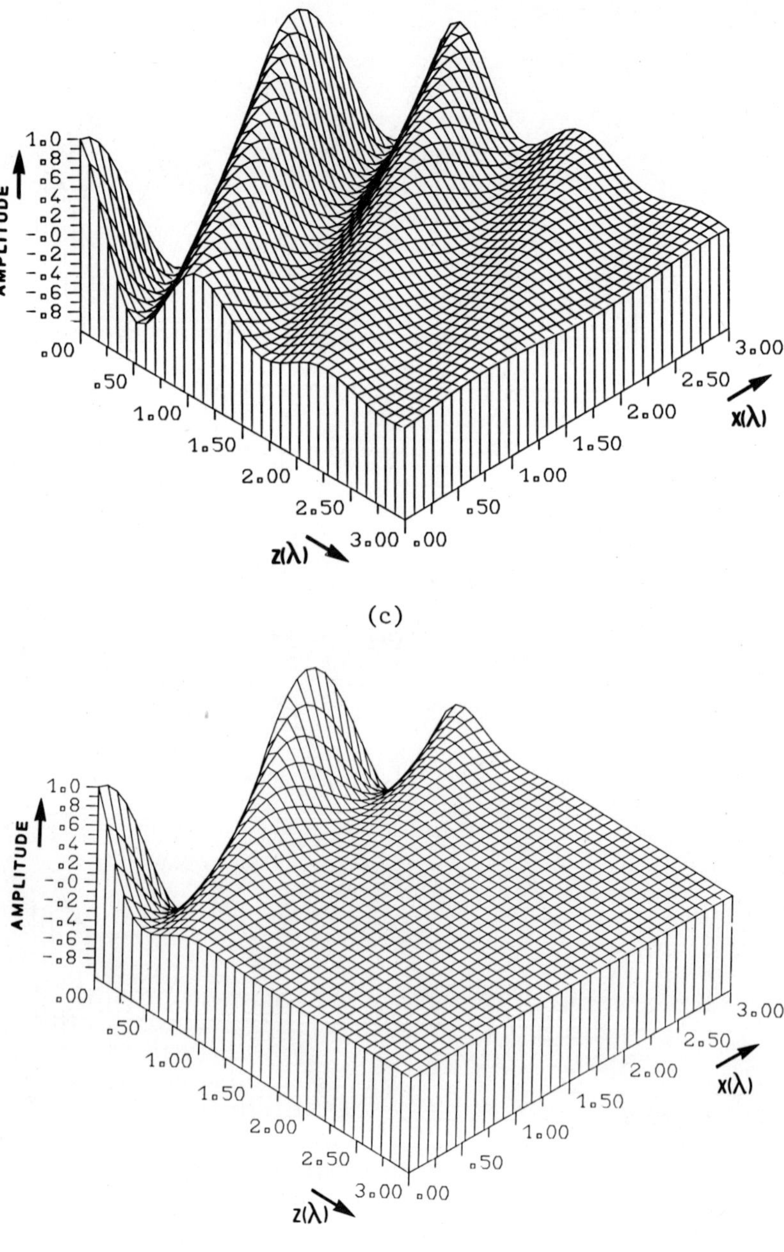

(d)

Fig. 2 (Continued).

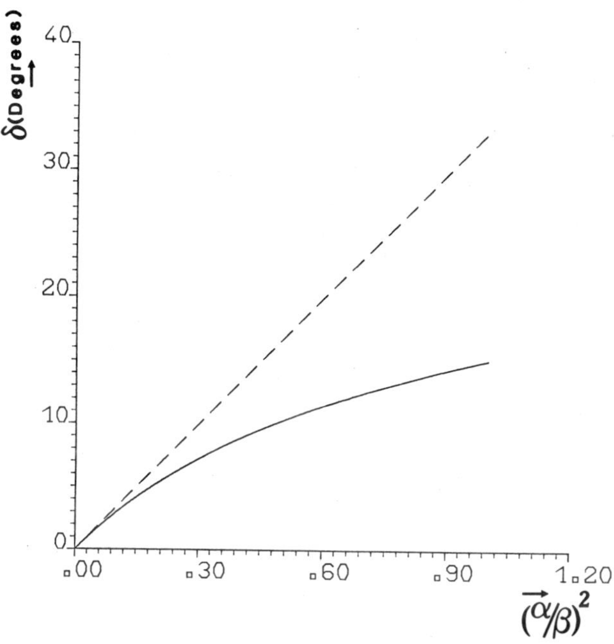

Fig.3. The angular deviation δ as a function of $(\frac{\alpha}{\beta})^2$. Dashed line, for analytical approximate case. Solid line, analytical exact case.

This deviation is shown in Figure 3, and the result is compared with the approximate analytical solution of equation 20. The approximation is seen to be reasonably valid for values of the loss factor $(\alpha/\beta)^2<0.1$, which covers most of the situations encountered in practice.

Low Loss Approximation for the Elementary Waves.

We can obtain very simple results for the low loss case $\alpha\ \beta<<$ Under these conditions, equation 17 shows that

$$\left.\begin{aligned} &\theta' \simeq \phi \\ \text{Also, } &\cosh\theta'' \simeq 1 \\ &\sinh\theta'' \simeq \tanh\theta'' = \frac{\alpha}{\beta}\tan\theta' \end{aligned}\right\} \quad (23)$$

Applying these approximations to equations 12 and 13, we find, to the first order in α/β,

$$k_x \simeq \beta \sin\theta'$$

$$k_z \simeq \beta(\cos\theta' - j(\frac{\alpha}{\beta})\sec\theta') \tag{24}$$

The elementary waves of equation 14, then take the form

$$dU(x,z) = T(f,0)\varepsilon^{-j\beta(x\sin\theta' + z\cos\theta')}\varepsilon^{-\alpha z/\cos\theta'} \tag{25}$$

With reference to Figure 1, we can then write the decay term in the form, $\exp(-\alpha \ell_2)$.

We can therefore conclude that this is the correct form for the attenuation term - in agreement with that used by Alais[4]. The controversy as to whether high spatial frequencies decay faster or slower than low spatial frequency components is therefore resolved. We conclude further that we would expect a loss of resolution when focussing a beam in an attenuating medium - a conclusion which is confirmed by the rigorous computations presented in the next section.

It remains to discover up to what values of parameter α/β, the results derived from the approximations of equation 23 remain valid. If we describe the attenuation by a term $\exp[-\alpha(\phi).z]$, we can compute $\alpha(\phi)/\alpha(0)$, as a function of ϕ. The results are shown in Figure 4 for three values of α/β. We see that for values of α/β

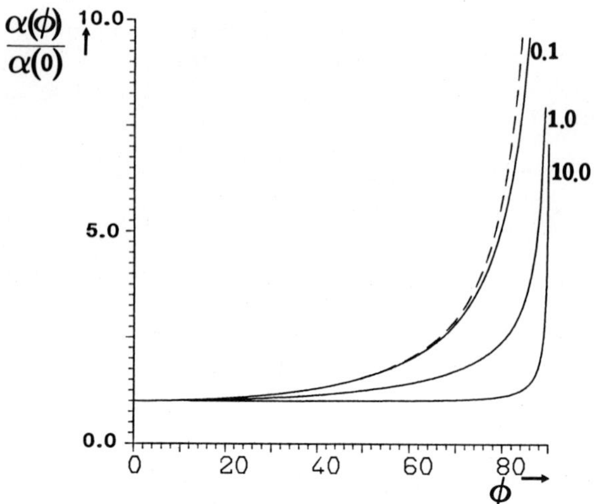

Fig.4. Variation of $\alpha(\phi)/\alpha(0)$ with ϕ for different values of α/β. Dashed line for analytical approximate case.

as large as 0.1, the approximation remains reasonably satisfactory over the whole range of ϕ. It is unlikely, at least in acoustic microscopy, that one will encounter much larger losses.

FOCUSSING IN HIGHLY ABSORPTIVE MEDIA

Our aim was to establish the primary effect of losses on the action of a lens in an absorptive medium. Using the formulation established, and representing the lens by an aperture illuminated by a uniform amplitude, parabolic phase waves , we can obtain the field distribution in the focussing region. Two typical cases are illustrated in Figure 5. In each case we have also presented the distribution in the paraxial focal plane, Figure 6. The most striking effect is, of course, the direct effect of the attenuation whereby the height at the focus is, for large attenuations, substantially less than that of the sidelobes of the distribution, further "upstream". Clearly, if one is looking at water-like objects, there will be a marked effect on the depth of field. In examining solid surfaces by the V(z) technique[5,6], one would also expect a marked change as a result of the coupling medium attenuation. We also note that the width of the main lobe changes very little, whilst there is a marked reduction in the sidelobe levels. The model we have used is that of a thin lens. This, we believe, is adequate to grasp the main features of focussing in lossy liquid media. However, if one is concerned with subsurface imaging in a solid, the situation can be entirely different. The path length of the peripheral rays in the liquid can be substantially shorter than for axial rays. This results in a marked apodisation which in extreme cases corresponds to focussing by means of a thin annular lens. Such situations can be grasped using a Fresnel-Kirchoff approach[7].

CONCLUSIONS

We have presented an approach which extends the methods of Fourier optics to the case of propagation in highly absorptive media. The elementary waves from which the solution is constructed are truncated inhomogeneous plane waves.

By examining the limiting case, we have shown that the correct way of taking losses into account in a perturbation approximation, leads to the conclusion that high spatial frequencies are preferentially attenuated with respect to low spatial frequencies. As a result, the main lobe of the field distribution in the focal plane of a lens will be broadened by the presence of losses. However, for losses encountered in high resolution acoustic microscopy, this effect is insignificant. The reduction in sidelobe level is, however, much more marked.

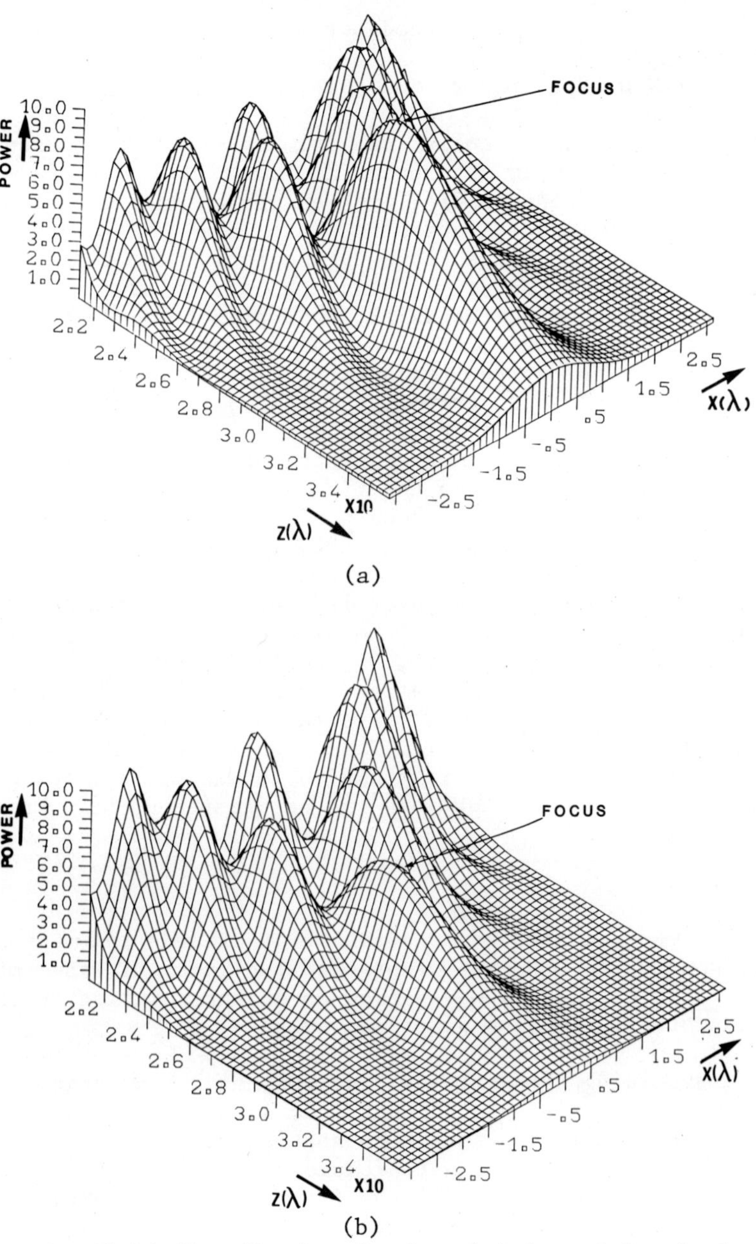

(a)

(b)

Fig.5. The field distribution in the vicinity of focal plane of a thin lens, (radius 30λ, aperture 52λ, velocity ratio 7.3 (a) α = 0.3 dB/λ and (b) α = 1.0 dB/λ. λ is the wavelength in the focussing medium.

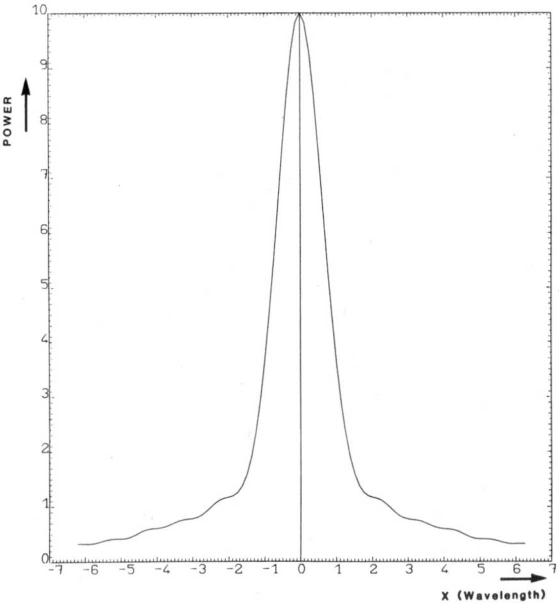

(a)

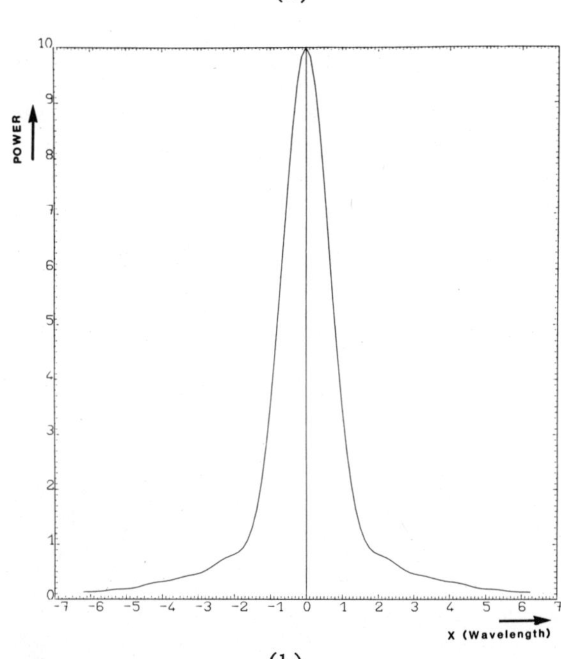

(b)

Fig.6. The field distribution at the paraxial focal plane of lens described in Figure 5 (a) $\alpha = 0.3\text{dB}/\lambda$ and (b) $\alpha = 1.0\text{dB}/\lambda$.

ACKNOWLEDGEMENTS

The authors would like to express their thanks to their colleagues, Dr Brian Davies and Dr Kumar Wickramasinghe, for a number of helpful discussions.

REFERENCES

1. H. K. Wickramasinghe, "Contrast and Image Performance in the Scanning Acoustic Microscope", J. Appl. Phys., 50 (2), pp 664-672 (1979).
2. P. Alais et P.Y. Hennion, "Etude par une methode de Fourier de l'interaction non linéaire de deux rayonnements acoustiques dans un fluide absorbant. Cas Particulier de l'emission Parametrique", Acoustica, 43 (1), (1979).
3. A. Rosencwaig, "Thermal-wave Imaging and Microscopy", Scanned Image Microscopy, Academic Press, pp 291-317 (1980).
4. P. Alais, P.Y. Hennion et M. Lagreve, "Théorie Fourier de la propagation linéaire et non-lineaire dans un fluide absorbant, applications à la transduction parametrique", Journal de Physique, 40 (11), (1979).
5. R. D. Weglein and R. G. Wilson, "Characteristic Material Signatures by Acoustic Microscopy", El. Letters, 14 (12), pp 352-354 (1978).
6. A. Atalar, "An Angular Spectrum Approach to Contrast in Reflection Acoustic Microscopy", J. Appl. Phys., 49 (10), (1978).
7. M. Islam, M. Nikoonahad and E. A. Ash, "Large Aperture Lens Focussing of Acoustic Waves in Highly Absorptive Media, using Fresnel Kirchoff Approach". Unpublished.

SCANNING PHOTOACOUSTIC MICROSCOPY AND DETECTION OF SUBSURFACE STRUCTURE

Zhang Shu-yi, Yu Chao, Miao Yong-zhi, Tang Zheng-yan and Gao Dun-tang

Institute of Acoustics, Department of Physics, Nanjing University, Nanjing, People's Republic of China

ABSTRACT

A scanning photoacoustic microscopy with a surface imaging resolution of about 2μm has been established. Some images of the subsurface structure of metals and layered materials have been obtained by measuring the amplitude or the phase angle of the photoacoustic signal. These preliminary results show potential applications of this instrument in nondestructive testing of materials.

INTRODUCTION

Several investigators [1-3] have studied the microstructure of solid surface and subsurface inhomogeneities by using photoacoustic techniques on a microscopic scale. In their methods, a focused laser beam is scanned over the surface of the sample in a raster-scan fashion, and the resultant photoacoustic signal is used to form an image on a synchronously scanned image tube. The scanning photoacoustic microscope (SPAM) has been developed into a valuable NDE technique for detecting flaws and inhomogeneities in solid surfaces or subsurfaces [4-5].

In this paper, we report a SPAM system which attains a resolution of about 2μm for photoacoustic imaging of integrated circuits. Some images of the subsurface structure of metals and layered materials have been obtained by measuring the amplitude or phase of the photoacoustic signal.

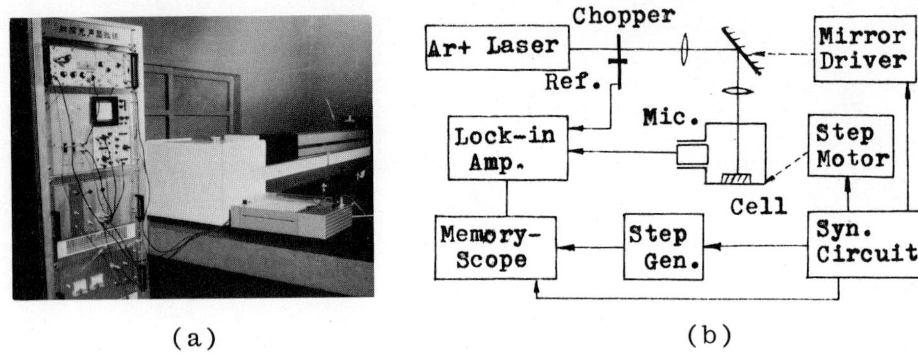

(a) (b)

Fig.1. Photograph (a) and block diagram (b) of SPAM

SCANNING PHOTOACOUSTIC MICROSCOPY SYSTEM

The system of our SPAM is given in Fig.1, (a) is the photograph of this instrument; (b) shows a block diagram of it. An Argon ion laser of 1W is modulated by a mechanical chopper at audio frequencies in the range of 25 - 1500 Hz. In order to obtain the focussing and X-scanning optical beam, a deflection vibration mirror is placed between two sets of convergent lenses. The waist of the focused optical beam has a diameter of about 3μm. The solid sample is placed in a brass photoacoustic cell (0.4c.c.). Y-scanning of the optical spot is accomplished by moving the cell with a stepper motor. A high sensitivity (<5 mv/μb) electret microphone is made specially and arranged to receive the photoacoustic signal. In-phase or quadrature components of the signal may be taken individually by lock-in amplifier to form the amplitude or phase angle photoacoustic image on the memory oscilloscope. The scanning circuit for the synchronous control of scanning optical spot as well as the electric beam of imaging tube is designed specially to improve the quality of imaging.

RESULTS AND DISCUSSION

By using this SPAM, some photoacoustic images of materials have been obtained.

Surface Structure of Solid Sample

A set of lines with width of 2μm of the integrated circuit was obtained as given in Fig.2. The frequency of chopper was 800 Hz. The width of line was much less than the thermal wavelength in the solid. It shows that the surface resolution of this system is mainly determined by the size of focused optical spot.

Fig.2. PA image of lines with width 2μm

Subsurface Structure of the Materials

Though SPAM imaging has been used for studying optical surface structures, its most useful applications are for detecting the subsurface structure of materials. In the later case, the imaging is dominated by thermal wave effects.

A. Samples of Metals

A brass sample was drilled with a hole of 1mm diameter parallel to the surface and at a distance of 0.2mm from the surface, as shown in Fig.3(a). The photoacoustic images of this sample is shown in Fig.3(b) and (c). Photograph (b) is the PA signal distribution pattern represented in curves and (c) is the structure image displayed according to the brightness of each image element. From these pictures, we also found a small defect near the drilled hole as indicated by the arrows.

An aluminium sample has been drilled with two 0.7mm diameter holes which cross each other and depart from the surface at 0.2mm and 0.5mm respectively. In addition, a black line was painted on the upper surface as shown in Fig.4(a). The structure images of this sample obtained by amplitude and phase angle imaging are shown in Fig.4(b) and (c). From these photographs, we can conclude as Busse[6] that the amplitude image (b) displays the structure of surface and very near subsurface; while the phase angle image (c) exhibits mainly the substructure. Therefore, the structure of either the surface or the subsurface can be distinguished by different imaging methods.

In the subsurface images, the modulation frequency was 70 Hz.

B. Layered Sample

The layered sample was composed of an integrated circuit of large power transistors with a thin silver layer covering.

Its profile structure is shown in Fig.5(a). The line width was 90μm with left and right intervals of 110μm and 170μm respectively. The thickness of Al electrode was about 1μm. The amorphous silicon film grown on the substrate was also about 1μm thick. The silver layer was about 10μm thick and was covered by chemical method. Consequently, the Al lines could not be observed by optical microscopy. However, the photoacoustic image of Al lines could be obtained using SPAM imaging as shown in Fig.5(b). This was obtained by phase angle detection of frequency of 1kHz.

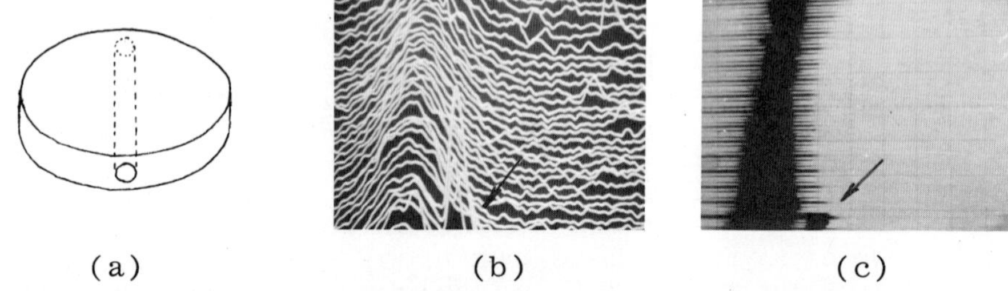

(a) (b) (c)

Fig.3. PA signal distribution pattern (b) and the structure image (c) of a brass sample (a)

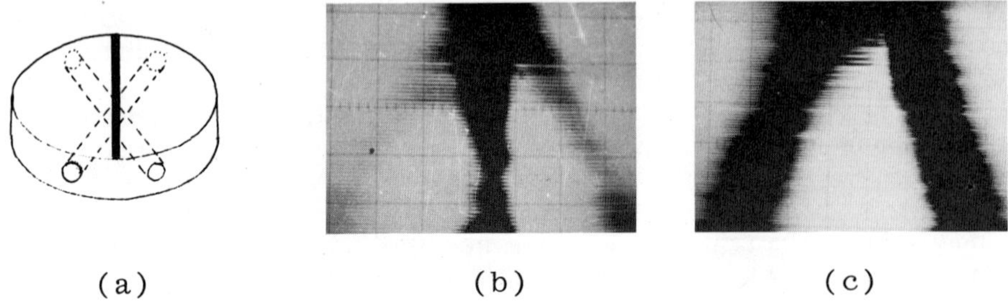

(a) (b) (c)

Fig.4. PA magnitude (b) and phase (c) image of an aluminium sample (a)

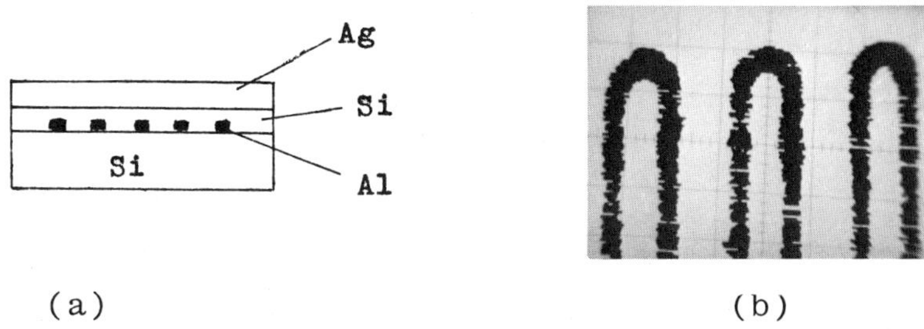

Fig.5. Substructure (b) of layered sample (a)

In conclusion, the technique described should provide a valuable tool for nondestructive testing of the subsurface structure of opaque materials.

ACKNOWLEDGEMENTS

The authors wish to thank Professor Wei Rong-jue and Wu Wen-qieu for their very interesting and helpful discussions.

REFERENCES

1. Y H Wong et al., Appl. Phys. Lett. 32:538 (1978).
2. H K Wickramasinghe, Appl. Phys. Lett 33:912 (1978).
3. M Luukkala & A Penttineu, Elect. Lett. 15:325 (1979).
4. L J Inglehart et al., J. NDE. 1:287 (1980).
5. R A McFarlane, Ultrasonic Symposium Proceedings 628 (1980).
6. G Busse, Ultrasonics Symposium Proceedings 622 (1980).

APPLICATIONS OF ACOUSTIC MICROSCOPY IN THE SEMICONDUCTOR INDUSTRY

A J Miller

The General Electric Company, p.l.c.
Hirst Research Centre
Wembley England

1 INTRODUCTION

The scanning acoustic microscope has now been under development for several years in university research groups. There has been some interest from the industrial and medical sectors, and commercial versions are starting to appear. This paper is concerned with initial experience with a scanning acoustic microscope in one industrial laboratory, and describes likely future requirements for such instrumentation.

2 THE MICROSCOPE

The instrument has been set up in a Device Diagnostics area because it was thought to be one of several potentially valuable new techniques for device and materials studies. The target of the design was a 1.5 GHz reflection mode system, giving a resolution of 1 μm with water as the coupling medium. This is a good compromise of lateral resolution and depth penetration for current semiconductor devices, and for other objects gives images which can easily be compared with optical microscopy.

The microwave electronics is most suited to the 1 GHz to 2 GHz range, and the images presented here were taken at 1 GHz. The details generally follow previous practice, using a PIN modulator to select the sample echo and a fast sample-and-hold circuit to form the video signal, which is stored in an analogue scan converter memory.

The sapphire lens used in the current studies is 120 μm in radius and the lens head is heated to 40°C. In the near future it is hoped to use a 65 μm lens at 1.5 GHz, employing a ZnO transducer deposited using in-house facilities. The fast scanning movement of the lens head at about 40 Hz is accomplished by loudspeaker coil and leaf springs, as in Stanford designs of about 1977, and the slow scan movement is achieved by a motor and belt-drive translation stage underneath the specimen stage.

The specimen stage has a size of 25 mm x 25 mm and is mounted on gimbals. There is no associated optical microscope, and as yet there is no sophisticated means of vibration isolation. Focussing in the z direction, which is vertical, is by differential micrometer and by piezoelectric pusher. The image displayed is of an object area approximately 300 μm by 230 μm although alternative ranges should soon be available. At present we are detecting amplitude images rather than phase images; facilities for recording V(z) curves will be added shortly.

The instrument was seen as a means for rapid, 'non-destructive' examination of subsurface features and as a method of monitoring elastic properties, which may be related to important electrical properties of devices and materials. After the initial development and adjustment period, a number of types of sample have been examined. The microscope has indeed given the expected lateral resolution of approximately 1.5 μm at 1 GHz.

3 EXAMPLE OF IMAGES OBTAINED

3.1 Devices

The earliest samples were bulk silicon CMOS integrated circuits. Simple parts of the circuit with known structure are of value in establishing the value of acoustic microscopy, and under this heading come markings representing different production stages

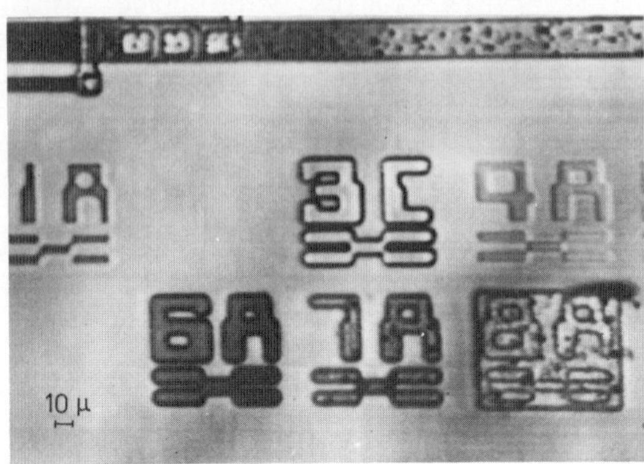

Fig 1a Acoustic image of processing marks on bulk silicon CMOS integrated circuits

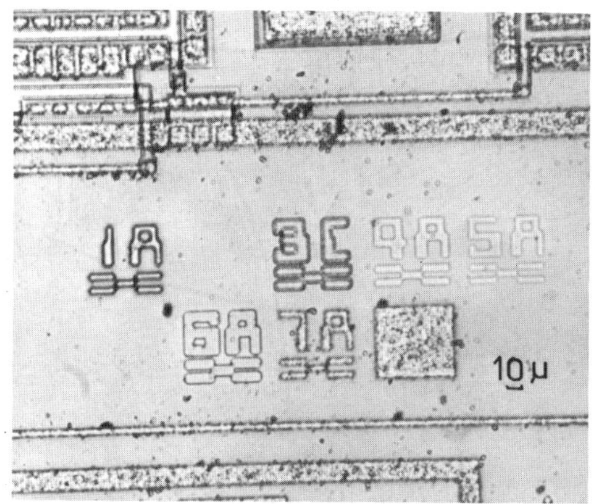

Fig 1b Optical image corresponding to 1a

(Figure 1) and those used for alignment purposes (Figure 2). Series of micrographs have been taken as a function of the z-coordinate and the classic reversals of contrast have been seen, although the differing nature and thicknesses of the surface layers can make contrast observation more complex. Known subsurface detail, such as polysilicon tracks underneath aluminium tracks, is easily seen. Consider however Figure 3. The tracks are obvious, but there are a few rectangular areas darker than the surroundings. It would be attractive to suppose that these areas were of different doping level, and indeed it has been proposed by Keyes (1982) that such an effect should be just observable only in the acoustic microscope, if the doping level was high ($>10^{19}$ cm^{-3}). It is, to digress briefly, thought more likely that electron thermal-wave imaging (Rosencwaig 1981) would be more useful here. However, in Figure 3, we are examining the surface of a finished device, which though macroscopically very flat has, after several processing steps, a highly contoured surface, and the dark areas must certainly be due to topographic features rather than material properties. Areas known to be ion-implanted have not shown acoustic distinguishability so far. It must also be mentioned that finished devices, as opposed to those investigated from steps in processing, have an overall passivation layer which must further complicate the unambiguous interpretation of acoustic micrographs.

The other major device technology under examination has been the silicon on sapphire (SOS) system. Slices are supplied with a layer of silicon grown on a sapphire surface, but the silicon not in the immediate area of active devices is removed. The quality of the silicon/sapphire interface is important in the electrical properties of devices, and it is thought that this should be amenable to acoustic examination as the silicon is less than a micron thick. Investigations by Wilson and Weglein (1980) have

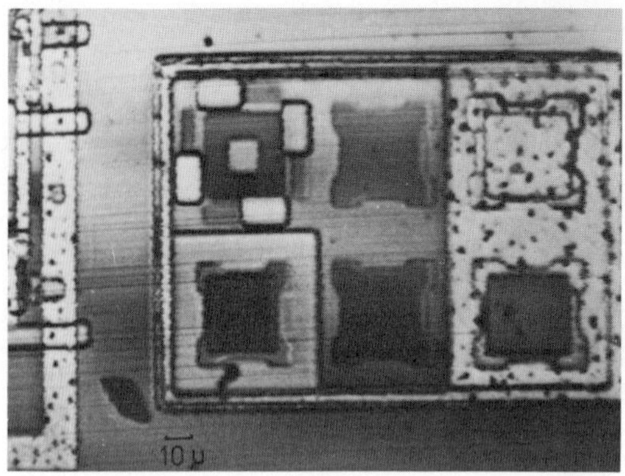

Fig 2a Acoustic image of alignment marks on bulk silicon CMOS integrated circuit, focussed at surface

Fig 2b Subsurface image corresponding to (a)

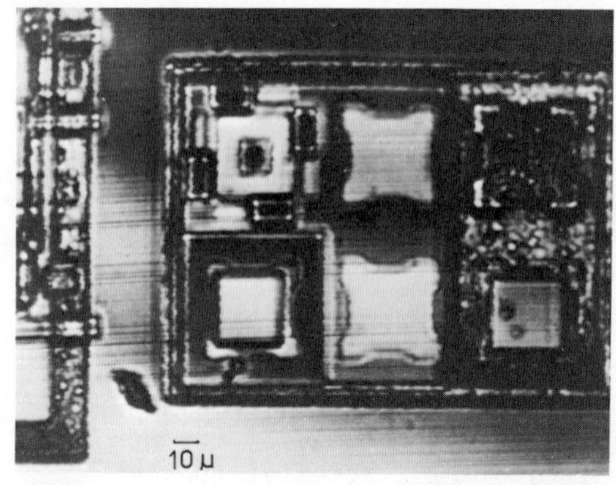

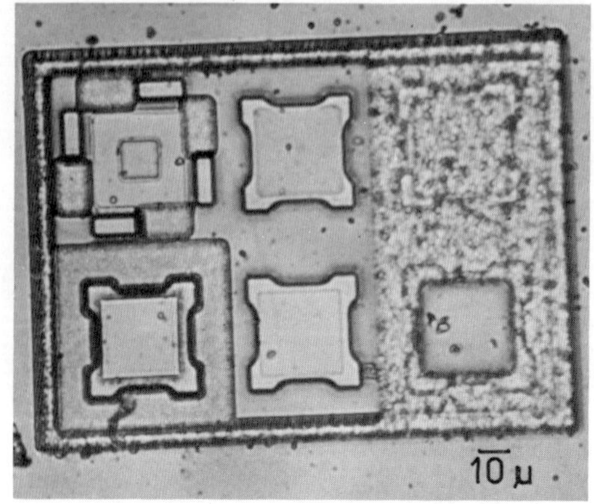

Fig 2c Optical comparison

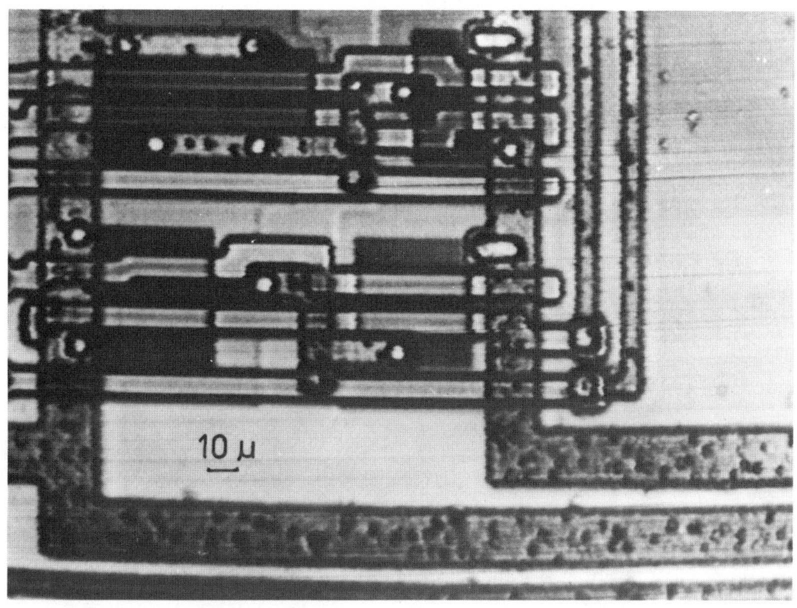

Fig 3 General area on CMOS integrated circuit

shown that for thin film systems such as this one, the relationship between the layer thickness and the acoustic wavelength may conspire to reveal or obscure interface detail and so this technique is not universally applicable. Examples of imaging a SOS test insert chip are shown in Figure 4. The first of these shows resolution of 1 µm features and the others show test devices. Successful operation of devices with minimum power consumption depends on leakage currents being low, and Jipson (1979) has observed acoustic differences in the gate regions of devices with normal and high leakage currents. Jipson's V(x,z) type of acoustic scan is to be implemented with a view to exploring such effects further. It would thus be of interest to examine the gate regions in depth, but it has been observed that fringes appear on SOS chip images in particular, near sharp changes in acoustic impedance at edges. This may be related to the very high contrast obtainable in such images, which has made them most valuable for testing the microscope itself, but is a confusing feature which is unlikely to lessen in its effect at higher resolutions. When focussing beneath the surface, all edges diffract and obscure real detail near them, and so very narrow gates tend to disappear completely.

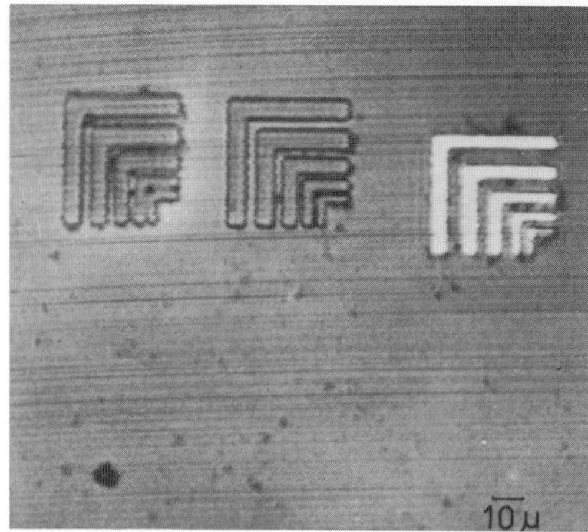

Fig 4a Lithography marks on SOS test insert. The smallest mark has a linewidth on 1 µm

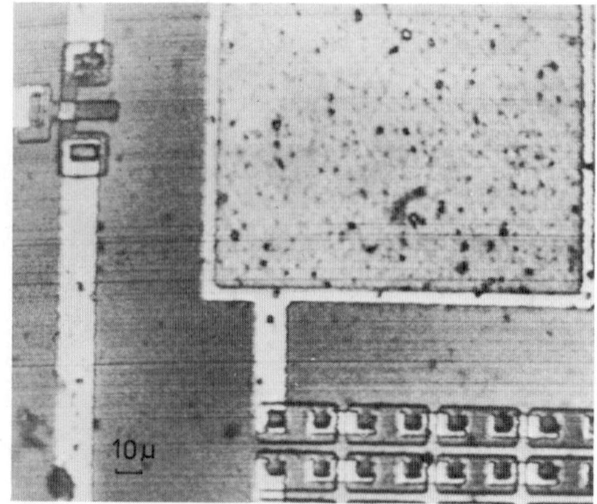

Fig 4b General area on SOS test insert

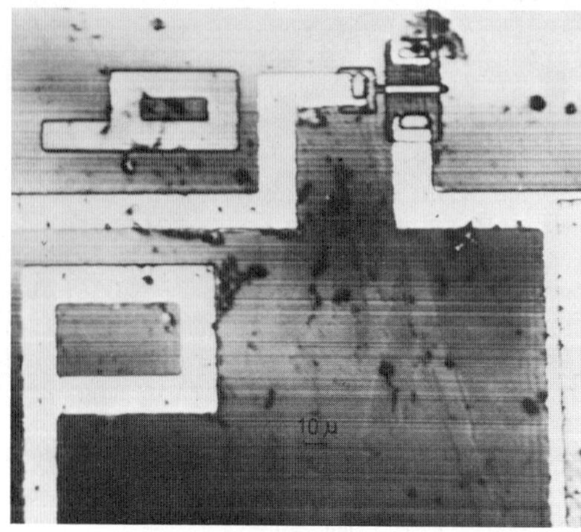

Fig 4c General area on SOS test insert, showing one test device structure

3.2 Metals

The usual materials use claimed for the scanning acoustic microscope is the acoustic analogue of the metallurgical microscope. It can be readily apparent that the sources of contrast are quite different in the acoustic and optical cases. We can see the grains, but their relative brightness or darkness has changed. While it has been shown that acoustic microscopy can see grains through a polished layer (or a thin surface layer of an acoustically isotropic material), and this has been tried, it is believed the main use will be on normally prepared surfaces which are difficult to observe optically.

Fig 5a Acoustic image of medium carbon steel showing ferrite and pearlite

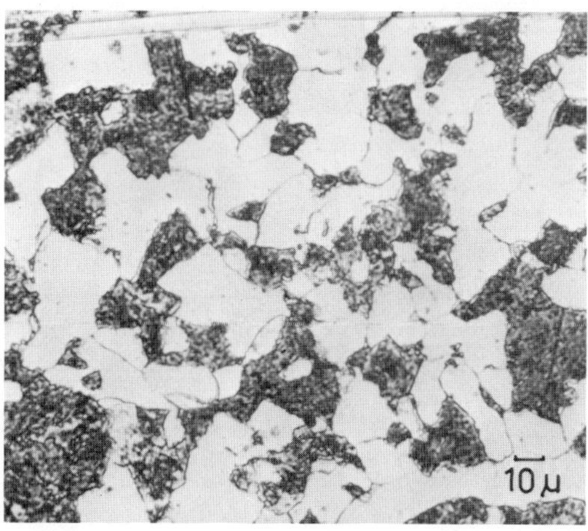

Fig 5b Corresponding optical image

Figure 5 shows an etched ferrite and pearlite surface, and Figure 6 shows another quite similar medium steel sample after heat treatment. In some metallurgical pictures the grain boundaries stand out rather more than in optical metallography, and this must be due to the leaky surface wave necessary for image formation in acoustic microscopy encountering the grain boundary and causing additional interference phenomena, in a similar way to the study by Yamanaka and Enomoto (1982) on cracks in surfaces. It appears that this is an acoustic effect rather than an artefact of the scanning or video electronics, as it does not change when the scan blanking is reversed i.e. the information is recorded on the opposite half cycles of the lens traverse in the X (fast) direction.

3.3 Langmuir-Blodgett films

There is considerable current interest in the deposition of successive monomolecular layers of organic molecules. These are known as Langmuir-Blodgett films and they are of potential use in devices as thin insulating layers, and in integrated optics as waveguides; one attraction is that films may be specifically tailored for their task by the deposition process.

Just as with visible light, one would expect these films to be very difficult to observe with the acoustic microscope since their thickness is usually less than 1000 Å. However, the acoustic microscope should be particularly sensitive to acoustic impedance

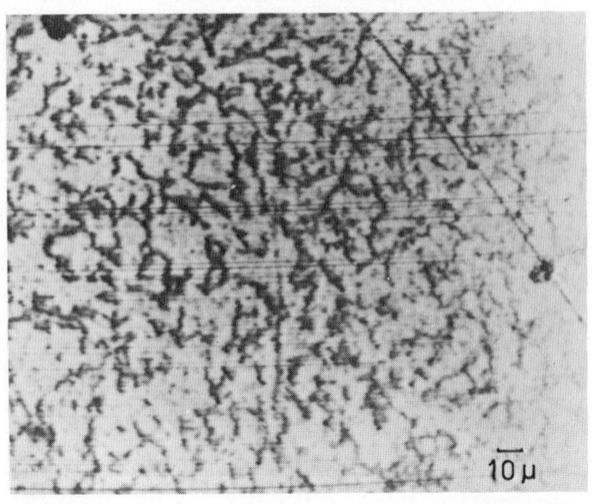

Fig 6 Similar medium carbon steel after heat treatment

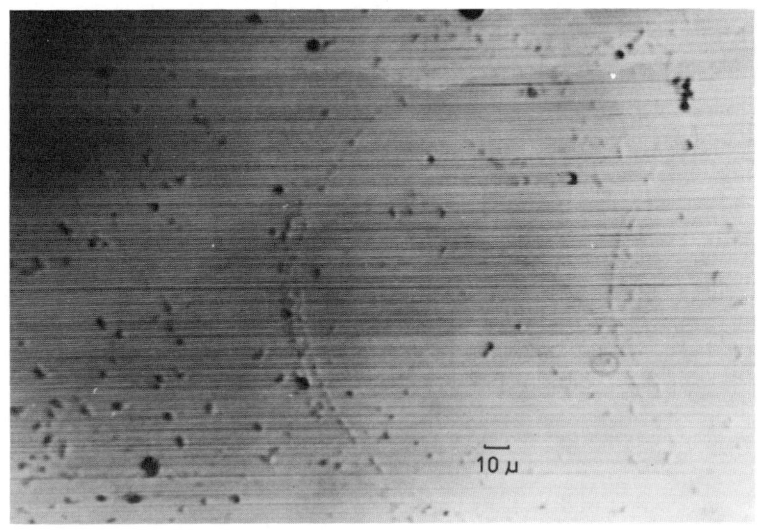

Fig 7 General area on Langmuir-Blodgett film

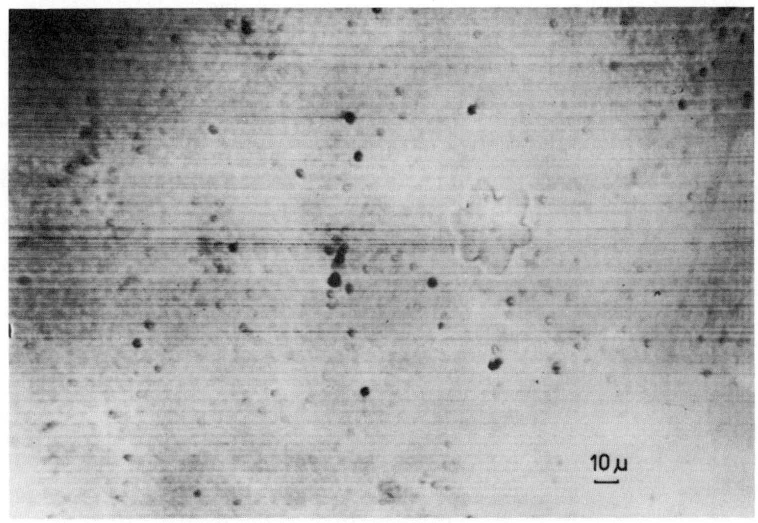

Fig 8 Unusual feature on polydiacetylene LB film revealed by acoustic microscopy

changes associated with thin layers. This was clearly shown by Bray et al (1980) in studies of the adhesion of thin films (1000 Å) of chromium on glass with a 2.6 GHz microscope in water.

The particular films of interest are polydiacetylene, where diacetylene is actually deposited on aluminium-coated glass slides and the films are subsequently polymerised. This is a system in which mechanical features of the films, caused by problems at the film deposition or polymer growth stages, might influence electrical properties in, for example, dielectric films. It was suspected that 'domains' of 1 μm to 300 μm in size might be observable. Optically, a good LB film is quite hard to observe : it is faulty films of various kinds that should be amenable to acoustic and optical examination.

This has turned out to be a different problem. No features of interest have been observed associated with the aluminium or glass, and one can take the microscope across a number of films without seeing any significant contrast variations. It was found that the focussing is quite critical, and that it is necessary to use the maximum enhancement of contrast currently available, which results in undue prominence of scanning lines across the image. Initial experiments showed high contrast features near the LB film edge, but this was a structure also having gold dots deposited over the LB film. More commonly, detail such as in Figure 7 is observed.

What we believe we are seeing is deposits of starting material globules or foreign particles which adhere particularly to the film edge area. Figure 8 shows an unusual feature, believed to be associated with the polymerisation process, as similar shapes have been observed optically. As an extreme example of acoustic thin film studies, it is expected to develop slowly in competition with other techniques such as photoacoustics and advanced optical methods.

4 ASSESSMENT OF FUTURE PROSPECTS

It is now comparatively easy to observe simple structures, such as the SOS chip already referred to, compare them with optical micrographs of various kinds, and think that the interpretation is trivial. It is however evident that the distinction between material and topography on some specimens may be troublesome. It could be argued that all the acoustic processes taking place are ultimately amenable to modelling, and that it should be possible to eliminate unwanted artefacts by a great deal of computing applied to a succession of frame-stored and depth-focussed images on an object consisting of materials whose elastic properties are entirely characterised, but this approach may be some years away, even if it could be justified. A clear visual impression possibly aided by colour-coding (Hammer and Hollis 1982) and supplemented by data from V(z) investigations of parts of the image is probably the right course in the immediate future.

Another point to be taken into consideration is specimen preparation. The limited depth of field means that, apart from any interesting features, the sample should be flat over the scanning area so that interference fringes are absent (though they may be occasionally useful for measurement purposes). This flatness and tilt requirement is particularly important in thin film interface studies, and gets worse at the highest frequencies, although there is little problem with integrated circuits.

Acoustic contrast in the images can be great or small for reasons already discussed, and in the present work it has been less easy to observe gold films on gallium arsenide than aluminium on sapphire, for example. There is probably a need for computer-enhanced contrast and intensity slicing to extract information that is present but not obviously available in the best 'average' image. This leads to a completely computer-driven system with image processing facilities, easily operable by someone other than the builder himself: commercial microscopes may achieve this within the next two years.

One of our investigations has highlighted the need for an intermediate resolution scanning acoustic microscope using, say 100 MHz waves and having greater depth penetration. In IMPATT devices, faults leading to device failure have been found by successive optical observation and polishing down. The position of what we want to see is known but completely out of range of the high-resolution instrument. A lower frequency microscope should also be very useful for the observation of the contact between a thin device and the heat-sinking material on which it is mounted. This is highly relevant to the problems associated with mounting and cooling power semiconductor devices, and some work on this topic, also employing phase imaging, has been published by Lee et al (1980).

Water has on balance been the preferred material for acoustic coupling, but this has caused corrosion problems with ferrous specimens; this must be seen to be a non-destructive method as far as possible. The microscope should be capable of accepting other working fluids if necessary.

5 CONCLUSIONS

From investigations so far carried out, the scanning acoustic microscope has enjoyed a wide measure of interest. For the solution of particular industrial problems, it seems there is a need for further work, some instrumental, but mainly to assist the useful interpretation of images. Theoretical work is necessary to develop further V(z) theory for more complicated layer structures, so that experience could be built up on what the image 'ought to look like'. Working in the opposite direction, given acoustic data, it should be possible to identify or otherwise assess substances from microscopic examination, after a normal image has indicated an area worthy of detailed examination.

ACKNOWLEDGEMENT

The author wishes to acknowledge the assistance of Dr H K Wickramasinghe, University College, London, in the design and construction of the scanning acoustic microscope.

REFERENCES

Bray, R C, Quate, C F, Calhoun, J and Koch, R: 1980, Film adhesion studies with the acoustic microscope, Thin Solid Films, 74:295-302

Hammer, R, and Hollis, R L: 1982, Enhancing micrographs obtained with a scanning acoustic microscope using false-colour encoding, Appl Phys Lett, 40:678-680

Keyes, R W: 1982, Device implications of the electronic effect in the elastic constants of silicon, IEEE Trans on Sonics and Ultrasonics, SU-29:99-103

Jipson, V B: 1979, Acoustic microscopy at optical wavelengths, PhD dissertation, Stanford University

Lee, C C, Wang, J K, Tsai, C S, Wang, S K and Hower, P: 1980, Detection and characterisation of alloy spikes in power transistors using transmission acoustic microscopy, in "Ultrasonic Materials Characterisation", H Berger and M Linzer, eds., NBS Publication 596, Gaithersburg : 387-391

Rosencwaig, A, and White, R M : 1981, Imaging of dopant regions in silicon with thermal-wave electron microscopy, Appl Phys Lett, 38:165-167

Wilson, R G, and Weglein, R D : 1980, Acoustic Material signatures using the reflection acoustic microscope, in: "Ultrasonic Materials Characteristion", H Berger, and M Linzer, eds., NBS Publication 596, Gaithersburg, 345-355

Yamanaka, K, and Enomoto, Y : 1982, Observation of surface cracks with the scanning acoustic microscope, J Appl Phys, 53:846-850

APPLICATION OF SCANNING ACOUSTIC MICROSCOPE TO THE STUDY OF FRACTURE AND WEAR

K.Yamanaka, Y. Enomoto and Y.Tsuya

Mechanical Engineering Laboratory
Namiki 1-2, Sakura-mura, Niihari-gun
Ibaraki-ken, 305 Japan

INTRODUCTION

The fracture toughness of structural materials are largely influenced by small flaws near the surface such as cracks, pores and inclusions. They had not been extensively investigated so far, because the observation technique has not been established. In this work, we describe two types of new applications of scanning acoustic microscope to the study of fracture and wear. First, detailed features of subsurface cracks in silicon nitride were observed and a new method to estimate the fracture toughness is proposed. It is also proposed that the measurement of attenuation of the leaky surface acoustic wave (leaky SAW) can be used to characterize the surface of steels with various heat treatment.

APPARATUS

The acoustic microscope used in this study was manufactured by Olympus Co. Ltd. The V(z) curve measurements were done by recording the peak value V of the detected video pulse while the lens is moved along the z axis perpendicular to the surface of specimens. A peak detector and a microcomputor connected to the microscope was used for this purpose. Signal averaging and correction of non linearity of amplifier gain was performed by the microcomputor.

CHARACTERIZATION OF SUBSURFACE CRACKS

Observation of Subsurface Cracks

A Vickers indentation was made on the lapped surface (Ra=0.02

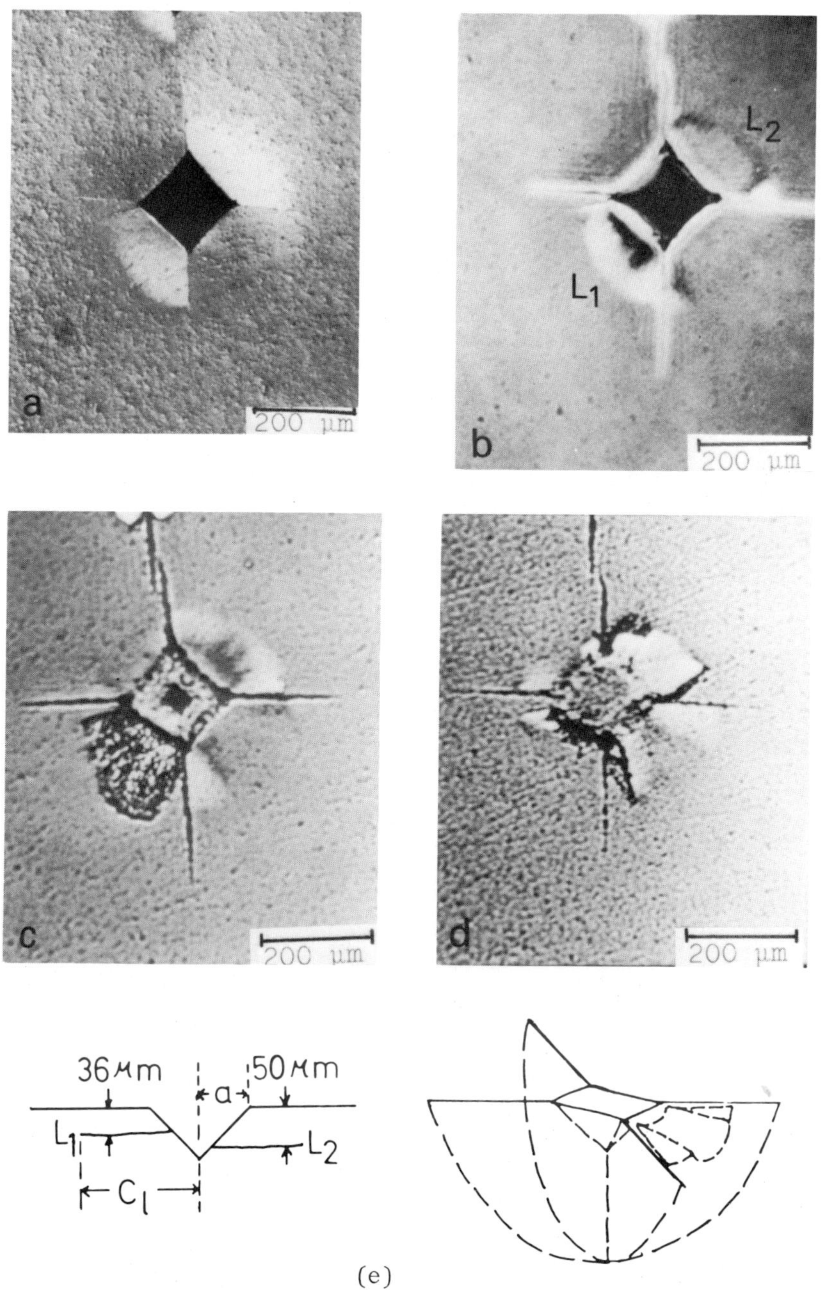

Fig. 1. Subsurface cracks in hot-pressed silicon nitride. (a) Optical image. (b)-(d) Acoustic Images (200 MHz, z=-30 µm). (b) As indented. (c) 36 µm, (d) 50 µm of surface layer removed. (e) Schematic representation of carcks.

where H is the hardness, a is the indent radius and ϕ is the ratio of H and the yield strength and f is an empirical function of C_1/a shown in Fig. 17 of their report. In our case, H and ϕ are known to be 18.9 GN/m^2 and 1.3, respectively. Since a and C_1 are determined from Fig. 1 (b) as 66 μm and 152 μm, C_1/a is 2.3 and f is 0.08. Thus, K_c is estimated with equation (1) to be 9.5 $MN/m^{3/2}$. This value is consistent in the order of magnitude with a reported value of 5 $MN/m^{3/2}$ (Evans and Wilshaw, 1976).

The lateral cracks are potential sites of material removal in solid particle erosion or abrasive wear. Therefore the K_c value eatimated above will provide a measure of erosion and wear resistance. The lateral cracks tend to propagate when an external stress is applied such as by grinding or lapping. Actually, the removal of surface layers by grinding and lapping slightly increased the length C_1 of the crack L_1 as seen by comparing Fig. 1 (b) and (c). Therefore the nondestructive measurement with acoustic images is necessary for the precise measurement of length of subsurface cracks.

CHARACTERIZATION OF SURFACE PHASE OF STEEL

Measurement of Attenuation of SAW

As is well known, the leaky SAW is generated by the incident converging beam, propagates on the surface, returns to the lens

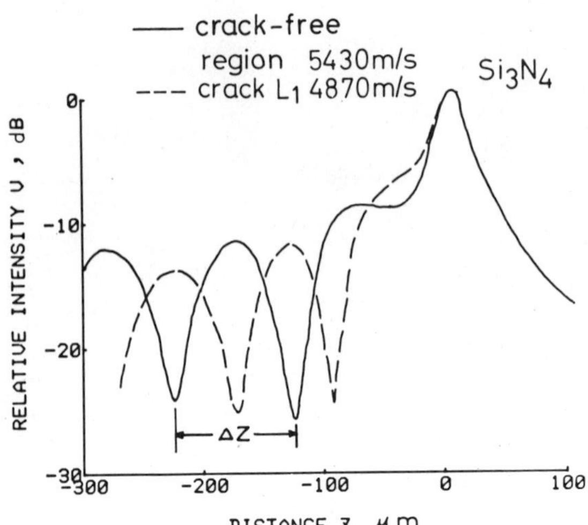

Fig. 2 V(z) curves of silicon nitride. The solid curve was taken on the crack-free region and the broken curve was taken on the region L_1 above the subsurface crack.

and interfere with the specularly reflected wave. This interference is responsible for the periodical dips in V(z) curves (Parmon and Bertoni, 1979). Then, if the attenuation of the leaky SAW due to the loss in a solid is high, the amplitude of the leaky SAW radiated to the lens will be small, as schematically shown in Fig. 3. In such cases the depth of dips in V(z) curves is expected to be small.

Though a ray model approach to estimate the attenuation of leaky SAW was reported in the literature (Weglein, 1982), it could not separate the attenuation due to the loss in solid from the attenuation due to the liquid loading. So, we proposed a method that can estimate the attenuation due to the loss in solid (Yamanaka, 1982), based on the angular spectrum approach (Atalar, 1978). The reflection coefficient R(θ) of a plane acoustic wave at a liquid-isotropic solid interface is given (Becker and Richardson, 1972) as

$$R(\theta)=\frac{\rho_1 v_l^*\cos^2 2\gamma_1/\cos\theta_1+\rho_1 v_t^*\sin_2 2\gamma_1/\cos\gamma_1-\rho v/\cos\theta}{\rho_1 v_l^*\cos^2 2\gamma_1/\cos\theta_1+\rho_1 v_t^*\sin^2 2\gamma_1/\cos\gamma_1+\rho v/\cos\theta} \qquad (2)$$

where ρ and ρ_1 are the density of liquid and solid, v, v_l* and v_t* are the velocity of longitudinal waves in the liquid, longitudinal waves in the solid and transverse waves in the solid, θ, θ_1 and γ_1 are the incident angle, the refraction angle of longtitudinal waves and the transeverse waves. When the solid attenuates acoustic waves, v_l* and v_t* become complex as

$$1/v_l^* = 1/v_l + i\,\alpha_s/\omega \quad , \quad 1/v_t^* = 1/v_t + i\,\alpha_s/\omega \qquad (3)$$

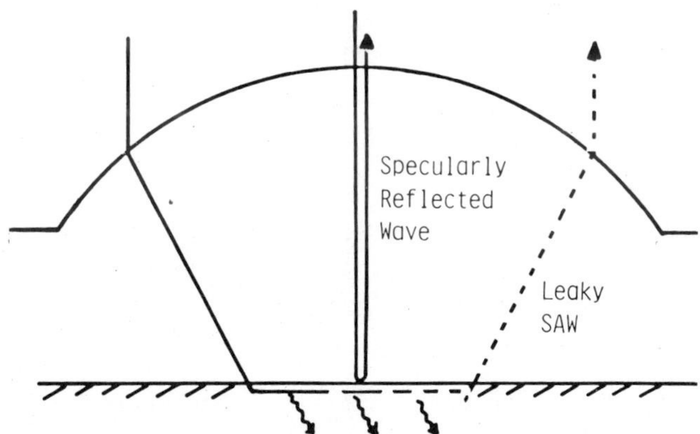

Fig. 3 A model for attenuation of leaky SAW (dotted line) due to loss in solid.

where α_S is the attenuation of acoustic waves in the solid and ω is the angular frequency. Using $R(\theta)$ given above, $V(z)$ curves are calculated with Atalar's formulation (Atalar, 1978).

The amplitude and the phase of $R(\theta)$ for SCM 3 steel were calculated and shown in Fig. 4 (a). The attenuation used in the calculation was 40 dB/cm, 350 dB/cm and 2000 dB/cm. The calculated $V(z)$ curves are shown in Fig. 4 (b). As expected, the depth of dips ΔV decreased as the attenuation increased. This is due to the decrease in the phase variation near the Rayleigh critical angle.

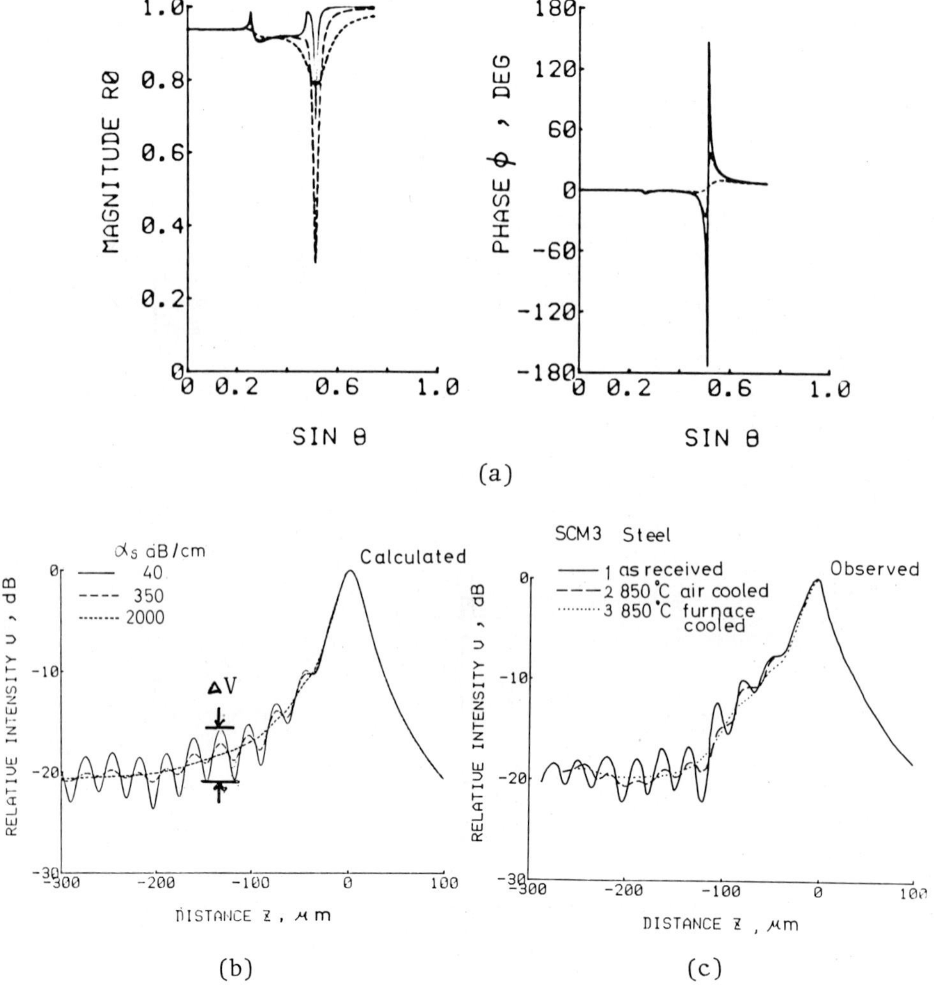

Fig. 4. (a) Reflectance function (b) caluculated and (c) observed V(z) curves of SCM3 steel with different heat treatment.

To examine the validity of above calculation, we measured V(z) curves of SCM 3 steels with three types of heat treatments. The specimen 1 was as received SCM 3 steel and the phase was martensite. The specimen 2 and 3 were heated in a furnace at 850°C for 2 hours. Then the specimen 2 was cooled in air whereas the specimen 3 was cooled slowly in the furnace. The phase of the specimen 2 was bainite or fine-grained ferrite plus pearlite whereas that of the specimen 3 was large grain ferrite plus pearlite. They were all polished before the V(z) curve measurement. The results are shown in Fig. 4 (c). The depth of dips decreased in the same order as the specimen type.

In the calculation of V(z) curves of Fig. 4 (b), the attenuation had been chosen to best fit the observed V(z) curves in Fig. 4 (c). The small change in v_l and v_t due to the heat treatments were neglected, because they are less than 1 % and do not have significant influence on the V(z) curves. The agreement between the calculated and observed V(z) curves was quite good. The attenuation of low frequency bulk waves in SAE 4150 steel with the similar composition to our specimens are reported in the literature (Papadakis, 1964). The high-frequency extrapolation of their data is 40 dB/cm for martensite, 200 dB/cm for bainite and 7000 dB/cm for ferrite plus pearlite phase at 200 MHz. Therefore the above calculation is accurate to the order of magnitude and can serve to nondestructively characterize the phase in the surface of steel.

Application to transformation hardened steel by eletron beam.

As an application of this method, the attenuation of S45C steel (containing 0.45 % carbon) was measured after and before the transformation hardening process by electron beam (Iwata,1982). This technique was recently developed to improve wear resistance of steel by hardening a localized area of the surface. When an annealed steel is heated by electron beam and rapidly cooled, the surface layer of 100 to 1000 μm thickness is transformed from ferrite plus pearlite phase to quenched martensite phase. It has an advantage of saving energy by hardening only necessary parts of machine elements.

A S45C steel specimen was first annealed, and then a part of the surface was transformation hardened by electron beam. Fig. 5 (a) shows the optical image and (b) the acoustic image (200 MHz, z=-60 μm) of boundary between the hardened and not-hardened areas. The hardened area is in the left. Though any sign of the boundary was not observed in the optical image, a large contrast was observed in the acoustic image.

The mechanism of the contrast in the acoustic image was

investigated by taking V(z) curves of each area. As shown in Fig. 6 the V(z) curve of not-hardened area was best fitted with attenuation of 1800 dB/cm, whereas the V(z) curve of hardened area was fitted with attenuation of 40 dB/cm. Therefore, the contrast in the acoustic image is attributed to the difference in attenuation of leaky SAW. The low attenuation is due to the small effective grain size in martensitic phase of the hardened area, whereas the high attenuation is due to the large grain size in ferrite plus pearlite phase of the not-hardened area.

The micro Vickers hardness H_v of the surface increased from 200 kg/mm^2 to 850 kg/mm^2 by the hardening (Iwata, 1982). As

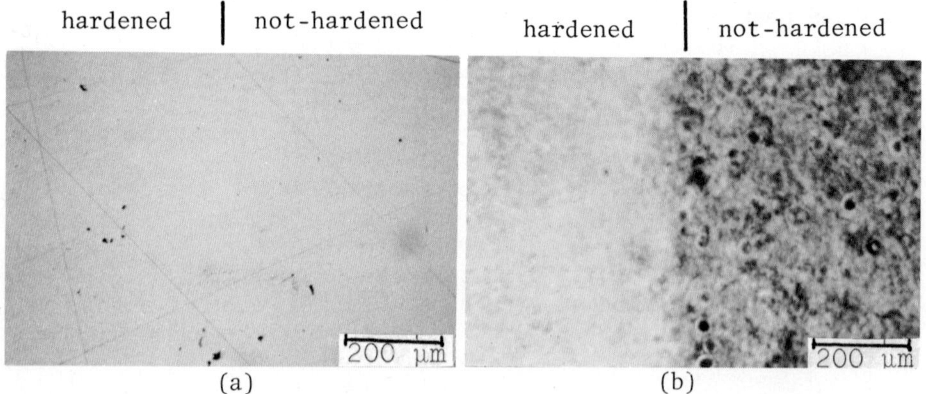

Fig. 5 Micrographs of S45C steel transformation hardened by electron beam. (a) optical image (b) acoustic image (200 MHz, z=-60 μm).

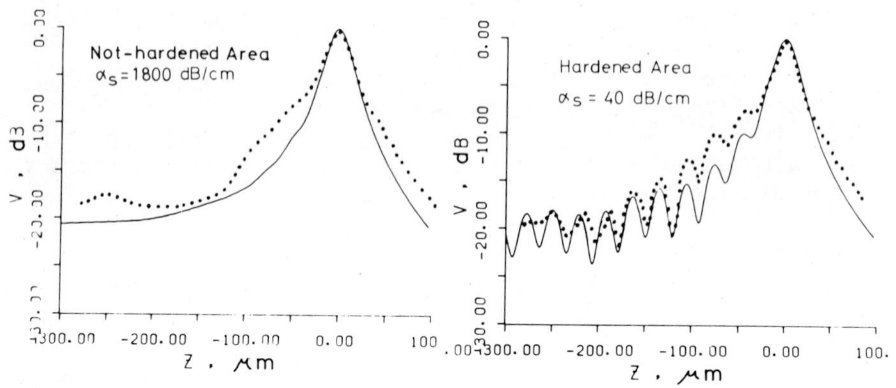

Fig. 6 V(z) curves of S45C steel (0.45 % carbon) transformation hardened by electron beam. The solid curves represent the calculated curves, whereas the dotted curves represent the observed ones.

is well known (Tetelman and Mcevily, 1967), the increase in hardness or yield stress has a close relation to the reduction in grain size. Therefore the attenuation measurement of leaky SAW from V(z) curves is particularly useful to nondestructively characterize the surface of steel. It has another advantage, of course, of the high spatial resolution. This is important when only a small area of the surface is hardened.

CONCLUSIONS

Subsurface cracks in silicon nitride was observed with an acoustic microscope. It enables an estimation of fracture toughness that can serve as a measure of erosion or wear resistance. An analysis on the attenuation measurement of leaky SAW was also proposed. This enables a characterization of surface of the transformation hardened steels.

ACKNOWLEDGEMENT

The authors thank Dr. A. Iwata, Mechanical engineering laboratory for providing the transformation hardened S45C steel specimen. The authors gratefully acknowledge Professor N. Chubachi and Dr. J. Kushibiki for discussions on attenuation of leaky SAW.

REFERENCES

Atalar, A., 1978, An Angular Spectrum Approach to Contrast in Reflection Acoustic Microscopy, J. Appl. Phys, 49 : 5130

Becker, F. L. and Richardson, R. L., 1972, Influence of Material Properties on Rayleigh Critical Angle Reflectivity, J. Acoust. Soc. Am. 51 : 1609.

Evans, A. G. and Wilshaw, T. R., 1976, Quasi- Static Solid Particle Damage in Brittle Solids I, Acta. Meta. 24 : 939.

Iwata, A, 1982, Transformation Hardening by Electron Beam, J., Jpn., Soc., Precision Engineering, 48 : 91.

Papadakis, E. P., 1964, Ultrasonic Attenuation and Velocity in Three Transformation Products in Steel, J. Appl. Phys, 35 : 1474.

Parmon, W. and Bertoni, H. L., 1979, Ray Interpretation of the Material Signature in the Acoustic Microscope, Electron. Lett. 15 : 684.

Tetelman, A. S. and Mcevily, A. J., 1967, Fracture of Structural Materials, Chap. 4, John Wiley and Sons Inc., New York.

Yamanaka, K., 1982, Analysis on SAW Attenuation Measurement Using Acoustic Microscopy, Electron. Lett.,18 to be published.

Weglein, R. D., 1982, Rayleigh Wave Absorption via Acoustic Microscopy, Electron. Lett. 18 : 20.

ACOUSTIC MICROSCOPY FOR MATERIALS STUDIES

G A D Briggs, C Ilett and M G Somekh

Department of Metallurgy & Science of Materials,
University of Oxford, Parks Road, Oxford OX1 3PH

INTRODUCTION

Over the past two years transmission and reflection scanning acoustic microscopes have been built at Oxford in collaboration with AERE Harwell. The purpose of our work is to discover applications in materials studies for which acoustic microscopy is suitable, and then to exploit the technique in tackling problems in those areas. The microscopes have been operating for a little under one year now, and in these Proceedings we summarise our progress in that time.

TRANSMISSION MICROSCOPY

Our transmission microscope works at 45 and 140MHz. It is designed for imaging within the bulk of a material, and can accept specimens up to 2mm thick. In passing from crystalline solids to water the refractive index is high, usually greater than four, and this ensures that spherical aberrations due to the lens are negligible. Unfortunately, the same fact causes the aberrations when trying to focus inside a solid with a plane surface to be very large. The resolution is optimised by using lenses of small numerical aperture (Pino et al., 1981), giving at 140MHz a resolution of 50μm, though features smaller than this can be detected.

When specimens of metals or alloys are imaged in the transmission microscope at this frequency the most prominent effect is a dappled appearance which is attributed to multiple scattering at grain boundaries. This has almost 100% contrast. This pattern may contain useful information about the microstructure of the specimen, but for the purpose of imaging other features it is a nuisance. It

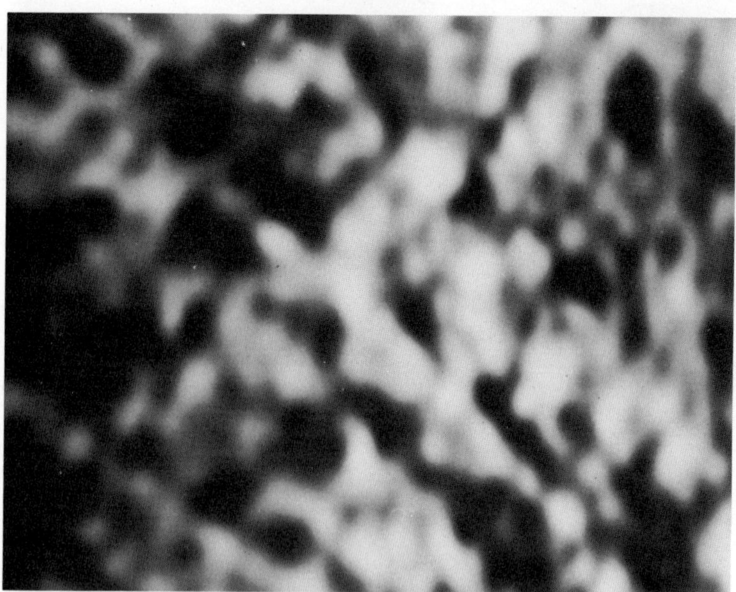

Fig.1. Transmission acoustic micrograph at 140MHz of a 3 minute diffusion bond in En8 steel. 100μm

can be greatly reduced by using a lower frequency (e.g. 45MHz), at the expense of resolution. Other methods of reducing this scattering, by special specimen preparation, are being investigated. Nevertheless, despite this result of grain boundary scattering we have had some success in imaging in transmission.

One problem of interest has been the development and growth of diffusion bonds. Images at 12MHz of sintered bonds have been shown by Sinclair and Ash (1980). Diffusion bonds are made by grinding the two surfaces to be joined, and then pressing them together with the grinding directions parallel at a temperature of 0.5-0.8 T_m (the absolute melting point) and at a pressure below that which would cause gross deformation. After small local plastic deformation the bond develops by diffusional processes over a period of up to an hour or so. In order to test a theoretical model of their development (Darby and Wallach, 1982), bonds which had been allowed to grow for different times were imaged in transmission. Because the surfaces have been brought together with their grinding directions parallel, long narrow voids are expected to appear corresponding to troughs in the ground surfaces. These might typically be 20μm wide and would appear dark in transmission. An image at 140MHz of a deliberately poor bond (bonded for only 3 minutes, at 900°C with a pressure of 20MPa) in En8 steel is shown in figure 1. The dappled appearance is due to the multiple grain boundary scattering. Also

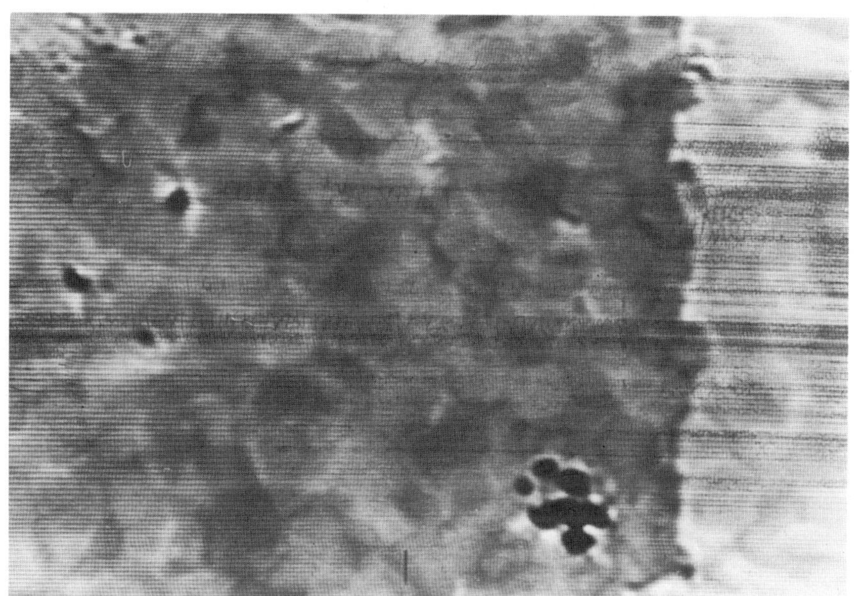

Fig.2. Reflection acoustic micrograph at 730MHz of a PLZT ceramic. The bright area at the right corresponds to a gold film on the surface. 10μm

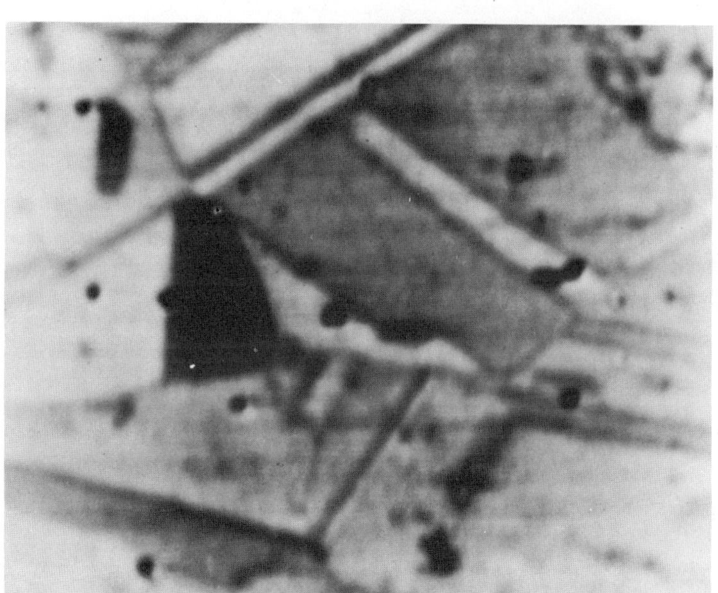

Fig.3. Acoustic micrograph of stainless steel, polished but not etched. The contrast depends on the amount of defocus; here it is z=-5μm. 20μm

visible are dark regions running diagonally across the picture (slight curvature is due to distortions in the temporary imaging electronics). It is believed that these correspond to features of the bond itself.

REFLECTION MICROSCOPY

Our reflection microscope currently operates at 730MHz, giving a surface resolution of 2μm and sampling a comparable depth of the specimen. Contrast is often enhanced by defocussing the specimen 5μm towards the lens. This should not be thought of as focussing under the surface; rather it is a result of interference between waves reflected at different angles to the normal (the V(z) effect).

The easiest material feature to image in reflection from a polished but unetched surface is the grain structure (Atalar et al., 1979). This is shown in figure 2 for PLZT, a transparent ferroelectric ceramic (Yin et al., 1982). The contrast arises because of the elastic crystal anisotropy, so that grains at different orientations present different elastic properties to the incident beam. The slightly brighter region to the right of the picture corresponds to a gold film 0.1μm thick on the surface. Grains are still acoustically visible through this film. An image of a stainless steel specimen is shown in figure 3. Once again the grains are visible, but in addition to contrast between one grain and another

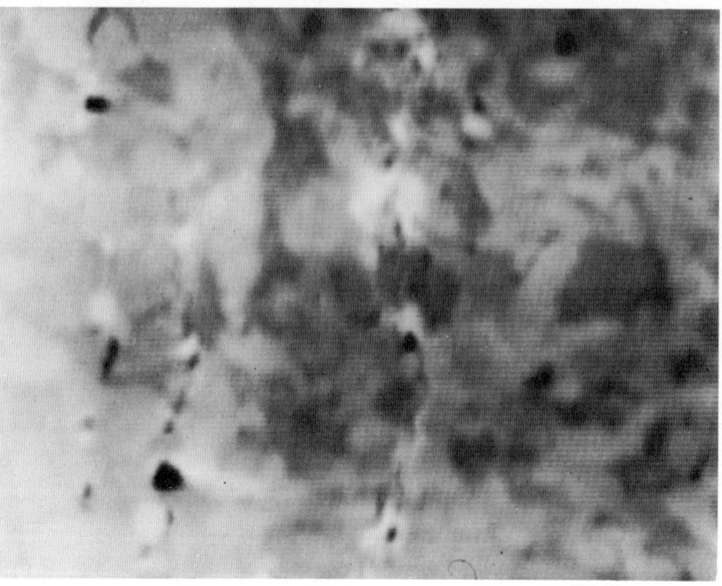

Fig.4. Acoustic micrograph of stress corrosion cracks in in pipeline steel. [scale bar: 20μm]

Fig.5a. Acoustic image of granite containing cracks. 20μm

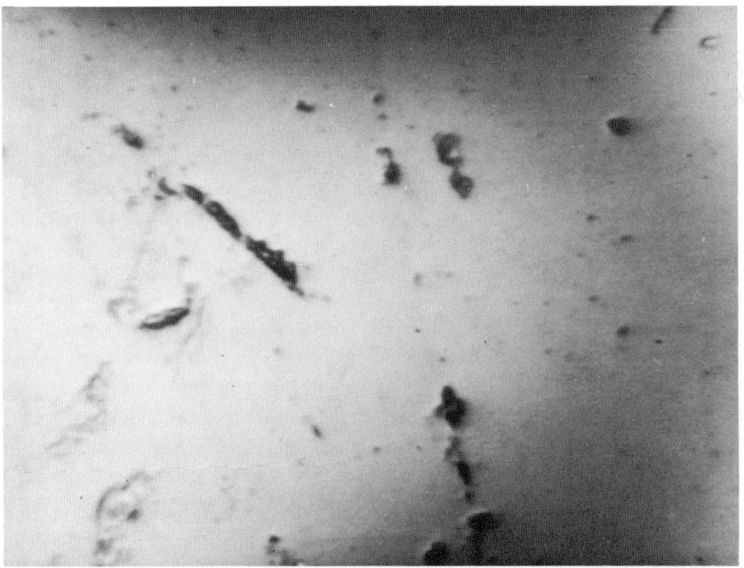

Fig.5b. Optical image of the same area as 5a.

there is contrast at the grain boundaries themselves (this effect is also noted by Miller, 1982, although his images are of etched specimens). This contrast undergoes reversals as the lens-object spacing is varied. A theory is being developed to account for this phenomenon and to clarify what information can be determined from it.

There is considerable interest in the study of cracks using the acoustic microscope. Figure 4 shows a specimen of pipeline steel which has been cyclically loaded for 7 days in a corrosive environment to produce stress corrosion cracking. The maximum stress was approximately $0.95\sigma_y$, with a stress ratio R=0.75 and a frequency of 10^{-5} Hz. The specimen was immersed in a 0.5M Na_2CO_3-1.0M $NaHCO_3$ solution, at a temperature of 80°C and a potential of -640mV relative to a standard calomel electrode. Stress corrosion cracks can be seen, and because of the grain contrast it is possible to see that in some places the cracking is intergranular, particularly at the initiation pitting sites, while in other places transgranular crack growth has occurred. Figure 5a is an acoustic image of granite. This contains porosity in the form of cracks which are important because the diffusion of radionucleides into these cracks contributes to retarding their transport in the pathway back to man. The crack in the centre of figure 5a, which runs approximately normal to the surface, is rendered particularly noticable by the interference fringes each side of it (Yamanaka and Enomoto, 1982). This may be compared with the optical image of the same area, figure 5b, in which the crack is scarcely visible.

Because the reflection acoustic microscope samples a thickness of about a wavelength, the technique has potential for examining thin films which are bonded to a substrate (Quate, 1980). Figure 6a is an optical micrograph of the surface of a cutting tool. The hard metal substrate is coated with a layer of TiN a few microns thick in order to increase wear resistance. During cooling in manufacture the coatings crack to allow stress relief. These cracks are so fine that they are scarcely visible optically, although lines of defects caused during metallographic polishing can be seen in figure 6a. The same defects are seen in the acoustic image of the same area in figure 6b. However in addition there is another network which appears brighter and broader acoustically, and which does not always coincide with the polishing damage. This is believed to be the network of stress relief cracks. Another problem, where the bonding of a surface film to a substrate is important, is in the growth of oxide films. In studies of oxide films on NiCrAl alloys it has been found that the adhesion of the film may be increased if small quantities of yttrium are added (Stott et al., 1979). The gross structures of the oxide film on samples with and without yttrium appear different optically, because the precipitation of yttrium at grain boundaries renders them visible in that alloy, but it is difficult to see any difference optically in the microstructure of the films. Differences can, however, be seen acoustically. This is illustrated in Figures 7a

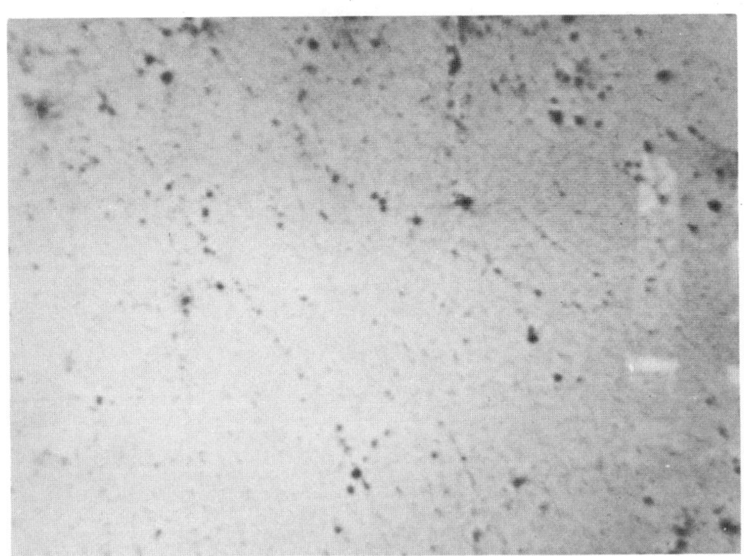

Fig.6a. Optical image of a hardmetal cutting tool. 20μm

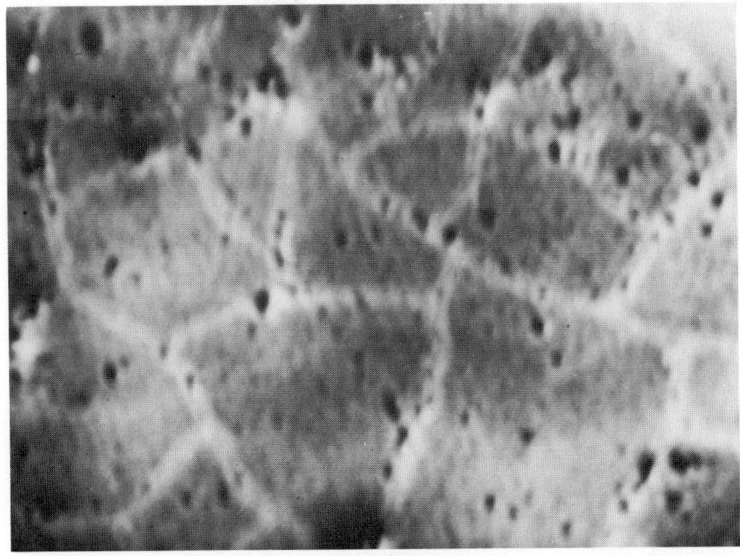

Fig.6b. Acoustic image of the same area as 6a.

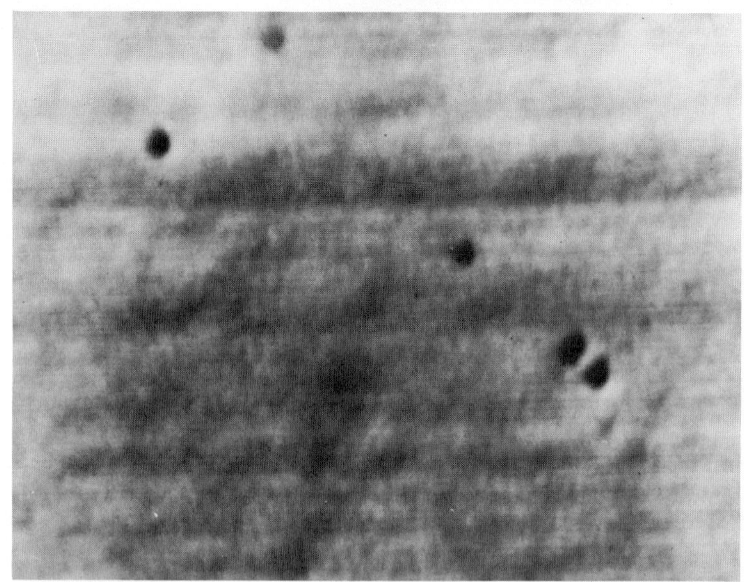

Fig.7a. Acoustic image of an oxide film grown on a NiCrAl alloy exposed to moist air for several hours at 850°C.

20μm

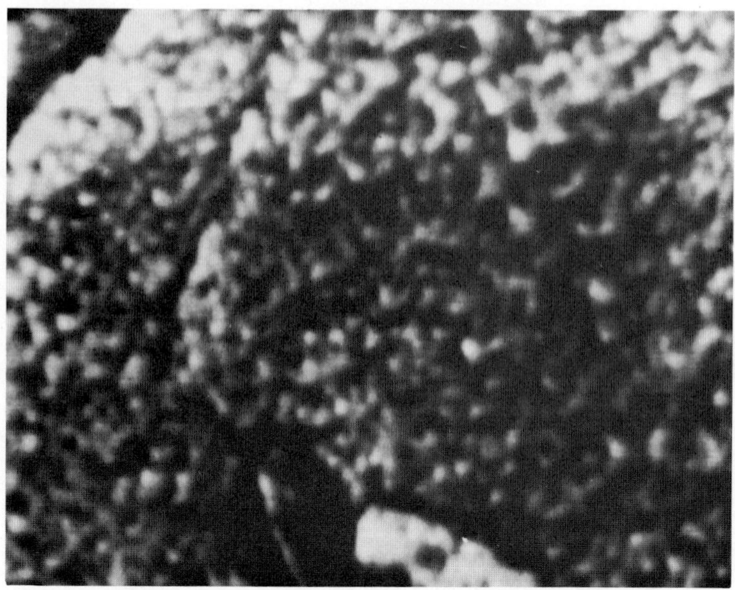

Fig.7b. As 7a, but with yttrium added to the alloy. Both images were taken with z=-5μm, and with identical settings of electronic contrast etc.

and b, which are acoustic images of oxide films grown on samples with and without yttrium respectively, by exposing to moist air at 850°C for several hours. The differences between these, and also between similar pairs of images, is now being investigated more quantitatively.

ELASTIC MICROPROBE

It is well established (e.g. Quate et al., 1979) that the variation in signal in reflection as the lens is scanned along its axis depends on the elastic properties of the surface being studied. The expression for V(z) may be written as a Fourier transform. For a lens with axial symmetry, a flat isotropic specimen, and small z

$$V(z) = \frac{A}{4k^2} \mathcal{F}[P^2(\frac{t}{k}) \cdot R(\frac{t}{k}) \cdot t]$$

where A is a normalizing constant
- k is the wave vector in the coupling fluid
- $t = 2k \cos \theta$
- θ is the angle which a given point on the lens subtends with the axis at the focal point
- $P(\frac{t}{k})$ is the pupil function of the lens (including aberration)
- $R(\frac{t}{k})$ is the reflection function of the substrate
- z,t are the transform pair.

This may be inverted to give

$$R(\frac{t}{k}) = \frac{4k^2}{A.P^2(\frac{t}{k}).t} \mathcal{F}^{-1}[V(z)]$$

Thus the elastic properties of a small region on a specimen may be deduced directly from measurements of V(z). As an example of how this might be used, figure 8 shows a specimen of a nickel based superalloy which had been held at 1050°C for 3000 hours to test its stability. This specimen has been polished but not etched, so that in conventional optical microscopy it would appear uniformly bright; the acoustic contrast is due entirely to variations in elastic properties. Various features can be seen in figure 8, such as precipitation along grain boundaries which is believed to be $M_{23}C_6$, (M is mainly Cr, with a little W), and some individual precipitates which are probably MC (M may be Hf or, Ti + Ta). When z is varied, changes in contrast are observed which differ at different regions of the image. By measuring V(z) at each region of interest and deducing the elastic properties, the information in such images can be made quantitative. This may also be of relevance in theoretical predictions of the scattering and attenuation which might be measured in lower frequency nondestructive characterisation of microstructures.

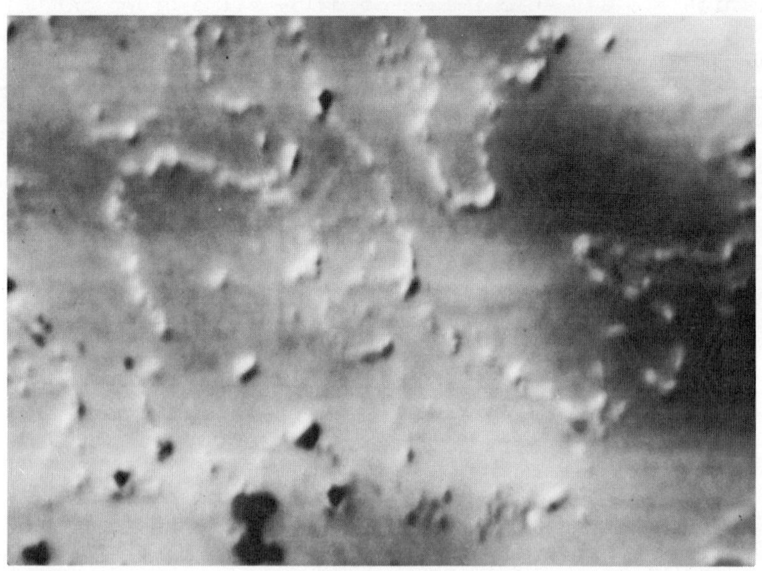

Fig.8. Acoustic micrograph of a nickel based super alloy which had been held at 1050°C for 3000 hours. The precipitates are various metal carbide compounds

50μm

CONCLUSIONS

A variety of problems in materials science have been studied in the scanning acoustic microscope. These studies have been of a preliminary nature, and much further work is needed in each case. Nevertheless, it is hoped that they indicate the kind of metallurgical problems for which the acoustic microscope may prove useful.

ACKNOWLEDGEMENTS

This work has been supported by AERE, SERC and NPL. We wish to express our thanks to Mr R Martin for extensive work on the electronics, to Dr H K Wickramasinghe for advice on the design and operation of the microscope and to Dr C J R Sheppard for discussions on imaging theory. The specimens have been studied in collaboration with the following: figure 1, Dr E R Wallach and Dr B Darby; figure 2, Mr Q R Yin; figure 4, Mr J Avila-Mendoza and Dr J M Sykes; figure 5, Dr A Atkinson; figure 6, Dr E A Almond; figure 7, Dr S R J Saunders; figure 8, Dr P N Quested. More detailed individual investigations of these specimens are in progress. Finally, we wish to thank Mr S F Pugh, Dr J A Champion and Professor Sir Peter Hirsch for support and advice throughout this project.

REFERENCES

Atalar, A., Jipson, V., Koch, R. and Quate, C.F., 1979, Acoustic microscopy with microwave frequencies, Ann. Rev. Mater. Sci., 9:255-81.

Darby, B. and Wallach, E.R., 1982, Theoretical model for diffusion bonding, Metal Science, 16:49-56.

Miller, A.J., 1982, Applications of acoustic microscopy in the semiconductor industry, in these Proceedings.

Pino, F., Sinclair, D.A. and Ash, E.A., 1981, New technique for sub-surface imaging using scanning acoustic microscopy, Ultrasonics International 81, IPC, Guildford, 193-198.

Quate, C.F., 1980, Microwaves, acoustics and scanning microscopy, in: "Scanned Image Microscopy", E.A. Ash, ed., Academic Press, London, 23-55.

Quate, C.F. Atalar, A. and Wickramasinge, H.K., 1979, Acoustic microscopy with mechanical scanning - a review, Proc.IEEE 67:1092-1114.

Sinclair, D.A. and Ash, E.A., 1980, Bond integrity evaluation using transmission scanning acoustic microscopy, Electronics Lett. 16:880-2.

Stott, F.H., Wood, G.C. and Golightly, F.A., 1979, The isothermal oxidation behaviour of Fe-Cr-Al and Fe-Cr-Al-Y alloys at 1200°C, Corrosion Sci., 19:869-887.

Yamanaka, K. and Enomoto, Y., 1982, Observation of surface cracks with scanning acoustic microscope, J.Appl.Phys., 53:846-850.

Yin, Q.R., Ilett, C. and Briggs, G.A.D., 1982, Acoustic microscopy of ferroelectric ceramics, J. Mat. Sci., 17:2449-52.

MATERIAL CHARACTERIZATION BY ACOUSTIC MICROSCOPE WITH LINE-FOCUS BEAM

Jun-ichi Kushibiki, Akira Ohkubo,
and Noriyoshi Chubachi

Department of Electrical Engineering
Faculty of Engineering, Tohoku University
Sendai 980, Japan

INTRODUCTION

The V(z) curves[1] have played a very important role in the recent rapid progress of development of the mechanically scanned acoustic microscope using a highly convergent beam. In the acoustical imaging measurements, the V(z) curves have been effectively employed for the interpretation of contrast mechanisms in acoustic images[1-4] obtained in scanning version and for the imaging signal processing for obtaining false-color micrographs[5]. Further, it has been found out that the V(z) curves are of particular importance in the quantitative measurements of acoustic properties of materials[6] because they are unique and characteristic of specific materials. Recently, for this latter case, a new acoustic line-focus beam[7] has been introduced with which the nonscanning reflection acoustic microscope can appropriately pick up acoustic properties of solid materials, including acoustic anisotropy.[8-10]

In this paper, the acoustic microscope system with the line-focus beam is discussed experimentally and theoretically from the point of view of establishing it as a useful measurement system for characterizing solid materials. The material characterization is carried out by measurements of the propagation characteristics of leaky surface acoustic waves (SAWs) whose phase velocities are determined through the V(z) curve measurements. Application of the system is made to a variety of solid materials for which the velocities of leaky SAWs range from 2600 to 6000 m/s. Experiments are described in which an acoustic sapphire lens of 1.0 mm radius, at a frequency around 200 MHz, was used and the measured results are compared with theoretical calculations.

V(Z) CURVES AND CHARACTERIZATION PRINCIPLE

The material characterization described here is determined quantitatively by V(z) curve measurements made with the nonscanning reflection acoustic microscope using a line-focus beam, as illustrated in Fig. 1. The acoustic line-focus beam is produced with a wedge-shaped structure, and formed by an acoustic sapphire lens having a cylindrical concave surface. The V(z) curves are the records of the piezoelectric transducer output as a function of distance between the acoustic probe and a sample, when the sample is moved along the z axis. A V(z) curve is schematically depicted in Fig. 1; the transducer output yields a maximum at the focal point, and dips (minima) appear periodically in the negative z region.

The mathematical representation for the V(z) curves, obtained by the acoustic line-focus-beam lens, is given briefly in the following[11], according to a theoretical analysis[12,13] based on Fourier optics. Figure 2 shows the cross-section geometry for analyzing the V(z) curves and for explaining the mechanism of operation. Here the coordinates x_0, x_1, x_2 and x_3 refer the transducer, the acoustic lens, the focal and the sample planes, respectively. The final form of the transducer output is determined, as follows, by assuming that the acoustic fields do not vary in the y direction,

$$V(z) = A\int_{-\infty}^{\infty} u_{3R}^{+}(x_3)\cdot u_3^{-}(x_3)\,dx_3 \;,$$
$$= A\int_{-\infty}^{\infty} [u_{3R}^{+}(x_3)\cdot F^{-1}\{R(k_x)\cdot F\{u_3^{+}(x_3)\}\}]\,dx_3 \;, \qquad (1)$$

where A is the arbitrary amplitude constant and $u_{3R}^{+}(x_3)$ is the reference field distribution defined as

$$u_{3R}^{+}(x_3) = (T^{-}(x_3)/T^{+}(x_3))\cdot u_3^{+}(x_3) \;. \qquad (2)$$

The quantities $u_3^{+}(x_3)$ and $u_3^{-}(x_3)$ are the incident and reflected field distributions at the sample plane, respectively. The quantities $T^{+}(x_1)$ and $T^{-}(x_1)$ are, respectively, the transfer coefficients from lens to water and from water to lens at the lens plane, including the effect of the acoustic antireflection coating layer. The function of $R(k_x)$ is the reflectance function of reflecting the elastic information of solid materials at the water/sample interface. The quantity of k_x is the wavenumber corresponding to the spatial frequency in the x direction. The distribution of $u_3^{+}(x_3)$ is given by

$$u_3^{+}(x_3) = F^{-1}\{F\{u_2^{+}(x_2)\}\cdot \exp(jk_z z)\} \;, \qquad (3)$$

where $u_2^{+}(x_2)$ is the incident field distribution at the focal plane and k_z is the z-component of the wavenumber k_l for longitudinal waves in water, defined as $k_z=(k_l^2-k_x^2)^{1/2}$. The F and F^{-1} denote the Fourier transform and the inverse Fourier transform, respectively.

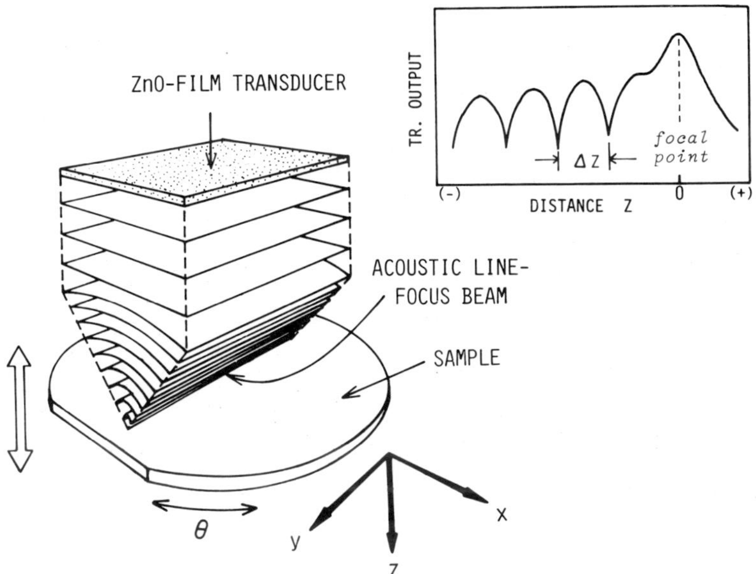

Fig. 1. Illustration of material characterization method by the nonscanning reflection acoustic microscope using a line-focus beam.

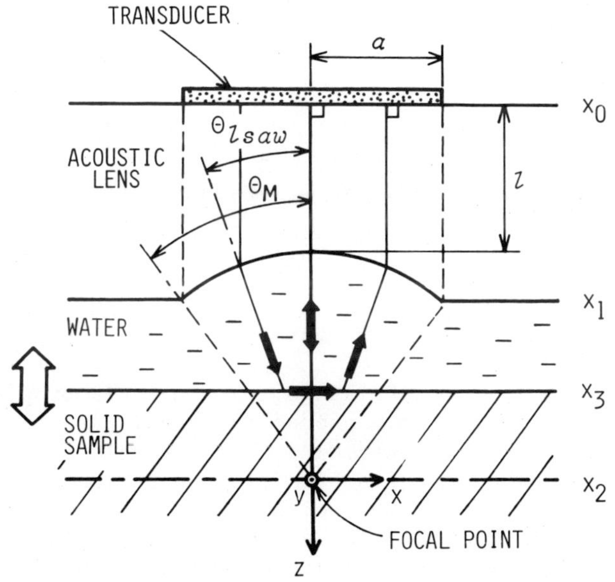

Fig. 2. Cross-section geometry for analyzing V(z) curves obtained by the acoustic line-focus-beam lens.

Thus, the transducer output of V(z) obtained with the line-focus beam is represented by both the reference field distribution $u_{3R}^{+}(x_3)$, determined uniquely by the employed acoustic line-focus-beam lens, and the reflected field distribution $u_3^{-}(x_3)$, containing the acoustic response of the materials to be measured. The shape of the V(z) curves is essentially dominated by the reflectance function $R(k_x)$ of the material. It also depends strongly on the wave propagation direction when the material is anisotropic about the z axis.

According to the recent studies[11,14] involving numerical calculations for determining the V(z) curves, it has been found that the leaky SAW parameters are the most important part of the reflectance function. The phase velocity of leaky SAWs makes a dominant contribution to the spatial interval of the dips (minima) in the V(z) curves, while the attenuation factor of leaky SAWs affects the shape of the V(z) curves.

The present material characterization is based on the fact that the periodicity of the dips appearing in the V(z) curves is closely related to the phase velocity of leaky SAWs propagating on the water/sample boundary. These periodic dips are caused by interference of two components of the acoustic waves detected by the transducer. As shown in Fig. 2, one component constitutes the waves near the z axis directly reflected from the sample, and the other is associated with those waves reradiated from the sample but excited on the boundary at the critical angle θ_{lsaw} for leaky SAWs.

The phase velocity v_{lsaw} of leaky SAWs can be determined from the interval Δz of the dips and is approximately given the relation[15,16]

$$\Delta z = v_l / 2f(1-\cos\theta_{lsaw}) \; , \qquad (4)$$

where $\theta_{lsaw}=\sin^{-1}(v_l/v_{lsaw})$, f is the acoustic frequency and v_l is the longitudinal velocity of water. Equation (4) can be represented also in terms of v_{lsaw} as

$$v_{lsaw} = v_l / (1-(1-v_l/2f\Delta z)^2)^{1/2} \; . \qquad (5)$$

Thus, by measuring the dip interval of the V(z) curve obtained with the acoustic microscope system, the phase velocity of leaky SAWs can be determined easily by calculation. For an anisotropic sample, by rotation through an angle of θ around the z axis and then repeating the similar V(z) curve measurements, the variations in phase velocities for leaky SAWs, as a function of propagation direction θ, can be obtained. In this way, the anisotropy of acoustic properties of materials around the z axis can be characterized by the nonscanning reflection acoustic microscope using the line-focus beam.

EXPERIMENTS AND CONSIDERATIONS

Experiments have been carried out using an X-cut rutile (TiO_2) crystal as the sample. The acoustic line-focus beam employed is formed from an acoustic sapphire lens with a cylindrical concave surface, with the following parameters: the radius of curvature R=1.0 mm, the aperture angle θ_M=60°, the transducer width $2a$=1.73 mm, and the distance from the transducer surface to the top surface of the lens l=12.0 mm, as shown in Fig. 2. This lens is suitable for the acoustic measurements around 200 MHz. A ZnO film transducer is formed on the flat surface of the lens rod to generate and detect acoustic longitudinal waves. A chalcogenide glass film is deposited on the lens surface as an acoustic antireflection coating layer with a quarter-wavelength thickness, for efficiently transmitting acoustic waves across the sapphire/water interface.[17] Acoustic plane waves radiated from the transducer are converted into the acoustic line-focus beam in water, the coupling medium between the lens and the sample, by the cylindrical concave acoustic lens. The acoustic output of the line-focus beam is so sensitive to the alignment of the beam and the sample that the measurements are usually carried out by placing the sample and the lens assembly on the microscope mechanical stage, which provides for high precision displacements along, and rotations about, each of the three orthogonal axes. The pulse mode measurement method[18] is used for the generation of RF pulses and detection of RF pulses reflected from the sample. The V(z) curves are automatically recorded employing motor-driven translation of the acoustic line-focus beam along the z axis.

Acoustic Velocity Measurement and Anisotropy Detection

Figure 3 shows an oscilloscope trace of a V(z) curve determined for Z-axis propagation direction (θ=90°) of leaky SAWs on the water/ X-cut rutile boundary, at the frequency of operation of 216.4 MHz. The dips appearing periodically in the negative z region have an interval Δz=72.55 μm, giving the phase velocity of the leaky SAW as 4907 m/s, using Eq. (5). In the same way, the leaky SAW velocities are measured for the other propagation directions. The experimental results are plotted as open circles in Fig. 4 where the variation of leaky SAW velocities extends from 4141 to 4907 m/s, depending on wave propagation direction.

To compare the experimental results with the theory, the exact numerical calculations are made for the propagation characteristics of leaky SAWs on the boundary between water and rutile according to the analytic procedure of Campbell and Jones[19]. The physical constants published by Wachtman et al.[20] are used for the rutile crystal in the calculations. For water, the longitudinal velocity of 1483 m/s and the density of 998.2 kg/m^3, at 20 °C, are used. The calculated results are shown by the solid line in Fig. 4. The calculated values of the phase velocities for the water/X-cut-rutile

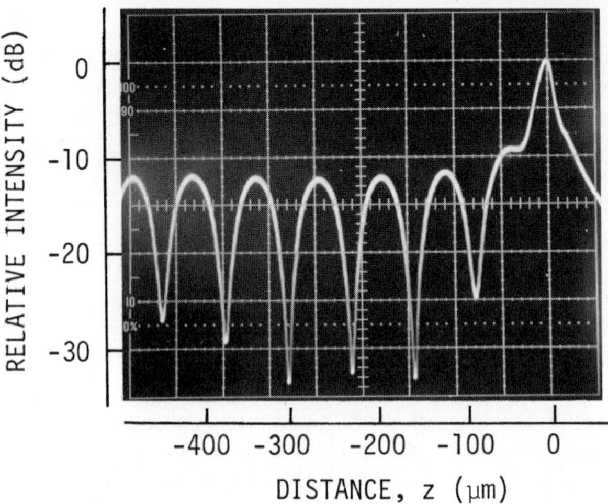

Fig. 3. V(z) curve for Z-axis propagation (θ=90°) of leaky SAW on water/X-cut-rutile boundary measured with acoustic line-focus beam at 216.4 MHz.

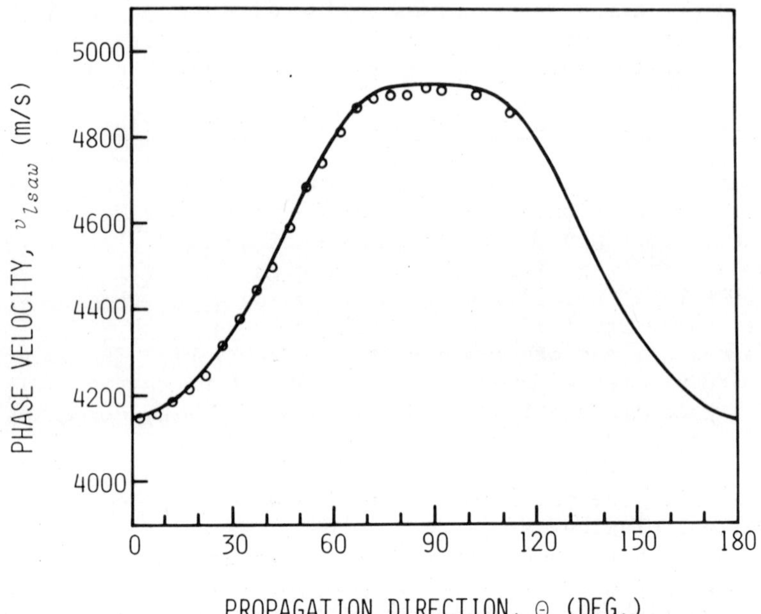

Fig. 4. Experimental and theoretical results of propagation properties of leaky SAWs on water/X-cut-rutile boundary. o; measured, ———; calculated

boundary are a few meters per second larger than those for SAWs on a free surface, as a water loading effect appears on the substrate.

The experimental results are seen to agree well with the theoretical calculations. The differences between the experimental and calculated values are within about 0.5 %, and the experimental errors are now believed to be less than ±0.2 %.

Numerical Considerations for V(z) Curve Measurements

Next, the V(z) curves obtained by the acoustic microscope with the line-focus beam are numerically analyzed in comparison with the experimental V(z) curve for the XZ rutile shown in Fig. 3. In the numerical calculations, the same lens parameters as used in the experiments are employed.

A calculated V(z) curve is shown in Fig. 5. It is seen that the experimental V(z) curve is, as a whole, explained well in the shape and the dip interval of the V(z) curve is in good agreement for both V(z) curves (compare Figs. 3 and 5). The intervals Δz of dips relating to the measurement of the leaky SAW velocity v_{lsaw} are given in Table 1. The values of v_{lsaw} in the table were calculated using Eq. (5). It is seen that the values of Δz and v_{lsaw} obtained by the numerical calculations for the V(z) curves agree well with the values obtained experimentally. Both the values of phase velocities determined by the calculation also agree well with the theoretical velocity value obtained by the exact numerical calculation of leaky SAWs propagating on the water/XZ-rutile boundary.

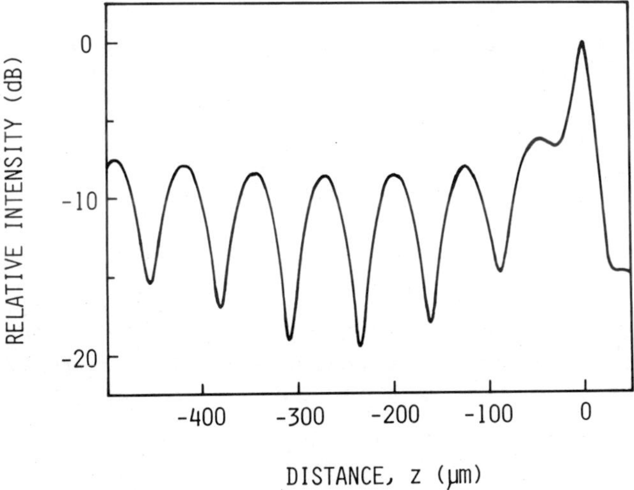

Fig. 5. V(z) curve numerically calculated for XZ rutile with acoustic line-focus-beam lens at 216.4 MHz.

Table 1. Comparison of theoretical calculations and experimental results relating to velocity measurement for leaky SAWs obtained by V(z) curves on XZ rutile.

	Dip interval Δz (μm)	Leaky SAW velocity v_{lsaw} (m/s)
Numerical analysis	72.83	4916
Experiment	72.55	4907
Exact analysis	—	4917

Thus, these results of the numerical analysis for the V(z) curve support theoretically that the nonscanning reflection acoustic microscope system with the line-focus beam can be appropriately applied to characterize solid materials via the measurements of leaky SAW velocity by using the relation (5) between the interval of dips in the V(z) curves and the leaky SAW velocity.

Applications to Various Materials

For the purpose of establishing the nonscanning reflection acoustic microscope system with the line-focus beam as a means of nondestructive (NDE) and noncontacting (NCE) evaluation for characterizing solid materials, the system was applied to a variety of materials. Glasses of fused quartz, Corning 7740 and Ohara E6 as isotropic materials, three nonpiezoelectric crystals of Z-cut sapphire, (111) Ge and (111) Si, and some piezoelectric crystals of Y-cut α-quartz, X-cut, Y-cut and Z-cut $LiNbO_3$ were examined.

In the same way as for X-cut rutile, the measurements of the propagation characteristics of leaky SAWs are made for each sample, so that the different dependences of leaky SAW velocities on the propagation direction are measured corresponding to the acoustic anisotropy of each sample around the z axis normal to the sample plane. The experimental values were determined directly by using relations (4) or (5) representing the principle of this material characterization method. The following normalized relation is introduced to obtain a universal representation for the characterization,

$$f\Delta z = v_l/2(1-\cos\theta_{lsaw}). \qquad (6)$$

Here it is seen that the product of $f\Delta z$ depends uniquely on the acoustic properties of the samples and the reference liquid, water. The experimental results observed above for each sample construct a

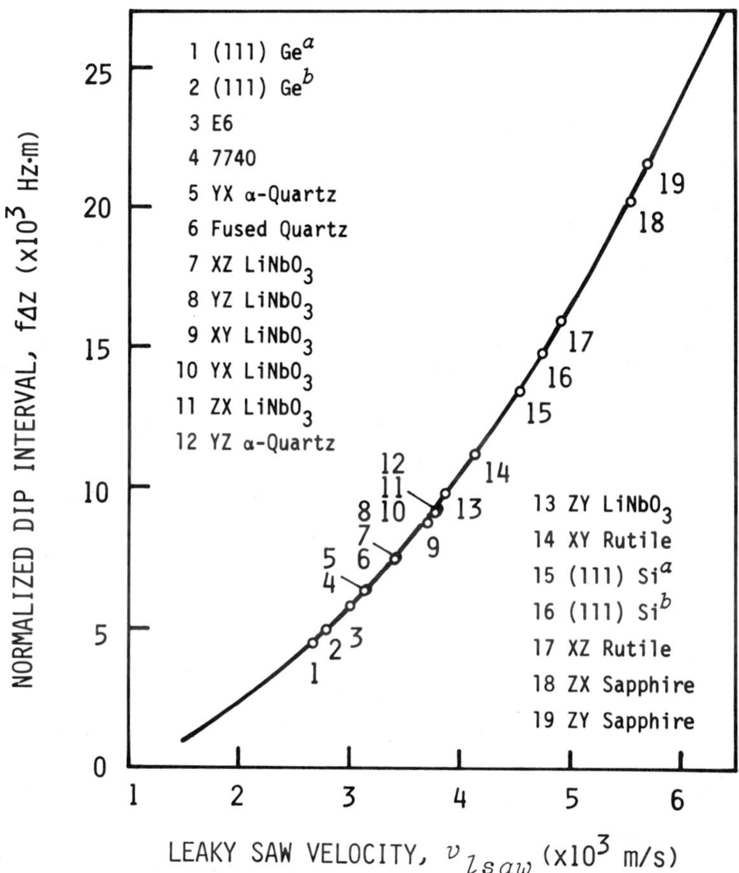

Fig. 6. Experimental results and theoretical calculations of material characterization curve.
——; experimental curve, o; theoretical values
a; [110], *b*; [112]

part of one experimental curve calculated from Eq. (6), and will now be called the material characterization curve. This curve is shown as a solid line in Fig. 6. In the same figure, the theoretical velocity values of leaky SAWs in the typical propagation directions for all the characterized materials are plotted as open circles, calculated using their published physical constants[21]. As seen from the comparison, the characterization curve is well explained by the theoretically calculated values, and the differences between the experimental and calculated values are less than 1.0 %.

From these results, it is satisfactorily demonstrated that the nonscanning reflection acoustic microscope system with the line-focus beam is a very useful system for material characterization. Quantitatively, the acoustic properties, including acoustic anisotropy, can be determined with a high measurement accuracy of about ±0.2 % for materials that are not only nonpiezoelectric but also strongly piezoelectric, for which leaky SAW velocities range from 2600 to 6000 m/s.

CONCLUSION

The material characterization method of determining acoustic properties using the nonscanning reflection acoustic microscope system with the line-focus beam, such as sound velocity and acoustic anisotropy of materials, both experimentally and theoretically has been investigated. The acoustic properties have been obtained as the propagation characteristics of leaky SAWs whose phase velocities are determined from the interval of dips appearing in the V(z) curves. As shown in the experiments, using an acoustic sapphire lens with a cylindrical concave surface of 1.0 mm radius around 200 MHz, the acoustic system has been satisfactorily applied to characterize solid materials for which the velocities of leaky SAWs are ranging widely from 2600 to 6000 m/s. The measured results agree well with the theoretical results obtained by the exact numerical calculations for leaky SAWs at the boundary of water/solid-materials. From these investigations it has been demonstrated that this system is useful for material characterization to measure quantitatively the acoustic properies for solid materials, both isotropic and anisotropic, with a high measurement accuracy of ±0.2 %.

ACKNOWLEDGMENTS

The authors are very grateful to K. Horii and H. Maehara for their helpful discussions and technical assistance on the experiments, and to M. Aihara for the X-ray analysis. We wish to thank Prof. F. Dunn for his critical reading of the manuscript. This work was supported in part by the Research Grant-in-Aids from the Ministry of Education, and the Toray Science & Technology Grants.

REFERENCES

1. A. Atalar, C. F. Quate, and H. K. Wickramasinghe, Phase imaging in reflection with the acoustic microscope, Appl. Phys. Lett. 31:791 (1977).
2. C. F. Quate, A. Atalar, and H. K. Wickramasinghe, Acoustic microscope with mechanical scanning — A review, Proc. IEEE 67: 1092 (1979).
3. R. D. Weglein and R. G. Wilson, Characteristic material signatures by acoustic microscopy, Electron. Lett. 14:352 (1978).
4. R. C. Bray, C. F. Quate, J. Calhoun, and R. Koch, Film adhesion studies with the acoustic microscope, Thin Solid Films 74:295 (1980).

5. R. Hammer and R. L. Hollis, Enhancing micrographs obtained with a scanning acoustic microscope using false-color encoding, Appl. Phys. Lett. 40:678 (1982).
6. R. D. Weglein, A model for predicting acoustic material signatures, Appl. Phys. Lett. 34:179 (1979).
7. J. Kushibiki, A. Ohkubo, and N. Chubachi, Linearly focused acoustic beams for acoustic microscopy, Electron. Lett. 17:520 (1981).
8. J. Kushibiki, A. Ohkubo, and N. Chubachi, Anisotropy detection in sapphire by acoustic microscope using line-focus beam, Electron. Lett. 17:534 (1981).
9. J. Kushibiki, A. Ohkubo, and N. Chubachi, Acoustic anisotropy detection of materials by acoustic microscope using line-focus beam, 1981 IEEE Ultrasonics Symp. Proc. pp.552-556 (1981).
10. J. Kushibiki, A. Ohkubo, and N. Chubachi, Propagation characteristics of leaky SAWs on water/$LiNbO_3$ boundary measured by acoustic microscope with line-focus beam, Electron. Lett. 18:6 (1982).
11. J. Kushibiki, A. Ohkubo and N. Chubachi, Theoretical analysis for V(z) curves obtained by acoustic microscope with line-focus beam, Electron. Lett. (1982), (in press).
12. A. Atalar, An angular spectrum approach to contrast in reflection acoustic microscope, J. Appl. Phys. 49:5130 (1978).
13. H. K. Wickramasinghe, Contrast in reflection acoustic microscopy, Electron. Lett. 14:305 (1978).
14. J. Kushibiki, A. Ohkubo, and N. Chubachi, Effect of leaky SAW parameters on V(z) curves obtained by acoustic microscopy, Electron. Lett. (1982) (in press).
15. W. Parmon and H. L. Bertoni, Ray interpretation of the material signature in the acoustic microscope, Electron. Lett. 15:684 (1979).
16. A. Atalar, A physical model for acoustic signatures, J. Appl. Phys. 50:8237 (1979).
17. J. Kushibiki, H. Maehara, and N. Chubachi, Acoustic properties of evaporated chalcogenide glass films, Electron. Lett. 17:322 (1981).
18. J. Kushibiki, T. Sannomiya, and N. Chubachi, A novel acoustic measurement system for pulse mode in VHF and UHF ranges, (unpublished).
19. J. J. Campbell and W. R. Jones, Propagation of surface waves at the boundary between a piezoelectric crystal and a fluid medium, IEEE Trans. SU-17:71 (1970).
20. J. B. Wachtman, Jr., W. E. Tefft, and D. G. Lam, Jr., Elastic constants of rutile (TiO_2), J. Res. Natl. Bur. Std. — A. Phys. & Chem. 66A:465 (1962).
21. A. J. Slobodnik, E. D. Conway, and R. T. Delmonico, Microwave Acoustic Handbook Vol. 1A. Surface wave velocities, (AFCRL-TR-73-0597, 1973).

NDE OF SOLIDS WITH A MECHANICALLY B-SCANNED ACOUSTIC MICROSCOPE

I.R. Smith, D.A. Sinclair and H.K. Wickramasinghe

Dept. of Electronic & Electrical Engineering
University College London,
Torrington Place, London WC1E 7JE

ABSTRACT

Acoustic microscopy of the interior of solids is hampered by spherical aberration introduced at the sample surface. The incorporation of a spherical coupling piece in front of a conventional acoustic lens radically reduces the aberration. This is so not only when the coupling piece is concentric with the lens, but when the lens is axially displaced. This means that diffraction limited focussing can be achieved over a wide range of depths, typically up to several hundred wavelengths. Thus a B-scan imaging system with mechanical scanning of the lens focus can be used to image a slice through a solid specimen. Alternatively, the lens may be focussed at a constant depth and transversely C-scanned in the normal way. This paper reports on theory and experiments using this system for non-destructive imaging of cylindrical specimens.

INTRODUCTION

Several problems are encountered when one attempts to image within the bulk of a solid using the scanning acoustic microscope (SAM)[1]. These problems can, in general, be related to the large impedance and velocity ratios that exist between the solid and liquid coupling medium. The large impedance mismatch greatly reduces the fraction of acoustic energy transmitted into the solid. This in itself does not represent a formidable problem except at the highest frequencies where signal to noise considerations become paramount. In this case, one can overcome the loss at the interface by applying a quarter wave matching layer. A much more serious problem arises due to the large velocity mismatch; converging spherical waves produced by the acoustic lens are severely distorted due to the spherical aberration at the liquid-solid interface. This distortion results in a focal spot diameter which is several times the diffraction limited value.

One way of overcoming this problem is to use a liquid such as gallium for the acoustic coupling medium; the longitudinal velocity in the liquid then closely matches the shear velocity in the solid. By resorting to the pulsed reflection mode and gating the shear waves within the solid, one can then achieve diffraction limited resolution[2]. Another possibility is to use a suitably shaped aspheric lens or even a reduced numerical aperture lens[3]. In the former case, if one wishes to image aluminium using water as a coupling medium, the required departure from sphericity is typically less than 10% in order to achieve a numerical aperture of 0.5 within the solid. The latter case is more interesting in that it can be shown that one could achieve diffraction limited performance within most solids using reduced aperture lenses provided the numerical aperture within the solid is chosen to be less than 0.5.

Reduced aperture lenses will only work for objects which have planar surfaces and aspheric lenses are very difficult to fabricate so as to phase match to anything other than a planar object surface. Although, in principle, the gallium coupling technique should work for objects with arbitrary surfaces, there are other problems such as wetting and object damage associated with its use.

In this paper, we study an alternative scheme for diffraction limited imaging within solids having an arbitrary surface. It has recently been shown[4] that by utilising one additional spherical surface, it is possible to totally eliminate spherical aberration. The schematic diagram of the system described in that article is shown in Figure 1. Its operation relies on the fact that, due to the gross refractive index change at the lens surface, the spherical waves generated by the lens are incident approximately normally on the convex spherical coupling surface. The material of the coupling piece is chosen so that its longitudinal velocity matches the longitudinal (or shear) velocity in the object. The object and coupling piece are acoustically coupled by a thin liquid layer. When the curvature of the phase front leaving the lens matches the curvature of the surface of the coupling piece a diffraction limited focus is obtained in the object. If the object is mechanically scanned in the X-Y plane, the points in a diffraction limited (C-scan) image can be obtained from a plane a fixed depth below the surface.

Although in Figure 1 the coupling element has been shown to have a planar bottom surface, it is important to realise that this surface can be chosen so that it matches the surface of the object being investigated. In this paper, we shall use the technique to image cylindrical objects at different focal depths within the object.

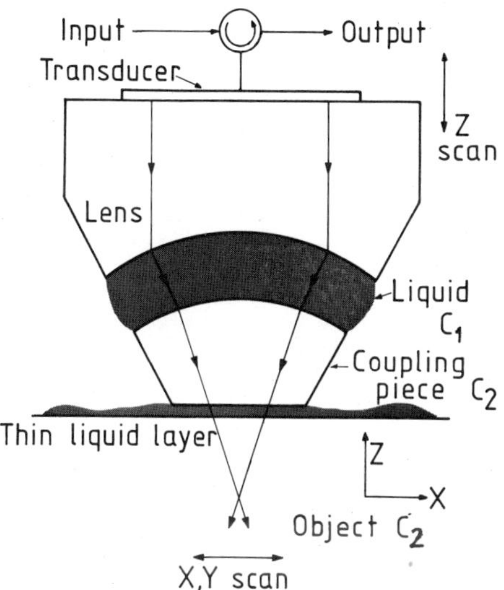

Fig. 1 Mechanical B-scan acoustic microscope for interior imaging of planar objects.

The theory developed in the following section shows that one can scan the lens in the Z direction and thereby scan the focal depth in the object over a large distance without seriously affecting the spherical aberration. It is, therefore, possible to build a B-scan system where the lens is scanned mechanically, say at 100Hz, in the Z direction and the object is scanned slowly in the orthogonal X or Y direction. The resulting image will be a cross-section view of the object.

THEORY

Let us now calculate the spherical aberration produced by our system when the spherical waves produced by the lens do not quite match the spherical surface on the coupling piece. We shall assume that the thin liquid layer between the coupling element and the object does not affect the spherical aberration so that we only need consider the liquid-solid interface at the spherical surface of the coupling element. This assumption is valid provided the liquid coupling film is thin enough; a simple ray tracing calculation shows that, in the situation where the coupling piece has the same acoustic properties as the object, the film thickness should be less than the diffraction limited spot diameter within the object.

The geometry we shall use is shown in Figure 2. The solid curve PA represents the spherical surface on the coupling piece with its centre of curvature at O. PQ is the incoming spherical wavefront from the liquid with centre of curvature at C. AX is the reference wavefront coming to a perfect focus at B. The spherical aberration $W(\theta)$ is obtained quite simply by invoking Fermat's principle:

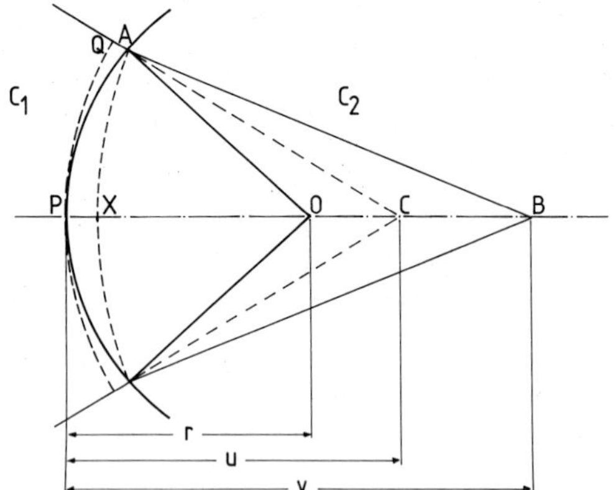

Fig. 2 Geometry for aberration calculation. r is the radius of the coupling piece, u is the radius of the incoming spherical wavefront and v is the radius of the perfectly focussed spherical wave within the coupling piece.

$$W(\theta) = PX - QA/\mu \tag{1}$$

where $\mu = C_1/C_2$ is the ratio of the longitudinal velocity in the liquid to the longitudinal (or shear) velocity in the solid. Now:

$$AB = 2r^2\left[(1-\cos\theta)-2vr(1-\cos\theta)+v^2\right]^{1/2} \tag{2}$$

and

$$AC = 2r^2\left[(1-\cos\theta)-2ur(1-\cos\theta)+u^2\right]^{1/2} \tag{3}$$

Also v can be obtained using the paraxial lens formula for a single surface:

$$v=\mu ur/\left[r+(\mu-1)u\right] \tag{4}$$

When the incident wavefront PQ matches the spherical surface on the coupling piece i.e. u = r and the spherical aberration is zero, equ. (4) yields:

$$v = u = r \tag{5}$$

We are interested in studying the situation where u is perturbed about its ideal value u = r. Let us take $u = r(1+\Delta)$. Then equ. (4) gives:

$$v=\mu(1+\Delta)r/\left[\mu+\Delta(\mu-1)\right] \tag{6}$$

The corresponding perturbation in v is given by:

$$\delta v = \Delta r / \left[\mu + \Delta(\mu - 1)\right] \tag{7}$$

Inserting equs. (2), (3), and (6) into equ. (1) yields an expression for the spherical aberration $W(\theta)$:

$$\frac{W(\theta)}{r} = (\Delta+1)\left(\frac{\mu}{\mu+(\mu-1)\Delta} - \frac{1}{\mu}\right) - \left(\frac{(\Delta+1)^2\mu^2}{\left[\mu+(\mu-1)\Delta\right]^2} - \frac{4\Delta\sin^2(\theta/2)}{\mu+(\mu-1)\Delta}\right)^{1/2} + \frac{1}{\mu}\left[(\Delta+1)^2 - 4\Delta\sin^2(\theta/2)\right]^{1/2} \tag{8}$$

As expected, equ. (8) predicts that $W(\theta) \to 0$ both as $\theta \to 0$ and as $\Delta \to 0$. Table 1 lists $W(\theta_{max})/r$ as a function of Δ for a water (or mercury)/aluminium interface (μ=0.25) where θ_{max} corresponds either to a numerical aperture of 0.25 or 0.5 in the object.

Table 1. Normalised spherical aberration as a function of Δ for a water/aluminium interface (μ=0.25)

Δ	$W(\theta_{max})/r$ (n.a. = 0.25)	$W(\theta_{max})/r$ (n.a. = 0.5)
1/50	2.1×10^{-6}	37.8×10^{-6}
1/40	3.2×10^{-6}	57.1×10^{-6}
1/30	5.3×10^{-6}	95.8×10^{-6}
1/20	10.5×10^{-6}	191.1×10^{-6}
1/10	27.5×10^{-6}	506.8×10^{-6}

For both cases considered in Table 1, it is clear that spherical aberration is negligible; in the worst case, $W(\theta_{max})/r$ is around 0.5×10^{-3} for Δ=0.1.

In our experiments, we have used an aluminium coupling element with a convex surface having a radius of 10 mm and the other surface was machined to fit the cylindrical object. The acoustic frequency used was 22 MHz so that the wavelength within the aluminium cylinder was approximately 300 microns. The results in Table 1 indicate that in the worst case $W(\theta)$ = 5 microns or less than a fiftieth of an acoustic wavelength. Equ. (7) predicts that one could scan the focus over a distance of approximately 9 mm and still be clearly diffraction limited. We can qualitatively see this in a ray tracing model of the scanner, Figure 3. This shows focussing at a) 1 mm b) 3 mm and c) 5 mm depth within the cylindrical specimen and in all cases, the circle of least confusion is substantially smaller than our expected diffraction limited spot size.

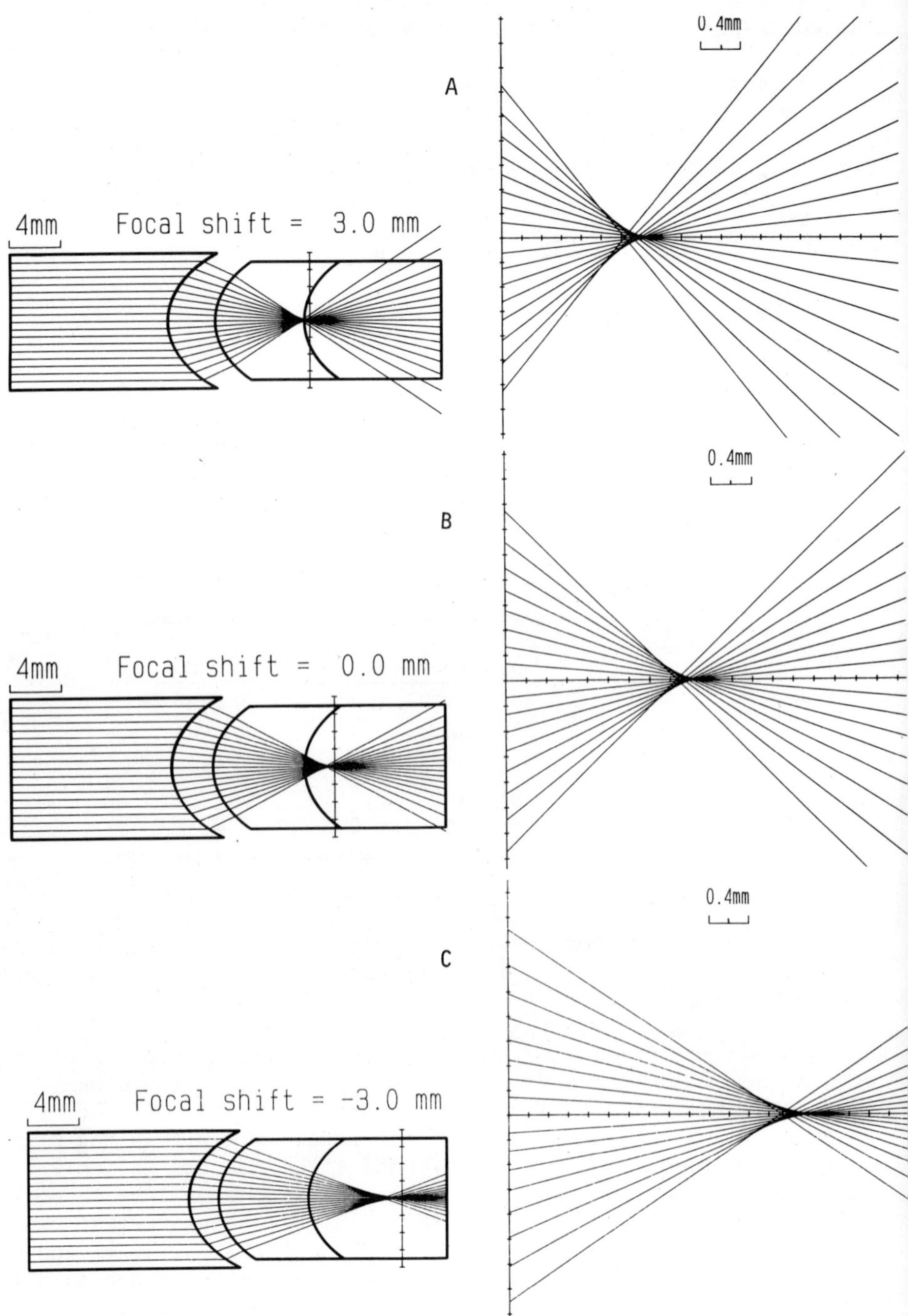

Fig. 3 Ray tracing model of B-scan microscope focussing at a) 1 mm b) 3 mm and c) 5 mm beneath the surface of a cylindrical specimen

B-SCAN MICROSCOPE FOR CYLINDRICAL SPECIMENS

We have built a B-scanning acoustic microscope for NDE applications, operating in pulsed mode at a centre frequency of 22 MHz, Figure 4. An important feature of our apparatus is that the coupling piece (element 2) is easily interchangeable. Our aberration theory predicts good imaging performance provided that the top surface of the coupling piece is spherical, that the coupling piece is acoustically identical to the specimen and that the lower surface is in intimate contact with the specimen - however, there is no specification as to the shape of the lower interface. In effect, the coupling piece makes an arbitrarily shaped specimen appear to the lens as having a spherical surface. We can use a range of coupling pieces to mate with almost any shape of specimen and still obtain diffraction limited images from the interior. We see this as the most powerful argument for the additional complexity of the B-scan microscope - it relaxes the requirements on the geometry of the specimen. Thus, in our experiments, where we have imaged within a cylindrical rod, we used a coupling piece with a convex spherical top surface and a concave cylindrical bottom surface.

The aluminium lens (radius 10 mm, half angle 45 degrees) slides within the microscope block and a micrometer is used to adjust the focal depth. The aluminium coupling piece mounts in the bottom of the block. In our experiments, we have used liquid mercury (impedance 19.7 MRayls) to acoustically couple the two elements because it gives a good impedance match to aluminium (impedance 17 MRayls). A reservoir of mercury relieves the pressure when the lens is focussed. The entire assembly presses the coupling piece onto the cylindrical aluminium specimen (radius 10 mm), and the coupling is improved and lubricated by a thin film of glycerin, loaded with aluminium powder. The specimen is rotated to give a Θ scan and a micrometer is used to drive the microscope in the Y-axis, along the length of the rod. Thus two imaging modes are possible. Firstly, an image may be recorded over a range of focal depths as the specimen is rotated - a cross-sectional image or B-scan. Alternatively, the scanner may be set to a particular focal depth and an image recorded both around and along the rod - a C-scan. Standard pulse-echo electronics are used to form the image, which is scanned and displayed by a microprocessor. In both cases, the receiver pulse window is set to 'range-gate' the focus. In practice, because of transducer bandwidth limitations, the depth of image field of our apparatus was set by diffraction rather than by the pulse duration.

The dimensions of the coupling piece are chosen such that when the focussed wavefront and the coupling piece are concentric (when $\triangleright$=0, a condition easily identified by a peak in the signal reflected from the coupling pieces surface) then the beam is focussed at 3 mm beneath the surface of the specimen. Figure 5 shows a rotation line scan over a 0.2 mm (.6λ) hole at 2 mm depth showing a beamwidth of

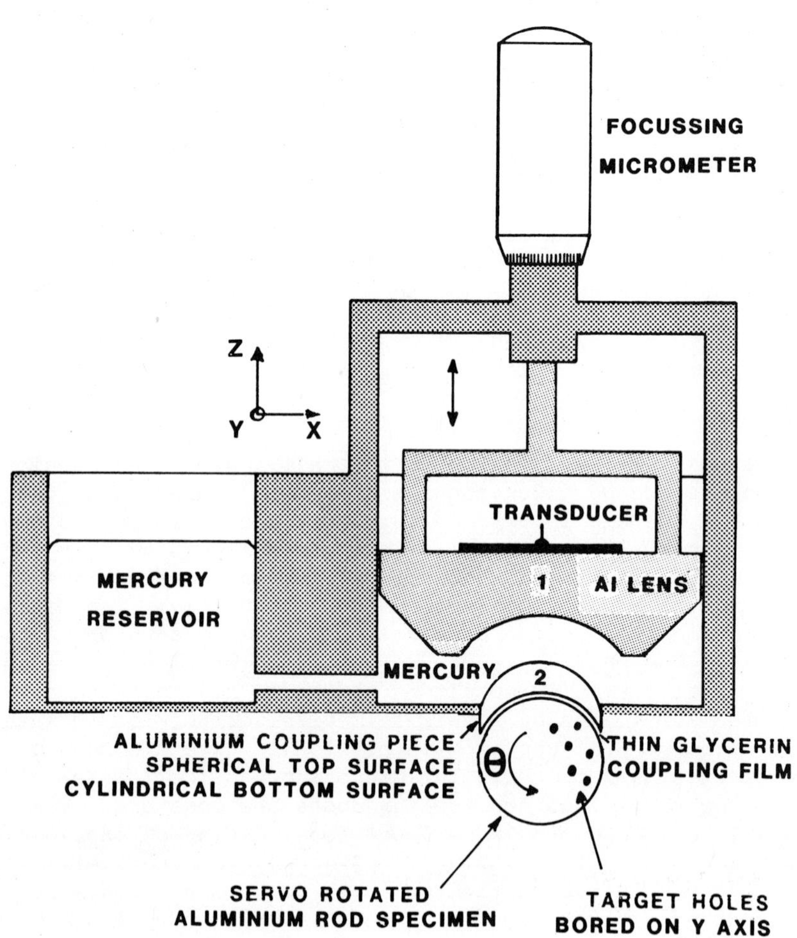

Fig. 4 Mechanically B-scanned acoustic microscope for interior imaging of cylindrical specimens

1.3λ, in good agreement with our expected point spread beamwidth of 1.25λ. The specimen has two sets of three holes drilled at depths of

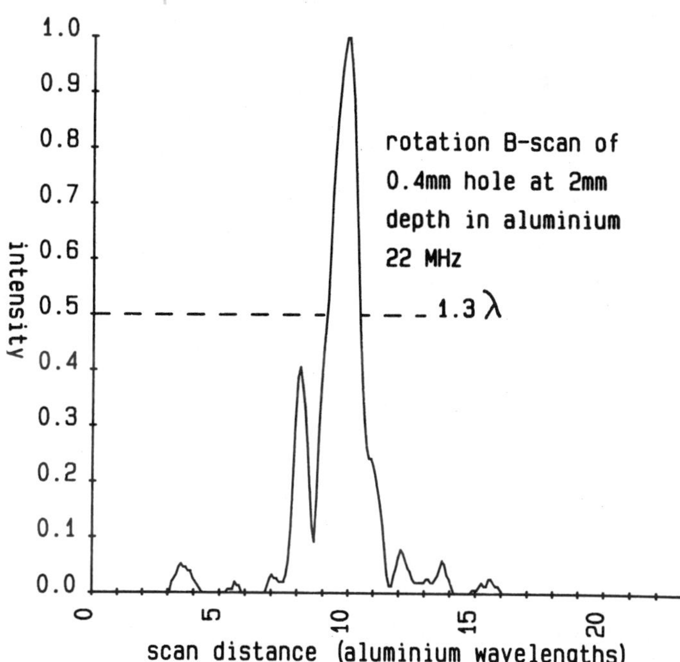

Fig. 5 Rotation scan of 0.2 mm hole at 2 mm depth inside aluminium cylinder. The point spread beamwidth is 1.3 wavelengths.

4 mm and 1 mm. The holes are drilled parallel to the axis of the rod with lengths differing by 2.5 mm. Figure 6 shows a C-scan of the holes at 4 mm depth over a specimen rotation angle of 210 degrees (corresponding to a circumferential scan distance of 21 mm). The differing hole lengths are easily determined. Whilst the sidelobe levels are low, the holes appear speckled. This may be attributed to roughness in both the hole and the contact between coupling piece and specimen - the familiar V(z) effect. Figure 7 shows a C-scan of the set of holes at 1 mm depth. The holes are masked by two effects. The pulse duration (0.25μs, limited by transducer bandwidth) does not permit discrimination of surface and subsurface echoes - these combine to form the image. In addition, the specimen surface is nearly in focus and so the image shows surface detail. In Figure 6 this was not apparent since the specimen surface was defocussed.

INSTRUMENTAL MODIFICATIONS

The coherent sensitivity to small depth variations in our images could be reduced by using swept frequency or noise transmissions[5].

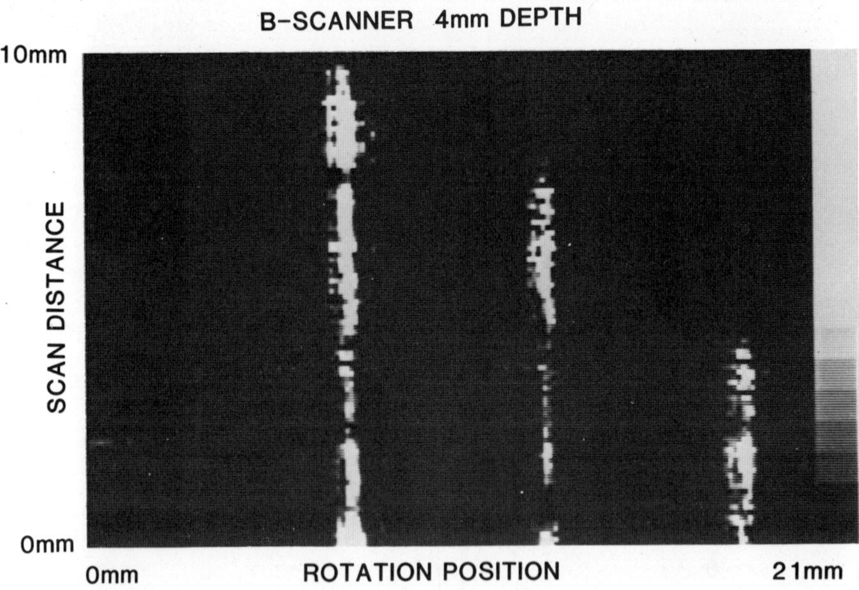

Fig. 6 C-scan of holes at 4 mm depth inside aluminium cylinder. The holes differ in length by about 2.5 mm.

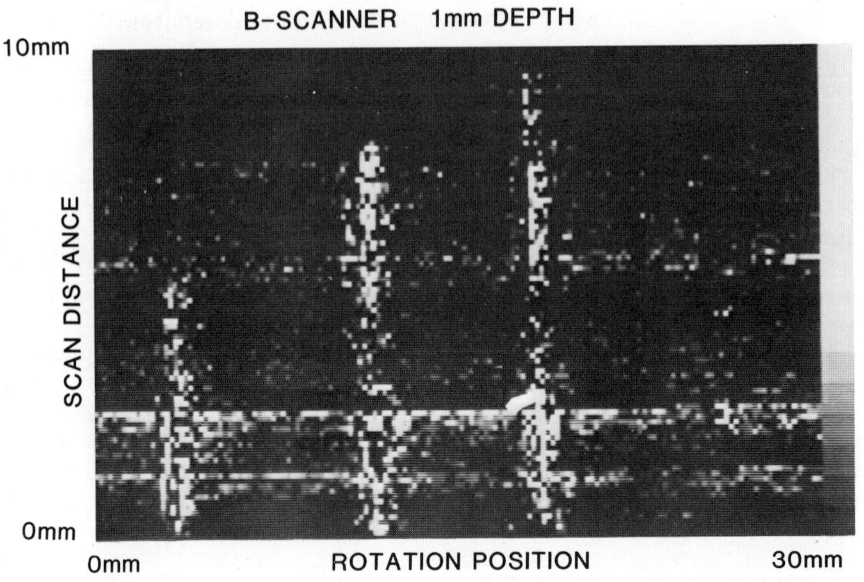

Fig 7 C-scan of holes at 1 mm depth inside aluminium cylinder. The holes differ in length by about 2.5 mm

In our present apparatus, we have been unable to image B-scans over a suitably wide range of depths because of the amount of energy contained within the lens - we cannot separate the focal echoes from reverberant ones over a continuous range. This problem can be simply overcome by a more careful choice of reverberation period within the lens rod and coupling piece.

CONCLUSIONS

We have shown how high (diffraction limited) resolution interior imaging may be achieved within arbitrarily shaped objects. An aberration theory shows that B-scanning over a wide depth range is possible, and this has been confirmed by a ray tracing model and by experiments on a cylindrical specimen. We shall use an improved form of our apparatus for NDE of ceramic components.

ACKNOWLEDGMENTS

The authors are grateful to E.A. Ash for helpful advice and to W. Raven, who skillfully constructed the apparatus. This work was supported by the Wolfson Unit for Micro-NDE.

REFERENCES

1. D. A. Sinclair, I. R. Smith and H. K. Wickramasinghe, "Recent Developments in Scanning Acoustic Microscopy", The Radio and Electronic Engineer,Vol. 52, No. 10, October 1982.

2. V. B. Jipson, "Acoustic Microscopy of Interior Planes", Appl. Phys. Lett., 35, pp. 385-387, 1979.

3. F. Pino, D. A. Sinclair and E. A. Ash, "New Technique for Sub-Surface Imaging Using Scanning Acoustic Microscopy", Ultrasonics International '81, Brighton, England, 30th June to 2nd July, 1981, IEE Press, 1981.

4. H. K. Wickramasinghe, "Mechanically Scanned B-Scan System for Acoustic Microscopy of Solids", Appl. Phys. Lett. 39(4), pp. 305-307, 15th August, 1981.

5. J. Attal, N. Truong-Quang,G. Cambon, J. M. Saurel and M. Rouzeyre, "Acoustic Microscopy with Noncoherent Source", Elec. Lett. 17, pp. 116-117, 1981

SCANNING ACOUSTIC MICROSCOPY INSIDE CERAMIC SAMPLES (*)

B. NONGAILLARD, J.M. ROUVAEN - Laboratoire O.A.E.

H. SAISSE - Laboratoire de Marcoussis Division Matériaux

Laboratoire O.A.E. - ERA 593 CNRS - Université de Valenciennes, 59326 Valenciennes Cedex, France / Laboratoire de Marcoussis - CGE Route de Nozay, 91460 Marcoussis Cedex, France.

Since a few years, new materials have been worked out to comply with the contraints of severe operating conditions. A good example is from the ceramic materials, which are potentially very interesting. These compounds are synthesised from very common and cheap raw materials, their mechanical stiffness and strength are high and their general physical and chemical properties make them wear resistant and very refractory. However, these materials are not ductile, so that very good quality materials must be produced and a good knowledge of the harmfulness of the defects is required. This harmfulness will be deduced from a comparison of the actual dimensions of the defect to the critical length a_c for that defect, corresponding to the fracture of the sample under a known stress field σ_r.

It is therefore capital to determine a maximum number of default parameters (geometry, nature, situation) by using non destructive testing methods. The small toughness of ceramics leads to particular features for the non destructive detection and characterization of their internal defects. The ceramic components are often used under very high stress conditions (up to 300 MPa). The critical length of a defect is then nearly equal to 0.2 mm, assuming a 5 MPa $\sqrt{m}$ value for the critical stress intensity factor K_{IC} (a figure valid for SiC ceramics).

(*) This work has been done with the collaboration of Laboratoires C.G.E. in Marcoussis (France) under the financial support of Direction des Recherches Etudes et Techniques (France).

This factor may be computed from the formula $K_{1C} = \sigma \sqrt{\Pi a}$ where σ stands for the stress leading to facture and 2a for the length of the crack. Very sophisticated non destructive testing techniques are therefore required for such ceramic materials, since the defects to be detected are nearly two order of magnitude smaller in length than those encountered for metallic materials.

The scanning acoustic microscopy in the reflection mode is a technique well suited to the non destructive investigation of ceramics. It provides not only means for the detection and sizing of defects but also for the display of a high resolution picture necessary for a better characterization. The acoustic microscopy which is already used for non destructive testing near the surface of electronic chips using an operating frequency in the gigahertz range (1, 2), may also be used here for exploring the bulk of ceramic samples using a frequency near 100 MHz. It must be noticed that the study of alumina samples is one of the most difficult in ultrasonic non destructive testing area. The ultrasonic waves propagate very fast in these materials (say 10,000 m/s), leading to a large acoustic wavelength in the operating frequency range and therefore to a decrease in the spatial resolution (with respect to slower materials). Moreover, the acoustic impedances of ceramics are generally very high (say 40 MRayls), so that only a small part of the incident acoustic energy (typically 7% if water is used as coupling medium) is transmitted to the sample, and the dynamic range of the image is so reduced, assuming a given sensitivity for the electronic receiver.

By solving the propagation equation for the velocity potential Φ, it has been shown possible to keep the acoustic focal spot nearly aberration-free after traversal of the sample surface(3). These results have also been verified for materials with a high acoustic velocity ($v \simeq 10{,}000$ m/s). Some results from this calculation are given in tab. 1 where λ_3 stands for the wavelength in the sample and Φ_f for the velocity potential taken at the focal spot of the lens (fig. 1).

Aperture of the lens	$\theta_m = 5°$	$\theta_m = 10°$	$\theta_m = 20°$
focal spot width for $\frac{\Phi}{\Phi_f} = 0.5$	$2.25\ \lambda_3$	$1.8\ \lambda_3$	$1.8\ \lambda_3$
length of focus for $\frac{\Phi}{\Phi_f} = 0.5$	$8.5\ \lambda_3$	$6\ \lambda_3$	$6\ \lambda_3$

Table 1

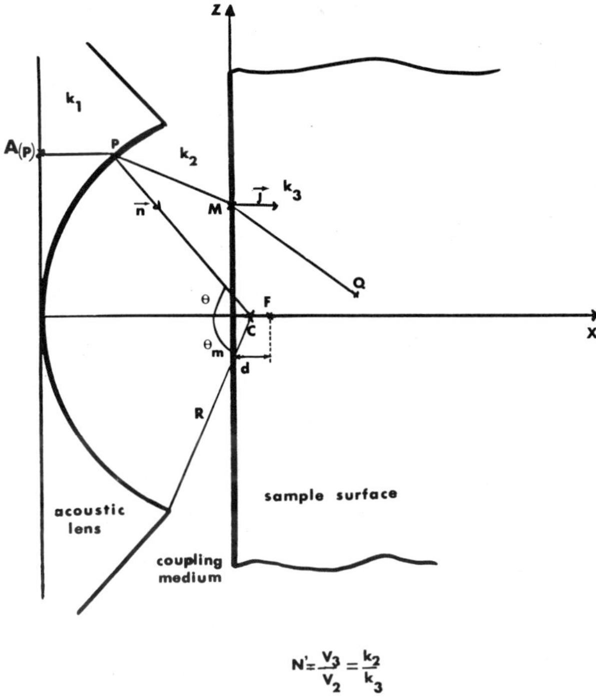

Fig. 1 - Notation used in the computer calculation

The results obtained for a lens aperture angle $\theta_m = 10°$ (see fig.1) are shown in figs 2 and 3.

The distance d between the focal plane of the lens and the sample surface has been taken equal to R/6, where R stands for the radius of curvature of the spherical lens.

This radius of curvature has been assumed equal to 1,000 times λ_2, the acoustic wavelength inside the acoustic coupling medium.

It may be noticed that the behaviour of the curves doesn't vary much if the aperture of the lens exceed the critical reflection angle (nearly 8 degrees for alumina). The position of the amplitude maximum may be fairly precisely given by the distance $z = d/n$ from the sample surface, where n stands for the relative index (velocity ratio) of the studied sample with respect to the coupling medium. (often water).

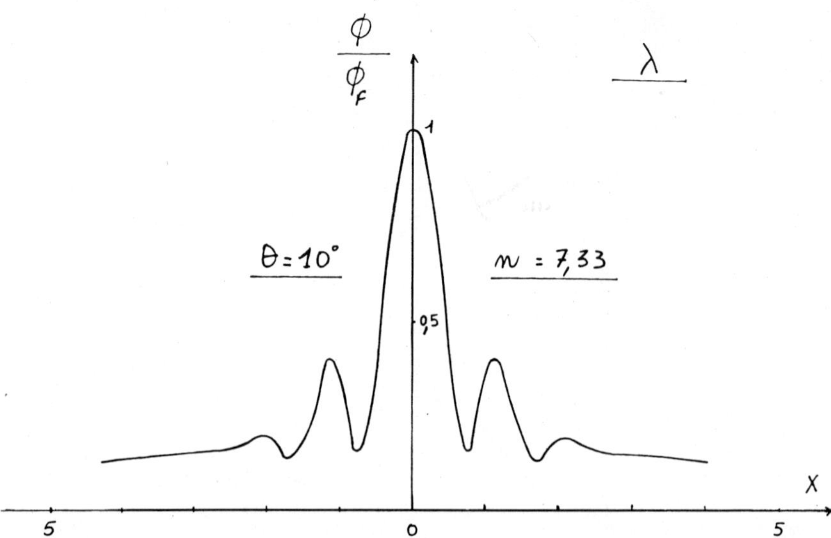

Fig. 2 - Transverse variation of the acoustic field in the focal plane

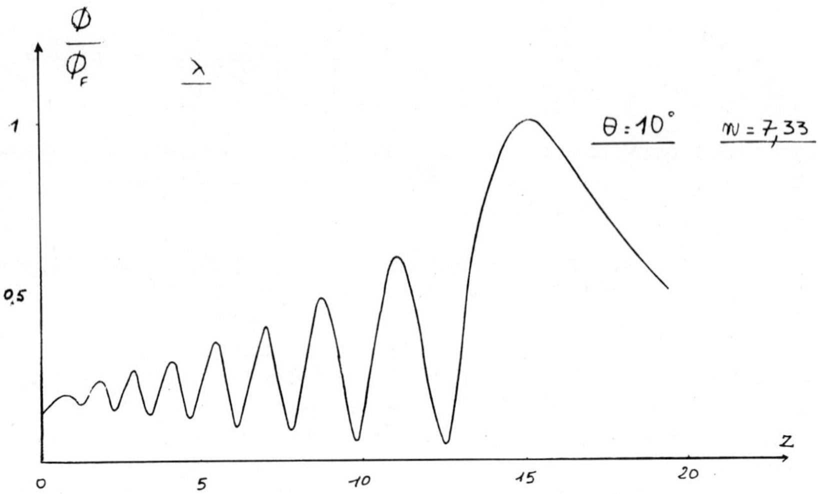

Fig. 3 - Axial variation of the acoustic field under the surface of the sample

From this approximate formula, it may be seen that a large radius of curvature is needed for an acoustic lens designed to study ceramic materials (for R = 15 mm, the maximal focusing depth is nearly equal to 2 mm inside alumina samples). It has been said that the minimum length of the defects to be detected is 0.1 mm. This implies the choice of an operating frequency near 100 MHz, where the spatial resolution is theoretically equal to 100 µm (nearly 0.9 λ).

At this particular frequency the intrinsic ultrasonic attenuation of ceramics enables a propagation of acoustic waves over several millimeters (for our alumina samples the attenuation was measured to be nearly 1dB/mm at 100 MHz, but this figure may vary a lot among samples from different sources and the compactness got during the manufacturing process).

For the purpose of studying the ceramic samples, an apparatus has been built (fig. 4) with an automatic magnifying power control (between 10 and 500), which drives the mechanical scanning over the sample and generates the electronic signals used for monitor display.

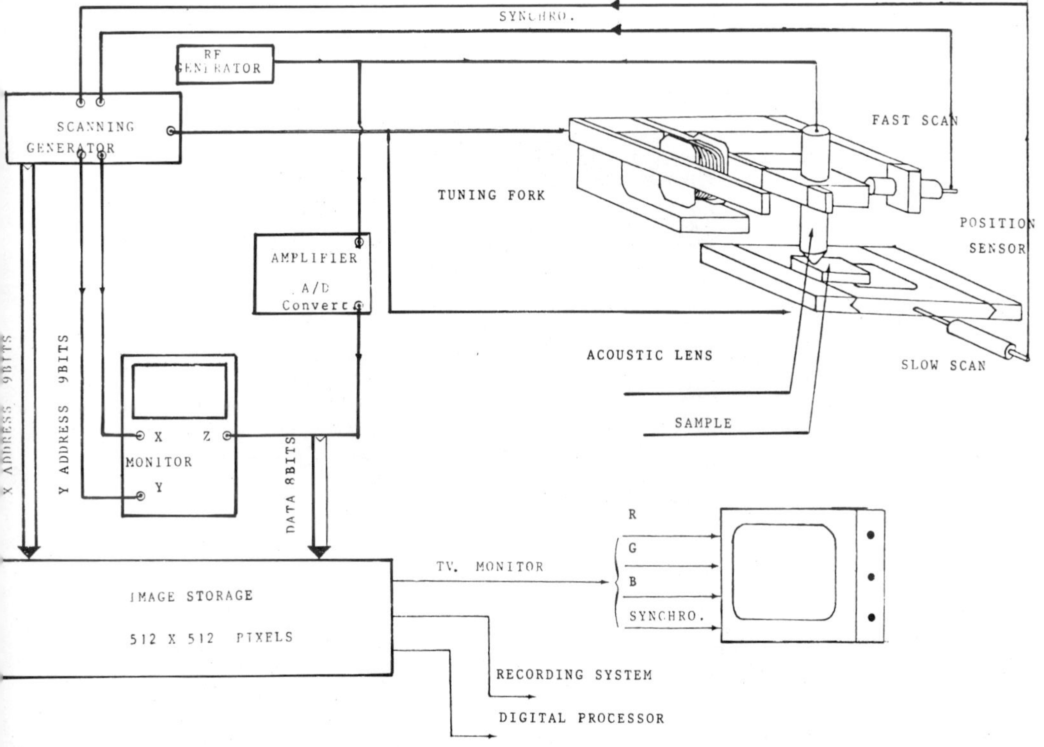

Fig. 4 - Synoptic scheme of the acoustic microscope

Moreover an image digitizing system has been added in order to enable the digital storage and processing of the image together with a good quality display using a TV set. The image rate is very low (typical frequency between 0.1 and 1 Hz) since it is limited by the faster mechanical scanning (a tuning fork here). After digitizing and storing, the image may be read at the TV rate, a screen with high remanence is no longer needed, which makes the observation more comfortable.

For the images to be shown late, the digitizing has been performed with a 8 bits precision, but only 6 bits are used in a false color encoding scheme. The effectiveness of the display system is shown in fig. 5, where the optical and acoustic images of a periodic grid with a 200 μm period are given.

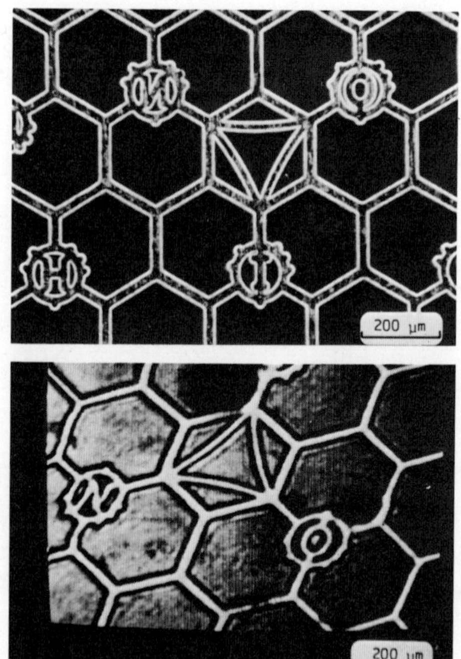

Fig. 5 - Optical (at the top) and acoustic images of a test grid

In order to evaluate the spatial resolution of the system under the surface of ceramic samples, the image of several gratings observed through the opposite surface have been taken. In fig.6, the acoustic image of a crossed grating with a period 0.8 mm, seen at a depth of 2 mm is shown. The acoustic beam is focused over the valley of the grooves. In fig.7, a linear grid is shown with a period of 300 μm at a depth of 1.5 mm. This image corroborates the theoretical figure of 100 μm for the spatial resolution.

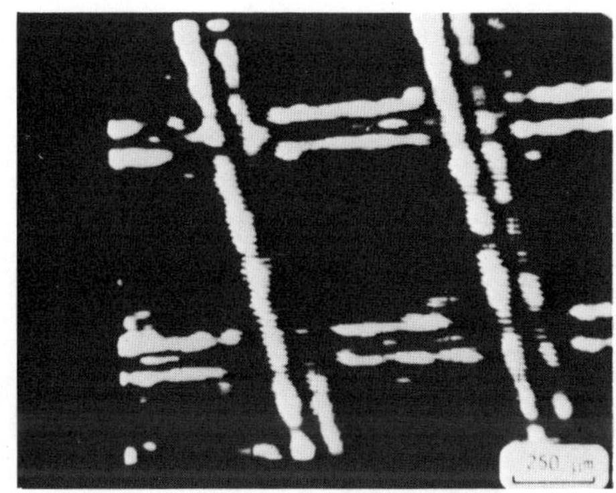

Fig. 6 - Acoustic image of a crossed grating at a depth of 2mm in alumina

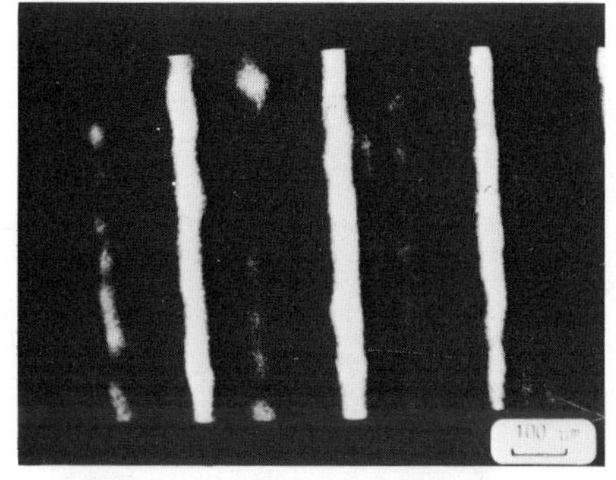

Fig. 7 - Linear grid at a depth of 1.5mm in alumina sample

It has not been technologically feasible to decrease the period of the gratings to 100 μm. All the previous images have been taken with a single lens, with a 16 mm radius of curvature.

These results lead to the following important conclusion concerning the observation of the bulk of ceramic samples using focused ultrasound. The spatial resolution may be kept constant over the exploration depth, since it is nearly given by the formula $\lambda d/a$ (where d stands for the focusing distance below the surface and a the diameter of the entry pupil of the lens) and d and a are proportional to the radius of curvature R. This phenomenon is exactly the same as that of the synthetic aperture focus system used in B scan mode ultrasonic echography.[4] It is therefore possible to explore at a larger depth, while keeping the spatial resolution constant only if the sample surface doesn't limit the entry pupil diameter.

Owing to the brittleness of ceramic materials, the ability to observe surface cracks and also those emerging at the surface is essential. This kind of defect has been simulated with controled length and depth, using an indentation method for alumina and glass samples. By focusing at a depth of 80 μm below the surface (fig. 8), the cracks are clearly visible at the apex of the indentation shape, but when focusing right over the surface, these cracks are no longer visible, because of the raw unpolished surface obtained after sintering.

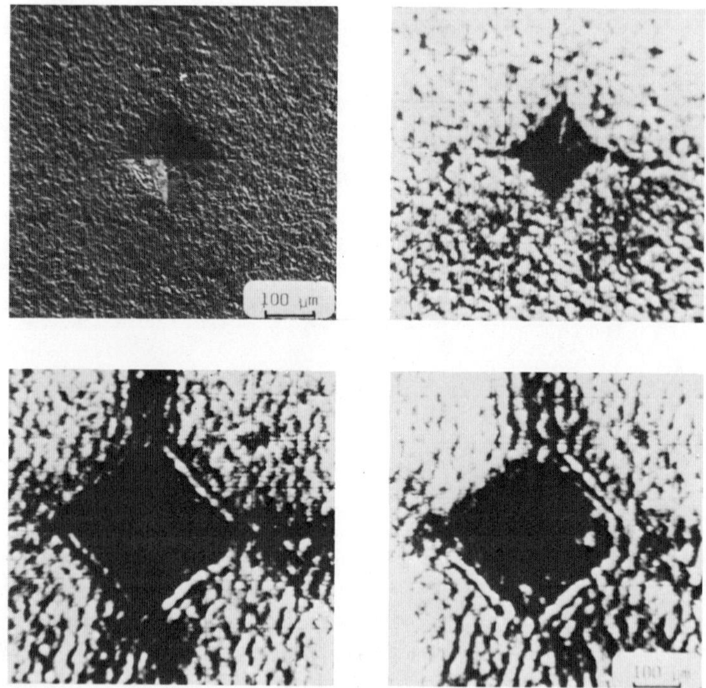

Fig. 8 - Indentation picture at different depth in alumina sample

By using glass samples, the perturbing influence of the unpolished surface has been eliminated (see fig. 9)

The fig.10 shows the image of the emerging part of a zircon inclusion using a scanning electron microscope. The acoustic image given at fig. 11 shows darkened the bottom of a broken glass, the emerging part of the inclusion appearing lightened. The inclusion is clearly visible for a focusing depth of 100 μm under the sample surface (fig. 12).

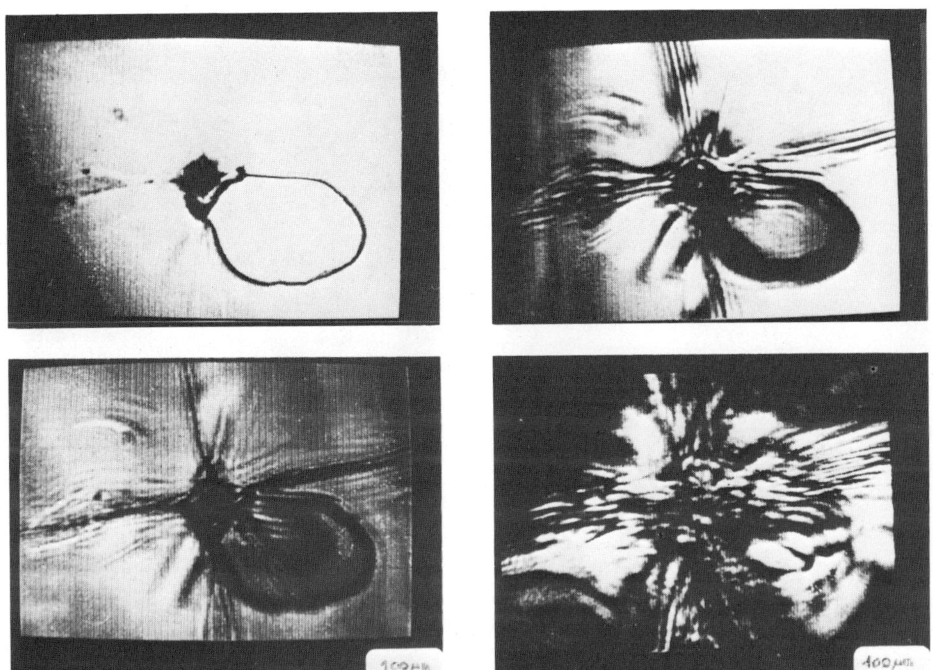

Fig. 9 - Indentation picture in glass sample acoustic image a) on the surface c) 50 μm depth
b) 10μm depth d) 110 μm depth

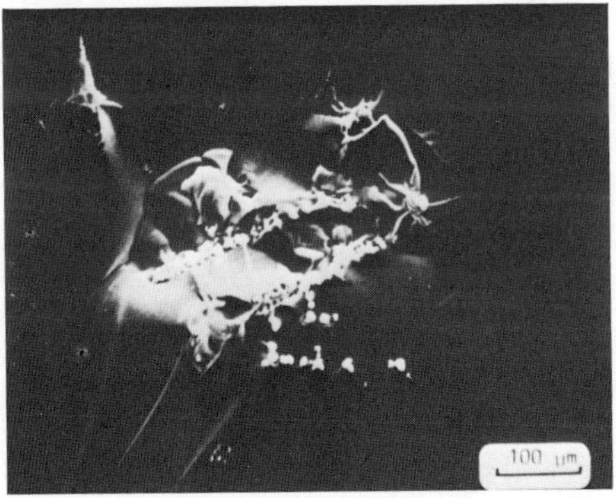

Fig. 10 - Scanning electron microscope image of the glass sample surface

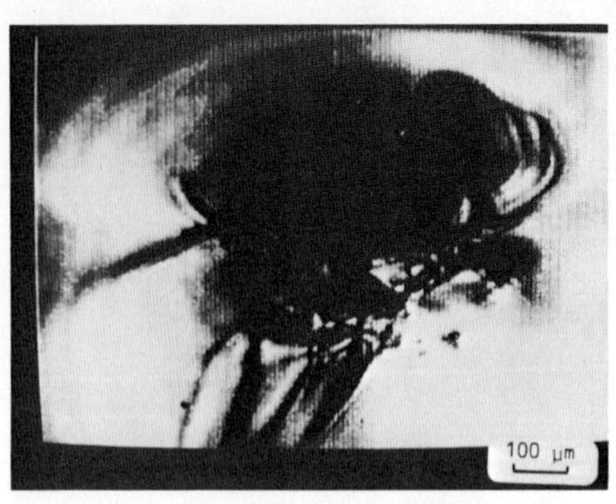

Fig. 11 - Acoustic image of the surface

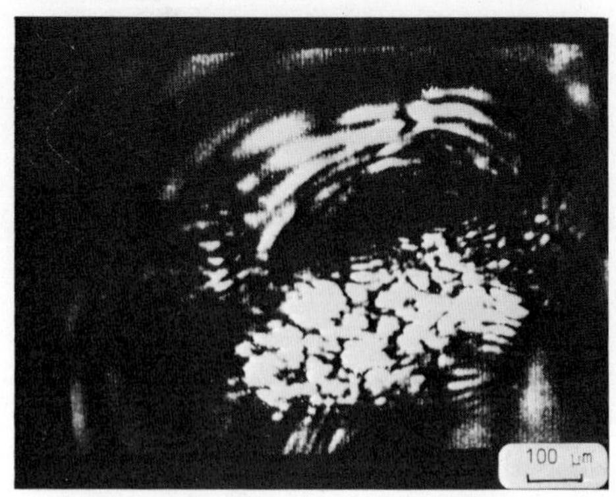

Fig. 12 - Acoustic image at a depth of 100 μm

These last images have been obtained using a lens with a 1mm radius of curvature. It may be seen that the structures under the surface are evidenced by associated structures (seeming like interferance fringes) whose interpretation is now in progress. Owing to the corresponding exploration depth, it is indeed not possible to discriminate in time the surface echo from that returned from the visualized structure and carried by a longitudinal or a transverse acoustic wave.

As a matter of fact; for a 100μm depth in alumina sample, these echoes are delayed only 20ns and 40 ns from the surface one, whereas the driving electrical pulse has a minimum duration of 50ns. The feasibility of applying acoustic microscopy to the high resolution non destructive testing of ceramics has been demonstrated here. Pictures of controlled defects lying at the sample surface or below it have been obtained. These results must, in the near future, be extended to the sizing and identification of the nature of the detected defects, in order to estimate their harmfulness in industrial applications.

REFERENCES:

1. Attal J and Cambon G.
Non destructive testing of electronic devices by acoustic microscopy.
Revue de Physique Appliquée , 13 ,815-819, 1978.
2. Hollis R.L and Hammer R.
Imaging techniques for acoustic microscopy of microelectronic circuits.
Acoustical Imaging, vol.10 ,817-827, 1980.
3. Nongaillard B and Rouvaen J.M.
Acoustic microscopy in non destructive testing.
Acoustical Imaging, vol.10 ,797-802, 1980.
4. Delannoy B,Torguet R ,Bruneel C ,Bridoux E ,Rouvaen J.-M ,and Lasota H.
Acoustical image reconstruction in parallel-processing analog electronic systems.
J.A.P,50,3153-3159, 1979.

ULTRASONIC-WAVE GENERATION BY SURFACE AND BULK HEATING IN MULTIMATERIAL STRUCTURES

Grover C. Wetsel, Jr.

Department of Physics
Southern Methodist University
Dallas, Texas 75275

INTRODUCTION

In many light-absorbing materials a substantial fraction of the annihilated photon's energy is transformed into heat. If the light is pulsed, then pulses of heat are produced which, in turn, generate pulsed elastic disturbances; this phenomenon is commonly known as the photoacoustic effect. Whereas the heat pulses are exponentially damped as they move away from the source, the elastic waves propagate with relatively little attenuation. The elastic waves so generated contain information about the thermal, elastic, and optical properties of the material in which they are produced as well as information about materials in contact with the generating material. The information carried by the elastic wave is more localized--that is, the resolution is greater-- the higher is the frequency of the light pulses; and if the frequency is increased to the microwave range, the resolution can approach or even surpass that of the optical microscope. The generation of high-frequency elastic waves by photothermal means is of current interest because of its relevance to several areas of physics, including: the development of the photoacoustic microscope[1], thermal-wave imaging[2], determination of thermoelastic material parameters, nondestructive evaluation of devices[3], and laser annealing and melting phenomena in semiconductors.

Since photothermally-generated ultrasonic waves carry information characteristic of the generating medium and adjacent media, one-dimensional theoretical models have been developed with the goal of understanding the basic material effects on the generation process. A bulk-heating model consists of a nonabsorbing backing material through which the incident light or particle beam propagates, an absorbing film, and a nonabsorbing sample. As a further aid to understanding the essential physics of the phenomenon, a surface-heating model with the absorbing film replaced by an infinitesimal, surface source has also been treated. Calculations of the temperature, elastic-displacement amplitude and phase, and ultrasonic intensity as functions of position, structure dimensions, and absorption coefficient have been made in the frequency domain for several material combinations. The results have been compared with available experimental results. Calculations in the time domain have also been performed for purposes of comparison with developing experiments.

In this paper the results of these calculations are discussed in terms of their relevance to efficiency of ultrasonic-wave generation and imaging. The results of initial experiments using a probe-laser beam to detect ultrasonic waves photothermally generated by 10 nsec pulses from a N_2 laser are also presented.

THEORY

The bulk-heating model provides for a semi-infinite backing ($x<0$), a light-absorbing film with optical absorption coefficient, β ($0 \leq x \leq d$), and a semi-infinite sample ($x > d$). The thermal-diffusion equation with negligible coupling to the elastic-wave equation was first solved in the frequency domain for all three regions. In the film, the heat source density was taken to be $H=(\beta I_0/2)\exp(-\beta x+j\omega t)$, where I_0 is the intensity of the light converted into heat at $x=o$ and ω is the pulse repetition or chopping frequency of the light.[4] Continuity of temperature and heat flux at $x=o$ and $x=d$ were used to determine the temperature $T(x), -\infty < x < \infty$. Then, the elastic-wave equation with losses was solved for all three regions using the previously-determined solutions to the thermal diffusion-equation. Continuity of particle velocity and stress at $x=0$ and $x=d$ were used to determine the elastic displacement, $u(x)$, $-\infty < x < \infty$. Using values of material parameters obtained from the literature, computer calculations of temperature, elastic displacement, and ultrasonic intensity were initially made for two cases of the multimaterial structure: (A) backing=sapphire, film=molybdenum, sample= fused quartz, and (B) backing=air, film=molybdenum, and sample=sapphire. The theoretical value of the ultrasonic amplitude at 20MHz for case (A) was about 42dB greater than that for case (B).[5,6] This compares very well with the corresponding experimental value of 40dB obtained by von Gutfeld and Melcher.[7] Temperature and elastic displacement distributions and displacement amplitude in the sample as a function of d, β, and ω have been previously presented for case (A), above.[5]

Computer-calculations of temperature and displacement distributions based on the bulk-heating model are shown in

(a)

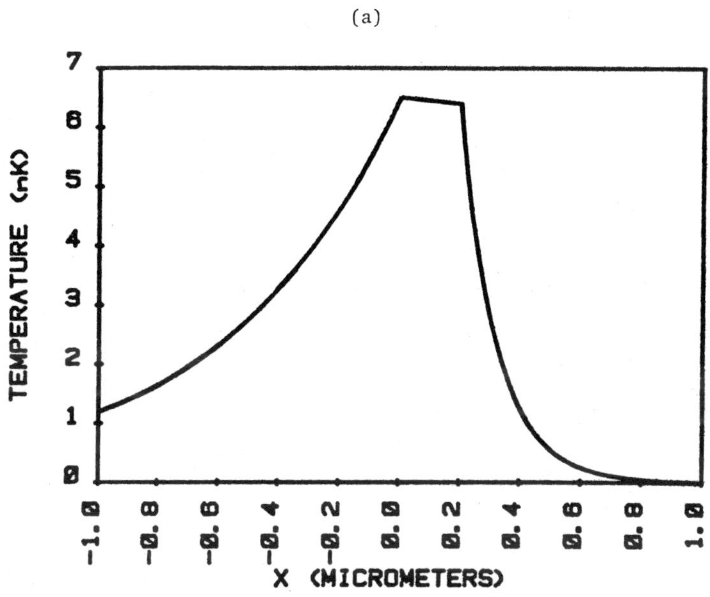

(b)

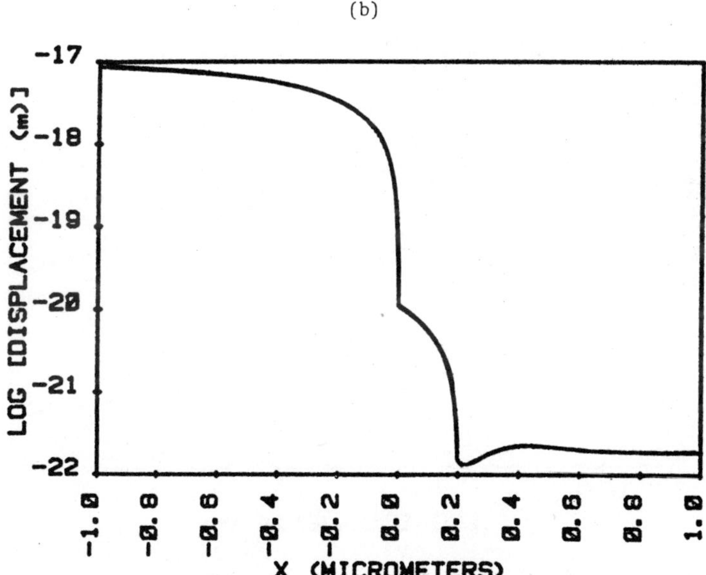

Fig. 1. Backing=air, film=Mo, sample=quartz; f=20MHz, $\beta=6.51\times10^{7}m^{-1}$, d=200nm. (a)T(x), (b)u(x).

Fig. 1 for the case: backing=air, film=molybdenum, and sample= fused quartz. Harmonic modulation of light of wavelength, 580 nm, at 20 MHz was assumed for these calculations. Under these conditions the optical-absorption length in Mo is 15.4 nm and the thermal-diffusion length in Mo is 920 nm. Since the film thickness is 200 nm, the Mo film is optically thick and thermally-thin in this case. It is particularly interesting to notice that an air backing -- even though it is a large acoustic-impedance mismatch to Mo -- has a nonnegligible effect on the ultrasonic-wave generation in the quartz.

The amplitude of the ultrasonic wave in the sample (quartz) for d=200 nm is shown in Fig. 2 as a function of frequency for air-molybdenum-quartz and sapphire-molybdenum-quartz. For purposes of comparison, calculations of the amplitude of the displacement in quartz based on the surface-heat-source model of White[8] are also shown in Fig. 2 for his cases of ideal-clamp backing and ideal-stress-free backing. One should notice that an air backing -- because of its thermal properties -- has a substantially greater effect on the ultrasonic-wave generation than an ideal stress-free backing. One should also notice that a sapphire backing gives a greater efficiency of photothermal generation than an ideal-clamp backing.[5,6]

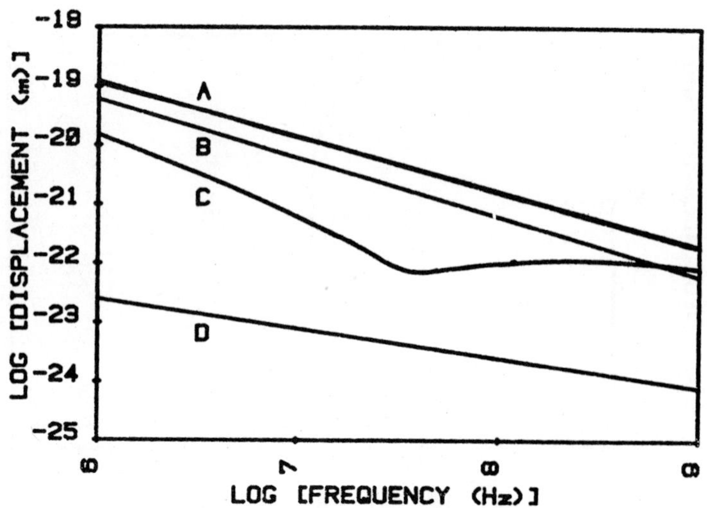

Fig. 2. $\tilde{u}(f)$ for surface heat density or I_0=1.0W/m^2. (A) Sapphire-Mo-quartz, $\beta=6.51\times10^7 m^{-1}$,d=200nm; (B) quartz with surface source and ideal clamp; (C) air-Mo-Quartz,$\beta=6.51\times10^7 m^{-1}$,d=200nm; (D) quartz with surface source and ideal stress-free surface.

The effects on the amplitude of photothermally-generated ultrasonic waves due to the thermoelastic parameters of the materials in the composite structure can best be understood using an analytical rather than a numerical solution to the problem. Such a solution can be accomplished using an extension of White's surface-heat-source model to include the effects of the thermoelastic parameters of the backing material.[9,6] According to this model, the most efficient photothermal ultrasonic-wave generation occurs when either sample or backing (or both) has a large value of $\alpha(D)^{1/2}$ and there is not a great acoustic-impedance mismatch of backing and sample, where α is effectively the thermal-expansion coefficient and D is the thermal diffusivity.

Computer calculations of the temperature at x=d as a function of time are shown in Fig.3 for air-molybdenum-quartz assuming that the film is heated by a sine-squared-shaped laser pulse with a 10 ns pulse width. The laser intensity is shown as a function of time for comparison. One should notice the characteristic delay of the temperature pulse and the long cooling tail. These calculations were made in support of the design of the experiment described below.

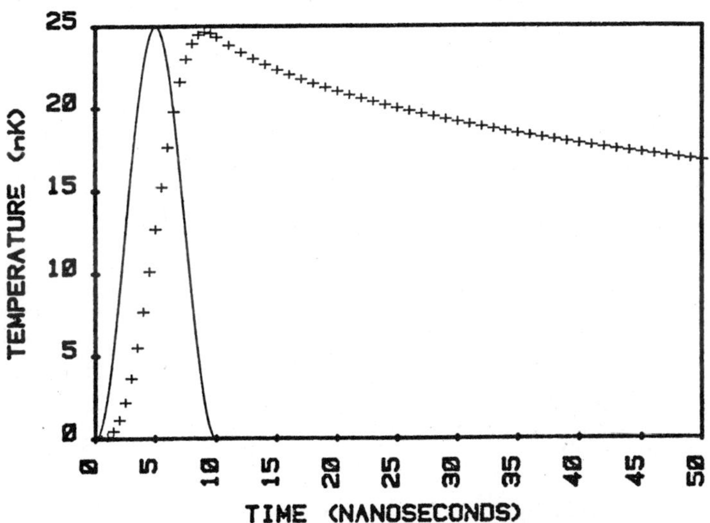

Fig. 3. Temperature vs. time (+) for air-Mo-quartz heated by a sine-squared laser pulse with I_0=1.0 W/m^2, β=6.51×10^{7}m^{-1}, x=d=200 nm; solid curve is normalized laser pulse.

EXPERIMENT

The most effective experimental evaluation of theoretical models of photothermal ultrasonic-wave generation would make use of the frequency domain. However, apparatus suitable for this purpose would involve tunable broad-band microwave-frequency modulation of light. The present impracticality of that approach and the availability of a pulsed laser have dictated experiments in the time domain. Photothermal generation was accomplished using 10ns, ∿ 400 kW pulses from a Molectron UV22 N_2 laser with a variable pulse-repetition frequency up to 100Hz. This "pump" laser was focused on solids and liquids having optical absorption at 337nm. Because of the rise time (∿3ns) of the laser pulse, it would be necessary to have fast-risetime transducers and electronics to detect the ultrasonic waves in the usual way. The requirement of broad brandwidth, the feature of integrating as opposed to localized detection, and the never-popular ultrasonic bonding problem argue against the use of electromechanical transducers. Therefore, the ultrasonic pulses were detected using a probe laser (Ar^+) beam directed perpendicular to the direction of wave propogation.[10] The probe beam is deflected by the elasto-optic (photoelastic) effect when the elastic wave travels through the probed region of the sample. A fast (∿0.2 ns risetime) photodiode placed behind an aperture or knife edge detected the deflection signal, which was connected to a Tektronix 7854 waveform-digitizing oscilloscope. The oscilloscope interfaced with a Hewlett-Packard 9825T computer and 9872B graphics plotter to produce a permanent record of the signals.

In the initial experiments, photothermally-generated elastic waves were generated by absorption of light in the glass wall of a cuvette containing water and by absorption in a solution of quinine sulfate in water (a good optical absorber at 337 nm) placed in the cuvette. The elasto-optic deflection signal in the quinine sulfate is shown in Fig. 4 for 5 positions of the probe beam. The time delay between the primary pulses is characteristic of the compressional-wave velocity in water. The time delay between the small pulses following each large pulse is characteristic of echoes in the glass wall of the cuvette. It is interesting to note the half width of the detected pulse is about 10 times that of the exciting laser pulse.

DISCUSSION OF RESULTS

The one-dimensional models provide physical insight into the essential features of photothermally-generated ultrasonic waves. The models predict that the elastic-displacement amplitude, A, in the sample increases with increasing β until essentially all the incident beam is absorbed within the film; A increases with d for

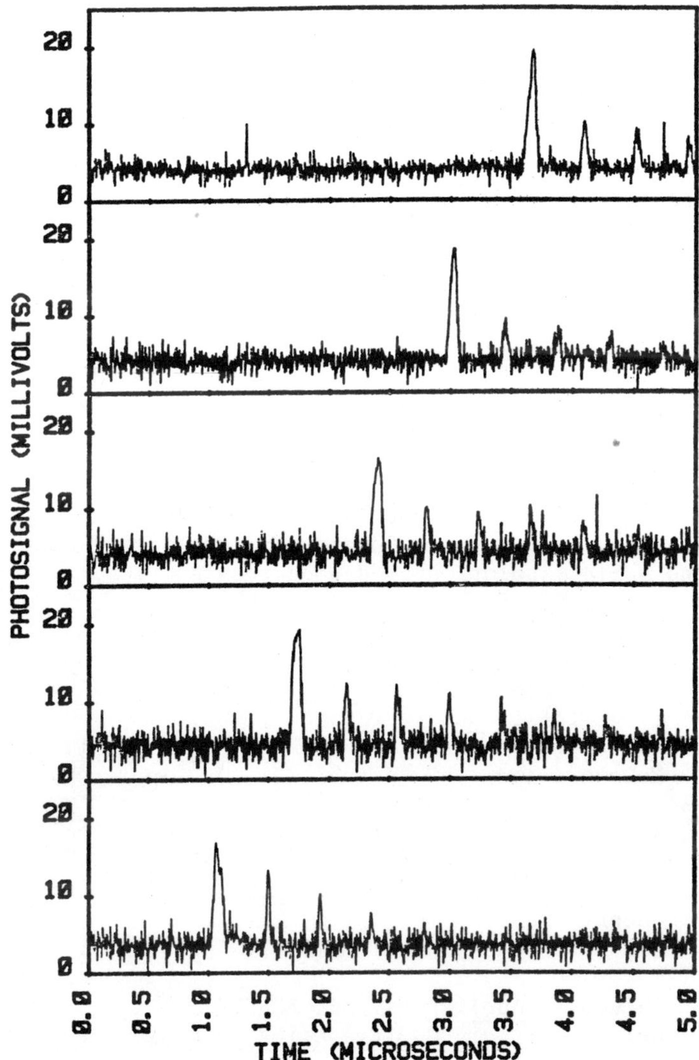

Fig. 4. Probe-laser deflection signal vs. time for 5 positions of the probe beam spaced 1.0 mm apart; pump-laser pulse width=10 ns; sample=glass cuvette containing quinine sulfate solution in .04 M HCl.

a given β until d is equal to about $5/\beta$, that is, about 5 absorption lengths. However, A may continue to increase with further increases in d if the film is thermally thin. The frequency dependences of the amplitude, phase, and intensity of the ultrasonic wave are functions of the material parameters of the backing, film, and sample. These predictions are in substantial agreement with what experimental evidence is available, [7,11-14] but experiments more appropriate for evaluation of details should be done--particularly those related to three dimensional effects.[15]

If the detected ultrasonic amplitude and phase are to provide information that is principally characteristic of the heated volume of the film or sample, as in thermal-wave imaging, then the backing material should have relatively small values of the relevant parameters, viz $\alpha(D)^{1/2}$. Thus, thermal-wave-imaging measurements might best be made with a vacuum backing.

For purposes of understanding the details of photothermal ultrasonic-wave generation, it appears that laser probing is the optimal experimental method; however, sensitivity can be a problem.

I gratefully acknowledge the contribution of Mr. Steven A. Stotts to the experimental results reported here. My appreciation is also extended to Dr. Henry Donato for identifying and supplying the quinine sulfate solution as a good test sample.

1. H. K. Wickramasinghe, et.al., Appl. Phys. Lett. 33, 923 (1978).
2. A. Rosencwaig and G. Busse, Appl. Phys. Lett. 36, 725 (1980); G. S. Cargill III, Nature 286, 69 (1980).
3. R. J. von Gutfeld and R. L. Melcher, Mater. Eval. 35, 97 (1977)
4. F. A. McDonald and G. C. Wetsel, Jr., J. Appl. Phys. 49, 2313 (1978)
5. G. C. Wetsel, Jr., 1981 Ultrasonics Symposium Proceedings, p.p. 810-814, publ. by IEEE, N. Y. (1981).
6. G. C. Wetsel, Jr., scheduled for publication in Appl. Phys. Lett., Sept.,1982.
7. R. J. von Gutfeld and R. L. Melcher, Appl. Phys. Lett. 30, 257 (1977).
8. R. M. White, J. Appl. Phys. 34, 3559 (1963).
9. G. C. Wetsel, Jr., 1980 Ultrasonics Symposium Proceedings, pp. 645-648, publ. by IEEE, N. Y. (1980).
10. The experimental configuration is similar to that reported by F. A. McDonald, G. C. Wetsel, Jr., and S. A. Stotts, "Scanned Photothermal Imaging of Subsurface Structure", also appearing in these Proceedings. Detection of compressional waves in liquids by this method has also been reported by W. Zapka and A. C. Tam, Appl. Phys. Lett. 40, 310 (1982).

11. R. M. White, J. Appl. Phys. 34, 2123 (1963).
12. M. J. Brienza and A. J. DeMaria, Appl. Phys. Lett. 11, 44 (1967).
13. G. Cachier, J. Acoust. Soc. Am. 49, 974 (1971).
14. R. J. von Gutfeld and H. F. Budd, Appl. Phys. Lett. 34, 617 (1979).
15. R. J. Dewhurst et al., J. Appl. Phys. 53, 4064 (1982).

SCANNED PHOTOTHERMAL IMAGING OF SUBSURFACE STRUCTURE

F. Alan McDonald, Grover C. Wetsel, Jr., and Steven A. Stotts

Physics Department
Southern Methodist University
Dallas, Texas 75275

INTRODUCTION

Photothermal imaging using laser beam deflection (PTLBD imaging) has recently shown considerable potential for nondestructive detection of subsurface structure in solids.[1] In this technique localized heating of a sample by a modulated laser beam produces a periodic temperature gradient in the fluid near the sample surface. The thermally induced gradient in the index of refraction of the fluid causes a periodic deflection of a laser-probe beam passing through the heated surface region. As the sample is translated relative to the beam intersection point, the magnitude and phase of the deflection signal give information about local variations in sample optical and thermal properties. Preliminary indication of the detection of subsurface structure with this technique has been given.[2,3] The PTLBD technique has proved successful in studying surface structure,[2-4] and has also been used in spectroscopic studies.[5] The potential advantages of PTLBD imaging over related photothermal approaches have already been identifieid. When compared to photoacoustic cell detection,[6] this technique has much greater potential sensitivity,[2] is not restricted to small sample size, and allows both excitation and detection to be spatially localized.[7] Thermal-wave microscopy[8] requires that a transducer be attached to the sample to detect the thermally-generated acoustic waves, so acoustic as well as thermal properties of the sample affect the signal, and localization occurs only in excitation.

In this paper we present PTLBD data on well-characterized subsurface structures, and on a thermal defect discovered in one of the samples. The extant theory[7] is compared to the data and found to be inconsistent in some cases. A generalization of the theory

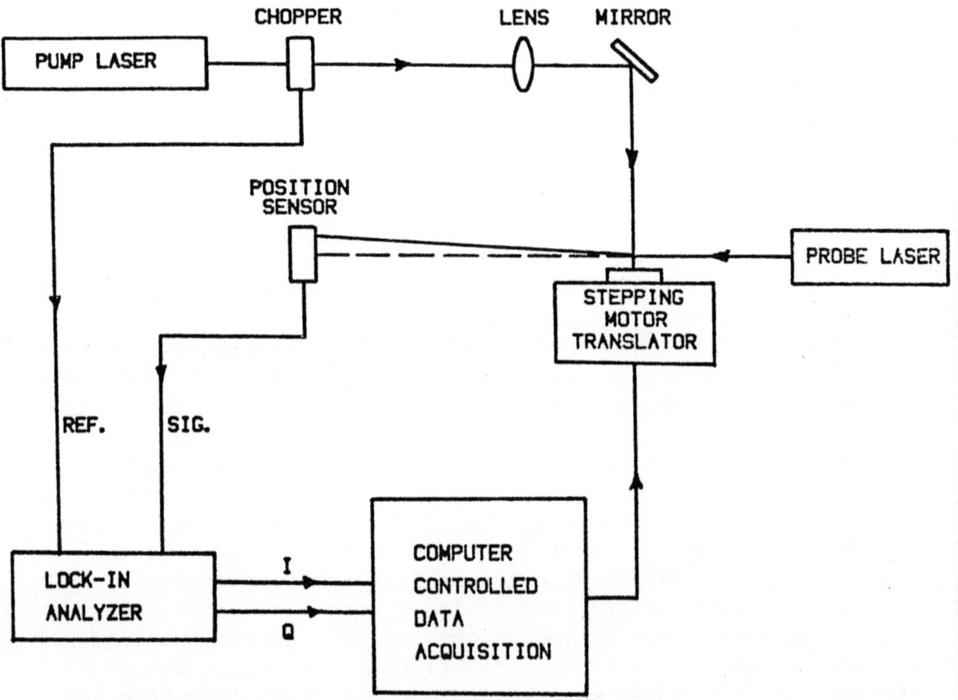

Fig. 1. Schematic diagram of photothermal laser-beam deflection imaging apparatus.

to include subsurface thermal contact resistance gives signal variations which agree in character with the experimental data on small subsurface structures.

EXPERIMENT

The PTLBD apparatus is shown schematically in Fig.1. An Ar^+ beam, modulated by a mechanical chopper, was focused onto the surface of the sample at normal incidence. A He-Ne probe beam was directed parallel to the sample surface and intersected the pump beam just above the sample. The periodic deflection of the probe beam was detected about 1.8 m from the sample by a United Detector Technology SC25 position sensor and 301B amplifier. The periodic position signal was connected to a P.A.R.C. 5204 lock-in analyzer; the output in-phase and phase quatrature signals were measured by a DVM. The sample position was determined by a stepping-motor-driven translator capable of 1.5 μm step size. The sample position, repetitive-data averaging, amplitude and phase calculations, and plotting were controlled by a Hewlett-Packard 9825T computer and associated peripherals.

The samples were machined aluminium alloys with top surfaces lapped with 600 grit emery paper. Samples #1 and #2 were 25mm x 25mm x 6.4mm blocks with 6.4mm wide channels milled on the underside; about 0.25mm of material was left between the top surface and the channel for sample #1 and about 0.11mm was left for sample #2. Sample #3 was a 25mm x 6.4mm x 3.2mm bar with a 0.30mm wide slot of varying depth cut on the underside with a wire saw. These samples were chosen to provide subsurface flaws of known character and to provide comparison with enclosed cell photoacoustic imaging.[6]

The amplitude and phase of the deflection signal are shown in Fig.2 as sample #1 is scanned (translated past the beam-intersection point). The channel region is clearly shown in each scan. The channel width can be deducted from the data (amplitude or phase) by determining the separation of points which lie halfway between the channel-center value and the base value (well away from the channel). The second-order variation in the signal is reproducible with at least part of it due to small variations in the depth of the machined channel. For later comparison with theory a useful experimental parameter is the ratio of deflection angle at channel center to the base value. This signal ratio decreases with increasing frequency for samples #1 and #2.

Phase scans for the narrow subsurface channel (#3,0.3mm) are shown in Fig.3. The phase increase over the channel region is an opposite change to that shown in Fig.2. Amplitude scans for this sample are similar to those in Fig.2, but with less of a "plateau" character, especially at lower frequencies. The channel width can

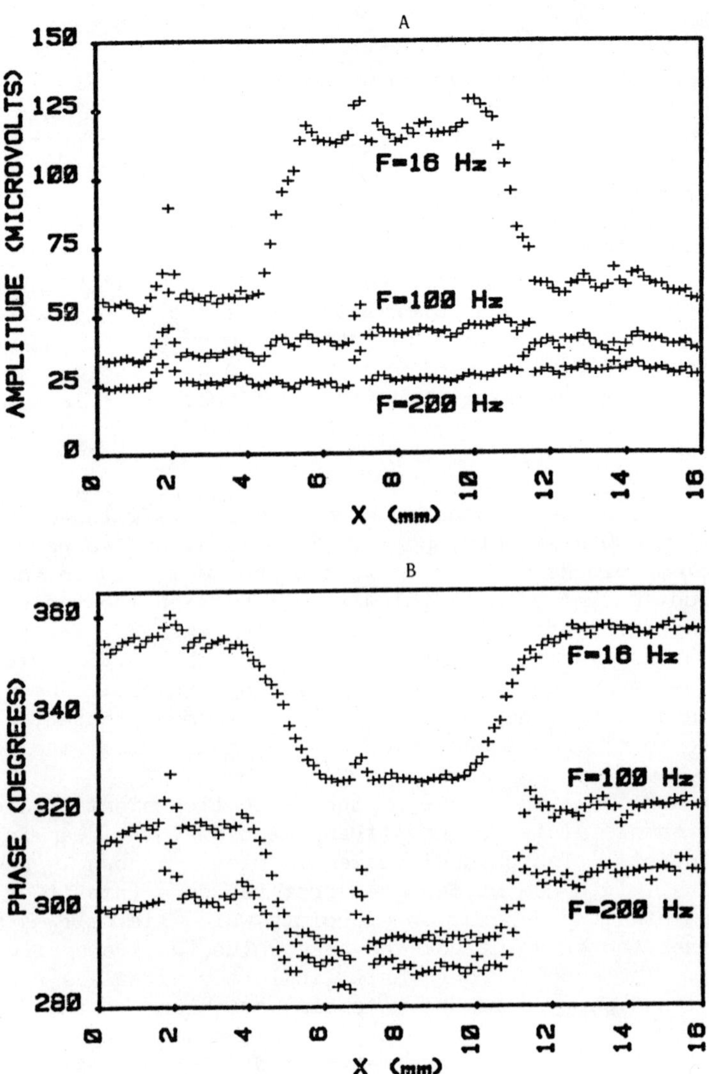

Fig. 2. Photothermal deflection signal vs. distance for Sample No. 1: (a) amplitude, (b) phase.

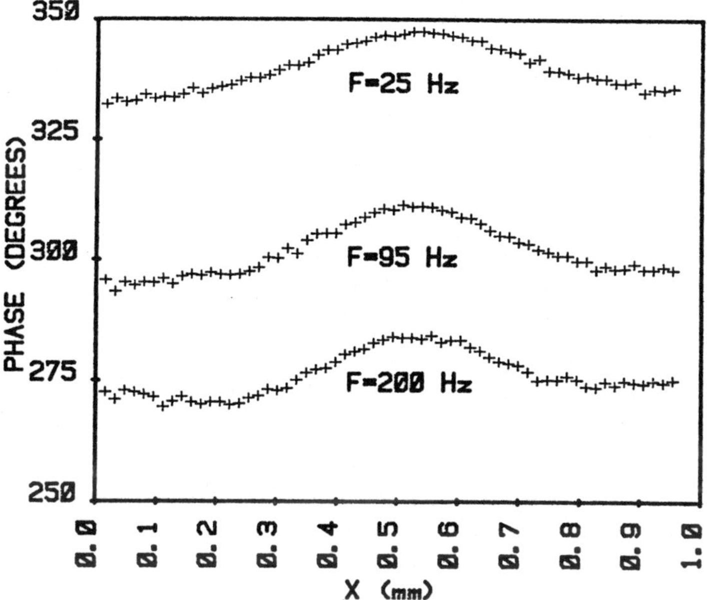

Fig. 3. Phase of photothermal deflection signal vs. distance for Sample No. 3.

again be deduced from the amplitude or phase data. A second difference for the narrow channel (compared to the wide channels) is that the signal ratio (channel center/base/ increases with frequency.

THEORY

A theoretical treatment of the probe-beam deflection has been given by Aamodt and Murphy[7] (also see ref.5), assuming that the sample is comprised of an infinite plate of thickness ℓ, with a semi-infinite medium (the "backing") in good thermal contact with the back surface of the sample. The temperature variation resulting from a periodic localised excitation decreases more rapidly than $\exp(-r/\mu)$,[9] where r is the distance from the source and μ is the thermal diffusion length ($\mu = \sqrt{2\alpha/\omega}$, where α is the thermal diffusivity and $\omega = 2\pi f$). Hence the assumed sample geometry need only be correct within a few diffusion lengths of the source, the region primarily responsible for the signal. If the beam is centered over the channels of samples #1 and #2, the channel sides are more than two diffusion lengths away (μ for aluminium at 25 Hz is $\sim$1.2mm), so that the theory of ref.7 should be a good approximation over the frequency range in use. The expression for the deflection angle is (for the present case, with air as backing):

$$\phi = 2T_0^{-1} k_s^{-1} \int_0^{\infty} h(\lambda) \coth(b_s \ell)(b_g/b_s)\exp(-b_g z)d\lambda \ , \qquad (1)$$

where $b_i = (\lambda^2 + 2j/\mu_i^2)$ [i = g(gas), s(aluminium sample)], ℓ is the sample thickness, z is the distance from sample surface to probe beam, k_s is the sample conductivity, and $T_0^{-1} = dn/ndT$ (relative change in gas index of refraction); for a Gaussian beam, $h(\lambda) = (I_0 R_G^2/2) \exp(-\lambda^2 R_G^2/4)$. We have calculated the ratio of deflection angles (channel center/base) and find[10] relatively good agreement with the experimental ratios for samples #1 and #2. The predicted phase decrease is about 20% smaller than the experimental phase decrease.

The contrast between the signal variations for wide and narrow channels indicates that a different model is required. A subsurface thermal defect (crack, inclusion, etc.) may be represented by introducing a thermal contact resistance between the sample and backing (the resistance R is the ratio of temperature difference across the defect to the heat flux, assumed continuous). Then the coth $(b_s \ell)$ in Eq. (1) is replaced by $[1+\Gamma \exp(-2b_s\ell)]/[1-\Gamma \exp(-2b_s\ell)]$, where $\Gamma = k_s b_s R\,(2+K_s b_s R)^{-1}$, and the "backing" is aluminium. The factor Γ is essentially a complex reflection coefficient for thermal waves at the interface at depth ℓ; the form of Eq. (1) corresponds to $R \rightarrow \infty$, so that $\Gamma \rightarrow 1$. This model can lead to signal variations having the character of those for the narrow subsurface channel: a phase <u>increase</u> and a signal ratio <u>increasing with frequency</u>. This occurs because Γ has a positive phase and increase with frequency (due to the factor b_s), provided ℓ is small enough that $\exp(-2b_s\ell) \sim 1$. Although the model is still an infinite plate model, work in similar thermal diffusion problems[11] suggests that the walls of a narrow channel can influence the heat flow (and thus the PTLBD signal) much as would an infinite backing place behind an air layer, so that the qualitative agreement between the model predictions and the narrow-channel data can be regarded as significant. Such signal variations can then be taken to indicate a near-subsurface thermal defect.

THERMAL DEFECT DETECTION

An unknown (and optically invisible) defect was discovered in the course of taking data on sample #1. The defect appears as a small signal variation in the scan of Fig.2, well away from the channel region. Study of this region showed that there was a phase increase and a signal ratio (channel center/base) which increased with frequency. The discussion above suggests that a thermal defect is located just under the surface of the sample at

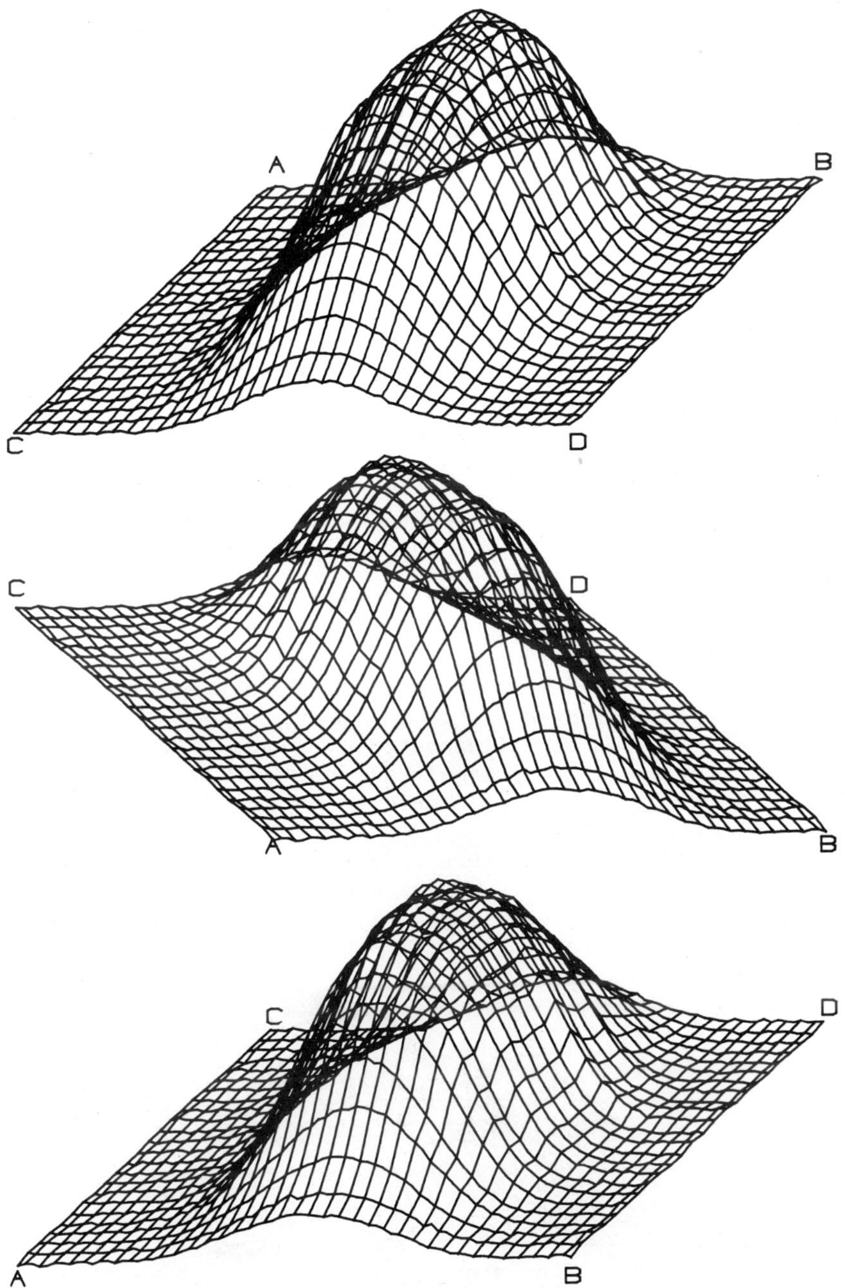

Fig. 4. Photothermal amplitude image of unknown defect, f = 95 Hz. The three images correspond to different "views" with fixed points on the sample labeled a,b,c,d. The distance ab (parallel to the x axis) is 476 µm. The distance ac (parallel to the y axis) is 230 µm.

this point. It is noteworthy that study of the sample by an electron microprobe detecting both backscattered electrons and x-rays showed no structure in this region, with a nominal detection depth of about 10μm. Thus the PTLBD technique has proved capable of detecting an invisible defect which would have been missed in a microprobe analysis.

By displacement of the sample along the "y" axis in 10μm steps between stepping-motor-driven scans along the "x" axis, a two dimensional photothermal image of the unknown defect was determined. Three views of the "amplitude" image are shown in Fig.4; the maximal value of the amplitude difference was about 80μV. A "phase" image is shown in Fig.5; the maximal value of the phase change is about 21°. As determined from the photothermal images, the defect is about 0.17 x 0.2mm. However, the estimated sizes of pump (.12mm) and probe (0.8mm) laser beams indicate that the actual size of the defect is probably smaller than the above estimate.

It would be very useful to be able to characterize the defect (e.g., ℓ and R) from the PTLBD data (signal ratio and phase). While it is possible to find values of ℓ and R which give the experimental ratio and phase at a single frequency, no single pair of values is consistent with the data for several frequencies. In fact it would have been surprising if the infinite plate model could give quantitative agreement for a small defect. Nevertheless, the range of ℓ values allowed within the model is 20-40 μm, and there may be some physical significance to these predications of defect depth. Further work on "thermal wave scattering" is underway in order to better characterize a defect from the PTLBD data.

The results presented here indicate that PTLBD imaging is a useful means of detecting subsurface structure, with definite advantages over other techniques. Further experimental and theoretical work is needed to define the ultimate capabilities of this method.

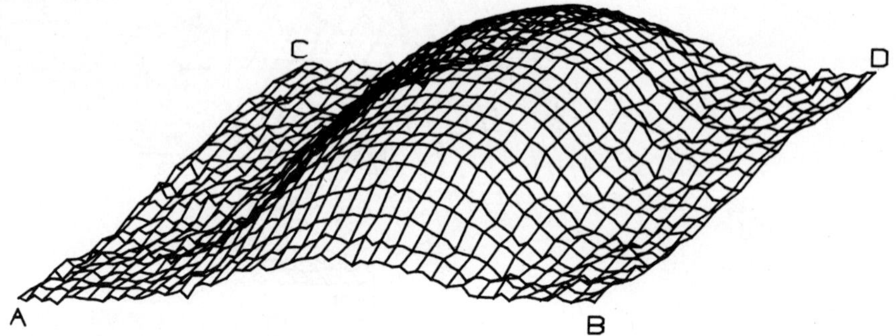

Fig. 5. Photothermal phase image of unknown defect, f = 95 Hz. The distance ab (parallel to the x) axis) is 476 μm. The distance ac (parallel to the y axis) is 230 μm.

REFERENCES

1. F. A. McDonald and G. C. Wetsel, Jr., Bull. Am. Phys. Soc. 27, 227 (1982).
2. D. Fournier and A. C. Bocarra, in Scanned Image Microscopy, E. A. Ash, Ed. (Academic, London, 1980); and also private communication.
3. J. C. Murphy and L. C. Aamodt, Appl. Phys. Lett. 38, 196 (1981)
4. J. C. Murphy and L. C. Aamodt, Appl. Phys. Lett. 39, 519 (1981)
5. W. B. Jackson, N. M. Amer, A. C. Boccara, D. Fournier, Appl. Opt. 20, 1333 (1981).
6. R. L. Thoms, J. J. Pouch, Y. H. Wong, L. D. Favro, P. K. Kuo, and A. Rosencwaig, J. Appl. Phys. 51, 1152 (1980).
7. L. C. Aamodt and J. C. Murphy, J. Apply. Phys. 52, 4903 (1981).
8. A. Rosencwaig and G. Busse, Appl. Phys. Lett. 36 , 725 (1980).
9. F. A. McDonald, Appl. Phys. Lett. 36, 123 (1980).
10. G. C. Wetsel, Jr., and F. A. McDonald, (unpublished).
11. F. A. McDonald, J., Appl. Phys. 52, 381 (1981).

ACOUSTICAL IMAGING OF NEAR SURFACE PROPERTIES AT THE RAYLEIGH CRITICAL ANGLE

G. L. Fitzpatrick, B. P. Hildebrand, and A. J. Boland

Spectron Development Laboratories, Inc.
1010 Industry Drive
Seattle, WA 98188

ABSTRACT

Surface waves excited on a liquid-solid boundary by an incident longitudinal wave in water can be used to examine flaws and other defects near the surface of a solid. The method involves a focused acoustic source producing longitudinal waves at an average angle of incidence equal to the so-called "Rayleigh" critical angle. At this angle, an incident longitudinal wave in water excites a surface wave along a liquid-solid boundary. This surface wave penetrates roughly one shear wavelength into the solid and then reradiates back into the water along the direction of a would-be specular reflection angle θ. A point-like receiver at a reflection angle θ is used to record the amplitude $R(\theta)$ and the phase $\phi(\theta)$ of this nonspecularly reflected signal. Because these signals are influenced by subsurface flaws, images of $R(\theta)$ and $\phi(\theta)$ obtained by scanning the detector in a plane parallel to the sample (or scanning the sample) holding $\theta = \theta(\text{critical})$, yield images of these flaws. The quality of these images is good and provides an excellent method of near surface flaw detection. However, certain quantitative aspects of the critical angle phenomenon remain unresolved, making detailed image interpretation difficult. If these problems can be solved, a new and useful tool for nondestructive examaination will become available.

INTRODUCTION AND BACKGROUND

It is well known that an ultrasonic wave propagating from one medium to another undergoes significant alteration at the boundary.[1-3] In particular, when a longitudinal wave propagates from a liquid into a solid, three new waves generally but not always appear; one reflected longitudinal wave, and two propagated waves (bulk shear and longitudinal). The propagated waves travel at angles defined by Snell's law, and the reflected wave obeys the law of reflection.[4] The direction of the propagated bulk longitudinal wave θ_{12} is given by

$$\text{Sin } \theta_{12} = \frac{V_{12}}{V_{11}} \text{ Sin } \theta_{11}, \tag{1}$$

where angles are measured from the normal to the boundary.

The direction of the propagated bulk shear wave θ_{22} is given by

$$\text{Sin } \theta_{22} = \frac{V_{22}}{V_{11}} \text{ Sin } \theta_{11}, \tag{2}$$

and the direction of the reflected longitudinal wave by

$$\theta_{1r} = \theta_{11} \tag{3}$$

where V_{11} = velocity of longitudinal waves in liquid
V_{12} = velocity of longitudinal waves in solid
V_{22} = velocity of shear waves in solid
and $\theta = \theta_{11}$ = angle of incidence of longitudinal waves in water.

These equations hold for ideal linear elastic media having no attenuation. When attenuation is present, Snell's law is complicated by the requirement that velocities and/or wavenumbers in the various media are complex quantities.[5]

Not every longitudinal wave incident from the liquid to the solid produces two bulk waves in the solid. Equations (1) and (2) predict (for θ real) that at some incident critical angle $\theta = \theta_L$ (θ_S), the angle of refraction of the longitudinal (shear) wave becomes 90° indicating that the bulk longitudinal (shear) wave propagates along a direction parallel to the boundary.

Snell's law predicts no more than this. In fact, it suggests that for $\theta > \theta_S$ the incident longitudinal wave is totally reflected, there being no bulk waves in the solid. Experiments show, however, that the behavior just beyond the shear critical angle θ_S ($\theta \geq \theta_S$) is considerably different than that expected.[6-8] The amplitude of the reflected longitudinal wave $R(\theta)$ drops sharply just beyond θ_S (typically 1° beyond) and its phase $\phi(\theta)$ undergoes a large phase shift (typically 360°).

These effects are attributed to a Rayleigh-type surface wave (not a bulk wave) which is excited on the liquid-solid boundary at an angle $\theta_R \geq \theta_S$. A number of investigators[9-23] have proposed and/or tested models which explain this behavior and allow one to compute the complex reflection coefficient ($R(\theta)$ and $\phi(\theta)$) for the incident longitudinal wave. Both spatially bounded and unbounded beams have been considered.

Although these models correctly describe experimental results under certain conditions, they may have to be modified in other situations. For example, these models typically assume among other things that: A) materials are homogeneous, uniform and isotropic, B) materials are linear, and C) Hooke's law (linear response) is to be supplemented by damping terms.

It has been argued[9] that without requirement C the surface waves would not be excited at the critical angle $\theta_R \geq \theta_S$. For example, if one derives the reflection coefficient using only Hooke's law without attenuation,[1] the surface wave excitation does not enter the analysis (is not required to fit the boundary conditions). This argument is somewhat circular, however, since neither damping or a mechanism for coupling the incident longitudinal waves to the respondent surface waves could be provided by a strictly linear theory (assumption B).

Nevertheless, in spite of such limitations, it is to be observed that both the models and experiment show that the reflected longitudinal wave at $\theta_R \geq \theta_S$ is modulated by the surface wave. Since this surface wave penetrates roughly one shear wavelength into the solid, information on near surface materials properties could in principle be obtained by examining this reflected wave alone.

In the following report, we develop an experimental approach using a focused acoustic source to excite surface waves and a point detector to form images of local near surface properties of materials.[24] We also point out that certain anomalies we have observed suggest that nonlinear materials response may be important at the Rayleigh critical angle.

APPLICATIONS USING A FOCUSED ACOUSTIC SOURCE

Applications of the critical angle phenomenon to problems in nondestructive evaluation (NDE) pose a number of practical and theoretical questions. Because specimens are seldom homogeneous, isotropic or uniform (assumption A in previous section), any successful NDE technique based on this phenomenon must be capable of obtaining sample information of a local nature.

A possible solution to this problem and the one employed in the present experiments, is illustrated in Fig. 1. An acoustic source is fitted with a lens and the sound is brought to a sharp focus on the liquid-solid boundary. A small point-like receiver samples a small portion (typically 10^{-5} steradians) of the scattered beam profile. In practice we find that this experimental arrangement produces a very significant amplitude reduction over a considerable range of frequencies at the critical angle for many materials.[25]

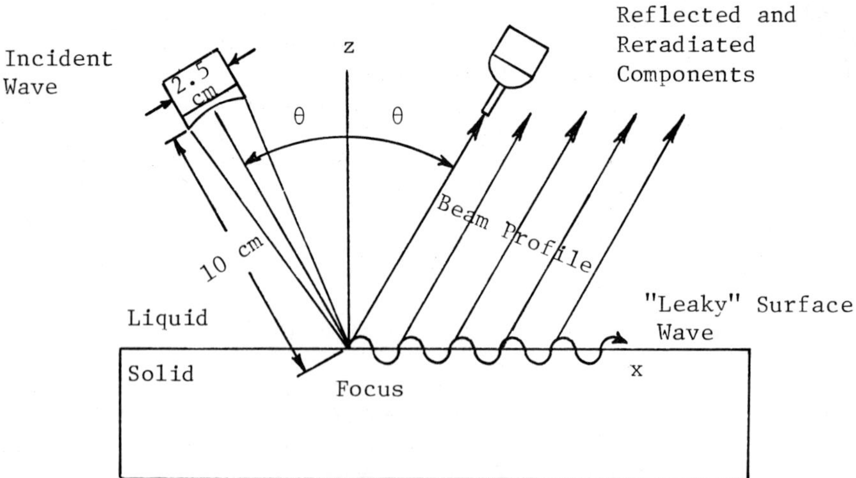

Fig. 1. Focused acoustic source and point receiver combination used in experiments. The source has an aperture of 2.54 cm and a 10 cm focal length. The transducer had a 5 MHz center frequency and is typically driven at 10 volts peak-to-peak (at various frequencies 1-10 MHz). Amplitudes recorded by the point detector are typically in the 1 volt range away from the critical angle while they are typically in the millivolt range at the critical angle.

MEASUREMENTS OF R(θ) AND ϕ(θ)

In Fig. 2 we illustrate the goniometer and associated apparatus used in obtaining graphs of reflected amplitude R(θ) and phase ϕ(θ) as a function of the incident angle θ. The goniometer is equipped with a computer controlled stepper motor that allows the operator to begin the experiment at any initial angle θ_i and proceed in arbitrary steps $\Delta\theta$ to some final angle θ_f. R(θ) and ϕ(θ) are measured using a phase meter; the output of which is digitized, recorded in disk memory, and displayed on a monitor. The computer operating system is an IBM personal computer.

A special gimbaled sample holder is provided so that samples of varying sizes and shapes (up to 5 cm in diameter but possessing at least one flat surface) can be examined. The water bath (tank not shown) used for acoustic coupling is maintained at a constant temperature (usually 20°C) in order to reduce effects due to temperature changes. Typically, a critical angle, of say 30°, measured in our apparatus increases by approximately 1° for every 5°C increase in temperature for temperatures in the range 15°-25°C.

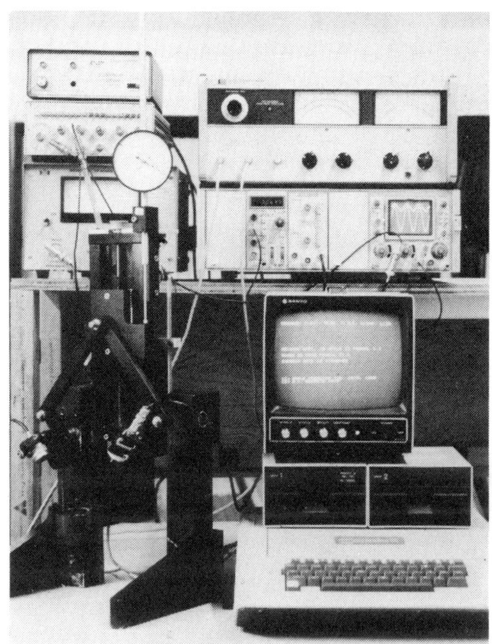

Fig. 2. Goniometer and associated electronic apparatus used in measuring critical angle data.

In Fig. 3 we illustrate R(θ) and ϕ(θ) graphs for a specimen showing a sharp amplitude reduction and a phase shift of 360°. This behavior is very similar to that predicted by the linear theories.[9-23] However, as we will point out later, this is not the only type of phase curve we have observed.

To demonstrate that our apparatus is functioning properly, we also illustrate graphs obtained on a thin air-backed mica sheet (see Fig. 4). As expected,[26] R(θ) and ϕ(θ) for a thin mica sheet are nearly constant, except for small variations due to temperature-density fluctuations in the circulating water bath.

In Table I we list measured critical angles for a variety of materials including metals, ceramics, and several naturally occurring rocks and minerals. One can also easily estimate the critical angle θ_R, using Eq. (1), and $\theta_R \simeq \theta_S$ ($\sin \theta_R \geq V_W/V_S$) given the shear wave velocity in the solid (V_S) and the longitudinal wave velocity in water (V_W). Values of the critical angle listed in Table I should be taken as a rough guide only since different alloys or chemical compositions, heat treatments, mechanical work histories, grain size, anisotropies, etc. lead to widely differing critical angles for the "same" material.

For example, crystals generally have different critical angles in different directions, and similarly, polycrystalline materials that are anisotropic have different critical angles in different directions. These effects are in fact useful in characterizing such anisotropy.[27]

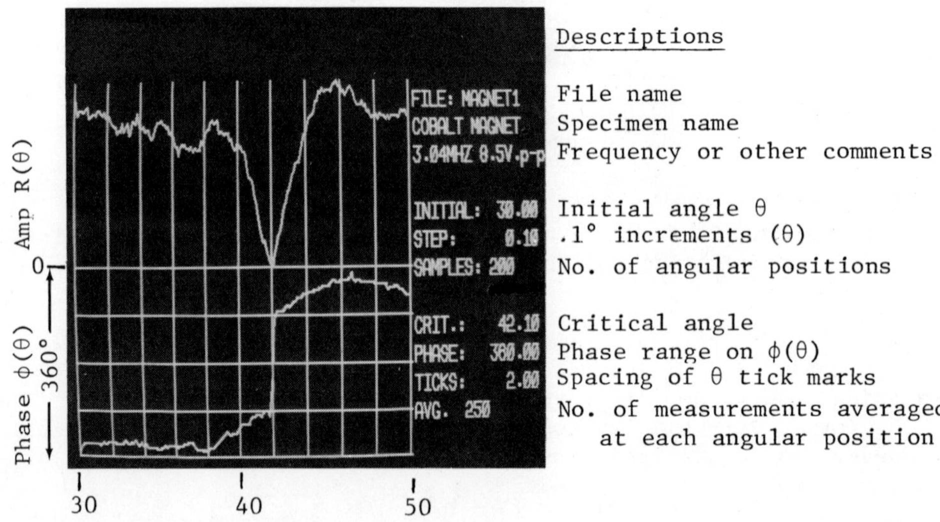

Fig. 3. Reflected amplitude $R(\theta)$ and phase $\phi(\theta)$ for a cobalt powder composite showing a "normal" phase curve $\phi(\theta)$ at 20°C at a frequency of 3 MHz. Note that the phase shift for these "normal" phase curves is typically 360°. The critical angle for this specimen is located at 42°.

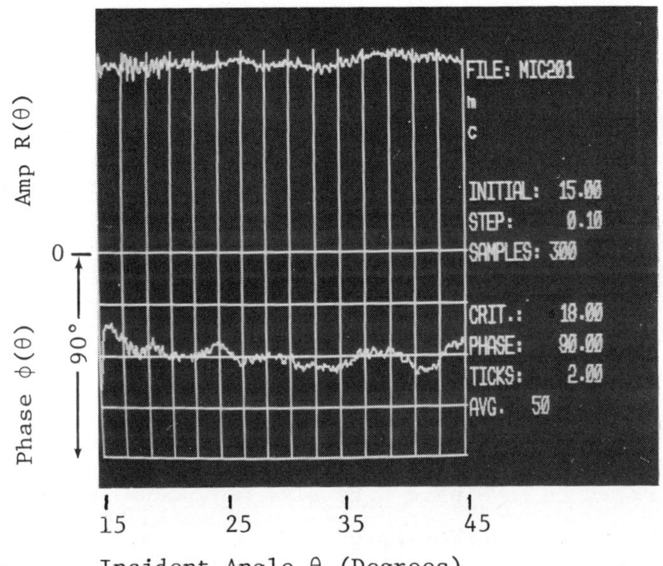

Fig. 4. Response of a thin (.05 mm) mica window. The incident angle varied from $\theta = 15°–45°$. The curves show the relative amplitude (arbitrary units) $R(\theta)$ and the phase $\phi(\theta)$ of the reflected wave at each angle θ (300 angle increments chosen) in the range 15°–45°. At each angular position, fifty measurements of $R(\theta)$ and $\phi(\theta)$ were made, and the average of each was plotted as shown on a cathode ray tube monitor. The frequency for this experiment was 3.1 MHz, and the transducer involved a 10 cm focal length lens focused on the mica sheet. Temperature was maintained constant at 20° during these measurements.

Table I. Typical Critical Angles for a Variety of Materials at 25°C

Materials	θ_R(°)	Frequency of Measurement (MHz)
	Metals and Alloys	
Brass	50.7	2.14
Copper	47.5	4.46
4340 Steel	33.8	2.50
Aluminum (7075)	33.5	4.29
Titanium	31.6	3.10
	Ceramics	
Tungsten Powder Composite	37.8	4.90
Coors Alumina	17.8	4.90
	Glasses	
Obsidian	26.9	2.62
Silica	26.5	2.55
	Complex Silicates	
Fine grained granite	26.2	4.00
"Tiger Eye"	25.2	4.00
Epidote	25.0	5.00
	Single Crystals (Not Oriented)	
Quartz	28.6	3.74
Fluorite	28.5	4.77
Silicon	21.1	4.45
Saphire	15.6	3.79

IMAGES OF R(θ) AND φ(θ) OVER A BOUNDARY SURFACE

If for any reason the critical angle should change as a function of position of the focal spot on the liquid-solid boundary, then both $R(\theta)$ and $\phi(\theta)$ will become complicated functions of position. Thus, one may set the incident angle equal to the average critical angle (measured or estimated) θ_R and scan the focus on the sample (or move the sample) in a raster pattern.

Any material variations encountered during the scanning that change θ_R will shift $R(\theta)$, $\phi(\theta)$ to new graphs $R'(\theta)$, $\phi'(\theta)$ characterized by a different critical angle θ'_R. By performing a sufficient number of such raster scans, one can image the reflected amplitude $R(\theta)$ and phase $\phi(\theta)$ over the boundary surface.

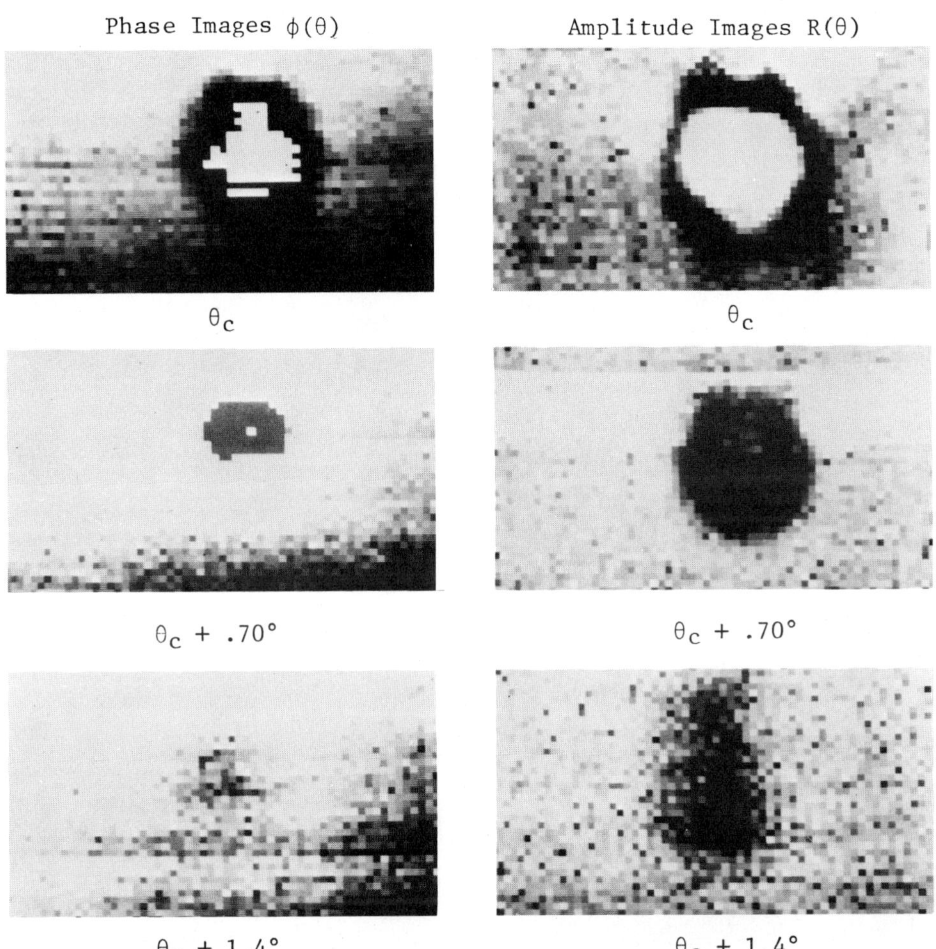

Fig. 5. Phase $\phi(\theta)$ and amplitude $R(\theta)$ images of a .6 cm diameter flat-bottomed drill hole .8 mm from the surface of an aluminum plate .9 cm thick. Images at the top of the figure were made at the critical angle for aluminum (≃31°) while those below were made at successively larger angles. Note that as θ increases, the conditions for surface wave generation are eventually ruined and the images grow steadily worse. The frequency for these experiments was 3.75 MHz. A lens and point receiver similar to that shown in Fig. 1 was employed. The hole images are somewhat eliptical in shape owing to the incorrect aspect ratio of the monitor.

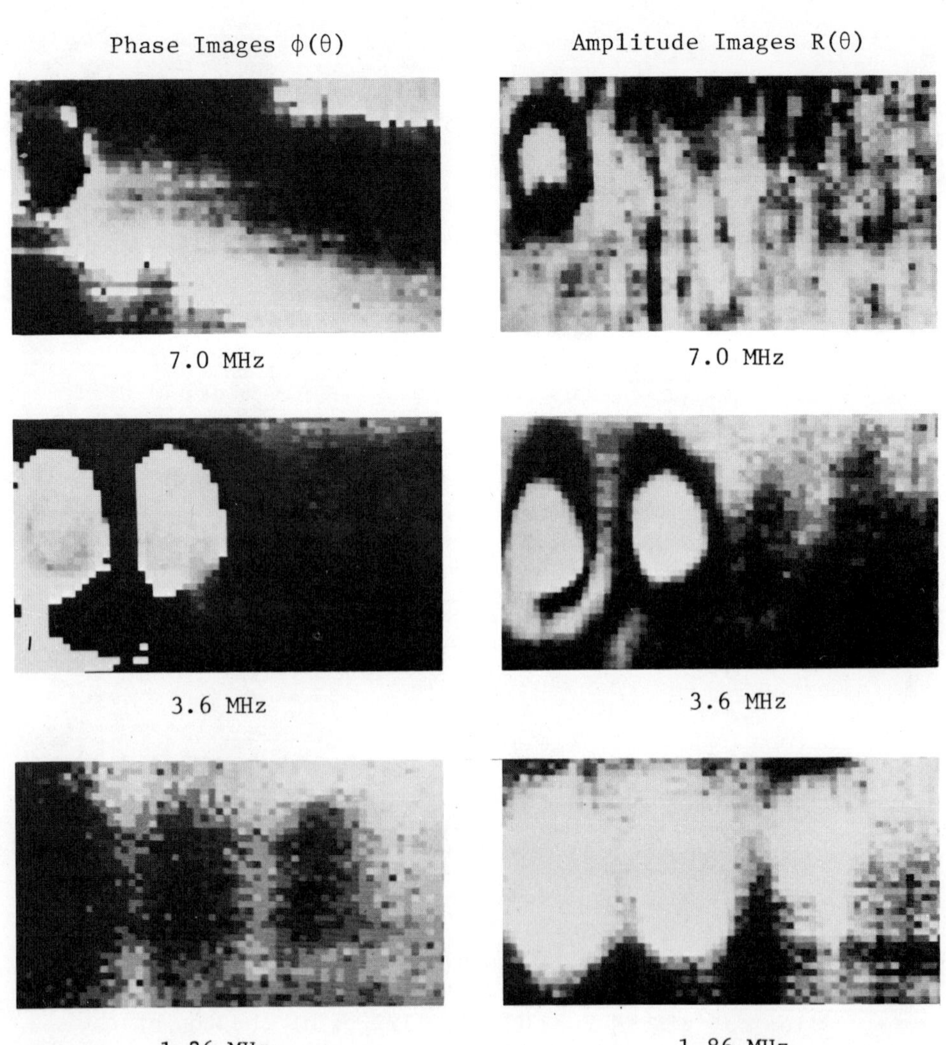

Fig. 6. Phase $\phi(\theta)$ and amplitude $R(\theta)$ images of a series of .6 cm diameter flat-bottomed drill holes at varying depths (.4, .8 and 1.6 mm) from the surface of an aluminum plate .9 cm thick. All images were made at the critical angle for aluminum (≃31°) and varying the frequency as shown. Note that as the frequency (f) increases, the depth of penetration of the surface waves ($\lambda_s \simeq V_s/f$) decreases, and for a frequency of 10 MHz (not shown) the hole at .4 mm depth is no longer visible for example.

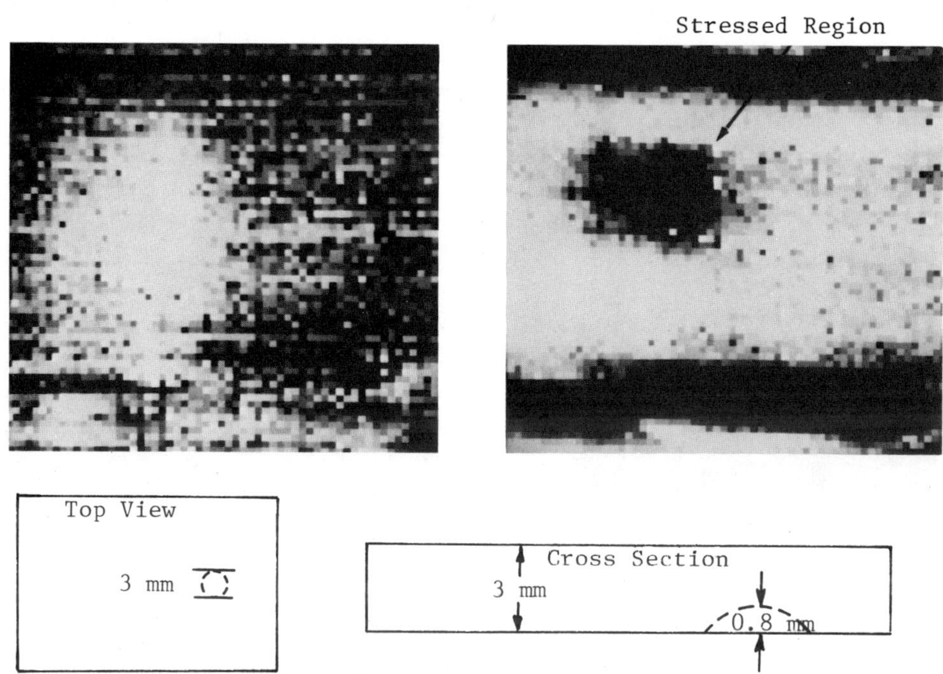

Fig. 7. Phase $\phi(\theta)$ and amplitude $R(\theta)$ images of a stress-dislocation field in a 3.2 mm thick aluminum plate in which a 4.8 mm steel ball bearing was pressed to a depth of .8 mm. The frequency chosen was 3.75 MHz but the critical angle (not recorded) was that of the flaw itself and not the aluminum (≈31°).

Because the reflected longitudinal wave is modulated by the surface wave and because the surface wave extends roughly one shear wavelength ($\lambda_s=V_s/f$) into the solid, these images provide information on the solid within λ_s of the liquid-solid boundary.

In Figs. 5 through 8, we illustrate images made by scanning the sample (using a computer controlled x-y scanner not shown in Fig. 2). Targets scanned ranged from flat-bottomed holes in aluminum to a piece of fine grained granite. Note, as in Fig. 5, that whenever the boundary conditions are insufficient to generate a surface wave ($\theta \neq \theta_R \geq \theta_s$) or the depth of the discontinuity is greater than the surface wave penetration (λ_s) as in Fig. 6, no discontinuity can be seen in the images.

Figure 7 illustrates the result of imaging a stress-dislocation field resulting from a steel ball bearing being pressed into an aluminum sheet.[28-30] Figure 8 illustrates images of a granite specimen showing what appears to be individual crystal grains.

Phase Image $\phi(\theta)$

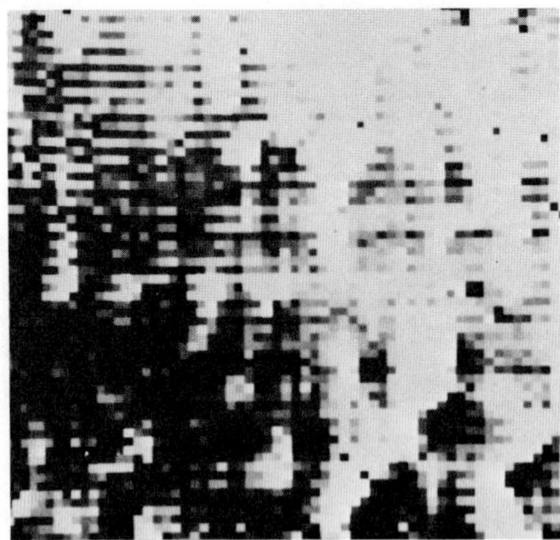

Amplitude Image $R(\theta)$

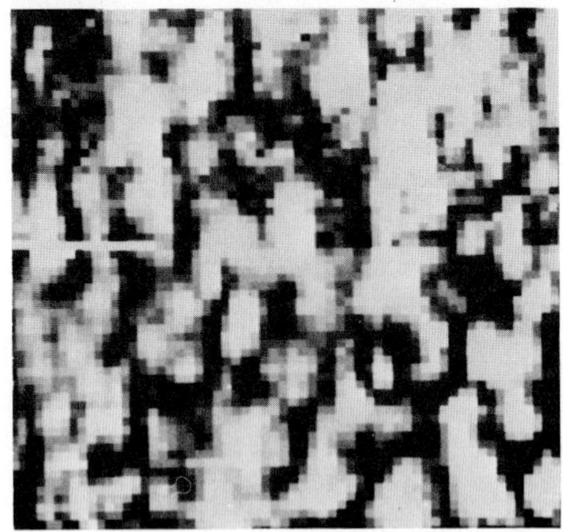

Fig. 8. Phase $\phi(\theta)$ and amplitude $R(\theta)$ images of a fine grained granite specimen (1-2 mm diameter grains) at a chosen incident angle of 29.5°. The frequency was 3.75 MHz. Both phase and amplitude images show what appear to be individual crystal grains (3.8 cm scan length).

DISCUSSION

From a qualitative standpoint, the foregoing results are encouraging. However, from a quantitative standpoint much remains to be done before a detailed understanding of such images can be achieved.

For example, the use of a focused acoustic source is a special case of a bounded beam so that eventually one must calculate the expected beam profile and the received signal at a point detector. Unfortunately, any such calculation is very sensitive to the model of the material and the equations of motion chosen. If these are not sufficient to describe a real sample, the quantitative interpretation of the measured data will not be possible.

In particular, a focused acoustic source may induce nonlinear materials response.[26,31] owing to 1) the increased intensity at the lens focus, and 2) the fact that a resonance condition is set up at the critical angle.

Two pieces of experimental evidence may be cited to suggest that nonlinear effects are important at the Rayleigh critical angle.

The first piece of evidence comes from the Fourier spectrum of the reflected wave. The sound waves incident on a titanium sample were observed to contain not only a fundamental at 3.1 MHz (the driving frequency of the source), but also some higher harmonics (6.2 and 9.3 MHz) due to the expected nonlinear response along the water path.[26,31] Higher harmonics were also present because the source-oscillator combination produces these at the source.

Now, the linear theory of the Rayleigh critical angle phenomenon[9-23] predicts that each of these incident frequency components (regardless of their actual origin) will undergo amplitude reductions at the critical angle. We expect, moreover, that because we are using a lens these reductions will be comparable (same percentage) for each incident frequency component.[25]

What one actually observes, however, is that the third harmonic seems to increase at the critical angle, at least, relative to the fundamental (see Fig. 9). This suggests that a nonlinear mechanism in the sample has compensated for the expected reduction of the third harmonic.

The second piece of evidence for nonlinearities is obtained from certain anomalous phase curves we have observed. If second and third harmonics are being generated in the sample by nonlinear interactions, then a certain amount of fundamental may also be generated by intermodulation. If the reflected signal at the point detector, as predicted by linear theory, is $R(\theta)\exp(i\phi_R(\theta))\cdot\exp(-i\omega_o t)$ and the contribution at the same frequency ω_o due to nonlinearities is $N(\theta)\exp(i\phi_N(\theta))\cdot\exp(-i\omega_o t)$, then the combined signals at the detector are (dropping $\exp(-i\omega_o t)$)

$$D(\theta)\exp(i\phi_D(\theta)) = R(\theta)\exp(i\phi_R(\theta)) + N(\theta)\exp(i\phi_N(\theta)). \tag{4}$$

Therefore, the resultant phase of the combined signals is

$$\phi_D(\theta) = \tan^{-1}\left\{\frac{R(\theta)\mathrm{Sin}\phi_R(\theta) + N(\theta)\mathrm{Sin}\phi_N(\theta)}{R(\theta)\mathrm{Cos}\phi_R(\theta) + N(\theta)\mathrm{Cos}\phi_N(\theta)}\right\}. \tag{5}$$

Note that unless the nonlinear contribution to the fundamental is large, ($N(\theta)$ is small relative to $R(\theta\simeq 0°)$ say) the total amplitude at the point detector $(R^2(\theta) + N^2(\theta))^{1/2}$ will always be small and approach a minimum at the critical angle as predicted by the linear theory. If on the other hand $N(\theta)$ and $R(\theta)$ are comparable at the critical angle, the phase $\phi_D(\theta)$ could, in general, be quite different from the linear theory prediction $\phi_R(\theta)$ since in that case ($N(\theta_R)\simeq R(\theta_R)$) and

$$\phi_D(\theta_R) \simeq \tan^{-1}\left\{\frac{\mathrm{Sin}\phi_R(\theta) + \mathrm{Sin}\phi_N(\theta)}{\mathrm{Cos}\phi_R(\theta) + \mathrm{Cos}\phi_N(\theta)}\right\}\Bigg|_{\theta=\theta_R}. \tag{6}$$

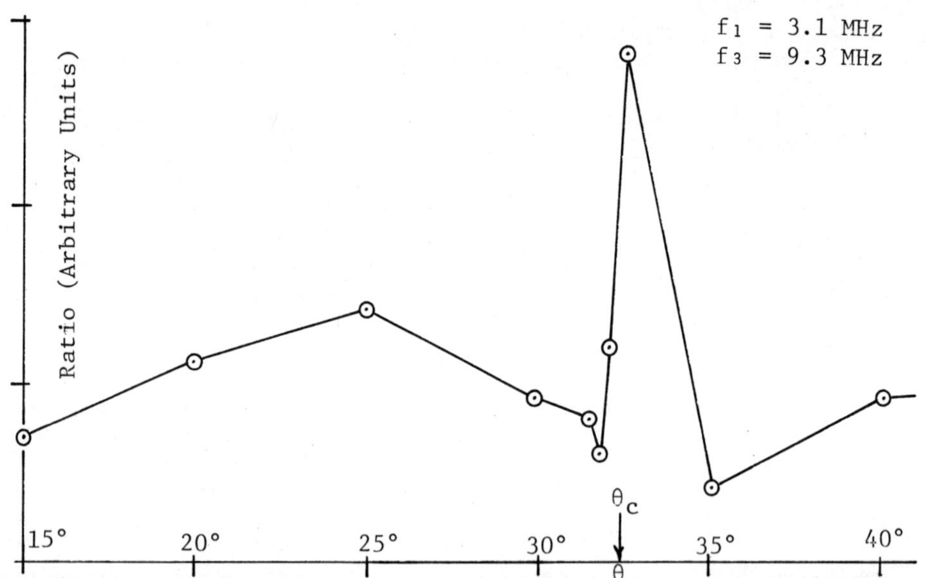

Fig. 9. Ratio of the third harmonic amplitude to the fundamental as a function of the incident angle θ for a titanium specimen (f=3.1 MHz, T=20°C). Note that there is a sizeable increase in the relative amount of the third harmonic at the critical angle θ_c=32.5°. A 3 MHz suppression filter (40 DB reduction of 3 MHz fundamental and little or no effect on 6 and 9 MHz harmonics) was used in obtaining this ratio.

Thus, the appearance of phase curves $\phi(\theta)$, which are distinctly different from their linear counterparts $\phi_R(\theta)$, could signal the presence of nonlinear interactions at $\theta \simeq \theta_R$. We have often noticed that when the minimum of the reflected amplitude curve is particularly low (millivolt range for source driving voltages of 10 V at 5 MHz), an anomalous phase curve of the type shown in Fig. 10 is obtained. This phase curve should be compared with a "normal" phase curve as predicted by linear theory (see Fig. 3).

For the foregoing reasons, we stress that the interpretation of phase images produced by a focused acoustic source (at $\theta=\theta_R$) may under certain circumstances require the introduction of a nonlinear materials response. Given that nonlinearities are important at $\theta=\theta_R$, other nonlinear phenomena such as, solitary acoustic surface waves (solitons)[32,33] may also play a role in understanding the phenomena.

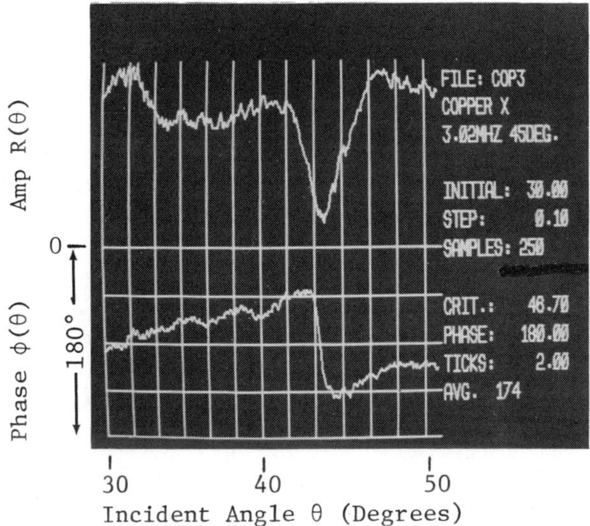

Fig. 10. Reflected amplitude R(θ) and phase φ(θ) for a Cu single crystal (100 plane) at 20°C and a frequency of 3.0 MHz. These curves (φ(θ)) illustrate the phase "reversal" effect. Compare with a "normal" type phase curve φ(θ) such as that illustrated in Fig. 3.

CONCLUSIONS

Practical application of the Rayleigh critical angle phenomenon to the problem of detecting and imaging local surface discontinuities requires the use of a focused acoustic source. Using such an approach, we have achieved excellent qualitative results including the formation of images of the reflected amplitude and phase of a focused longitudinal wave incident at the average critical anlge on a solid material. These images clearly show the presence of any discontinuity that changes the critical angle locally. A variety of practical applications (not discussed here) which take advantage of these qualitative successes can easily be imagined. However, from a quantitative standpoint, the method requires more work. Besides a better understanding of the effects produced by the bounded beam (the focused acoustic beam) we must consider the possibility that the increased intensitites at the lens focus can excite a nonlinear material response. While small nonlinear effects would not be expected to significantly alter curves of reflected amplitude R(θ), they could significantly affect the observed phase curve φ(θ). Thus, the quantitative interpretation of images, especially the phase images, may eventually require the introduction of nonlinear elasticity theory.

A realization that the Rayleigh critical angle phenomenon may involve nonlinear materials response (especially when lenses are employed) could lead to techniques for local measurements of such things as the third order elastic constants. Nonlinearities at the critical angle also open the possibility that phenomena such as, solitary surface waves [32,33] are involved.

ACKNOWLEDGMENTS

The authors thank Y. Pipkin for editing and typing the manuscript. We thank T. J. Davis for numerous helpful suggestions regarding our electronic apparatus and measurement procedures. We also thank both R. L. Silta and M. C. Cochran for constructing experimental apparatus and for performing experiments. We are especially indebted to R. L. Richardson for many helpful discussions regarding this phenomenon. The research work leading to this paper was sponsored by the National Science Foundation under Contract No. DMR-8116240 and the Air Force Office of Scientific Research under Contrac No. F49620-81-C-0040.

REFERENCES

1. L. M. Brekhovskikh, Waves in Layered Media, Academic Press, New York (1980).
2. J. D. Achenbach, Wave Propagation in Elastic Solids, North-Holland Publishing, New York (1973).
3. B. A. Auld, Acoustic Fields and Waves in Solids, Vol. I & II, John Wiley, New York (1973).
4. D. H. Towne, Wave Phenomena, Addison Wesley, Reading, MA (1967).
5. M. Newlands, J. Acoust. Soc. Am., 26, 434-448 (1954).
6. A. Schoch, Acustica 2, 18 (1952).
7. L. M. Brekhovskikh, Waves in Layered Media, Academic Press (1960)
8. F. R. Rollins, Jr., "Ultrasonic Reflectivity at a Liquid-Solid Interface Near the Angle of Incidence for Total Reflection," Appl. Phys. Lett., Vol. 7, No. 8 (1965).
9. C. E. Fitch and R. L. Richardson, "Ultrasonic Wave Models for Nondestructive Testing Interfaces with Attenuation," Progress in Applied Materials Research, edited by E. G. Stanford et. al. Iliffe Books, London, Vol. 8, 79-120 (1967).
10. F. L. Becker and R. L. Richardson, "Influence of Material Properties on Rayleigh Critical Angle Reflectivity," J. Acoust. Soc. Am., 51, 1609 (1972).
11. F. L. Becker, J. Appl. Phys., 42, 199-202 (1971).
12. C. E. Fitch, Acoust. Soc. Am., 40, 989-997 (1966).
13. F. L. Becker and R. L. Richardson, "Ultrasonic Critical Angle Reflectivity," Research Techniques in Nondestructive Testing, edited by R. S. Sharpe, Academic Press, London, 91-130 (1970).
14. F. L. Becker, C. E. Fitch, and R. L. Richardson, "Ultrasonic Reflection and Transmission Factors for Materials with Attenuation," Battelle Memorial Inst. Report, BNWL-1283 (1970) (Unpublished).
15. V. M. Merkulova, Sov. Phys. Acoust., 15, 404 (1970).

16. F. L. Becker and R. L. Richardson, J. Acoust. Soc. Am., 51, 1609-1617 (1971).
17. H. L. Bertoni and T. Tamir, "Unified Theory of Rayleigh Angle Phenomena for Acoustic Beams at Liquid-Solid Interfaces," Appl. Phys. 2, 157-172 (1973).
18. H. L. Bertoni and Y. L. Hou, "Effects of Damping in a Solid on Acoustic Beams Reflected at the Rayleigh Critical Angle," in Proceedings of the 10th Symposium on Nondestructive Testing (NDE), San Antonio, 136, 142 (1975).
19. T. Tamir and H. L. Bertoni, "Lateral Displacement of Optical Beams at Multilayered and Periodic Structures," J. Opt. Soc. Am., Vol. 61, No. 10 (1971).
20. L. E. Pitts, "A Unified Theoretical Description of Ultrasonic Beam Reflections from a Solid Plate in a Liquid," Ph.D. Thesis, Georgetown University, Washington, D.C. (1976).
21. L. E. Pitts and T. J. Plona, "Theory of Nonspecular Reflection Effects for an Ultrasonic Beam Incident on a Solid Plate in a Liquid," IEEE Trans. Sonics Ultrason., Vol. SU-24, No. 2 (1977).
22. T. D. K. Ngoc and Walter G. Mayer, "Ultrasonic Nonspecular Reflectivity Near Longitudinal Critical Angle," J. Appl. Phys., 50 (12) (1979).
23. T. D. K. Ngoc and W. G. Mayer, "Numerical Integration Method for Reflected Beam Profiles Near Rayleigh Critical Angle," J. Acoust. Soc. Am., Vol. 67, 1149-1152 (1980).
24. B. P. Hildebrand and F. L. Becker, "Ultrasonic Holography at the Critical Angle," J. Acoust. Soc. Am., 56, 459-462 (1974).
25. Werner G. Neubauer, "Observation of Acoustic Radiation from Plane and Curved Surfaces," Physical Acoustics, edited by Warren P. Mason and R. N. Thurston, Academic Press, New York, Vol. X, 61-126 (1973). (We have observed that when a focused acoustic source is used to generate surface waves the amplitude reduction is very significant at many different frequencies. The phenomenon of the so-called frequency of least reflection (FLR) is not obvious in our data. Neubauer has argued that the FLR is more a function of the beam geometry than a material property. Our experiments with a sharply focused beam are apparently in qualitative agreement with his observations.)
26. A. L. Van Buren and M. A. Breazeale, "Reflection of Finite-Amplitude Ultrasonic Waves," J. Acoust. Soc. Am., Vol. 44, No. 4, 1014-1020 (1968).
27. O. I. Diachok and W. G. Mayer, "Crystal Surface Orientation by Ultrasonic Beam Displacement," Acustica, Vol. 26, 267-269 (1972).
28. B. G. Martin and F. L. Becker, "The Effect of Near-Surface Metallic Property Gradients in Ultrasonic Critical Angle Reflectivity," Materials Evaluation (1980).
29. F. L. Becker, "Ultrasonic Determination of Residual Stress," Battelle Pacific Northwest Laboratories (1973) (Unpublished).

30. B. G. Martin, "Theory of the Effect of Stress on Ultrasonic Plane-Wave Reflectivity from a Water-Metal Interface," McDonnell Douglas Corp., Paper #6514 (1978) (Unpublished).
31. M. A. Breazeale, "Ultrasonic Studies of the Nonlinear Properties of Solids," International Journal of Nondestructive Testing, Vol. 4, Gorden and Breach, Great Britain (1972).
32. J. F. Ewen, R. L. Gumshor, and V. H. Weston, "An Analysis of Solitons in Surface Wave Devices," Purdue University (Aug. 1981) (Unpublished).
33. M. J. Ablowitz and H. Segur, Solitons and the Inverse Scattering Transform, SIAM, Philadelphia (1981).

AN EXPERIMENTAL INVESTIGATION OF THE 'HOSEPIPE' TECHNIQUE OF REAL-TIME C-SCANNING

C.P. Oates
T.A. Whittingham

Regional Medical Physics Department
Newcastle General Hospital
Newcastle upon Tyne NE4 6BE

INTRODUCTION

This paper describes an in-vitro experimental study of the 'hosepipe' real-time C-scanning technique proposed by Whittingham (1982) at the 10th International Symposium on Acoustical Imaging. This technique was proposed as a means of overcoming the 'go and return' time redundancy inherent in conventional pulse-echo C-scanning without recourse to parallel processing, as used in the real-time C-scanning system of Green et al.(1974).

The method can best be described by the analogy of rapidly turning the nozzle of a hosepipe through an angle so as to send out the jet of water in an oblique arc across a sector of a circle. Any field point in the sector experiences a "splash" as the arc of water crosses it. Thus, although the transmission is continuous,each field point experiences a discrete pulse. Putting this in terms of an ultrasonic system, Figure 1.shows that if we transmit a c.w. beam and electronically steer it across a sector we produce a transmitted beam forming an oblique arc across the sector in the same manner as with the jet of water. Looking along a line of constant range R, as shown, it can be seen that the transmitted beam crosses that line at only one point at a time. If we then consider the echoes arriving back at the transducer from targets lying along the line of constant range, we will only receive echoes from one direction at a time. We therefore arrange for our receive transducer to be sensitive to the direction in which echoes from the scan depth R are returning. This means we must steer the receiving beam to follow behind the transmitted beam,with a time delay between the two of

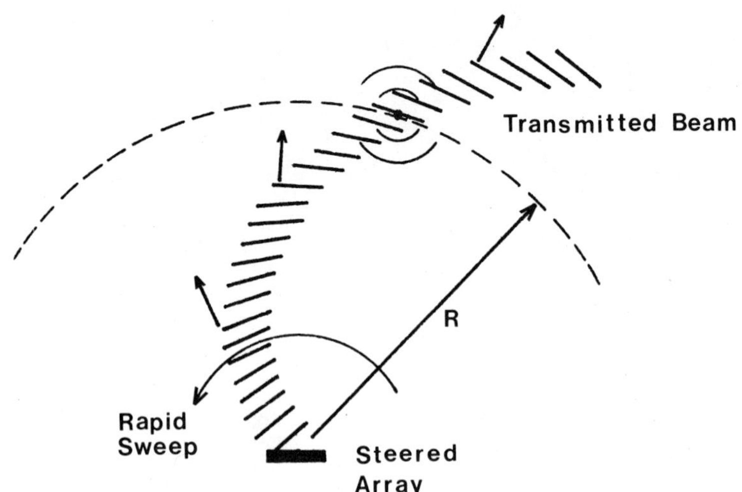

Fig 1 Principle of the'hosepipe'technique in steering a beam across a sector

$\frac{2R}{c}$, where c is the speed of sound.

Although the method was described as a c.w. technique, the principle may be applied to a quasi-continous wave of closely spaced discrete pulses, each several cycles long, with the beam deflection being advanced between pulses. This modification was adopted in this study for experimental convenience.

SAMPLE VOLUME

Whittingham showed that, for the c.w. case at moderate sweep speeds, the transmitted beam has a sinc function profile. Fig 2a. shows the amplitude distribution within the transmitted beam more clearly. For each direction, the transducer transmits a beam with the usual sinc function profile. By drawing in lines of constant amplitude through these profiles at a given instant in time, the arc of the transmitted beam is made apparent. The profile of the receiver beam is simply that of the stationary transducer, since the transducer is stationary whilst receiving echoes from a particular bearing. Fig 2b.shows how the arc-shaped transmitted beam combines with the conventional sinc profile of the receiving beam to define a crossover region from which echoes must have originated. The dashed arc represents the line of constant range from which echoes are required. By sweeping the receiving beam behind the transmitted beam with a time delay equal to the "go and return" time of this range, the crossover region will travel along this line and thus interrogate one line of the scan. The

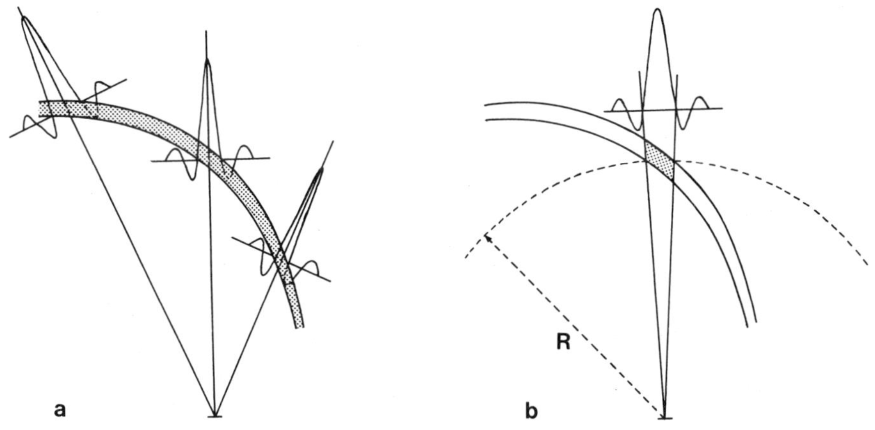

Fig 2a) The transmitted beam showing its amplitude profile for a clockwise sweep and b) the transmitted beam together with the receiving beam showing the sample volume.

size and shape of the crossover region determines the lateral and axial resolving power of the system; it will be referred to subsequently as the sample volume.

The size of the sample volume depends on the beamwidth of the transducer when stationary and the rate of sweep across the sector. Fig 3a.shows the condition of zero sweep speed. Here lateral resolution is described by a sinc^2 profile but there is no axial resolution. Fig 3b.shows the sample volume for a slow sweep speed. As the sweep speed is increased (Fig 3c) the transmitted beam gets narrower and there is a corresponding improvement in axial resolution. However, the increased sweep speed leads to a decrease in lateral resolution due to undersampling of the sector within the sweep. In the limit of an infinite sweep speed (Fig 3d) the sinc profiles for every direction in the transmitted beam lie along a line of constant range. This gives perfect axial resolution. However, at infinite sweep speed the sampling within the sweep is effectively zero so there is no lateral resolution. In practice therefore a trade-off must be made between a high sweep speed, giving better axial resolution at the expense of lateral resolution, and a lower sweep speed giving improved lateral resolution at the expense of poor axial resolution. Of course for any particular range the size of the sample volume can be further improved by focusing the beam at that range. Indeed the ability to use strong focusing is a principal advantage of any C-scan system.

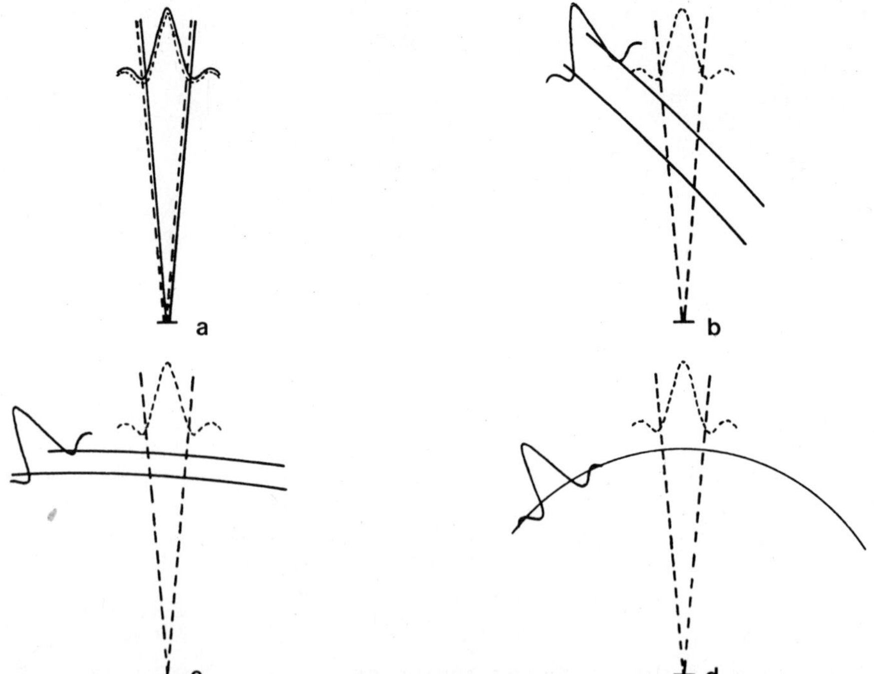

Fig 3 The variation in sample volume shape with increasing sweep speed a) static c.w., b) slow sweep speed, c) fast sweep speed, and d) infinite sweep speed.

EXPERIMENTAL ARRANGEMENT

The hosepipe technique enables one complete line of C-scan to be imaged in a few hundred microseconds. In order to scan a plane it is necessary to sweep the beam relatively slowly in a direction orthogonal to the fast line sweep. Whittingham proposed achieving this by mechanically rocking the transducer as is presently done in mechanical sector scanners (Fig 4). By this means some tens of fields could be swept in a second to give a real-time C-scan image. In the present experiment the mechanical field sweep has been achieved by a slow hand driven mechanism, taking about 15 seconds to image a single field. However, individual line sweeps are completed in under 700 μs so that real-time operation could be achieved.

In transmission, the required deflection of the beam is effected by a progressive time delay across the transducer array of the drives to each element, in the normal manner of electronic sector

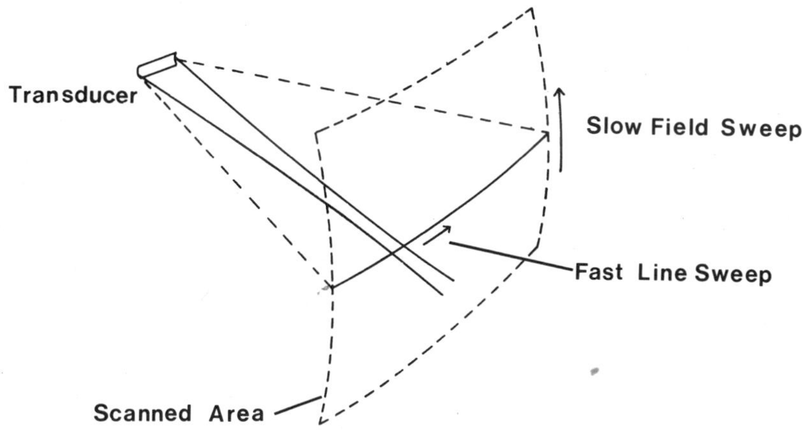

Fig 4 Showing how the constant depth plane is scanned by the combination of fast electronic line sweeps and a slow mechanical field sweep.

scanning. A voltage ramp and comparator method is used to provide the delays, which may be easily altered to change the angle advanced by the beam between pulses, the angle covered in each sector, and the sweep rate, and also to impose focusing on the beam.

In receive we steer the transducer beam using a technique involving cross correlation of the echo signal to each array element with a reference signal. For each element a simulation of the transmitted pulse is generated at a time when an echo from the required range is expected. These simulated pulses then form the reference signals for their respective elements and are mixed in double balanced mixers with the returning echoes. The output of a mixer can be expressed as the combination of the sum and difference frequencies of the two input frequencies:

$$A_1 \cos(w_1 t + \phi_1) . A_2 \cos(w_2 t + \phi_2) =$$

$$\frac{A_1 A_2}{2} \cos((w_1 + w_2)t + (\phi_1 + \phi_2)) + \frac{A_1 A_2}{2} \cos((w_1 - w_2)t + (\phi_1 - \phi_2))$$

where arbitrary phases ϕ_1, ϕ_2 and amplitudes A_1, A_2 are included. Since, in our application, both inputs are of the same frequency, the difference term reduces to a d.c. level whose amplitude depends on the cosine of the phase angle between the inputs. This d.c. component is easily recovered by low pass filtering. With the subscripts R and E for the reference and echo parameters respectively, the detection process may be summarised as:

$A_R \cos(w_o t + \phi_R)$

$A_E \cos(w_o t + \phi_E)$

$\rightarrow \frac{A_R A_E}{2} \cos(\phi_R - \phi_E)$

If either input has no signal present the output will also be zero. The mixer for each element will therefore only give a non-zero output when the reference signal has a pulse on it, and then it will give an output whose amplitude depends on the phase difference between the echo and the reference; i.e. the output is a measure of the correlation between the two.

Consider the echo arriving back from a target at the scan depth and lying in the direction from which we wish to receive. This is shown in Fig 5a. for two adjacent elements of the array. This echo will arrive at each of the mixers at exactly the same moment as does the reference pulse for that direction, and thus there will be zero phase difference between them.

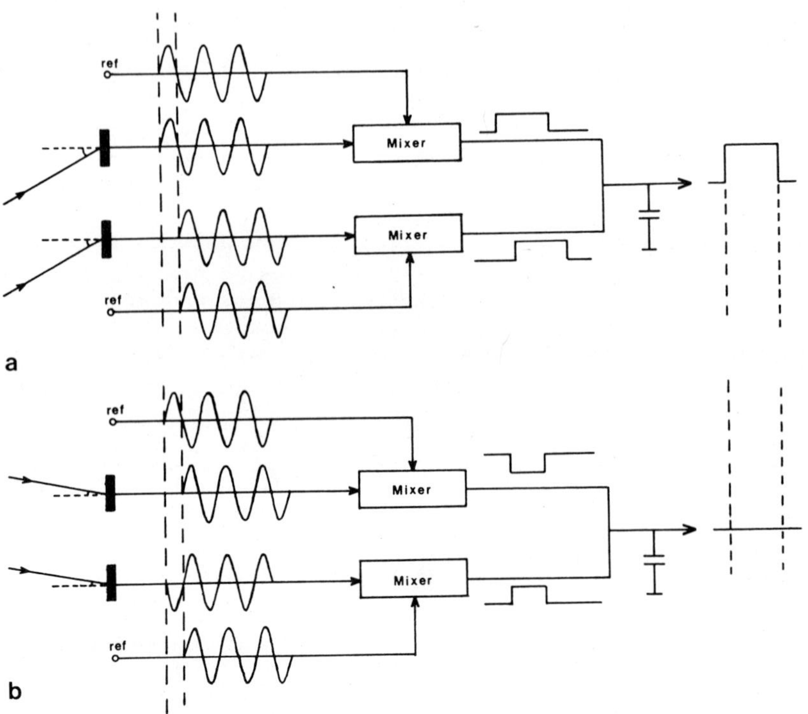

Fig 5 Showing the signals in the detection circuit a) for a target at the correct range and bearing and b) for some other target.

The filtered output from each mixer will then be a positive going pulse. Summing the output from all the mixers together will give a large output response from a target lying in the intended receive direction. If we now consider an echo arriving from some other target not lying in the intended direction (Fig 5b) differences in path length between the target and the array elements will mean that the echo will arrive back at the mixers either before or after the reference pulse. Summing the mixer outputs then gives a smaller response than for a target lying in the intended receive direction. Since an echo from a given range and bearing reaches the different elements of the array at different times, this summing is achieved by integrating the mixer outputs over the period of echo arrival across the array. The angular dependence of the summed response $v(\theta, \theta')$ can be shown to be:

$$v(\theta,\theta') = a A_R A_E . \operatorname{sinc}\left(\frac{aw}{c}\left(\sin\theta - \sin\theta'\right)\right)$$

where a is the half-width of the array, θ is the receive angle and θ' is the bearing of a scattering target. This shows that this method of detection gives the usual sinc function dependence on bearing.

This correlation method of detection is phase sensitive, so that a change of a quarter wavelength in the range of a target would produce a change in sign of the mixer output. Such undesirable range sensitive fluctuations in the mixer output could be eliminated by squaring and adding the outputs of two mixers having quadrature reference signals; in the present study, however, the effect of these fluctuations was simply reduced by full-wave rectification. Two further simplifications of the experimental system were made which imposed limits on the image quality we could expect to obtain from it. The first is that the same transducer was used for both transmit and receive. This imposed a minimum depth on the scan plane since echoes could not be received until transmission right across the sector had been completed. The second further simplification was that an array with only 16 elements was used. The transducer used was a Toshiba 32 element phased array with a width of 15mm and a frequency of 2.6 MHz. The elements were paired together to give 16 active elements. The format used in the images shown below was 62 pulses across a 30° sector with the transmit and receive phase each taking 330 µs. This gave minimum scan depth in water of 24.5cm. Note that even with a line scan time of 660 µs it would be possible to produce a 100 line image 15 times per second. Having to work at a range of 24.5 cm unfortunately limits the resolution capabilities. The 16 element array has a main lobe beam width of 1.9 cm at the range and the sample volume is 17 times the equivalent pulse-echo sample volume. Furthermore, because this range is deep in the far field, no focusing can usefully be applied. The large sample volume of the experimental system meant that considerable

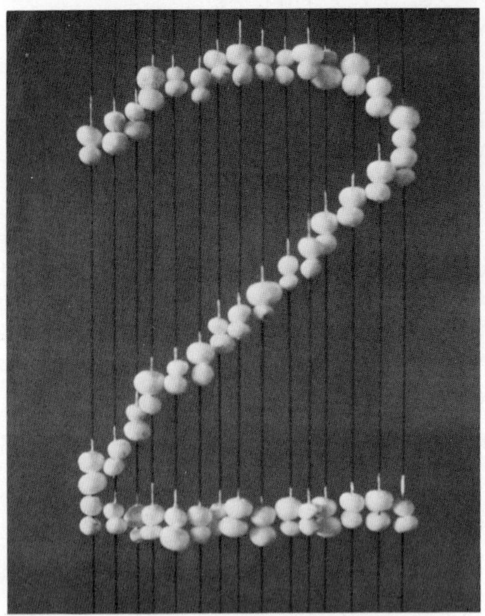

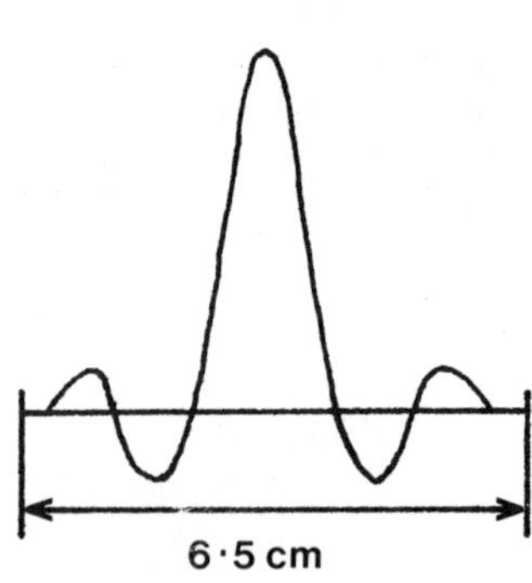

Fig 6.The "figure 2" target with the receiver beam profile at a range of 24.5cm,drawn to scale.

speckle could be predicted when multiple targets were scanned. The targets used to make the images were a "figure 2" and a "figure 4" (dimensions 6.5 x 10.5cm) made of expanded polystyrene pellets suspended on cotton threads. These gave a good signal to noise ratio and form what is essentially an array of point targets.

RESULTS

An image of the "figure 2" target at the correct scan range of 24.5 cm is shown in Fig 7a. At this range the lateral resolution is too poor to show the individual target points and speckle in the image is very apparent. The vertical lines in the image correspond to the 62 pulses transmitted across the sector for each scan line. There are many hundreds of scan lines in the field because of the slow hand driven mechanism used to produce the field sweep. Fig 7b shows a similar image obtained with a target that was agitated vertically whilst it was being scanned, thus smoothing the speckle pattern. Fig 7c shows an image with the "figure 2" target at a range of 19.5 cm, i.e. 5 cm in front of the scan range of 24.5cm. An image formed with a pair of targets is shown in Fig 7d. It was formed with the "figure 2" at the correct range of 24.5cm and the"figure 4" at 28cm. The relatively poor range resolution demonstrated here was expected from the slow sweep speed used and the beam width at this depth.

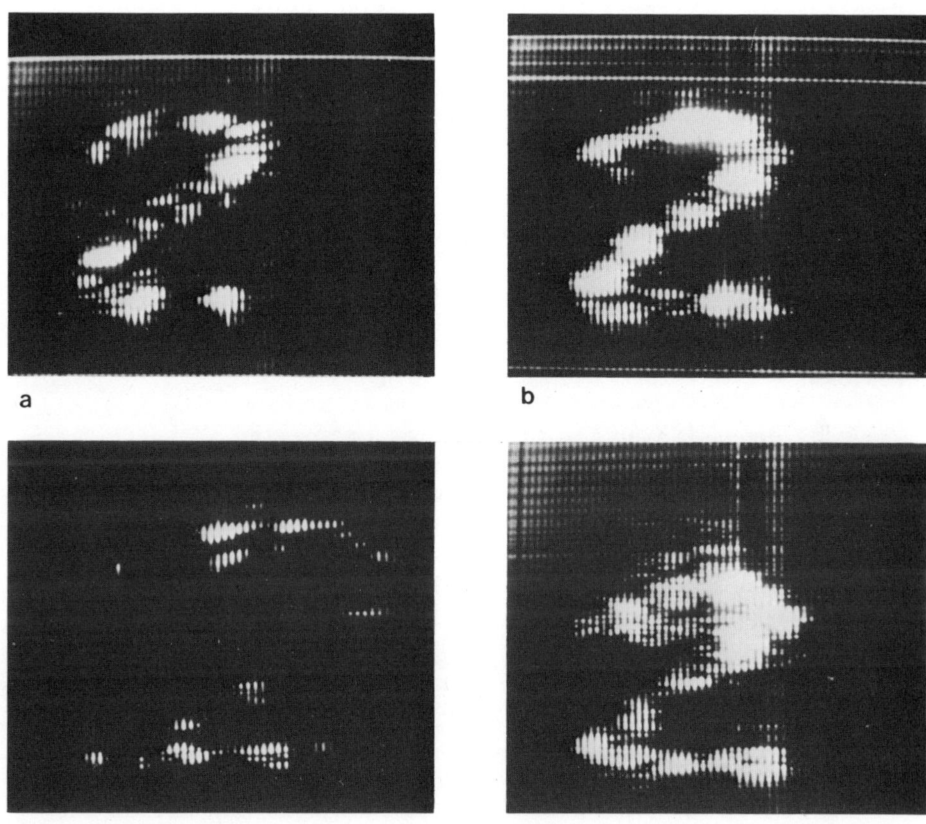

Fig 7. Images formed with the experimental system:
a) the "figure 2" target at the correct scan depth range of 24.5cm
b) as for(a) with vertical agitation of the target.
c) the "figure 2" target 5cm in front of the correct scan depth,ie. at 19.5cm
d) two targets; the "figure 2" at the correct range of 24.5 cm with the"figure 4" at 28cm.

DISCUSSION

The images presented here have relatively poor resolution and prominent speckle. Some of the apparent speckle is simply due to the phase sensitivity of the detection system and could be eliminated as discussed above or by employing a phase insensitive detection system. The remaining speckle is due to interference between echoes from multiple targets within the large sample volume.

The large width of the sample volume is mainly due to the large target range, resulting from the use of the same transducer for transmission and reception. Separate transmission and reception transducers would allow the range to be set to a clinically more realistic depth, with a consequent reduction in sample volume width. A clinical system would also use larger, strongly focussed transducers.

The depth of the sampling volume could be reduced by increasing the sweep speed. The accompanying degradation of lateral resolution due to undersampling could be corrected, at the expense of scanning rate, by interlacing repeat scans of each scan line.

CONCLUSIONS

We have demonstrated the possibility of forming a C-scan image at real-time rates, using a single channel phased array transducer in the 'hosepipe' mode. The limited resolution and prominent speckle (poor contrast) of the images we have obtained could be significantly improved by the use of separate transmission and receiver transducers, having larger apertures and strong focussing.

ACKNOWLEDGEMENTS

We would like to thank Professor K. Boddy for supporting this work, and Dr.S.Hunter of the Department of Paediatric Cardiology of Freeman Road Hospital, Newcastle upon Tyne, for the use of the Toshiba phased array transducer.

REFERENCES

Green, P.S., Schaefer, L.F., Jones, E.D., and Sharez, J.R., 1974, A new high-performance ultrasonic camera, in: "Acoustical Holography and Imaging, Vol.5", P.S.Green, ed., Plenum Press, New York.

Whittingham, T.A., 1982, Real-time constant depth scanning with phased arrays, in: "Proc 10th Int.Symp.Acoustical Imaging, Cannes, 1980, Acoustical Imaging, Vol.10", Plenum Press, New York(pp 1-16).

A STOCHASTICAL IMAGING PROCEDURE

R. H. T. Bates and B. S. Robinson

Electrical Engineering Department
University of Canterbury
Christchurch, New Zealand

INTRODUCTION

Following Labeyrie's (1970,1976) lead, we have for several years been devising means, both optical (Gough and Bates, 1974; Bates and Gough, 1975; Bates and Milner, 1979; Cady and Bates, 1980; Bates and Fright, 1982) and ultrasonic (Bates and Robinson, 1981), for restoring images of objects viewed through distorting media. Our concern here is with a new extension to the shift-and-add technique (Bates and Cady, 1980).

When the distortions of the medium are time-varying and sufficiently pronounced, "speckle processing" allows faithful estimates of true images to be formed by appropriate averaging of sequentially recorded distorted images (Bates and Robinson, 1981; Bates and Fright, 1982). In ultrasonic applications, although typical propagation media introduce severe distortion, their temporal fluctuations are rarely rapid enough for our purposes. We have found, however, that we obtain sufficiently "independent" distortions by transmitting wideband signals (1.7 to 3.4 MHz for the experiments reported here) which are separated after reception into many (128 here) different narrow bands (Robinson and Bates, 1980).

The shift-and-add principle stems from an approach we introduced some years ago (Bates, 1976). The essential idea is that a blurred image is likely to be least distorted where it is brightest (or most intense). So, each of the images (in a set, all of the same object, but distorted differently) are <u>shifted</u> such that their "brightest" points are superimposed <u>and</u> they are then <u>added</u> together, and averaged.

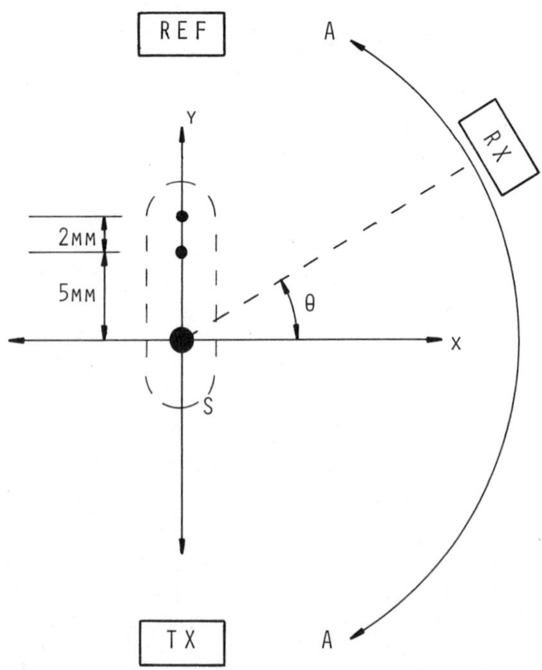

Fig. 1. The experimental arrangement. *TX* and *REF* were both 12 cm, and *RX* was 25 cm, from the origin of the x,y-coordinates. The arc *A*-*A* spanned the interval $|\theta| < 60°$. The three filled circles on the y-axis are the cross-sections in the x,y-plane of three cylindrical scatterers (constituting a linear array of scatterers) with their axes perpendicular to that plane. The larger filled circle represents a 0.183 mm diameter wire. The two smaller filled circles represent 0.125 mm diameter nylon monofilaments. The scattering amplitude of the wire was very close to being three times as large as that of each monofilament.

EXPERIMENTAL DETAILS

Fig. 1 shows the essentials of our experimental apparatus, which is described in detail elsewhere (Robinson and Bates, 1980). A transmitting transducer *TX* illuminates the scattering region *S*, which is part of the y-axis. The field scattered from objects within *S* is sensed by the receiving transducer *RX*, which is scanned over the arc *A-A*. The transducers and scattering objects are all immersed in a water tank. Our main interest has been in imaging with animal tissue placed between *S* and *A-A*, but we have noticed (Bates and Robinson, 1981) that artefacts (i.e. spurious responses) in images formed conventionally tend to be appreciable even when the tank contains water, which is about as ideal an ultrasonic propagation medium as can be found in the real world. We concentrate here on the improvements obtainable by speckle processing of images "viewed" through water (rather than animal tissue) because we feel that the results are likely to suggest a wider range of promising applications.

We used an array of cylindrical scatterers (see Fig. 1) in order to have close control over the experimental conditions. The scattering geometry is thus effectively two-dimensional. Throughout the band of transmitted frequencies the distance from *S* to *A-A* was large enough that *RX* was in the far field (Fraunhofer region) of the scatterers. In early experiments we placed a third transducer *REF* directly in line with *TX* (see Fig. 1), so as to provide a physical phase reference, whose sole purpose was to compensate for "phase drift". Thus, *REF* is not to be accorded any "holographic" connotation. We have found that our speckle processing makes *REF* superfluous, which is potentially of practical technical significance.

Inspection of Fig. 1 shows that the ideal image, denoted by $f(y)$, of the array of scatterers consists of three isolated peaks (Fig. 2a shows the magnitude of this image), whose widths vary inversely with the angular extent of the arc *A-A* (120° in our experiments) and whose heights are proportional to the scattering amplitudes of the individual cylinders.

TIME-DELAY IMAGING

The wideband signals sensed by *RX* can of course be transformed into short pulses of the kind employed in conventional medical B-scan systems. So, we are able to form a time-delay image, denoted here by $f(t,y;\theta)$, when *RX* is positioned at any angle θ. Such an image is best resolved when $\theta = -90°$, which corresponds to "back-scattering". However, the resolution for $\theta = -60°$ is not much worse, and it is the best we can do with our apparatus, because of mechanical constraints. Fig. 2b shows $|f(t,y;-60°)|$. While the three actual scatterers can be recognised easily enough, the artefact level is sufficiently high to suggest that at least two others (one larger and one smaller) might be present. We have not found any satisfactory means of pro-

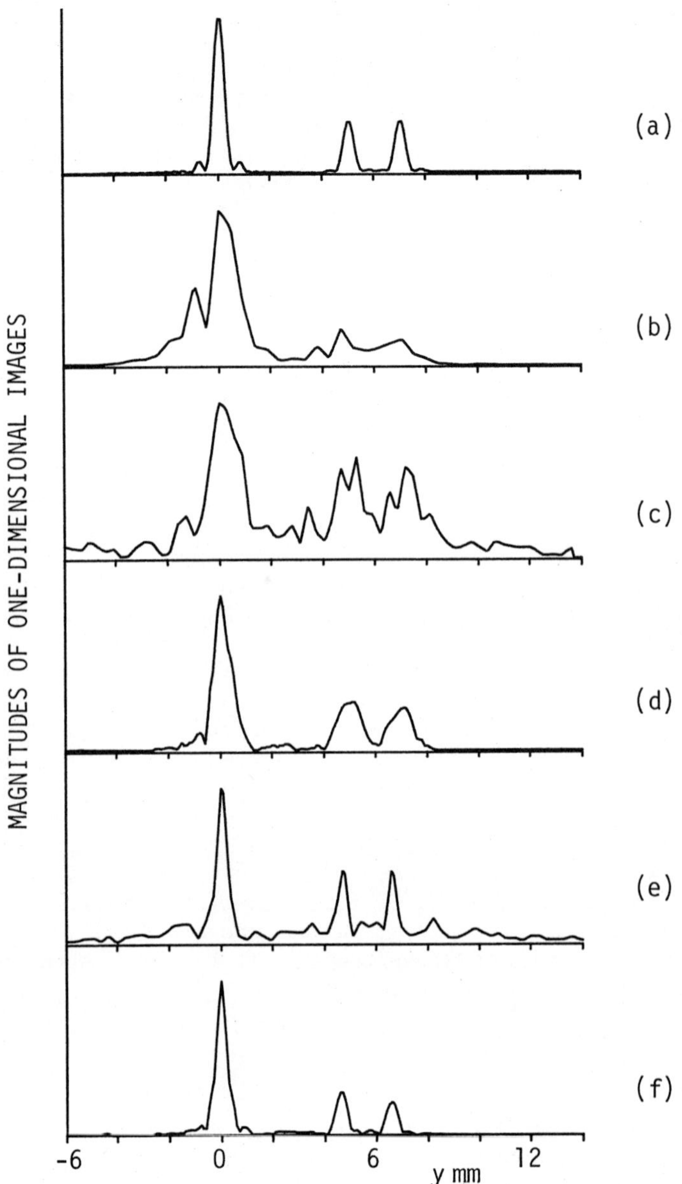

Fig. 2. Magnitudes (maxima all normalised to the same arbitrary value) of one-dimensional images of the array of three cylinders depicted in Fig. 1: (a) $|f(y)|$; (b) $|f(t,y;-60°)|$; (c) $|f(75;y)|$; (d) $|fsa(y)|$; (e) $|g(75;y)|$; (f) $|g(y)|$.

cessing $f(t,y;\theta)$ over the arc A-A to produce a final image with an artefact level significantly lower than that apparent in Fig. 2b.

SHIFT-AND-ADD PROCESSING

We use the integer m to label quantities relating to the 128 narrow bands. We introduce the Fourier variables $\alpha(m)=(sin(\theta))/\lambda(m)$, where $\lambda(m)$ is the ultrasonic wavelength in water at the centre of the mth narrow band. The (complex) amplitude $F(m;\alpha(m))$ of the field measured by RX within the mth band is, when appropriately normalised, the Fourier transform of the (complex) amplitude of the density of equivalent re-radiating sources induced in the scatterers by the signal emanating from TX. We call such a source density a direct narrow band image and we write it as $f(m;y)$. Fig. 2c shows a typical $|f(m;y)|$ - the one for $m=75$. Provided one knows what to look for(!) the three cylinders can be recognised, but the artefact level is considerably higher than in Fig. 2b.

The direct narrow band images are complex quantities, which means that they can be subjected to our coherent shift-and-add procedure the details of which - and the differences between it and our incoherent optical shift-and-add technique (Bates and Cady, 1980) - can be found elsewhere (Bates and Robinson, 1981). We write this shifted-and-added image as $fsa(y)$ - its magnitude is shown in Fig. 2d. The artefact level is very low, but there is some "smudging" of the image, in that the peaks are wider than we might expect.

STOCHASTIC SHIFT-AND-ADD

We now describe what we call our "stochastical imaging procedure". From each $F(m;\alpha(m))$ we form 100 degraded narrow band fields $F(n;m;\alpha(m))$ by the following operation (the integer n runs from 1 to 100):

$$F(n;m;\alpha(m)) = F(m;\alpha(m))\ exp\left(i\ \phi(n;m;\alpha(m))\right)$$

where each $\phi(n;m;\alpha(m))$ is generated within the computer using a pseudo-random number program. The Fourier transforms of the degraded narrow band fields are here called the quasi-speckle images $f(n;m;y)$. For any m, all 100 of the $f(n;m;y)$ are shifted-and-added coherently to produce what we call the processed narrow band image $g(m;y)$. Fig. 2e shows a typical $|g(m;y)|$ - the one for $m=75$. While the artefact level is noticeably higher than in Fig. 2d, there is appreciably less smudging. Remember also that Fig. 2e was formed from data measured within only one of the narrow bands.

All 128 of the $g(m;y)$ are complex quantities, which can be coherently shifted-and-added to produce the processed wide band image $g(y)$. Fig. 2f shows $|g(y)|$. The artefact level is lower than in Fig. 2e, there is less smudging than in Fig. 2d and, in fact, Fig.

2f is very similar to the ideal image shown in Fig. 2a.

CONCLUSIONS

We were at first tempted to interpret the superiority of Fig. 2f over Fig. 2d as an improvement in resolution. However, we have recently imaged scatterers distributed over an area, which has forced us to form two-dimensional images (Bates et al., 1982). Our results suggest that, in fact, the resolution of $fsa(\cdot)$ is usually comparable with that of $g(\cdot)$. However, the artefact level of the latter tends to be considerably reduced. This of course makes it all the more worthwhile to try and find further potentially promising extensions of the stochastical approach introduced here.

The most significant aspect of what we have reported here is that a well-resolved image of a collection of discrete scatterers can be formed from a single narrow band scattering pattern, even when the overall system distortion is so severe that the number and relative amplitudes of the scatterers cannot be easily ascertained from a conventionally generated image. All that is necessary is to subject the measured field to simple-minded stochastical operations in a computer!

The shift-and-add procedure only gives a (moderately) faithful image if one of the scatterers is significantly stronger than the others. We assert nevertheless that our idea has plenty of potential applications. One need only consider the many situations in radar, sonar, acoustics, microscopy, non-destructive testing and medical ultrasonics where active beacons or strong isolated reflectors either already occur or could be introduced easily enough.

REFERENCES

Bates, R. H. T., 1976, A stochastic image restoration procedure, Opt. Commun., 19:240.

Bates, R. H. T., and Cady, F. M., 1980, Towards true imaging by wide-band speckle interferometry, Opt. Commun., 32:365.

Bates, R. H. T., and Fright, W. R., 1982, Towards imaging with a speckle-interferometric optical synthesis telescope, Mon. Not. R. astr. Soc., 198:1017.

Bates, R. H. T., and Gough, P. T., 1975, New outlook on processing radiation received from objects viewed through randomly fluctuating media, IEEE Trans. Computers, C-24:449.

Bates, R. H. T., and Milner, M. O., 1979, Towards imaging of star clusters by speckle interferometry, in:"Image Formation from Coherence Functions in Astronomy," C. van Schooneveld, ed., D. Reidel Publishing Co, Dordrecht, Holland.

Bates, R. H. T., and Robinson, B. S., 1981, Ultrasonic transmission speckle imaging, Ultrasonic Imaging, 3:378.

Bates, R. H. T., Robinson, B. S., Fright, W. R., Gough, P. T., and Hunt, B. R., 1982, Aspects of speckle interferometric imaging, in:"Electronic Image Processing," Proceedings International Conference, York, UK, 26 - 28 July (I.E.E. London).

Cady, F. M., and Bates, R. H. T., 1980, Speckle processing gives diffraction-limited true images from severely aberrated instruments, Opt. Lett., 5:438.

Gough, P. T., and Bates, R. H. T., 1974, Speckle holography, Optica Acta, 21:243.

Labeyrie, A., 1970, Attainment of diffraction limited resolution in large telescopes by Fourier analysing speckle patterns in star images, Astron. & Astrophys., 6:85.

Labeyrie, A., 1976, High resolution techniques in optical astronomy, in:"Progress in Optics, vol. XIV," E. Wolf, ed., North-Holland, Amsterdam.

Robinson, B. S., and Bates, R. H. T., 1980, Wideband ultrasonic diffraction measurements, Australasian Physical & Engineering Sciences in Medicine, 3:233.

LATERAL INVERSE FILTERING OF ULTRASONIC B-SCAN IMAGES

W. Vollmann, H. Schomberg, G. Mahnke

Philips GmbH Forschungslaboratorium Hamburg

D-2000 Hamburg 54, Vogt-Koelln-Str. 30, FRG

INTRODUCTION

The ultrasonic B-scan technique is a well-established diagnostic tool in medicine. In the past the image quality was improved by a steady development of the ultrasonic equipment. Nevertheless, a still existing limitation is caused by the finite width of the ultrasonic beam limiting the lateral resolution of B-scan images. There have been attempts to improve the lateral resolution by means of digital signal processing. So far, several authors have applied inverse filtering techniques to simulated B-scan images or to real images of overidealized objects, e.g. wires in a water tank [1,2,3,4], whereas tests on clinical B-scan images seemingly have not yet been reported. In this paper we try to answer the question of whether inverse filtering can improve clinical B-scan images.

In a B-scan image, a point reflector appears as a little streak perpendicular to the ultrasonic beam. Therefore, in a system with parallel lines of sight, the process of image formation can be discussed rowwise. The measured image intensity in a row, $m(y)$, may be written as a convolution of a lateral point-spread function, $f(y)$, and a reflection coefficient, $r(y)$.

$$m(y) = \int f(y-y')r(y')dy' + n(y) \quad . \tag{1}$$

The noise function $n(y)$ takes the experimental and modelling errors into account. In principle, each row of the image may have a different point-spread function $f(y)$. If it does not, the point-spread function is called depth independent. In the examples studied below the assumption of a depth independent point-spread function is sufficiently satisfied.

Lateral inverse filtering can be defined as any method that tries to reconstruct r(y) from m(y), given f(y) and using equation (1). Lateral inverse filtering thus amounts to a deconvolution. Such a deconvolution can be performed in the frequency domain or in the spatial domain. We shall discuss one method of either approach. Then we shall apply the two methods to two B-scan images, one of a tissue mimicking phantom and one of a human liver.

One criterion for the quality of a filtering procedure is the resolution enhancement, R. This is the factor by which the width of the image of a point reflector is decreased. In reference [5] the following relation between the achievable resolution enhancement, R, and the signal-to-noise ratio of an image, SNR, has been established:

$$R = \sqrt{\ln \mathrm{SNR}} \quad . \tag{2}$$

This relation has been derived for the deconvolution technique used in [1] and for the case of a Gaussian point-spread function. However, one may expect that all filtering techniques have comparable upper limits for the resolution enhancement. So for our clinical B-scan images with a signal-to-noise ratio of about 15 we may expect a resolution enhancement in the order of 1.6.

DECONVOLUTION USING THE FOURIER TRANSFORM TECHNIQUE

Deconvolution techniques based on Fourier transforms are standard in the field of image processing [6]. The mathematical idea is as follows. Defining the Fourier transform of m(y), M(k), by

$$M(k) = \int_{-\infty}^{+\infty} e^{iky} m(y)\, dy \tag{3}$$

one obtaines from equation (1) the simple relation

$$M(k) = F(k)R(k) + N(k) \quad . \tag{4}$$

The reflection coefficient r(y) can be expressed as

$$r(y) = \frac{1}{2\pi} \int_{-\infty}^{+\infty} e^{-iky} \frac{M(k) - N(k)}{F(k)}\, dk \quad . \tag{5}$$

The noise function n(y) and its Fourier transform N(k) are unknown functions. So N(k) must not appear in a reconstruction formula. The step from equation (5) to a reconstruction formula for r(y) cannot be done by a simple elimination of N(k) because M(k)/F(k) generally is not defined at the zeros of $F(k)$. A suitable modification is achieved by the replacement

$$\frac{1}{F(k)} \rightarrow \frac{F^*(k)}{|F(k)|^2 + p^2|F(0)|^2} \tag{6}$$

with a positive parameter p. For $|F(k)| \gg p|F(0)|$, the two expressions in (6) are approximately equal. The new expression is well-behaved for all values of k.

The reconstruction formula then reads

$$\tilde{r}(y) = \frac{1}{2\pi} \int_{-\infty}^{+\infty} e^{-iky} M(k) \frac{F^*(k)}{|F(k)|^2 + p^2|F(0)|^2} dk \ . \tag{7}$$

This filtering procedure is a simplified version of Wiener filtering. For a given B-scan system, an optimum value for p can be found empirically.

Formula (7) differs from the exact equation (5) and so can never yield the true reflection coefficient $r(y)$, but only an estimate $\tilde{r}(y)$. The calculation of $\tilde{r}(y)$ is done numerically using fast Fourier transforms. The true reflection coefficient $r(y)$ is a positive function. If $\tilde{r}(y)$ attains negative values, they can be simply replaced by zero.

DECONVOLUTION USING THE KACZMARZ METHOD

In this section we describe a deconvolution which is performed in the spatial domain. The idea is to approximate equation (1) by a discrete, linear system of equations. Then this system is solved numerically to yield a discrete version of the unknown reflection coefficient.

Let $y_1 < y_2 < \ldots < y_I$ denote the lateral coordinates in a specified row of the underlying B-scan image. In our case $I = 123$. Also we have

$$y_i = y_1 + (i-1)\Delta \ , \ i = 1,\ldots,I \quad , \tag{8}$$

for some $\Delta > 0$ independent of i. As discrete version of equation (1) we choose:

$$m(y_i) = \sum_{j=i-K}^{i+K} f(y_i - y_j)\, r(y_j)\Delta + \tilde{n}(y_i), \ i=1,\ldots,I. \tag{9}$$

Here the integer K is such that $f(y)$ is zero for $|y| > K\,\Delta$. The noise terms $\tilde{n}(y_i)$ now also incorporate the errors due to replacing the integration by summation. We drop these error terms and replace (9) by the approximating system

$$\sum_{j=i-K}^{i+K} f(y_i - y_j)\Delta u_j = m(y_i) \quad , \; i = 1,\dots,I. \tag{10}$$

This is now a linear system of I equations in I+2K unknowns $u_{1-K},\dots,u_0,u_1,\dots,u_I,u_{I+1},\dots,u_{I+K}$. Although not absolutely necessary, we decide to reduce the number of unknowns to the number of equations by requiring that

$$u_{1-K} = \dots = u_0 = m(y_1) \quad , \tag{11}$$

$$u_{I+1} = \dots = u_{I+K} = m(y_I) \; .$$

This reflects the belief that m(y) is already a good guess for r(y) and that r(y) is constant outside the interval $[y_1,y_I]$. Equations (10) and (11) now make up a linear system of I equations, in I unknowns, $u_1,\dots,u_I$. In obvious matrix-vector notation, the system reads

$$Au = b \quad . \tag{12}$$

The existence of a unique solution to (12) is guaranteed only when the determinant of A is nonzero. But in practice it is very unlikely that this determinant is <u>exactly</u> zero so that we may well assume the existence of a unique solution, which we call $u^* = (u_1^*,\dots,u_I^*)$. It follows from our derivation of (12) that the vector $r^* = (r(y_1),\dots,r(y_I))$ satisfies the equation

$$Ar^* = b + e \quad , \tag{13}$$

where e is a small, but unknown error term. It is therefore tempting to compute u* and to take it as an approximation to the wanted r*. Such an approach would fail, however. The reason is the "ill-posed" character of the underlying convolution equation (1) which makes the system (12) extremely "ill-conditioned". As a result, slight changes in the right hand side of (12) will cause drastic changes in the solution, and in particular u* will have nothing to do with r*. The remedy in such a situation consists in computing an approximate solution of (12) which, in addition, is "regular" in the sense that it does not oscillate too much.

There are various ways to compute an approximate, regular solution of (12). Here we suggest to proceed iteratively: Suppose we had an iterative method which, when applied to (12), produces a sequence of vectors $u^0,u^1,u^2,\dots$, converging to the exact solution

u^*, for each initial guess u^0. Then we might try to start with a smooth initial guess and stop the iteration before the iterates begin to exhibit the unwanted irregular features of the exact solution. This is the basic idea. Of course, one should start with the best smooth initial guess available. In our case this is

$$u^0 = (m(y_1),\ldots,m(y_I)). \qquad (14)$$

As iterative method we suggest to use a method due to Kaczmarz [7]. The typical iterative step of this method consists in selecting one equation and to update the current iterate in a deliberate way so as to satisfy the equation selected. The equations are swept through cyclically. One complete sweep through all equations is called a cycle. Applied to (12) and with $A = (a_{ij})$, $b = (b_1,\ldots,b_I)$, $u^i = (u_1^i,\ldots,u_I^i)$, the first cycle of the Kaczmarz method reads:

For $i = 1,\ldots, I$ do:

$$\alpha_i = \left(\sum_{j=1}^{I} a_{ij}\, u_j^{i-1} - b_i\right) \Big/ \sum_{j=1}^{I} a_{ij}^2 \quad ;$$

For $j=1,\ldots, I$ do:

$$u_j^i = u_j^{i-1} - \alpha_i\, a_{ij} \quad ; \qquad (15)$$

end;

end.

The last iterate of the first cycle may then serve as the initial guess of a second cycle, and so on. But in our case it was found experimentally that a single cycle gave already sufficient improvement when starting with (14) as initial guess. Occasional negative components were again set back to zero.

RESULTS

The two presented methods of lateral inverse filtering are now applied to two B-scan images which have been made with a 3.5 MHz unfocussed, linear array consisting of 64 transducer elements. Each ultrasonic beam is generated by a simultaneous excitation of either 3 or 4 neighbouring transducer elements. An alternating change between 3 and 4 elements produces 123 equidistant ultrasonic beams, which add up to an image width of 15.25 cm. The number of rows is 320. As usual, the ultrasonic signal is compressed logarithmically. The output of the system is a non-negative 8-bit number for each picture element.

The first B-scan image is shown in figure 1. It has been obtained from a tissue mimicking phantom. The white streaks at the bottom of the image originate from wires inside the phantom. Figure 2 is a section of figure 1 consisting of 60 rows and showing some wires. The point-spread function can be easily obtained from figure 2.

Figure 3 shows a filtered version of figure 2, obtained by Wiener filtering with p = 0.16. The resolution enhancement is about 1.4. Another somewhat vague criterion of image quality is the appearance of image degrading artifacts. The speckle structure in the main part of the phantom seems to be enhanced by artifacts. The filtered version of figure 2, obtained by Kaczmarz filtering, can be seen in figure 4. The resolution enhancement is about 1.5, whereas the artifacts are relatively small. One additional difference in the preparation of figures 3 and 4 has to be mentioned. The Wiener filtering has been directly applied to the original, logarithmically compressed data (figure 3), whereas the Kaczmarz filtering has been applied to digitally decompressed data. In figure 4, the result has been compressed again. By these procedures, we found the best results.

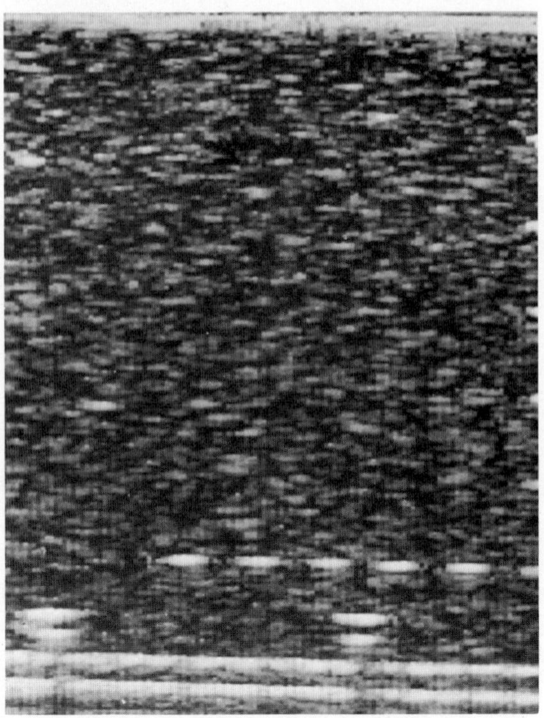

Fig. 1. B-scan image of a tissue mimicking phantom.

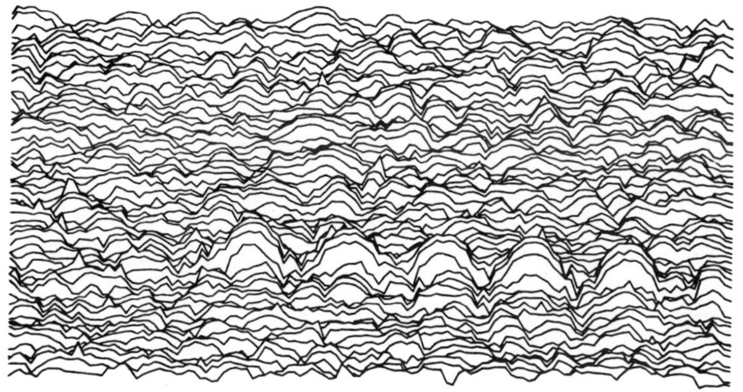

Fig. 2. Section of fig. 1, original.

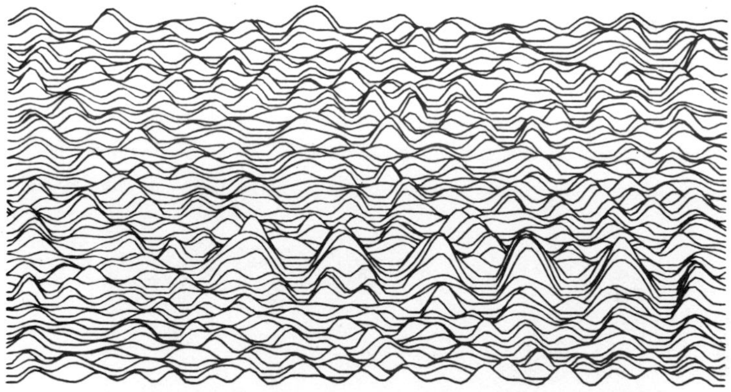

Fig. 3. Wiener filtering.

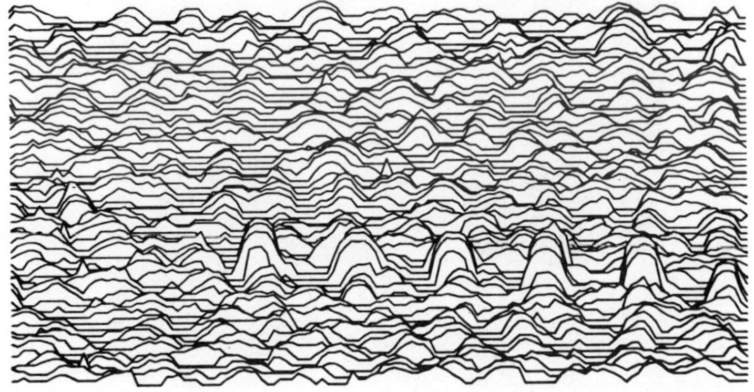

Fig. 4. Kaczmarz filtering.

Figure 5 shows an image of a human liver. It was obtained in the same manner as the image of the wire phantom in figure 1. Figures 6 and 7 are filtered versions of figure 5. Each image was filtered under the same conditions that turned out to be optimum for the wire phantom. The filtered images show a slight enhancement of the resolution and a different artifact pattern each. We find that they contain neither more nor less information than the original image of figure 5. To rank them is a matter of taste. We like the original image the most.

The question arises whether other approaches to lateral inverse filtering might perform better than the ones presented here. We believe they cannot, because it is the signal-to-noise ratio together with (2) which limits the resolution enhancement. In order to improve the resolution enhancement by a factor of 2, the signal-to-noise ratio has to be improved by a factor of 100. Such an improvement cannot be expected, because unavoidable modelling errors,

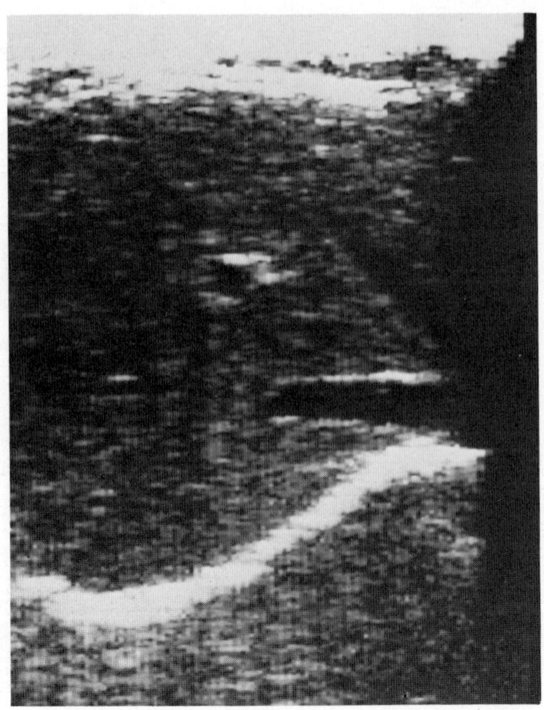

Fig. 5. Image of a human liver.

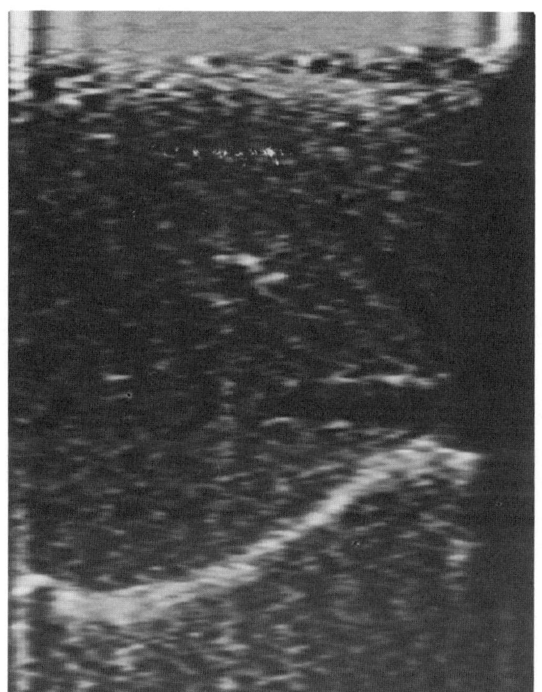

Fig. 6.
Wiener filtering.

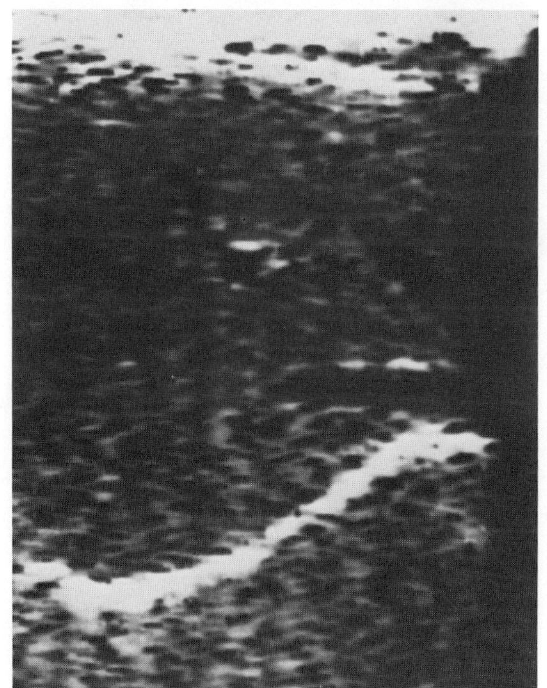

Fig. 7.
Kaczmarz filtering.

which contribute to the noise, keep down the signal-to-noise ratio. One source of a modelling error is the macroscopic inhomogeneity of tissue which makes the point-spread function to depend on the position. Another source of a modelling error is the microscopic inhomogeneity of tissue which is responsible for the appearance of speckles. Every deconvolution algorithm will interpret a speckle as a point reflector and will enhance the undesired specular structure. Concluding, the noise level is largely determined by the tissue itself.

[1] E.E. Hundt and E.A. Trautenberg, "Digital Processing of Ultrasonic Data by Deconvolution", IEEE Transactions on Sonics and Ultrasonics SU-27 : 249 (1980).

[2] R.B. Kuc, "Application of Kalman Filtering Techniques to Diagnostic Ultrasound", Ultrasonic Imaging 1 : 105 (1979).

[3] M. Fatemi and A.C. Kak, "Ultrasonic B-Scan Imaging: Theory of Image Formation and a Technique for Restoration", Ultrasonic Imaging 2 : 1 (1980).

[4] W.v. Seelen, A. Gaca, E. Loch, W. Scheiding and G. Wessels, "Recognition of Patterns in Ultrasonic Sectional Pictures of the Prostate for Tumor Diagnosis", in : "Ultrasonic Tissue Characterization II", M. Linzer, ed., National Bureau of Standards, Spec. Publ. 525, Washington, D.C., 1979.

[5] W. Vollmann, "Resolution Enhancement of Ultrasonic B-Scan Images by Deconvolution", IEEE Transactions on Sonics and Ultrasonics, SU-29 : 78 (1982).

[6] T.S. Huang, W.F. Schreiber and O.J. Tretiak, "Image Processing", Proceedings of the IEEE 59 : 1586 (1971).

[7] S. Kaczmarz, "Angenäherte Auflösung von Systemen linearer Gleichungen", Bull, Acad. Polon. Sci., Lett. A 35 : 355 (1937).

TEXTURE CLASSIFICATION OF B-SCAN ULTRASOUND IMAGES:

AN ASSESSMENT USING TISSUE MODELS

D.K. Nassiri, D. Nicholas and C.R. Hill

Institute of Cancer Research:Royal Marsden Hospital
Sutton
Surrey, U.K.

ABSTRACT

Image texture analysis procedures have been carried out on B-scans of a set of different tissue phantoms, acquired under a variety of different imaging conditions in respect of the parameters: system gain, TGC rate, acoustic beam profile.

Quantitative measures of all of the features of image texture that have been examined are found to be strongly sensitive to variations in the above parameters of the imaging system. It is concluded that successful application of texture analysis and related procedures to *in vivo* tissue classification will call for very rigorous control of acquisition conditions, and that the present findings may account for some apparent discrepancies between the reports of different research groups who are attempting to apply such techniques.

Keywords : tissue characterization, ultrasound, B-scan, image analysis

INTRODUCTION

Several reports have been made of attempts to classify tissue *in vivo* on the basis of quantitative or statistical features of pulse echo signals recorded in either A-mode or B-mode (eg.1-9). Included in these reports, for investigation of liver diseases, are those of our own group for the B-mode approach (1,6,7) and those of Lerski et al for A-mode (2,3) where, in each case, appreciable success in classification has been achieved. More recently, however, Gonzalez et al (10) have reported that they were not able to achieve classifications on the lines reported by Lerski (2,3).

We have, in particular, been aware that many of the quantitative features employed in these studies might be expected to be sensitive to changes in various parameters of the acquisition procedure, including : the beam profile, frequency dependent acoustic attenuation, and system gain operating on non-linearities in the acquisition/recording system (including deliberate gain compression). In a preliminary attempt to investigate the practical importance of these variables we have therefore carried out a study of the effect on the accuracy of the classification procedure of deliberate, controlled changes in acquisition parameters. Initially we have used a set of simple tissue phantoms, with progressively changing characteristics, which are the basis for this study.

METHODS

Tissue models were prepared as pseudo-random suspensions of spherical particles in gelatin and gelatin-alginate gel designed to have acoustic properties similar to those of soft tissues (11). Specifications of the individual phantoms were as follows :

Phantom	Diameter(μm)	Particle Density(cm^{-1})
Sephadex-gelatine	10-40	8000, 4000, 1000
	100-300	667, 64, 8
Polyethylene-gel. algenate	100-300	4000, 64, 8

In addition, measurements were carried out using a domestic plastic sponge embedded in gelatin after having previously been degassed.

B-scans were acquired in a conventional manner using a 3.5 MHz nominal frequency narrow band-width circular transducer of aperture 13 mm, spherically focussed over the range 40-100 mm. Scans were performed manually through a waterbath, with the beam directed vertically downward as the transducer was moved laterally to produce a rectilinear scan (Fig. 1, left).

Signal processing was by conventional rf. amplification, demodulation and compression, following which B-scan data sets were temporarily stored in an analogue scan convertor from which a selected portion of the data (3 x 3 cm in real space, see Fig. 1, right) were digitised to 64 x 8 bits for subsequent analysis. Conventional means for control of system gain (dB) and TGC rate (dB cm^{-1}) were provided in the analogue circuitry.

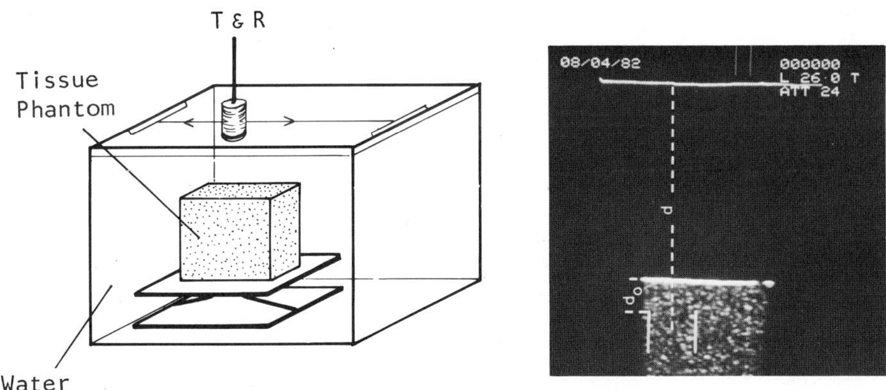

Fig. 1. Geometry for B-scan data acquisition.

Using this system, B-scans were acquired, and relevant portions digitised with high precision, corresponding to several different selected regions in a given tissue phantom. In particular, it was possible to acquire data sets under a variety of combinations of gain and TGC rate settings.

In subsequent analysis of the digitised portions of the images the following features, or groups of features, were computed :

(a) first order statistics of signal amplitude;
(b) spatial density of peaks in signal amplitude;
(c) spatial gradients of signal amplitude;
(d) Fourier space counterparts of the above features;
(e) signal level co-occurrence features (statistical measures of the spatial relationships of signal amplitude values occurring at various points in the image portions).

From the above groups we evaluated a total of 120 individual features, although the effective number of independent features is considerably less than this, since many pairs of features are appreciably correlated. Thus, in the present study, attention was concentrated on a limited number of features which, in previous more empirical work, had been found to be useful as tissue classifiers either when taken alone or when combined in discriminant analysis.

RESULTS

The values of the features were computed for the selected

portions of B-scans acquired under a variety of different imaging conditions in respect of the parameters: system gain, acoustic beam profile and TGC rate. Although the effect of these variables are not independent, in this preliminary report we have tried to demonstrate their effects separately.

a) System Gain

Examples of computed values of some of the features, plotted as functions of relative system gain, are shown in Fig. 2. The deviation in the values of any particular feature, calculated for different sites in the tissue phantom (at the same depth, when the system settings have not been changed) seems to be small, while the changes in the values of the features are strongly correlated with the relative gain. The non-linearity of the system may cause some changes in the form of gain dependence at very high or very low gain. This effect, combined with the fact that, in calculation of the co-occurrence matrix, the 64 x 64 original image is normalised to its minimum and maximum gray levels, results in a very 'dark', (low mean level) original image, giving a very noisy normalised image. A similar form of variation may be seen in very 'bright' images.

b) Acoustic Beam Profile

Variation of the features were studied as a function of distance of the scattering volume from the transducer. This study was facilitated by keeping the effect of the attenuation of the beam constant. In these experiments the value of d_o was kept constant, and no time-gain compensation was applied. The depth dependence of the arithmetic mean of the amplitude, and 'contrast' (a second-order statistic) are shown in Fig. 3 a and b).

The unexpected variation of the arithmetic mean as a function of depth may be explained by theoretical evaluation of the averaged scattering signal received by the transducer (12,13). Such a variation can introduce changes of the features as a function of depth. Since these measurements have been made as a function of system gain, as well as of depth, one may plot the features of the images which have the same arithmetic means for the signal amplitude. Such a corrected form of variation of contrast with depth is plotted in Fig. 4. At far distances from the transducer the depth dependence of this feature seems to be negligible, but it becomes more apparent in the focal region (i.e. 4-10 cm) and in the near field. Reasons for these effects are not fully understood but theoretical investigations are being pursued to aid in these investigations.

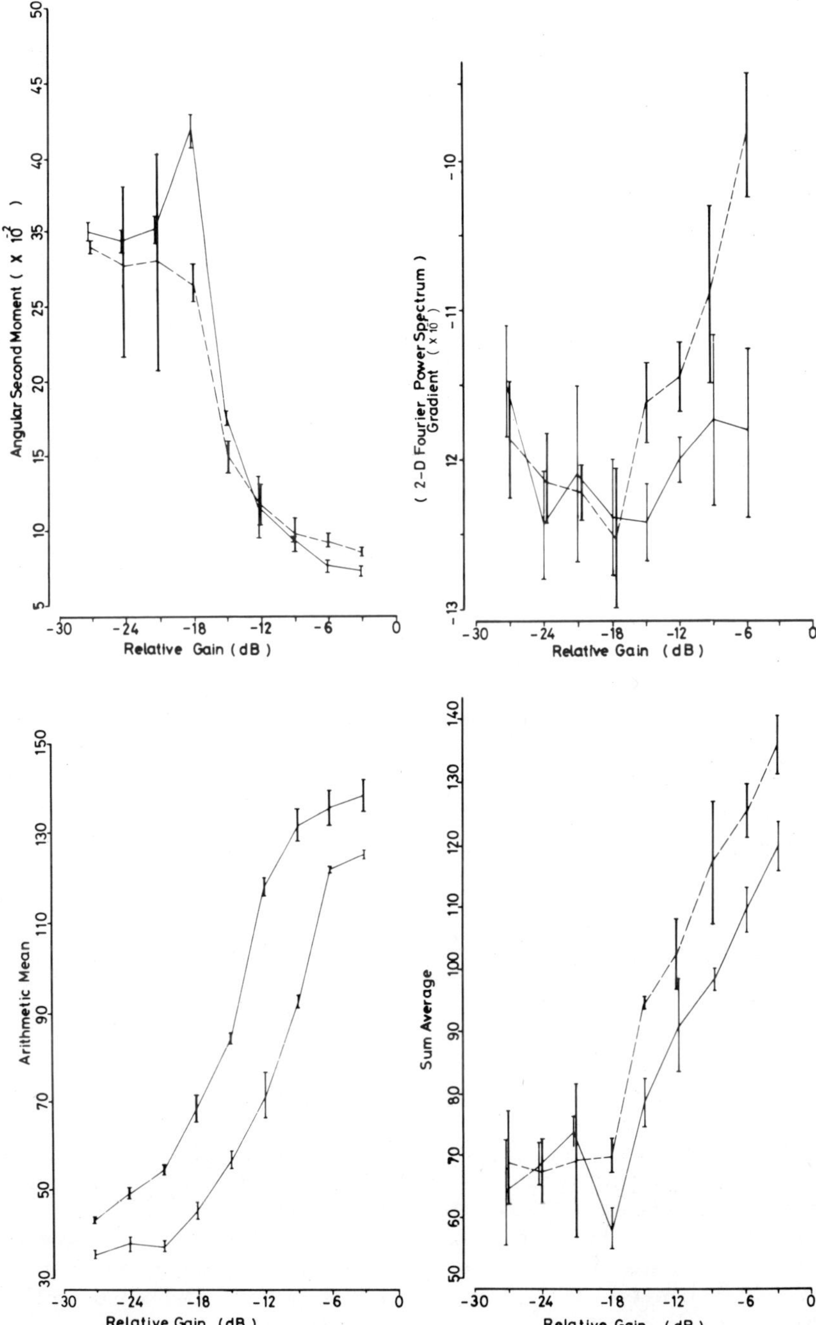

Fig. 2. Examples of computed values (± s.d.) for four selected features, plotted as functions of relative system gain.

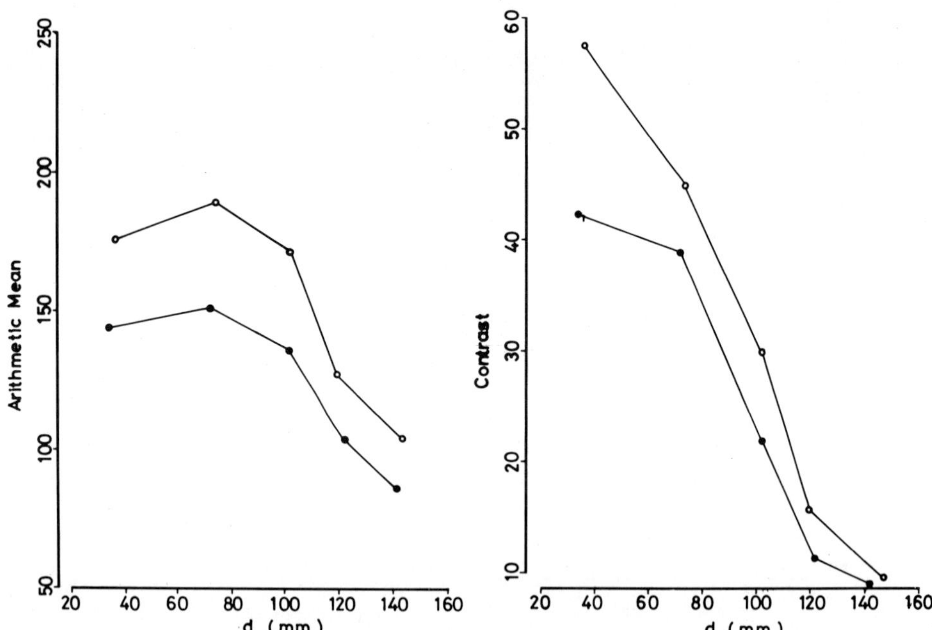

Fig. 3. Variation in the values of two selected features as a function of distance, d, of the scattering volume from the transducer.

c) TGC Rate Dependence

The B-scans were acquired under a variety of combinations of gain and TGC settings; for example by varying TGC rate and applying a compensating gain correction to maintain an effective constant gain at the centre of each box or at a plane indicated by the line AB in Fig. 5. With knowledge of the effect of gain and depth, one may correct the feature values and evaluate the effect of the changes due solely to TGC rate. However in practice variation of the features due to $\pm$ 1 dB cm^{-1} TGC changes were not significant. This is partly due to the large deviation of the features, calculated at various positions within the specimen. Further measurements to complete these findings are currently being pursued.

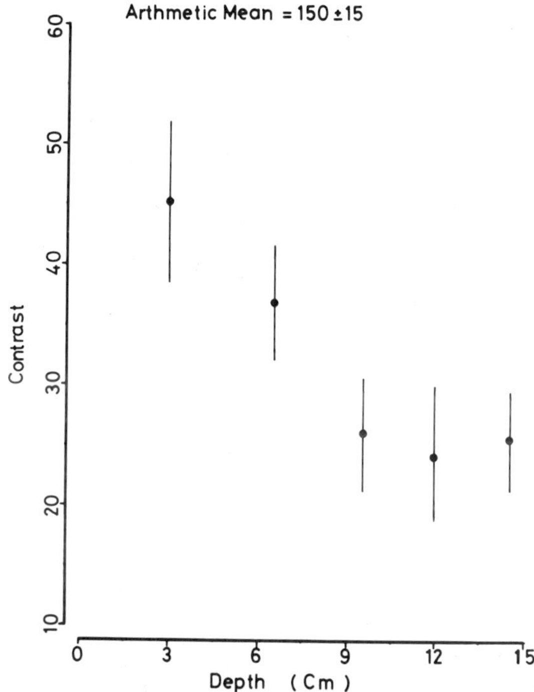

Fig. 4. Variation with depth of the value of "contrast" for images normalized to equal arithmetic mean.

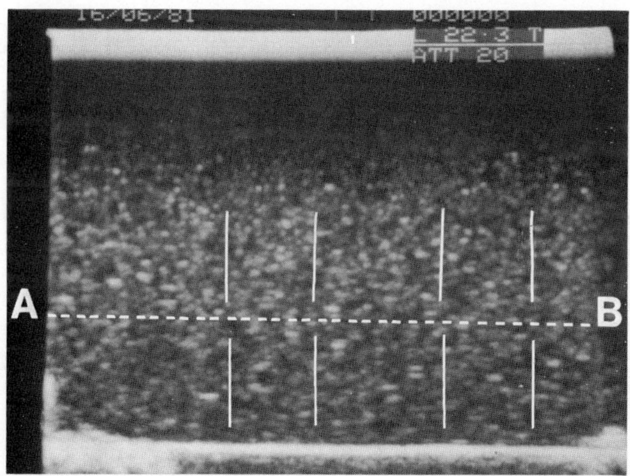

Fig. 5. Acquisition geometry used for investigation of the dependence of feature values on TGC rate.

SUMMARY AND DISCUSSION

Both A-mode and B-mode texture analysis techniques exploit statistical features which are sensitive to variations in both the amplitude and the spatial frequency content of the data. In this paper we have demonstrated experimentally the extent to which some of these features are sensitive to relative gain setting: a 3 dB change in gain, for example, being found generally to give rise to appreciable changes in numerical values of features of a normalized image. This sensitivity to gain appears to be enhanced by the presence of non-linearities in the data acquisition system.

One possible conclusion from these findings may be that the particular features shown to have such gain dependence may, in practice, be useful ones, in the sense that they are sensitive to some conditions which could be tissue specific. At the same time, however, the results suggest that optimum practical performance in tissue differentiation will call for very careful control and standardisation of the conditions of data acquisition.

The findings also suggest an explanation for the apparent disagreement between the previously published results of different groups in this field, using comparable techniques, who have been able to achieve tissue differentiation.

Our findings reported here are preliminary only and we believe that there is a need for much further work in order to demonstrate the situation more completely. An additional, and very desirable, outcome of such work which may be anticipated would be the ability to interrelate and intercalibrate image texture data acquired using a somewhat different data acquisition system. This would have the immense value of enabling "learning" sets of data acquired on one system to be accessible for general use on other systems.

REFERENCES

1. D. Nicholas, A. Barrett, J.M.G. Chu, D.O. Cosgrove, P. Garbutt, J. Green, S. Pussell, and C.R. Hill, Computer analysis of grey scale tomograms, in: "Acoustical Imaging" Vol.8, A. Metherell, ed., Plenum Press, New York (1980).
2. R.A. Lerski, E. Barnett, P. Morley, P.R. Mills, G. Watkinson, and R.N.M. MacSween, Computer analysis of ultrasonic signals in diffuse liver disease, Ultrasound Med. Biol. 5:341 (1979).
3. R.A. Lerski, M.J. Smith, P. Morley, E. Barnett, P.R. Mills, G. Watkinson, and R.N.M. MacSween, Discriminant analysis of ultrasonic texture data in diffuse alcoholic liver disease: 1. Fatty liver and cirrhosis, Ultrasonic Imaging 3:164 (1981).

4. G. Wessels, W.V. Seelen, and U. Scheiding, The application of pattern recognition in ultrasonic sectional picture of prostate (B-mode analysis), in: "Ultrasonic Tissue Characterization: Clinical achievements and technological potentials", J.M. Thijssen, ed., Stafleu's, The Netherlands (1980).

5. W.J. Lorenz, H. Bihl, G. van Kaick, A. Lorenz, and M. Geissler, Methods of Image Analysis and Enhancement, in: Medical Ultrasonic Images. Formation, Display, Recording and Perception, C.R. Hill and A. Kratochwil, eds., Excerpta Medica, Amsterdam (1981).

6. D. Nicholas, M. Bamber, C. Bossi, P. Garbutt, J. Hinton, D. Nassiri, and S. Pussell, Quantitative image analysis for diffuse liver disease (Abstract), Ultrasonic Imaging 3:202 (Proc. of 6th Internat. Symp. on Ultrasonic Tissue Characterisation) (1981).

7. D.K. Nassiri, D. Nicholas, C. Bossi, P. Garbutt, M. Bamber, J. Hinton, S. Pussell, D.O. Cosgrove and C.R. Hill, B-scan texture classification of diffuse liver disorders, (Abstract), Ultrasonic Imaging 4:182 (Proc. of 7th Internat. Symp. on Ultrasonic Tissue Characterisation) (1982).

8. S.I. Finette, A.R. Bleier, K. Haber, W. Swindell, M. Good, and B.B. Goldberg, Analysis of digitized A-scans from breast tumours using pattern recognition techniques. (Abstract) Ultrasonic Imaging 4:181 (Proc. of 7th Internat. Symp. on Ultrasonic Tissue Characterisation) (1982).

9. Ultrasonography in Ophthalmology, (Proceedings of 8th SIDUO Congress) J.M. Thijssen and A.M. Verbeek, eds. Junk, The Hague (1981).

10. V. Gonzalez, J. Jones, L. Ferrari, and M. Behrens, The application of a one-dimensional texture algorithm to A-mode ultrasound waveforms recording *in vivo*, (Abstract), Ultrasonic Imaging 4:181 (Proc. of 7th Internat. Symp. on Ultrasonic Tissue Characterization) (1982).

11. N. L. Bush, and C.R. Hill, Improvements in stable, acoustically tissue-equivalent materials. Ultrasound in Med. and Biol. (submitted).

12. D.K. Nassiri, PhD Thesis, University of London (1983).

13. D. Nicholas, C.R. Hill, and D.K. Nassiri, Evaluation of back-scattering coefficients for excised human tissues: principles and techniques, Ultrasound in Med. and Biol. 8:7 (1982)

REDUCTION OF SPECKLE IN ULTRASOUND B-SCANS BY DIGITAL PROCESSING

R.J. Dickinson

The General Electric p.l.c.
Hirst Research Centre
Wembley, Middlesex, U.K.

INTRODUCTION

Ultrasonic B-scans of regions of tissue whose inhomogeneities are small compared to the acoustic wavelength have a textured appearance with large fluctuations in echo amplitude; termed speckle. This texture pattern is caused by random interference between scattering regions too fine to be resolved, and is a consequence of the coherent nature of ultrasonic waves. As such it is similar in properties to laser speckle and has been investigated by Buckhardt (1), and Abbott and Thurstone (2). If a simple model of speckle is considered, where the scatterers are weak, randomly positioned, and numerous, so that there are a large number within each resolution cell, then the B-scan image of such a scattering ensemble will have the following properties. The amplitude distribution will always have the same form, a Rayleigh distribution. The mean amplitude is determined by the scatterer strength, and the standard deviation is half the mean. The spatial properties are determined solely by the interrogating pulse; the auto-correlation function of the speckle pattern is dependent on the pulse shape and beam-width of the pulse.

The properties of speckle mean that little information concerning the tissue is contained in ultrasonic B-scans of tissue with small random inhomogeneities. Both the first and second order statistics of the image are independent of the tissue properties, so long as the tissue structure cannot be resolved. Work on visual perception has shown that if the first and second order statistics of a texture are identical, then they cannot be differentiated by eye (3). So a visual display of a speckle pattern, as presented in a conventional B-scan, will not allow different scattering ensembles to be differentiated, and computer techniques of image analysis are called for. The one property of a speckle pattern that is characteristic of the scattering ensemble is the mean echo level. However small differences in mean level will be obscured by the large standard deviation of the speckle pattern, and hence regions with slightly differing mean echo levels will be difficult to discern. This paper examines ways of reducing speckle and improving the information content of the B-scan.

One method of reducing speckle is to reduce the standard deviation of the speckle while retaining the mean level, and there are a number of methods that achieve this. The most common is to combine N uncorrelated scans, which achieves a $\sqrt{N}$ reduction in the standard deviation of the speckle pattern if the scans are incoherently combined such as by averaging intensities. The uncorrelated scans can be at different angles, as in compound scanning (1), or different frequencies (2). The reduction is limited to the number of uncorrelated scans that can be obtained, and hence by the acoustic window or bandwidth available. While successful in reducing speckle they have the drawback that more time is required to obtain a full scan, and the spatial resolution is also degraded.

A simple method of speckle reduction is to use a low-pass filter. This will reduce the standard deviation of the speckle pattern, but will also degrade the resolution of other components of the scan such as edges. The next section describes a non-linear smoothing filter which reduces speckle while leaving edges unchanged.

Another method is to process A-scans prior to display to prevent interference occurring. Speckle is a consequence of the non-linear signal processing used in ultrasonic scanners which displays echo intensity and discards phase. The A-scans can be processed to give echo amplitude without interference effects by using a Wiener filter. This converts the imaging process into a linear convolution with no need for demodulation, and is described in a later section.

SPECKLE REDUCTION USING AN UNSHARP MASK FILTER

Method

The filter used is an unsharp mask filter (4), and has the form

$$x'_{ij} = \bar{x}_{ij} + c(x_{ij} - \bar{x}_{ij}) \quad (1)$$

where x_{ij} is the original image, $\bar{x}_{ij}$ is the local mean, and x'_{ij} is the processed image, c is a parameter which determines the effect of the filter. If c = 0.0 it is a moving average filter, if $0.0 < c < 1.0$ it is a low pass filter of varying severity; if c = 1.0 the filter has no effect, and if $c > 1.0$ it is a high pass filter.

As it stands the filter is linear, and has the same effect on all parts of the image. A useful filter for speckle reduction would smooth regions of speckle while leaving other regions such as edges unprocessed or even high pass filtered. This can be done by making c image dependent, and hence the filter becomes non-linear. One approach would be to use an edge detection algorithm which would differentiate regions of the image corresponding to edges from regions corresponding to speckle, and to set c high in the former and low in the latter. Such edge detection algorithms are available, but the approach taken here uses the fact that in ultrasonic B-scans regions corresponding to speckle are low in amplitude, and edges have high amplitudes, so c can be made amplitude dependent. This criteria is very simple, and there will be exceptions, but it can be easily implemented.

The images processed are shown in Figure 1, and are taken with an in-house prototype linear array scanner. One frame is digitised to 128 x 128 x 8 bit resolution, then stored and processed on a PDP 11/60 computer. The most significant 4 bits of original and processed images are then displayed on a Barco high resolution graphics monitor. The processing algorithms, written in FORTRAN, are used on the stored images. For each pixel, the local mean of a 4 x 4 block of nearest neighbours centred on that pixel is calculated to give $\bar{x}_{ij}$, and then the new pixel value, x'_{ij} is calculated from equation (1).

Results

Figure 1 also shows the result of processing the images of 1(a) and (b) with an unsharp mask filter, where c is proportional to the local mean. It can be seen that low level echoes are smoothed, while high level echoes are edge enhanced, since c is greater than 1.0. There is a range of smoothing and edge enhancement, depending upon the echo level.

Discussion

The non-linear unsharp mask filter used here is capable of smoothing speckle without blurring edges. These results could be improved by using a more sophisticated edge detector. There is a further advantage in preferentially high pass filtering high amplitude echoes. The effect of the non-linear compression on the echoes performed in ultrasonic B-scanners is to degrade the resolution of high amplitude echoes, giving an amplitude dependent effective resolution (5). This algorithm has the opposite effect, amd thus can compensate for this undesirable side-effect of dynamic range compression.

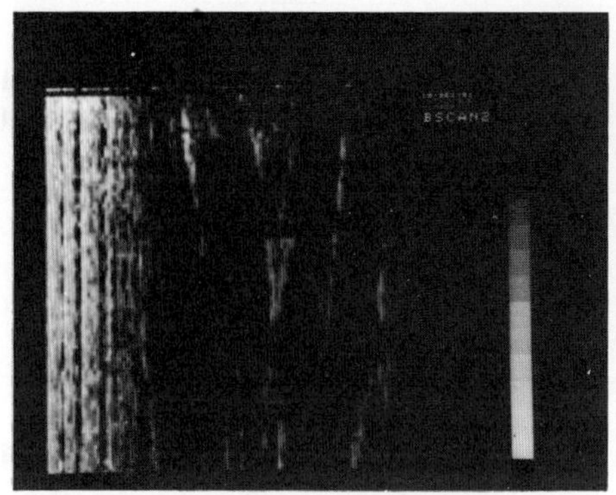

(a) Original 7 MHz carotid scan

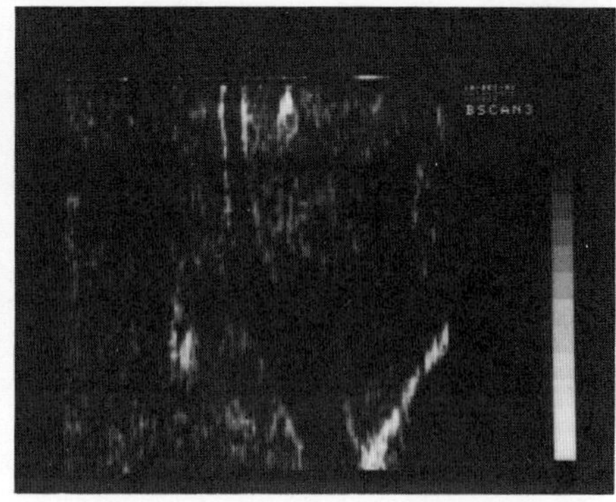

(b) Original 3.5 MHz liver scan

(c) $c = \overline{x}/64$

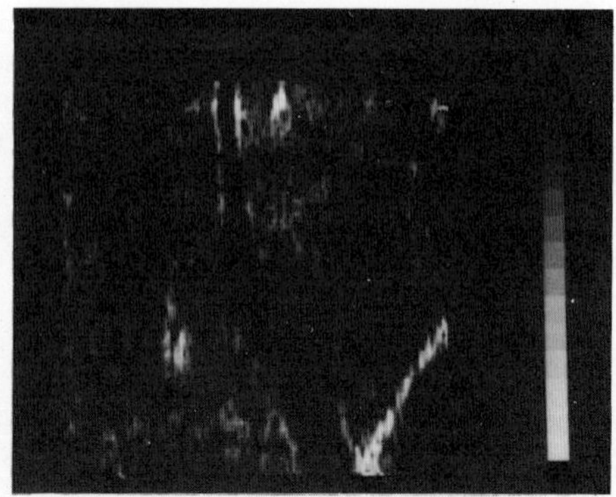

(d) $c = \overline{x}/96$

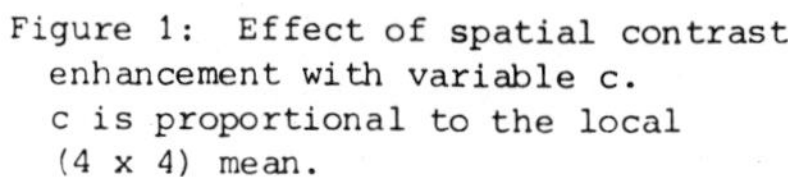
Figure 1: Effect of spatial contrast enhancement with variable c. c is proportional to the local (4 x 4) mean.

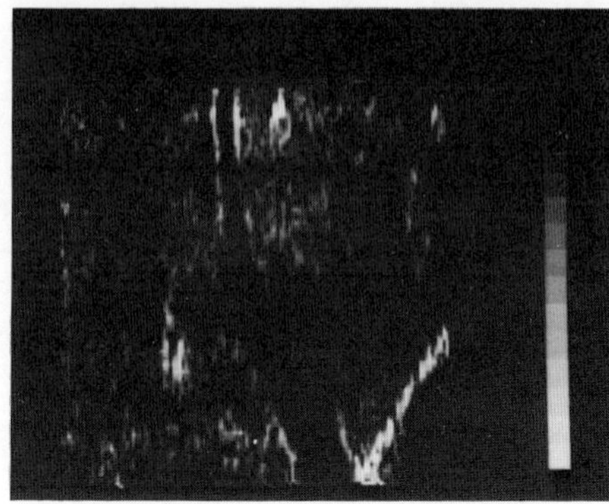

(e) $c = \overline{x}/64$

SPECKLE REDUCTION BY DECONVOLUTION

Introduction and Method

Deconvolution is used in image processing to remove the effect of the point spread function from an image, giving an optimum estimate of the original object. As such it improves the high contrast resolution, or the ability to resolve point targets. A limiting factor in ultrasound B-scanning is the low contrast resolution, or the ability to resolve small regions of differing scattering levels in the presence of speckle. This section examines the effect of deconvolution on the low contrast resolution.

It has been shown theoretically (6) that the process of B-scan image formation can be modelled by a convolution process followed by demodulation, and thus it is valid to use a deconvolution algorithm to process the image. The investigation uses simulated images generated by a computer model of B-scanning reported by Bamber and Dickinson (6). This simulation generates a tissue model, which may be one or two point scatterers or an ensemble of randomly positioned scatterers, and then convolves with a pulse whose wavelength, envelope length and width are specified. Finally the simulated echoes are digitally demodulated. The resultant image has many of the properties of the B-scan image. The simulation is useful because the original scatterer distribution is known, and so the success of a particular deconvolution routine at retrieving it can be judged. This is not true for real B-scan images.

The method of deconvolution used is Wiener filtering. If an object F is convolved with a point spread function H, with additive noise, then in Fourier space the image G can be represented by

$$G(\underline{k}) = H(\underline{k})\,F(\underline{k}) + N(\underline{k}) \tag{2}$$

($N(\underline{k})$ is the noise spectrum).

It can be shown (7) that an optimum estimate of F from G is given by F', where

$$F'(\underline{k}) = \frac{H^* G}{|H|^2 + |N|^2/|F|^2} \tag{3}$$

The value of $|F|^2$ in the denominator must be estimated, and here it is approximated by $|G|^2$. White noise is added to the simulated image, so the noise spectrum is flat. The filter finally used is

$$F'(k) = \frac{H^* G}{|H|^2 + \alpha/|G|^2} \tag{4}$$

α is the signal to noise ratio.

(a) Undemodulated point spread function. wavelength = 6.0, Beam width = 6.0, Pulse length = 6.0.

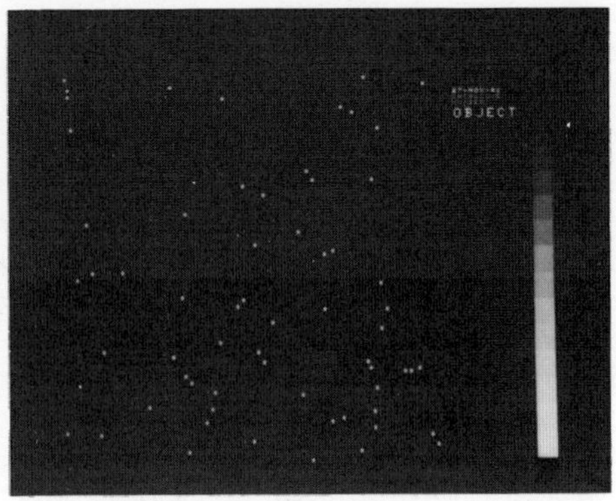

(b) Scattering object mean separation = 16.0.

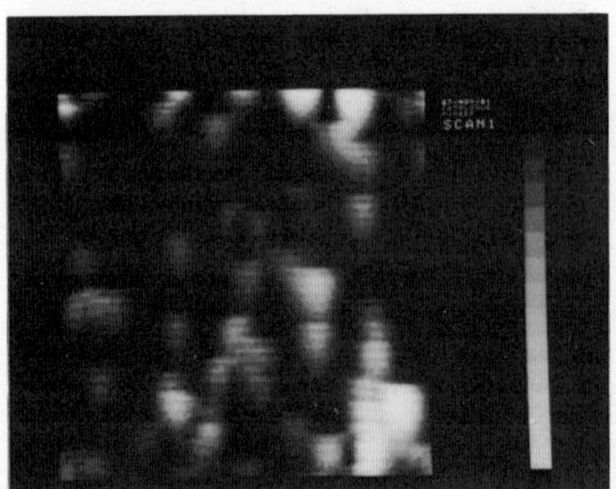

(c) Simulated B-scan.

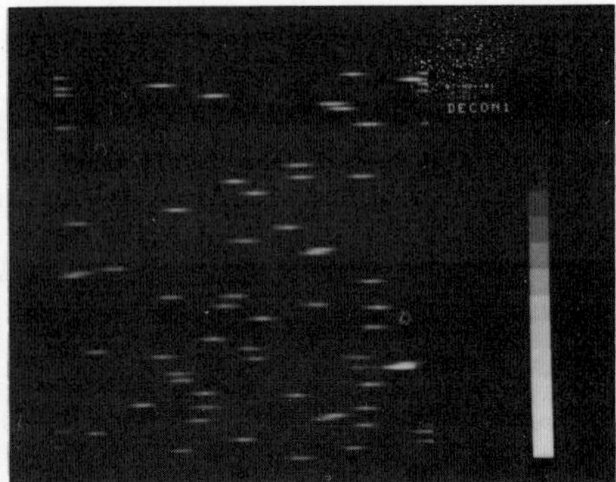

(d) Scan deconvolved before demodulation.

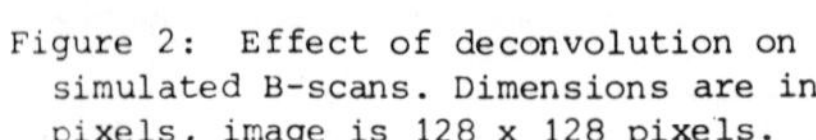
Figure 2: Effect of deconvolution on simulated B-scans. Dimensions are in pixels, image is 128 x 128 pixels.

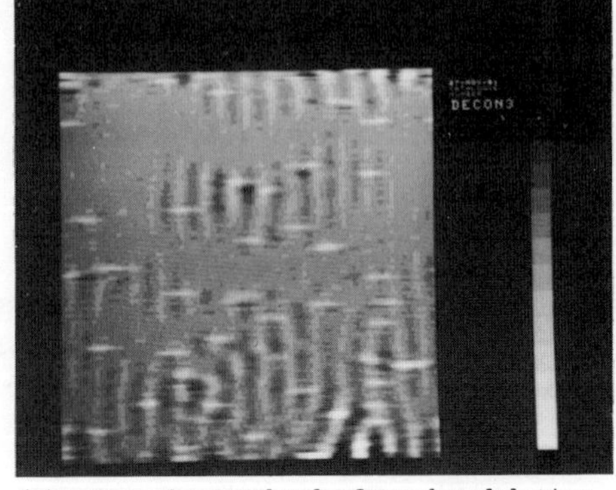

(e) Scan deconvolved after demodulation.

Results

Figure 2 shows how the algorithm operates. The point spread function 2(a) is in the top left hand corner by convention, and use of the discrete Fourier transform means that it is wrapped round the edges of the image. The scattering object 2(b) is an ensemble of randomly positioned scatterers, and the simulated B-scan is shown in 2(c). Because the object is sparse, individual points can in some parts be resolved, but in others cannot. Figure 2(d) shows the result of deconvolution, where the image before demodulation is deconvolved with the undemodulated point spread function. All the original points are resolved due to the improvement in high contrast resolution. If the demodulated B-scan is deconvolved with a demodulated point spread function, then the result 2(e) contains interference effects. This is because deconvolution is no longer valid if a non-linear process such as demodulation is used to form the image, and emphasises the importance of processing the r-f scans. If a denser array of scatterers is used as an object, then the equivalent of 2(e) is completely dominated by interference effects.

Figure 3(a) shows a tissue model of a 'cyst'. The demodulated scan, 3(b), distorts the size and shape of the cyst, and the exact nature of the object is equivocal. The deconvolved scan is shown in Figure 3(c), and the size and shape of the cyst is more clearly defined. Some further improvement is obtained if the deconvolved scan is smoothed slightly with a local standard deviation filter. This is like a moving average filter, but instead of displaying the local average, the local standard deviation is displayed, and is particularly effective on these images. Figure 3(d) shows the effect of applying this to 3(c), and the cyst is clearly displayed. The standard deviation filter has no beneficial effect on 3(b).

An object which tests the effect on low contrast resolution is Figure 5(a), modelling a 'lesion' with a lower scattering level to the surround. The effect of the standard deviation filter on this class of object is seen in Figure 4(b). The B-scan is shown in 4(c), and the 'lesion' is not displayed unambiguously. The result of deconvolving and processing with a standard deviation filter is shown in 4(d). It is possible to estimate the scattering level of the lesion, and differentiate it from a cyst.

The effect of deconvolution in one direction only can be seen in Figure 5. Figure 5(a) is identical to 3(c), and shows the result of deconvolving in both directions. If only the axial pulse shape is deconvolved, then the result, 5(b), although blurred in the lateral direction, does display the 'cyst'. There is no need to demodulate this picture because the effect of deconvolution is to remove the r.f. If the original r.f. scan is deconvolved in the lateral direction then it still needs to be demodulated, to remove the r.f. fluctuation, and the result of the two processes is shown

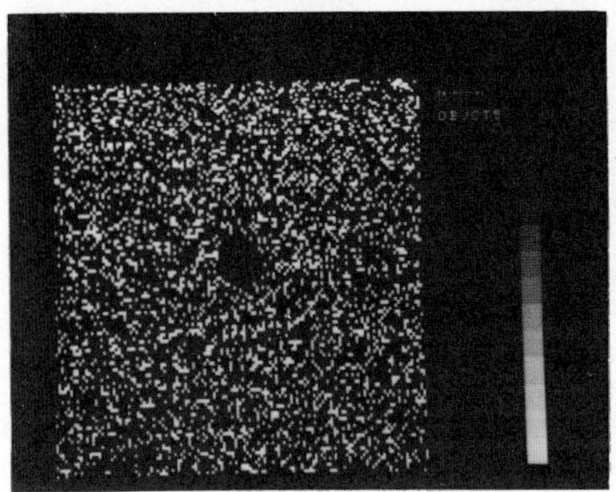

(a) Scattering Object
Mean separation = 2.0,
target diameter = 15.0.

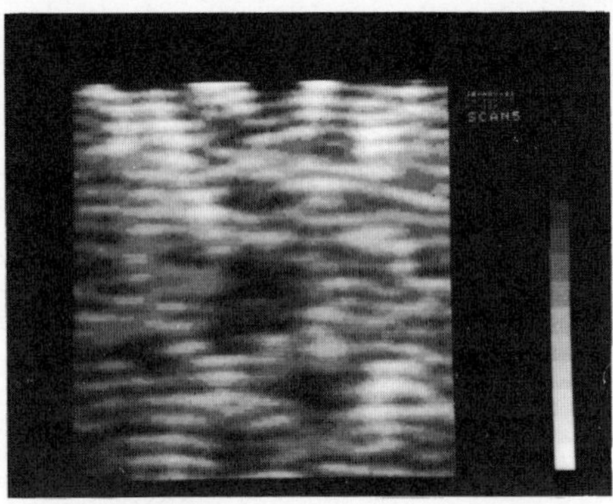

(b) Demodulated scan formed with pulse 2(a)

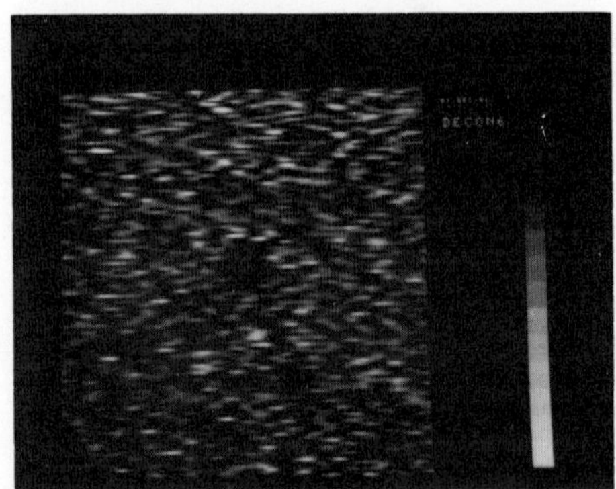

(c) Scan deconvolved before demodulation.

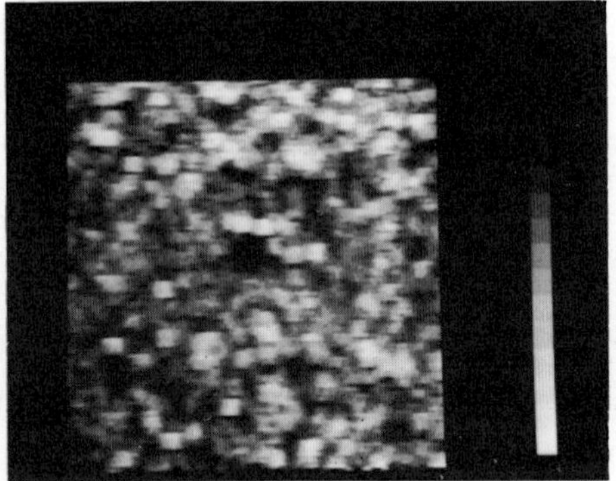

(d) Deconvolved scan smoothed with a 4 x 4 standard deviation filter.

Figure 3. Effect of deconvolution on detectability of cysts.

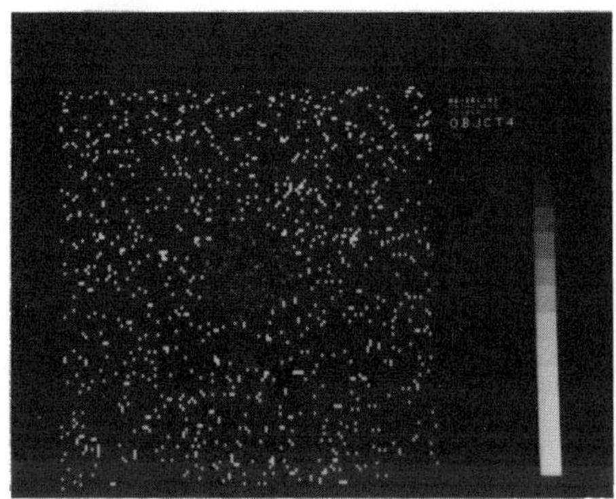

(a) Scattering object. Mean Separation = 4.0. Diameter of target = 30.0, Level = 50%.

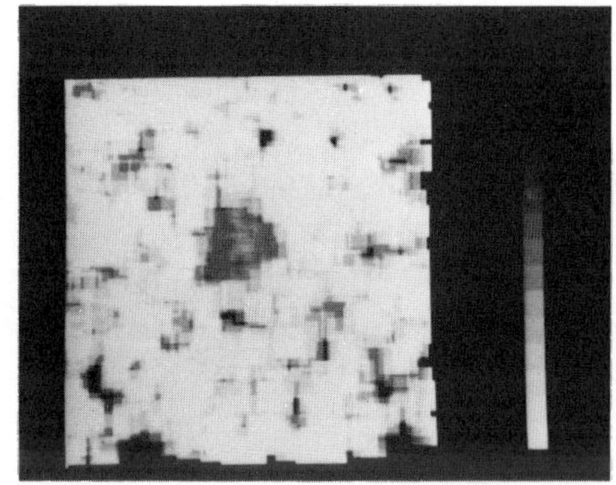

(b) Effect of 8 x 8 moving standard deviation filter on (a).

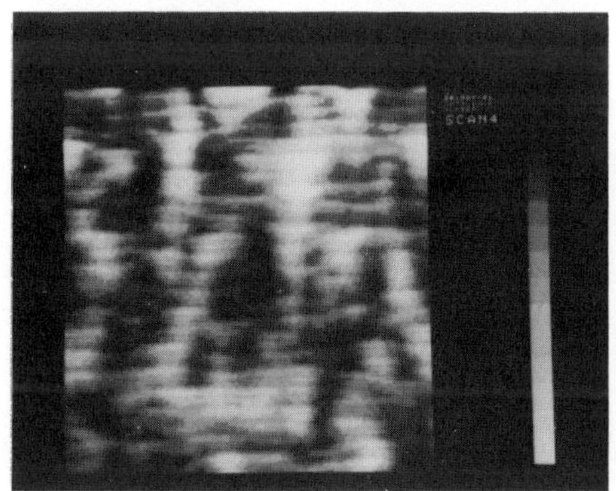

(c) Demodulated scan of (a) formed with pulse 2(a).

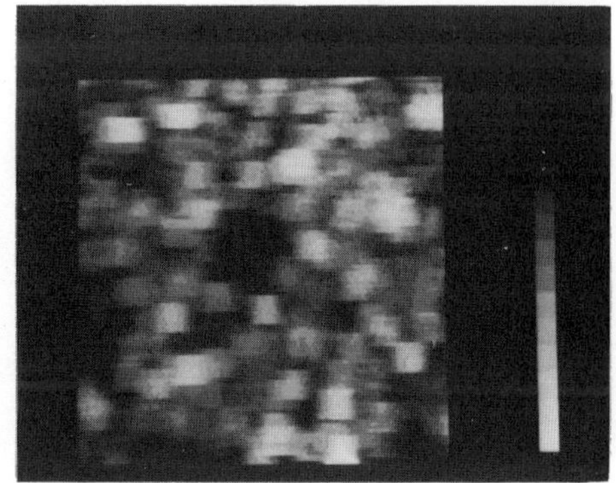

(d) Result of deconvolution followed by 8 x 8 moving standard deviation filter.

Figure 4. Effect of deconvolution on low contrast resolution.

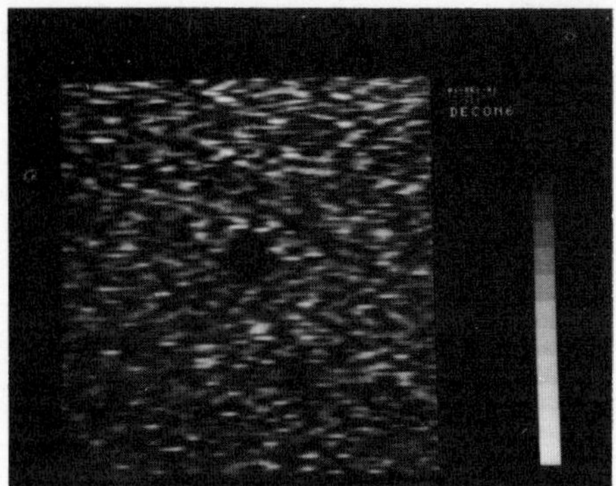

(a) Scan of 3(a) deconvolved in both directions = 3(c)

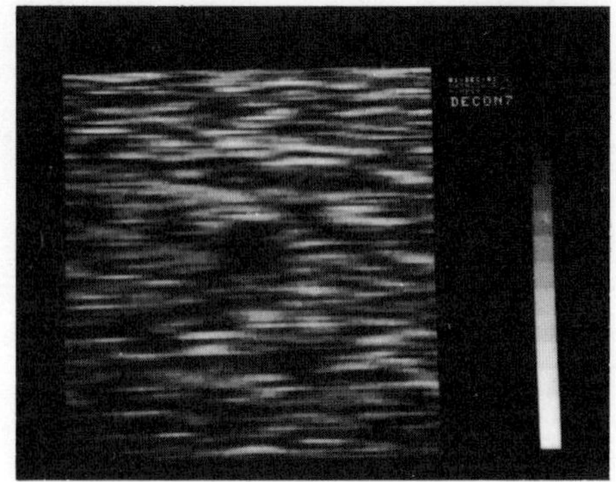

(b) Scan of 3(a) deconvolved in axial direction only.

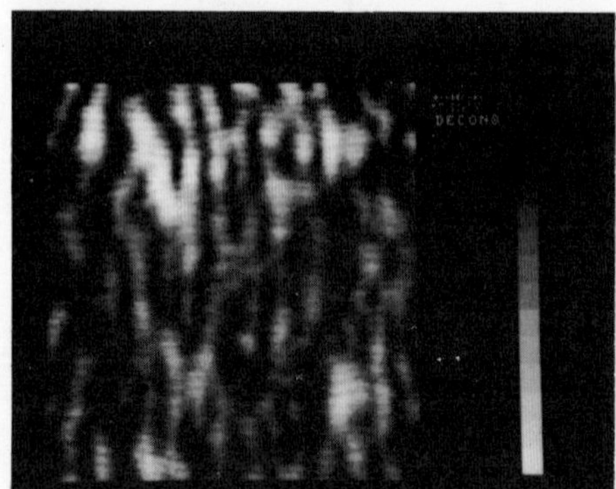

(c) Scan of 3(a) deconvolved in lateral direction and then demodulated.

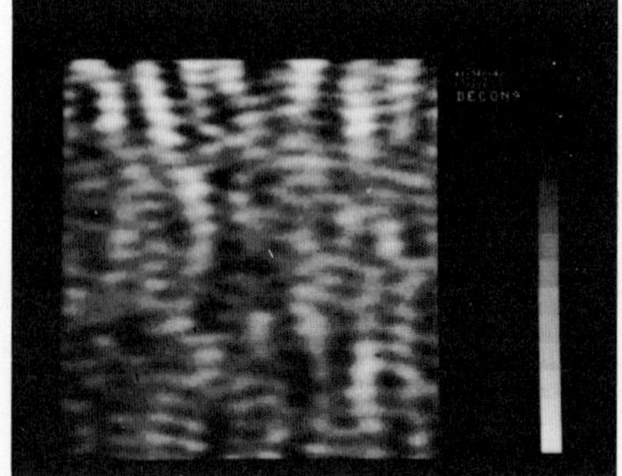

(d) Demodulated scan 3(b) deconvolved by lateral beam profile.

Figure 5. The effect of deconvolution in one direction only.

in 5(c). Finally, it is possible to deconvolve the beam profile from the demodulated scan, as shown in 5(d). Neither of the latter two images successfully display the cyst, so it can be concluded that deconvolving the axial pulse shape alone is worthwhile, whereas deconvolving the lateral beam profile alone is not. This is found to be true for a range of pulse lengths and beam profiles.

Conclusions

Deconvolution does then give a moderate degree of improvement in the low contrast resolution as well as the expected improvement in the high contrast resolution, and so does reduce the effect of speckle. Deconvolution of an r.f. image is a linear process, unlike demodulation which is non-linear, and it is the replacement of a non-linear process by a linear process which is responsible for the reduction in speckle. This is best demonstrated in Figure 5, where linear deconvolution of the axial pulse is worthwhile, whereas non-linear demodulation combined with deconvolution of the beam profile is not. In fact the deconvolution of the bipolar pulse can be considered a form of linear demodulation, giving a monopolar pulse whose sign is determined by the phase of the original pulse, as opposed to demodulation which gives a monopolar pulse of positive sign irrespective of the phase of the original.

DISCUSSION

Two methods have been described which reduce speckle. The first smooths out the low-level speckle, and gives some reduction in the speckle. To fully evaluate the routine, it needs to be implemented on a real-time scanner so that it can be used in a normal clinical environment. To do this in real-time would be relatively simple, the algorithm works on a pixel-by-pixel basis using nearest neighbours, and only a small temporary store is required.

Deconvolution is more complicated, since the whole image is required if the calculation is done using Fourier transforms. There are recursive methods of deconvolution which could be more readily implemented in real-time (8). Alternatively, it has been shown that deconvolution in the axial direction alone is worthwhile, so it could be performed one A-scan at a time. A major problem is that r.f. data is required, which results in data rates in excess of 10 M bytes/sec, which with current digital technology would preclude the implementation of deconvolution in a real-time scanner.

One possible problem in implementing the deconvolution routine in real-time is that the point spread function (p.s.f.) will vary throughout tissue from that obtained in a water bath. Using the simulation it is found that if the scan is deconvolved with a p.s.f.

larger than that used to form the scan, the deconvolution is unsuccessful, whereas if the p.s.f. used in the deconvolution is smaller, the deconvolution is successful although the resolution is worse than if the correct p.s.f. is used. This latter case is more likely in tissue, where the effect of tissue attenuation, inhomogeneities and beam diffraction is to make the pulse larger than that in water.

In conclusion, deconvolution applied to ultrasound B-scans is a promising technique for removing speckle degradation, and warrants further investigation.

REFERENCES

1. C.B. Burckhardt, Speckle in ultrasound B-mode scan. IEEE Trans. Sonics & Ultrasonics, SU-25, 1-6 (1978).
2. J.C. Abbott and F.L. Thurstone, Acoustic speckle: Theory and Experimental Analysis. Ultrasonic Imaging, 1: 303-324 (1979).
3. B. Julesz, E.N. Gilbert, L.A. Shepp and M.L. Frisch. Inability of humans to discriminate between visual textures that agree in second-order statistics-revisited. Perception, 2: 391-405 (1973).
4. B. Pratt, Digital Image Restoration. (Wiley: New York) (1978).
5. J.C. Bamber and J.V. Phelps, The effective directivity characteristic of a pulsed ultrasound transducer and its measurement of semi-automatic means. Ultrasonics, 15: 169-174 (1977).
6. J.C. Bamber and R.J. Dickinson, Ultrasonic B-scanning: A Computer Simulation. Ultrasound in Med. & Biol. 25: 463-479 (1980).
7. H.C. Andrews and B.R. Hunt, Ch. 7 in 'Digital Image Restoration' (Prentice Hall, Englewood Cliffs). (1977).
8. R.B. Kuc, Application of Kalman filtering techniques to diagnostic ultrasound. Ultrasonic Imaging, 1, 105-120 (1979).

OPTICAL PROCESSING OF LINEAR ARRAY ULTRASONIC IMAGES

Giorgio Rizzatto
Dept. of Radiology
General Hospital
Gorizia (Italy)

Paolo Sirotti
Inst. of Electronics
University
Trieste (Italy)

INTRODUCTION

Recent advances in dynamic scanners resulted in construction of high resolution linear array transducers that have rapidly become the most popular instruments in ultrasound examination. Nevertheless the apparent resolution of the resulting scans is degraded due to the structure of the displayed image which is formed by separated scan lines. In this paper we propose a new system of optical processing which allows both to enhance the effective resolution and to suppress the raster.

THEORETICAL APPROACH

We suppose that from an image $f(x,y)$ it should be obtained a second image $g(x',y')$ related to $f(x,y)$ by the equation

$$G(p,q) = F(p,q)\ W(p,q) \qquad (1)$$

where: - $G(p,q)$, $F(p,q)$ are the bidimensional Fourier transforms of $f(x,y)$ and $g(x',y')$;
- $W(p,q)$ is the transfer function of a linear, shift-invariant system;
- p and q are the spatial frequencies.

It is well known[1] that this operation (1) can be performed by a coherent optical system like that shown in fig. 1. On the back focal plane P_2 the lens L_1 operates the Fourier transform of $f(x,y)$. As in our research $f(x,y)$ may be a transparency; it must be under-

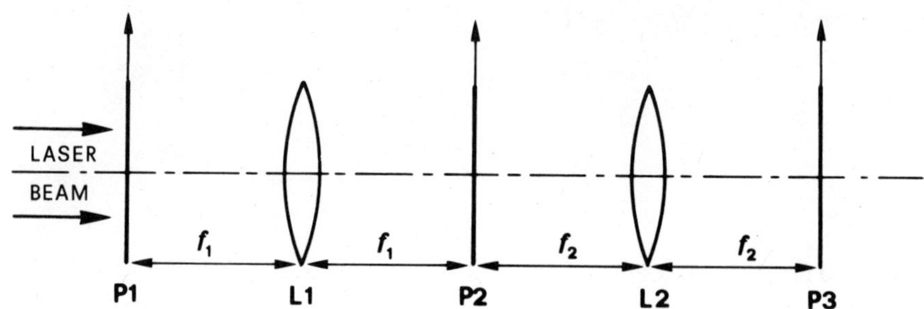

Fig. 1. Coherent optical system - schematic view.

lined that different real time methods allow to transfer an image to a device that can be read out by a laser beam. On the P_2 plane may be located a filter whose transfer function is W(p,q). The lens L_2 operates a further transform and reconstructs the image g(x',y') on the output plane P_3. Being α , β (mm) the coordinates of the P_2 plane, the spatial frequencies p,q (lines/mm) are

$$p = \frac{\alpha}{\lambda f_1} \quad ; \quad q = \frac{\beta}{\lambda f_1} \qquad (2)$$

where: - f_1 is the focal distance of L_1
- λ is the laser wavelenght (He-Ne/λ = 632.8 nm).

Other optical systems may perform operation (1). For instance, a single lens system was used in our research: the input image is placed at a distance d = f+s from the lens. The Fourier transform is found on the back focal plane; it is affected by a quadratic phase error while the correctly reconstructed image is read out at a distance $l = f + f^2/s$. The various quantities are illustrated in fig. 2.

Any real or complex W(p,q) may be realized. Without considering the filtering systems devoted to pattern recognition, we mention the case in which

$$W(p,q) = H^{-1}(p,q) \qquad (3)$$

where H(p,q) is the transfer function of the systems which produces the image to be processed f(x,y); in that case W(p,q) is usually a complex function and must be realized by holographic methods.

A general solution[2,3] of this problem can be found when W(p,q) is a real and positive function, with a circular or directional symmetry. In our proposed method the filter rotates on the transform

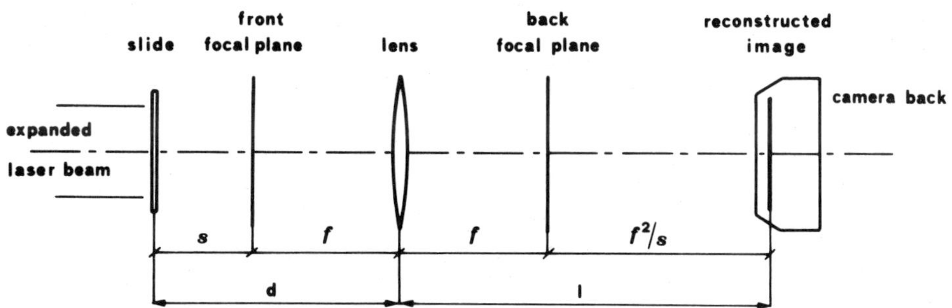

Fig. 2. Single lens system.

plane; it is formed by opaque sectors realized on photomechanical film or by photolithographic etching on copper sheets.

The symmetric circular filters are usually arranged as a single sector rotating around the optical axis. The sector profile, on polar coordinates r, ϑ is

$$\vartheta(r) = 2\pi\, T(r) \tag{4}$$

where T(r) is the transmittance to be realized.

The directional filters are more easily realized by an array of sectors drawn on an annulus at the periphery of a rotating disc. The transform is centered on the median circle of the annulus. The filtering direction is that joining the rotation center with the origin of the transform plane. The profile of every single sector in the array, whose period D is arbitrarily choosen, is:

$$y(x) = \frac{D}{2}\, T(x) \tag{5}$$

where T(x) is the requested transmittance.

The filtering direction can be practically defined by the orientation of the input image on its own plane.

LINEAR ARRAY ULTRASONIC IMAGES AND FILTERING PROCESS

Coherent optical processing was applied by dynamic scans produced by dynamic scanners with linear array transducers and dynamic focusing. Basically the quality of the displayed image is limited as it concerns the effective spatial resolution. Axial resolution for all linear array transducers is around 1 mm and lateral resolution varies between 1.5 and 6 mm.[4]

Moreover these images are formed by a raster of individual scan lines which degrades the apparent resolution and gives rise to a pattern of replicas of the basic spectrum on the transform plane (fig. 3).

A global approach to the optical processing of these images would have been the inverse filtering related to the equation (3), using holographic filters. Nevertheless, as the echographic system is not shift-invariant along the whole scanning field, only a limited area of the original image would be processed. Therefore we have preferred to deal separately with the raster suppression and the resolution enhancement.

Partial or complete raster suppression may be achieved using a low-pass filter which cuts out the spectrum replicas (fig. 4B). Similar results may be obtained by electronic filtering, often operating in the available scanning system. Nevertheless this filtering may deteriorate the contrast of the displayed image. On the contrary a derivative filter acting in the scanning direction can easily suppress the raster; a proper selection of its continuous transmittance preserves the original contrast (fig. 4 C).

As suggested by Gore and Leeman[5] a similar filter could also improve the effective axial resolution of the system. Basing on the same report an inverse gaussian filter could theoretically improve lateral resolution.

The transmittance of the derivative filter was assumed to be:

$$T(x) = 0.2 + \frac{0.8}{B} x \qquad (6)$$

where: - x is the filtering direction
- B is the bandwidth in the scanning direction

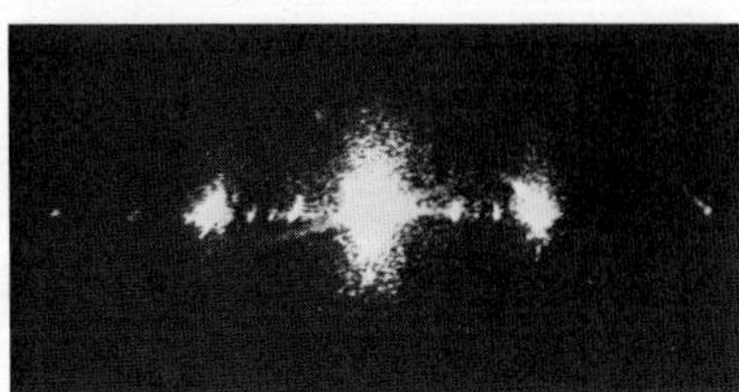

Fig. 3. Spectrum of ultrasonic linear array image (1 mm = —— = = 4.16 1/mm).

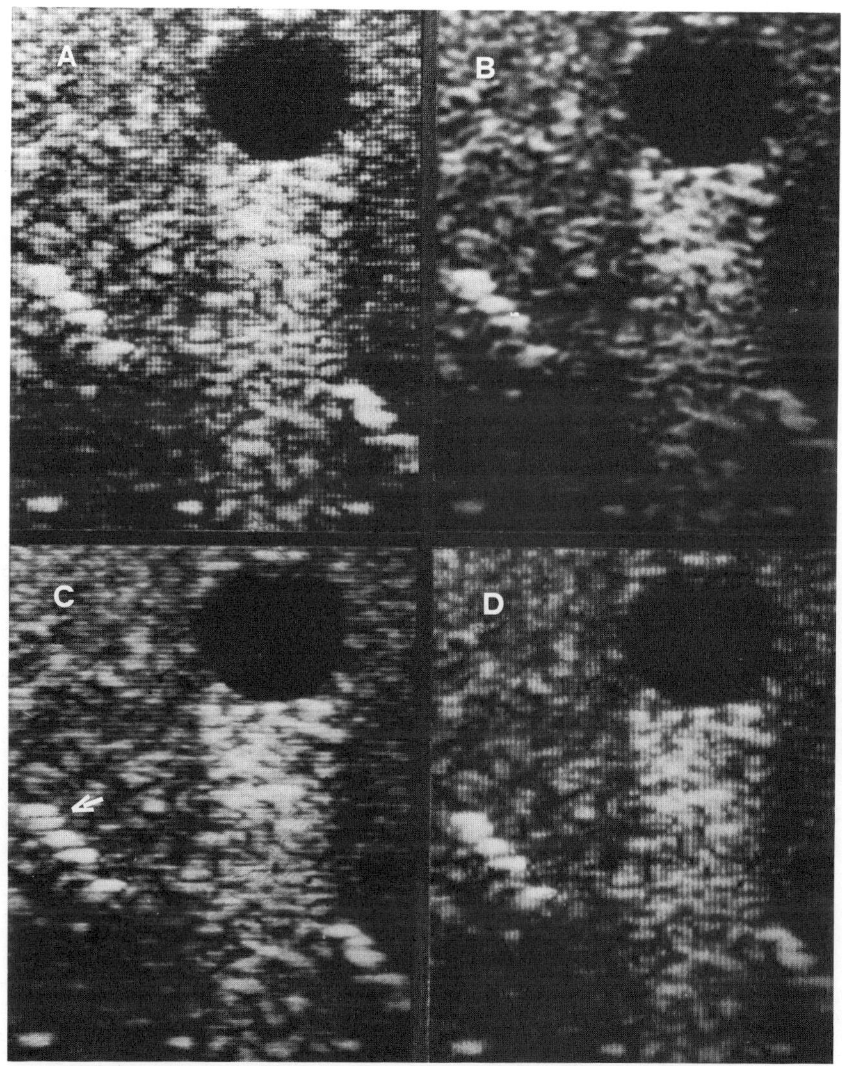

Fig. 4. Tissue equivalent phantom.
A) Original image; B) Low-pass filtering;
C) Derivative filtering; D) Inverse gaussian filtering

The value of 20% was the best filter continuous transmittance to maintain a proper contrast.

The transmittance of the inverse gaussian filter was assumed

$$T(x) = e^{k^2(\tau^2 x^2 - 1)} \tag{7}$$

In a first approximation τ (mm) was assumed to be equal to the scanning line width; thus, expression (7) approximates the Fourier transform in lateral direction of the single scanning line. k is a numerical constant whose value was estimated to be 1.7.

The rotating filters realizing (6) and (7), according to the equation (5) are shown in figures 5 and 6.

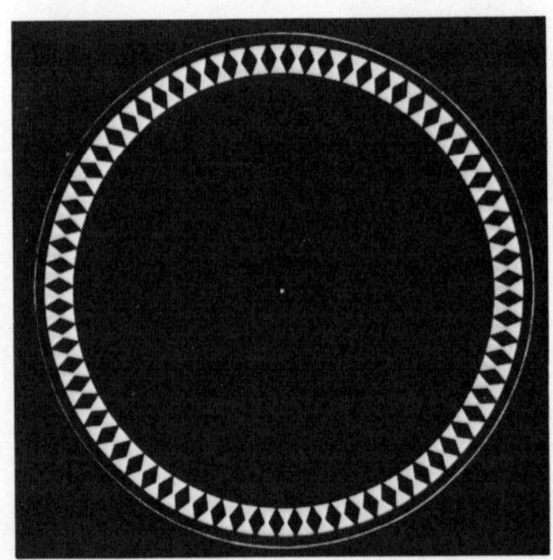

Fig. 5. Derivative directional filter.

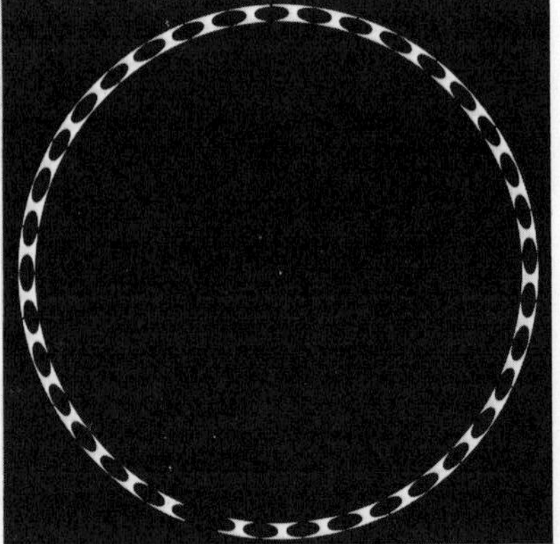

Fig. 6. Inverse gaussian filter.

RESULTS

The actual effectiveness of the filtering process was tested using both tissue equivalent phantom and micrometric calipers.

The use of the derivative filter was always profitable in removing the vertical raster due to the scan lines separation (fig. 4C); the horizontal raster due to the TV monitor remains as in the original image. At the same time this filter acted also on the axial resolution that improved from 0.7 to 0.5 mm. This enhancement was also evident with the tissue equivalent phantom; the nylon targets on the left side of the image and with axial spacing of 0.5 and 0.8 mm were better resolved (◄—) like those on the right side down (fig. 4C). Less impressive results were obtained with the inverse gaussian filter; the lateral resolution only enhanced from 1.5 to 1.4 mm.

CLINICAL APPLICATION

Experimental results on tissue equivalent tests suggest that the proposed optical processing improves both resolution and perception of boundaries. Based on these considerations we have extended the filtering process to clinical work-up according to definite suggestions:

- walls of fluid filled cavities
- normal and abnormal vessels
- fetal structures and measurements
- small focal lesions within parenchymas.

The original ultrasonic images were obtained using a dynamic scanner with high resolution linear array probes (3.5 and 5 MHz); as the results of the inverse gaussian filter were not impressive, only the derivative filter has been employed. Some examples of clinical application are shown in figures 7 and 8. The better quality of the processed images is easily appreciated; while other electronic methods for raster suppression give rise to contrast enhancement, in all the cases the separation between the different grey levels keeps very similar to the original scans allowing good echo pattern definition. Besides the better visualization of normal anatomy and pathological lesions the improved perception of boundaries allows more precise measurements; this is specially true for structures that originally cannot be well defined as they present an echo pattern very similar to the surrounding structures (focal intraparenchymal lesions, fetal kidneys, etc.). Although some results have been achieved the evaluation of the clinical effectiveness of the method is still critical. A basic problem concerns the proper image selection as the ultrasonographer is inclined to pick out only those scans in which some information is already present. In all the other cases when filtering process adds new information these are often questionable as it is difficult to obtain equal pathological specimens; this is linked to the tomographic nature of the ultrasonic images and to the difference between an in vivo or excised organ. Anyhow optical processing of small breast carcinomas allowed to visualize marginal uneveness that strictly correlated to tumoural digitations shown on mammography.

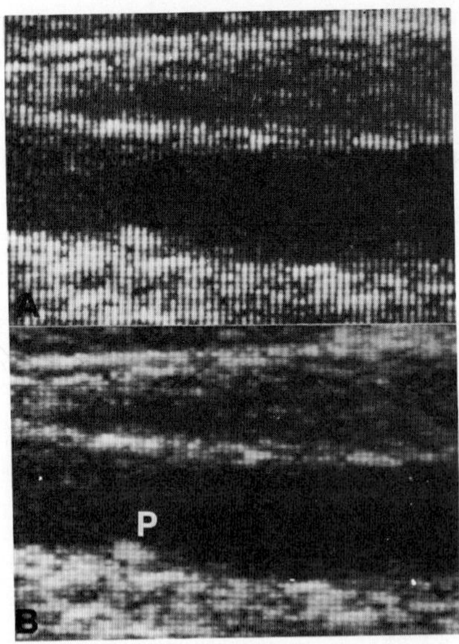

Fig.7. Longitudinal scan through the right common carotid artery (5 MHz).
A - original image
B - processed image
P - atherosclerotic plaque (diameter 3.5 mm).

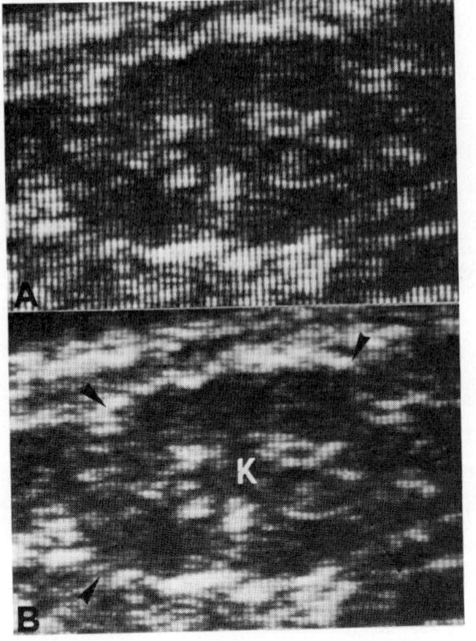

Fig.8. Longitudinal scan through a right fetal kidney - 36 weeks (3.5 MHz).
A - original image
B - processed image
K - kidney (longitudinal diameter 4 cm).

REFERENCES

1. D.C. Champeney, "Fourier Transforms and their Physical Applications". Academic Press, London (1973)

2. R. Mottola and P. Sirotti, "Rotating Amplitude Filters for Image Optical Processing". Alta Frequenza N.3 vol. XLVII (1978)

3. P. Sirotti, G. Rizzatto and F. Beltrame, "Elaborazione di immagini ecografiche con luce coerente". LXXXI Riunione Annuale AEI. Trieste (1980)

4. R.L. Deter and J.C. Hobbins, "A Survey of Abdominal Ultrasound Scanners: the Clinician's Point of View". Proc. IEEE vol. 67, 664:671 (1979)

5. J.C. Gore and S. Leeman, "New Criteria for the Assessment of the Resolution of Ultrasonic Scanners". In E. Kazner, M. de Vlieger, H.R. Müller and V.R. McCready, Eds. Ultrasonics in Medicine. Excerpta Medica, Amsterdam, 1975, pag. 197-203

INVERSE FILTERING TO MINIMIZE THE EFFECT OF MECHANICAL ABERRATION IN FOCUSING PIEZOELECTRIC TRANSDUCERS

Mostafa Mortezaie
Glen Wade

Department of Electrical and Computer Engineering
University of California, Santa Barbara, California

ABSTRACT

The resolution of acoustic images from a "positively scanning transmitter" system is limited by the size of the focal spot of the acoustic beam. Frequently a spatial focusing electrode pattern, such as a Fresnel zone plate, is used to excite the piezoelectric transducer of such a system. However, the transformation of that electrode pattern to the acoustic pattern actually generated is not exact. This effect, called mechanical aberration, is a result of wave generation and propagation in the piezoelectric plate and can be represented by a spatial transfer function. The aberration enlarges the focal spot and decreases the resolving power of the instrument.

The resolution can be improved by inverse filtering. The approach is to assume that a desired acoustic pattern has already been generated by an unknown electrode pattern. The electrode pattern is found by inverse filtering. If this electrode pattern is then used to excite the piezoelectric plate, it will produce very nearly the desired acoustic pattern. The actual pattern will usually not be exactly the same as the desired pattern because frequency cutoff in the transducer may prevent the propagation of high spatial frequencies necessary for the two patterns to be exactly the same.

INTRODUCTION

In a "positively scanning transmitter" system, the resolution of acoustic images is limited by the size of the focal spot of the acoustic beam [1]. An electrode pattern, such as a Fresnel Zone plate, which is capable of producing spatial focusing is frequently used to excite the piezoelectric transducer of such a system [2]. However, the electrode pattern and the acoustic pattern generated by it are not exactly the same [3]. This is due to an effect called mechanical aberration and results from unwanted wave generation and propagation in the piezoelectric plate. The effect can be represented by a spatial transfer function. Mechanical aberration enlarges the focal spot and decreases the resolving power of the instrument [4].

Inverse filtering can be applied to improve the resolution. The approach is to assume that a desired acoustic pattern, such as one that will produce a highly-focused spot, has already been generated by an unknown electrode pattern. From knowing the spatial transfer function, the electrode pattern can be found by inverse filtering. If this electrode pattern is then used to excite the piezoelectric plate, it will produce very nearly the desired acoustic pattern. In general, the actual pattern and the desired pattern will not be precisely the same due to the fact that frequency cutoff in the transducer may eliminate the high spatial frequencies necessary for the two patterns to be exactly the same.

We model the transducer as a linear, spatial system [5]. The transducer is fully electroded on the water side and has a focusing electrode pattern on the air side. The electrode pattern on the air side is the input and the particle displacement just in front of transducer in the water is the output.

This is a valid model because the piezoelectric effect and elastic wave propagation are linear phenomena over a wide range of amplitude. To simplify the analysis we assume a one-dimensional variation of the electrode pattern in the x direction. Generalization to two dimensions is self evident and may be obtained by simple mathematical extension [6]. Figure 1 shows the block diagram of the system. f_x is the spatial frequency in the x direction and $H(f_x)$ is the spatial transfer function of the plate.

For notational convenience let us define the following terms:

$g_i(x)$ is the electrode pattern;

$g_0(x)$ is the acoustic output;

$h(x)$ is the point spread function for the system;

$G_i(f_x)$, $G_0(f_x)$ and $H(f_x)$ are the Fourier transforms of $g_i(x)$, $g_0(x)$, and $h(x)$.

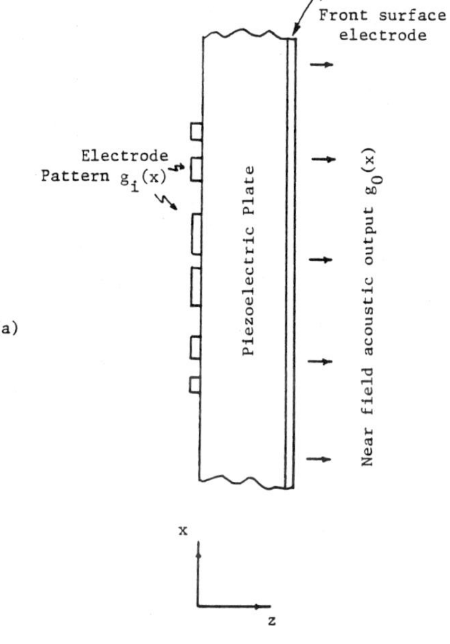

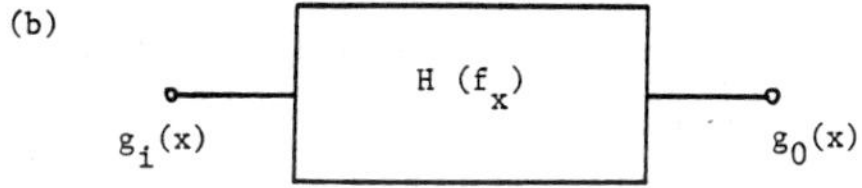

Fig. 1 (a) Physical structure of the transducer

(b) Linear System Model of (a)

The system is assumed to be spatially invariant and noise has been neglected. With these assumptions we can write

$$g_0(x) = \int_{-\infty}^{\infty} h(x-x')g_i(x')dx' \tag{1}$$

or alternatively by the inverse transform relation

$$g_o(x) = \int_{-\infty}^{\infty} H(f_x)G_i(f_x)\, e^{j2\pi f_x x}\, df_x \tag{2}$$

With this technique it is possible to adjust the amplitude and phase of the input to get a desired output. This is done by assuming that a desired acoustic pattern $g_0(x)$ is generated by an unknown electrode pattern $g_i(x)$. This electrode pattern is found by inverse transforming the relation

$$G_i(f_x) = H^{-1}(f_x)G_0(f_x) \tag{3}$$

where $H^{-1}(f_x)$ is the inverse filter for $H(f_x)$. Care must be used in regions where $H(f_x)$ goes to zero. In these regions $H^{-1}(f_x)$ is infinite and the inverse filter is not realizable. However, approximations can be used to find a proper inverse filter by replacing infinities in $H^{-1}(f_x)$ with large values.

The electrode pattern $g_i(x)$ found by this method, if used to excite the piezoelectric plate, will generate a near-field acoustic pattern $\hat{g}_0(x)$ which may not exactly duplicate the desired pattern $g_0(x)$. This is because waves whose frequencies are above the spatial frequency cut-off for the system are not present in the output. However, in the passband, $\hat{G}_0(f_x)$, the Fourier transform of $\hat{g}_0(x)$, is exactly the same as $G_0(f_x)$. Therefore $\hat{g}_0(x)$ will be very nearly the desired pattern, much more so, in general, than the output of a transducer using the desired pattern as the electrode pattern.

ABERRATION LIMITED TRANSFER FUNCTION

$H(f_x)$ characterizes the effect of wave generation and propagation inside the piezoelectric plate. It may be represented by

$$H(f_x) = |H| \exp\{j \angle H\} \tag{4}$$

where $|H|$ and $\angle H$ are the amplitude and phase of $H(f_x)$ and are plotted in Figures 2(a) and 2(b) as a function of the normalized spatial frequency, $\nu_x = f_x\lambda$, where λ is the acoustic wavelength in water. To obtain the result in Figure 2, only compressional waves have been taken into account. The contribution due to shear waves has been neglected, because shear waves are weakly coupled to compressional wave in this case.

The normalized cut-off spatial frequency ν_{xc} is given by [5]:

$$\nu_{xc} = f_{xc}\lambda = \frac{v}{(C_{11}^E/\rho)^{1/2}} \tag{5}$$

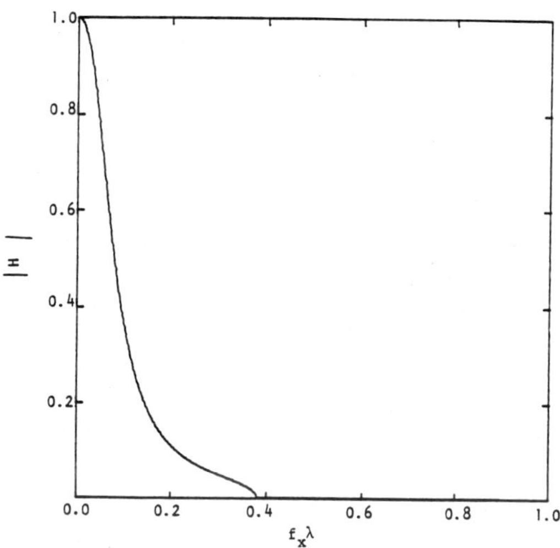

Fig. 2 (a)
Magnitude of the abberation-limited transfer function versus the normalized spatial frequency.

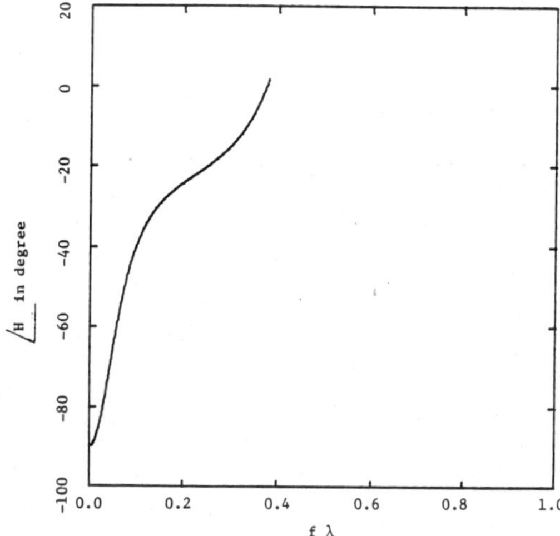

Fig. 2 (b)
Phase of the aberration-limited transfer function versus the normalized spatial frequency.

where v is the sound velocity in water, ρ is the density of the piezoelectric material, and C_{11}^{E} is one of its elastic constants. The plot in Figure 2 has been obtained for a PZT-5A transducer resonant at 3.4 MHz, air backed and in contact with water.

INVERSE FILTER OF ABERRATION-LIMITED TRANSFER FUNCTION

The inverse filter of the aberration-limited transfer function is given by

$$H^{-1}(f_x) = \frac{1}{H(f_x)} = \left| H^{-1}(f_x) \right| \exp\{j \angle H^{-1}(f_x)\} \tag{6}$$

This filter is not realizable because it possesses infinities in regions of f_x where $H(f_x)$ is zero. Fortunately, these regions are in the stop band with frequencies larger than f_{xc}. We can obtain a realizable approximation to the filter by replacing the infinities with large numbers.

This can be done by approximating $H(f_x)$ with $\hat{H}(f_x)$, whose values are very small but not zero in regions of f_x where $H(f_x)$ is zero. The approximated inverse filter is given by $\hat{H}^{-1}(f_x)$. Amplitude of $\hat{H}(f_x)$ and $\hat{H}^{-1}(f_x)$ are shown in Figure 3(a) and 3(b).

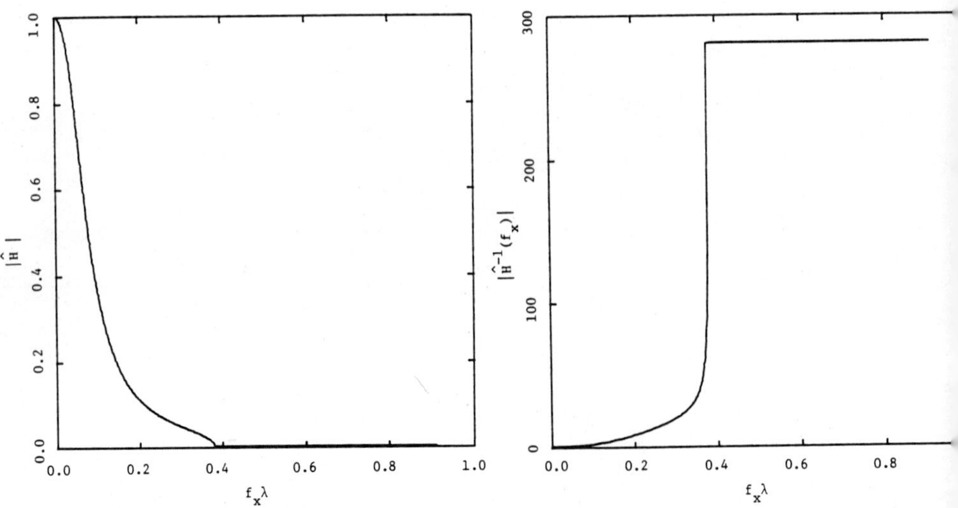

Fig. 3 (a)
Amplitude of the approximated aberration-limited transfer function versus the normalized spatial frequency.

Fig. 3 (b)
Amplitude of the approximated inverse filter of the aberration-limited transfer functio[n] versus the normalized frequen[cy]

FRESNEL ZONE PLATE PATTERN

The transmission function $g_0(x)$ of a one-dimensional Fresnel zone plate (FZP) pattern is shown in Figure 4. The parameters given in that figure are related to λ, and the FZP focal length, f, by the following equations [4]:

$$x_n = [nf\lambda + (n\lambda/2)^2]^{1/2} \quad \text{(a)}$$

$$x_{(n)} = (x_{n-1} + x_n)/2 \quad \text{(b)}$$

$$\Delta x_{(n)} = x_n - x_{n-1} \quad \text{(c)} \qquad (7)$$

$$D = 2x_n \quad \text{(d)}$$

$$N = -2(f/\lambda) + [4(f/\lambda)^2 + (d/\lambda)^2]^{1/2} \quad \text{(e)}$$

where N is the number of zones and D is width of the zone plate.

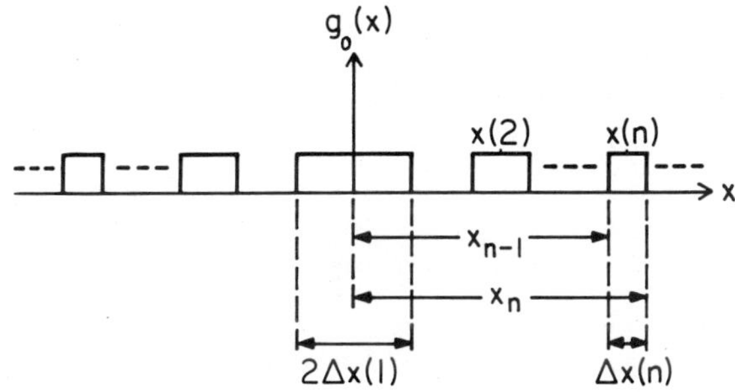

Fig. 4 Transmission function $g_o(x)$ of a one-dimensional FZP Pattern.

Each zone consists of a pair of electrodes situated symmetrically about the x = 0 line. Using equations (7), the Fourier transform of $g_0(x)$ can be written as:

$$G_0(f_x) = \sum_{n=1}^{2N-1} [2\Delta x(n)\cos(2\pi x(n)f_x)][\frac{\sin(\pi\Delta x(n)f_x}{(\pi\Delta x(n)f_x)}]$$

Figure 5 shows the magnitude of $G_0(f_x)$ as a function of normalized spatial frequency for three values of N with f = 4.0 cm = 90.66 λ and frequency of 3.4 MHz.

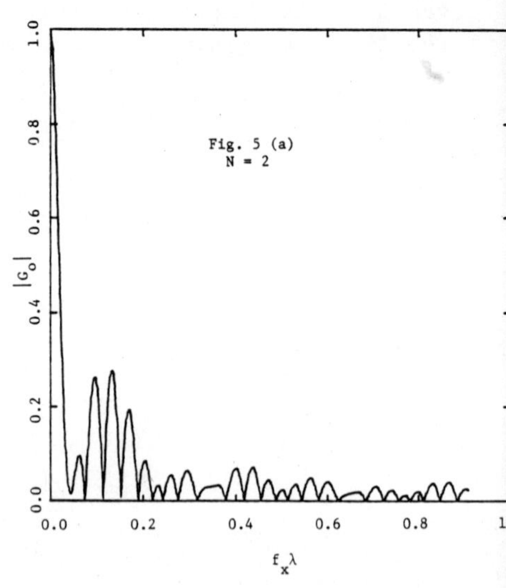

Fig. 5
Spectral content of an FZP pattern for (a) N = 2; (b) N = 4; (c) N = 6.

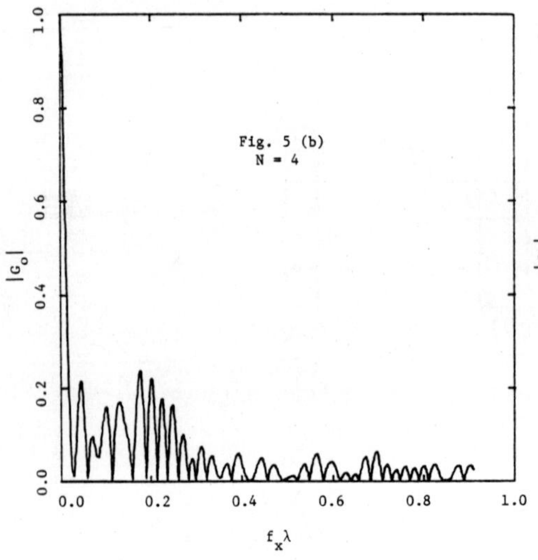

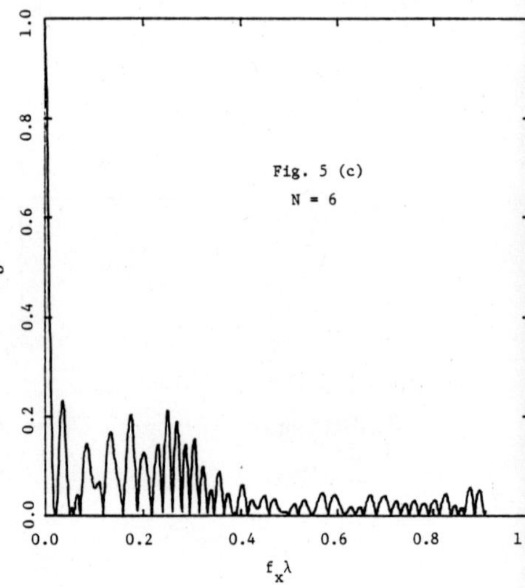

INVERSE FILTERING TO MINIMIZE ABERRATION

$G_i(f_x)$ is found from the relation

$$G_i(f_x) = \hat{H}^{-1}(f_x)G_0(f_x)$$

Figure 6 shows the magnitude of $G_i(f_x)$ as a function of normalized spatial frequency for three values of N. The high spatial-frequency content is amplified whereas the low spatial-frequency content is decreased due to inverse filtering.

The electrode pattern $g_i(x)$ can be found by inverse transforming $G_i(f_x)$. The amplitude of $g_i(f_x)$ is plotted in Figure 7.

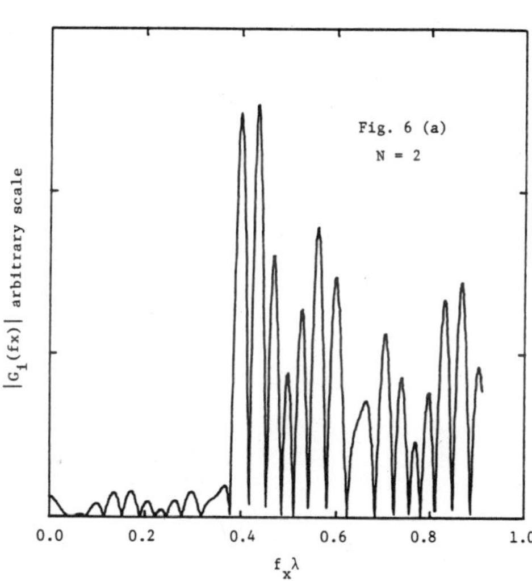

Fig. 6

Magnitude of $G_i(f_x)$ as a function of normalized frequency for (a) N = 2 (b) N = 4 (c) N = 6

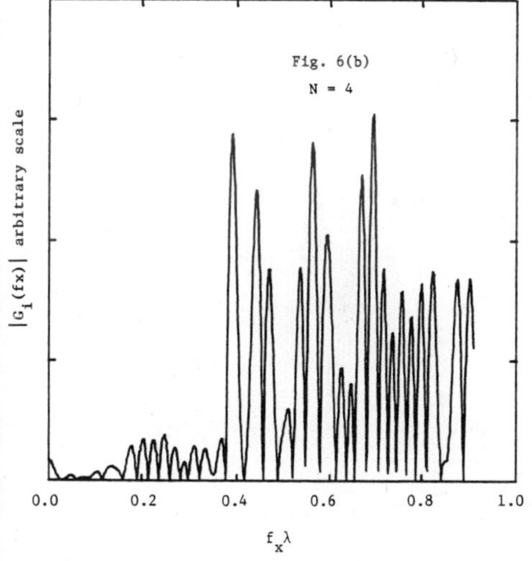

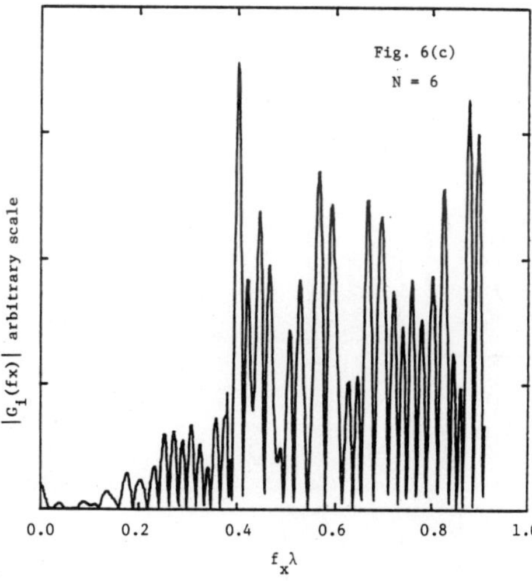

Fabrication of this electrode pattern with metal electrodes is certainly not simple and may be impossible. However, in optically activited transducers like the opto-acoustic transducer (OAT) [7] the amplitude of the input signal can be carefully controlled [8] and the production of $g_i(x)$ might be feasible in such transducers.

This electrode pattern was used as the input to the piezo-electric model. The Fourier transform of acoustic output is shown in Figure 8. We see that $\hat{G}_0(f_x)$ in the passband coincides with the $G_0(f_x)$ shown in Figure 5. The output acoustic intensity $\hat{g}_0(x)$ is shown in Figure 9.

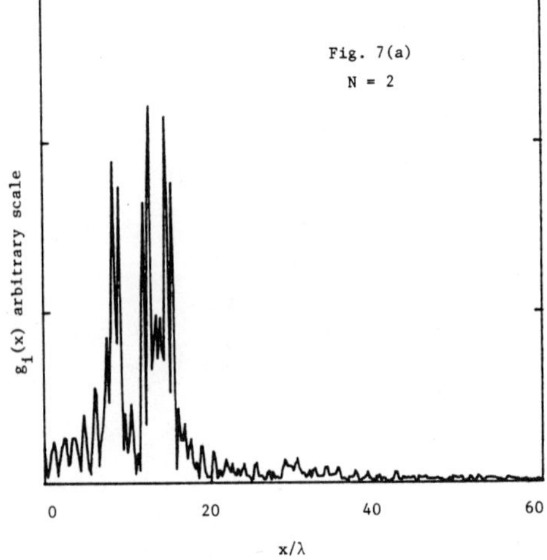

Fig. 7
Amplitude of the electrode pattern as a function of lateral distance x/λ found in inverse filtering from an acoustic FZP acoustic pattern for (a) N = 2 (b) N = 4 (c) N = 6

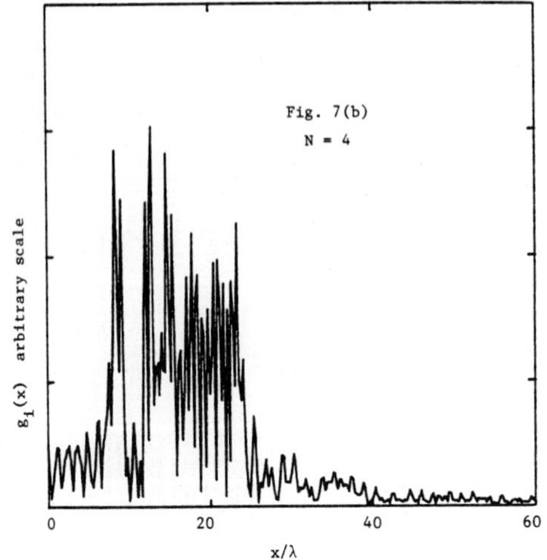

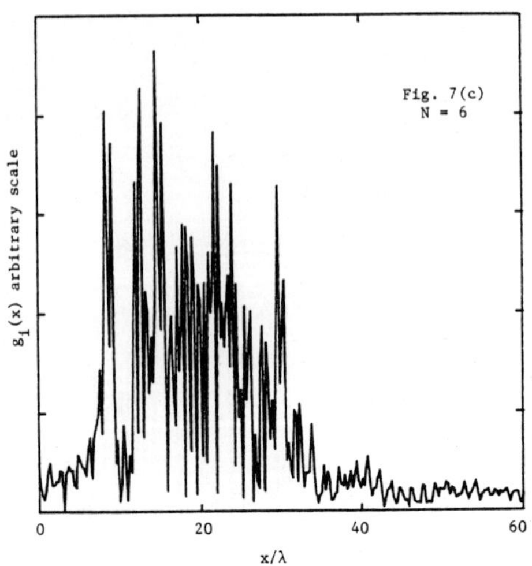

Fig. 8
Spatial frequency content of restored acoustic output $\hat{G}_o(f_x)$ for (a) N = 2 (b) N = 4 (c) N = 6

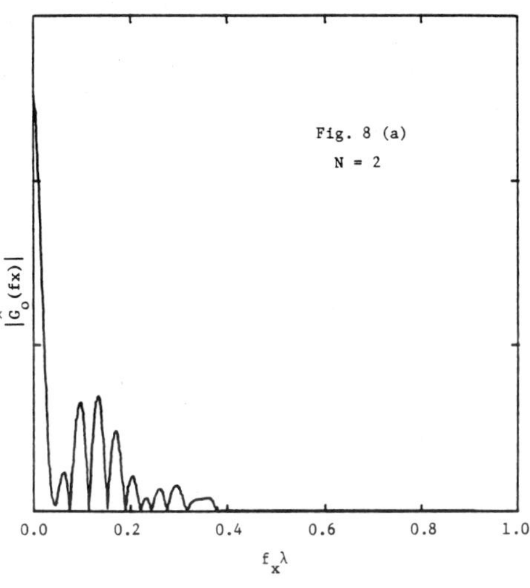

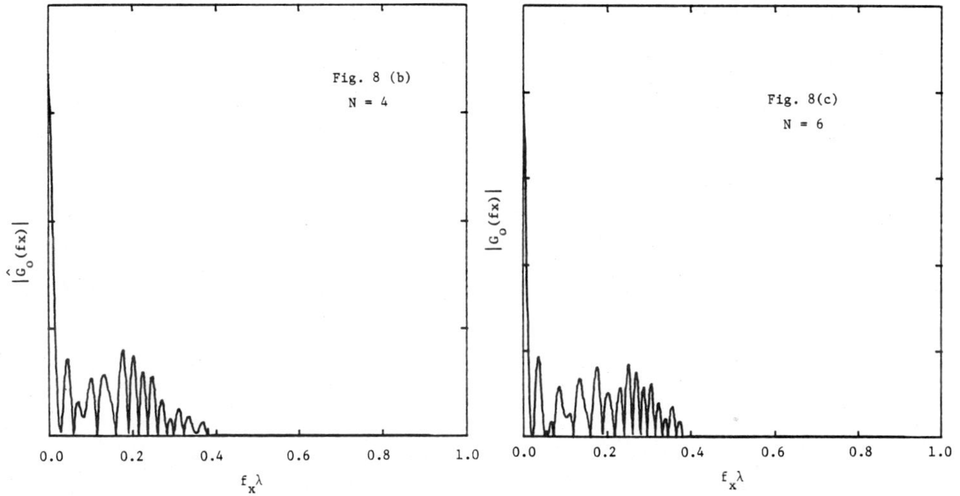

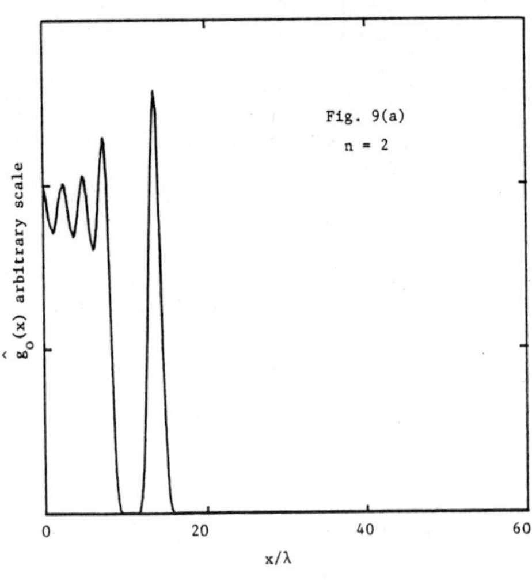

Fig. 9
Restored acoustic output $g_o(x)$ as a function of lateral distance x/λ for (a) N = 2 (b) N = 4 (c) N = 6

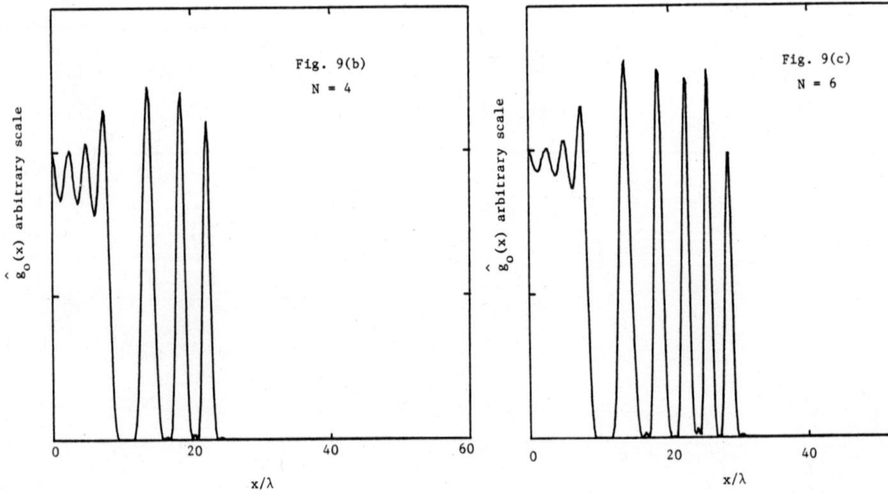

DISCUSSION

Note the similarity of the restored acoustic output of Figure 9 to the FZP pattern. Comparing this figure with the acoustic output of the FZP electroded transducer [3], reveals the following:

(1) The reduced amplitude in the region of the outer zones of FZP electroded transducer [3] does not show up in Figure 9. The amplitude corresponding to the transmitting region of the FZP in Figure 9 is quite uniform.
(2) The distinction between different zones in the FZP electroded transducer [3] is not very clear. However, different zones are clearly distinguishable in Figure 9.
(3) The sharp edges of the FZP pattern have not been completely reproduced and the output pattern has rounded edges.

The sharp edges of the FZP pattern are responsible for the high spatial frequency content of that pattern. The FZP pattern has higher frequency content than does a pattern with rounded edges, like a Gabor zone plate (GZP) pattern. The spatial spectrum of a GZP pattern has more information in the passband of the transfer function than does that for an FZP pattern. Consequently during restoration more information can be retrieved for the GZP pattern than for the FZP pattern. Our technique therefore works better for the GZP pattern.

CONCLUSION

We have presented a way to minimize aberration introduced by the piezoelectric plate. A spatial linear piezoelectric model enables us to find an approximate inverse filter for the system. Input electric excitation for a known acoustic output can be found by using an inverse filter. However, this technique is an approximation. The input electric excitation applied to the system to produce a given acoustic output will be filtered to some extent by the transfer function. Therefore, in general, aberration cannot be completely eliminated by this method.

ACKNOWLEDGMENT

We would like to thank Michael Galindo and Sherri Corr for the typing of this manuscript. This work was supported by the National Science Foundation under Grant NSF ECS-79-18779.

REFERENCES

1. K. Wang, and G. Wade, "Threshold Contrasts for Various Acoustical Imaging Systems," Acoustical Holography, V. 4, G. Wade Ed., Plenum Press, N.Y., 1972.

2. S.A. Farnow, and B.A. Auld, "Acoustic Frensnel Zone Plate Transducers," App. Phys. Lett., Vol. 25, pp. 681-682. 1974.

3. M. Mortezaie, G. Wade, and B. Noorbehesht, "Near Field Acoustic Pattern of Fresnel Zone Plate Transducers," International Workshop on Physics and Engineering in Medical Imaging, 1982. IEEE Trans. on Computer Science (in press).

4. B. Noorbehesht, M. Mortezaie, G. Wade, and C. Schueler, "Effect of Mechanical Aberration on Resolution of Fresnel Zone Plate Transducers," Acoustical Imaging, Vol. 11, J. Powers, Ed., Plenum Press, New York, 1981.

5. B. Noorbehesht, G. Flesher, and G. Wade, "Spatial Response of Arbitrarily Electroded Piezoelectric Plates by Plane Wave Decomposition," Ultrasonic Imaging, Vol. 2, 1980, pp. 102-121.

6. Generalization to two dimension may be achieved by taking advantage of the crystallographic symmetry of the piezoelectric ceramics. The two dimensional transfer function $H(f_x,f_y)$ or $H(f_r,f_\theta)$ (in the polar coordinates) may be obtained from $H(f_x)$ by a rotation of $H(f_x)$ around the $H(f_x)$ axis. This is mathematically equivalent to replacing f_x in $H(f_x)$ by f_r. Due to the symmetry, there is no dependence on f_θ. It should be noted that $h(r,\theta)$ cannot be obtained from $h(x)$ by simple rotation of $h(x)$ about the $h(x)$ axis.

7. K. Wang and G. Wade, "A High-Sensitivity Real-Time Acoustic Imaging System for Medical Diagnosis," Proc. IEEE, Vol. 62, 1974, pp. 650-651.

8. K. Bates and K. Wang, "PEOATS and ESOATS," 1979 IEEE Ultrasonics Symposium Proceedings, pp. 189-193, (1979).

THE APPLICATION OF MAXIMUM ENTROPY TO THE PROCESSING OF ULTRASONIC IMAGES OF NUCLEAR REACTOR COMPONENTS IMMERSED IN LIQUID SODIUM

S.F. Burch[1] and J.A. McKnight[2]

1. Atomic Energy Research Establishment, Harwell, England

2. Risley Nuclear Power Development Laboratories, UKAEA, England

ABSTRACT

Under sodium images of the top of the core of the Prototype Fast Reactor at Dounreay have been obtained by an ultrasonic pulse-echo scanning technique. Restoration of these sparsely sampled images requires non-linear processing. The maximum entropy method has been used to remove the visually confusing scan arcs, whilst simultaneously improving the image resolution. The maximum entropy principle has also provided the basis for a novel method to determine automatically the central positions of reactor components which are imaged as rings.

1. INTRODUCTION

Nuclear Fast Reactors use liquid sodium as coolant. Whilst examination of the internal structure of a nuclear reactor is never an easy task, the opacity of sodium makes optical methods impossible. The idea of using ultrasonic radiation to obtain internal information from the reactor is not new. An ultrasonic device has been installed in the French PHENIX reactor[1], a 'viewer' has been installed in the American FFTF facility[2] and work in Japan[3] has been reported. The intentions of the UKAEA for under sodium viewing have been described and progress on various aspects of the work has been published[4-8].

Recently, under sodium images have been obtained of the top of the core of the Prototype Fast Reactor (PFR) at Dounreay. It is believed that these are the first such images to be demonstrated in a working reactor. The images obtained on-line can be used directly for identification and measurement of the distortion of core components, but for extraction of maximum information further processing of the data is necessary. This paper discusses the application of the non-linear maximum entropy principle to the two-dimensional core images.

2. USE OF ULTRASONICS UNDER SODIUM

Ultrasonics technology for under sodium viewing usually depends on the pulse-echo principle. In this, an ultrasonic pulse emitted by a transducer produces an echo from a target, and the echo is then detected by the same transducer. The time difference between transmission and reception of the pulse is a measure of the distance between target and transducer. To know where the target is, however, it is necessary to know the position of the transducer and its orientation precisely.

The detection of an echo defines the position of a single point of the target only. To produce an image, it is necessary that the transducer is moved, to 'scan' over the target so that thousands of echoes can be assembled to produce the image.

Available technology limits the scanning for ultrasonics to mechanical movements. (Scanning of the ultrasonic beam using phase-control of a static transducer array has yet to be demonstrated in a hot sodium environment[4]). The scanning pattern is chosen principally for engineering convenience.

In the PFR viewer[6], eight transducers are mounted in an assembly so that they transmit downwards, and are suspended a few centimetres above the reactor core. The assembly may be rotated about its own axis, and the reactor top shield in which it is mounted may also rotate. The combined action of the two rotations is to cause the transducers to sweep out an annular area over the core. The width of the annulus is 18 cm, and its mean diameter varies depending on the location of the device in the rotating shield. This double rotation scan does not cover the whole of the core in one sweep, the viewer being re-loaded in a different position if needed.

A substantial computer assembly is used with the viewer to collect, and record the echo data and to display the composed images. This has been described elsewhere[7].

The transducers used are piezo-electric, using lead zirconate elements lead-bonded within a nickel encapsulation by a special technique[5]. This results in a uniformity of performance that is needed for this particular application. When set up the transducer sensitivities agree within 2 dB and the ultrasonic beam axes are truly normal to the transducer faces to within $\pm 0.08^\circ$. Adequate performance is maintained through many temperature cycles up to 280°C[4].

The most obvious way of obtaining good resolution with a simple point-to-point scan is to use focussed transducers which limit the insonified area with each pulse. For the PFR such a technique was not adopted, because in the general application of the viewer, the target distance cannot be guaranteed. As a consequence, imaging outside the focal plane can be anticipated and this would be worse than with an unfocussed transducer because of increased flare of the ultrasonic beam[4,8].

Unfocussed transducers have therefore been used, but the spatial resolution of the data has been improved using the maximum entropy method for image restoration.

3. IMAGE RESTORATION BY MAXIMUM ENTROPY

The spatial resolution of ultrasonic images of targets in the far-field is limited by the beam spread of the transducers, which becomes more important at large ranges. For the purposes of this paper, we model the formation of the observed ultrasonic amplitude image by a convolution equation (written for simplicity in one-dimension, but with a straightforward extension into two-dimensions):

$$g(i) = \sum_j f(j)\, h(i - j) + n(i) \qquad (1)$$

where the amplitudes $g(i)$ are the observed transducer amplitudes at point i, and $f(j)$ is the ideal target image (i.e. the image that _would_ be observed with _zero_ beam spread). The blur due to the _actual_ beam spread is modelled by a spatially invariant, but range dependent, point-spread function (PSF) h, and the random noise, introduced in our case principally by the receiver electronics, is represented by $n(i)$.

If the targets in the image to be processed have closely similar ranges, then the PSF in equation (1) is independent of position. Provided this (constant) PSF is known reliably, image restoration methods can, in principle, be used to invert equation (1) and hence estimate $f(j)$ from the observed data $g(i)$.

Many image restoration methods have been proposed in the literature[9] and linear methods such as Wiener and constrained least squares filters[10] are relatively straightforward to implement. However, these methods can only be applied to complete images, such as optical photographs, in which every data point has a known amplitude or intensity. The circular scanning procedure used for the PFR viewer gives highly incomplete or sparsely sampled images when the recorded data are mapped onto a

2-D grid of points. Measurement of every point in the grid could not be achieved in the available scan time. Linear restoration methods are therefore inappropriate, but the more powerful non-linear method of maximum entropy[11,12] can be applied.

Details of the maximum entropy method have been presented elsewhere[12,13], but the basis of the method is as follows. The restored image is represented by the amplitudes $\hat{f}(j)$ ($j = 0, 1 \ldots N$) at each point in a square grid (the suffices are written in one-dimension, for simplicity). For the examples in this paper $N = 512 \times 512$. If the amplitudes $\hat{f}(j)$ correctly represent the ideal target, $f(j)$, then equation (1) shows that, in the absence of noise, the transducer amplitudes that would be observed are given by

$$r(i) = \sum_{j=0}^{N-1} \hat{f}(j)\, h(i - j) \tag{2}$$

The misfit between the actual (noisy) data $g(i)$ and the restoration $\hat{f}(j)$ is then measured using a chi-squared statistic

$$\chi^2 = \sum_{i=0}^{N-1} [g(i) - r(i)]^2/\sigma^2(i) \tag{3}$$

which allows for noise through the variance $\sigma^2(i)$ of datum i. Nothing in the maximum entropy method requires all the $\sigma^2(i)$ to be the same, and thus the unmeasured samples within a sparsely sampled image can be allowed for by setting $1/\sigma^2(i) = 0$ at the appropriate points. For large N, the plausible values of χ^2 are all close to N. The condition $\chi^2 \simeq N$ therefore determines the large set of restorations which is consistent with the actual data, $g(i)$. On purely experimental grounds, any of the consistent restorations is acceptable, but to distinguish between them the configurational entropy[14] is calculated

$$S = -\sum_{j=0}^{N-1} p(j) \log p(j) \qquad p(j) = \hat{f}(j)/\Sigma\hat{f} \tag{4}$$

The unconstrained maximisation of S gives a flat restoration, but the maximum entropy method selects the (unique) restoration which has maximum S subject to the constraint that $\chi^2 = N$. In information theory terms, S is (minus) the information content of the shape of the image, so the maximum entropy criterion selects that particular restoration which has least configurational information. Maximum entropy restorations therefore contain the minimum of false detail (artefacts such as "ringing", and noise), and have the best resolution allowed by the quality of the observed data.

When the observed image contains large connected areas of

unmeasured points, the maximisation of S becomes locally unconstrained so that in the restoration these areas will be flat, with a chosen 'default' amplitude. However, in regions containing a mixture of measured and unmeasured points, the maximum entropy method selects a restoration which fits the measured points, within the limits imposed by the noise in the data. The unmeasured points merely exert a weak influence towards the default amplitude level.

Because S is a non-linear function, an iterative scheme is needed for its constrained maximisation[13], and substantial computing time is required. For example, with N = 512 x 512, the computations needed to obtain a restoration take about 15 hr on a PDP 11/60 with a hardware floating point unit.

4. PROCESSING THE PFR DATA

The data from the PFR viewer recorded by the on-line computer includes time-sequential information on echo range and amplitude, together with the appropriate transducer (x,y) positions. To achieve the required scan rates, data were recorded only when the echo amplitude exceeded a threshold value.

The PFR viewer was used to scan one complete annular area at the outer perimeter of the core. The amplitude data from targets at the top of the core (identified from the range data) were then mapped onto a 2-D grid, using the off-line computer. The resultant image (Fig. 1) shows the annulus covered by the viewer, and the tops of the reflector rod assemblies are visible as small circles. The grid mesh interval was 1.25 mm so that a 2048 x 2048 matrix covered the complete core.

An expanded view of a section of the core (512 x 512 points), shown in Fig. 2, clearly demonstrates the sparse nature of the data. For detailed inspection work the circular scan arcs can confuse the interpretation of the data. Restoration by maximum entropy (ME) is therefore valuable because the sampling pattern can be removed and the resolution of the image simultaneously improved. Fig. 3 shows the ME restoration of Fig. 2, using a Gaussian point-spread function (PSF) as a first approximation to the blur from beam spread. A low default amplitude (Section 3) was used since the majority of unmeasured points had amplitudes below the threshold. It is evident that ME has effectively removed the sampling pattern, and improved the resolution of the image. Significant amplitude variations around the circumferences of the reflector rod assemblies are clearly visible, and, on careful inspection, there is some evidence for most of these in the original data. ME processing has, however, enhanced these features, which are due to the non-uniformity of the reflecting surfaces.

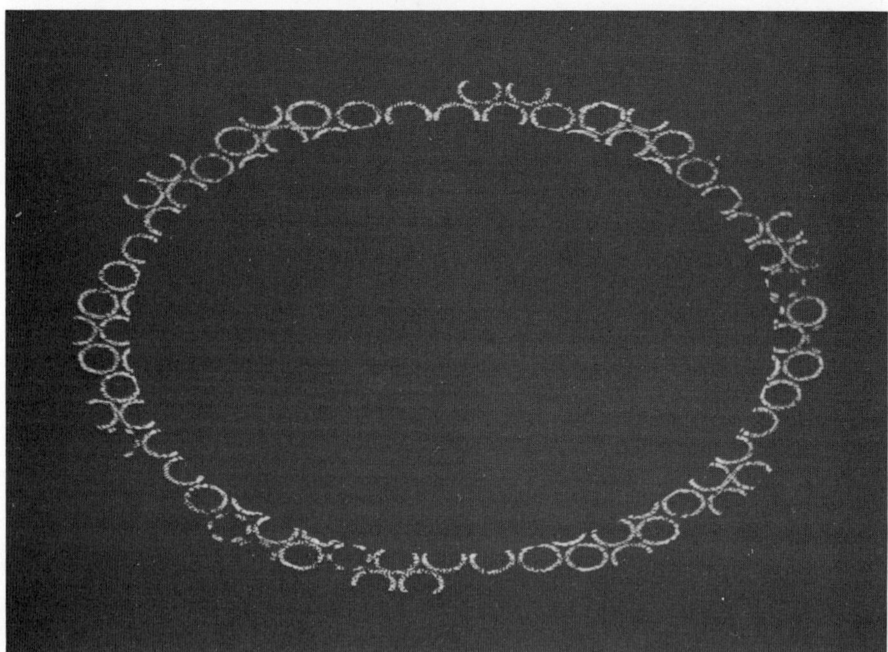

Fig 1. Ultrasonic echo amplitude data from targets at the top of the PFR core, as scanned by one complete rotation of the viewer.

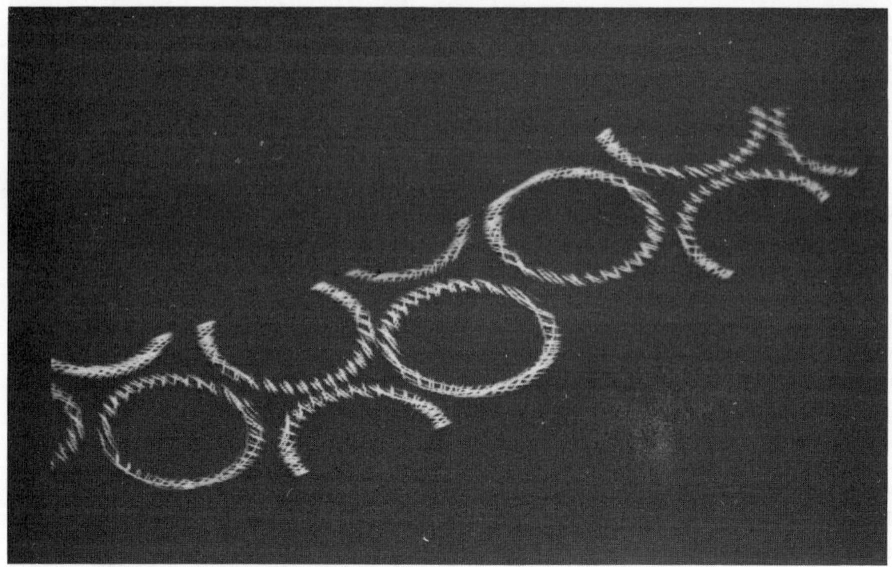

Fig 2. Expanded view of one section from Fig 1.

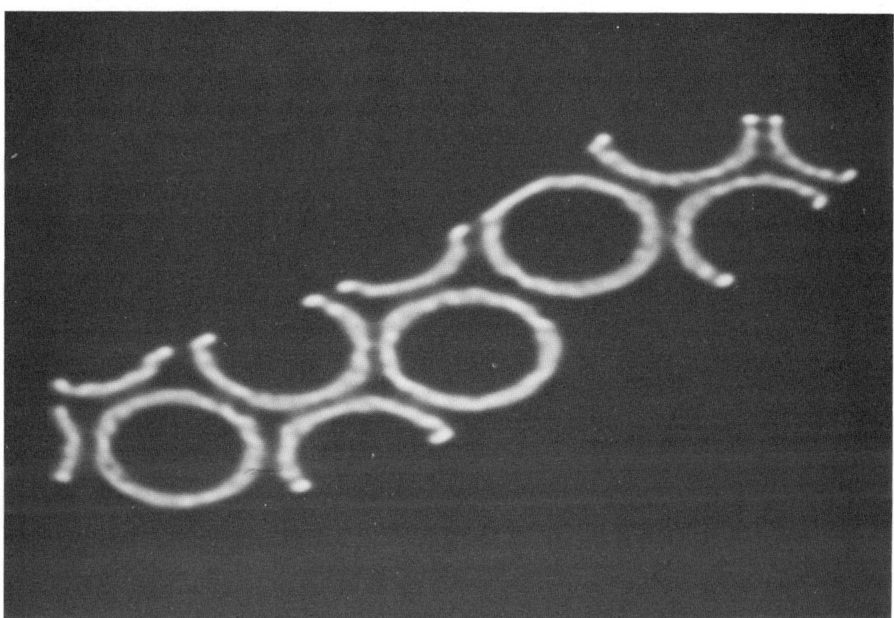

Fig 3. Maximum entropy restoration of Fig 2,using a Gaussian point-spread function.

Fig 4. Maximum entropy reconstruction of Fig 2, using a ring point-spread function to facilitate positional determination of reflector rod assembly centres.

One of the main aims of the PFR viewer is to allow the positions of the centres of the assemblies to be determined. Whilst conventional ME restoration is useful in improving image quality, it does not, in itself, greatly facilitate positional measurements. However, an adaptation of this method is directly relevant to this problem. This relies on the simplicity of the target shapes which are essentially narrow rings, with a constant diameter. It is then possible to obtain a restoration consistent with the observed data using a PSF equal to a narrow ring. By reversing the definitions of f and h in equation (1) it is evident that this restoration would contain the beam spread function situated at each of the assembly centres, provided the targets acted as simple ring reflectors. To improve measurement accuracy the widths of the central responses can be reduced by using a PSF consisting of a ring convolved with a Gaussian having a width less than the beam-spread function.

In practice, the quality of the data does not permit a substantial sharpening of the central responses. Nevertheless, the resultant maximum entropy reconstruction (Fig. 4) shows bright central maxima, clearly distinguishable from much fainter diffuse background features. Some of the central maxima depart significantly from circular symmetry, due to the amplitude variations around the circumferences of the observed rings. However, an automated determination of the positions of the assembly centres can be made. The accuracy of these positional determinations is currently under investigation, but is probably better than 1 mm for an assembly diameter of ≃125 mm.

5. CONCLUSIONS

The initial application of non-linear maximum entropy processing to ultrasonic image data from the Prototype Fast Reactor at Dounreay has been encouraging. The main benefits are (1) removal of visually confusing scan arcs, and (2) the development of an automated procedure to determine the central positions of reactor components.

REFERENCES

1. N. Lions, et al. Special Instrumentation of Phenix. pp 525-535 of Fast Reactor Power Stations, Thomas Telford Ltd. London, (1974).
2. N.C. Hoitink, et al. Under Sodium Viewing Development for FFTF. Second International Conference on Liquid Metal Technology in Energy Production, pp 4-37 to 4-43 USDOE, CONF-800401-P1 (April 1980).
3. J. Taguchi et al. An ultrasonic Viewing System in Liquid Sodium. IEEE transactions on Nuclear Science. Vol NS-27, No 1. (February 1980).

4. J.A. McKnight, J. Bishop, D.K. Cartwright, and W.R. Diggle. The Application of Ultrasonic Technology Under Sodium. Second International Conference on Liquid Metal Technology in Energy Production, Hanford, USA, 20-23 April 1980.
5. J. Bishop, British Patents 1,525,950, 2029,306.
6. J.R. Fothergill, J.A. McKnight and L.M. Barrett, Ultrasonic Imaging in LMFBRs Using Digital Techniques. UKAEA Report ND-R-581(R). UKAEA, Risley Nuclear Power Development Establishment, Risley, Warrington, WA3 6AT, UK (1980).
7. J.A. McKnight and J.A. Parker, Data Collection Instrumentation for Ultrasonic Imaging Under Sodium. UKAEA Report No MD R 616(R). UKAEA, Risley Nuclear Power Development Establishment, Risley, Warrington, WA3 6AT, UK (1980).
8. E.J. Burton, D. Firth, I.D. Macleod and J.A. McKnight, Advances in signal processing for LMFBR diagnostic techniques, in "Progress in Nuclear Energy Vol 9". (Reactor Noise - SMORN III) Pergamon Press (1982).
9. A. Rosenfeld and A.C. Kak, "Digital Picture Processing", Academic Press, New York (1976).
10. B.R. Hunt, The application of constrained least squares estimation to image restoration by digital computer, IEEE Trans Computers, C22:805 (1973).
11. B.R. Frieden, Restoring with maximum likelihood and maximum entropy, J Opt Soc Amer 62:511 (1972).
12. S.F. Gull and G.J. Daniell, Image reconstruction from incomplete and noisy data, Nature 272:686 (1978).
13. J. Skilling, Algorithms and applications, Workshop on maximum entropy estimation and data analysis, University of Wyoming, Laramie, Wyoming (1981).
14. C.E. Shannon, A mathematical thoery of communication, Bell System Technical Journal, 29:379 and 623 (1948).

A SELF-FOCUSING ULTRASONIC IMAGE RECONSTRUCTION TECHNIQUE

S. Sepehr, J.M. Reeves, S.O. Harrold

Portsmouth Polytechnic
Department of Electrical and Electronic Engineering
Portsmouth, Hants., U.K.

1. INTRODUCTION

Near field imaging systems require a knowledge of the range of the object plane in order to obtain a focused image from the measured aperture data. This range information is often not available, and considerable effort has been made (Powers, 1973) to provide an acceptable answer to the problem in this case. Given that the recorded wavefront can be back propagated to any plane, Powers developed a feedback technique to determine if the computed diffraction field corresponds to an object in that plane on the assumption that a focused computed image will have sharp changes of intensity at the boundaries of an object. The eye is focused by these means using eye-brain interaction, but although this highly complex process is not presently amenable to computer simulation, nevertheless Powers' computerised approach combining the backward wave propagation method and an automatic edge detection scheme has been successful. The method involves programming a computer to calculate the field intensity across a number of different planes by backward propagation of the recorded wavefront and then to search for a focused image. The image which possesses the greatest rate of change of intensity at its boundary is then taken to be the focused result. Not only is this technique time consuming but also it can only be applied to a restricted class of objects with edge boundaries.

A mathematical solution to the problem of self-focussing has been developed by the authors at Portsmouth Polytechnic (Sepehr, 1980) which enables the reconstruction of the image of an object to be obtained solely from a measurement of the phase and amplitude of the wavefront in a plane at an unknown distance from the object.

2. MATHEMATICAL BASIS OF SELF-FOCUSED RECONSTRUCTION

The mathematics of focusing, given range information as distinct from self-focusing, is well established. For example, it can be shown, referring to Figure 1, that in reconstructing an object by backward propagation (Goodman, 1968, Sondhi, 1969) using the spatial frequency domain approach, that the Fourier transform of the pressure pattern $p_s(u,v)$, observed at the sampling plane, may be related to the Fourier transform of the scattered pressure wave $p_o(x,y)$ by the equation:

$$P_s(f_x, f_y) = P_o(f_x, f_y) \exp\left[j \frac{2\pi z}{\lambda} \sqrt{1 - (\lambda f_x)^2 - (\lambda f_y)^2}\right] \quad (1)$$

where: λ is the wavelength, f_x and f_y are spatial frequencies, and $P_s(f_x, f_y)$, $P_o(f_x, f_y)$ are the Fourier transforms of $p_s(u,v)$, $p_o(x,y)$.

The term $\exp\left[j \frac{2\pi z}{\lambda} \sqrt{1 - (\lambda f_x)^2 - (\lambda f_y)^2}\right]$ is the transfer function of the system. By reverse or back propagation the object wave $p_o(x,y)$ may be obtained in terms of the observed or sampled wavefront and the sampling plane object plane distance z as:

$$p_o(x,y) = F^{-1}\left[P_s(f_x, f_y) \exp\left[-j \frac{2\pi z}{\lambda} \sqrt{1 - (\lambda f_x)^2 - (\lambda f_y)^2}\right]\right] \quad (2)$$

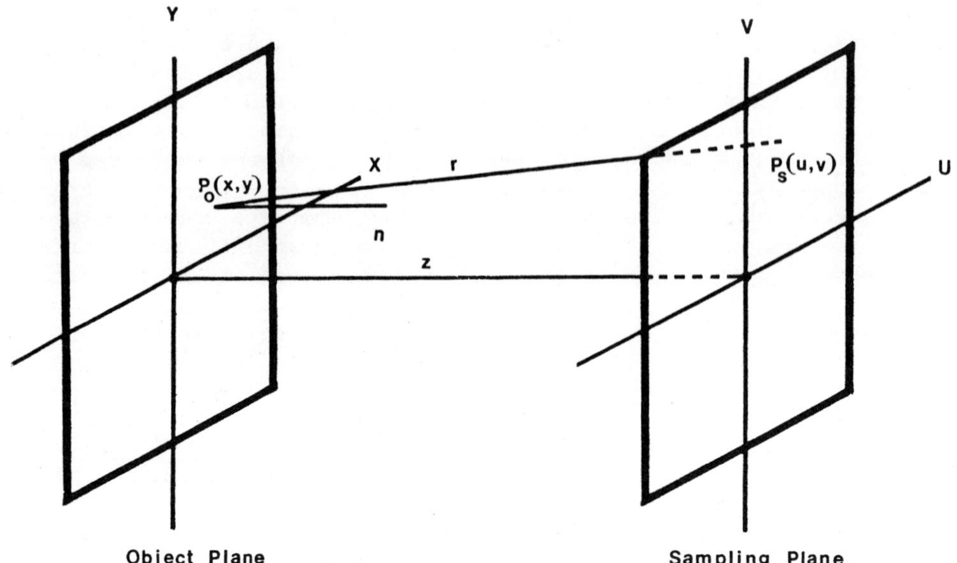

Figure 1. Object plane - sampling plane geometry.

Thus the image may be computed from the wavefront data providing that the wavefront is recorded at a known distance z from the object plane.

For self-focused reconstruction, however an equation is required which enables the object wave to be computed without the knowledge of z. The pressure wave $p_s(u,v)$ at the sampling plane may be expressed by the Fresnel integral with the obliquity factor omitted:

$$p_s(u,v) = \frac{1}{j\lambda} \iint_{-\infty}^{\infty} p_o(x,y) \frac{\exp(jkr)}{r} \, dx \, dy \tag{3}$$

where: $r = \sqrt{z^2 + (u-x)^2 + (v-y)^2}$ and $k = \frac{2\pi}{\lambda}$

The object wave $p_o(x,y)$ is in general a complex quantity, but in order to obtain a mathematical solution to the self-focused reconstruction problem it is necessary for the object wave to be a real quantity. This implies a uniform phase distribution across the object plane, limiting the approach to planar objects, but it can be demonstrated that the effect on image definition of a small non-uniformity in phase distribution is noticeable, but acceptable. If however, this limitation is applied, equation (3) may be expanded into the real part p_1 and the imaginary part p_2 as:

$$p_1 = \iint_{-\infty}^{\infty} p_o(x,y) \frac{\cos\left(k\sqrt{z^2 + (u-x)^2 + (v-y)^2} - \frac{\pi}{2}\right)}{\lambda\sqrt{z^2 + (u-x)^2 + (v-y)^2}} \, dx \, dy \tag{4}$$

and

$$p_2 = \iint_{-\infty}^{\infty} p_o(x,y) \frac{\sin\left(k\sqrt{z^2 + (u-x)^2 + (v-y)^2} - \frac{\pi}{2}\right)}{\lambda\sqrt{z^2 + (u-x)^2 + (v-y)^2}} \, dx \, dy \tag{5}$$

Both the above equations may be recognised as two dimensional convolution integrals and may then be written as:

$$p_1 = p_o(x,y) * c(x,y) \tag{6}$$

$$p_2 = p_o(x,y) * s(x,y) \tag{7}$$

where: $$c(x,y) = \frac{\cos\left(k\sqrt{z^2 + x^2 + y^2} - \frac{\pi}{2}\right)}{\lambda\sqrt{z^2 + x^2 + y^2}} \tag{8}$$

and $$s(x,y) = \frac{\sin\left(k\sqrt{z^2 + x^2 + y^2} - \frac{\pi}{2}\right)}{\lambda\sqrt{z^2 + x^2 + y^2}} \tag{9}$$

Equations (6) and (7) may now be transformed into the spatial frequency domain. Thus if $P_1(f_x,f_y)$ and $P_2(f_x,f_y)$ represent the

Fourier transform of p_1 and p_2, and $P_o(f_x,f_y)$ represents the Fourier transform of the object wave, the diffraction equations in the spatial frequency domain will be :

$$P_1(f_x,f_y) = P_o(f_x,f_y)\ C(f_x,f_y) \tag{10}$$

$$P_2(f_x,f_y) = P_o(f_x,f_y)\ S(f_x,f_y) \tag{11}$$

where $C(f_x,f_y)$ and $S(f_x,f_y)$ are the Fourier transforms of the cosine and sine propagation terms. The sum of the squares of equation (10) and (11) gives:

$$P_1(f_x,f_y)^2 + P_2(f_x,f_y)^2 = P_o(f_x,f_y)^2 \left[S(f_x,f_y)^2 + C(f_x,f_y)^2\right] \tag{12}$$

It has been shown (Sepehr, 1980) that at low spatial frequencies the Fourier sum $\left[S(f_x,f_y)^2 + C(f_x,f_y)^2\right]$ is unity. Since most of the information about a finite duration function is contained within the low frequency coefficients of its Fourier transform this limitation on bandwidth does not in practice have a serious effect on image quality, although some loss of fine detail must occur. Equation (16) may therefore be approximated at low spatial frequencies to:

$$P_o(f_x,f_y)^2 = P_1(f_x,f_y)^2 + P_2(f_x,f_y)^2 \tag{13}$$

The object wave will then be given by:

$$p_o(x,y) = F^{-1} \left[P_1(f_x,f_y)^2 + P_2(f_x,f_y)^2\right]^{\frac{1}{2}} \tag{14}$$

Equation (14) may be regarded as a mathematical solution to the self-focused reconstruction problem within the given limitations.

The following algorithm may be used for the self-focused reconstruction of ultrasonic images by means of Equation (14).

(a) Obtain the complex wavefront $p_s(u,v)$ at an arbitrary sampling plane U-V.

(b) Calculate the real part p_1 and imaginary part p_2 of the wavefront.

(c) Perform Fourier transformations on p_1 and p_2 to obtain the Fourer functions $P_1(f_x,f_y)$ and $P_2(f_x,f_y)$.

(d) Calculate the sum of the squares of $P_1(f_x,f_y)$ and $P_2(f_x,f_y)$ to obtain the object wave $p_o(x,y)$.

(e) Calculate the square root of this sum to obtain $P_o(f_x,f_y)$.

(f) Take the inverse Fourier transform of $P_o(f_x,f_y)$ to obtain the object wave $p_o(x,y)$.

(g) Calculate the square of the modulus of $p_o(x,y)$ to obtain the image.

Thus the algorithm may be implemented using the F.F.T. three times.

It should be noted that step (e) introduces a phase ambiguity of π radians. This problem has been overcome by developing a special software routine which employs a 'rotating phasor' technique (Sepehr, 1980).

3. SIMULATION STUDIES

To simulate the reconstruction process on the computer, a simple object, letter H, with an arm size of 8 mm was chosen (see Figure 2(a)). The diffraction pattern was calculated using the Fresnel integral over a 6 x 6 cm^2 observation plane positioned at a distance of 10 cm in front of the object using 3600 complex data points. Figure 2(b) shows the magnitude of the computed field. In this calculation all points in the object were in phase and the wavelength of 1·5 mm was assumed which corresponds to the wavelength of a 1 MHz wave frequency in water. The simulated data was then processed on PDP8/E according to the self-focused reconstruction algorithm outlined in Section 2. The resulting computed image is shown in Figure 2(c). This clearly demonstrates the accuracy of the self-focused reconstruction. Due to the discrete Fourier calculation the reconstruction time was three hours and ten minutes. Further calculations have been carried out at 5 and 15 cm ranges with similar results.

The simulation has been extended to indicate the likely effect of non-uniform phase distribution in the object plane. A good degree of tolerance has been shown giving recognisable reconstruction providing that the extent of non-uniform phase distribution does not exceed that with which the sign ambiguity program can cope.

4. EXPERIMENTAL SET-UP

The transmitter, test object and the sampling probe are all suspended in a water filled standard glass fibre water-storage tank with approximate dimensions: 120 cm x 60 cm x 60 cm. For all tests through transmission is used, the transmitter unit (3 x 3 mm) being cut from a 1 MHz thickness mode resonance PZT-5 transducer disc, and driven by a gated sinewave burst of 20 cycles duration with an amplitude of 60 volts in order to improve the signal to noise ratio of the received signal. A minimum time gap of one millisecond between successive transmitted bursts is chosen in order to allow

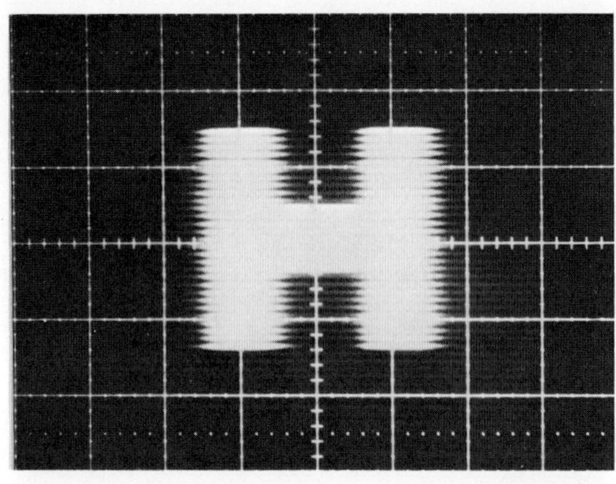

Figure 2(a)
Amplitude Distribution of object function

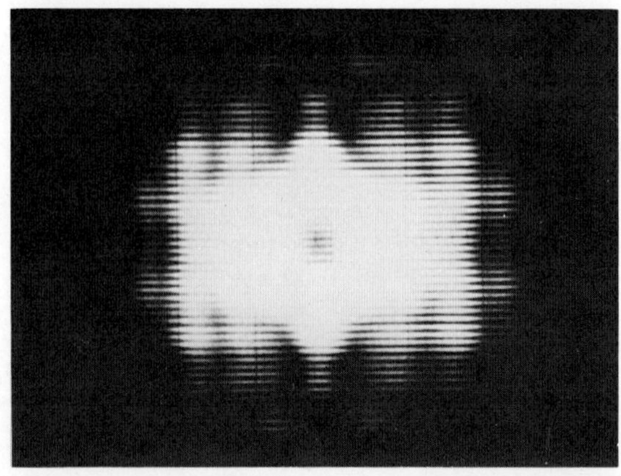

Figure 2(b)
Computed diffraction field

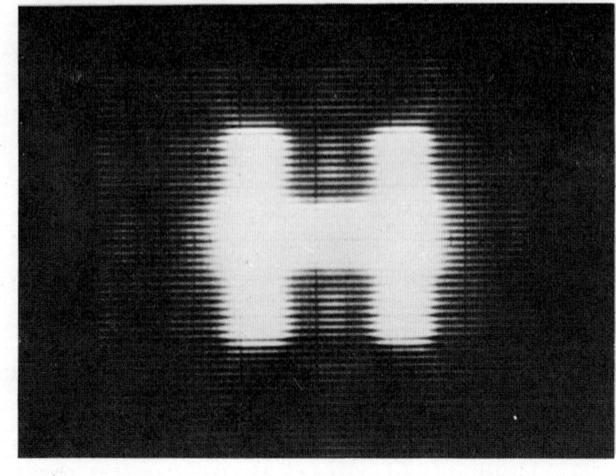

Figure 2(c)
Reconstructed image

time for the reflections from the sides of the tank to die away to an insignificant level before the next burst is transmitted. This avoids the creation of standing waves within the test tank. A test object is placed in the far field region of the transmitter unit (approximately ½ metre from the transmitter), effectively giving an object wave of uniform phase distribution. Figure 3 shows in a schematic form the acoustic section of the system.

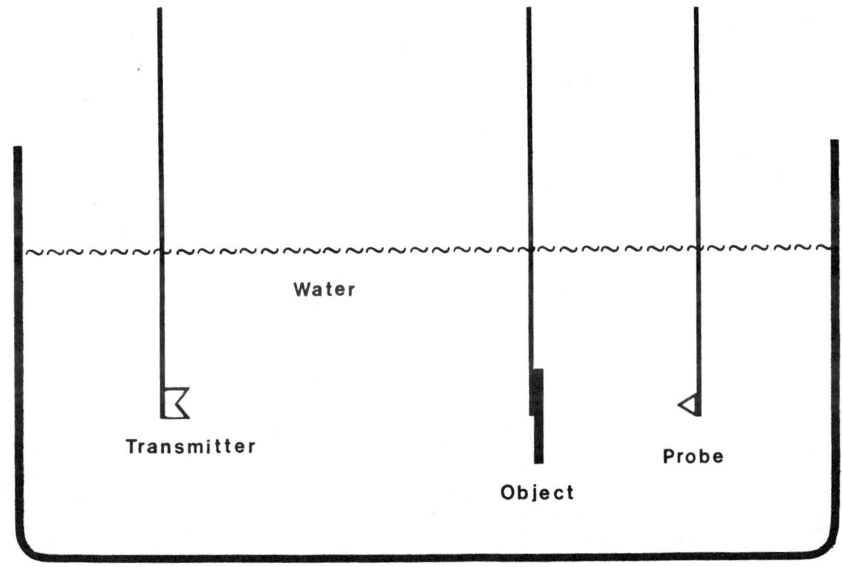

Figure 3. The acoustic test tank arrangement.

The propagated wave on reaching the measurement plane is sampled by a single manually scanned probe to perform point-by-point sampling, over 60 x 60 sampling points spaced 1 mm apart.

Since the phase is significant, the recording plane must be flat to within a small fraction of a wavelength. At the experimental frequency of 1 MHz ($\lambda = 1{\cdot}5$ mm), this implies a stability and accuracy of movement in the order of 0•15 mm. However, the movement accuracy and the flatness of the mechanical scanner is at least three times worse than this figure and it s major source of error in the current experimental system.

The instantaneous value of the pressure signal at each sample point is recorded twice with a time interval of a quarter period to specify the signal. This has the advantage over phase and amplitude detection that it gives the real and imaginary parts directly. The inherent loss in signal to noise performance is compensated for by averaging over 100 recordings.

The detected signals are digitized on-line and delivered to the computer memory. System timing is controlled by a programmable timing unit driven by a 1 MHz crystal clock generator. This unit delivers the necessary timing and command signals to the other units in the system. The reference time signal to the sample-hold amplifier and the A/D clock pulse for the conversion of the detected signal into a binary signal, are provided by the programmable timing unit. The transmit and receive cycle is also controlled and programmed by this unit. The operation of the computer for reading the binary signal is synchronised to the timing sequence of the programmable timing unit.

5. EXPERIMENTAL RESULTS

Several test objects of different shape were used, Figure 4(a) is a photograph of one of the objects used, an aluminium plate 2 mm thick in which five 7 mm holes were drilled, so that the test piece represented a high contrast object. The magnitude of the recorded field in the sampling plane for this object is shown in Figure 4(b). The reconstructed image using the self-focusing algorithm is shown in Figure 5(a).

For comparison the same data, together with the range, were used to produce a reconstructed image by backward propagation, the result is shown in Figure 5(b).

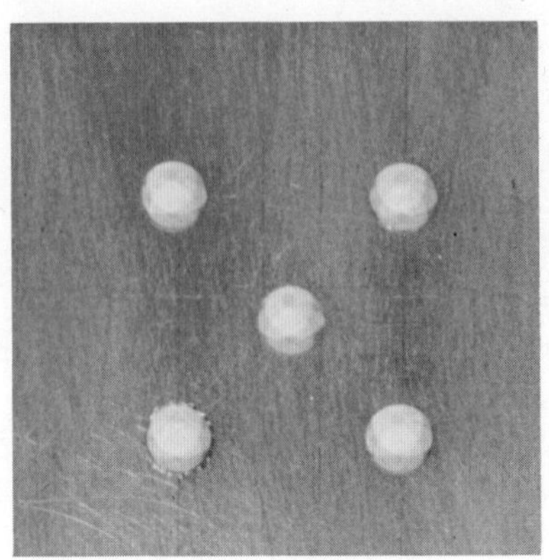

Figure 4(a)
Test Object

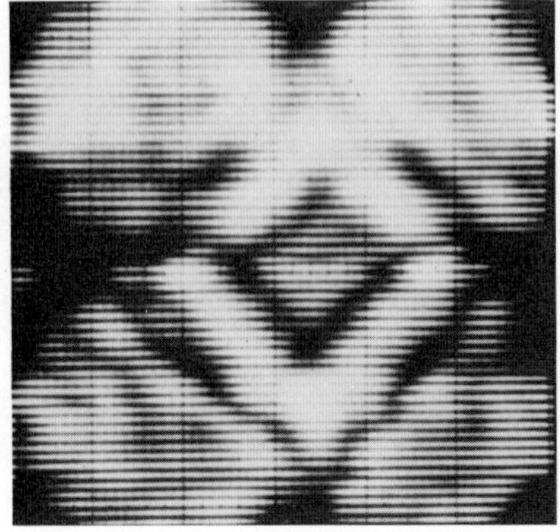

Figure 4(b)
Diffraction field

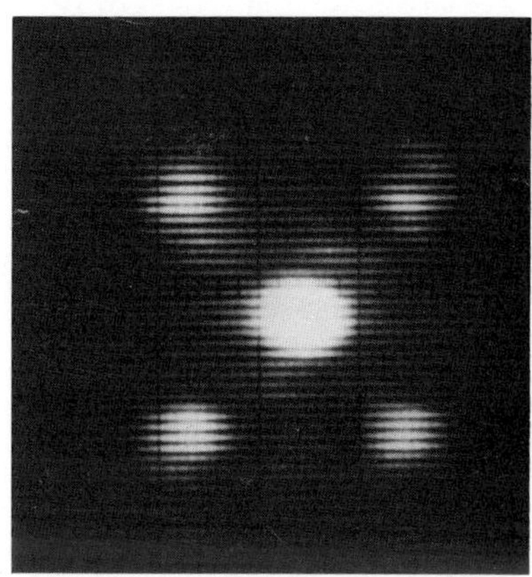

Figure 5(a)
Self-focused image

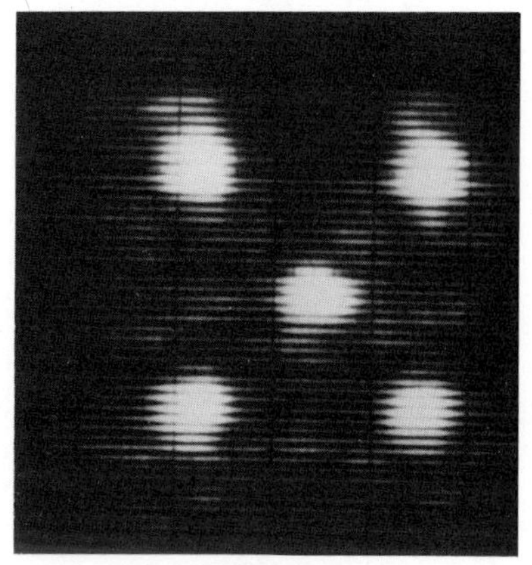

Figure 5(b)
Backward Propagation Image

A further computation on the same data and range was carried out using the computer modelled lens method (Harrold, 1974), the result was indistinguishable from that shown in Figure 4(d). However the computation time was an order of magnitude greater as the fast Fourier transform could not be used.

6. CONCLUSIONS AND FURTHER WORK

A solution to the problem of self-focusing has been developed and demonstrated practically. Initial results although obtained in the presence of quite substantial data errors, adequately confirm the theoretical basis on which the technique stands.

The work is by no means complete and further investigations are necessary. Those presently considered important are listed below:

(i) The limitations imposed on the range of f_x and f_y by the requirement for $S(f_x,f_y)^2 + C(f_x,f_y)^2 = 1$ need further consideration, and methods of compensation outside this range need to be developed.

It has been stated that the near unity value of the sum $S(f_x,f_y)^2 + C(f_x,f_y)^2$ is limited to low spatial frequencies but rise above unity (by as much as 100%) for high spatial frequencies. This magnifies the high frequency components in the Fourier spectrum of the reconstructed image. This effect is inherent in the reconstruction, and its correction would greatly improve the fine detail of the results obtained.

It has been tentatively shown that the sum $S(f_x,f_y)^2 + C(f_x,f_y)^2$ is virtually independent of propagation distance. The significance of this is that it allows for the possibility of correcting the Fourier spectrum of the image by dividing by stored estimates of the sum $S(f_x,f_y)^2 + C(f_x,f_y)^2$.

(ii) Simulation studies have shown that the effect of excessive non-uniformity of the object wave phase distribution is to cause serious image deterioration. A technique by which the effect of this may be reduced is an important subject for further work, since uniform phase insonification is of limited practical value.

(iii) To properly investigate the self-focused reconstruction technique a highly stable and accurate mechanical system is needed. An improved system should include a computer controlled stepping drive facility in order to speed up the scanning of the field.

REFERENCES

Goodman, J.W. (1968). 'Introduction to Fourier Optics'. McGraw-Hill, New York.

Harrold, S.O. (1974). 'Solid State Ultrasonic Imaging'. PhD Thesis, University of Warwick.

Powers, J.P. (1973). In 'Acoustical Holography', Vol. 5, pp 527-537, Plenum Press, New York.

Sepehr, S. (1980). 'Ultrasonic Imaging by Means of Frequency Domain Processing'. PhD Thesis, C.N.A.A.

Sondhi, M.M. (1969). J. Acoust. Soc. Am. 46, 5, 1158-1164.

NEW POSSIBILITIES IN DATA MEASUREMENT, SIGNAL PROCESSING AND INFORMATION EXTRACTION: PHILOSOPHY AND RESULTS

A.J. Berkhout, J. Ridder, M.P. de Graaff

Delft University of Technology
Dept. of Applied Physics, Group of Acoustics
P.O.B. 5046, 2600 GA Delft, The Netherlands

INTRODUCTION

Recently,the acoustics group of the Delft University has introduced a theoretical framework which significantly improves the insight in and understanding of fundamental problems in acoustical echo techniques.In this theoretical framework wave theory and systems theory have been combined,resulting in an accurate and concise description of the basic physical processes in acoustical echo experiments.

WAVE-THEORETICAL MODEL

In acoustic echo techniques forward travelling pressure waves occur in two ways (Fig. 1):

a. Waves travelling <u>downward</u> from the surface to the inhomogeneities of interest,and
b. Waves travelling <u>upward</u> from the inhomogeneities back to the surface.

The physical process of the two-way wave propagation and reflection has been quantified by us with the aid of the following matrix equation

$$\vec{P}(z_0) = D(z_0)\left[\sum_m W(z_m,z_0)\, R(z_m)\, W(z_0,z_m)\right]\vec{S}(z_0)\ , \qquad (1)$$

where $\vec{S}(z_0)$ = Source vector representing the source configuration at the surface (z_0)

$W(z_0,z_m)$ = downward propagation matrix representing the propagation properties from the surface (z_0) to depth level z_m

$R(z_m)$ = scattering matrix representing the scattering properties of the inhomogeneities at depth level z_m

$W(z_m,z_0)$ = upward propagation matrix representing the propagation properties from depth level z_m to the surface (z_0). It can be shown that for a time-invariant medium $W(z_m,z_0) = W^T(z_0,z_m)$, where T denotes a matrix transposition.*

$D(z_0)$ = detector matrix representing the detector configuration at the surface (z_0)

$\vec{P}(z_0)$ = data vector representing the single-scattered echo data at the surface (z_0) due to source $\vec{S}(z_0)$.

In expression (1) W is defined by the wave equation for pressure waves in inhomogeneous absorptive fluids, R is defined by the elastic boundary conditions and S,D are defined by the geometrical and the acoustical properties of the tranducer in emission and in reception respectively.

In a simulation procedure R has to be specified and the effects of propagation (W,W^T) for a given data acquisition technique ($\vec{S}$,D) are evaluated. In an inversion procedure (focussing, deconvolution) the effects of propagation for a given data acquisition technique have to be eliminated from the data ($\vec{P}$) to yield an estimate of R.

To show the basic principle of our inversion technique, let us write

$$\vec{S}(z_m) = W(z_0,z_m)\ \vec{S}(z_0) \tag{2a}$$

and

$$\vec{P}(z_m) = \left[D(z_0)\ W(z_m,z_m)\right]^{-1} \vec{P}(z_0) \tag{2b}$$

* See also Berkhout, 1980, chapter 6.

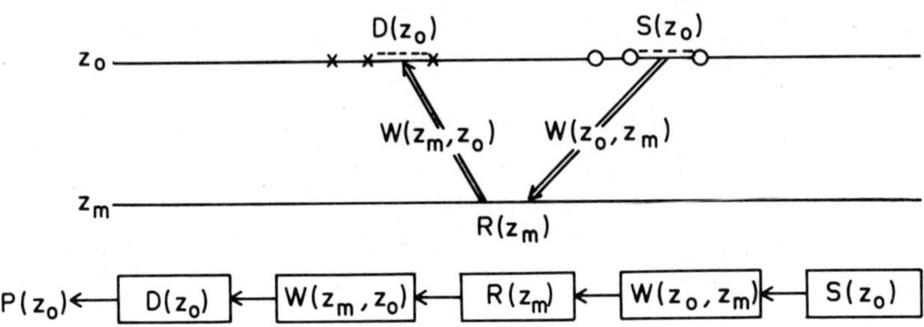

Fig.1.: Basic acoustic model.

then, according to (1),

$$\vec{P}(z_m) = R(z_m)\, \vec{S}(z_m) \,, \tag{3}$$

where $\vec{S}(z_m)$ represents the propagated source wave field at depth z_m and $\vec{P}(z_m)$ represents the reflected wave field at depth z_m. The latter has been estimated by matrix inversion (2b).
Hence, if R represents a diagonal matrix then (3) can easily be solved. Note from (3) that for a stable inversion the elements of $\vec{S}(z_m)$ should be large, i.e. all inhomogeneities at depth level z_m should be well illuminated.
In practical situations high quality results can only be obtained if echo data due to several sources are used with different source positions:

$$(\vec{P}_1(z_m), \vec{P}_2(z_m), \ldots\, \vec{P}_N(z_m)) = R(z_m)\; (\vec{S}_1(z_m), \vec{S}_2(z_m), \ldots\, \vec{S}_N(z_m)) \tag{4a}$$

$$P(z_m) = R(z_m)\; S(z_m). \tag{4b}$$

In (4b) the columns of S and P contain the data from different physical experiments. Note that if R represents a diagonal matrix then (4) defines an overdetermined system.

From the foregoing it follows that the best inversion results may be expected if omni-directional transducers are used, since

a) every inhomogeneity in the medium will be illuminated by all individual transducer elements;
b) every transducer element receives the reflected wave fields from all inhomogeneities.

By inverting matrix equation (4b) all contributions for a given image point are collected from the measured data matrix and combined in phase with optimum weighting factors.

Conventional focussing techniques are formulated in our theoretical model by (3) where the lateral extension of $\vec{S}(z_m)$ is made as small as possible by using a narrow source beam and the inversion for $D(z_0)\, W(z_m, z_0)$ is carried out by in-phase addition along the receiver array.
We will see that the spatial resolution as obtained by properly inverting for (4) is many times better than the lateral resolution of conventional focussing results.

In our approach inversion for the matrices

$$D(z_0)\, W(z_m, z_0) \quad \text{and} \quad W(z_0, z_m)\, S(z_0)$$

plays a central role. According to us, any texture analysis technique should not be applied on P but it should be applied on an estimate of R and,therefore.texture analysis should occur after inversion (Mesdag et al.,1982).

AN EXAMPLE

The wave-theoretical model will be illustrated by means of an example.A very simple model is chosen which consists of one point diffractor at depth $z = z_m$,while a single source with omnidirectional sensitivity and an array of omnidirectional detectors are located at the surface (Fig. 2).Because there is only one single source all elements of the source vector $\vec{S}(z_0)$ are equal to zero,except for one element:

$$\vec{S}(z_0) = (0, \ldots ,0,S_n(\omega),0, \ldots ,0) \tag{5}$$

After transmission of the ultrasound pulse the acoustical energy propagates from the surface $z = z_0$ to the subsurface $z = z_m$ (Fig.3):

$$\vec{P}(z_m) = W(z_0,z_m)\, \vec{S}(z_0) \ , \tag{6}$$

where the elements of the propagation matrix $W(z_0,z_m)$ are defined by the wave-equation for pressure waves in (in)homogeneous media. In figure 3b $\vec{P}(z_m)$ is given as a function of time for all lateral positions at $z = z_m$.Note that for the position directly below the source $\vec{P}(z_m)$ is maximal and the transition time is minimal,while at the left boundary of the subsurface $z = z_m$ $\vec{P}(z_m)$ is minimal and the transition time is maximal.
At the subsurface $z = z_m$ the acoustical energy is reflected by the point diffractor.In figure 3b the lateral position of the point diffractor is indicated by an arrow.Part of the reflected energy propagates back to the surface $z = z_0$ (Fig. 4):

$$\vec{P}(z_0) = W(z_m ,z_0)\, R(z_m)\, W(z_0,z_m)\, \vec{S}(z_0) \ . \tag{7}$$

For an ideal point diffractor the scattering matrix $R(z_m)$ is a diagonal matrix with just one element unequal to zero (spatial delta pulse).Note from figure 4b that the measured pressure distribution $\vec{P}(z_0)$ is symmetric around the position of the point diffractor with respect to both transition time and the amplitude.In figure 4b the position of the detector array is indicated by arrows.So $\vec{P}(z_0)$ is only registered over the very limited aperture bounded by these two arrows.The outputs of the detector array are defined by equation (1). Because this model can be described by just one subsurface the summation in equation (1) can be omitted.

To obtain an image of the point diffractor (i.e. an estimation of the scattering matrix $R(z_m)$)we have to evaluate equation (3).So the focussing procedure (i.e. the inversion technique) starts with the calculation of the forward extrapolated source function at depth

$z = z_m$ $\vec{S}(z_m)$, and the backward extrapolated detector responses at depth $z = z_m$ $\vec{P}(z_m)$ (equations 2a and 2b). Figures 5a and 5b show $\vec{S}(z_m)$ and $\vec{P}(z_m)$ as a function of depth. Now $\vec{S}(z_m)$ and $\vec{P}(z_m)$ are known $R(z_m)$ (i.e. the image of the point diffractor) can be calculated by means of equation (3). The result is shown in figures 5c and 5d. In figure 5d the maximum amplitude of $R(z_m)$ is given for all lateral positions at $z = z_m$. The beam width is indicated at the -6,-10 and -20dB levels. For a perfect inversion technique figure 5d would have shown a spatial delta pulse. Because the reconstruction is based on just one physical experiment and because a detector array is used with a very limited aperture the energy of the reconstructed delta pulse is smeared out over a large area. So for more physical experiments or application of a detector array with a large aperture area a much better result can be expected. For example figure 6 shows the result of zero-offset data acquisition. During zero-offset data acquisition a sound pulse is emitted at a certain position of the aperture by a (omnidirectional) transducer and the echos are registered by means of a (omnidirectional) receiver located at the same position as the source. These experiments are repeated for a large number of positions within the aperture area. So zero-offset data acquisition consists of a large number of experiments. Figure 6a shows the result of the different physical experiments, while figures 6b and 6c show the calculated estimation of $R(z_m)$ (i.e. the image of the point diffractor). Figure 6c resembles a spatial delta pulse much better than figure 5c.

PARAMETER ESTIMATION

The acoustic parameters of interest in tissue characterization can be subdivided in two categories:

1. Bulk parameters which define the propagation velocity and absorption;
2. Local parameters which define the reflectivity distribution.

Synthetic focussing plays a central role in our method to determine these parameters. For velocity analysis the synthetic focussing process is applied for a range of velocities around an expected value

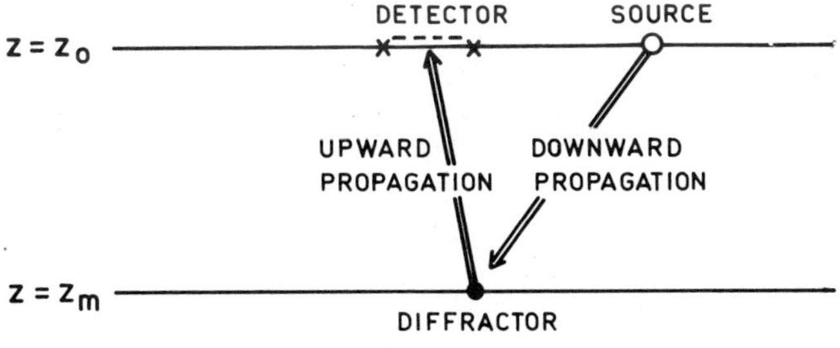

Fig. 2.: Model.

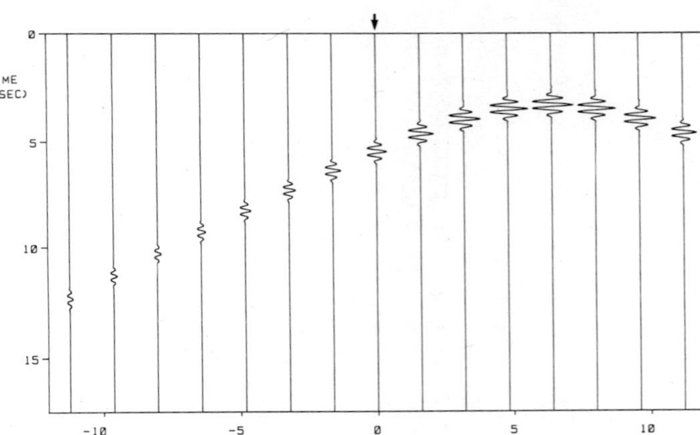

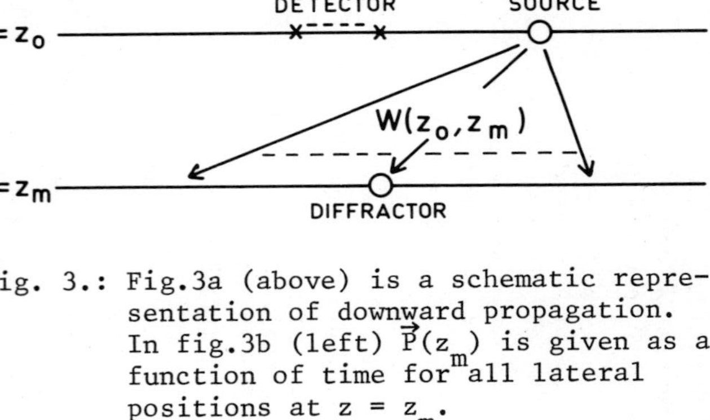

Fig. 3.: Fig.3a (above) is a schematic representation of downward propagation. In fig.3b (left) $\vec{P}(z_m)$ is given as a function of time for all lateral positions at $z = z_m$.

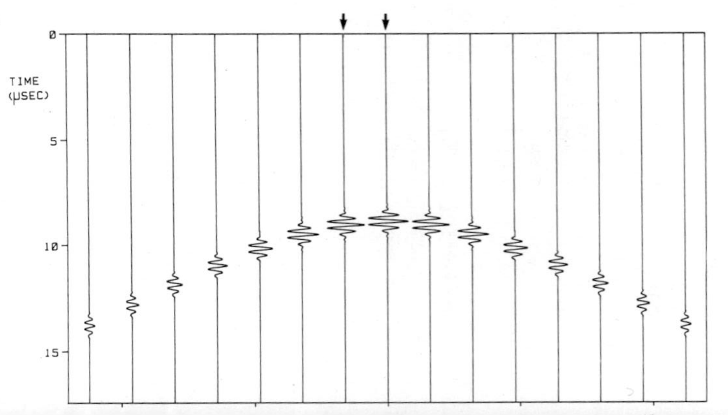

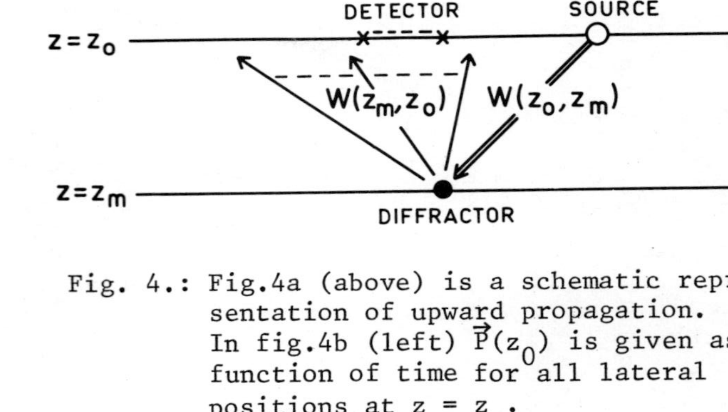

Fig. 4.: Fig.4a (above) is a schematic representation of upward propagation. In fig.4b (left) $\vec{P}(z_0)$ is given as a function of time for all lateral positions at $z = z_m$.

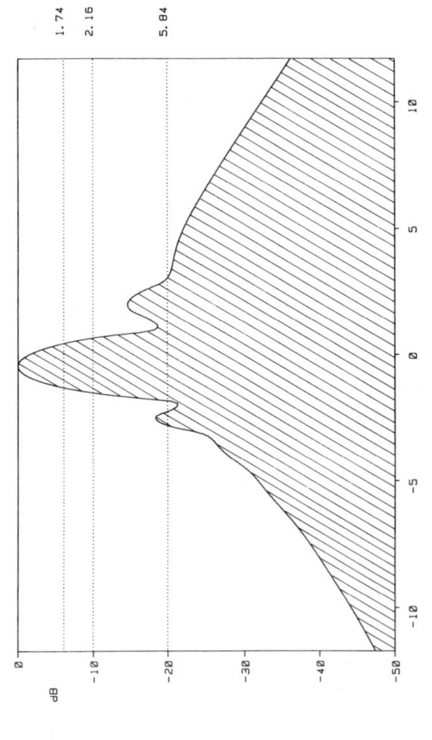

Fig. 5.: In fig.5a (above left) the forward extrapolated source function is given as a function of the propagation velocity times the transition time (c*t). Fig.5b (middle left), same as fig.5a but for the backward extrapolated detector responses. Fig.5c (below left) and fig.5d (above) show the calculated estimation of $R(z_m)$. In fig.5d the maximum amplitude of $R(z_m)$ is given for all lateral positions at $z = z_m$. The beamwidth (in mm) is indicated at the -6, -10 and -20 dB levels.

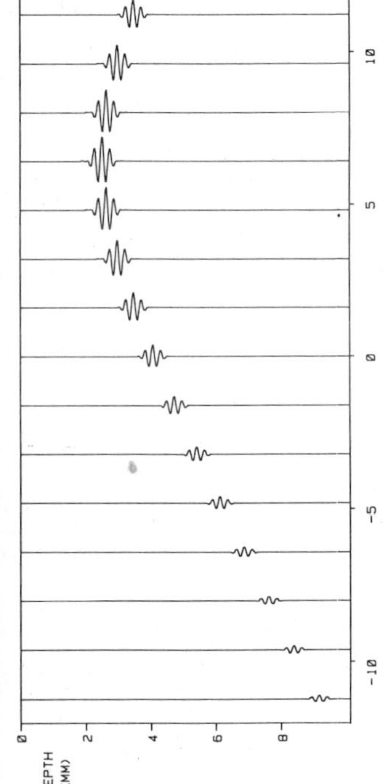

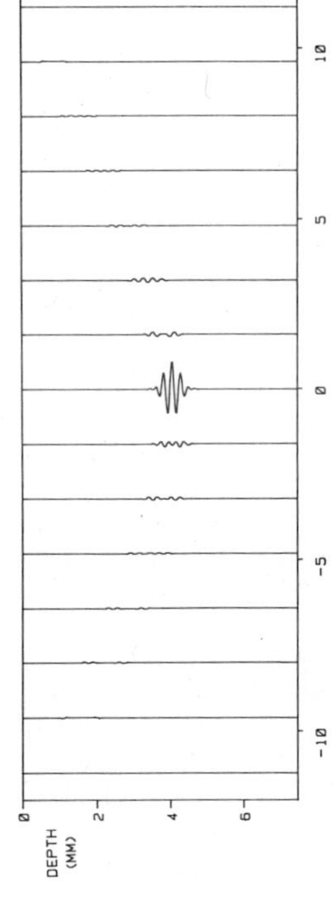

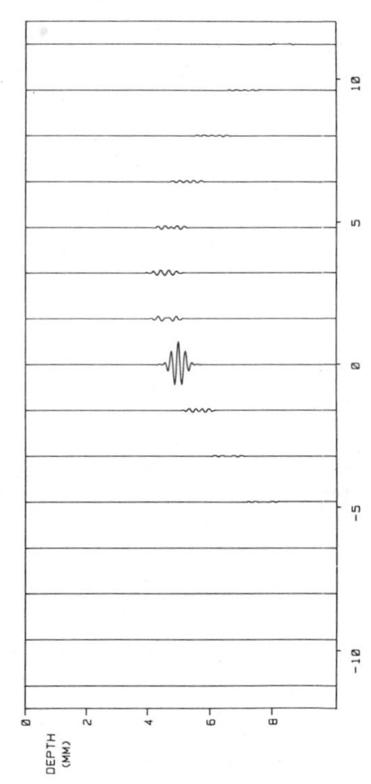

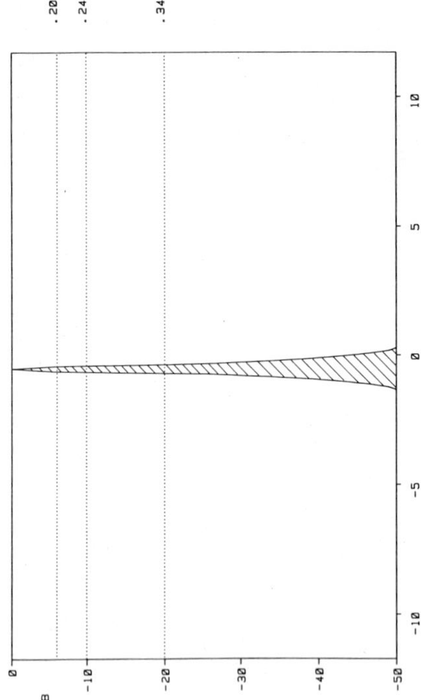

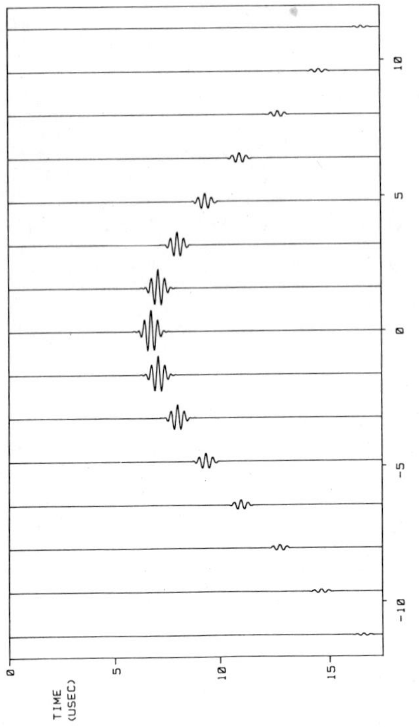

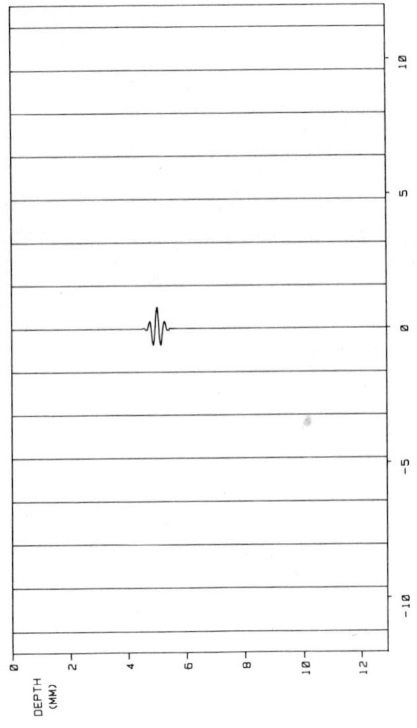

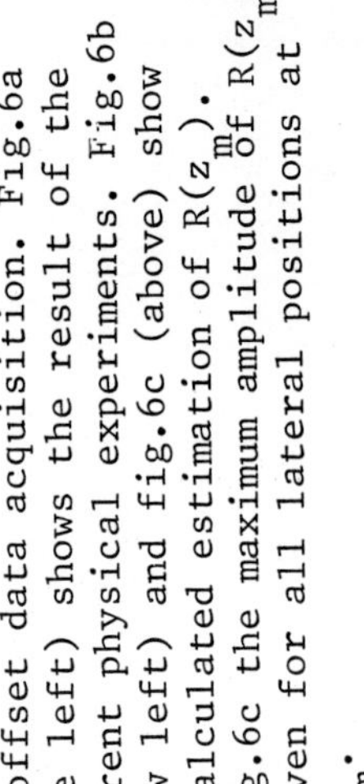

Fig. 6.: Zero-offset data acquisition. Fig.6a (above left) shows the result of the different physical experiments. Fig.6b (below left) and fig.6c (above) show the calculated estimation of $R(z_m)$. In fig.6c the maximum amplitude of $R(z_m)$ is given for all lateral positions at $z = z_m$.

and the results are inspected by a minimum-entropy criterion (De Vries et al.,1982).
For the determination of local parameters the acoustic energy must first be concentrated to the image points of the inhomogeneities. This is done by our synthetic focussing process.Then the area of interest is selected,isolated from the total data set,and our amplitude and texture analysis techniques are applied (Mesdag et al.,1982).

SOME RESULTS OF WATERTANK EXPERIMENTS

So far all results shown were based on computer simulations.In this section we will present some results of our inversion technique applied on measured data.All measurements were done with a zero-offset data acquisition technique.

In figure 7 the resolution is given as a function of depth. A thin steel wire (ϕ = 100 μm) is used as a target.Data is acquired over an aperture area of 51.2 mm.As transducer one element of a multi-element transducer is used.This element has an aperture of .3 mm and a bandwidth of 2 - 4 MHz (-20 dB).Also shown in figure 7 is the resolution as a function of depth calculated from computer simulations.Measured and calculated values are in good agreement from a depth of about 30 mm.
The increase of the beam width nearer the transducer is mainly caused by the fairly long recovery time (± 35 μsec) of the applied detector pre-amplifier.Immediately after transmission of a sound pulse this pre-amplifier is completely saturated by the transmission pulse.So

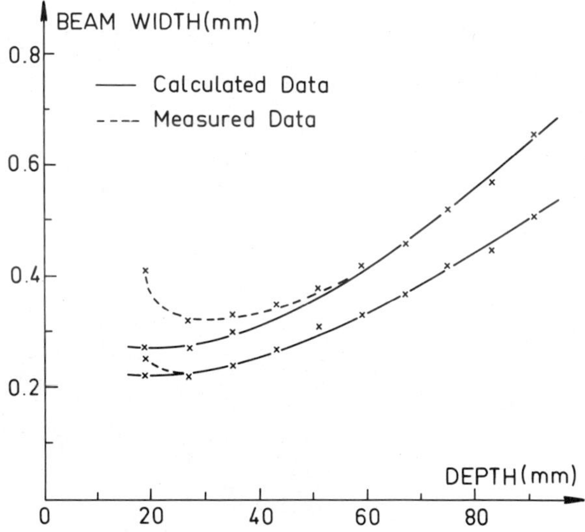

Fig.7.: -6 dB & -10 dB Beam width as a function of depth for synthetically focussed data.

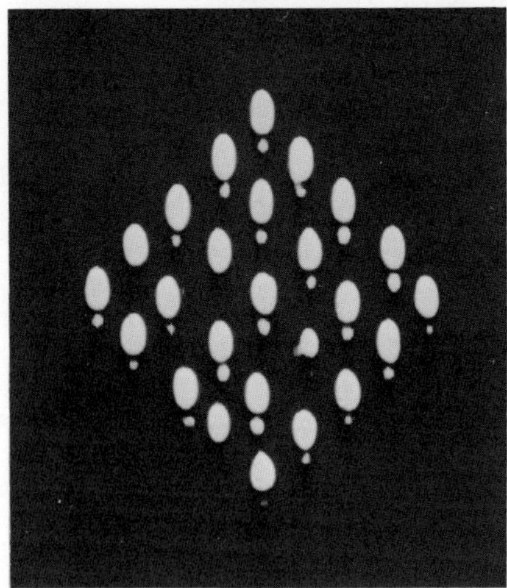

Fig.8.: Image of an array of scatterers.The transducer is situated at the top of the image.

the first 35 μsec or 25 mm the measured values are inaccurate. Figure 8 shows an image of an array of thin steel wires (ϕ = 100 μm). The minimal distance between the wires is equal to 1.4 mm.The midpoint of the array of wires is situated at a depth of 80 mm.Here also the data is acquired over an aperture of 51.2 mm and the same transducer is used as in the previous experiment.All individual wires can easily be distinguished from each other.Note that the lateral resolution in this image is better than the axial resolution!Behind each wire a second response can be seen.This second response is caused by an echo from the backing of the transducer.

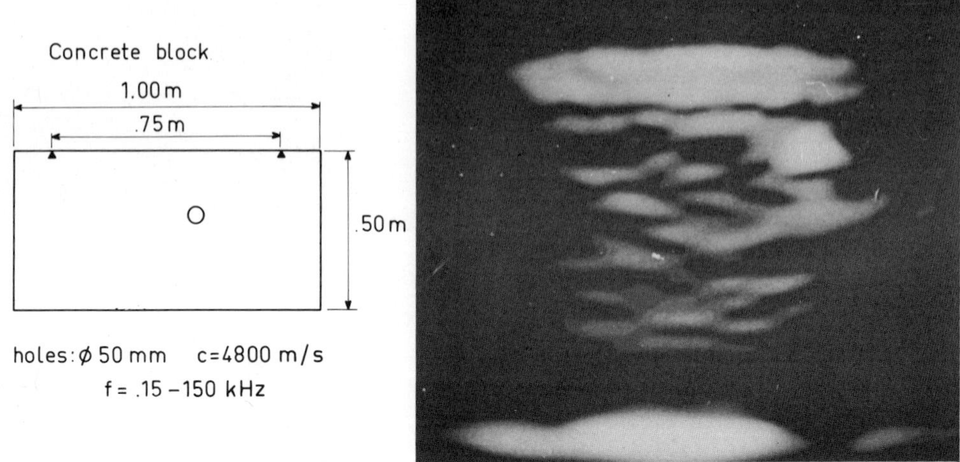

Fig.9.: Schematic drawing of a concrete block and its ultrasonic image.

The last result we want to show is the image of a concrete block with a hole in it (Fig. 9).Imaging of concrete blocks is difficult because of the high scattering level caused by local inhomogeneities in the concrete (e.g. pebbles) and because of the relatively low attenuation (multiples).In figure 9a a schematic drawing of the block is given.In figure 9b the hole and the bottom of the block are visualized.Note the shadowing effect of the hole on the response from the bottom.The diameter of the hole measured from the image matches the actual value very accurately.The horizontally shaped response above the hole is caused by a Rayleigh wave over the surface of the block. For this image data was acquired over an aperture of 750 mm.

CONCLUSIONS

1. A discussion is given on a wave-theoretical framework which describes the propagation of primary pressure waves in inhomogeneous media in terms of matrices.

2. The focussing problem has been presented in terms of matrix inversion. It is shown that for good inversion results the data of many different experiments should be combined and/or a large aperture detector array should be applied.

3. The by computer simulations predicted resolution is in good agreement with the results of our inversion technique obtained by application on measured data.

4. For zero-offset data acquisition over an aperture area of 51.2 mm and a transducer with a bandwidth of 2 - 4 MHz (-20 dB) at a depth of 80 mm the lateral resolution has proven to be better than the axial resolution. In fig. 8 the lateral resolution is about .55 mm while the axial resolution is about 1.2 mm (both values at the -10 dB level).

5. With this technique even from strong scattering media like concrete good images can be obtained.

REFERENCES

Berkhout,A.J.,"Seismic Migration",1982,Elsevier,Amsterdam-Oxford-New York.

Mesdag,P.R.,De Vries,D.,Berkhout,A.J.,"An Approach to Tissue Characterization Based on Wave Theory Using a New Velocity Analysis Technique",1982,Acoustical Imaging,Vol 12,Plenum Press.

De Vries,D,Berkhout,A.J.,"Influence of Velocity Errors on Computerized Acoustic Imaging",1982,Submitted for publication in J.Acost.Soc.Am.

A METHOD FOR IMPROVING ACOUSTIC SWITCHING RATIO IN OPTO-ACOUSTIC TRANSDUCERS

B. Noorbehesht and V. Arat

Electrical Engineering Department
University of Houston
Houston, Texas 77004

ABSTRACT

Previous theoretical and experimental work on Opto-Acoustic Transducers (OATs) has shown their switching ratios to be lower than that necessary for most imaging applications. We will show that this problem can be overcome by applying intensity-modulated light to the OAT instead of constant intensity light, as suggested previously.

The conductivity of the photoconductive layer of the OAT follows the same time variations as the intensity of the light applied to it. Under these conditions the OAT may be considered as a linear system with a time-varying component. It will be shown, using an equivalent circuit, that sinusoidally intensity-modulated light applied to the OAT gives rise to sideband signals in the illuminated region which do not exist in the dark region. Therefore, in principle, infinite switching ratios may be realized by allowing one of these sidebands to occur at the resonant frequency of the OAT.

There is an additional advantage to using intensity-modulated light. When a Fresnel zone pattern is projected onto the OAT the acoustic signals from the illuminated and dark regions, being at different frequencies, focus at different depths thus improving the signal to noise ratio at the desired focal point.

INTRODUCTION

An opto-acoustic transducer (OAT) is a device which transforms an optical pattern into a corresponding acoustic pattern

[1]. When light carrying a focus-inducing pattern (e.g., a zone-plate pattern) is incident on the OAT, part of the generated sound beam converges to a point. Scanning of the focused beam can be accomplished by setting the projected light pattern into motion.

Figure 1 shows the physical structure of a positive opto-acoustic transducer (P-OAT). For a detailed description of the P-OAT structure see Ref. 1.

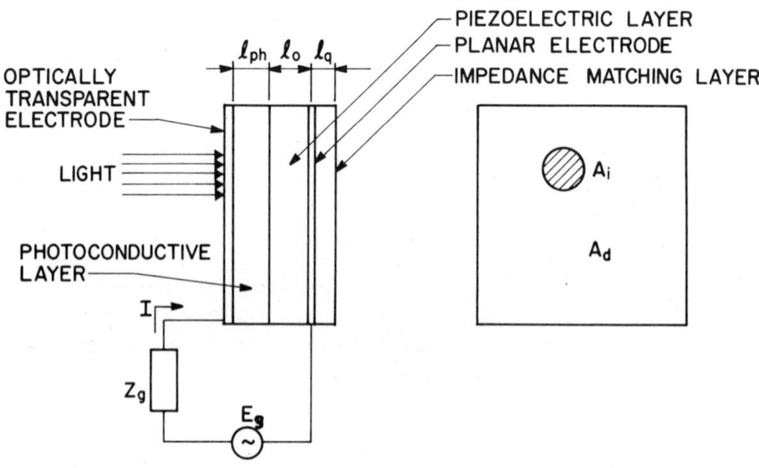

Fig. 1 Physical structure of the OAT

The amplitude of the acoustic signal generated at any point on the device is proportional to the voltage across the piezo-electric layer at that point. Thus the acoustic amplitude generated in the illuminated region is stronger than that in the dark region. The usefulness of such a device depends on its ability to transform a spatial optical intensity distribution into a corresponding spatial acoustical pressure distribution. This, in turn, depends on such characteristics as the switching ratio, $K_{sw}(f)$, defined as the ratio of the acoustic pressure at the illuminated region, $P_i(f)$, to that at the dark region, $P_d(f)$:

$$K_{sw}(f) = P_i(f)/P_d(f) \tag{1}$$

In the above equation f denotes the acoustic frequency. $K_{sw}(f)$ is a complex quantity and it contains important phase information.

An equivalent circuit for a P-OAT has been used to study the frequency-dependent nature of the switching ratio [2,3]. The predictions of this model and subsequent experimental work, however, have shown the switching ratio to be lower than that necessary for most imaging applications. One way to increase the switching ratio is to apply intensity-modulated light to the photoconductor instead of a constant intensity optical input [4]. This results in a proportional variation in the conductivity of the photoconductor in the illuminated region. Under these conditions the OAT model is still a linear one but it now contains a time-varying conductance.

We have developed a technique for studying the OAT under these conditions. It will be shown that when the device is driven by an electrical signal at a frequency f_g, and the optical input is intensity modulated at a frequency f_m, sidebands are generated in the illuminated region whose frequencies are given by $f_g \pm nf_m$ with n a positive integer. In the dark region, however, only one frequency component at f_g exists. By choosing any one of the sideband frequencies to coincide with the resonant frequency, f_{res}, of the device, in principle, an infinite switching ratio may be obtained. This is because the sideband frequencies only exist in the illuminated region. The unwanted sidebands can be reduced by utilizing the filtering action of the device. Choosing f_g to be far away from the resonant frequency, the dark signal can be made to be very small. Then f_m can be adjusted so as to put the strongest sideband (usually one of the first sidebands at $f_g \pm f_m$) at f_{res}. This would result in a large acoustic output at that frequency.

If the transducer's filtering action does not suppress the dark signal adequately, one may use a band pass electrical filter at the receiving end to extract the desired sideband. This filter must have its center frequency equal to f_{res}. This technique has been successfully applied to receiver OATs at low modulating frequencies [5].

There is an additional advantage to using intensity-modulated light. When a Fresnel zone plate pattern is projected onto the OAT to focus the acoustic output, signals from the illuminated and dark regions, being at different frequencies, focus at different distances thus improving the signal to noise ratio at the desired focal point.

THE METHOD

The OAT of Fig. 1 may be represented by the two-port model shown in Fig. 2 [2].

To simplify the analysis we will assume that only the conductivity of the photoconductor varies with light intensity, i.e., we will neglect any variations of the permittivity. This is a good approximation for many photoconductive materials.

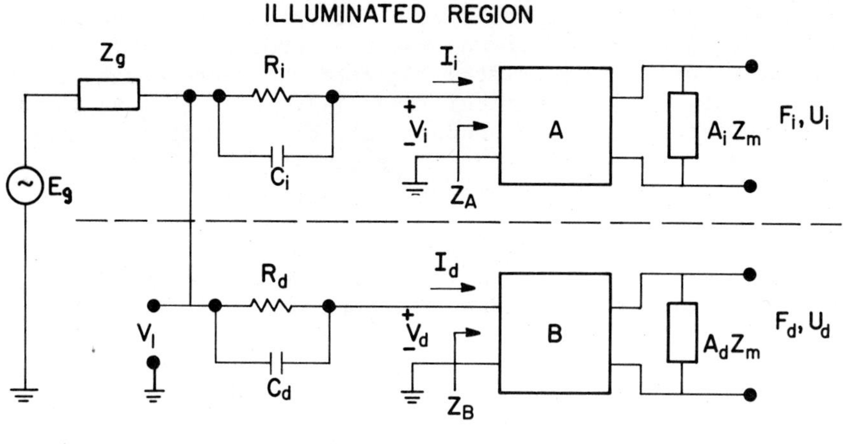

Fig. 2 Two-port model of the OAT

The circuit in Fig. 2 can be redrawn in a more convenient form as shown in Fig. 3(a). In that figure E_g is the independent voltage source connected across the input port (port 1), and $G_i(f)$ is the Fourier transform of the time-varying conductance $g_i(t)$ connected across the output port (port 2). We now define $H(f)$ to be the short-circuit driving admittance at port 2, and $I_g(f)$ to be the short-circuit current at that part. Figure 3(a) can then be represented by its Norton equivalent shown in Fig. 3(b). Our aim is to find $V_r(f)$ from which $V_i(f)$ and $V_d(f)$ can be calculated. Here we will only present the key equations in the derivation of these voltages, the details of the analysis will be published later [6].

Assuming sinusoidally intensity modulated light to be incident on the photoconductor, we can write

$$g_i(t) = [\frac{A_i}{\ell_{ph}}][\frac{\sigma_{max}+\sigma_{min}}{2} + \frac{\sigma_{max}-\sigma_{min}}{2} \text{Cos } (2\pi f_m t)] \tag{2}$$

and A_i is the cross sectional area of the illuminated region of the OAT, ℓ_{ph} is the thickness of the photoconductive layer, σ_{max} and σ_{min} are the maximum and minimum conductivities of the photoconductive material, respectively. Modulation index of the conductivity is defined by

$$m = \frac{\sigma_{max}-\sigma_{min}}{\sigma_{max}+\sigma_{min}} \leq 1 \tag{3}$$

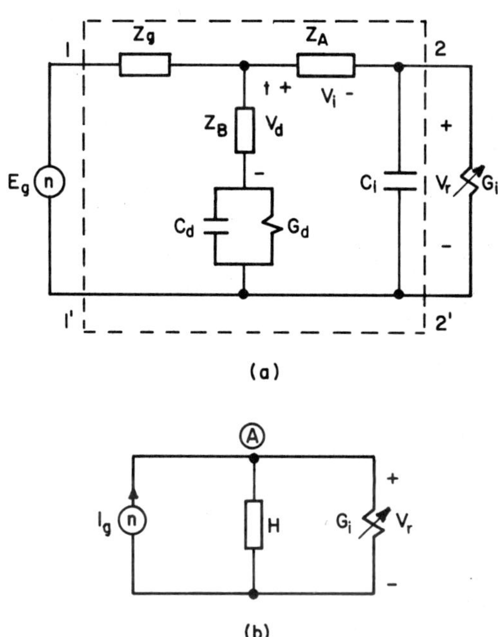

Fig. 3 (a) Simplified two-port model of the OAT
(b) Norton equivalent of 3(a)

Kirchhoff's current law at node A in Fig. 3(b) gives

$$I_g(f) = V_r(f)H(f) + \int_{-\infty}^{+\infty} V_r(f')G_i(f-f')df' \tag{4}$$

The above equation can be solved to give

$$V_r(f) = \sum_{n=-\infty}^{+\infty} X_n \delta(f-f_g-nf_m) \tag{5}$$

or equivalently in the time-domain

$$V_r(t) = \sum_{n=-\infty}^{+\infty} X_n \exp[j2\pi(f_g+nf_m)t] \tag{6}$$

Thus, it can be seen that when the OAT is excited by intensity-modulated light, the voltage across the photoconductor consists of frequency components at the applied electrical frequency, f_g, and an infinite number of sideband frequencies. The magnitude and phase of these components are given by the complex coefficients X_n, which can be found from a recurrence relation. It can be shown that in practice, under certain conditions [7], only a finite number of sidebands have appreciable magnitudes and the others may be neglected.

DISCUSSION

To determine the effectiveness of the modulation technique as a means of improving the switching ratio one must consider several factors.

First is the effect of truncation of the infinite series in Eq. (6) on the accuracy of the calculated output pressure values. To study this effect, output pressure magnitudes at three sideband frequencies were calculated as a function of the number of sidebands assumed to be nonzero. Figure 4 shows the results, where we have plotted the pressure magnitudes at the three sidebands versus n (number of sidebands = 2n). It can be seen that after n=7 the graphs seem to reach a steady state, i.e., including more sidebands

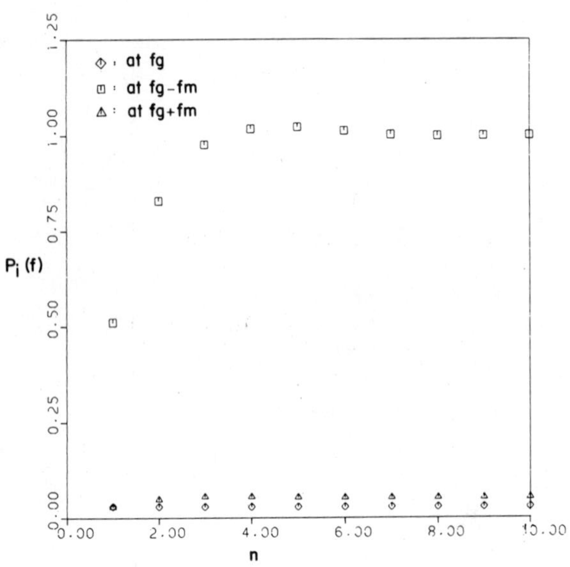

Fig. 4 Magnitude of the acoustic pressures at three different sidebands versus the number of sidebands assumed nonzero on each side of the operating frequency. Assumptions: E_g = 20V; ℓ_{ph} = 10 μm; f_g = 6.112 MHz; f_m = 3.592 MHz.

into the calculations does not appreciably increase the accuracy of the results.

Another factor is the choice of the applied frequency, f_g, and the optical modulation frequency, f_m. These frequencies should be chosen so as to maximize the desired sideband at f_{res} with respect to all other sidebands. To illustrate the effect of f_g and f_m on the switching ratio, we have obtained plots of the normalized pressure magnitudes as a function of the applied electrical frequency for two values of the photoconductive layer thickness, ℓ_{ph}. The results are shown in Figs. 5 and 6. The emergence of additional peaks is due to the presence of the photoconductive layer. These results can be used to determine the value of f_g which minimizes the dark signal. Next f_m can be chosen such that the signal at f_g-f_m or f_g+f_m coincides with the main peak for maximum illuminated signal generation.

To calculate these frequencies a computer program was written to compute and compare sideband magnitudes iteratively. For this analysis we assumed the piezoelectric layer of the OAT to be a PZT-5A ceramic, vibrating in the thickness expander mode, resonant at f_0 = 3.0 MHz. The photoconductive layer was assumed to be cadmium sulphide with its dark conductivity, σ_d, equal to $5\times10^{-6}(\Omega\cdot m)^{-1}$ and its average illuminated conductivity, σ_i, equal to $0.2(\Omega-m)^{-1}$. Under these conditions maximum value of the desired sideband at f_{res} was obtained when f_g was chosen to be 6.11 MHz and f_m 3.59 MHz. Figure 7 shows the normalized magnitude spectra of the acoustic pressure generated in the illuminated and the dark regions for this pair of f_g and f_m. It can be seen that at the resonant frequency of the OAT ($f_{res} \simeq 2.52$ MHz) there is a large sideband generated in the illuminated region while there is no signal in the dark region at that frequency. As an example, Fig. 8 shows a case where f_g and f_m are not at their optimal values.

The modulation index of the conductivity of the photoconductive layer, m, is also a major factor to be considered. Until now we have assumed the modulation index to be at its maximum (i.e., the value corresponding to a 100% intensity modulated light input). To study the effect of lower modulation indices, we have plotted pressure magnitudes at the resonant frequency as a function of the modulation index. The result is given in Fig. 9. It can be seen that the output pressure falls off rapidly in magnitude as the modulation index is reduced below one. This result indicates that a strong light source is needed to completely switch the photoconductor in order to generate a strong illuminated signal at the resonant frequency. Equivalently, one could use a highly sensitive photoconductor with very low dark conductivity.

The above results indicate that under certain conditions modulation of the optical input to the OAT would result in a very

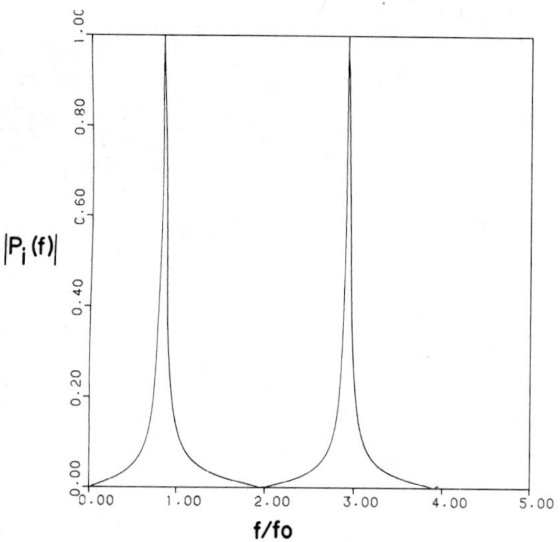

Fig. 5 Magnitude of the acoustic pressure as a function of normalized applied electrical frequency for ℓ_{ph} = 10 μm.

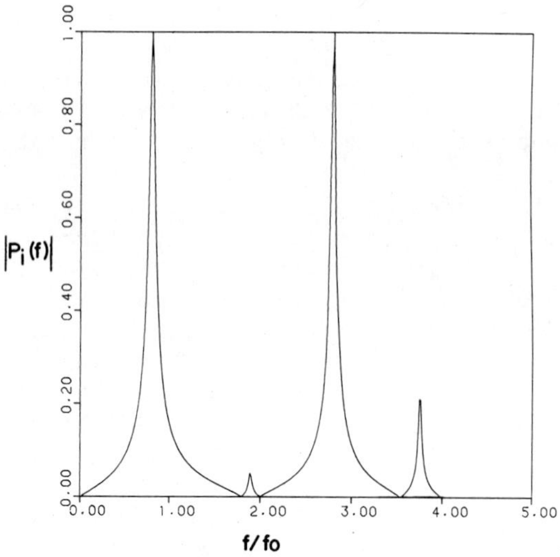

Fig. 6 Magnitude of the acoustic pressure as a function of normalized applied electrical frequency for ℓ_{ph} = 100 μm.

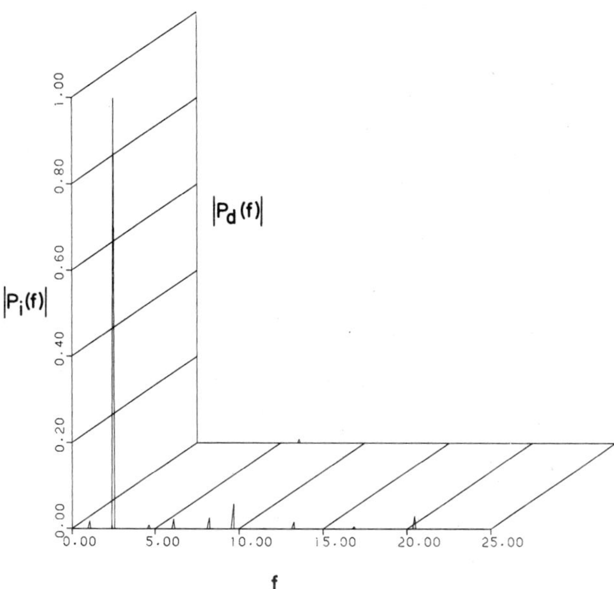

Fig. 7 Magnitude spectra of the acoustic pressures in the illuminated (foreground) and dark (background) regions with f_g = 6.112 MHz; f_m = 3.592 MHz.

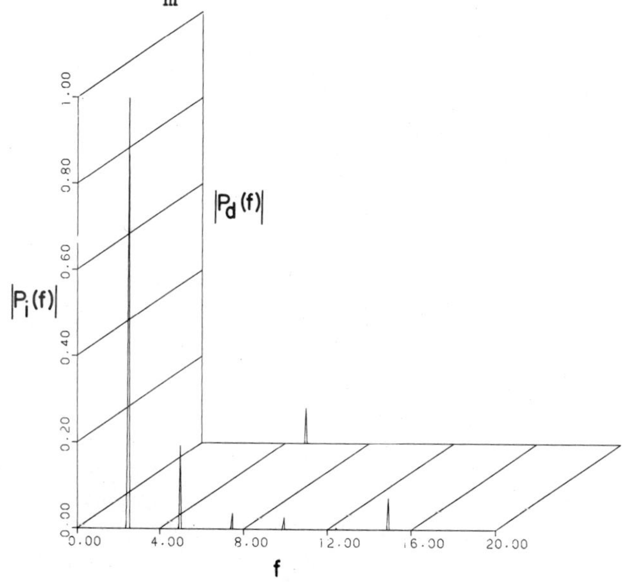

Fig. 8 Magnitude spectra of the acoustic pressures in the illuminated and dark regions, with f_g = 5.5 MHz; f_m = 2.48 MHz; ℓ_{ph} = 10 μm.

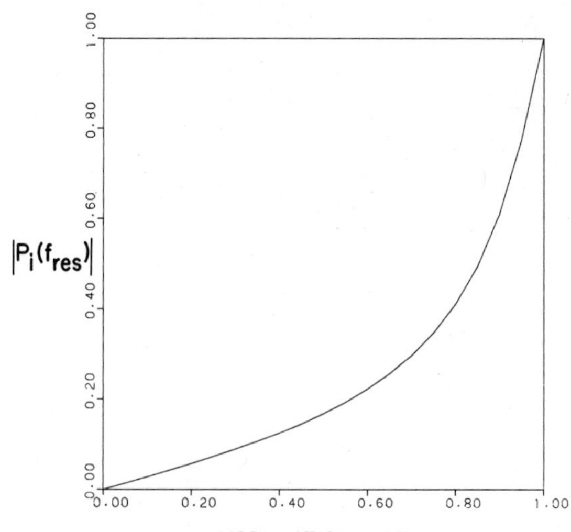

Fig. 9 Normalized magnitude of the acoustic pressure at f_{res} versus modulation index of the conductivity of the photoconductive layer, with ℓ_{ph} = 10 μm; f_g = 6.112 MHz; f_m = 3.59 MHz.

high switching ratio. We are now in the process of performing experiments to verify the above theoretical predictions. We are also developing experimental techniques to determine material properties and other pertinent device characteristics which an OAT must have to function efficiently when subjected to an intensity modulated optical input.

ACKNOWLEDGMENT

The authors wish to thank the National Science Foundation's Automation, Bioengineering, and Sensing Systems Section for its generous support of this work under Grant No. ECS-7913259.

REFERENCES

1. K. Wang and G. Wade, "A Scanning Focused Beam System for Real-Time Diagnostic Imaging", Acoustical Holography, Vol. 6, N. Booth Ed., Plenum Press, New York, pp. 213-228, 1975.
2. S. Elliott, V. Domarkas and G. Wade, "Frequency Characteristics of Opto-Acoustic Transducers", IEEE Trans. Sonics and Ultrasonics, SU-25, pp. 346-353, Nov. 1978.
3. B. Noorbehesht and G. Wade, "An Improved Equivalent Circuit for Opto-Acoustic Transducers", 1979 Ultrasonics Symposium Proceedings, pp. 195-199.

4. B. Noorbehesht, "A Modified Equivalent Circuit for Opto-Acoustic Transducers", IEEE Trans. on Sonics and Ultrasonics, 1982. (In Press)
5. C. W. Turner, and S. O. Ishrak, "Comparison of Different Piezoelectric Transducer Materials for Optically Scanned Acoustic Imaging", Acoustical Imaging, Vol. 10, P. Alais and A. F. Metherell Eds., Plenum Press, New York, pp. 761-778, 1980.
6. B. Noorbehesht, and V. Arat, "Opto-Acoustic Transducers with Intensity Modulated Optical Input," IEEE Trans. on Sonics and Ultrasonics, 1982. (Under Review)
7. V. Arat, M. S. Thesis, University of Houston, Houston, Texas, 1982.

THE DESIGN OF ELECTRIC EXCITATIONS FOR THE FORMATION OF DESIRED TEMPORAL RESPONSES OF HIGHLY EFFICIENT TRANSDUCERS

R. Y. Liu

ADR Ultrasound
Tempe, Arizona 85282-9990
U.S.A.

Abstract

A new approach in designing highly efficient transducers with the desired temporal responses is proposed. In this method, the efficiency of the transducer is achieved acoustically by air backing and quarter-wave matching, and the shape of the transducer response is controlled electrically by electrical driving signals which are computed by deconvolving the desired transducer responses by the transducer impulse response. The obtained digitized driving waveforms are stored in memory and serve as the basis for generating the desired driving patterns through an ultrafast digital-to-analog converter and a pulse driver. A broadband tuning network, which has been taken into account in determining the transducer impulse response, is added between the transducer and the driver. Design examples based on theoretical investigations are presented in this paper. Effects of transducer responses to the formation of spatial beam profiles are also addressed.

1. INTRODUCTION

It has long been recognized (1-2) in the field of ultrasound imaging that range resolution as well as spatial resolution and accompanying sidelobe levels are affected by the transducer response (the instantaneous normal particle velocity at the transducer front surface). This is because, under certain approximations (3), the instantaneous pressure of a transducer at any spatial point in a homogeneous but attenuating dispersive medium can be computed by convolving the derivative of the transducer response with the aperture impulse response (a function of transducer geometry) and the medium impulse response (a function of medium attenuation). Therefore, apodization for a single element transducer, which usually cannot be performed in the spatial domain, can be obtained by shaping the transducer response either acoustically or electrically in the temporal domain.

Standard approaches for controlling transducer responses are divided into two categories. The acoustic approach (4-8), methods of attaching proper matching or/and backing layers to piezoelectric materials, is the most widely adopted method. Another useful technique is the electric approach (9-11), where either electric matching networks are used or special driving waveforms are applied. Recent efforts (12) have been directed at shaping transducer responses by applying both electric and acoustic methods simultaneously. Good improvements have been achieved, but so far only part of the parameters are controllable.

This paper proposes a new approach in designing highly efficient transducers with desired transducer responses for ultrasound imaging applications. Acoustic spatial beam profiles obtained from different shapes of transducer responses for a disk transducer are compared theoretically. A two-and-a-half cycle Gaussian modulated cosine response is selected because of its characteristics in affecting axial resolution, spatial resolution and sidelobe levels. The waveform of the driving electrical signal is computed by deconvolving this selected transducer response by the transducer impulse response which is computed by considering both the transducer and the broadband tuning network. The obtained driving waveform is stored in memory and serves as the basis for generating the desired driving pattern through an ultrafast digital-to-analog converter and a pulse driver. In this way, the efficiency of the transducer can be achieved acoustically by air backing and quarter-wave matching, and the shape of the transducer response can be controlled electrically by driving waveforms.

2. TEMPORAL APODIZATION

The effects of the shape and the duration of the transducer response on the transducer spatial resolution and the sidelobe level are investigated as follows. A 10 mm spherical disk transducer, whose radius of curvature is 60 mm, is chosen for study. Temporal apodization effects can be observed first by applying two types of transducer response, a gated cosine and a Gaussian modulated gated cosine, to the transducer. Figure 1-(b) shows a 2.5 cycle Gaussian modulated gated cosine transducer response which is formed by applying a temporal apodization function (Gaussian) to a gated cosine response (Figure 1-(a)). Spectrums of these two types of transducer response are shown in Figure 2. A 3.5 MHz center frequency is used for all simulated transducer responses throughout the paper.

Transmitting transducer beam profiles at a distance 30 mm away from the transducer are shown in Figure 3, where (a) is the profile obtained by having a gated cosine transducer response and (b) is the profile formed by applying a Gaussian modulated gated cosine response. Clearly, the improvement of spatial resolution and the reduction of sidelobe level have been achieved simultaneously in this example by applying Gaussian temporal apodization to a gated cosine response. This is not the case in applying spatial apodizations, where a compromise between the spatial resolution and the sidelobe level must be made. The beam profile simulation program has taken a 0.7dB/CM/MHz attenuating dispersive property of the propagating medium into account.

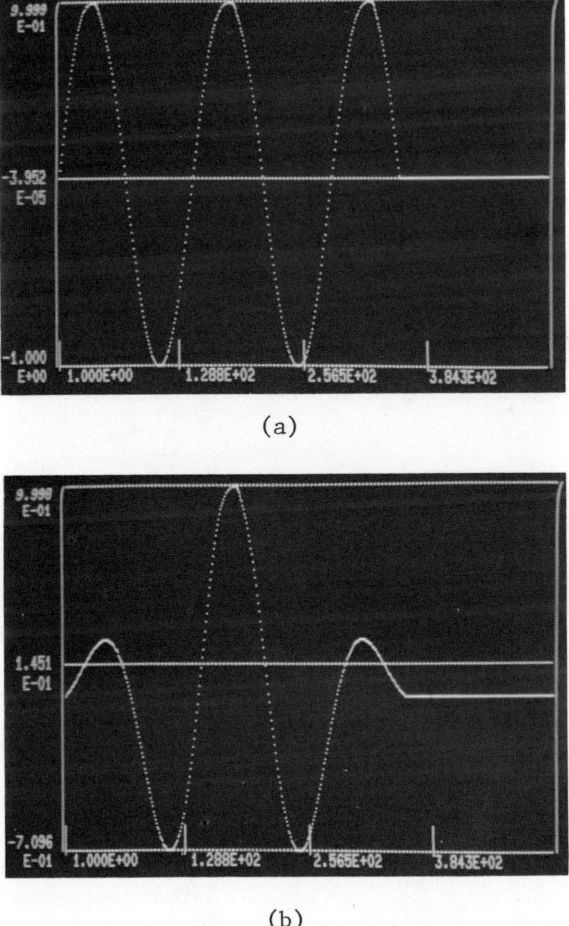

(a)

(b)

Figure 1 - (a) A 2.5 cycle cosine transducer response.
(b) A 2.5 cycle Gaussian modulated cosine transducer response.

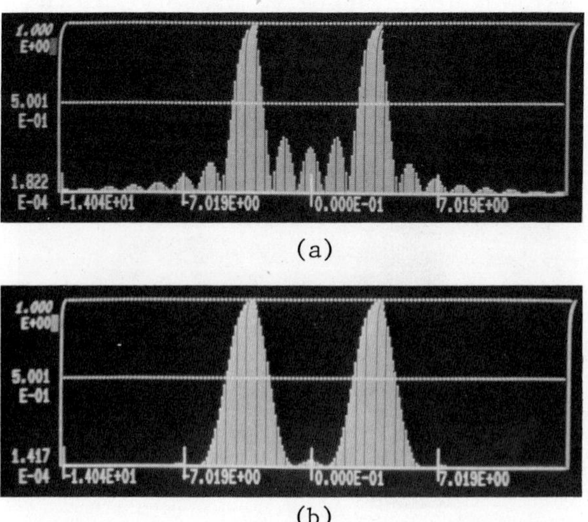

Figure 2 - (a) Spectrum (amplitude) of a gated cosine transducer response.
(b) Spectrum (amplitude) of a Gaussian modulated gated cosine transducer response. Both have a 3.5 MHz center frequency and a 2.5 cycle duration. The horizontal scale is in MHz.

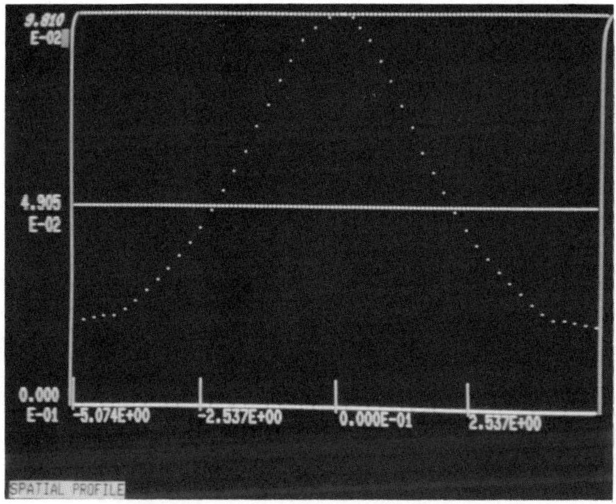

(a)

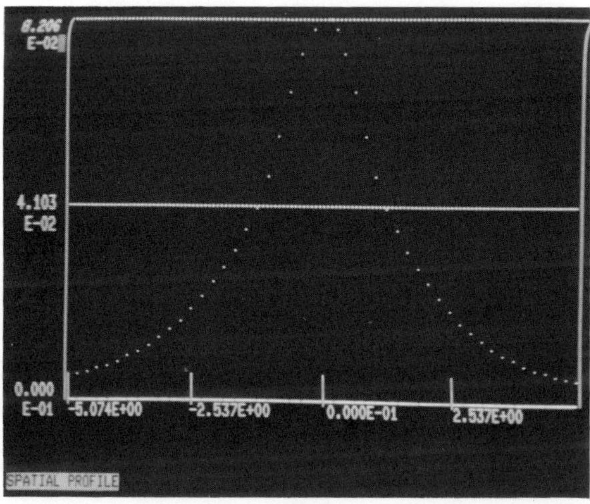

(b)

Figure 3 - Transmitting spatial beam profiles at a distance 30 mm away from a 10 mm spherical disk transducer with a 60 mm radius of curvature. (a) is the profile obtained by assuming a transducer response shown in Fig. 1-(a), and (b) is the profile obtained by having a transducer response shown in Fig. 1-(b). The center frequencies of both responses are 3.5 MHz, the horizontal scale is in millimeters.

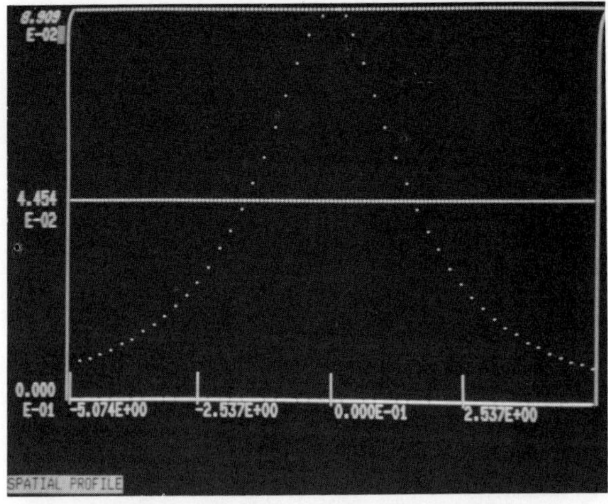

(a)

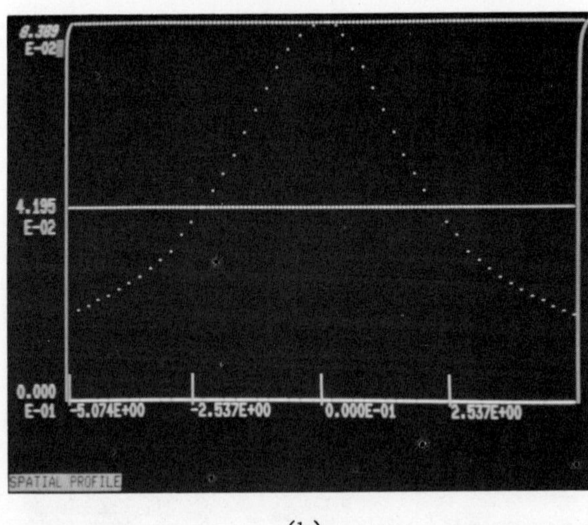

(b)

Figure 4 - Transmitting spatial beam profiles (Z = 30 mm) of a 10 mm spherical disk transducer with a 60 mm radius of curvature. (a) shows a profile obtained by applying a 1.5 cycle Gaussian modulated cosine response, and (b) shows a profile obtained by having a 0.5 cycle Gaussian modulated cosine response. The center frequencies of both responses are 3.5 MHz, and the horizontal scale is in millimeters.

The duration of the transducer response also plays an important role in the formation of spatial beam profiles. Figure 4 shows two transmitting spatial beam profiles formed with a 1.5 cycle and a 0.5 cycle Gaussian modulated cosine transducer response respectively. Trying to improve the axial resolution by reducing the duration of the transducer response has resulted in the deterioration of the spatial resolution. Therefore, a 2.5 cycle Gaussian modulated cosine response is selected as the desired transducer response for the following studies due to its good characteristics in affecting both axial and spatial resolutions.

3. EXCITATION WAVEFORMS

The impulse response of the 10 mm PZT 5A disk transducer is calculated by using the KLM equivalent circuit for the piezoelectric crystal and the transmission line theory for air backing, water loading and quarter-wave matching (Z = 4.5). The calculated complex electric input impedance of this transducer is shown in Figure 5-(a). With the consideration of a 50 Ohm source impedance and adding a series 2.2 μH tuning inductor, the two-way insertion loss and the impulse response of the transducer are shown in Figure 5-(b). For practical applications, the design of broadband (four element) tuning networks for transducers may be necessary in some cases. The excitation waveform is computed by deconvolving a selected 3.5 cycle Gaussian modulated cosine response (results for the desired 2.5 cycle will be given later) by the transducer impulse response. The spectrum of the calculated result is shown in Figure 6-(a). By windowing the spectrum (Figure 6-(b)) and taking the inverse FFT, the driving waveform (Figure 7-(a)) is obtained. Figure 7-(b) shows the transducer response produced by driving the transducer with a waveform shown in Figure 7-(a). The computed results for the desired 2.5 cycle Gaussian modulated cosine transducer response is shown in Figure 8.

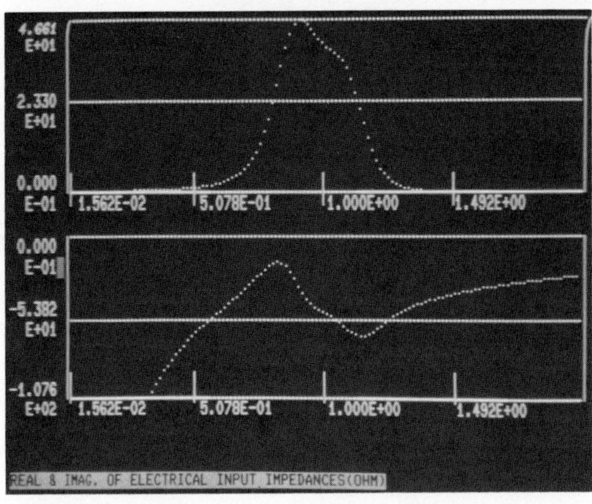

(a)

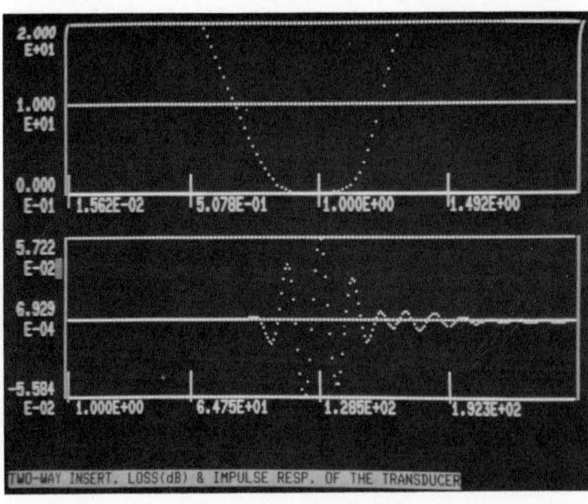

(b)

Figure 5 - (a) The complex electric input impedance (vertical scale in Ohm) of a 10 mm, 3.5 MHz, PZT 5A disk transducer. This transducer has an air backing, water loading and a quarter-wave matching layer (Z = 4.5).

(b) The two way insertion loss (vertical scale in dB) and the impulse response of the transducer with the considerations of a 50 Ohm source impedance and adding a series 2.2 μH inductor.

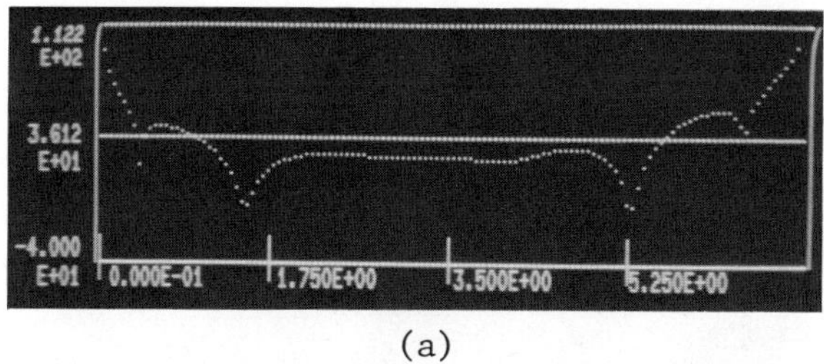

(a)

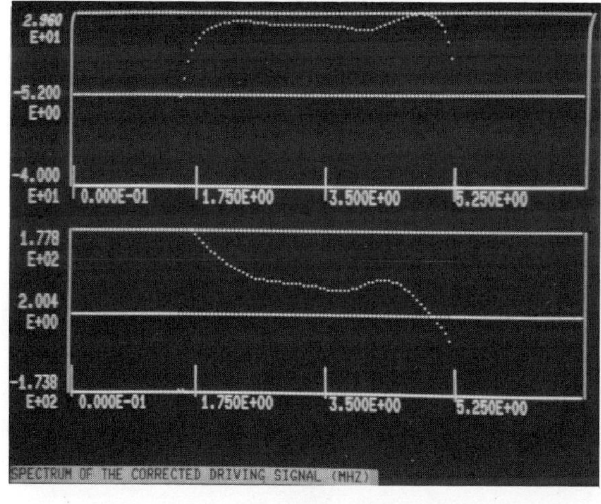

(b)

Figure 6 - (a) The computed amplitude spectrum (vertical scale in dB) after deconvolving a 3.5 cycle Gaussian modulated cosine response by the transducer impulse response shown in Fig. 5-(b).
(b) The windowing spectrum of (a). A rectangular window covering the spectrum between 1.58 and 5.25 MHz is applied.

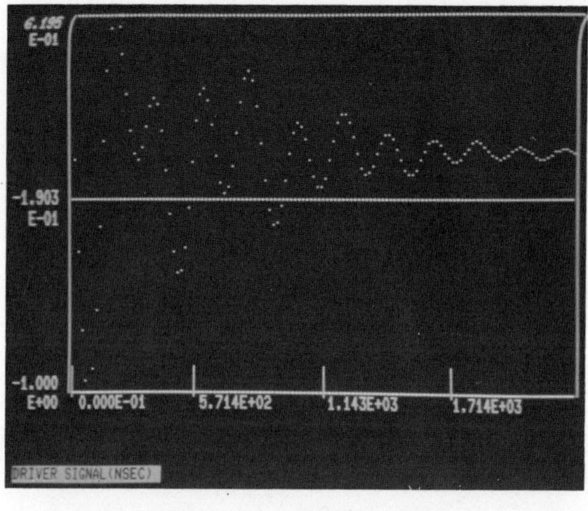

(a)

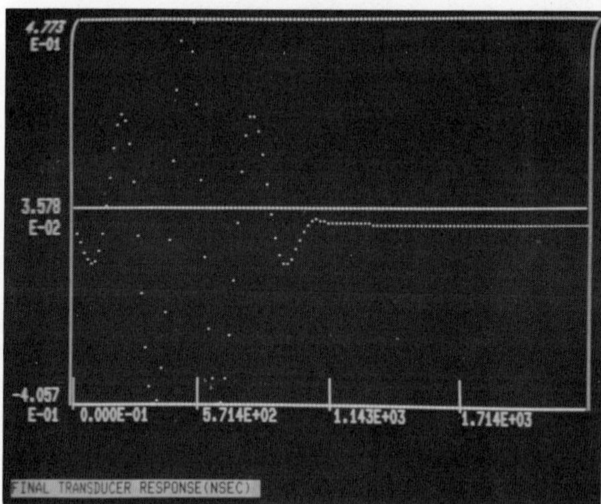

(b)

Figure 7 - (a) The excitation waveform (horizontal scale in nanosec) obtained by taking the inverse FFT of the spectrum shown in Fig. 6-(b).
(b) Transducer response produced by driving the transducer with an excitation waveform shown in (a).

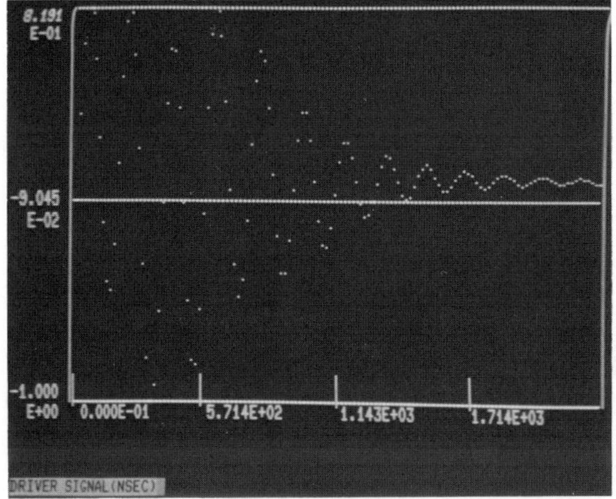

(a)

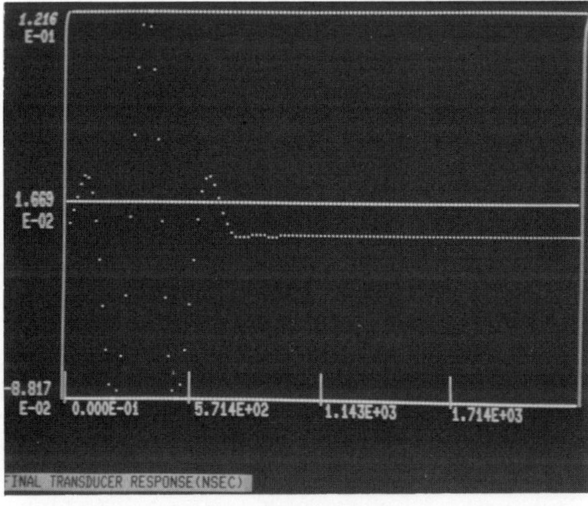

(b)

Figure 8 - (a) The excitation waveform (horizontal scale in Nsec) for the formation of a 2.5 cycle Gaussian modulated cosine transducer response.
(b) Transducer response produced by driving the transducer with an excitation waveform shown in (a).

The upper 3-dB frequency is around 5 MHz for the driving signal shown in Figure 7-(a), and around 6 MHz for the driving signal shown in Figure 8-(a). Therefore, an operating bandwidth of up to 20 MHz or higher (four times rule) might be needed for practical applications. The block diagram of a wideband arbitrary waveform generator/driver is shown in Figure 9. With state-of-the-art high speed electronic technology, the design of such a driver is feasible.

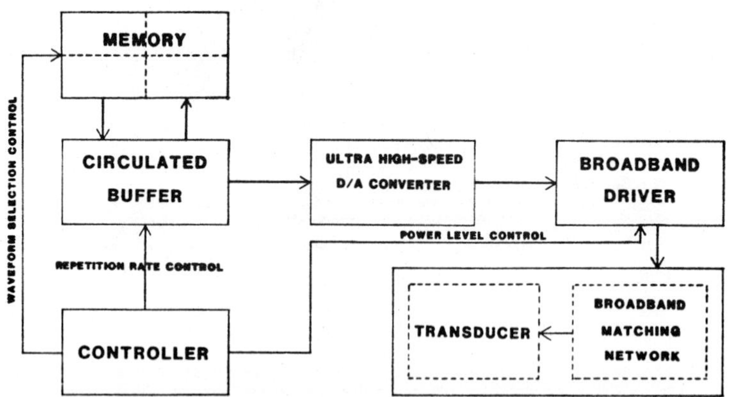

Figure 9 - The block diagram of a wideband arbitrary waveform generator/driver.

4. DISCUSSION

The proposed method have combined acoustic matching, electric tuning and pulse shaping all together to try to obtain a highly efficient transducer with desired transducer response. For applications where acoustic loads are unchanged, only one driving waveform is necessary for each transducer. This is the case in medical ultrasound imagings. The drawback of this approach, however, is the variation of transducer impulse responses due to material and manufacturing processes. The empirical determination of the transducer impulse response for each transducer might be necessary for calculating the driving waveform.

References

1. W. L. Beaver, "Sonic Nearfields of A Pulsed Piston Radiator," J. Acoust. Soc. Am., Vol. 56, No. 4, 1974, pp. 1043-1048.
2. B. P. Hildebrand, "An Analysis of Pulsed Ultrasonic Arrays," Acoustical Imaging, Vol. 8, A. F. Metherell, Editor, Plenum Press, New York, 1980, pp. 165-185.

3. M. Fink, "Theoretical Study of Pulsed Echographic Focusing Procedures," Acoustical Imaging, Vol. 10, P. Alais and A. F. Metherell, Editor, Plenum Press, New York, 1982, pp. 437-453.
4. G. Kossoff, "The Effects of Backing and Matching on the Performance of Piezoelectric Ceramic Transducers," IEEE Trans. Sonics Ultrasonics, SU-13, No. 1, 1966, pp. 20-30.
5. E. K. Sittig, "Transmission Parameters of Thickness-Driven Piezoelectric Transducers Arranged in Multilayer Configurations," IEEE Trans. Sonics Ultrasonics, SU-14, No. 4, 1967, pp. 167-174.
6. E. K. Sittig, "Effects of Bonding and Electrode Layers on the Transmission Parameters of Piezoelectric Transducers Used in Ultrasonic Digital Delay Lines, " IEEE Trans. Sonics Ultrasonics, SU-16, No. 1, 1969, pp 167-174.
7. J. H. Goll and B. A. Auld, "Multilayer Impedance Matching Schemes for Broadbanding of Water Loaded Piezoelectric Transducers and High Q Electric Resonators,", IEEE Trans. Sonics Ultrasonics, SU-22, No. 1, 1975, pp. 52-53.
8. C. S. Desilets, J. D. Fraser and G. S. Kino, "The Design of Efficient Broad-Band Piezoelectric Transducers," IEEE Trans. Sonics Ultrasonics, SU-25, No. 3, 1978, pp. 115-125.
9. L. J. Augustine and J. Anderson, "An Algorithm for the Design of Transformerless Broadband Equalizers of Ultrasonic Transducers," J. Acoust. Soc. Am., Vol. 66, No. 3, 1979, pp. 629-635.
10. C. H. Chou, J. E. Bowers, A. R. Selfridge, B. T. Khuri-Yakub, and G. S. Kino, "The Design of Broadband and Efficient Acoustic Wave Transducers," IEEE Ultrasonic Symposium Proceedings, 1980, pp. 984-988.
11. H. W. Persson, "Electric Excitation of Ultrasound Transducers for Short Pulse Generation," Ultrasound in Med. & Biol., Vol. 7, 1981, pp. 285-291.
12. A. R. Selfridge, R. Baer, B. T. Khuri-Yakub, and G. S. Kino, "Computer-Optimized Design of Quarter-Wave Acoustic Matching and Electrical Matching Networks for Acoustic Transducers," IEEE Ultrasonic Symposium Proceedings, 1981, pp. 644-648.

A LINEAR MONOLITHIC RECEIVING ARRAY
OF PVDF TRANSDUCERS FOR TRANSMISSION CAMERAS

Bernd Granz

Forschungslaboratorien der Siemens AG
D 8520 Erlangen, Federal Republic of Germany

ABSTRACT

Linear receiving arrays with ten transducer elements made of the piezoelectric polymer PVDF were constructed and the performance measured. The single transducers in the monolithic array are defined by standard photolithography and selective poling the remaining metallized 0.7x0.8 mm² spots. The 25µm PVDF foil is glued on to a hard backing to get increased sensitivity. The result we found was a sensitivity of $2 \cdot 10^{-6}$ Vm²/N and a broad band behaviour from 2 to 12 MHz. The interelement coupling was found to be about -35 dB. The angular acceptance was nearly ideal except narrow minima depending on the backing material.

INTRODUCTION

In pulse echo systems like B-scan imagers ultrasonic transducer arrays made of piezoelectric ceramics, like PZT, are widely used. In order to get broad band behaviour necessary for good axial resolution sophisticated constructions have to be used. Especially good matching to the backing and elaborated width-to-thickness ratios have to be selected. In addition, because of the large impedance missmatch of ceramics to tissue of $30 \cdot 10^6$ to $1.5 \cdot 10^6$ kg/m²s, a quarter wave matching layer within narrow thickness and impedance tolerances has to be chosen. These steps are all taken because of the PZT's good properties to transmitting ultrasound.
On the other side, according to Callerame's formula /1/, materials with low dielectric constants are better suited for pure receiving. Such a material, e.g. PVDF with an ϵ_r of 8 (2 MHz)

can be used in an ultrasonic transmission camera in which transmitting and receiving is well separated.
We started to construct receiving arrays capable of reconstructing the diffraction limited resolution offered by the camera. In transmission cameras with f/1.5 lenses a resolution of 1.5 mm at 2 MHz is achievable. So we chose the size and the number of the transducers of our arrays to fit in a camera like the SRI camera /2/. This leads to an individual area of the transducers of 0.7x0.8 mm, spaced by 0.1 mm, which is sufficient to sample the offered resolution. Because of the low capacity of only 1.6 pF for a single transducer, additional stress has to be layed on the preamplifier electronics.

To find a less difficult way to construct arrays beside slotting ceramics, a monolithic construction with the piezoelectric polymer PVDF has been chosen. Monolithic means that the piezoelectric foil as a whole is glued to a backing. The foil is covered on one side with the pattern of the transducer, on the other side with the common ground. A slotted foil is conceivable as well but results in a double number of electrical connections. With the 10-transducer-arrays of different construction we measured the receiving properties. Properties to characterize a transducer are absolute sensitivity, broad band behaviour and angular acceptance. Transducers in an array are characterized by their mutual cross-coupling and their homogeneity in sensitivity.

CONSTRUCTION

We used commercial 25 μm and 12 μm biaxially drawn capacitor-grade PVDF film as transducer material. This film was vacuum coated by Cr - Ag with a total thickness of 100 nm. After wet-etching the patterns of the transducers, they were selectively poled with an electrical field of about 1 MV/cm for one hour at 100°C and then were brought back to room temperature under voltage applied. With a quasistatic measurement /3/ the piezoelectric pressure constants d_h, d_{31} , d_{32} were determined and thus the piezoelectric constant can be calculated by

$$g_{33} = \frac{1}{\epsilon_{33}} (d_h - d_{31} - d_{32})$$

assuming a relative dielectric constant ϵ_r of 8 (2 MHz). One type of transducers was metallized with Cr-Ag on both sides, then patterned with 0.7x0,8mm electrodes on one and a 0.8mm strip on the other side (type A, Fig. 1). The electrical gold wire contacts were bonded by conductive adhesive. After this the poling was done. Another type of transducers was constructed by selectively poling a 0.8 mm broad metallized strip - Cr-Ag on one side, Al

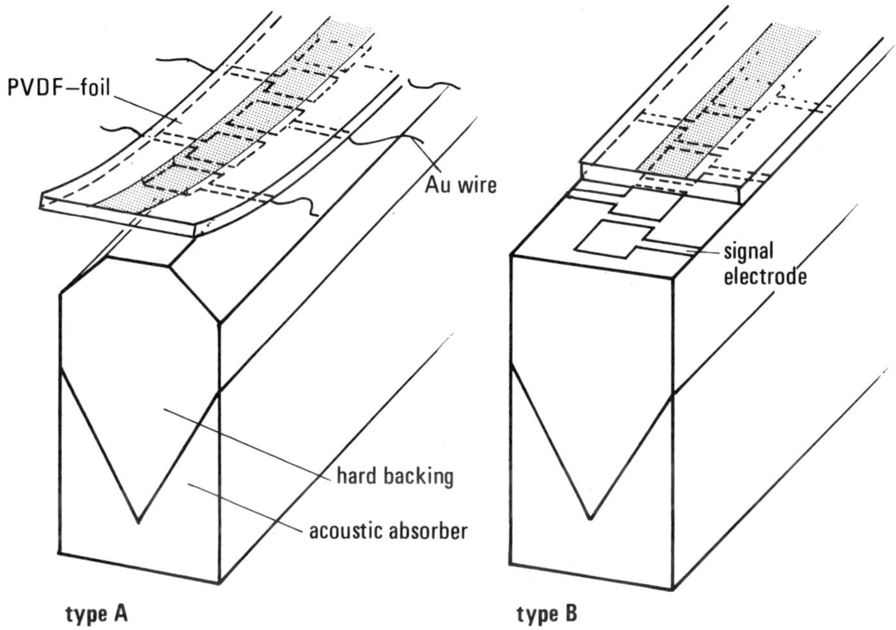

Figure 1. Construction of two different arrays with PVDF.

on the other side. After the poling procedure the Al strip was totally etched away and the foil was glued with this side on a backing containing on its surface the transducer patterns and the connection strips (type B, Fig. 1). The signal from the transducer was capacitively coupled to the pattern on the backing. This construction has the advantage of the connections being easily soldered to the backing. Because of the PVDF's low dielectric constant the glue with a thickness of 5 µm and relative dielectric constant $\epsilon_r = 4$ produces a signal reduction of less than 30 %.

Ultrasonic waves travelling in water and hitting a wall are reflected according to the impedance mismatch between water and the wall material. The reflected wave superposes the incident wave and this results in doubling the amplitudes near the wall for an ideal hard wall. PVDF with its impedance of $3.5 \cdot 10^6$ kg/m²s is acoustically well matched to water with $1.5 \cdot 10^6$ kg/m²s and thus well suited for measuring ultrasound in water. A PVDF transducer with a thickness very much smaller than the wavelength (25 µm to 1000 µm at 2 MHz) hardly disturbs the superposition explained above. On the other, wedge shaped, side of the backing an absorber of a tungsten/araldite mixture reduces reflection from there to -35 dB.

Compared to PVDF a backing of pure Al_2O_3 (95%) with an impedance of $35 \cdot 10^6$ kg/m²s reaches the ideal of an infinite hard backing. Al_2O_3 leads to an amplitude reflection coefficient of 0.92, another backing material used was quartz (Herasil ®) with an impedance of $12.5 \cdot 10^6$ kg/m²s leading to a coefficient of 0.79.

PERFORMANCE

Physical properties to characterize the performance of the transducers in the array are sensitivity, homogeneity, frequency response, cross coupling and angular acceptance. Because the transducers will be used in the camera at 2 MHz we measured most of the properties at this frequency.

SENSITIVITY

Sensitivity is measured by comparing the output voltage of the PVDF transducer with the signal of a calibrated hydrophone. For a transmitter we used a commercial 2 MHz transducer excited with a 10 µsec sine burst. The hydrophone was placed just behind the last axial maximum of the beam pattern. The hydrophone (Medisonics MK II) had been calibrated by the Physikalisch-Technische Bundesanstalt Braunschweig. The transmitter was then placed in another water tank in such a way that its last axial maximum was at the place of a 100 µm mylar window in the tank wall. The PVDF array touched the outside of the window by means of a thin film of aquasonics. Thus the array and the electronics could be used in air and the measurement could be made for a realistic setup. As preamplifier we used a 1:1 active probe (Tektronix P6201 with R_i = 100 kΩ and C_i = 3pF). For a pressure amplitude of $3 \cdot 10^4$ N/m² we got typical voltages of 70 mV which leads to a sensitivity of $2.2 \cdot 10^{-6}$ Vm²/N for a transducer of 25 µm thickness in this special arrangement. For a 12 µm transducer we found a sensitivity of $1 \cdot 10^{-6}$ Vm²/N showing the dependance on transducer thickness. These results include uncertainties of about $\pm$ 3 dB. This number is composed by $\pm$ 1.5 dB coming from the calibration of the hydrophone and the other part from the measurement technique.

The noise coming from the dielectric loss of a 0.56 mm² PVDF transducer of 25 µm thickness was calculated to 14 µV with bandlimits 1.5 to 2.5 MHz. Because the noise is reduced by a similar network as the signal e.g. the input capacitance of the preamplifier, we can compare it to the theoretical sensitivity defined by the measured piezoconstants and the condition of the hard backing. This leads to a minimum detectable intensity of about $1 \cdot 10^{-10}$ W/cm². In this calculation we assumed the transducer noise to be superior to the amplifier noise.

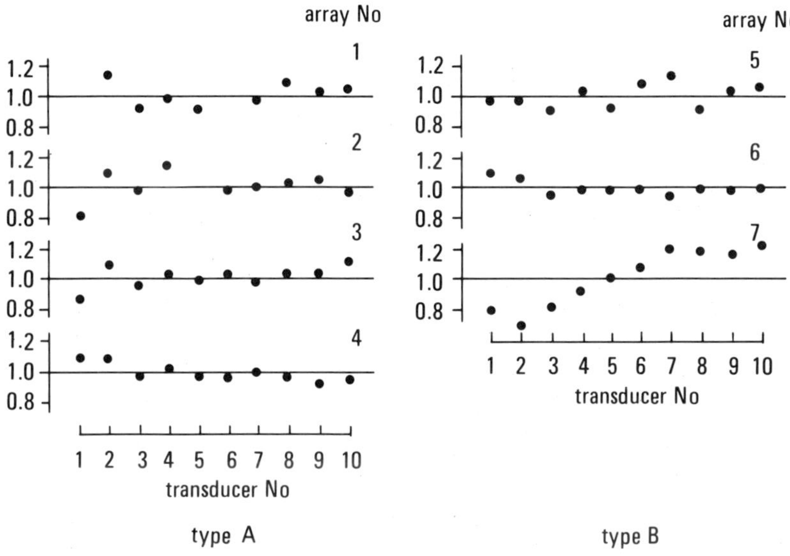

Figure 2. Relative receiving transducer efficiency for 7 arrays.

HOMOGENEITY OF SENSITIVITY

We measured the sensitivity of each transducer in an array having the neighbor receivers short-circuited. The values of the seven arrays normalized to the average value of every array are shown in Fig. 2. The variance of the sensitivity within one array is always less than 10%, except array No 7 which has a variance of 18.7%.

We think this inhomogenity comes from slightly different parasitic capacitances in the signal line, reducing the signal of the 1.6 pF electrical source with a different factor.

FREQUENCY RESPONSE

The frequency response is measured in the same way as the sensitivity by comparing the response of a PVDF transducer to the response of the calibrated hydrophone. To this end both were located one after the other in a water tank at the same place of the beam pattern of resonant transmitters with different center frequencies. The result of the relative frequency response of the PVDF receiver is shown in Fig. 3. There is a rise in sensitivity of about a factor 2 between 2 and 12 MHz, which is due partial-

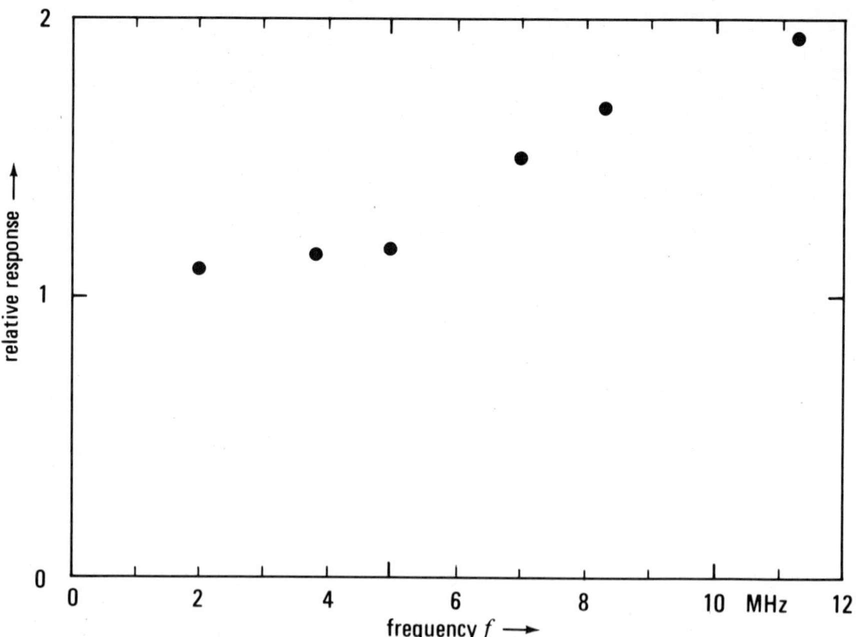

Figure 3. Relative ratio of output voltage to input stress vs frequency for a 25 µm PVDF transducer on quartz backing, Type B.

ly to the resonance predicted by the Mason model at 17 MHz for a 25 µm PVDF foil glued to a hard backing, and partially to our measuring network. On the other hand the relative dielectric constant of PVDF diminishes from 8 to 6 between the frequency borders of the measurement, and there is an unknown frequency dependance of the piezoelectric constant d_{33}.
The uncertainty is the same as for the sensitivity. The additional uncertainty caused by the transmitter's band widths of 25% does not affect the result, because of the extreme wide band width of the receiving transducer.

CROSS COUPLING

The cross coupling of the receiving transducers to their respective neighbors results in an effective broadening of the transducer size and thus reduces their angular acceptance and the spatial resolution to be reproduced. From the ratios of capacitances to the next neighbor and to the next but one neighbor of 0.06 pF and 0.04 pF resp. and to the capacitance of 1.6 pF of the transducer itself, an electrical cross coupling of -28 dB

and -32 dB is expected for an Al_2O_3 backing. We measured the actual cross-coupling by mechanically and electrically exciting the PVDF transducers in the monolithic array.
First we touched one receiver with a plexiglas rod of 0.8 mm diameter glued to a 2 MHz ultrasound transmitter, excited at its resonance. The rod was then moved away from the receiver to one neighbor at intervals of 50 µm. We measured the output voltage of the first receiver as well as the output voltage of its two neighbors. The relative voltage of these three transducers normalized to the sum voltage is shown in Fig. 4. Because of the low voltage values which are near the noise, Fig. 4 shows the dependence on the rod's position (only qualitatively).

Another way of measuring the cross coupling is by exciting one receiver with a known electrical signal e.g. a 2 MHz pulse of definied amplitude, and by comparing it with the voltage signal received by the neighbors. A third way is more realistic by exciting the receiving transducers with ultrasound. The 10 transducer array touches the outside of the mylar window of the water tank as explained above. The distance of the ultrasonic transmitter is chosen to result in a 3 dB pressure spot of 16 mm diameter at the window. The voltage output of one center transducer is measured and then the neighbors are successively short-circuited.

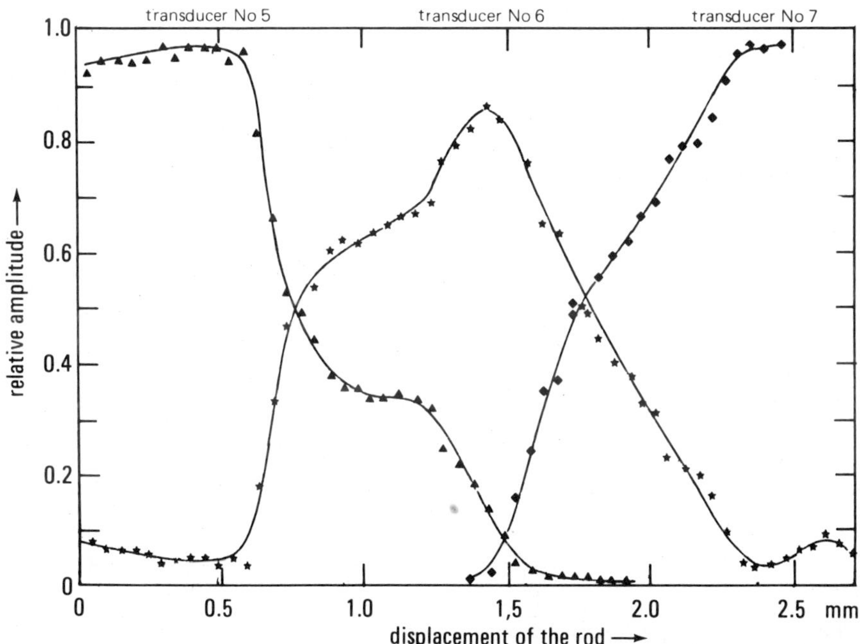

Figure 4. Relative amplitudes of neighboring transducers by mechanical excitation normalized to the sum amplitude vs rod displacement.

Table 1. Cross-coupling in the monolithic PVDF array.
Result in dB for pulses with 2 MHz

ultrasonic excitation		electrical excitation		type of construction
next neighbor	next but one	next neighbor	next but one	
		-38	-26	quartz backing, type B
-33	-19	-34	-27	Al_2O_3 backing, type B
-35	-26	-34	-28	Al_2O_3 backing, type A
-40	-23	-38	-26	Al_2O_3 backing slotted, type A (slots $0.1 \times 0.6 mm^2$ width x depth)

By the difference of the pulse amplitudes the additional effect of the neighbors is demonstrated. The result of the two last cross coupling measurements is shown in Table 1.

The higher cross-coupling to the next but one neighbors shows that a part of the cross coupling comes from the connection leads of the alternatively connected transducer elements. The cross-coupling to the next neighbors is about 6 dB less than expected and shows the influence of the additional series capacitance of the glue between the PVDF transducer and the backing. The differences between the various values in table 1 are due to backing material and construction. Lowest values are expected for slotted backings with low dielectric constants.

ANGULAR ACCEPTANCE

The angular acceptance of a single transducer is a sensitive indication for interelement cross-coupling and frequency response /4/. The angular acceptance is measured by turning the array in the water tank in the far field on an ultrasound beam of a 2 MHz transmitter. Axes of rotation are the length of the array (x-axis) and an axis perpendicular to it and to the ultrasonic beam (y-axis). To exclude the electrical cross coupling from the leads of the neighbors through the water, all neighbors are short-circuited.

The angular acceptance patterns as functions of the angle of rotation of the transducers of the different arrays are shown in Fig. 5a-c. Elementary calculations lead to a sin ξ/ξ distribution, where $\xi = \pi d/\lambda \sin$ and d is the width of the transducer perpendicular to the axis of rotation, λ is the wavelength of the ultrasound and ϑ is the angle of incidence. Thus the acceptance pattern is determined only by the transducer size and the wavelength. This pattern is shown in Fig. 5 as a line. The measured dots, however, show pronounced minima at ± 28° (Fig. 5a) and ± 18° (Fig. 5b and Fig. 5c). The differences in construction are given by different backing materials: quarz (5a) and Al_2O_3 (5b and 5c), and different PVDF thickness: 25 μm (5a and 5b) and 12 μm (5c).

The minima are at the same angular position for turning round the x-axis or the y-axis. They do not depend significantly on the frequency between 1 and 2.5 MHz and on the PVDF thickness. However, they depend on the backing material and are found near the angle ϑ_R with $\sin\vartheta_R = v_W/v_R$, where v_W/v_R is the ratio of the sound velocity in water to the Rayleigh wave velocity on the backing surface. At this angle Schoch showed the existence of a lateral displacement of the reflected beam /5/. Swartz /6/ measured minima at ± 22° with PVDF on a silicon backing. This angle is again near the angle ϑ_R of silicon.

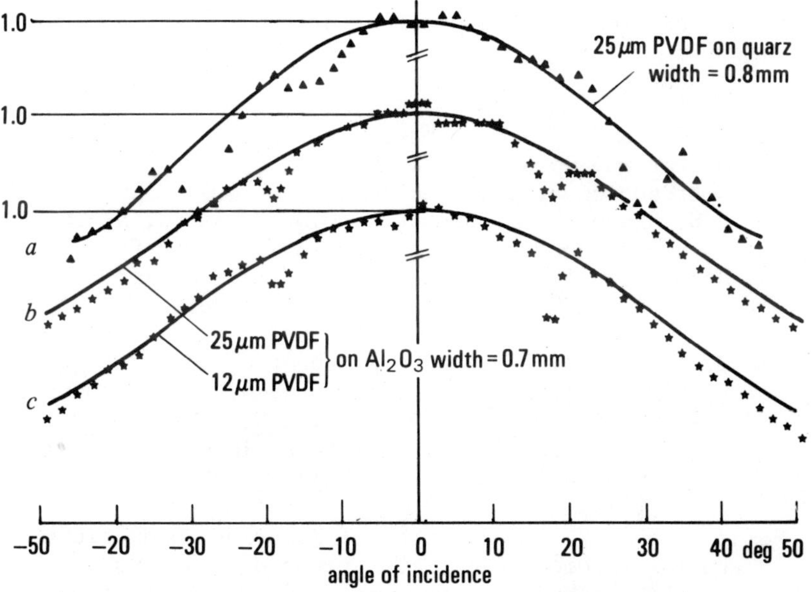

Figure 5. Angular acceptance of different PVDF transducers

CONCLUSION

Linear monolithic arrays each with 10 transducers, have been constructed with PVDF on hard backing. They convince by the ease of construction and the promising results dependent on the inherent properties of the polymer PVDF.

The most promising results are expected from the array type B (Fig. 1) with the signal electrodes on the backing combined with the signal detection by capacitive through-coupling to this electrode. The low cross-coupling allows the monolithic construction. The homogeneity of sensitivity allows the transducers to be used in an array. The number of transducers can be produced merely by reproduction technique of photolithography. The good angular acceptance enables the transducer to be used in a transmission camera to reproduce the spatial resolution offered. With the frequency response measured from 2 MHz to 12 MHz the PVDF receiver array is superior to arrays made of conventional transducer materials like PZT. Smaller PVDF transducers manufactured by photolithography and selective poling are conceivable. The limit in size of this construction, however, is set by the low capacitance of the transducers and the electronic network to be matched to them.

ACKNOWLEDGMENT

This work has been supported under the technological program of the Federal Department of Research and Technology of the FRG. The author alone is responsible for the contents.

REFERENCES

/1/ J. Callerame, R.H. Tancrell and D.T. Wilson, "Transmitter and Receivers for Medical Ultrasonics", 1979 Ultrasonics Symposium Proceedings, 407. (IEEE, London).

/2/ P.S. Green, L. F. Schaefer, E.D. Jones, J. R. Suarez, "A new high Performance Ultrasonic Camera System", Acoustic Holography, Vol. 5, N.Y. Plenum Press 1974.

/3/ H. Schewe, "Measurements of the piezoelectric Constants of PVDF", unpublished.

/4/ B.A. Auld, M.E. Drake, C.G. Roberts, "Monolithic acoustic imaging transducer structures with high spatial resolution", Appl. Phys. Lett., 25, (1974), 478.

/5/ A. Schoch, "Seitliche Versetzung eines total reflektierten Strahls bei Ultraschallwellen", Acoustica 2, (1952), 17.

/6/ R.G. Swartz, "Application of polyvinylidene fluoride to monolithic silicon/PVDF transducer arrays", SEL-79-013, Technical report No. G561-1, Stanford University 1979.

PHYSICAL LIMITATIONS OF OPTICALLY-SCANNED ACOUSTIC IMAGING TRANSDUCERS AT ULTRASONIC FREQUENCIES ABOVE 10MHz

C.W. Turner, A. Ayoola, S.O. Ishrak* and M. Salahi

Department of Electronic & Electrical Engineering
King's College, London, England
*Present address: Philips Ultrasound Inc., Santa Ana, California, U.S.A.

INTRODUCTION

Optically-scanned piezoelectric-semiconductor sandwich transducers using photoconductive switching provide an effective means of achieving amplitude and phase imaging of acoustic field distributions at low megahertz frequencies[1]. The particular practical realization discussed here is capable of producing C-scan images using a brightness-modulated flying-spot scanner tube as the optical source and a separated-medium transducer as shown in Fig. 1. The relative simplicity of this type of imaging transducer, in comparison with, for example, electrically-addressed two-dimensional mosaic transducers[2] or laser-based systems[3], suggests that it might be the most suitable approach to near-field orthographic imaging for VHF ultrasonic frequencies below the practical limit of the Quate or the SLAM instruments.

Experimental and theoretical results obtained in the range of 2 to 5 MHz have been reported previously[1]. In this paper, further results are presented for frequencies around 10MHz and the feasibility of operation at higher frequencies is considered. The principal physical limitations of the present design are discussed and the prospects of improved performance are reviewed in the light of those factors.

THE OPTICALLY-SCANNED PVF_2-SILICON TRANSDUCER AT 3MHz AND 10MHz

The PVF_2-Silicon composite transducer has been demonstrated to operate satisfactorily at 3MHz as an imaging (receiver) transducer[1], an optically-scanned point-source transmitter[4], and a high

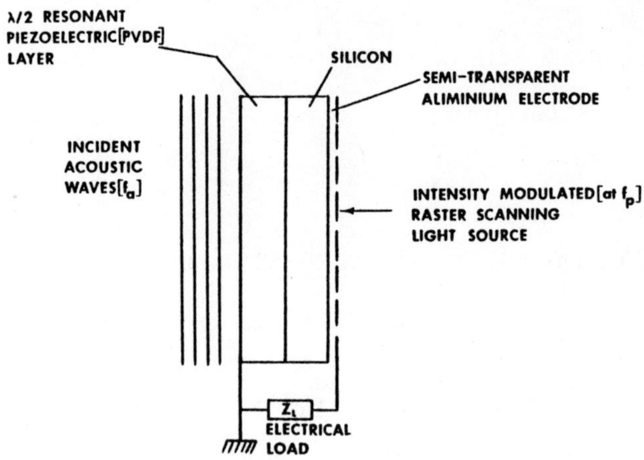

Fig. 1 The optically scanned transducer

resolution acoustic probe. PVF_2 films, although less active piezoelectrically than, say, PZT ceramic transducers, have important advantages in imaging applications[5], notably in their good impedance and velocity match to water. The improved spatial frequency response of the PVF_2 films over that obtainable from either PZT or $LiNbO_3$ transducers gives an image resolution close to the diffraction limit to be obtained at 3MHz. It should be emphasized that the piezoelectric element in this type of optically-scanned transducer is used in a very low-Q mode (typically a loaded Q of about 5) and is not mechanically coupled to the semiconductor layers. Reviewing

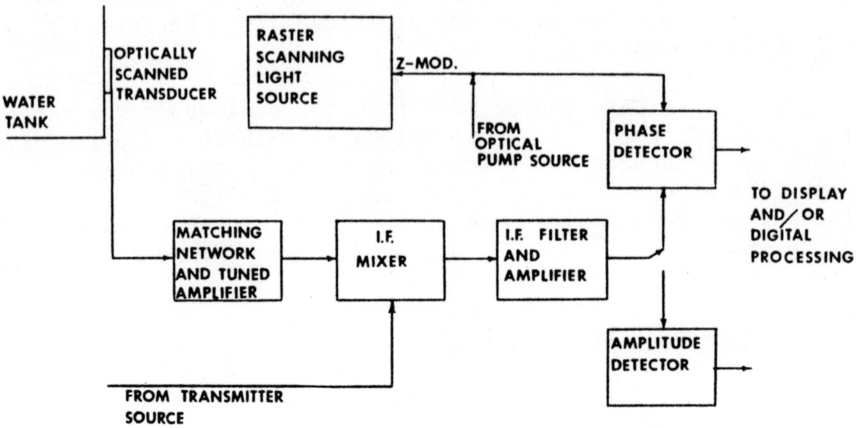

Fig. 2 System block diagram

the results obtained at 3MHz, the main constraints are imposed by the semiconductor parameters, namely the silicon photocarrier lifetime (which limits the maximum usable optical pump frequency), the resistivity of the silicon (the dielectric relaxation frequency should be in the region of the acoustic frequency), the presence of the unwanted and relatively large amplitude 'carrier' signal, and the optical efficiency and sheet resistance of the integrating electrode on the silicon wafer.

At 3MHz the transducer can be used in either a rapid-scan mode giving quasi-real-time transmission images of planar objects or in a slow-scan mode as a sensitive high resolution acoustic field probe for the evaluation of transducer radiation patterns. Both amplitude and phase images are obtained using the experimental arrangement shown in Fig. 2.

Similar imaging experiments have recently been carried out at 10MHz, using the same 0.35mm thick PVF_2 film in view of its broadband response. In addition, gold-doped silicon with appreciably reduced photocarrier lifetime has been used, permitting a higher optical pump frequency of 100kHz and therefore somewhat easier electronic filtering of the sideband, carrying the imaging information, from the acoustic carrier signal, as shown in the spectrum of Fig. 3. The anticipated improvement in spatial resolution was confirmed, Fig. 4, at the expense of about 3dB deterioration in sensitivity. As at 3MHz the performance of the transducer is thermal-noise limited. Additional losses are attributable to increased attenuation in the water medium, increased conversion loss of the non-resonant PVF_2 film and the lower photoconductive switching-ratio of the gold-doped silicon which is lifetime-dependent.

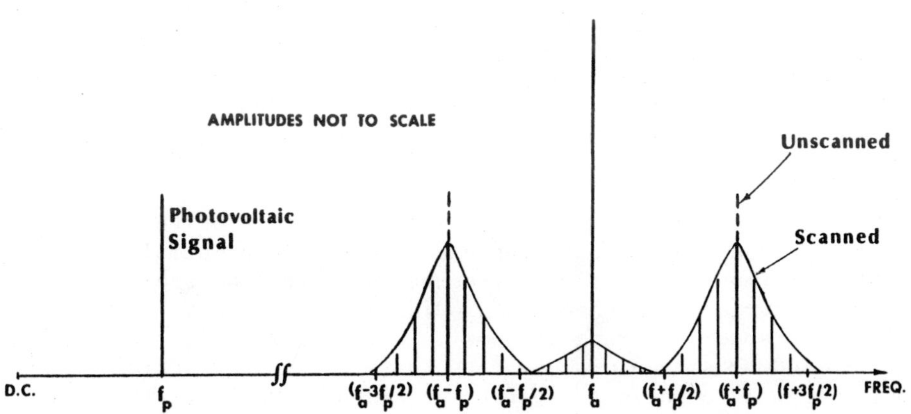

Fig. 3 Frequency content of output signal from a scanned transducer

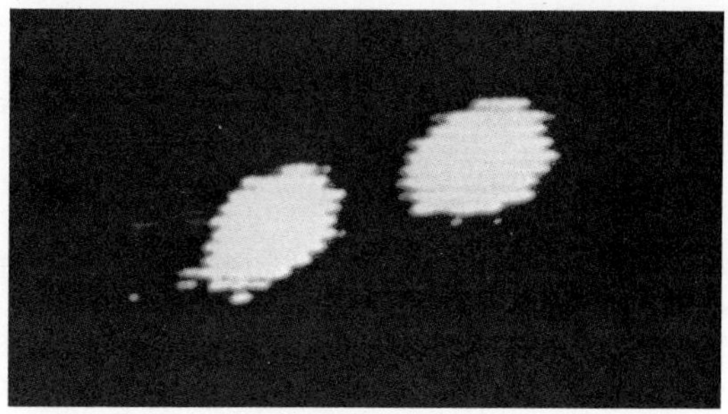

Fig. 4 Image of two 0.8mm holes drilled with a separation of 0.3mm (f=10MHz)

None of these factors is considered to be sufficiently influential to cause the performance of a suitably optimized transducer sandwich to be any less satisfactory at 10MHz. As the frequency is increased beyond 10MHz, however, the optimization of the piezoelectric film-semiconductor layer sandwich imposes severe constraints on its fabrication and the physical limitations of this type of transducer degrade the performance.

PHYSICAL PARAMETERS CONTROLLING THE ULTIMATE FREQUENCY OF OPERATION

A simplified equivalent circuit model that has given good agreement with experimental results is shown in Fig. 5, where the piezoelectric film and semiconductor layer are represented by lumped circuit elements for each image cell. For this model to be valid the scanning light spot must be no larger than the mean cell diameter and the scan time across each cell must be at least one period of the pump oscillation. In practice, the external load impedance is designed to give a near short-circuit at the acoustic 'carrier' frequency to maximize the rejection of the carrier in the output circuit (it carries negligible imaging information) and to minimize cross-talk between the 60 x 60 image cells. The load additionally presents a conjugate match at the upper sideband frequency.

A useful estimate of the relative sensitivity of the transducer over a wide range of frequencies can be obtained by simulating the acoustic carrier with a directly connected signal source as shown in Fig. 6, avoiding the need to provide suitable piezoelectric elements to cover the range. Some loss in sensitivity occurs from 10MHz to 100MHz as measured by both the sideband-to-carrier ratio and the signal-to-noise ratio for the sideband signal. It should be noted, however, that this simulation does not reveal how the imaging capability behaves as a function of spatial frequency and, secondly, that no optimization of the physical parameters of the piezoelectric-semiconductor sandwich has been attempted.

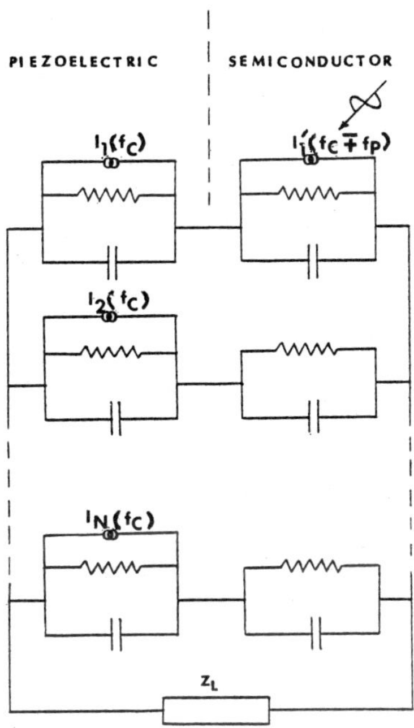

Fig. 5 Equivalent circuit

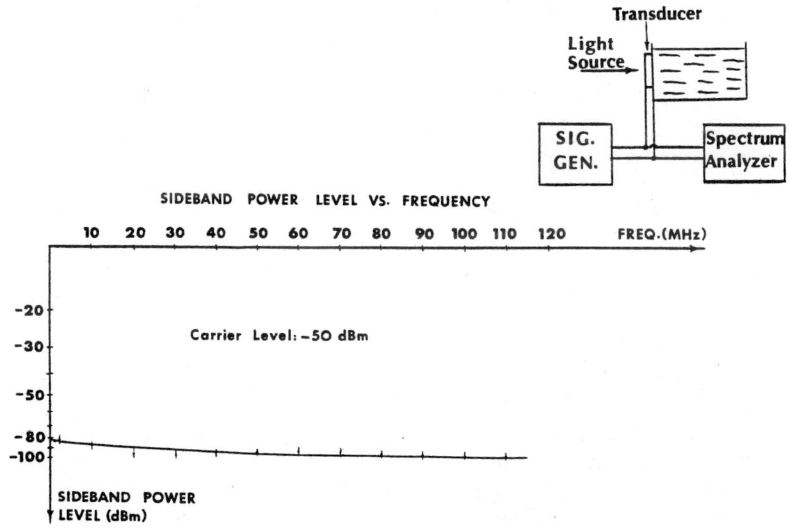

Fig. 6 Simulated response of transducer up to 110MHz

The sideband current generator in Fig. 5 is strongly dependent upon the spatial frequency k because of the finite thickness d of the silicon wafer. The major causes of this dependence are lateral photocarrier diffusion and flux fringing. Analytical results show that the amplitude response falls off sharply for spatial frequencies $k>1/d$. At 10MHz the semiconductor should be etched or mechanically polished to no greater than 50μm thickness: above 10MHz, because of the large area of the wafer, this requirement leads to mechanical difficulties until at 100MHz where thin film techniques are applicable.

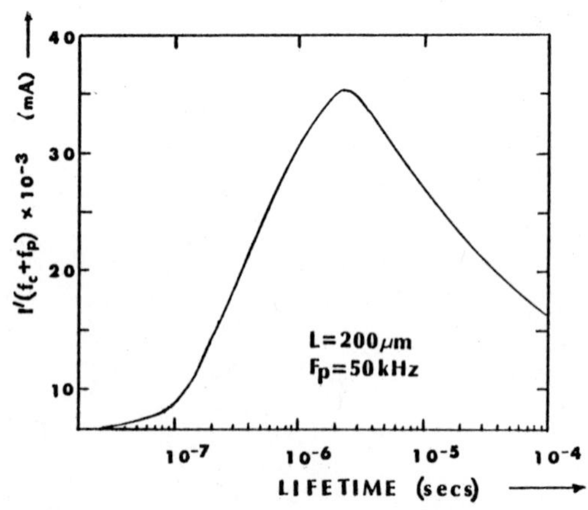

Fig. 7 Sideband current generator amplitude as a function of photocarrier lifetime τ_p for f_p=50kHz

For a given optical modulation angular frequency ω_p the optimum photocarrier lifetime τ_p is in the region of $1/\omega_p$ (Fig. 7). Although there are some signal processing advantages in reducing τ_p to the sub-microsecond range the characteristic loss of photoconductive response must be considered, together with the requirement that the photocarrier diffusion length L_D remains greater than d (Fig. 8). A more effective approach is to remove the strong acoustic carrier signal by arranging the integrating electrode in two halves, driving a differential amplifier. The desired image sideband is relatively little affected whilst the background signal is strongly suppressed. The optical and electrical efficiency of the integrating electrode can be considerably improved by the use of an evaporated ITO film which can provide over 80% light transmission with a sheet resistance of only a few ohms per square. A flying-spot scanner tube is much cheaper and simpler than mechanically-scanned or acousto-optically deflected laser sources, but the minimum spot size for adequate brightness level is about 25μm. Above 50MHz, therefore, scanned laser sources would be essential for high resolution imaging.

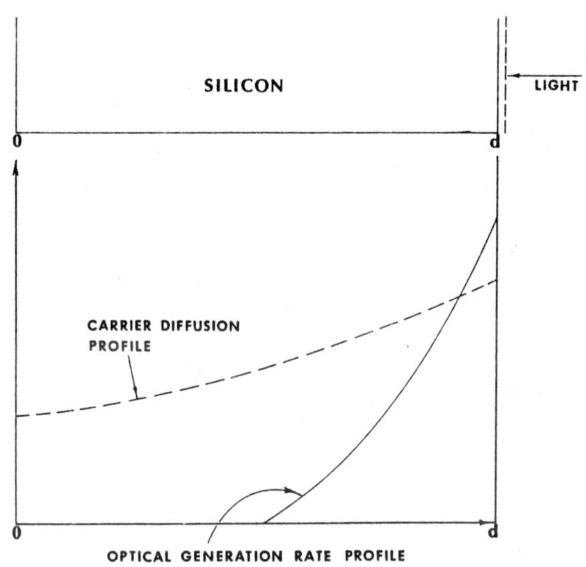

Fig. 8 Photocarrier profile for $L_D > d$

The choice of semiconductor material for imaging applications is dictated by the need for a comparatively large area wafer, typically $10^3 mm^2$, that can be photoconductively switched by visible light (the output from a fast response flying-spot scanner tube is mainly in the blue wavelengths) modulated at about 100kHz. High resistivity silicon gives the best performance: gold-doping or similar treatment is usually required to reduce the lifetime to the microsecond range. The output impedance of the sandwich becomes dominated by the wafer capacitive reactance which falls as $1/f^2$ because the thickness d needs to be scaled to preserve the spatial frequency response. However, it is unlikely that the same aperture size would be required at frequencies above 10MHz. If N^2, the number of image cells, is maintained constant instead, the output impedance would not change appreciably with frequency, so that matching to the output load impedance would not become difficult.

Although the fractional change in conductivity, $\Delta\sigma/\sigma$, optically induced in the wafer by the scanning light beam is small (typically below 1 per cent) and is constrained by the lifetime, the image sideband signal-to-noise ratio gives adequate dynamic range for amplitude and phase imaging at 10MHz. Nevertheless, the obvious alternative to bulk photoconductors, an array of photodiodes, offers the possibility of a greater dynamic range and faster response. Unfortunately this can only be achieved under reverse bias, an unwelcome complication to the fabrication of the sandwich since each photodiode must be isolated electrically for imaging purposes. The monolithic piezoelectric semiconductor transducer has been considered before at frequencies up to 1GHz (e.g. GdS, ZnO) but suffers from an excessively long lifetime, limiting the scan rate to about 100 elements per second.

FUTURE DEVELOPMENTS

Optically-scanned transducers using photoconductive switching should be capable of achieving satisfactory amplitude and phase imaging of acoustic field distributions in water up to frequencies around 50MHz, where the acoustic loss in water becomes excessive. Above these frequencies modulated laser sources would be needed for photoconductive switching and alternative forms of semiconductor geometry become necessary, since the thickness will need to be less than 10μm to maintain good spatial frequency response. The Debye space charge region will then extend throughout the semiconductor, whereas at 10MHz it is negligible in comparison to the length of the bulk (neutral) region, so that the optical switching mechanism will change character. In addition, the semiconductor would have to be in mechanical contact with the piezoelectric to minimize coupling loss across the air gap for high spatial frequencies. There is a close analogy between this behaviour and the electric coupling in air-gap surface wave devices (Fig. 9).

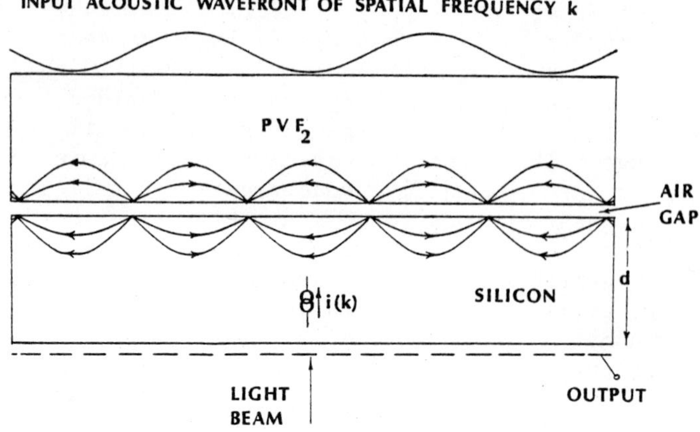

Fig. 9 Surface wave analogy of air-gap coupling loss for high spatial frequencies

PVF_2 films give good spatial frequency response without the need for matching layers when used in water-based imaging, against which must be set the relatively small coupling constant and high electrical and mechanical losses. For solid-based imaging, $LiNbO_3$ would probably be preferred although matching layers would, in general, need to be used.

At 100MHz it is arguable that the Quate microscope or the SLAM device would have significant advantages over the optically-scanned transducer for imaging, but acoustic-field probing may still be attractive if a modulated laser source were used.

CONCLUSIONS

In conclusion, the optically-scanned transducer is a strong candidate both for near-field imaging and acoustic field probing above the range where electrically-addressed arrays are feasible and below the practical limit of acoustic microscopes. The physical processes whereby the image sideband signal is generated have been shown to remain effective in the frequency range 10-100MHz provided that practical techniques for thinning large area silicon wafers are available.

ACKNOWLEDGEMENTS

Grateful acknowledgement is made of the support of the U.K. Science and Engineering Research Council. PVF_2 films were kindly supplied by Thorn-E.M.I. Central Research Laboratories.

REFERENCES

1. S.O. Ishrak and C.W. Turner, "Two-dimensional Imaging with a High Resolution PVF_2/Si Optically-scanned Receiving Transducer," Acoustical Imaging, Vol. 11, J.P. Powers, Ed., Plenum Press, New York, 1981

2. R.G. Swartz and J.D. Plummer, "Monolithic Silicon-PVF_2 Piezoelectric Arrays for Ultrasonic Imaging", Acoustical Imaging, Vol. 8, A.F. Metherll, Ed., Plenum Press, pp. 69-95, New York,1978.

3. S. Ellior, V. Domarkas and G. Wade, "Frequency Characteristics of Opto-acoustic Transducers", IEEE Trans. Sonics and Ultrasonics, SU-25, pp. 346-353, Nov. 1978.

4. A. Ayoola and C.W. Turner, "The Properties and Performance of a 'Si-PVF_2' Optically-controlled Point Source," Acoustical Imaging, Vol. 11, J.P. Powers, Ed., Plenum Press, New York, 1981.

5. C.W. Turner and S.O. Ishrak, "Comparison of Different Piezo-electric Transducer Materials for Optically-scanned Acoustic Imaging, Vol. 10, P. Alais and A. Metherell, Eds., Plenum Press, New York, pp. 761-778, 1980.

PARTICLE VELOCITY AND DISPLACEMENT PATTERNS OF DISC TRANSDUCERS WITH AMPLITUDE SHADING

D.A. Hutchins and J.A. Archer-Hall

Department of Physics, Queen's University, Kingston, Canada K7L 3N6, and Department of Physics, University of Aston, Birmingham B4 7ET U.K.

INTRODUCTION

The radiated fields from ultrasonic transducers have been investigated theoretically using many approaches, and both near-fields and farfields have been examined. Some treat the continuous-wave case[1-3], whereas others have examined pulsed behaviour[4]. The references cited investigate principally the uniformly vibrating disc, where the front face of the transducer is assumed to vibrate in phase, with a constant amplitude. The transducer is usually assumed to be mounted in an infinite baffle, a perfectly-reflecting barrier preventing modification of the field by waves radiated from the back surface. Various other theoretical approaches have been presented, which have treated variations in baffle configuration[5,6].

The above approaches, treating the rigid planar piston case, have been extended by analysis of nonuniform amplitude distributions across the disc face. The form of motion has usually been assumed to be axisymmetric, being only a function of the radial coordinate r. This is of importance, as nonuniform behaviour of transducers is often met experimentally e.g. the difference in behaviour between simply supported and clamped edge radiators.

As an experimental verification of the importance of this phenomenon, Lockhart and Miller[7], investigated the directivity patterns of transducers and demonstrated a reduction in side lobes by using a line shading. This effect had been noted earlier by Martin and Breazeale[8] and Papadakis[9]. Theoretically, Szabo[10] predicted the beamshaping and diffraction from tapered amplitude distributions, using theory derived from optics, but the various acoustic parameters were not identified. However, a Gaussian distribution across the face was seen to lead to a uniform radiated beam. Truncated Gaussian profiles were also effective in this respect.

In an extension of the theory for a piston radiator, Greenspan[11] examined the radiated field of a transducer with a Guassian particle velocity distribution across its face. In agreement with Szabo[10], axial nearfield intensity variations were smoothed out, when a comparison was made to those of a uniform vibrator. Recent advances have been made by both Tjøtta and Tjøtta[12], and Harris[13], who evaluated the transient field of pulsed planar pistons with nonuniform velocity distributions. The former authors derived expressions to describe the field for several amplitude distributions, whereas Harris presented impulse and pulse response functions for different axisymmetric source velocity distributions and pulse shapes.

The above methods are extremely useful, but it was felt by the authors that a more simple approach was needed which could investigate a wide range of velocity distributions across disc transducers. Such an approach will now be described.

THE APPROACH

General Discussion

All problems to be treated involve cylindrical symmetry, and so the wave equation to be solved, expressed in cylindrical polar coordinates (r, Z, θ), is

$$\frac{1}{r}\frac{\partial}{\partial r}\left(r\frac{\partial\phi}{\partial r}\right) + \frac{\partial^2\phi}{\partial Z^2} + k^2\phi = 0 \qquad (1)$$

where variation with θ is zero, Z is an axial coordinate and r a radial coordinate. A single frequency of excitation is assumed, described by the wavenumber k.

The expression

$$\phi = \frac{AJ_0(Cr)e^{j(k^2-C^2)^{1/2}Z}}{j(k^2-C^2)^{1/2}} \qquad (2)$$

is a solution of Equation 1; such solutions are well known and have been used previously in integral transform approaches[1].

Consider the following infinite series:

$$\phi = \sum_{n=1}^{\infty} \frac{A_n J_o (C_n r)}{j(k^2-C_n^{\;2})^{1/2}} e^{j(k^2-C_n^{\;2})^{1/2} Z} \tag{3}$$

in which each term is characterized by its own value of A_n and C_n, but which is still a solution of (1). Provided sufficient convergence occurs, a finite number of terms m may be used leading to an expression for the acoustic pressure P of the form

$$\begin{aligned} P &= -\rho \frac{\partial \phi}{\partial t} = -jk\rho c\phi \\ &= k\rho c \sum_{n=1}^{m} \frac{A_n J_o(C_n r)}{(k^2-C_n^{\;2})^{1/2}} e^{j(k^2-C_n^{\;2})^{1/2} Z} \end{aligned} \tag{4}$$

where c is the velocity of longitudinal waves in the medium of density ρ. Two component vectors of the particle velocity, q_z and q_r, may also be derived:

$$q_z = \frac{\partial \phi}{\partial Z} = \sum_{n=1}^{m} A_n J_o(C_n r) \; e^{j(k^2-C_n^{\;2})^{1/2} Z} \tag{5}$$

$$q_r = \frac{\partial \phi}{\partial r} = \sum_{n=1}^{m} \frac{C_n A_n J_1 (C_n r)}{j(k^2-C_n^{\;2})^{1/2}} e^{j(k^2-C_n^{\;2})^{1/2} Z} \tag{6}$$

The form of these series depends on the values of the C_n coefficients. If the C_n's are integers, a Schölmilch series results which is usually valid and convergent over the range r = 0 to π. This range may be extended, however, once boundary conditions are applied at the plane Z = 0 to find appropriate values for the A_n coefficients.

Boundary Conditions and their Application

The boundary conditions to be applied exist at the plane Z = 0, containing the face of the transducer. The distribution of q_z across this face is the parameter to be studied here; for a uniformly vibrating disc, q_z is constant across the face. Outside the transducer radius a, i.e. at radial distances r/a > 1, the boundary conditions at Z = 0 will be dictated by the presence or absence of a baffle. If no baffle is present, the boundary condition P = 0 exists for r/a > 1. If a perfectly-reflecting baffle is present, however, then the boundary condition becomes

$q_z = 0$ over this region.

At the plane Z = 0, expressions for P and q may be obtained from Equations 4 and 5 giving

$$P = k\rho c \sum_{n=1}^{m} \frac{A_n J_o(C_n r)}{(k^2 - C_n{}^2)^{1/2}} \tag{7}$$

and

$$q_z = \sum_{n=1}^{m} A_n J_o(C_n r) \tag{8}$$

Boundary conditions may be applied to these expressions by a simultaneous equation technique, which is illustrated in Fig. 1 for the case of a uniformly vibrating disc surrounded by an infinite baffle. The amplitude of q_z is constant at unity over the disc face, but is zero elsewhere. Equally spaced values of r between 0 and π are selected, and at each Equation 8 is equated to the appropriate value of q_z. In Fig. 1, 32 values of r are defined, leading to a 32 term series (m = 32). For the first ten values of r, the series is equated to unity; for the subsequent 22 values, it is equated to zero. The result is 32 simultaneous equations, with 32 unknowns (the values of $A_1 - A_{32}$) which may be found if the C_n's are already chosen. The resultant coefficients may then be used in Equations 4,5 and 6 to plot the radiated field.

A sufficient number of equations are required to provide adequate convergence of the series and to accurately represent the boundary conditions. In the above case, the use of 105 terms

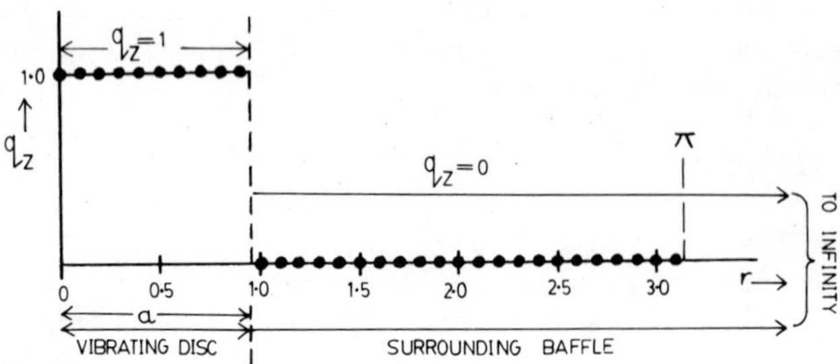

Figure 1: Illustration of application of boundary conditions for infinitely-baffled disc.

provided a good representation, as illustrated in Fig. 2, plotted using the resultant values of $A_1 - A_{105}$. The uniformly vibrating disc and surrounding baffle are well defined. Slight departures from ideal behaviour are noted at the disc edge, an effect analogous to the Gibbs phenomenon in Fourier series, and at $r = \pi$; these are not serious however.

Note that pressure may also be defined across the Z = 0 plane by this technique, so that in principle unbaffled and partially-baffled discs may be investigated also.

Comparison to Surface Integrals

The radiated field may be plotted as either pressure, particle velocity or particle displacement distributions. Velocity and displacement magnitudes will follow the same variations and only velocity distributions are presented here.

The field may be plotted radially, axially, or may be represented as a 3D plot through a section. In all cases presented, disc transducers of radius a are examined, surrounded by an infinite baffle. The frequency of excitation is such that a = 5 λ, where λ is the radiated wavelength. Fig. 3 shows a radial pressure distribution at a small distance from the uniformly-vibrating transducer face, and predictions of surface integral[6] and series formulations are compared. Reasonable agreement is observed. An axial plot of q_z is presented in Fig. 4 for the same transducer. Note the increasingly marked variations in amplitude with distance from the transducer in the nearfield region. Farfield oscillations of the series approach are due to the exponential term in Z, and are reduced with an increased number of terms in the series. Note that $q_r \mp 0$ on axis due to symmetry.

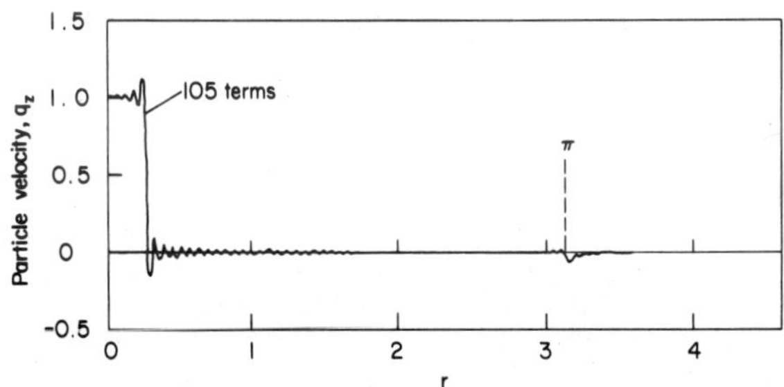

Figure 2: Representation of infinitely-baffled disc at plane z = 0, predicted by Equation (8) following the application of boundary conditions.

RADIATED PARTICLE VELOCITY DISTRIBUTIONS

In this section, radiated nearfield particle velocity distributions will be presented for discs with various q_z characteristic across their face. The frequency is such that $a = 5\lambda$ throughout.

For the uniformly vibrating disc, the nearfield normal and radial components of particle velocity are presented in Fig. 5(a) and (b) respectively. The normal component q_z exhibits marked variations on axis (see Fig. 4), and at a distance of Z = 3a from the front face it is possible to see significant side lobe formation. The radial component q_r is zero on axis, and in general is of smaller amplitude than q_z. Significant values are encountered in specific

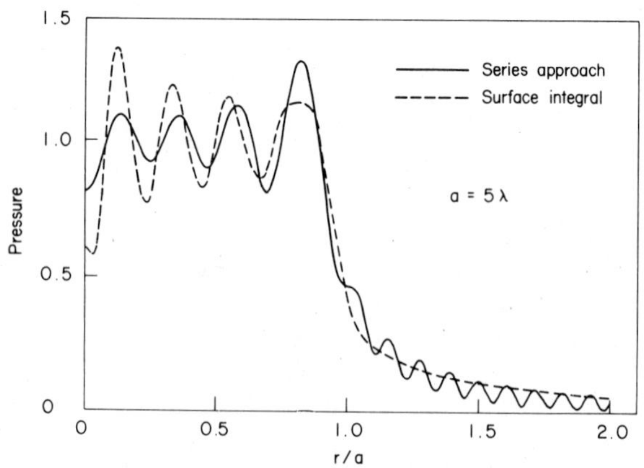

Figure 3: Radial pressure distributions for a uniformly vibrating disc at Z = a/5.

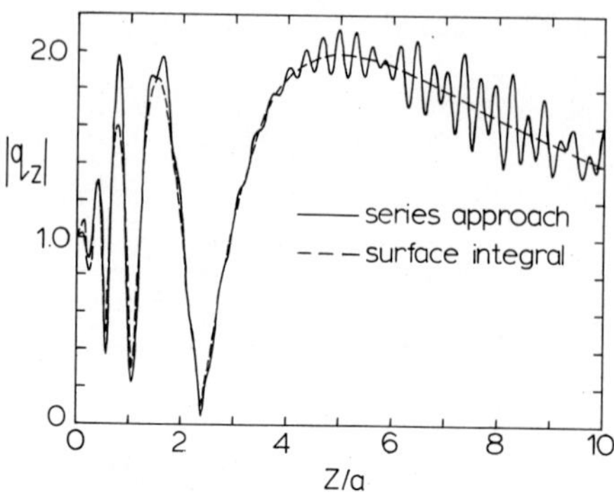

Figure 4: Axial variations of q_z for a uniformly vibrating disc.

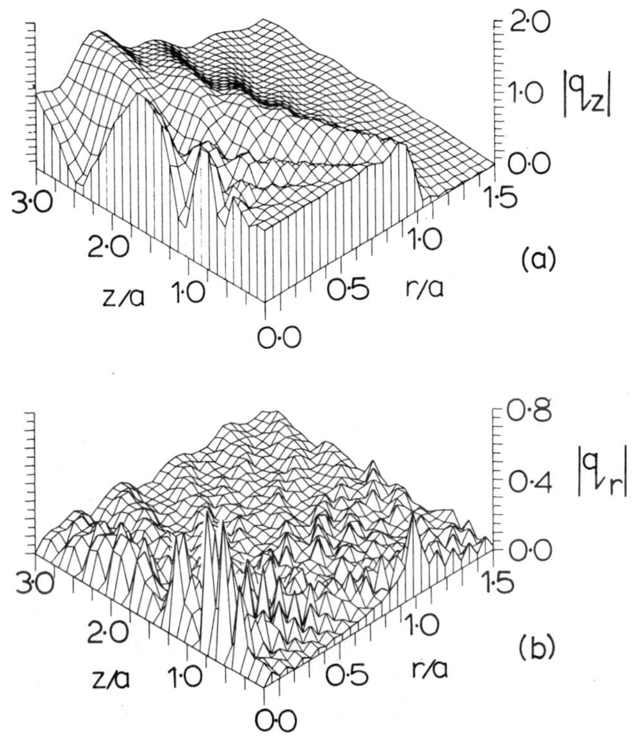

Figure 5: Radiated particle velocity nearfield distributions for a uniformly vibrating disc. (a) normal component, (b) radial component.

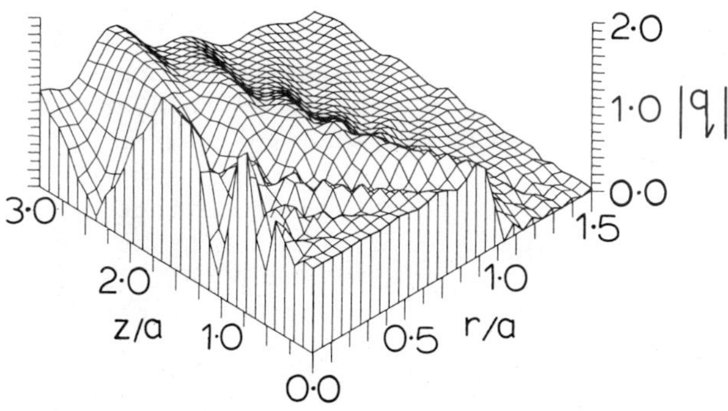

Figure 6: Resultant magnitude of particle velocity from two components of Figure 5.

locations, however, especially just off axis, and at the disc's edge at small z values.

The theory described evaluates the two components of particle velocity (or displacement), both acoustical parameters being vector quantities. Finding the resultant magnitude and direction at a particular position is complicated, in that the two spatial components in r and z are complex and not in phase. The means for finding the resultant is treated in publication by Archer-Hall and Gee[14], and details will not be given here. Fig. 6, however, shows the resultant magnitude of particle velocity, derived from the two spatial components for the uniformly vibrating disc. It differs in only minor detail from the variations in q_z due to the small q_r contribution.

Assuming cylindrical symmetry, the theory may be used to study radiated distributions for cases where the normal component of particle velocity q_z across the disc face is not uniform. The overall magnitude of particle velocity q is plotted for brevity, although they will closely resemble those of q_z in each case due to the minimal values of q_r at most positions.

The situation most investigated previously is the Guassian shading, with q_z being greatest at the disc centre. The predicted particle velocity radiated distribution is presented in Fig. 7. Evident is the smoothing out of nearfield amplitude variations, with the reduction of side lobe generation. This result agrees well with that expected from the investigations of previous authors, who were involved primarily in the prediction of pressure variations. There was a small deviation from an ideal Guassian distribution in the application of boundary conditions, which accounts for the small axial variations which would be expected to be absent.

It is found that the suppression of side lobes and nearfield amplitude variations may also be achieved by causing a linear decrease in q_z amplitude to exist over the transducer face, and to a slightly lesser degree by causing q_z to decrease with a sin dependence, as shown in Fig. 8(a) and (b) respectively.

In marked contrast to the above, the effect of increasing q_z linearly from zero at the centre, to unity at the disc edge, was seen to be detrimental. Fig. 9 shows that in this case substantial nearfield amplitude variations in particle velocity are observed, with a marked tendency for side lobe generation.

Other examples of amplitude shading are important in some imaging systems. The first is the case of concentric annuli, where the signal amplitude to each is decreased as a function of

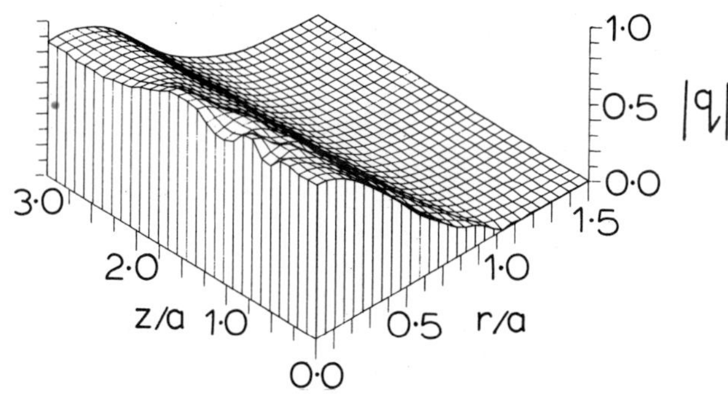

Figure 7: Particle velocity distribution from a disc with a Gaussian amplitude shading.

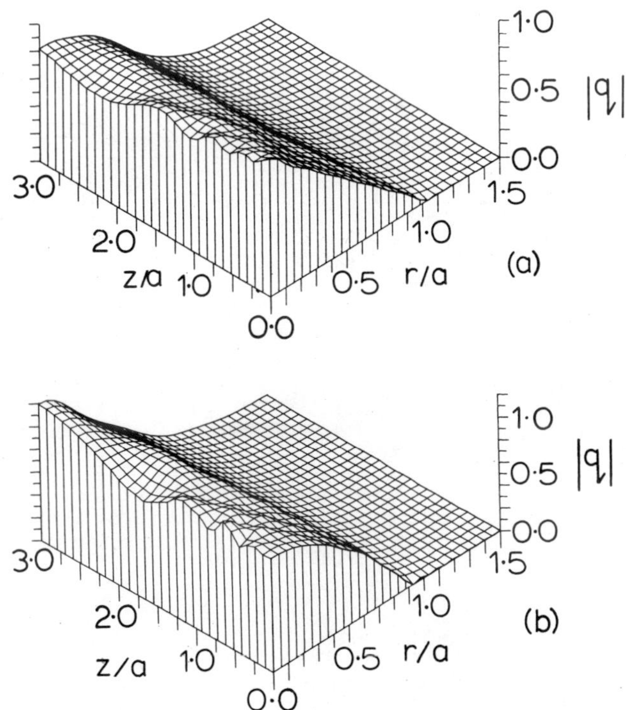

Figure 8: Particle velocity distributions from discs with tapered amplitude shading of q_z. (a) linear decrease , (b) sinusoidal decrease with distance from centre.

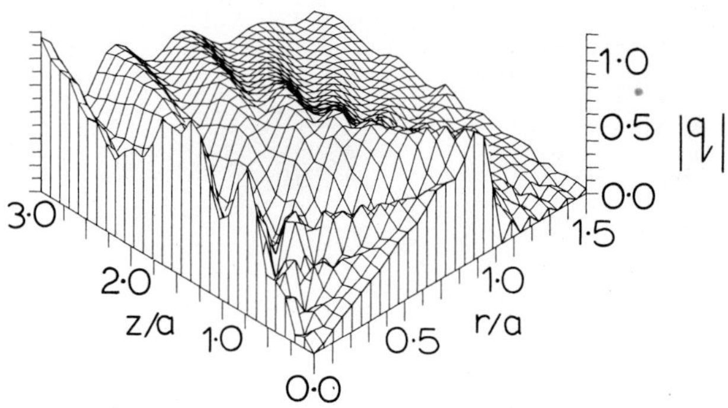

Figure 9: Particle velocity distribution from disc with linear increase in q_z amplitude with distance from its centre.

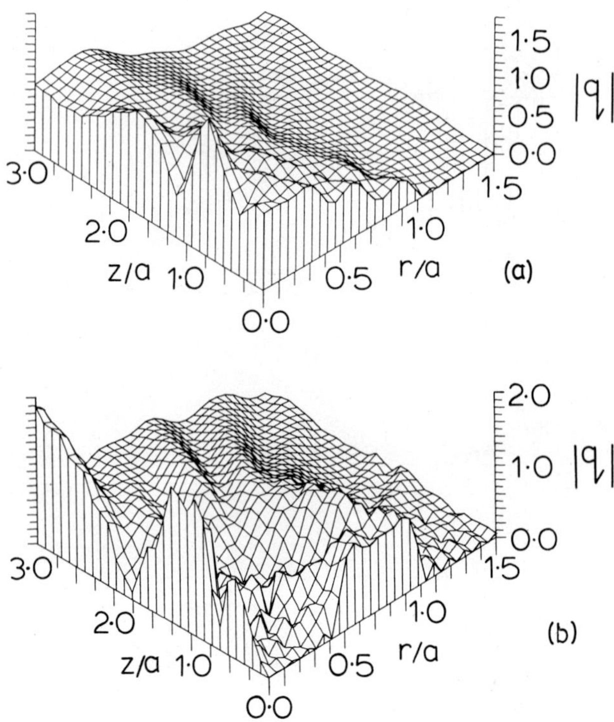

Figure 10: Particle velocity distributions from (a) concentric annuli with a central disc, and decreasing amplitude as shown, and (b) a single annulus.

distance from the axis, there being a central disc vibrating at the maximum amplitude. This situation was modelled as two closely-spaced outer rings, surrounding an inner central disc, each annulus vibrating at 2/3 and 1/3 of the central amplitude (unity) respectively. The predicted nearfield distribution is shown in Fig. 10(a). Comparison to Fig. 6, that for the uniformly-vibrating disc, indicates that an improvement in nearfield uniformity and side lobe suppression occurs.

The second case is that of a single annular transducer. In our case the annulus was of width a/2, with a central baffle of the same radius. The magnitude of q_z was unity over the face of the annulus and zero over the enclosed baffle. The radiated distribution is shown in Fig. 10(b). Of interest is the large on-axis amplitude as the nearfield/farfield transition region is approached. This phenomenon has been predicted previously for the pressure distribution from such transducers[6].

DISCUSSION AND CONCLUSIONS

The above has demonstrated that desirable features of radiated distributions may be enhanced by decreasing the value of q_z radially, with a maximum at the disc's centre. This agrees with the work of previous authors. Of interest is the fact that a linear decrease is very effective, but a sinusoidal decrease less so. A guassian distribution is the most desirable. Conversely, a linear increase in q_z from zero at the centre of the disc face results in a deterioration in nearfield uniformity. Other interesting cases include the possible use of concentric annuli to enhance desirable features, and the tendency for axial focusing from a single annulus transducer.

REFERENCES

1. L.V. King, On the acoustic radiation field of the piezo-electric oscillator and the effect of viscosity on transmission, Can. J. Res. 11:135 (1934).
2. J. Zemanek, Beam behaviour within the nearfield of a vibrating piston, J. Acoust. Soc. Am. 49:181 (1970).
3. J.C. Lockwood and J.G. Willette, High speed method for computing the exact solution for the pressure variations in the nearfield of a baffled piston, J. Acoust. Soc. Am. 53:735 (1973).
4. G.R. Harris, Review of transient field theory for a baffled planar piston, J. Acoust. Soc. Am. 70:10 (1981).
5. R.V. DeVore, B.D. Hodge and R.G. Kouyoumjian, Radiation by finite circular pistons imbedded in a rigid circular baffle. 1. Eigenfunction Solution, J. Acoust. Soc. Am. 48:1128 (1970).

6. J.A. Archer-Hall and D. Gee, A single integral computer method for axisymmetric transducers with various boundary conditions, NDT Int. 13:95 (1980).
7. C.M. Lockhart and M.K. Miller, Generalized shading formula from a given line shading, J. Acoust. Soc. Am. 68:1142 (1980).
8. F.D. Martin and M.A. Breazeale, A simple way to eliminate diffraction lobes emitted by ultrasonic transducers, J. Acoust. Soc. Am. 49:1668 (1971).
9. E.P. Papadakis, Effects of input amplitude profile upon diffraction loss and phase change in a pulse echo system, J. Acoust. Soc. Am. 49:166 (1971).
10. T.L. Szabo, Acoustic beamshaping and diffraction from tapered amplitude distributions, Proc. 1975 IEEE Ultrasonics Symp.:166 (1975).
11. M. Greenspan, Piston radiator: some extensions of the theory, J. Acoust. Soc. Am. 65:608 (1979).
12. J.N. Tjøtta and S. Tjøtta, Nearfield and farfield of pulsed acoustic radiators, J. Acoust. Soc. Am. 71:824 (1982).
13. G.R. Harris, Transient field of a baffled planar piston having an arbitrary vibration amplitude distribution, J. Acoust. Soc. Am. 70:186 (1981).
14. J.A. Archer-Hall and D. Gee, to be published in Ultrasonics.

RECENT DEVELOPMENTS IN AXICON IMAGING

[*]R.L. Clarke, J.C. Bamber[+], C.R.Hill[+] and P.F. Wankling[+]

[*]Physics Department, Carlton University, Ottawa , Canada
[+]Physics Division, Institute of Cancer Research, Sutton, Surrey, U.K.

INTRODUCTION

In the formation of ultrasound images for medical diagnostic purposes, it has been recognized that some considerable improvement in image quality is possible in principle through the use of wide aperture imaging devices. Merely increasing the aperture of a single spherically focussed source, however, results in a reduction of the range over which good lateral resolution is maintained. Focal scanning, either by switching between multiple fixed-focus sources (Dick et al., 1979) or by the electronically-phased focussing of a multiple-element source (Melton and Thurstone, 1978; Arditi et al., 1982), has been exploited to overcome this problem, which is particularly important when imaging the female breast. Synthetic aperture (computer reconstruction) techniques also aim to solve this problem, though perhaps in a more versatile way (see the section on reconstruction tomography in this volume). An alternative scheme has been, by physical means, to provide a wave front which simultaneously converges towards all points down the imaging axis. The name axicon has been applied to this system, which was introduced as a new optical element by McLeod (1954) and is now well known in optics. The principle of generating a line focus in this way was utilized in the "scatter scanners" of Foster et al., (1980), so named because a separate receiving transducer was aimed along the focal line to collect ultrasound scattered at angles other than 180°. Considerable promise was shown by a system consisting of a 45° cone transmitter and an f5.4[‡] receiver (Foster et al., 1981), which generated a lateral point response with a full width half maximum (FWHM) that remained about 0.7λ[‡] over a 5 cm depth of field. Very

‡ The "f-number" is the ratio focal length/aperture diameter. λ is the wavelength of sound.

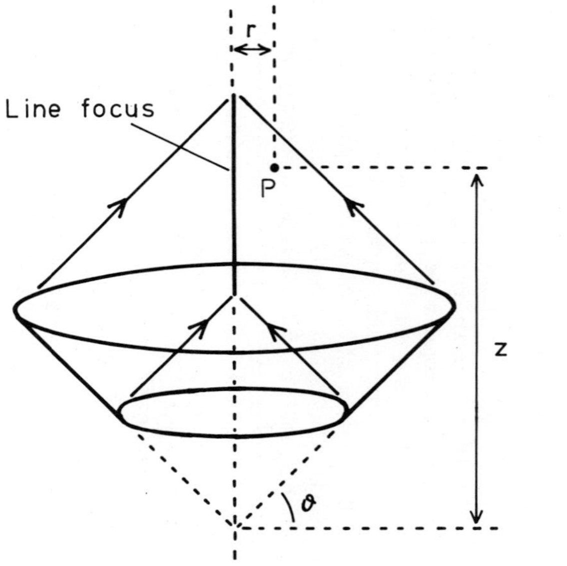

Fig. 1. Schematic diagram of a conically converging wave front showing the coordinate system for defining a field point P, and the convergence angle Θ.

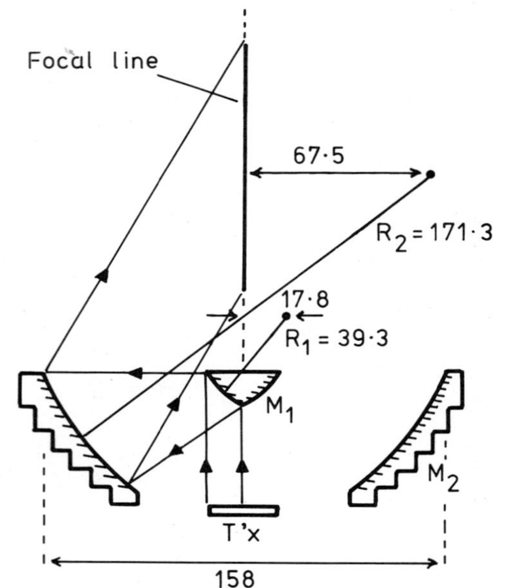

Fig. 2. Schematic diagram of the mirrors, M_1 and M_2, used in the present experiment. Dimensions shown are in mm, R_1 and R_2 are the radii of curvature of mirrors M_1 and M_2 respectively.

high quality images of excised breast tissue were demonstrated at a frequency of 4 MHz.

Axicons represent a family of devices which are figures of revolution and from which rays converge towards a common focal line on the axis of revolution. A variety of methods are in principle available for fabricating acoustic axicon sources, including the shaped piezoelectric plastic sheet used by Foster et al. (1981), a single annular transducer (Burckhardt et al., 1975; Weight and Brown, 1982) or phased annular arrays. Existing plane or other-shaped waves might also be converted to axicon form by systems of lenses, diffraction gratings or mirrors; all of which have been used in optical axicons. In an experimental ultrasound axicon Burckhardt et al. (1973) used a single plane transducer followed by a strongly focussing perspex lens and a small conical reflector, placed beyond the focal point of the lens, to spread the sound beam onto a large converging reflector which then formed the conical wave front returning to the axis. Using the same system for both transmission and reception encouraging results were achieved, but the attempt to focus over a depth range of some 20 cm using a convergence angle of only 7° meant that a lateral resolution of about 3λ only was attained. For the present work it was decided to build as simple an axicon mirror system as possible, having a large convergence angle and having the particular feature of utilizing existing B-scan equipment and transducers as the sound source and receiver. The time available for this study was limited, so that the necessary investigations of the physical characteristics, limitations and tolerance of the system were only preliminary.

THEORY

The relevant continuous wave theory, to a good first approximation for small values of the convergence angle and for points close to the axis, was given for the optical axicon by Fujiwara (1962) and was later refined by Lit and Brannen (1970). Patterson and Foster (1981) have recently provided a more complete analysis for conical acoustic radiators through the impulse response approach. Fujiwara's approximate expression for the axicon intensity distribution I(r,z), which we shall use as an approximate representation of the pulse-echo signal amplitude for a point target, may be written in terms of the geometrical parameters shown in Fig. 1 as

$$I_{axicon} = C_1 \frac{\sin^2\theta}{\lambda z} J_o^2(rk \sin\theta) \qquad (1)$$

where C_1 is a constant, $k = 2\pi/\lambda$ and J_o is the Bessel function of order zero. This is actually the result for a toroidal wavefront, but it differs from that for the cone only in the value of the constant and in that the on-axis intensity, I_o, falls off inversely with distance, z, whereas for the cone I_o is directly proportional to z. The latter distribution is in fact the more useful one for imaging in an attenuating medium.

An essential feature of Equ. 1 is that the J_o^2 function is constant with z, i.e. the lateral resolution is independent of distance within the focal zone. The intensity falls off initially very rapidly with the off-axis position, r, the rate of fall off depending strongly on the convergence angle for small values of θ, but not so strongly for large angles. In particular, for $\theta \gtrsim 30^o$ the improvement in lateral resolution due to increasing θ becomes much less marked. The lateral resolution of the axicon is defined by

$$FWHM_{axicon} \simeq \frac{0.41\ \lambda}{\sin\theta} \qquad (2)$$

This rapid fall off is followed by a series of diffraction minima and maxima. These define the side lobes of the device which are somewhat smeared to form a "skirt" to the lateral resolution curve in pulsed operation. For comparison, the intensity profile in the plane of focus of a spherically converging wave is

$$I_{sphere} = C_2 \frac{J_1^2(kr\ \sin\theta)}{(kr\ \sin\theta)^2} \qquad (3)$$

where J_1 is the Bessel function of the first order. The lateral resolution is then given by

$$FWHM_{sphere} \simeq \frac{0.55\ \lambda}{\sin\theta} \qquad (4)$$

Thus the resolution of the axicon is somewhat better, for a given aperture, than in the conventional focussing process. However, about 84% of the total power is contained within the main peak of Equ. 3 whereas the majority of the power in an axicon wave is carried in the side lobe regions. The J_o^2 function has a 1/r envelope for large values of the argument and the envelope of the function on the right of Equ. 3 varies as $1/r^3$. The axicon should therefore be useful for imaging an object composed of a few discrete points, but it is anticipated that when used within an extended scattering object the strong off-axis contributions will lead to a loss of imaging contrast, i.e. it could be said that the axicon system suffers from an inherently poor dynamic range.

As pointed out by Clarke (1981), multiple reflections within a sound beam may contribute additional (incoherent) noise to the image, which is a function of the effective beam diameter. The low f-number devices under discussion have very wide beams and so the signal-to-incoherent noise ratio is expected to be less than for a conventional system. Furthermore, since only a fraction of the total power in the wave is used to form the image of a given object point, we expect a lower sensitivity to point targets but a high sensitivity to large interfaces of a similar shape to the wavefront.

MIRROR DESIGN AND CONSTRUCTION

The geometrical design used for the mirrors (Fig. 2) is simple and offers considerable flexibility; the detailed parameters of a particular system may be chosen to accommodate a large variety of plane or focussed transducers, cone apertures, convergence angles and focal depths. The two mirror surfaces are each a figure of revolution derived from the rotation of a segment of a circle around the acoustic axis, but with the centre of curvature of the mirrors located as shown offset from the acoustic axis. Mirror M_1 causes the wave from the transducer to diverge in the plane of the figure, with rotational symmetry about the acoustic axis, and M_2 changes this diverging wave front to the conical form required.

A prototype system was designed with the application of breast scanning (with the patient lying in the prone position) in mind. It was decided to use plane transducers, and to fix the depth of focus and convergence angles at 10 cm and 30° respectively. With these parameters and the diameter of the transducer fixed, the dimensions of the mirrors shown are readily derived. It was assumed in this design that the small mirror is located in the far field of the transducer and that the wave from the transducer is plane. In fact, when used, the system was found not to be very sensitive to variations in the distance between the transducer and the small mirror.

The mirrors were made of brass, with the surfaces polished to an optical finish. Tolerances on radii were about 0.1 mm, where possible. Relative positioning was achieved by cutting the final surfaces after assembly. The system was so made that it could be used in the tank being developed for breast scanning, with any one of several standard transducers. The mirrors could be readily dismounted so that the breast scanner could be used in its normal mode for direct comparison. The mirrors are shown in Fig. 3, and again schematically, mounted in the breast tank, in Fig. 4.

RESULTS

Tests of the system were carried out with a 3.7 MHz, 7.5 mm diameter plane transducer having a pulse length of about 3 cycles, and a 3.6 MHz, 40 mm diameter bowl transducer focussed at 145 mm and having a pulse length of about 6 cycles. Fig. 5a shows the ray plot for the former of these transducers. In spite of the fact that the mirrors were not designed to be used with a focussed input wave the ray plot of Fig. 5b suggests that quite good conical wave fronts might still be expected from this arrangement. For both arrangements a 0.3 mm diameter nylon filament was scanned at various positions within the expected focal range by linear translation of the complete mirror/transducer assembly. Examples of the resulting lateral pulse-echo profiles, obtained in the manner described by Bamber and Phelps (1977), are provided in Figs 6a and 6b. Many of

Fig. 3. The subassembly of the mirrors used showing M_1 on the left with a rubber aperture stop in place, and M_2 on the right with the screws used for adjusting the orientation with respect to the incoming sound beam.

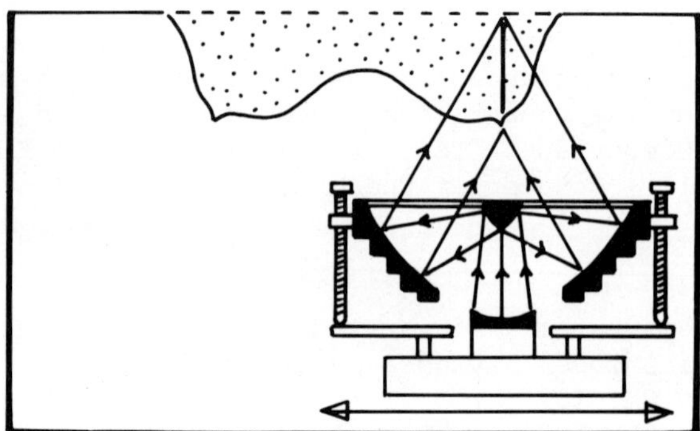

Fig. 4. Schematic diagram illustrating the proposed manner of use of the mirrors in the breast scanning tank.

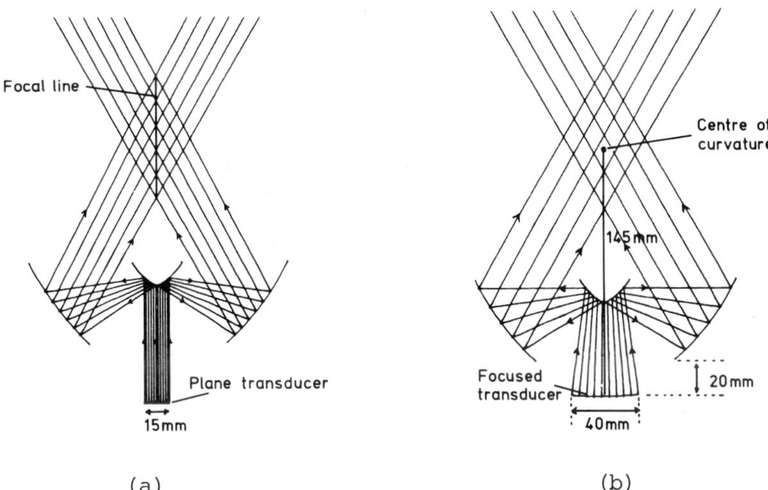

(a) (b)

Fig. 5. Computer ray plots for the mirror system in use with a plane transducer (a) and a focussed bowl (b).

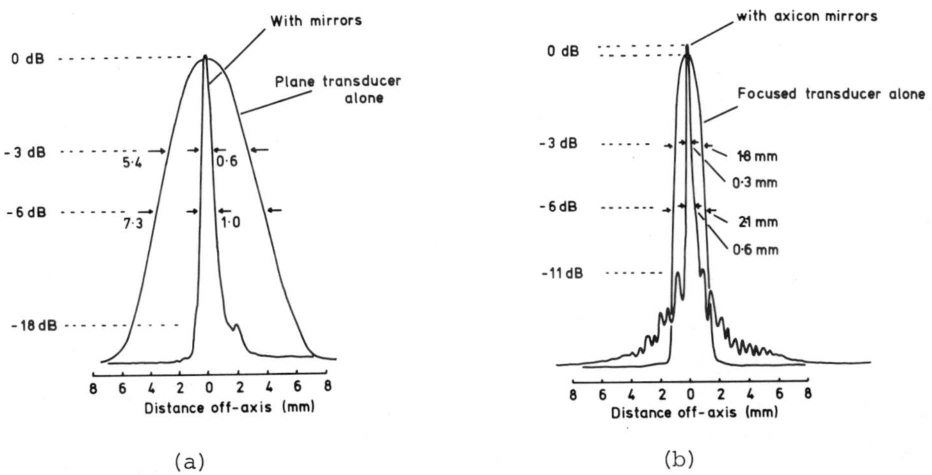

(a) (b)

Fig. 6. Lateral pulse-echo profiles for a line target corresponding to the arrangements (a) and (b) in Fig. 5. Positional accuracy was $\pm$ 0.1 mm and the axial distance for the target in (b) was 158 mm.

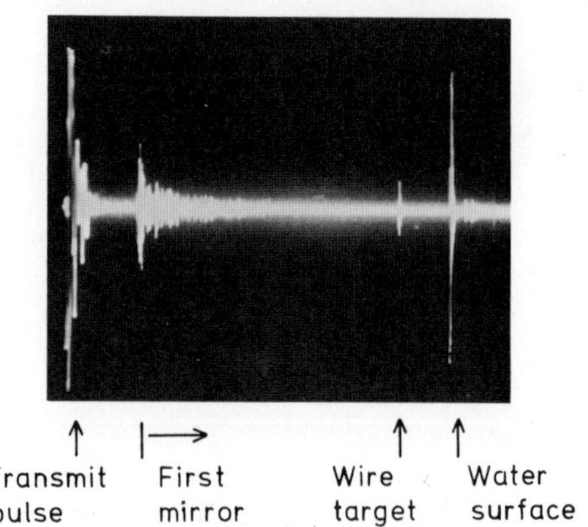

Fig. 7. A-scan from the axicon. The reverberations from the interior of the mirrors are clearly visible.

these plots were not completely symmetrical; this is presumed to be due to lack of perfect alignment between all the components of the system (including the target). It may be seen that over the central region a resolution (-6 dB width, which is the FWHM) of about 1.4λ was achieved (Equ. 2 would predict 0.85λ for a point target). As expected, smoothed side lobes and high intensity levels far from the central maximum were also observed. The use of a line, rather than a point, target tends to emphasize this region and to broaden the maximum.

Although, as expected, the sensitivity (i.e. echo height) does not drop off rapidly with distance down the axis, spurious reflections, arising predominantly from reverberations within mirror M_1, interfered with the ability to make detailed measurements over a large part of the expected focal region. These are clearly visible in the A-scan in Fig. 7. It was possible to note, however, that the full width at the -3 dB level remained approximately constant for at least 10 cm of range.

The images presented in Figs 8 and 9 are the results of scanning a piece of ordinary cellulose "sponge", which had been thoroughly degassed and had a hole, approximately 18 mm diameter, cut in it. This particular type of sponge has a wide range of sizes of cells (~ 0.05 to 3 mm) which can act as scattering structures. It will be seen from the figures that images of good quality are possible with the axicon mirror system. The improvement in image quality, over that obtained using the conventional transducers, is largely due to the reduced "speckle", which is a consequence of the large aperture of the axicon. The lateral resolution is clearly much improved

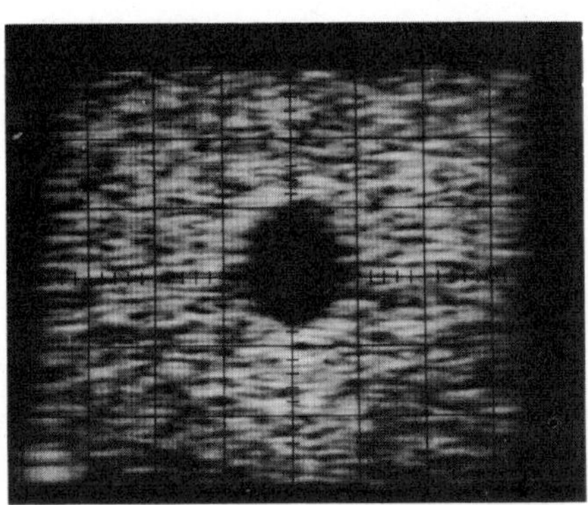

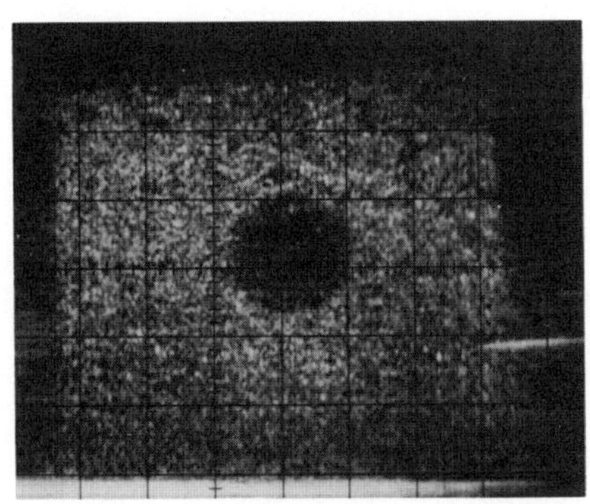

(a) (b)

Fig. 8. B-scans of the cellulose sponge taken with the plane transducer alone (a), and then with the mirrors in place (b). For the conventional image (a) the graticule is calibrated 1 cm/div. The sponge was scanned left-right and the direction of sound propagation is top to bottom.

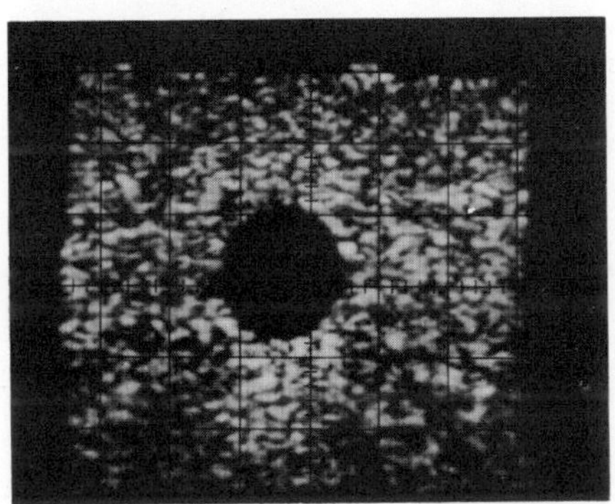

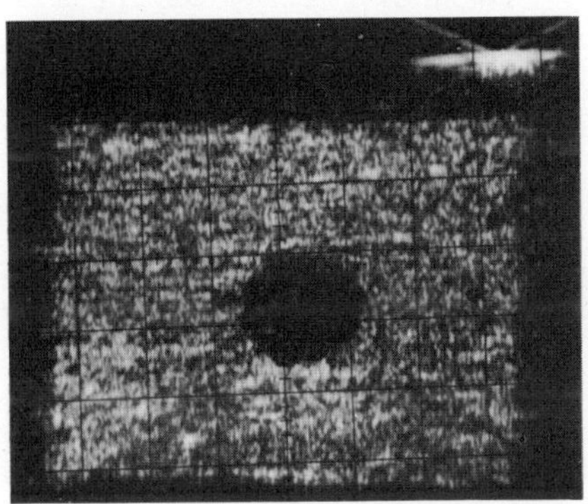

(a) (b)

Fig. 9. As for Fig. 8 but the images were created with the focussed transducer alone (a), and with the focussed transducer plus the axicon mirrors (b). The section on results contains the transducer details.

but the speckle will also tend to be smoothed somewhat by an averaging of the returned pressure impulse over a range of scattering angles, from 180° to 120°. It is interesting that the effective beam width has been reduced so much that for Fig. 9b, where the sound pulse is fairly long, the streaks which make up the speckle pattern are in the axial direction rather than the lateral direction which is more usual. Note also that, whereas the image quality falls off with distance in 8a and 9a, it is constant with depth in 8b and 9b.

Figs 8 and 9 also demonstrate very well the trade-off between point resolution and image contrast predicted by the axicon theory. In the top right corner of Fig. 9b there appears a prominant star shaped artefact which is known to arise from the strong reflection off the wire frame used to support the sponge. The star shape is a direct consequence of the conical shape of the wavefront and the relatively high level of the skirts of the resolution curve of Fig. 6. Presumably all parts of the axicon images have this star pattern associated with them but only in the case of strongly reflecting point objects is the pattern visible. For a distributed scattering object like the sponge the result is a general reduction of image contrast between the sponge and the 18 mm hole, as compared to the images made using the conventional transducers. Additional artefacts present in Figs 8b and 9b, seen as a faint series of lines crossing the images, are due to the mirror reverberations described earlier.

The sensitivity of the mirror system, judged by the ratio of the echo amplitude for a target at a given position to the amplitude of the same echo using the same transducer without the mirrors in place, was seen to vary with the shape of the target and with the transducer used. For the sponge echoes the mirror sensitivity was -15 dB when the focussed transducer was used and -6 dB with the plane transducer. A plane perspex surface placed normal to the beam at the focal plane of the focussed transducer, however, yielded a mirror sensitivity of about -7 dB.

CONCLUSIONS

The potential for high resolution over a large depth of focus and excellent reduction of the coherent speckle artefact has been shown to apply as well as can be expected in the presence of the artefacts and noise inherent in the present form of axicon. The system of mirrors used to produce the wide aperture conical wave is both simple and highly flexible; mirrors may be designed for almost any convergence angle or depth of focus, and for use with most easily available B-scanning transducers.

The present low sensitivity and noise due to reverberations in the mirrors have prevented us from obtaining clinically useful breast images. Nevertheless, the results of scanning volunteers have demonstrated most of the features mentioned above. Clearly, however, for the device to realize its potential usefulness the difficulties already discussed must be overcome. The reverberation noise can be reduced by making the back surfaces of the mirrors non-reflecting. Preliminary experiments have shown that a random pattern of small holes drilled in the back face, or backing the mirrors with lead, can be successful in this regard. A better choice of reflector material must be considered, which may also help to improve the sensitivity.

The star artefacts and poor imaging dynamic range associated with the high skirt (i.e. side lobe) level may prove more difficult to overcome. It is likely that some form of computer image treatment would enable the obvious star-shaped artefacts to be removed, but the poor contrast for distributed scattering media seems to be a sort of fundamental limitation of the axicon and may be serious. At present it seems sensible to limit the depth of focus, and hence the width of the conical beam, as much as possible. One method of achieving this in a variable manner is to use a transducer divided into annular zones which can be actuated individually or together in any combination. Some form of compromise between the width of the central peak and the height of the skirt, such as the use of a separate focussed receiver as suggested by Patterson and Foster (1981), may ultimately be necessary. It also seems likely that, for best results, it will be found that a particular design of axicon will be needed for each target organ.

It has been commented that, at least for the particular application of breast imaging, refraction at the tissue surface may degrade the image quality considerably. The results taken up to the present do not give any clear indication of how serious this problem will be, but it may constitute a further limitation on the resolution and depth of focus achievable.

Finally, we conclude that, in spite of some problems, the axicon principle has great potential for ultrasound imaging, and well merits further development. The original organ of interest, the breast, remains a most promising region for investigation.

ACKNOWLEDGEMENTS

It is a pleasure to acknowledge the efforts of Mr P. Collins, who fabricated the mirrors and their mountings to the high degree of precision required in a very short time. One of us (RLC) would like to acknowledge the hospitality and support of the Institute of Cancer Research while this work was carried out. He would also like to thank Dr J. Hunt for several useful conversations.

REFERENCES

Arditi, M., Taylor, W.B., Foster, F. S. and Hunt, J. W., 1982, An annular array system for high resolution breast echography, Ultrasonic Imaging, 4:1.

Bamber, J. C. and Phelps, J. V., 1977, The effective directivity characteristic of a pulsed ultrasound transducer and its measurement by semi-automatic means, Ultrasonics, 14:169.

Burckhardt, C., Hoffmann, H. and Grandchamp, P. A., 1973, Ultrasound axicon: a device for focussing over a large depth, J. Acoust. Soc. Amer., 54:1628.

Burckhardt, C. B., Grandchamp, P. A. and Hoffmann, H., 1975, Focussing ultrasound over a large depth with an annular transducer - an alternative method, IEEE Trans. Sonics and Ultrasonics, SU-22:11.

Clarke, R. L., 1981, The contribution of multiple scattering to the observed backscattered intensity from an ultrasound beam. Acoustics Letters, 4:224.

Dick, D. E., Elliott, R. D., Metz, R. L. and Rojohn, D. S., 1979, A new automated, high resolution ultrasound breast scanner, Ultrasonic Imaging, 1:368.

Foster, F. S., Arditi, M. and Hunt, J. W., 1980, Scatter imaging with cylindrical and conical transducers (abstract from Ultrasonic Imaging and Tissue Characterization Symposium), Ultrasonic Imaging 2:180.

Foster, F. S., Patterson, M. S., Arditi, M. and Hunt, J. W., 1981, The conical scanner: a two transducer ultrasound scatter imaging technique, Ultrasonic Imaging, 3:62.

Fujiwara, S., 1962, Optical properties of conic surfaces. I. Reflecting cone, J. Opt. Soc. Amer., 48:287.

Lit, J. W. Y. and Brannen,E., 1970, Optical properties of a reflecting cone, J. Opt. Soc. Amer., 60:370.

Melton, H. E. and Thurstone, F. L., 1978, Annular array design and logarithmic processing for ultrasonic imaging, Ultrasound Med. Biol., 4:1.

McLeod, J. H., 1954, The axicon: a new type of optical element, J. Opt. Soc. Amer., 44:592.

Patterson, M. S. and Foster, F. S., 1981, Acoustic fields of conical radiators, IEEE Trans. Sonics and Ultrasonics, SU-29:83.

Weight, J. P. and Brown, A. F., 1982, Improved resolution ultrasonic transducers, in: "Proceedings of the 10th World Congress on NDT", Moscow.

ENHANCED TRAPPED ENERGY MODE ARRAY TRANSDUCER USING THICKNESS OVERTONES*

D. V. Shick, H. F. Tiersten, R. P. Kraft,
J. F. McDonald, and P. K. Das

Rensselaer Polytechnic Institute
Troy, New York 12181

ABSTRACT

A trapped energy mode of thickness-extensional (TE) vibrations is shown to exist at certain overtones in various materials. Utilization of such a mode, which results in improved isolation between elements in a transducer array, is demonstrated. Knowledge of the dispersion curves (frequency vs. lateral wave number) for unelectroded and fully electroded plates, along with an understanding of the manner in which they determine trapped energy modes, is essential in order to choose a suitable overtone: one which, first, will exhibit TE energy trapping, and second, will allow for large imaginary wave numbers in the unelectroded region of the plate. Dispersion curves for PZT-7A and Z-cut lithium tantalate are examined and found to possess suitable overtones. Improved trapped energy mode spatial confinement is confirmed.

INTRODUCTION

The properties of trapped energy mode transducer arrays have been studied for several years by the authors [1,2,3,4,5,6,7,8] and imaging systems have been constructed using these arrays [9,10]. A monolithic mosaic transducer utilizing an essentially trapped energy thickness-extensional mode possesses certain fabrication advantages and avoids certain difficulties inherent in other procedures used to obtain acoustic isolation between elements [11,12].

*Partially supported by NSF under Grant No. ECS-8024297.

To date emphasis has been placed on the use of the dispersion curves in the vicinity of the fundamental TE branch and the lowest order complex branch [6,8,13]. However, the dispersion curves actually contain a large number of other branches which were not discussed in previous papers on TE trapped energy arrays. These correspond to various overtones and additional complex branches in the dispersion plots. The shape of these curves is clearly of importance in predicting the degree of spatial confinement achievable with energy trapping utilizing these overtones. These plots in turn must be computed from available material constants [14]. Some interesting results have been found for the fifth overtone of PZT-7A and most especially for the third overtone of Z-cut lithium tantalate which suggest that a greater degree of spatial energy confinement is feasible with these materials. Furthermore, the potential bandwidth over which trapping can occur for lithium tantalate is as large as that for PZT-7A operated in the fundamental mode.

CONDITIONS FOR TRAPPING WITH OVERTONES

A detailed mathematical formulation of the appropriate analysis for energy trapping has appeared in most of the previous papers by the authors on this subject. This formulation can readily be extended to handle trapping with overtones by including terms in the approximate eigensolutions corresponding to the branches of the relevant dispersion curves for the infinite electroded and unelectroded plate which are significant when operating in the vicinity of an overtone resonance frequency. The relevant geometrical considerations as in the past treatments are summarized in Fig. 1.

The desired isolation is achieved by employing the frequency lowering which results from the mass loading and electrical shorting of the electrodes [15,16,17]. Trapping is achieved by operating in a frequency range between the frequencies of thickness vibration of the electroded and unelectroded regions. This causes the dominant component of the mode in the unelectroded region to decay with distance from the electrode. It should be noted that this decay is not the result of dissipation of energy. Furthermore, while energy

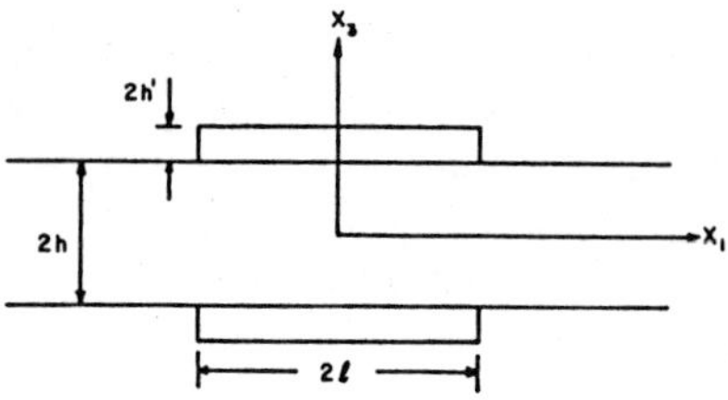

Fig. 1 Geometry of the trapped energy mode device.

can be stored in this dominant evanescent mode, during trapping little energy escapes by propagation along the lateral directions of the plate.

One of the important considerations in many applications is the fabrication of a large thickness-extensional transducer array that has effective transducer elements as small as possible so that the radiation lobe will be as wide as possible for each element. In the conventional trapped energy thickness-extensional mode transducer, which utilizes the fundamental thickness-extensional mode of a PZT-7A plate, the effective width of the transducer is limited by the spatial attenuation of the mode in the unelectroded region of the plate even for very small electrodes. The shape of the dispersion curves for suitable overtones can significantly affect this attenuation.

Dispersion curves for the piezoelectric ceramic PZT-7A poled in the thickness direction are shown in Fig. 2a for the unelectroded plate and in Fig. 2b for the fully electroded plate. These curves give the frequency as a function of the wave number for symmetric modes, calculated as in Ref. [5].

These are illustrated for the fundamental through the fifth overtone. The frequency range in the vicinity of the fifth overtone is shown in Fig. 3. When the dispersion curves are on the left of the vertical axis, the wave number is purely imaginary and the associated displacement field decays with lateral distance. It is important to note that in these figures, frequency and wave number

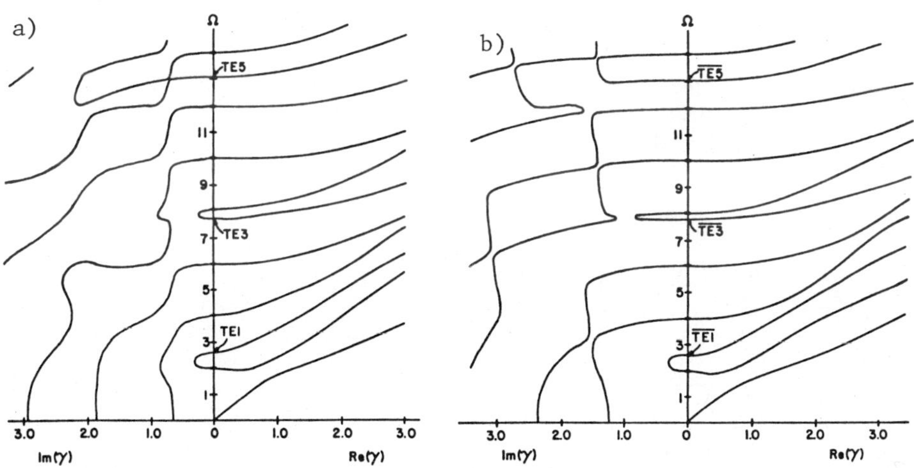

Fig. 2 Dispersion curves for a) unelectroded PZT-7A plate and b) fully electroded PZT-7A plate.

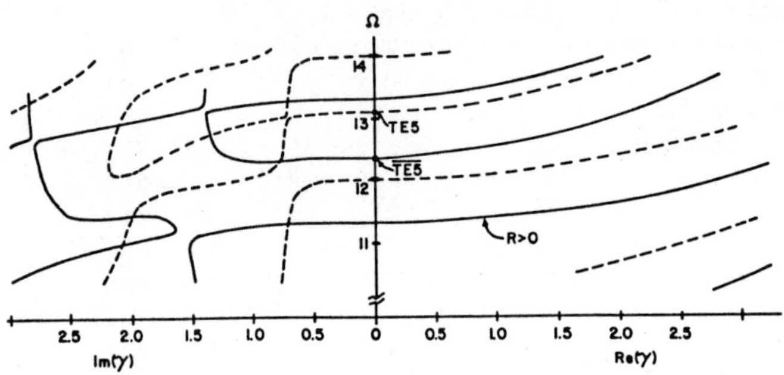

Fig. 3 Dispersion curves in the vicinity of the fifth thickness-extensional overtone for PZT-7A. Solid lines are for the electroded plate including effect of mass loading; dashed lines are for the unelectroded plate.

have been normalized with respect to the first thickness-shear frequency. That is, $\Omega=\omega/\omega_1^{TS}$ and $\gamma=2h\xi/\pi$ where Ω is the normalized frequency, ω is the actual (circular) frequency, ω_1^{TS} is given by Eq. (1a) below with n=1, γ is the normalized lateral wave number, ξ is the actual lateral wave number, and 2h is the plate thickness. The shape and position of these curves are critical in determining the conditions under which energy trapping can be realized.

The critical thickness frequencies (with $\xi = 0$ or zero longitudinal wave number) in the unelectroded plate are given by

$$\omega_n^{TS} = \frac{n\pi}{2h}\left(\frac{c_{55}}{\rho}\right)^{1/2} \qquad n = 2, 4, 6 \ldots, \tag{1a}$$

$$\omega_m^{TE} = \frac{m\pi}{2h}\left(\frac{\bar{c}_{33}}{\rho}\right)^{1/2} \qquad m = 1, 3, 5 \ldots, \tag{1b}$$

and in the electroded plate by

$$\bar{\omega}_n^{TS} = \frac{n\pi}{2h}\left(\frac{c_{55}}{\rho}\right)^{1/2}(1-R) \qquad n = 2, 4, 6 \ldots, \tag{2a}$$

$$\bar{\omega}_m^{TE} = \bar{\eta}_m\left(\frac{\bar{c}_{33}}{\rho}\right)^{1/2} \qquad m = 1, 3, 5 \ldots, \tag{2b}$$

where ω_n^{TS} and $\bar{\omega}_n^{TS}$ are the thickness-shear frequencies of the unelectroded and electroded plates, respectively, and where ω_m^{TE}

and $\bar{\omega}_m^{TE}$ are the thickness-extensional frequencies of the unelectroded and electroded plate. Also, $\bar{\eta}_m h$ is the mth root of

$$\tan(\bar{\eta}_m h) = \bar{\eta}_m h/(k_{33}^2 + R\bar{\eta}_m^2 h^2) , \tag{3}$$

where

$$k_{33}^2 = e_{33}^2/(\bar{c}_{33}\ \varepsilon_{33}) , \tag{4a}$$

$$R = 2\rho' h'/(\rho h) , \tag{4b}$$

$$\bar{c}_{33} = c_{33} + e_{33}^2/\varepsilon_{33} , \tag{4c}$$

in which c_{pq}, e_{33}, ε_{33}, ρ and ρ' denote the elastic constants, a piezoelectric constant, a dielectric constant, the mass density of the plate, and the mass density of the electrode. The quantities 2h and 2h' are the thicknesses of the plate and one of the electrodes, respectively.

To determine which thickness-extensional overtones are suitable for energy trapping one must compute the complete dispersion curves (Figs. 2-5) for the materials considered. Trapping useful for extensional wave imaging applications can be achieved by operating at a frequency between the TE frequencies of the unelectroded and electroded plates such that the longitudinal wave number ξ is imaginary (corresponding to exponential decay with distance from the electrode) in the unelectroded region and real (corresponding to a standing wave solutions) in the electroded region.

a)

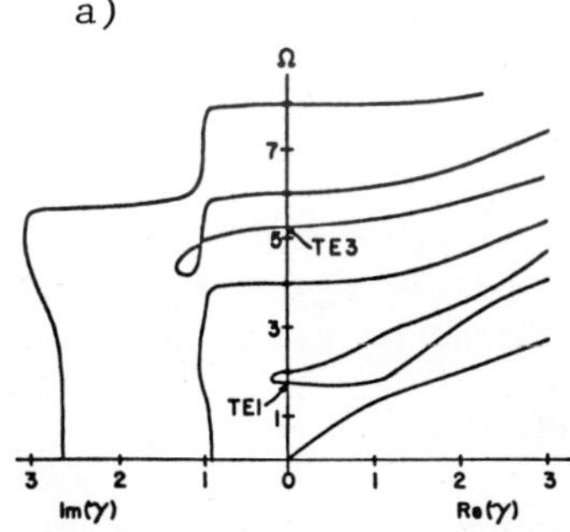

b)

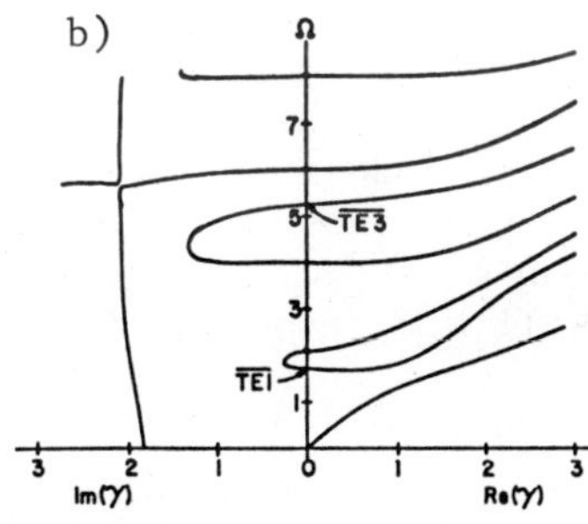

Fig. 4 Dispersion curves for a) unelectroded Z-cut $LiTaO_3$ plate and b) fully electroded Z-cut $LiTaO_3$ plate.

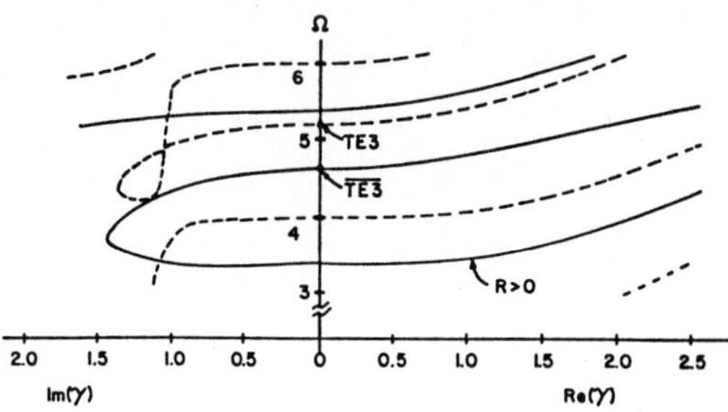

Fig. 5 Dispersion curves in the vicinity of the third thickness-extensional overtone for $LiTaO_3$. Solid lines are for electroded plate including effect of mass loading; dashed lines are for the unelectroded plate.

Since the thickness frequency for an extensional mode is always less for the electroded plate than for the unelectroded plate (see Eqs. (1b), (2b), and (3)) the primary concern is that the imaginary branch of the unelectroded plate's dispersion curve should curve down from the $\xi = 0$ thickness-extensional frequency.

Furthermore, since in the unelectroded region the lateral dependence of the (dominant) displacement in the x_3 direction behaves as $\exp(-|\xi|x_1)$ where x_1 is 0 at the electrode edge (see Eq. (8) of Ref. [8]), it is desirable to have $|\xi|$ as large as possible in order that the displacement of the plate surface be confined as closely as possible to the electroded area. This suggests that the downward curve of the dispersion plot should make an excursion leftward in the $Im(\xi)$ direction as far as possible.

Finally, the potential bandwidth over which TE trapping might occur is partially dictated by the frequency range over which the TE branch persists in curving downward.

Referring to Fig. 2, it is evident that the fundamental and fifth overtone thickness extensional modes (i.e. ω_1^{TE} and ω_5^{TE}, marked TE1 and TE5 in the figure) are appropriate in PZT-7A, whereas the third overtone (ω_3^{TE} or TE3) is not. Furthermore, the fifth overtone curve extends quite far in the $Im(\xi)$ region offering the possibility of enhanced isolation when compared with the fundamental mode. The potential bandwidth for trapping, however, is limited by the presence of the lower branches of the dispersion plot.

On the other hand, in Fig. 4 are shown dispersion curves for unelectroded and electroded Z-cut $LiTaO_3$ from which it is clear that, while trapping is impossible for the fundamental TE mode, the third overtone offers the possibility of excellent trapping indeed. Furthermore, the potential bandwidth for trapping is quite good.

Nevertheless, for higher-order modes than the fundamental the difference between thickness-extensional frequencies ω_n^{TE} and $\bar{\omega}_n^{TE}$ due to the shorting effect alone is diminished, since from Eq. (3) with R = 0, $\bar{\eta}_m h \rightarrow m\pi/2$ as m increases (c.f. Eqs. (1b), (2b)).

Therefore, to realize energy trapping at the desired frequency the thickness-extensional frequency must be lowered by introducing mass loading. This can be achieved by using thicker or denser electrodes. Increasing the thickness greatly, however, affects the aspect ratio h'/ℓ of the electrode. For very small electrode widths 2ℓ, this would create the possibility of lateral vibrations within the electrode itself which could increase the coupling to other modes. These, in turn, could either propagate or excite additional evanescent modes in the vincinity of the edge of the electroded region contributing to the difficulty of achieving isolation for small ℓ. Hence improvements in R demand use of extremely dense metals such as gold or tungsten (sp. gr. 19.3), lead (sp. gr. 11.3) or platinum (sp. gr. 21.4).

Another factor which contributes to the excitation of evanescent modes in the vicinity of the electrode edge is the width of the electrode. This can affect the way in which the electroded and unelectroded solutions "fit" together at the electrode edge. A complete calculation has been performed for PZT-7A including traveling-wave modes [8]. A typical mode shape is shown in Fig. 6.

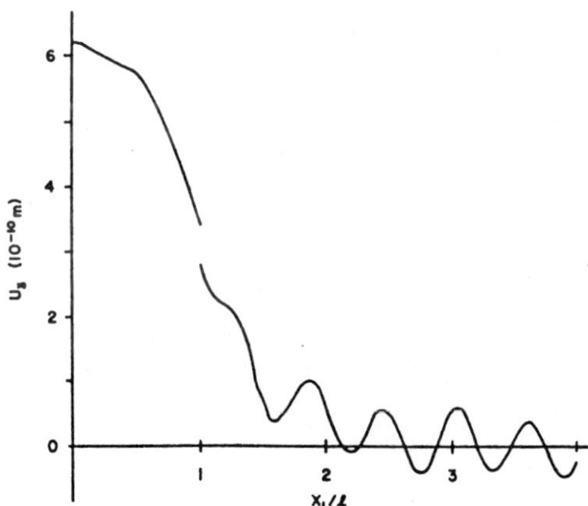

Fig. 6 Calculated thickness-displacement modeshape [Ref. 8] showing partial lack of energy trapping.

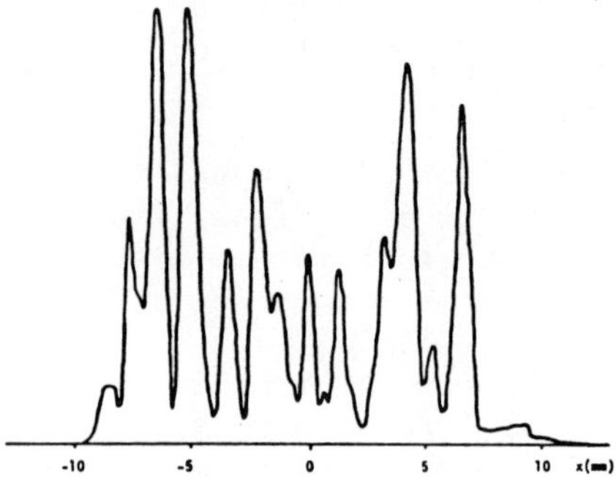

Fig. 7 Measured thickness-displacement modeshape showing absence of trapping.

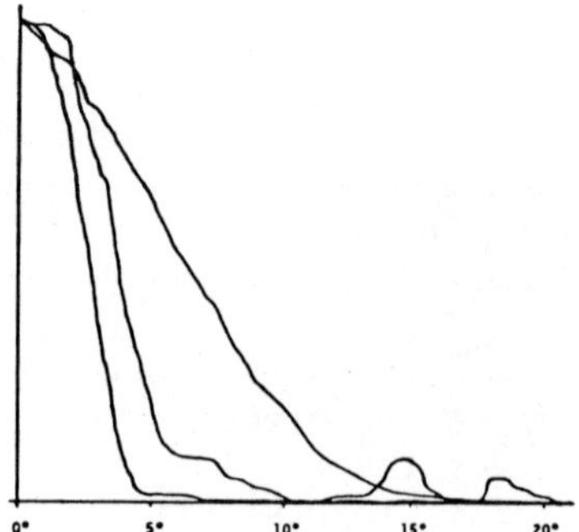

Fig. 8 Measured farfield radiation plot for the fundamental and fifth-overtone TE modes in PZT-7A and the third-overtone TE mode in $LiTaO_3$.

An obvious disadvantage of using overtones is the increasing multiplicity of lower order traveling wave modes which can couple in the unelectroded region to the dominant mode in the electroded region. Fig. 7 shows a typical surface measured radiation plot when operating outside the valid frequency range for trapping for the third overtone for lithium tantalate.

MEASUREMENTS WITH OVERTONES

Figure 8 shows the measured farfield radiation plots when operating at the optimum frequency for spatial trapping for PZT-7A and $LiTaO_3$. The corresponding farfield half beamwidth angles (to the first null) of 5° and 10° were measured for the fundamental and fifth overtone in PZT-7A. The angle of 15° was measured for the optimum frequency of the third overtone for the $LiTaO_3$ plate.

CONCLUSION

Trapped energy mode transducer arrays utilizing the fifth overtone of PZT-7A and the third overtone of $LiTaO_3$ are feasible. The potential advantage of operating with these branches of the dispersion curves is the additional spatial isolation due to their larger excursions leftward below the corresponding TE frequencies at zero longitudinal wave number. A potential disadvantage is the existence of a larger number of dispersion curve branches corresponding to laterally propagating modes which can be excited by coupling. This coupling is affected by electrode width, height, and mass density.

ACKNOWLEDGEMENT

The authors wish to express their gratitude for partial financial support from the National Science Foundation under Grant ECS-8024297. Many thanks are also due to Paul Cohen for processing of many of the plates used in this work.

REFERENCES

1. H. F. Tiersten, J. F. McDonald, and P. K. Das, "Monolithic Mosaic Transducer Utilizing Trapped Energy Modes", Appl. Phys. Lett. 29 (12), 761-763, (1976).

2. H. F. Tiersten, J. F. McDonald, and P. K. Das, "Two-Dimensional Monolithic Transducer Array", in Proc. 1977 IEEE Ultrasonics Symposium, 1977, pp. 408-412.

3. H. F. Tiersten, B. K. Sinha, J. F. McDonald, and P. K. Das, "On the Influence of a Tuning Inductor on the Bandwidth of Extensional Trapped Energy Mode Transducers", Proc. 1978 IEEE Ultrasonics Symposium, 1978, pp. 163-166.

4. P. Das, G. A. White, B. K. Sinha, C. Lanzl, H. F. Tiersten, and J. F. McDonald, "Ultrasonic Imaging Using Monolithic Mosaic Transducer Utilizing Trapped Energy Modes", in Acoustical Imaging, edited by A. F. Metherell (Plenum, New York, 1978), Vol. 8, pp. 119-135.

5. H. F. Tiersten and B. K. Sinha, "An Analysis of Extensional Modes in High Coupling Trapped Energy Resonators", Proc. 1978 IEEE Ultrasonics Symposium, 1978, pp. 167-171.

6. P. K. Das, S. Talley, H. F. Tiersten, and J. F. McDonald, "Increased Bandwidth and Mode Shapes of the Thickness-Extensional Trapped Energy Mode Transducer Array", Proc. 1979 IEEE Ultrasonics Symposium, 1979, pp. 148-152.

7. H. F. Tiersten and B. K. Sinha, "Mode Coupling in Thickness Extensional Trapped Energy Resonators", Proc. 1979 IEEE Ultrasonics Symposium, 1979, pp. 142-147.

8. D. V. Shick, H. F. Tiersten, and B. K. Sinha, "Forced Thickness Extensional Trapped Energy Vibrations of Piezoelectric Plates", Proc. 1981 IEEE Ultrasonics Symposium, 1981, pp. 452-457.

9. P. Das, S. Talley, R. Kraft, H. F. Tiersten, and J. F. McDonald, "Ultrasonic Imaging Using Trapped Energy Mode Fresnel Lens Transducers" in Acoustical Imaging, edited by K. Y. Yang, (Plenum, New York, 1980) Vol. 9, pp. 75-92.

10. N. Saiga and T. Suzuki, "Dependence of Local Sound Vibration on Time Frequency in a Monolithic Array Transducer", Appl. Phys. Lett. 40 (3), 220-222 (1982).

11. S. W. Smith, O. T. Von Ramm, M. E. Haran and F. L. Thurstone, "Angular Response of Piezoelectric Elements in Phased Array Ultrasound Scanners," IEEE Trans. on Sonics and Ultrasonics, Vol. SU-26 (#3) May 1979, pp. 185-191.

12. M. G. Magninness, J. D. Plummer, and J. D. Meindl, in Acoustical Holography (Plenum, New York, 1974), Vol. 5, p. 619.

13. H. Watanabe, K. Nakamura, and H. Shimizu, "A New Type of Energy Trapping Caused by Contributions from the Complex Branches of Dispersion Curves", 1980 IEEE Ultrasonics Symposium, 1980, pp. 825-828.

14. H. Jaffe and D. A. Berlincourt, "Piezoelectric Transducer Materials", Proc. IEEE 53, 1372 (1965).

15. W. Shockley, D. R. Curran, And D. J. Koneval, J. Acoust. Soc. Am. 41 (4) pp. 991-993 (1967).

16. a) M. Onoe and H. Junonji, Electronics and Comm. Eng. (Japan) 84 (1965); b) M. Onoe, H. Jumonji, and N. Kobori, in Proc. 20th Annual Symposium on Frequency Control (U.S. Army Signal Engineering Laboratory, Fort Monmouth, N.J., 1966) p. 266.

17. R. A. Sykes and W. D. Beaver, in Ref. 16b, p. 288.

A THEORETICAL STUDY OF THE TRANSIENT BEHAVIOUR OF ULTRASONIC TRANSDUCERS IN LINEAR ARRAYS

M. Th. Larmande, P. Alais

Laboratoire de Mécanique Physique
E.R.A. C.N.R.S. 537
Université Pierre et Marie Curie
78210 Saint-Cyr-l'Ecole, France

INTRODUCTION

It is well recognized that in classical echographic systems using transducers arrays, the behaviour of the elementary transducer inside the array is the most unknown, or at least not understood characteristics of the system. Even if for linear arrays a bidimensional model may be retained describing the transducers, the transient behaviour is affected in a very complex manner by the number of excited modes and the coupling effects between adjacent transducers. Recently, several authors {1,2,3} have pointed the importance of the width to thickness ratio for each individual block of ceramics used inside the array. However, a theoretical study of the transducer taking in account its real environment and specially coupling effects inside the array and through the propagative medium is difficult to achieve. We had presented at the 10th I.S.A.I. a tentative of analog electrical simulation of the problem {4}. But we met a lot of difficulties in the realization of adequate transformers and we preferred to go on this research work using numerical analysis. We want to show here the interest of our relatively simple model to attain the harmonic or transient behaviour of complex bidimensional structures.

THE PHYSICAL MODEL AND THE ASSOCIATED FINITE ELEMENT METHOD

We have retained the same physical approach as in {4}, i.e. assumed that the deformation is exclusively plane and described by (x_1,x_3) variables and also that the electrical induction exercised through the plane electrodes orthogonal to x_3 (Fig. 1) remains quasi parallel to x_3, $\vec{D} = D_3(x_1)\,\vec{x}_3$. The classical matricial piezo-

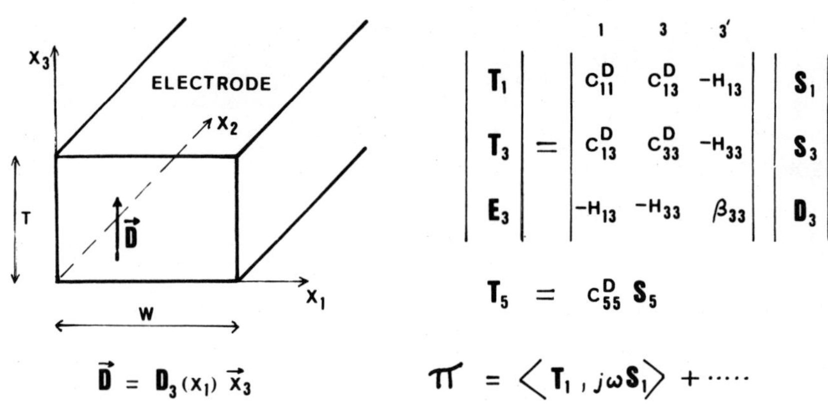

Figure 1 - The simplified physical model for a 2-D transducer.

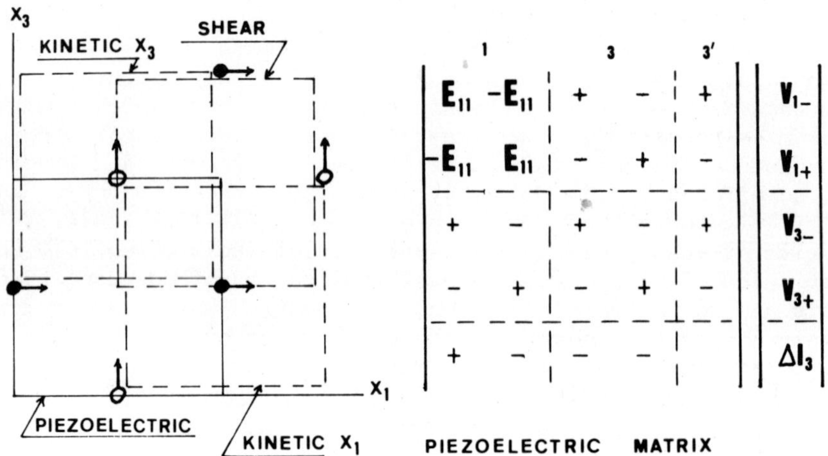

Figure 2 - Volume splitting in terms of the different involved energies - Example of the piezoelectric matrix.

electric relation is then reduced ($S_2 = S_4 = S_6 = 0$, $D_1 = D_2 = 0$) to uncoupled relations involving (S_1, S_3, D_3) and S_5 respectively. A correct description of the transducer may be attained by splitting in elementary volumes $L\Delta x_1 \Delta x_3$, and retaining as a correct set of (extensive) mechanical variables describing the harmonic oscillation at a pulsation ω, the complex amplitudes V_1 and V_3 of the velocity according to the x_1, x_3 directions at the level of a finite number of points associated to the splitting as indicated on figure 2. The electrical variables completing correctly this extensive set are the intensities $\Delta I_3 = j\omega D_3 L\Delta x$, feeding each column of width Δx_1. Then, we may write sequentially elementary matricial impedances associated to the different involved energies stored in elementary volumes as shown in figure 2, just by writing the involved power : $\pi = [B,A] = 1/2\, B_i A_i^*$, where (B_i) is the generalized force conjugate to the extensive vector (A_i). As an example, the piezoelectric energy will be attained for an elementary volume $L\Delta x_1 \Delta x_3$ from the power density :

$$\pi = T_1 \dot{S}_1 + T_3 \dot{S}_3 + E_3 \dot{D}_3 \ ,$$

associated to the (1,3,3') relation of figure 1, through a 5×5 matrix involving (V_{1-}, V_{1+}, V_{3-},V_{3+}, $\Delta_1 I_3$) , where the 2×2 matrix associated to V_{1-}, V_{1+} will be derived from the first diagonal coefficient of the (1,3,3') relation according to :

$$\left[c_{11}^D S_1, j\omega S_1\right] L\Delta x_1 \Delta x_3 = \left[\frac{c_{11}^D}{j\omega}\frac{L\Delta x_3}{\Delta x_1}\Delta_1 V_1, \Delta_1 V_1\right]$$

$$= \left(\Gamma_{11} \begin{vmatrix} 1 & -1 \\ -1 & 1 \end{vmatrix} \begin{vmatrix} V_{1-} \\ V_{1+} \end{vmatrix} , \begin{vmatrix} V_{1-} \\ V_{1+} \end{vmatrix}\right) , \Gamma_{11} = \frac{c_{11}^D}{j\omega} L \frac{\Delta x_3}{\Delta x_1}$$

All these elementary matricial impedances are then inserted into a unique matrix representative of the studied system on the whole retained set of extensive variables. We have to take in account, with the piezoelectric energy, the shear energy ($T_5 \dot{S}_5$) and the kinetic energies in the x_1 and x_3 directions.

Obviously, for an isotropic elastic non piezoelectric material, the electrical variables disappear and the relation (1,3,3') reduces to the isotropic relation :

$$\begin{vmatrix} T_1 \\ T_3 \end{vmatrix} = \rho c^2 \begin{vmatrix} 1 & h \\ h & 1 \end{vmatrix} \begin{vmatrix} S_1 \\ S_3 \end{vmatrix} , \quad h = \frac{\nu}{1-\nu}$$

where ρ, c and ν are the density, the sonic velocity and the Poisson coefficient of the material.

Problems are encountered at boundaries for the shear energies which are associated to volumes centered on the boundary. When the boundary connects a solid material to a liquid (or vacuum) like material, we have decided to neglect the shear energy of the half elementary volume.

In a first approach, we neglect the radiative coupling, i.e. the diffraction effects on the acoustical impedance presented locally by the backing or the propagative medium at the boundary of the studied structure, so that we may write in both cases the radiated power :

$$\pi_R = \left[ZV_3, V_3\right]$$

as if it was for a plane wave, Z being the characteristic impedance $Z = \rho c$ of the backing material or the propagative medium. If we include this power in the system described by the whole matrix, the external excitation reduces (for the emitter case) to the electrical excitation :

$$\pi_E = \sum_{\ell} \left[\Phi^{\ell}, \sum_{m} \Delta I^{\ell}_{m}\right]$$

where Φ^{ℓ} is the electrical potential exercised on the electrode ℓ and $I^{\ell} = \sum_{m} \Delta I^{\ell}_{m}$ the total intensity injected through this electrode.

In these conditions, the problem is correctly set by the equations associating, through the obtained matrix, the unknown vector $(V_{1i}, V_{3j}, \Delta I^{\ell}_{3m})$ with the data vector $(0,0,\Phi^{\ell})$.

We have developed a subroutine which permits to write automatically the matrix from simple physical data, i.e. the geometry of the structure, the nature of components and the retained splitting.

In the following part, we describe results obtained for a simple ceramic block which may be compared to those already published {1} and also results interesting more complex structures, which is our real purpose.

DYNAMIC BEHAVIOUR OF A FREE RECTANGULAR SECTION CERAMIC BLOCK

We have effected this study essentially for testing our computation technique and evaluating the optimal splitting. Due to the double symmetry of this structure, it is possible to reduce the representation to a quarter and to check the effects of the splitting complexity on the obtained accuracy.

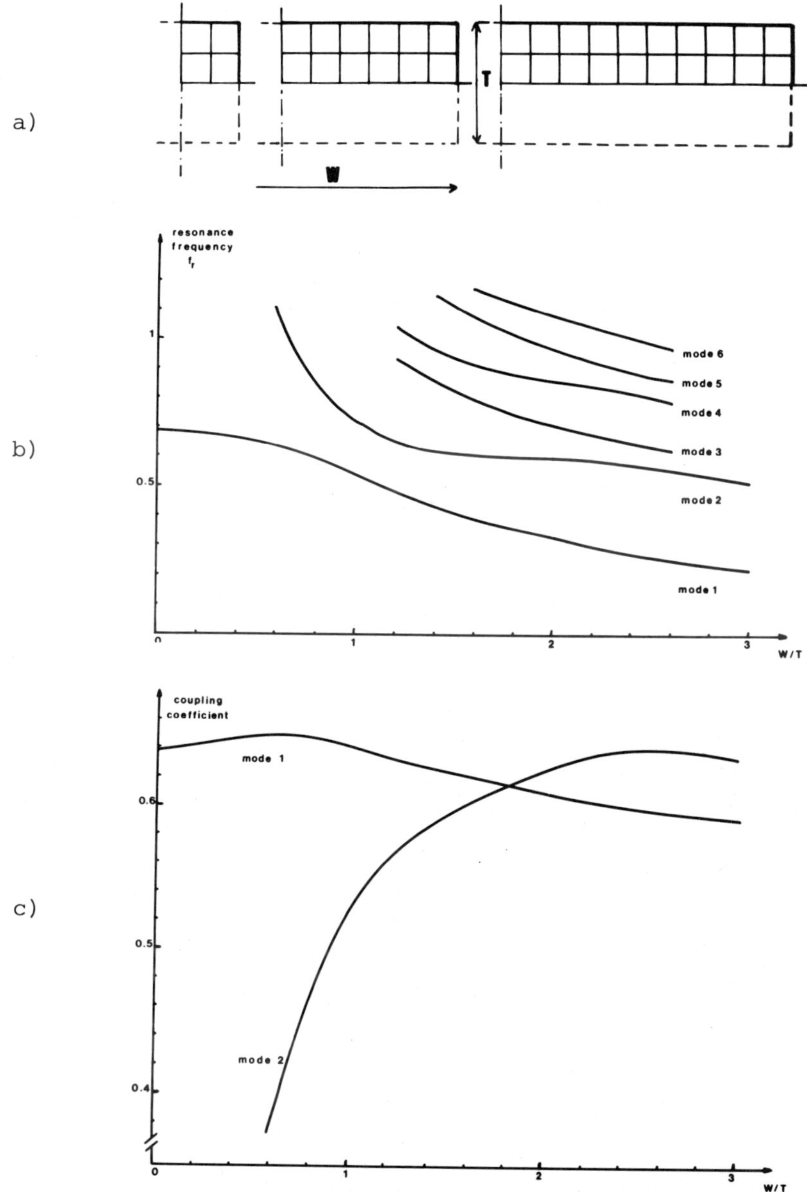

Figure 3 - Dynamical behaviour of a free rectangular section ceramic block.

a) Examples of retained splittings.
b) Resonance frequencies versus the width to thickness ratio.
c) Dynamical coupling coefficient for modes 1 and 2.

The figure 3-a shows three retained splittings for a study of the resonance modes observed for $0 < W/T < 3$. By comparison with computations effected with thinner meshes, we know that the obtained accuracy in the evaluation of the resonance frequencies shown by figure 3-b for the ceramic PZT-5A is about 10^{-2} for the first modes 1 and 2 ; it is a little rougher for higher modes. The one-dimensional thickness mode antiresonance frequency has been chosen as unit.

It is easy to check the frequency limit of the mode 1 when $W/T \to 0$. Assuming that $T_1 = 0$ in the (1,3,3') relation of figure 1 gives immediately {2} :

$$\begin{vmatrix} T_3 \\ \\ E_3 \end{vmatrix} = \begin{vmatrix} C'^D_{33} & -h'_{33} \\ \\ -h'_{33} & \beta^S_{33} \end{vmatrix} \begin{vmatrix} S_3 \\ \\ D_3 \end{vmatrix}$$

where $C'^D_{33} = C^D_{33} - (C^D_{13})^2/C^D_{11}$, $h'_{33} = h_{33} - h_{13}\ C^D_{13}/C^D_{11}$.

In the case of the PZT-5A ceramic for which all these coefficients are tabulated {5}, the velocity and the impedance of the material are reduced for the beam mode in comparison with the thickness mode by the ratio $v'_D/v_D = Z'_D/Z_D = (C'^D_{33}/C^D_{33})^{1/2} = f'_a = 0.878$, which is the adimensional value of the antiresonance frequency for $W/T \to 0$. Taking into account the value of the static beam coupling coefficient $k'_{33} = 0.66$ {5}, the resonance frequency f'_r is solution of :

$$\frac{\pi f'_r}{2f'_a} = (k'_{33})^2 \ \mathrm{tg}\ \frac{\pi f'_r}{2f'_a}$$

i.e. $f'_r = 0.688$, in excellent agreement with our computed value of $f'_r = 0.684$ for $W/T = 0.1$ (Fig. 3-b).

The figure 3-c shows the evolution of the dynamical coupling coefficient performed in the same conditions for the modes 1 and 2. This coefficient is obtained from its classical definition : $k' = E_m/(E_e E_d)^{1/2}$, where E_e, E_d and E_m are respectively the pure elastic energy, the pure dielectric energy and the electromechanical energy. Our results confirm the first observation by SATO {1} of a maximum of the coupling coefficient of the mode 1 in the neighbourhood of $W/T = 0.6$. It is interesting to note that the limit obtained for $W/T \to 0$ is not the static coefficient k'_{33}, but should be the dynamic one $(k'_{33})_{dyn} = \alpha k'_{33}$, where the coefficient α is obtained from an elementary calculation :

$$\alpha = \sqrt{2}\ \frac{\sin(\varphi/2)}{(\varphi/2)}\ \left(1 + \frac{\sin\varphi}{\varphi}\right)^{-1/2} , \quad \varphi = \pi\ f'_r/f'_a ,$$

taking into account the sinusoidal spatial variation of S, whereas

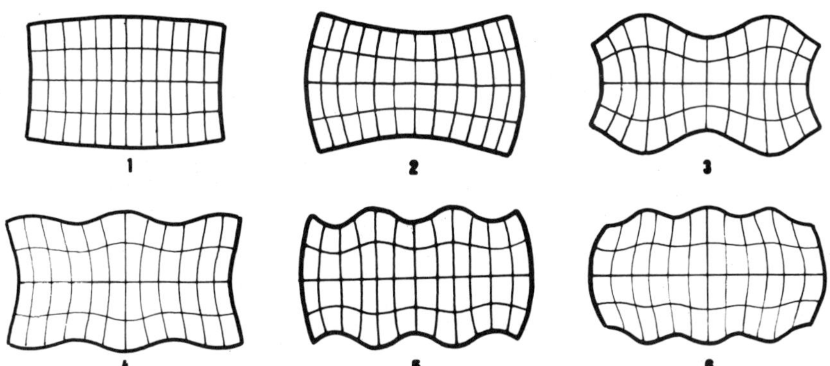

Figure 4 - Polarization of the first 6 modes observed at W/T = 2.

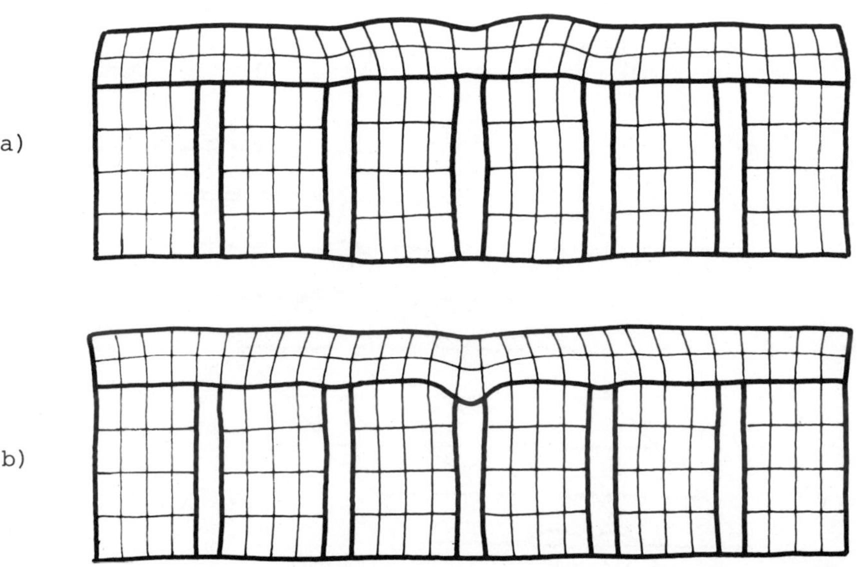

Figure 5 - Harmonic behaviour velocity of a set of 3 elementary transducers, the middle one being excited alone, in phase (a) and in quadrature of phase (b) with the excitation (intensity).

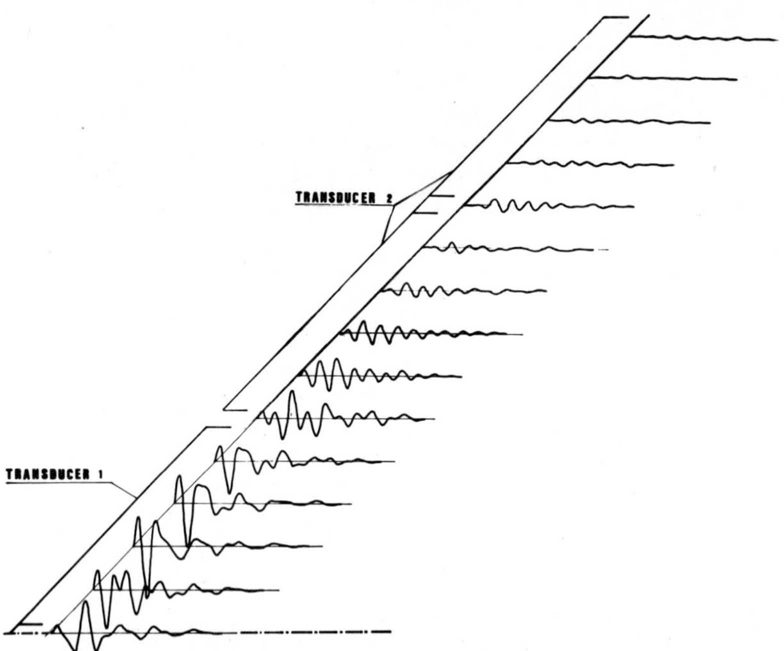

Figure 6 - Transient response in velocity of the radiating interface to an impulse in intensity exciting the transducer 1.

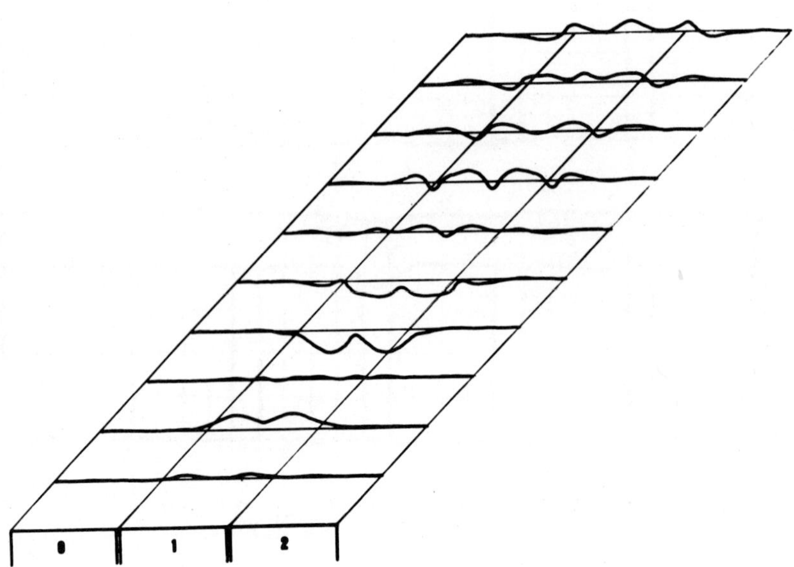

Figure 7 - Instantaneous profiles of deformation (velocity) of the 3 transducers array in the same conditions.

D remains spatially constant. Hence, $(k'_{33})_{dyn} = 0.967 \times 0.66 = 0.638$, in good agreement with our computed values, for W/T = 0.1 , k' = 0.646 for a $\lambda/8$ splitting in the x_3 direction and k' = 0.641 for a $\lambda/16$ splitting.

The figure 4 shows the polarization of the different modes observed at the value W/T = 2.

THE DYNAMIC BEHAVIOUR OF THE ELEMENTARY TRANSDUCER STRUCTURE IN A LINEAR ARRAY

As a first example, we have studied the symmetrical behaviour of a structure composed of an elementary transducer and its two neighbours, each transducer being composed of two PZT-5A ceramic blocks (W/T = 0,6), ($Z_C = 34\ 10^6$), mechanically coupled by the backing material ($Z_B = 4\ 10^6$) and by an adaptative $\lambda/4$ thick front plate ($Z_A = 4\ 10^6$) radiating in a liquid ($Z_E = 1.5\ 10^6$) (Fig. 5).

It is necessary to take into account the electrical adaptation due to the fact that a zero external excitation (I = 0) implies a non-zero intensity injected in the concerned ceramic blocks from the adaptative circuit through the potential Φ induced by coupling effects. We have retained as a simple adaptative circuit for each element the parallel connection of an inductance L matching the imaginary admittance at the resonant frequency ω_f and the resistance R verifying $Q = R/L\omega_f = 5$.

The harmonic response to an excitation I_1 of the central transducer, the adjacent ones being in open circuit ($I_2 = 0$), is obtained in two steps superposing the responses to the potential excitations ($\Phi_1, \Phi_2 = 0$) and ($\Phi_1 = 0, \Phi_2$). The figures 5-a and 5-b show the deformation observed at f = 0.58 in phase and in quadrature of phase with the intensity delivered to the central transducer.

The figure 6 shows the transient response in velocity of the radiating interface to a Dirac impulse in intensity, obtained by Fast Fourier Transform from the harmonic response computed in the bandwidth f = 0 to 1.1 .

The figure 7 gives profiles of velocity of deformation for the considered set of three adjacent transducers.

CONCLUSION

Our relatively crude approach has delivered information with an acceptable accuracy about the behaviour of a free rectangular section block of piezoelectric ceramic. Our first computation of a more complex structure modelizing an elementary transducer inside a linear array remains a first step. Before delivering the corresponding radiating diagram, which may be computed straight forward from the deformation of the radiating interface, we want to correct these results taking into account the radiative coupling inside the liquid, which should have strong effects for large angles of incidence.

REFERENCES

{1} J. SATO, M. KAWABUCHI and A. FUKUMOTO - "Dependence of the electromechanical coupling coefficient on the width-to-thickness ratio of plank-shaped piezoelectric transducers used for electronically scanned ultrasound diagnostic systems", J.A.S.A., 66, 6, (1979).

{2} G.S. KINO and C.S. DESILETS - "Design of slotted transducer arrays with matched backings", Ultrasonic Imaging, 1, pp. 189-209 (1979).

{3} B. DELANNOY, C. BRUNEEL, H. LASOTA and G. GHAZALEH - "Theoretical and experimental study of the Lamb wave eigenmodes of vibration in terms of the transducer thickness to width ratio", Jl of Appl. Phys., 52, p. 7433 (1981).

{4} P. ALAIS, Z. HOUCHANGNIA and M.Th. LARMANDE - "Analog electrical simulation of the transient behaviour of piezoelectric transducers", Acoustical Imaging, 10, pp. 731-750 (1980).

{5} D.A. BERLINCOURT, D.R. CURRAN and H. JAFFE - "Piezoelectric and piezomagnetic materials and their function in transducers", Physical Acoustics, Volume I, Part A, pp. 169-270 (1964).

AN EXPERIMENTAL METHOD

FOR CHARACTERIZING ULTRASONIC TRANSDUCERS

P. Alais, P. Cervenka, Z. Houchangnia, C. Kammoun

Laboratoire de Mécanique Physique
E.R.A. C.N.R.S. 537
Université Pierre et Marie Curie
78210 Saint-Cyr-l'Ecole, France

INTRODUCTION

It is well known that linear systems, when described through convenient variables, exhibit symmetrical properties in their matricial internal relations which may be interpreted in terms of reciprocal physical properties. For example, the efficiency of an emitting transducer may be related to its receiving transfer function. These symmetrical properties are in fact of thermodynamical origin and an in-depth view shows that they are related to Maxwell relations for conservative systems and both to Maxwell and Onsager relations for dissipative systems. We want here to recall briefly the rules for choosing adequate variables, giving the simple example of the unidimensional piston-like transducer. Afterwards, an extension will be given for a complex real transducer and we shall show that through the obtained symmetrical relations, it is possible to elaborate an experimental procedure permitting to obtain the complete description of the transducer as an emitter or as a receiver at any frequency, taking in account both its electrical and mechanical environments. This procedure may be of special interest for the study of the elementary transducer included in a linear or a matricial array used in acoustical imaging. A detailed illustration of the case of the linear array is given here.

I - CHOOSING GOOD VARIABLES - THE UNIDIMENSIONAL TRANSDUCER CASE

We consider that the linear system is a black box which exchanges energy with the external world in a way which may be correctly described through a set (q_i) of independent variables. If the system is in forced harmonic oscillation $(q_i = Q_i e^{j\omega t})$, the energy exchange will be attained through generalized forces $f_i = F_i e^{j\omega t}$ linearly related to the q_i or the associated generalized velocities $V_i = j\omega Q_i$ through linear relations :

$$F_i = Z_{ij} V_j$$

where Z_{ij} is the matricial impedance of the system in the description (q_i). The power received by the system from the external world may be explicited by :

$$\pi = \langle F_i, V_i \rangle = \pi' + j\pi''$$

where $\langle A,B \rangle$ means the product $1/2(A B^*)$, and π' is the real accepted power and π'' the reactive power.

For electromechanical systems, the q_i may be extensive variables, i.e. generalized displacements or charges, or intensive variables, i.e. generalized mechanical forces or electrical potentials. What we know from thermodynamics is that if the choice of the (q_i) set is homogeneous, i.e. constituted of extensive (or intensive) variables exclusively, the matricial generalized impedance Z_{ij} will be symmetrical. If not, the non diagonal impedance elements associated with a couple of variables of different nature, q_ℓ and q_m for example, will verify $Z_{\ell m} = - Z_{m\ell}$, i.e. a partial antisymmetry.

As an example, which will be useful for the following generalization in Part II, we give here the illustration of the unidimensional piston-like transducer mechanically and electrically adapted (Fig. 1). If we retain extensive variables, i.e. the mechanical velocities V_1, V_2 , V_3 of the three interfaces, and the intensity I feeding the transducer as generalized velocities V_i, it is simple to check that the generalized forces associated to the set (V_1, V_2, V_3, I) will be $(0, 0, -p_-, U)$, where U is the applied electrical voltage (emitter case) and p_- the pressure induced by the incident acoustic wave pressure (receiver case), if we define correctly the black box, i.e. if we say that the system including not only the electromechanical energy inside the ceramics and the front plate, but also the energy inside the adaptative (linear and passive) circuit and the acoustical energy radiated in the backing, and also emitted by the front plate in the emitter case, or re-radiated in the receiver case. In these conditions, the energy exchange is reduced to :

$$\pi = \langle -p_-, V_3 \rangle + \langle U, I \rangle \tag{1}$$

and the associated matricial relation (2) is symmetrical and may be simply explicited. Obviously, the emitted (or re-radiated energy) is obtained from the emitting part V_+ of the vibration V_3 of the front plate $V_3 = V_+ + V_-$, where V_- is the known term : $V_- = -p_-/\rho c$.

$$(2)\quad \begin{vmatrix} 0 \\ 0 \\ -p_- \\ U \end{vmatrix} = \begin{vmatrix} Z_b + \dfrac{Z_c}{j\,tg\varphi_c}, & -\dfrac{Z_c}{j\sin\varphi_c}, & 0 & \dfrac{h}{j\omega} \\ -\dfrac{Z_c}{j\sin\varphi_c}, & \dfrac{Z_c}{j\,tg\varphi_c} + \dfrac{Z_a}{j\,tg\varphi_a}, & -\dfrac{Z_a}{j\sin\varphi_a}, & -\dfrac{h}{j\omega} \\ 0 & -\dfrac{Z_a}{j\sin\varphi_a}, & \dfrac{Z_a}{j\,tg\varphi_a} + Z_e, & 0 \\ \dfrac{h}{j\omega} & -\dfrac{h}{j\omega} & 0 & \dfrac{\beta e_c}{j\omega} + Z \end{vmatrix} \cdot \begin{vmatrix} V_1 \\ V_2 \\ V_3 \\ I \end{vmatrix},$$

$$\varphi_c = \omega e_c/C_c \ , \quad \varphi_a = \omega e_a/c_a$$

Naturally solving the relation (2) permits to rewrite the useful relation between the variables $(-p_-, I)$ representing the external excitation of our system and the conjugate apparent variables (V_3, U). As the set $(-p_-, I)$ is inhomogeneous, the relation will be antisymmetrical :

$$(3)\quad \begin{vmatrix} V_3 \\ U \end{vmatrix} = \begin{vmatrix} \alpha & E \\ -E & \beta \end{vmatrix} \begin{vmatrix} -p_- \\ I \end{vmatrix}$$

and it is easy to predict the reciprocal property of the adapted transducer, i.e. :

Emission efficiency : $\dfrac{V_+}{I} = \dfrac{V_3}{I} = E(\nu)$, $p_- = 0$

Receiving efficiency: (open circuit) $\dfrac{U}{-p_-} = -E(\nu)$, $I = 0$

If the transducer is used in an echographic measurement, where the emitted signal is reflected by a perfect mirror, we may write the echographic efficiency neglecting absorption and diffraction effects in the propagating liquid (i.e. $-p_- = \rho c V_+$) according to :

Echographic efficiency : $\dfrac{U_{echo}}{I_{emis}} = \dfrac{E^2}{\rho_c}$

We have computed the spectrum $E(\nu)$ and the Fourier transform of the echographic efficiency which is just the impulse echographic response of the transducer in several cases of transducers with one or several adaptative front plates. The figure 1 gives an example of the obtained results.

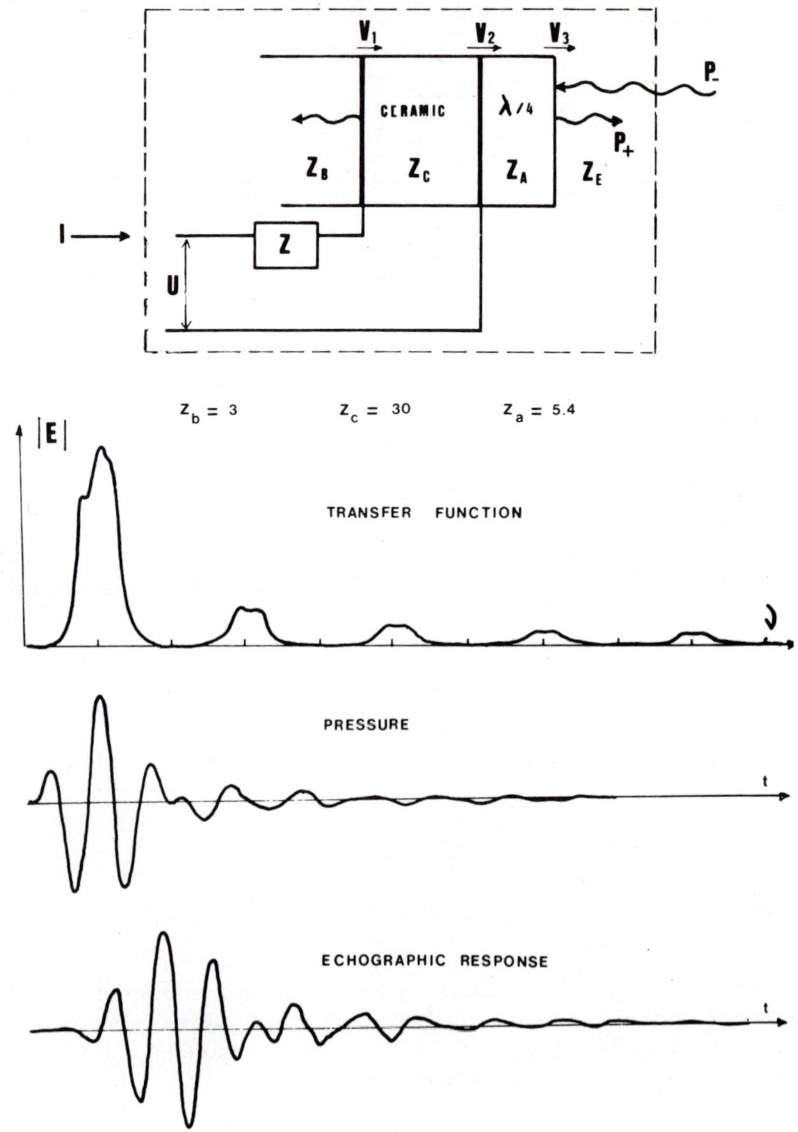

Figure 1 - The unidimensional transducer

Results obtained for the case ($Z_b, Z_c, Z_a, Z_e = 3, 30, 5.4, 1.5 \cdot 10^6$). Transfer function $|E(\nu)|$, transient response to a step function $i = I_o H(t)$ at emission (pressure) and in an echographic experiment with a perfect mirror.

II - ECHOGRAPHIC CHARACTERIZATION OF A REAL TRANSDUCER

We still define the black box as the linear system including the transducer, the mechanical and electrical environment and acoustical waves radiated backwards into the backing material, forwards into the propagating medium and eventually sidewards inside the array (Fig. 2). This system is receiving energy electrically and mechanically through the incident acoustical radiation. A correct description of the incident wave will be given by a Fourier spectrum associated to a plane Π_o which will be chosen as the emitting interface or very near the front plate if this one is not plane, according to :

$$p_-(\vec{m})\ e^{-j\omega t} = \{ \int P_-(\vec{f})\ e^{2j\pi \vec{f}\cdot\vec{m}}\ d\vec{f} \}\ e^{-j\omega t}$$

where the spectrum $P_-(\vec{f})$ is the 2-D Fourier transform of the complex amplitude of the pressure P_- delivered by the incident radiation in Π_o. Defining similarly the spectrum $V_n(\vec{f})$ of the normal velocity $V_n(\vec{f})$ of the medium at the level of Π_o, the energy exchange of our system will be correctly written as :

$$(4) \quad \pi = \langle U,I \rangle + \int \langle -P_-(\vec{f}), V_n(\vec{f}) \rangle\, d\vec{f}$$

which gives evidence of conjugate mechanical variables $V_n(\vec{f})$ and $-P_-(\vec{f})$, which in this case are a continuous distribution in the Fourier space. If we choose, as in the preceding example, to write the unknown variables (V_n,U) in terms of excitation variables $(-P_-,I)$, which are intensive and extensive respectively, we may predict that the solution will be antisymmetrical for the electromechanical coefficients :

$$(5) \quad \begin{cases} V_n(\vec{f}) = \int W(\vec{f},\vec{f}')\left[-P_-(\vec{f}')\right] d\vec{f}' + h(\vec{f})\, I \\ U = \int -h(\vec{f}')\left[-P_-(\vec{f}')\right] d\vec{f}' + ZI \end{cases}$$

and symmetrical for the pure mechanical coefficients :

$$W(\vec{f},\vec{f}') = W(\vec{f}',\vec{f})$$

This last property may be interpreted as a reciprocal property concerning the diffusive properties of a linear target.

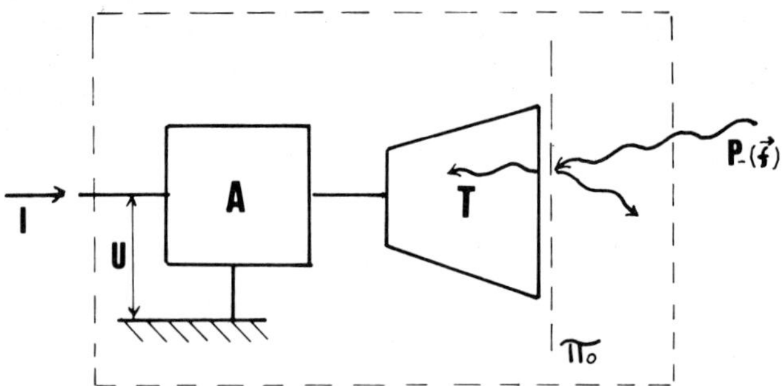

Figure 2 - The real transducer

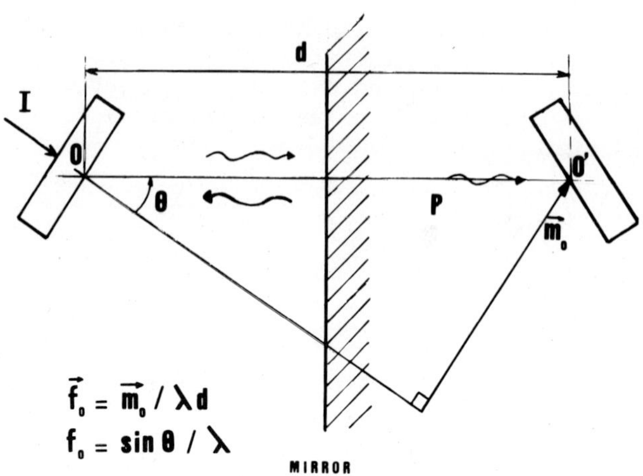

Figure 3 - The echographic experiment

Our purpose is to show that the electromechanical spatial spectrum $h(\vec{f})$ may be attained experimentally at any frequency $\omega = 2\pi\nu$ through a simple echographic experience using a plane perfect mirror set so that the "receiving transducer" image is at the distance d in the Fraunhofer field of "the emitting transducer" (Fig. 3).

If the transducer is excited during a sufficient time by a quasi-harmonic intensity $I\,e^{j\omega t}$, the vibration induced in Π_o will be attained through $V_n(\vec{f}) = h(\vec{f})I$, $(P_- = 0)$. A straightforward simple calculation shows that the acoustical velocity potential induced at the level of the receiver will be :

$$\varphi \cong - \frac{e^{jkd}}{2\pi d} V_n(\vec{f}_o)\; e^{-j\omega t},$$

where $\vec{f}_o = \vec{m}_o/\lambda d$ is the spectral frequency selected by the orientation of the mirror. This wave is equivalent for the receiving image transducer to the excitation :

$$P_-(\vec{f}) = P_o \delta(\vec{f} - \vec{f}_o),$$

where : $$P_o = j\omega\rho\,\varphi = \frac{\rho\nu}{jd}\, e^{jkd}\; V_n(\vec{f}_o)$$

and the harmonic echographic response measured at open circuit $(I = 0)$:

$$U = \int - h(\vec{f}')\left[- P_-(\vec{f}')\right] d\vec{f}'$$

(6)

$$= h(\vec{f}_o)\; P_o = \frac{\rho\nu}{jd}\, e^{jkd}\; h^2(\vec{f}_o,\nu)\; I$$

In fact, the experimental procedure is to submit the transducer to a wide band excitation $I(\nu)$ and to register the temporal delayed echographic response, $u(t) = u'(t - d/c)$, where :

$$u(t) = FT^{-1}\left[\frac{\rho\nu}{jd}\, h^2(\vec{f}_o,\nu)\; I(\nu)\right]$$

Naturally, the spatial frequency $\vec{f}_o$ must be interpreted in terms of ν :

$$\vec{f}_o = \frac{\vec{m}_o}{\lambda d} = \nu\,\frac{\vec{m}_o}{cd}$$

For transducers inserted in a linear array, we may separate the characteristics in the tomographic plane and reduce the spatial examination of the transducer to a rotation θ_o of the mirror

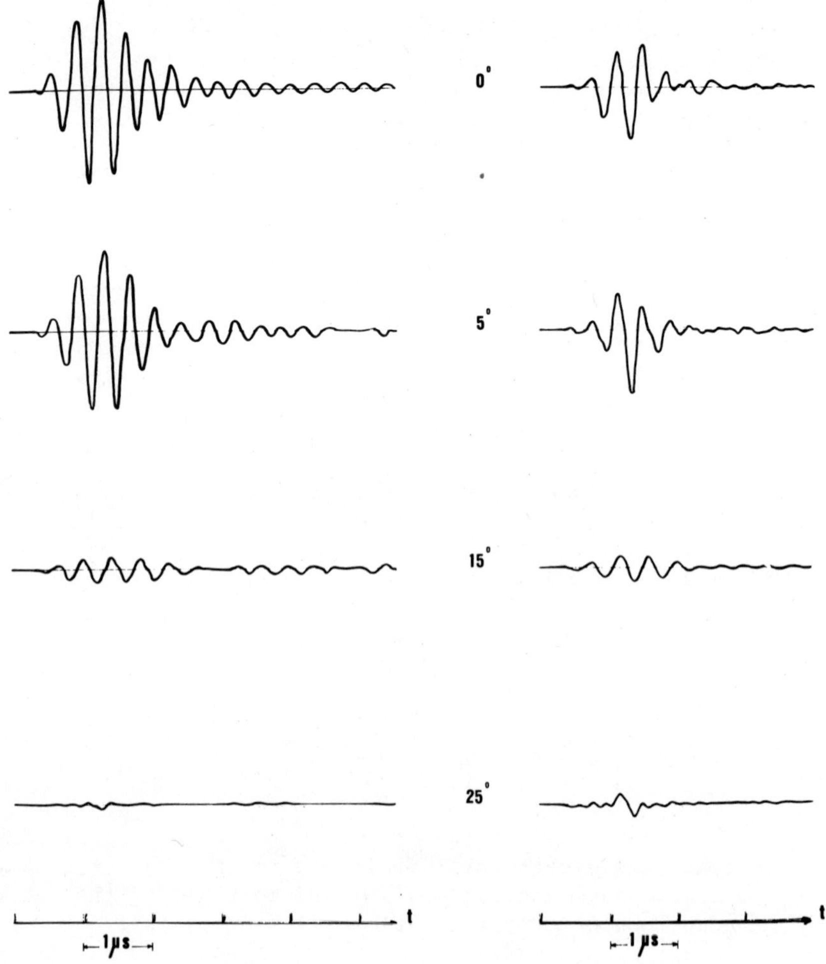

Figure 4 - Echographic data (left) and computed corresponding pressure transients (right) for different angles of incidence $\theta = 0.5°$, $15°$, $25°$, in response to a step function of intensity $i(t) = I_o H(t)$.

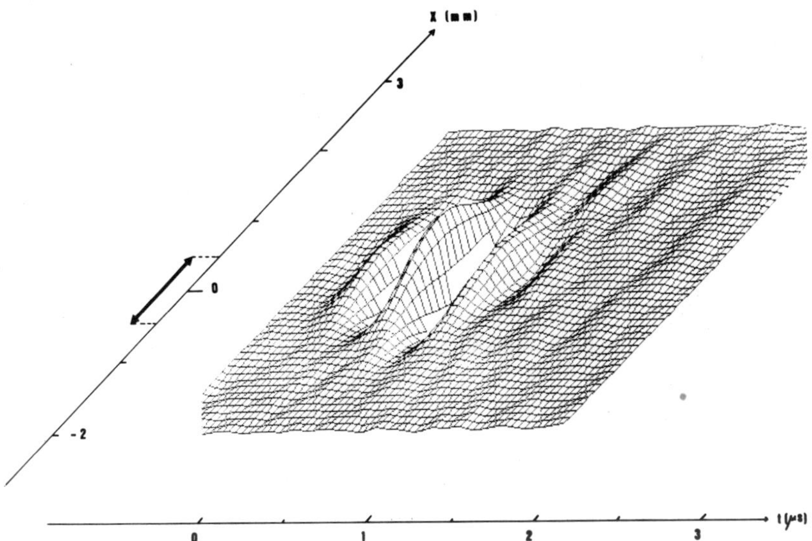

Figure 5 - Computed transient deformation of the array (velocity). The arrows show the location and the width of the transducer excited by a step function $i(t) = I_o H(t)$.

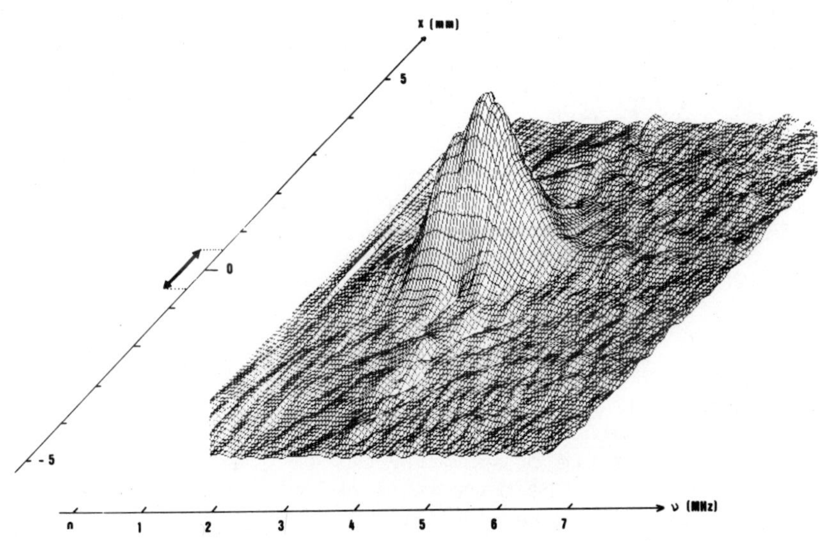

Figure 6 - Representation of the evolution (versus x) of the spectrum of the transient response obtained in the same conditions along the array. It gives a good idea of the location of the induced mechanical energy.

in this plane so that $f_o = \nu \sin\theta_o/c$. In that case, the sequence of operations is reduced to :

- recording of n echographic responses $u(\nu)$ for n θ_o ;
- determination of the F.T. $U(\nu;\theta_o)$.

From the data $I(\nu)$ and $U(\theta_o;\nu)$, it is now possible to compute:

$$(7)\qquad \begin{aligned} h(\nu, f = \nu\sin\theta/c) &= \left(\frac{jd}{\rho\nu}\,\frac{V}{I}\right)^{1/2} \\ V_n(\nu, f = \nu\sin\theta/c) &= \left(\frac{jd}{\rho\nu}\,UI\right)^{1/2} \\ P(\nu,\theta) &= \left(\frac{\rho\nu}{jd}\,UI\right)^{1/2} \end{aligned}$$

Then, it is possible through different $(F.T)^{-1}$ to attain :

- the temporal evolution of the pressure emitted in the $_o$ direction (in the tomographic plane) :

$$p(t,\theta) = \underset{\nu\to t}{FT^{-1}}\left[P(\nu,\theta)\right]$$

- the spatio-temporal evolution of the deformation of the array :

$$v(x,t) = \underset{f\to x,\nu\to t}{FT^{-1}}\left[V_n(\nu,f)\right]$$

These results concerning a 3 MHz medical echographic array transducer of 1 mm pitch are displayed by Figs. 4 and 5 respectively. The figure 6 shows an intermediate result, i.e. the spectrum $V_n(\nu,x)$, which gives a good idea of the localisation of the transient deformation of the array.

CONCLUSION

We have developed a technique for characterizing linear ultrasonic transducers, both at emission and at reception in a large bandwidth. It should be emphasized that absolute results in terms of acoustical pressure or velocity may be obtained from pure electrical measurements. The mechanical and electronical environment of the transducer is taken in account. This technique is specially interesting for characterizing array transducers or transducers of weak directivity. Corrections taking into account the attenuation of the propagative liquid and the normal reflexion coefficient of the mirror may be easily introduced in the computation.

A TWO-DIMENSIONAL PHASED ARRAY WITH AN EXTENDED DEPTH OF FOCUS: SOME PRELIMINARY RESULTS

D.R. Fox* and R.E. Reilly

Department of Electronic and Electrical Engineering
King's College London
Strand, London WC2, UK

INTRODUCTION

Many pulse-echo acoustic imaging procedures include near-field operation and hence demand some kind of focusing. The medical imaging field, for instance fetal imaging, is but one example where near-field operation is desirable. Acoustic lenses can be used, but these generally result in either a narrow depth of focus, or a small aperture. Alternatively, for imaging systems employing multiple elements, focusing can be achieved by electronic means (Welsby, 1968), a technique which can be extended to permit dynamic focusing (Thurstone and von Ramm, 1973). Yet another approach is the use of annular apertures (Vilkomerson, 1973; Burckhardt et al, 1973, 1975), which can offer a highly directive field pattern in two dimensions extending over a long axial distance. This can be achieved without any kind of physical lens or its electronic analogue, and is a result of the equivalence (apart from amplitude factors) of the field pattern in the Fresnel and Fraunhofer regions of the annular aperture. An annular array has the additional advantage of permitting high speed electronic scanning in two dimensions, resulting in greater versatility and allowing unconventional scan modes. It is the purpose of this paper to describe some results of simulations and experimental work on multi-element annular arrays, operating in pulse-echo at a centre frequency of 2MHz.

*Present address: Sonics Department,
Thorn EMI Central Research Laboratories,
Trevor Road, Hayes, UB3 1HH, UK.

ANNULAR APERTURES

It can be shown that a uniform annular array (that is, equi-angularly spaced elements on the circumference of a circle) has a field pattern with no grating lobes, provided 3-, 4-, or 6-element arrays are avoided. This is in marked contrast to the conventional uniform linear array, where a trade-off between system complexity and lateral resolution becomes necessary as a result of the grating lobes exhibited in the field pattern of the array. The annular array therefore allows the use of widely spaced elements, resulting in reduced inter-element coupling, and a consequent simplification in array fabrication. This leads to an additional advantage that the total number of elements in the array may be reduced without compromising the lateral resolution which is primarily a function of the aperture. In other words, the aperture can be spatially undersampled. The price to be paid for these advantages is the poor side-lobe structure (distinct from true grating lobes), which can reach peak levels approaching that of the main peak. It is, however, precisely because the side-lobes are not true grating lobes that techniques can be applied to re-distribute the energy in them. Such techniques include J^2-synthesis (Wild, 1965), employed by Vilkomerson (1973), and Burckhardt et al (1973, 1975), but the inherent high speed imaging capability of multi-element arrays is compromised by the requirement for signal storage. By shifting the side-lobe suppression problem away from the signal processing domain towards the array structure proper, the high speed capabilities of the array can be exploited to the full.

The field patterns for a continuous annular aperture, and a circular source follow the familiar zeroth and first order Bessel functions respectively. The oscillatory nature of Bessel functions result in the generation of side-lobes, and it is well known that the annular source has a poorer side-lobe level when compared with the circular source. Matters, however, can be improved by employing a second concentric annular source, the diameter of which can be adjusted to achieve the best overall side-lobe level. Thus, if the radius of the inner annulus is some fraction γ of the outer radius a, the overall directivity (in the one-way mode of operation) is given by

$$D = J_o\,(ka \sin\theta)\; J_o\,(k\gamma a \sin\theta)$$

Since a simple asymptotic approximation for the product of two Bessel functions does not exist, even when the orders are equal (Watson, 1966), a graphical approach was used, as shown in Fig.1. The improvement in the peak side-lobe level for intermediate values of the radius parameter results from a kind of "incoherence" in the angular position of the side-lobes for the individual rings, and the product is a smoothing out of these side-lobes. It will be seen that the best side-lobe level that can be achieved by this method is around -21 dB with respect to the main peak.

A practical array will be spatially sampled to enable steering to be implemented. This results, on the one hand in a much poorer peak side-lobe level (when compared with the continuous annulus), approaching 90% of the main peak for a ten element array. On the other hand, the freedom to orient the inner array with respect to the outer can be used to minimise the side-lobe level. This has an effect similar to that resulting from the employment of two concentric rings, that is, the side-lobes generated by one array do not coincide with those resulting from the other.

It is common practice in manually scanned systems to employ the same transducer for transmitting and receiving, by including a T/R switch. As far as such systems are concerned, there is nothing to be gained, apart from the elimination of the T/R switch, by using a separate transmitter and receiver. When considering arrays,

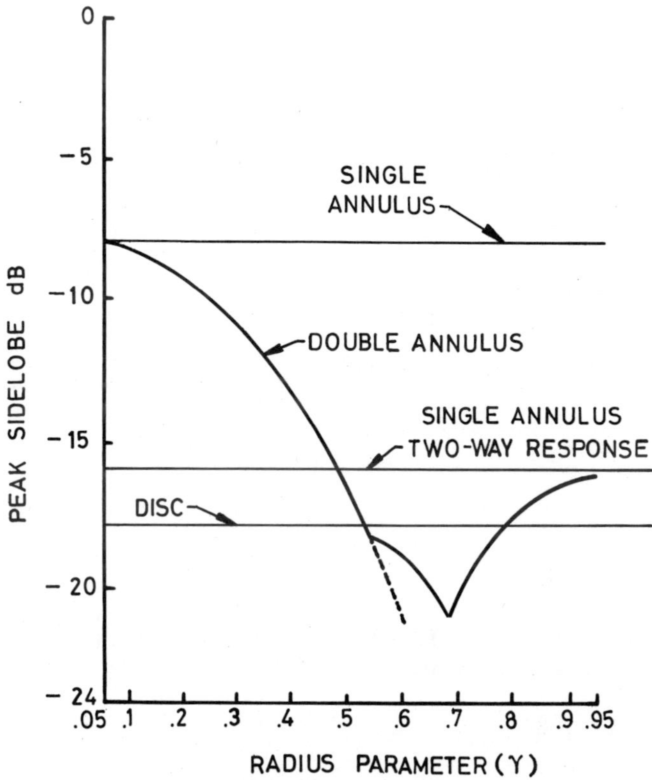

Fig.1. Peak side-lobe levels for various sources.

a separate transmitter and receiver. When considering arrays, however, an improvement in the side-lobe structure can be made, together with some simplification of the hardware, by the elimination of a T/R switch for each element, a not insignificant factor when a large number of elements are used. By employing separate transmit and receive elements, the orientation of the transmitter array with respect to the receiver array can be adjusted so that as far as possible, the side-lobes due to the receiver are not illuminated by those of the transmitter.

A single, n-element annular array of radius a, has a field pattern given by

$$J_o\ (ka\ \sin\theta) + 2J_n\ (ka\ \sin\theta)\cos n\emptyset + 2J_{2n}\ (ka\ \sin\theta)\cos 2n\emptyset + \ldots$$

where θ, $\emptyset$ are the elevation and azimuth angles respectively, and k is the wave number. For the sake of clarity, the arguments of the Bessel functions will be omitted in the following. If the expression above represents a transmit array, and if a similar receive array is oriented at an angle ξ to the transmit array, then the overall pattern will be given by the product,

$$(J_o + 2J_n\cos n\emptyset + 2J_{2n}\cos 2n\emptyset + \ldots)(J_o + 2J_n\cos n(\emptyset + \xi) + \ldots.)$$

In addition to the natural side-lobes arising from the oscillatory nature of the Bessel functions, a complex set of side-lobes is generated by the cross-product terms. Omitting all but the first two terms in each bracket simplifies the problem, and is equivalent to restricting the analysis to small elevation angles. Thus, for the interleaved array

$$D = J_o^2 + 4J_n^2\cos n\emptyset\cos n(\emptyset+\xi) + 4J_oJ_n\cos n(\emptyset+\frac{\xi}{2})\cos\frac{n\xi}{2} \quad \text{---(1)}$$

Now for an array employing coincident transmit and receive elements $\xi=0$ and

$$D = J_o^2 + 4J_n^2\cos^2 n\emptyset + 4J_oJ_n\cos n\emptyset$$

but for $\xi=\frac{\pi}{n}$

$$D = J_o^2 - 4J_n^2\cos^2 n\emptyset$$

Thus, the cross-product term in equation (1) can be made zero resulting in the elimination of a complete set of side-lobes.

COMPUTER SIMULATIONS

A wide variety of annular array configurations based on the approach outlined above were studied using computer simulations.

The simulations calculate the round-trip directivity, that is, the product of the transmit and receive directivities for the various array configurations, and plots them on a pseudo-three dimensional format with elevation and azimuth angles as the independent variables. The response as a function of azimuth angle is now required because, unlike the continuous annulus, the annular arrays do not possess full rotational symmetry.

All the array configurations investigated employed a total of forty elements; twenty transmit and twenty receive. However, in order to emphasise the effect of physically separating the transmit and receive elements, a ten channel system was simulated (Fig.2), where each element acts as a transmitter and a receiver. It is immediately apparent that there are a large number of very large side-lobes. It will be noted, however, that it is possible to locate azimuth angles where the side-lobes appear both small in number and in amplitude, for instance around the five degree line. This is where a "three-dimensional" plot proves invaluable where

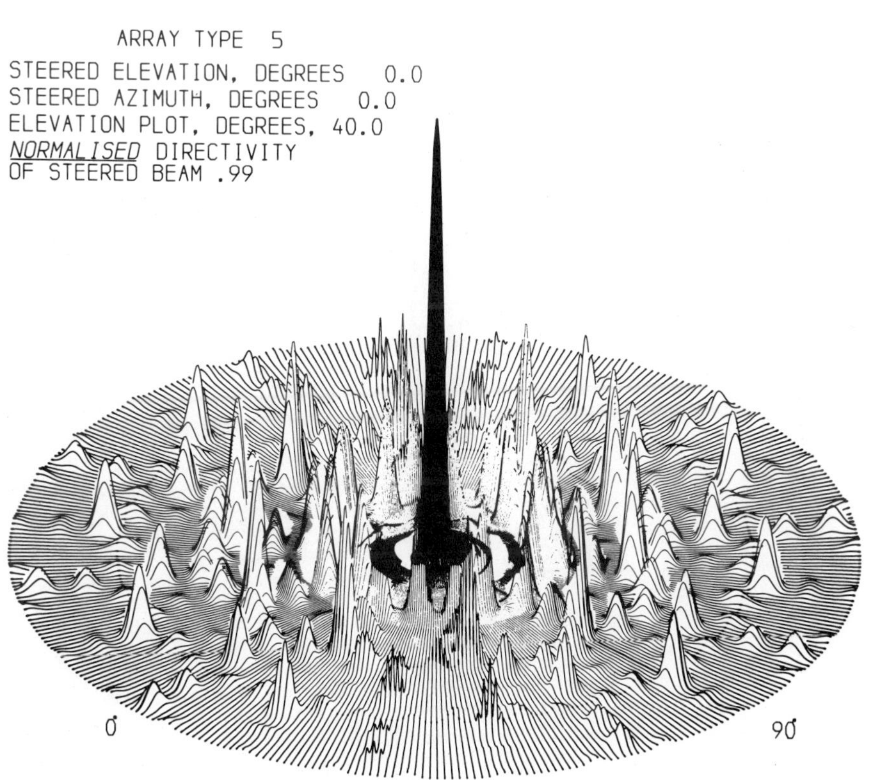

Fig.2. 10 element, 10 channel, single ring system.

a two-dimensional display could present a false impression of the side-lobe structure. Fig.3 shows the result of a similar array with ten transmit elements interleaved with ten receive elements. A considerable reduction in the number and level of the side-lobes has been achieved without any increase in hardware. Both configurations are ten channel systems, and in fact the second array would be marginally simpler to implement, since it dispenses with ten T/R switches that would otherwise be necessary.

Of the extensive studies performed on a large number of array configurations, several were chosen for further study, based on their low side-lobe levels. Fig.4 shows a result for one particular 40-element, 20-channel two ring system. The results of these, and other simulations suggest that the overall performance of selected double annular arrays may offer attractive alternatives to more conventional systems. Thus, Fig.4 shows a peak side-lobe level of about -25 dB, which stands favourable comparison with that of a linear array of the same aperture.

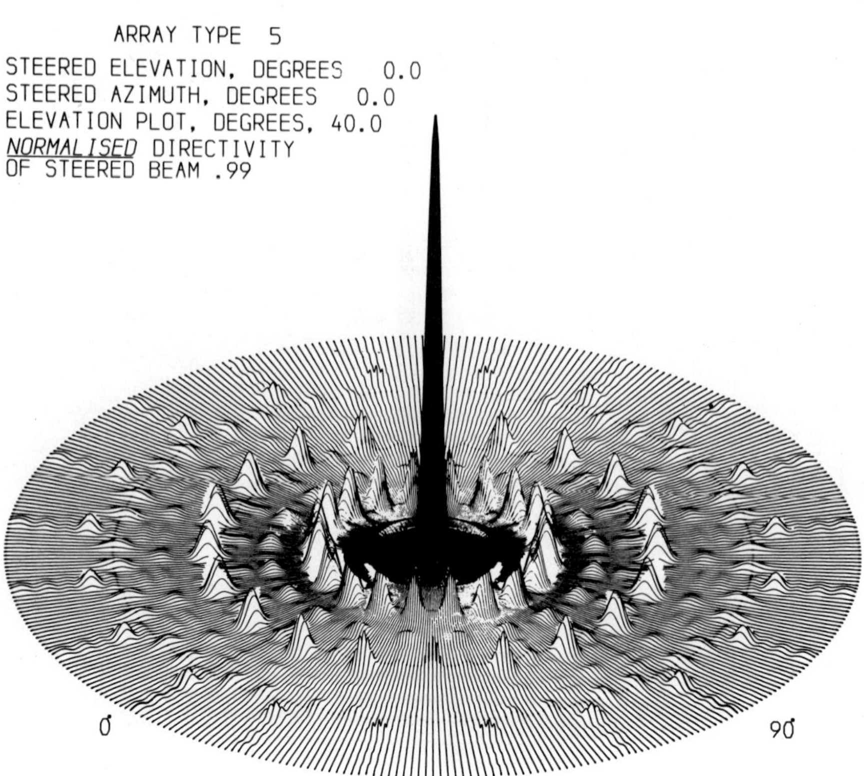

Fig.3. 20 element, 10 channel, single ring system.

EXPERIMENTAL RESULTS

A prototype array was fabricated as a monolithic structure from PZT-5A piezoelectric ceramic, by selective etching of the electrode pattern chosen on the basis of the simulations. The monolithic structure proved less than ideal since it resulted in a rather high element directivity. Nevertheless, some preliminary measurements of the reflection response were made and these are discussed below.

Fig.5 shows a composite plot of the reflection response measured for twelve azimuth angles, over a restricted elevation span of $\pm 7^{o}$. This compression of the data into a single diagram enables a better visualisation of the overall reflection directivity of the array, although it cannot show the detailed structure revealed in the computer simulations. The measurements were made at a target range of 113 mm, and it will be seen that a well-defined, uniform beam profile is obtained.

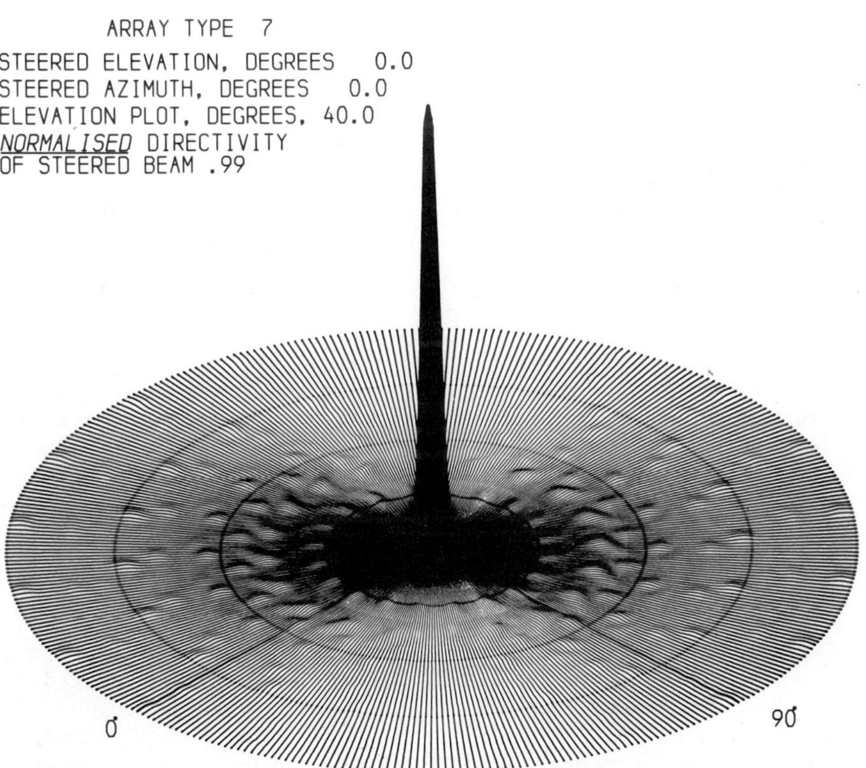

Fig.4. Optimised, 20 channel, two ring system; diameters 15mm and 10.2mm; elements 0.4mm square.

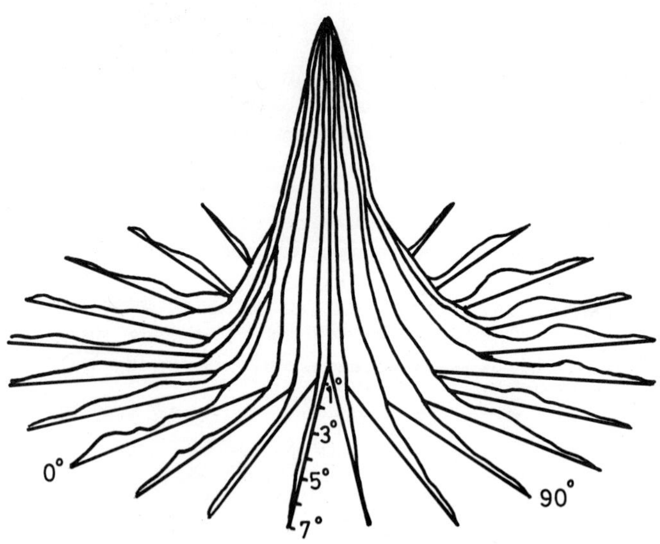

Fig.5. Composite plot of reflection response.

Measurements of the beamwidth were made over a range of azimuth angles and for several axial ranges encompassing a two-to-one ratio. Fig.6 shows the result of these measurements, each experimental point shown being the mean over all the azimuth angles. Beamwidths of 2.3 mm at 75 mm range, rising to 3.4 mm at 150 mm were obtained representing 3 and 4.5 wavelengths respectively, a significant improvement over conventional unfocused systems.

Finally, Fig.7 shows a typical plot of the side-lobe level measured at an axial range of 113 mm, and along an azimuth angle of 75°. The mean side-lobe level, averaged over all measured azimuthal directions was -34 dB, whilst the peak side-lobe level was found to be -24 dB. This is some 10 dB worse than a conventional unfocused disc transducer, and 3 dB worse than a linear phased array employing the same number of elements. On the other hand, the present array offers much greater versatility, and further techniques such as apodisation can be employed to make additional improvements.

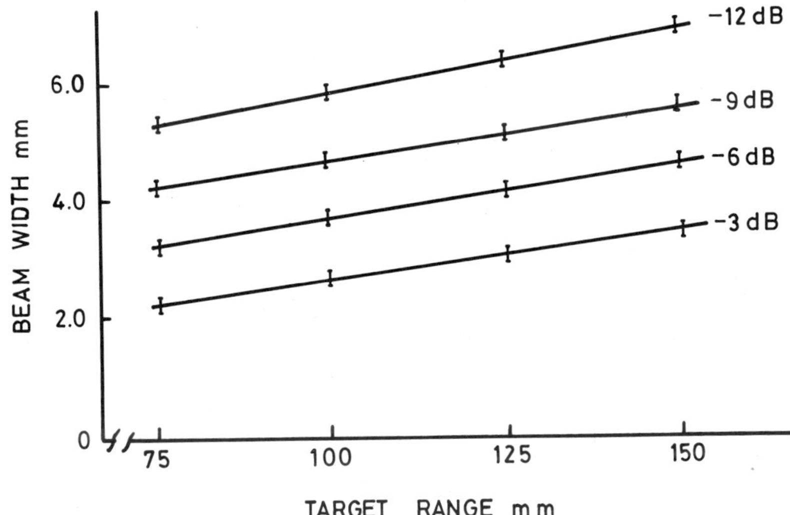

Fig.6. Measured beamwidth.

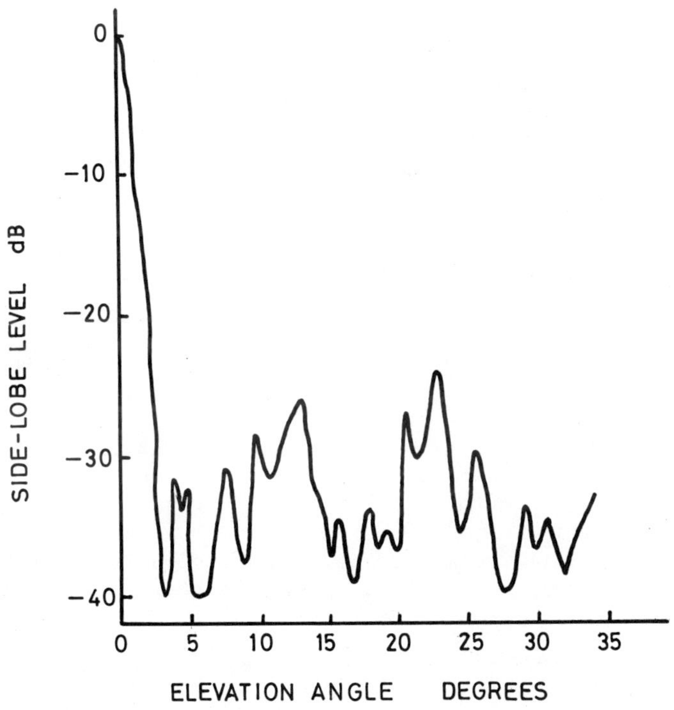

Fig.7. Side-lobe structure of reflection system.

SUMMARY

The development of a two-dimensional phased array has been described together with examples of computer simulations which enabled a particular configuration to be selected for experimental investigation. The simulations indicated that improvements in side-lobe level could be made by suitable juxtaposition of the elements of the array. The experimental results have indicated that a satisfactory two-dimensional resolution can be achieved, with a moderate side-lobe level, maintained over a substantial axial range, and without resort to electronic methods of resolution enhancement which tend to compromise the overall rate of image formation.

These are only very preliminary results and further work is necessary to characterise the array over various elevation angles. Nevertheless, they do point to the development of such an array as an effective basis for a high resolution imaging system.

REFERENCES

Burckhardt, Grandchamp, and Hoffman, 1973, Methods for increasing the lateral resolution of B-scan, _in_: "Acoustical Holography", vol 5, P.S. Green, ed., Plenum, NY.

Burckhardt, Grandchamp, and Hoffman, 1975, Focusing ultrasound over a large depth with an annular aperture - an alternative method, _IEEE Trans Sonics Ultrason._, SU-22: 11.

Thurstone and von Ramm, 1973, New ultrasound imaging technique employing two-dimensional electronic beam steering, _in_: "Acoustical Holography", vol 5, P.S. Green, ed., Plenum, NY.

Vilkomerson, 1973, Acoustic imaging with thin annular apertures, _in_: "Acoustical Holography", vol 5, P.S. Green, ed., Plenum, NY.

Watson, 1966, "A Treatise on the Theory of Bessel Functions", Cambridge University Press, Cambridge.

Welsby, 1968, Electronic scanning of focused arrays, _J Sound Vib._, 8: 390.

Wild, 1965, A new method of image formation with annular apertures and an application in radio astronomy, _Proc Roy Soc A._, 286: 499.

ESTIMATION OF ECHO SCATTERED FROM STRONGLY SCATTERING MEDIUM

Mitsuhiro Ueda

Tokyo Institute of Technology
Research Laboratory of Precision Machinery and Electronics
Nagatsuta, Midori-ku, Yokohama 227 Japan

The estimation of echo signal has been succeeded only for plane and spherical incident waves, but echoes are usually reflected in near field of an ultrasonic transducer in most ultrasonic pulse echo systems. Lately we have proposed the new estimation algorithm for echo signal which is based on a discrete expression of the medium. This algorithm can be applied for any incident waves if the medium can be assumed as weakly scattering one. In this paper it is modified to the case of strongly scattering medium and then echoes reflected from a wire of nylon are calculated. The comparison with experiments shows good agreement.

INTRODUCTION

Recently, an ultrasonic pulse echo method has been widely used in medical diagnosis and nondestructive testing. In these areas echo signals have been used to indicate the locations of discontinuity of medium. Recently there have been reported a few attempts to get quantitative information about acoustic characteristics of the medium from the echo signal.[1~3] The basic requirement for such signal processing is the ability of estimating the waveform of echo signal reflected from a scatterer whose geometrical and acoustic characteristics are specified beforehand. The estimation of the echo waveform has been successful only for plane and spherical incident waves.[4,5] But echoes are usually reflected in near field of an ultrasonic transducer in the conventional usage of pulse echo systems. Consequently it becomes necessary to develop a new estimation algorithm which works for arbitrary incident waves and arbitrary scatterers.

Lately we have proposed the estimation algorithm which is based on a discrete expression of the medium.[6] It can be applied for any incident waves if the scattering of the medium is weak. In this paper it is modified to the case of strongly scattering medium and echoes reflected from a wire of nylon are calculated. The comparison with echoes which are obtained experimentally shows good agreement.

PRINCIPLES

Echoes reflected from weakly scattering medium

According to the analysis of an echo signal which makes use of a discrete model of the medium,[6] the echo reflected from weakly scattering medium can be expressed as follows

$$E(\omega) = j\omega\rho_0 G(\omega)F(\omega) \tag{1}$$

where $E(\omega)$ shows the Fourier transform of the echo signal $e(t)$, ρ_0 shows the mean density of the medium, $G(\omega)$ shows an electrical characteristics of an ultrasonic transducer, and it is assumed that the same transducer is used as a transmitter and a receiver. $F(\omega)$ shows a frequency response of the medium and it is given by

$$F(\omega) = \int_V \Delta z(\mathbb{r})(j\omega\Phi(\omega,\mathbb{r})/c_0)^2/(2\pi z_0)dv \tag{2}$$

where V shows a scattering volume, c_0 shows the mean velocity of the medium, z_0 shows the mean specific acoustic impedance of the medium ($z_0 = c_0\rho_0$), $\mathbb{r}$ shows a position vector, dv shows a volume element, $z_0 + \Delta z(\mathbb{r})$ shows the specific acoustic impedance of the scatterer and it is assumed that $z_0 >> |\Delta z(\mathbb{r})|$. (Fig. 1). $\Phi(\omega,\mathbb{r})$ shows a velocity potential and is given by

$$\Phi(\omega,\mathbb{r}) = \int_\sigma \exp(-j\omega|\mathbb{r}-\mathbb{r}'|/c_0)/(2\pi|\mathbb{r}-\mathbb{r}'|)ds \tag{3}$$

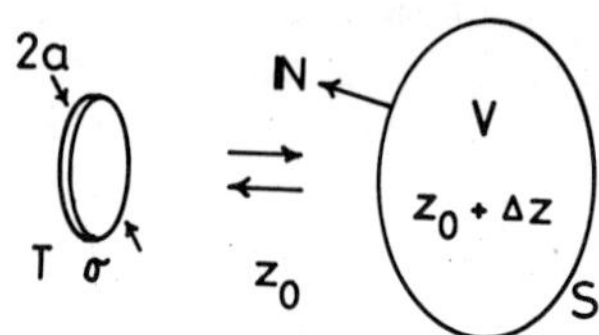

Fig. 1 Geometry of a transducer(T) and a scatterer.

where σ shows a surface of the transducer, ds shows a surface element, and $\mathbb{r}'$ shows a position vector which is concerned with the integration. If the inside of thescattereris uniform, Eq.(2) can be expressed by a surface integral on the scattering volume as follows

$$F(\omega) = (\Delta z/4\pi z_0)\int_S \Phi(\omega,\mathbb{r})\mathbb{N}\cdot\nabla\Phi(\omega,\mathbb{r})ds \quad (4)$$

where S shows a surface of the scatterer and $\mathbb{N}$ shows an outer normal to S. If the distance between the transducer and the scatterer becomes less than a radius of the transducer, echoes due to the velocity fluctuation and the density fluctuation must be dealt with separately. Consequently it is assumed that the distance between the transducer and the scatterer is larger than the radius of the transducer throughout this paper.

Echoes reflected from strongly scattering medium

Eqs.(2) and (4) have been derived from the wave equation for inhomogeneous medium under the condition of weak scattering. Most scatterers except soft biological tissues have large variation of acoustic impedance. Consequently these relations cannot be applied to them directly.

Let us consider a simple case. A small transducer is placed at the center of the cylindrical coordinates system (y,r,θ) as shown in Fig.2. A plane plate of thickness d is placed at a distance y_0 from the transducer. Then the frequency response of the plate is given by

$$F(\omega) = (\Delta z/z_0)\int_{y_0}^{y_0+d}\int_0^{\infty}(j\omega\Phi(\omega,\mathbb{r})/c_0)^2 rdrdy \quad (5)$$

where Eq.(2) is used in the derivation of Eq.(5), z_1 shows the specific acoustic impedance of the plate and $\Delta z = z_1 - z_0$. If the size of the transducer is sufficiently small, Eq.(3) becomes

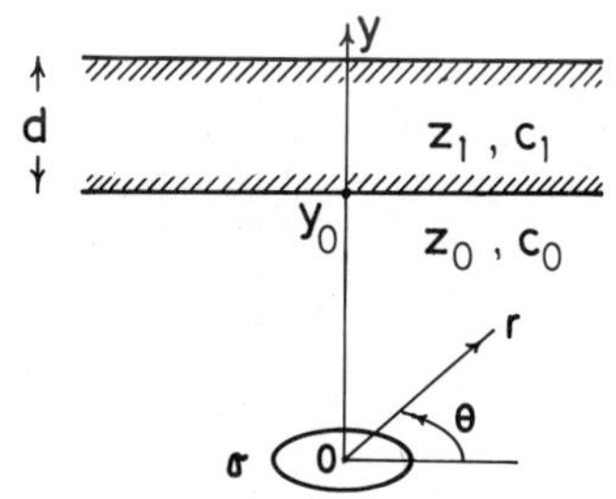

Fig. 2 Geometry of a transducer and a plate.

$$\Phi(\omega,\mathbf{r}) \doteqdot (\sigma/(2\pi|\mathbf{r}|))\exp(-j\omega|\mathbf{r}|/c_0) \tag{6}$$

where $|\mathbf{r}|^2 = (y^2 + r^2)$ and σ shows the area of the transducer. Substituting Eq.(6) into Eq.(5) and performing the integral under the condition of $\omega y_0/c_0 >> 1$, Eq.(5) becomes

$$F(\omega) = (\sigma^2/(4\pi^2(2y_0+d)))\cdot(\Delta z/2z_0)\cdot(\exp(-j\omega 2y_0/c_0) - \exp(-j2\omega(y_0+d)/c_0)) \tag{7}$$

where the first brackets show the contribution of the area of transducer and the distance between the transducer and the scatterer, the second brackets show the contribution of impedance change, and the third brackets show the delay time of echoes.

Since the plate is far from the transducer, the amplitude of echoes must be proportional to a reflection factor of normal incidence R, that is,

$$R = (z_1 - z_0)/(z_1 + z_0). \tag{8}$$

But in the case of small fluctuation Eq.(8) can be rewritten as $R = \Delta z/2z_0$. Consequently it may be reasonable to replace $\Delta z(r)/2z_0$ in Eqs.(2) and (4) by $R(\mathbf{r})$ in the case of large fluctuation.

As seen from Eq.(7), the echo reflected from the front surface of the plate returns after the time lapse of $2y_0/c_0$ and it coincides with the geometrical consideration. The echo reflected from the rear surface is supposed to return after the time lapse of $2(y_0+d)/c_0$ by Eq.(7). But according to the geometrical consideration it must return after the time lapse of

$$2y_0/c_0 + 2d/c_1$$

where c_1 shows the sound velocity of the plate. In the case of small fluctuation these two delay times are the same, since $c_1 = c_0$. But in the case of large fluctuation it may be necessary to change the thickness of the plate as

$$d' = dc_0/c_1 \tag{9}$$

where d' shows the modified thickness of the plate. Consequently in the case of strongly scattering medium it becomes necessary to deform the boundary of scatterers by using the relation (9).

As a result of above considerations it may be reasonable to modify Eq.(2) as follows in the case of large fluctuation.

$$F(\omega) = (1/\pi)\int_{V'} R(\mathbf{r})(j\omega\Phi(\omega,\mathbf{r})/c_0)^2 dv \tag{10}$$

And Eq.(4) becomes

$$F(\omega) = (R/2\pi)\int_{S'} \Phi(\omega,\mathbf{r})\mathbf{N}\cdot\nabla\Phi(\omega,\mathbf{r})ds \quad (11)$$

where V' and S' show the deformed scattering volume and surface respectively.

Estimation of echo signal

In order to estimate the waveform of echoes reflected from arbitrary objects, it is necessary to know the electrical characteristics of the transducer. Let us denote an echo reflected from a plane reflector by ep(t) and its Fourier transform by $E_P(\omega)$. The frequency response of the plane reflector $F_P(\omega)$ can be calculated by using Eq.(11) as follows

$$F_P(\omega) = -R_P\int_0^\infty \Phi(\omega,y_0,r)(\partial\Phi(\omega,y_0,r)/\partial y)rdr \quad (12)$$

where the plane reflector is located at $y = y_0$ and R_P shows a reflection factor of the reflector. Then the electrical characteristics of the transducer is given by

$$j\omega\rho_0 G(\omega) = E_P(\omega)/F_P(\omega) \quad (13)$$

Consequently the echo reflected from an arbitrary object can be estimated as follows

$$e(t) = F^{-1}[F(\omega)E_P(\omega)/F_P(\omega)] \quad (14)$$

where F^{-1} shows Fourier inverse transformation and $F(\omega)$ shows the frequency response of the object.

Calculation of frequency response

In order to calculate the frequency response by using either Eq.(10) or Eq.(11), the velocity potential $\Phi(\omega,\mathbf{r})$ must be calculated first. A time-dependent velocity potential $\phi(t,\mathbf{r})$, which is given by an inverse Fourier transform of $\Phi(\omega,\mathbf{r})$, has been derived analytically[7] for a plane circular transducer and is given as follows, where the geometry of Fig. 2 is used,

$$\phi(t,\mathbf{r}) = c_0 s(t,\mathbf{r}) + c_0 h(t,\mathbf{r}) \quad , a > r ,$$

$$= c_0 h(t,\mathbf{r}) \quad , a < r , \quad (15)$$

where $s(t,\mathbf{r}) = 1 \quad , \quad y < c_0 t < (y^2+(a-r)^2)^{\frac{1}{2}},$

$= 0 \quad , \quad$ otherwise,

and $h(t,\mathbf{r}) = (1/\pi)\cos^{-1}(((c_0t)^2-y^2+r^2-a^2)/(2r((c_0t)^2-y^2)^{\frac{1}{2}})),$

$(y^2+(a-r)^2)^{\frac{1}{2}} < c_0 t < (y^2+(a+r)^2)^{\frac{1}{2}},$

$= 0 \quad , \quad$ otherwise.

An impulse response of the scatterer $f(t)$, which is given by an inverse Fourier transform of $F(\omega)$, is represented by (Eq.(10))

$$f(t) = \int_{V'}(R(r)/\pi)(\phi_t(t,\mathbf{r})/c_0)*(\phi_t(t,\mathbf{r})/c_0)dv \tag{16}$$

where ϕ_t means $\partial\phi/\partial t$ and $*$ shows convolution operation. Let us consider the case of $a>r$ first, then ϕ_t/c_0 is given by

$$\phi_t(t,\mathbf{r})/c_0 = \delta(t-y/c_0) + h_t(t,\mathbf{r}) \ , \tag{17}$$

where $\delta(t)$ shows a delta function and it should be noted that $h_t(t,\mathbf{r})$ is a continuous function of time. Then Eq.(16) becomes

$$f(t) = \int_{V'}(R(r)/\pi)(\delta(t-2y/c_0)+2h_t(t-y/c_0,\mathbf{r})+h_t(t,\mathbf{r})*h_t(t,\mathbf{r}))dv \tag{18}$$

Since the integrand of Eq.(18) contains a delta function, let us define the sampled value of $f(t)$ as follows

$$\hat{f}(m) = \int_{(m-0.5)\Delta}^{(m+0.5)\Delta} f(t)dt \tag{19}$$

where $\hat{f}(m)$ shows the sampled value of $f(t)$ at $t = m\Delta$ and Δ shows a sampling interval. Then Eq.(18) becomes

$$\hat{f}(m) = \int_{V'}(R(r)/\pi)(\hat{\delta}(m,\mathbf{r})+2\hat{h}_{t1}(m,\mathbf{r})+\hat{h}_t(m,\mathbf{r}) * \hat{h}_t(m,\mathbf{r}))dv \tag{20}$$

where $\hat{\delta}(m,\mathbf{r}) = 1 \quad , \quad (m-0.5)\Delta < t-2y/c_0 \leq (m+0.5)\Delta \ ,$

$$= 0 \quad , \quad \text{otherwise} \quad , \tag{21}$$

$$\hat{h}_{t1}(m,\mathbf{r}) = h((m+0.5)\Delta-y/c_0,r) - h((m-0.5)\Delta-y/c_0,r) \ , \tag{22}$$

and $$\hat{h}_t(m,\mathbf{r}) = h((m+0.5)\Delta,r) - h((m-0.5)\Delta,r) \ . \tag{23}$$

Eqs.(22) and (23) correspond to sampled versions of $h_t(t-2y/c_0,\mathbf{r})$ and $h_t(t,\mathbf{r})$ respectively and * shows digital convolution operation. For the case a<r, Eq.(20) becomes

$$\hat{f}(m) = \int_{V'} (R(r)/\pi)\hat{h}_t(m,\mathbf{r}) * \hat{h}_t(m,\mathbf{r})dv . \tag{24}$$

As to the calculation of impulse response by using Eq.(11), Eq.(16) becomes

$$f(t) = (R/2\pi)\int_{S'} \phi(t,\mathbf{r}) * (n_y\phi_y(t,\mathbf{r}) + n_r\phi_r(t,\mathbf{r}))ds \tag{25}$$

where $N = (n_y,n_r,n_\theta)$ and ϕ_y and ϕ_r mean $\partial\phi/\partial y$ and $\partial\phi/\partial r$ respectively. Since the following relations hold for $\phi_y(t,\mathbf{r})$

$$\begin{aligned}\phi_y(t,\mathbf{r}) &= -\delta(t-y/c_0) + c_0h_y(t,\mathbf{r}) \quad , \quad a > r \quad , \\ &= c_0h_y(t,\mathbf{r}) \quad , \quad a < r \quad , \end{aligned} \tag{26}$$

the sampled value of impulse response is given by

$$\begin{aligned}\hat{f}(m) &= (R/2\pi)\int_{S'} (-\hat{\phi}_1(m,\mathbf{r})+\hat{\phi}(m,\mathbf{r})*(n_yc_0\hat{h}_y(m,\mathbf{r})+n_r\hat{\phi}_r(m,\mathbf{r}))ds, \ a>r, \\ &= (R/2\pi)\int_{S'} \hat{\phi}(m,\mathbf{r})*(n_yc_0\hat{h}_y(m,\mathbf{r})+n_r\hat{\phi}_r(m,\mathbf{r}))ds, \ a<r, \end{aligned} \tag{27}$$

where $\hat{\phi}_1$, $\hat{\phi}$, and $\hat{h}_y$ are the sampled versions of $\phi(t-y/c_0,\mathbf{r})$, $\phi(t,\mathbf{r})$, and $h_y(t,\mathbf{r})$ respectively and are given by

$$\begin{aligned}\hat{\phi}_1(m,\mathbf{r}) &= \int_{(m-0.5)\Delta}^{(m+0.5)\Delta} \phi(t-y/c_0,\mathbf{r})dt \\ &\doteqdot (\phi((m+0.5)\Delta-y/c_0,\mathbf{r})+\phi((m-0.5)\Delta-y/c_0,\mathbf{r}))\Delta/2 \ , \end{aligned} \tag{28}$$

$$\hat{\phi}(m,\mathbf{r}) \doteqdot (\phi((m+0.5)\Delta,\mathbf{r})+\phi((m-0.5)\Delta,\mathbf{r}))\Delta/2 \ , \tag{29}$$

$$c_0\hat{h}_y(m,\mathbf{r}) = -(y/(c_0m\Delta))\hat{h}_t(m,\mathbf{r}), \tag{30}$$

where the following relation is used for the derivation of Eq.(30).

$$c_0h_t(t,\mathbf{r}) = -(y/(c_0t))h_t(t,\mathbf{r}) \tag{31}$$

$\hat{\phi}_r$ is calculated by using the following difference relation.

$$\hat{\phi}_r(m,\mathbf{r}) = (\hat{\phi}(m,y,r+\Delta r)-\hat{\phi}(m,y,r))/\Delta r \tag{32}$$

where $\Delta r = c_0\Delta$.

The size of volume or surface elements is determined by a rule that the start and end times of impulse response for the points inside the element do not fluctuate more than a sampling interval.

Fig.3 Experimental arrangement. T shows a transducer(25.4mmϕ) and W shows a wire of nylon(94μmϕ).

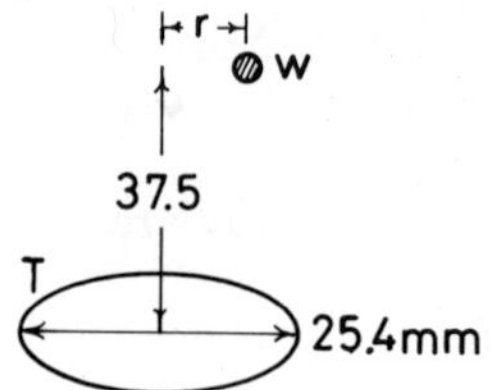

EXPERIMENTAL RESULTS

The echo signals reflected from a wire of nylon (ρ_0=1.1,c_0=2680 m/s, and R=0.325) are estimated and compared with the echoes obtained experimentally. The experimental arrangement is shown in Fig. 3, where a plane circular transducer of 25.4 mmϕ and 2.25 MHz is used. The echoes are calculated by changing a distance r between the wire and an axis of the transducer while the depth of the wire is kept at 37.5 mm. Impulse responses of the wire at r=0,4,8 mm are shown in Fig. 4, where they are calculated at every 100 ns by using Eqs.(20) and (24). The length of the wire is assumed to be 36 mm in the calculation, but if it is a little longer than the diameter of the transducer, almost the same waveform is obtained irrespective of the length.

In order to estimate the electrical characteristics of the transducer, an echo reflected from a plane reflector made of acrylic acid resin (R_p=0.36) is recorded and is shown in Fig.5(a). The frequency response of the plane reflector is computed by using Eqs.(27) and is shown in Fig.5(b). The diameter of the plane reflector is taken as 36 mm. The echo signals reflected from the wire are computed by using Eq.(14) at every 100 ns and are shown in Fig.6(b).

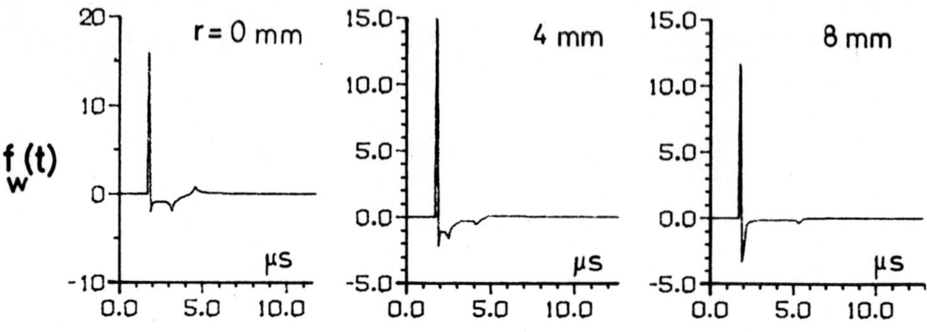

Fig. 4 Impulse responses of the wire. The ordinates are plotted by an arbitrary unit.

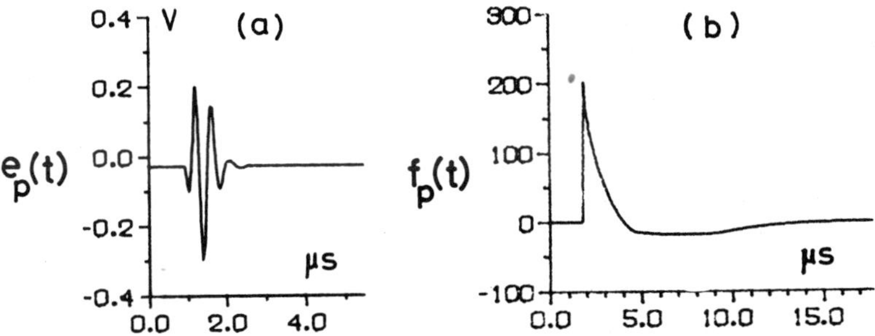

Fig. 5 An echo signal reflected from the plane (a) and its impulse response (b). The ordinate of the response is arbitrary.

The experimentally obtained echoes are shown in Fig.6(a). Since the sampling rate is not sufficiently close that some sharp peaks are collapsed in Fig. 6. It is interesting to note the amplitude of the third positive peaks which are indicated by arrows in Fig.6. According to the theoretical estimation the amplitude of the third peak decreases suddenly at r=6 mm and recover again at r=8 mm. We can see the same behavior of the amplitude in the experimentally obtained echoes. Consequently it may be concluded that our estimation algorithm works rather well.

The amplitudes of echoes are plotted in Fig.7 as a function of the distance between the wire and the axis of the transducer, where the peak to peak amplitudes relative to that of an echo reflected from a perfect reflector which is located at the same distance from the transducer are shown. As seen from Fig.7, when the wire is located near the center of the transducer, the amplitude of the estimated echoes are a few dB larger than those of the actually obtained

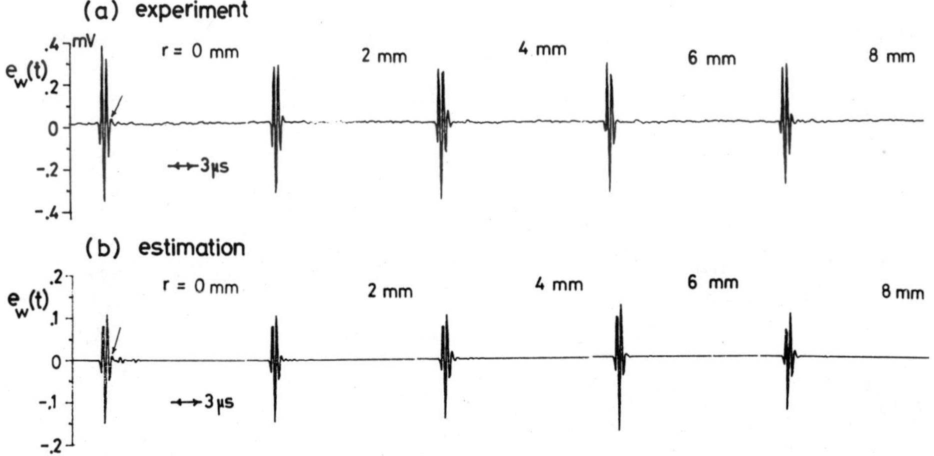

Fig. 6 Echoes reflected from the wire obtained experimentally (a) and theoretically (b). The ordinates of the calculated echoes are arbitrary.

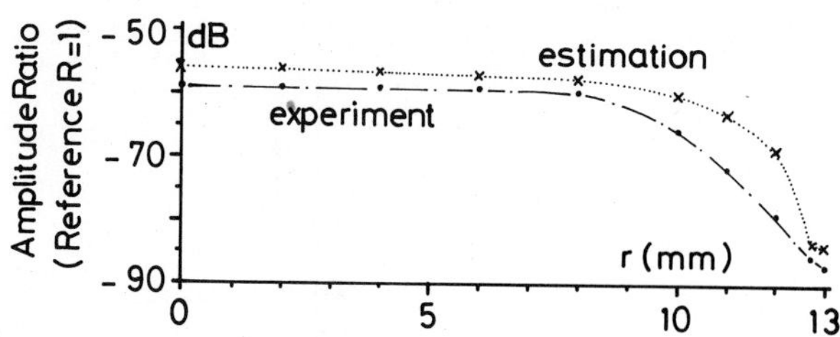

Fig. 7 Amplitude of echo reflected from the wire(e_W) relative to that reflected from a perfect plane reflector(e_P) located at the same distance

echoes. This difference may come from discarding the effects of transmission coefficient and refraction in this algorithm. When the wire approaches to the edge of the transducer, the difference of amplitude becomes larger. The reason for this increase is not clear at this stage.

CONCLUSIONS

It becomes clear that the new expressions for echo signals work rather well for strongly scattering medium. They are, however, deduced from the considerations for the simple case. Consequently it may be desirable to derive more rigorous relations from the wave equation for inhomogeneous medium. It will be also necessary to extend the estimation algorithm for concave transducers, since most practical systems use them.

REFERENCES

1. J.P. Jones, "Recent Advances in Ultrasound in Biomedicine" Research Studies Press, Oregon, vol.1, p.131 (1977).
2. M. Ueda and Y. Igarashi, Oyo Buturi, 49:1049 (1980)(in Japanese).
3. J.E. Barger, IEEE Trans. Sonics & Ultrason., SU-28:311 (1981).
4. A. Freedman, Acustica, 12:61 (1963).
5. D. M. Johnson, J. Acoust. Soc. Am., 59:1319(1976).
6. M. Ueda and H. Ichikawa, J. Acoust. Soc. Am. 70:1768 (1981).
7. P.R. Stepanishen, J. Acoust. Soc. Am., 49:1629 (1971).

DIRECTIVITY PATTERNS IN INHOMOGENEOUS ACOUSTIC MEDIA

Henryk Lasota*, Bernard Delannoy, Michel Moriamez

- C.N.R.S., ERA 593, Valenciennes
- Institut Industriel du Nord B.P. 48
  59651 Villeneuve d'Ascq Cédex - France

INTRODUCTION

Acoustic sources directivity patterns in homogeneous liquid media where the waves propagation velocity is constant everywhere, are determined by the well know diffraction formulae[1]. In such a medium it is possible to focus the wave as well as to reconstruct an image based on delay or phase processing of radiated or received waves. However, most of the media in which acoustical imaging is performed are, in fact, inhomogeneous. The propagation velocity varies from point to point in human tissues, under water as well as in soil. It leads, in some cases, to phase perturbations of propagating wave that can deform directivity patterns as well as limit focussing possibilities and, in the case of strong inhomogeneities, make impossible any image reconstruction of objects submerged in the medium.

A rising interest can be observed concerning inhomogeneous media problems, acoustic energy backscattering in dependence on the nature, size and distribution of inserted inhomogeneities being largely studied in a context of acoustic speckles[2] or medium characterisation[3]. However, there is no, or little, interest in the influence of inhomogeneities on directivity pattern or focus formation.

The paper presents a theoretical approach to the above problem making use of a modified Rayleigh formula where the homogeneous space Green's function is replaced by a random space one. A medium with randomly distributed wave propagation velocity is assumed, the velo-

* Permanent affiliation: Institute of Telecommunication, Technical University of Gdansk, 80952 Gdansk, POLAND

city fluctuations being a small amplitude continuous random function. A qualitative criterion is proposed that sets limits on the possibility of unperturbed directivity pattern creation and focussing, the parameters of the medium and the wavelength being taken into consideration. An experimental setup is presented that allows simultaneous automatic recording of both amplitude and phase of directivity patterns. The results of some preliminary measurements are given for illustrating the problem.

DIRECTIVITY PATTERNS

The Rayleigh diffraction formula, largely used in the calculations of acoustic sources directivity patterns in homogeneous media[4], can be presented in a form showing explicitly the homogeneous space Green's function

$$p(\vec{r}) = 2\,\rho \int_{S_o} a_n(\vec{r}_o)\; G(|\vec{r}-\vec{r}_o|)\; dS(\vec{r}_o) \tag{1}$$

where $p(\vec{r})$ is the acoustic pressure amplitude in a field point $\vec{r}$, $a_n(\vec{r}_o)$ is the normal component of acceleration on the source surface S_o, ρ is the medium density. The Green's function $G(|\vec{r}-\vec{r}_o|)$, that is given by :

$$G(|\vec{r}-\vec{r}_o|) = \frac{e^{jk|\vec{r}-\vec{r}_o|}}{4\pi|\vec{r}-\vec{r}_o|} \tag{2}$$

with $k = 2\pi/\lambda$ being the wave number, represents a spherical wave generated in the source point $\vec{r}_o$, with phase term in the numerator and amplitude term in the denominator. In homogeneous, isotropic medium both phase and amplitude are functions only of the distance $|\vec{r}-\vec{r}_o|$ between field and source points.

In inhomogeneous media the Green's function has, in general, a complex form, both phase and amplitude terms can be strongly modified by scattering effects that lead to attenuation and general deformation of the wavefront[5]. For simplification purposes, we consider here only weak inhomogeneities being small amplitude space fluctuations of the wavenumber (wave velocity). In such a case, we can assume that :

a) only phase-type perturbations are introduced into the propagating wave,
b) there is no scattering in the medium (no acoustic impedance fluctuations),
c) wavelets propagate along straight lines (no refraction of acoustic rays).

These assumptions lead to a slightly modified Green's function

$$G(\vec{r}/\vec{r}_o) = \frac{e^{j\phi(\vec{r}/\vec{r}_o)}}{4\pi|\vec{r}-\vec{r}_o|} \tag{3}$$

where the phase $\phi(\vec{r}/\vec{r}_o)$ depends now on both source and field points positions and not only on their distance. The form of the amplitude term remains unchanged, that means the wave-front is, from the energetic point of view, practically spherical.

The phase term is given in an integral form :

$$\phi(\vec{r}/\vec{r}_o) = \int_{\ell(\vec{r}/\vec{r}_o)} k(\vec{r})\, d\ell \tag{4}$$

where the wave number $k(\vec{r})$ is integrated along a straight trajectory $\ell(\vec{r}/\vec{r}_o)$ linking the source and field points.

In our study, we consider a medium the wave number function of which is the sum of its mean and fluctuating parts :

$$k(\vec{r}) = k_o\left[1 + \varepsilon f(\vec{r})\right] \tag{5}$$

where k_o is the mean value of the wave number, ε is the fluctuations amplitude coefficient, and $f(\vec{r})$ is a nondimensional centered, normalised random function that describes the inhomogeneities in the medium[5], i.e. :

$$\langle f(\vec{r})\rangle = 0 \ ; \ \langle f^2(\vec{r})\rangle = 1$$

The phase can be presented now as the sum of a mean term and a fluctuating one :

$$\begin{aligned}\phi(\vec{r}/\vec{r}_o) &= \bar{\phi}\,(|\vec{r}-\vec{r}_o|) + \tilde{\phi}\,(\vec{r}/\vec{r}_o)\\ &= k_o\,|\vec{r}-\vec{r}_o| + k_o\, s\,(\vec{r}/\vec{r}_o)\end{aligned} \tag{6}$$

The mean term has the same form as in case of homogeneous media. In the second term, $s(\vec{r}/\vec{r}_o)$ is a random function being the result of rectilinear integration of wave number fluctuations along the $\ell(\vec{r}/\vec{r}_o)$ trajectory :

$$s(\vec{r}/\vec{r}_o) = \varepsilon \int_{\ell(\vec{r}/\vec{r}_o)} f(\vec{r})\, d\ell \tag{7}$$

This function having a dimension of length has the sense of distance correction. We call it "perturbation length".

Finally, the Green's function (3) takes the following form :

$$G\,(\vec{r}/\vec{r}_o) = \frac{e^{jk\left[|\vec{r}-\vec{r}_o| + s\,(\vec{r}/\vec{r}_o)\right]}}{4\pi\,|\vec{r}-\vec{r}_o|} = G\,(|\vec{r}-\vec{r}_o|).e^{j\,k_o\,s\,(\vec{r}/\vec{r}_o)} \tag{8}$$

Let us study the influence of the phase fluctuations on the directivity pattern of a linear source being, more precisely, a narrow baffled piston of width $w \ll \lambda$ and length L (Fig. 1). The source symmetry allows to limit the analysis to the $y = 0$ plan. The source surface acceleration is defined as $a(\vec{r}_o) = a_o.A(x)$, where a_o is the acceleration amplitude and A(x) is a dimensionless acceleration distribution defined on the limited interval $x \in \langle -L/2\ ,\ L/2 \rangle$.

In these conditions, substitution of the inhomogeneous medium Green's function (8) to the Rayleigh formula (1) gives, in the far field, the following result :

$$p(\vec{r}) = \frac{\rho w a_o}{2\pi r}\,e^{j\,k_o\,r}\int_{-L/2}^{L/2} A(x).e^{j\,k_o\,s(\vec{r}/x)}.e^{-j\,k_o\,x\,\sin\theta}.dx = p_o(r).\mathcal{F}_{k_o\,\sin\theta}\left(A(x).e^{j\,k_o\,s(\vec{r}/x)}\right) = p_o(r).P(r,\theta) \tag{9}$$

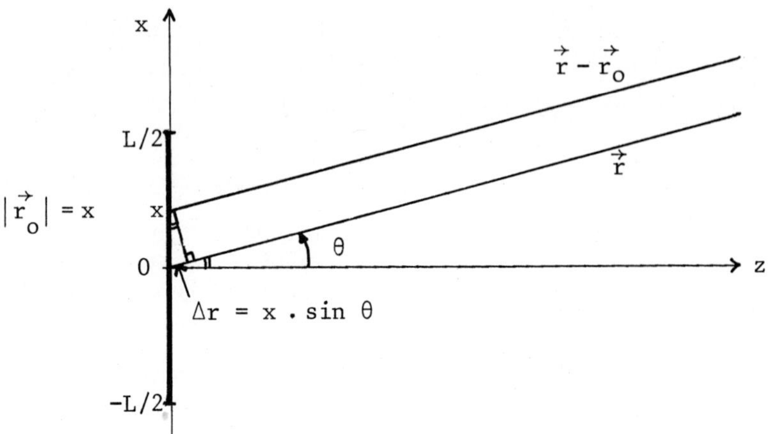

Fig. 1 - Linear source geometry

where $p_o(r) = \frac{\rho w L a_o}{2\pi r} \cdot e^{j k_o r}$ is the function of distance r only and $\mathcal{F}_{k_o \sin\theta}(\cdot) = P(r, \theta)$ is a Fourier transform with $k_o \sin\theta$ playing the role of spatial angular frequency. We have made use, above, of the usual Fraunhofer approximations, i.e. $|\vec{r} - \vec{r}_o| \cong r$ for amplitudes and $|\vec{r} - \vec{r}_o| \cong r - x \sin\theta$ for phases.

In comparison with the classic directivity pattern formula, the phase perturbation term has appeared as an additional phase modulation of the input acceleration distribution, the effective distribution being now $A(x).\exp\left[j k_o s(\vec{r}/x)\right]$. As the modulation changes from one field point to another, it is not possible to separate the angle and distance dependence of acoustic pressure. The notion of directivity pattern has changed its usual sense, being now dependent on the distance. The Fourier transform (9) must be computed separately for each field point $\vec{r}$.

In order to understand better the phase perturbations influence, we can imagine a field point $\vec{r}$ in which the perturbation length $s(\vec{r}/x)$ is not random and has a linear dependence on x. The pressure amplitude in this point is the same as it would be in case of a phased linear array in an homogeneous medium. Similary, a parabolic dependence of $s(\vec{r}/x)$ means the same calculus as for a focused array. In the first case, the overall influence of inhomogeneities on the resulting field in the considered point corresponds to the inclusion of a prism between the point and the source. The second case corresponds to a lens-type inclusion.

The general case of random fluctuations needs a detailed analysis that would link stochastic properties of :

a) inhomogeneities distribution $f(\vec{r})$,
b) perturbation length $s(\vec{r}/\vec{r}_o)$,
c) Green's function $G(\vec{r}/\vec{r}_o)$ and
d) pressure angular distribution $P(r, \theta)$.

Leaving this study to a later term, it is possible, however, to draw just now some practical conclusions concerning the limits of the directivity patterns formation in inhomogeneous media.

It seems natural to presume that the directivity pattern will not be much deformed if the phase perturbation is much smaller than π, it means when :

$$|s\ (\vec{r}/x)| \ll \frac{\pi}{k_o} = \frac{\lambda}{2} \qquad (10)$$

The above condition shows that the same medium that is, from the field formation point of view, an homogeneous one at low frequencies becomes inhomogeneous at higher frequencies.

It is worth remembering here the proportional dependence of the perturbation length on the fluctuations amplitude ε (7).

In many practical cases, the phase of directivity pattern is of no importance and a less stringent condition can be introduced. As a matter of fact, in a fixed field point $\vec{r}$, we can decompose the perturbation length $s(\vec{r}/x)$, as it varies in the integration limits $x \in \langle -L/2\ ,\ L/2 \rangle$, to the mean and fluctuating components :

$$s\ (\vec{r}/x) = \overline{s}\ (\vec{r}) + \tilde{s}\ (\vec{r}/x) \tag{11}$$

Substituting (11) into (9), we obtain :

$$p(\vec{r}) = p_o(r).e^{j\, k_o\, \overline{s}(\vec{r})} . \mathcal{F}_{k_o \sin\theta} \{A(x).e^{j\, k_o\, \tilde{s}(\vec{r}/x)}\} \tag{12}$$

that means that the amplitude of the directivity pattern remains unperturbed if :

$$|\tilde{s}\ (\vec{r}/x)| \ll \frac{\lambda}{2} \tag{13}$$

The mean component $\overline{s}(\vec{r})$, being a function of direction, causes fluctuations of the phase of directivity pattern. Having usually no practical importance, these fluctuations should not be neglected in the analysis of the synthetic aperture field.

FOCUSING IN INHOMOGENEOUS MEDIA

Having estimated the influence of medium inhomogeneity on far-field insonification, we can apply the same approach to estimate this influence on near-field focusing.

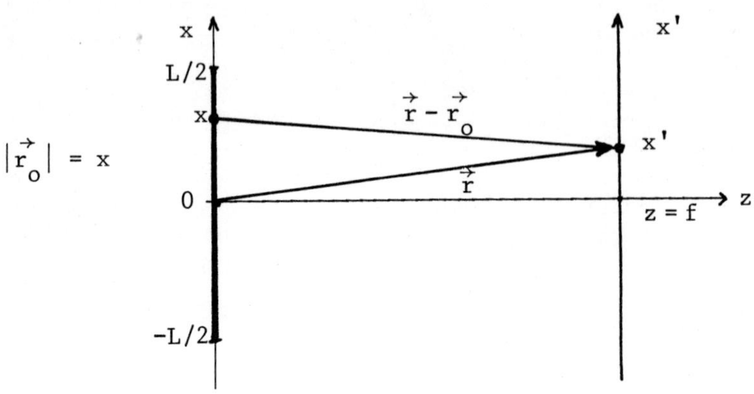

Fig. 2 - Source and focal planes

For this purpose, we consider the linear source having an additional parabolic phase distribution that corresponds to the output of a lens of a focal length f (Fig. 2) :

$$A_f(x) = A(x).e^{-j\frac{kx^2}{2f}} \tag{14}$$

The analysis concerns now the focal plane $z = f$, the distances between source points $\vec{r}_o = (x\ ,\ 0)$ and $\vec{r} = (x'\ ,\ z)$ being in the Fresnel zone :

$$|\vec{r} - \vec{r}_o| \cong z + \frac{(x - x')^2}{2z}$$

The phase integral has the form :

$$\phi(\vec{r} - \vec{r}_o) \cong k_o\ z + k_o\ \frac{(x' - x)^2}{2z} + k_o\ s\ (x'/x) \tag{15}$$

It is easy to show that the pressure amplitude distribution in the focal plane takes the following form :

$$p(x'\ ,\ f) = p_o(f).e^{j\frac{k_o x'^2}{2f}}.\mathcal{F}_{\frac{k_o x'}{f}}\{A(x).e^{j\,k_o\,s(x'/x)}\} \tag{16}$$

This result differs from the classic one[6] in the same manner as in the previous case. The perturbation length $s(x'/x)$ introduces the additional random modulation of phase. The same remarks concerning the conditions in which the pressure distribution is not perturbed remain valuable. Particularly, it is essential to have for a fixed x' small fluctuations of the perturbation length, i.e. :

$$|\tilde{s}\ (x'/x)| \ll \frac{\lambda}{2} \tag{17}$$

When this condition is fulfilled there is no change in the imaging systems resolution in such a medium.

PRELIMINARY MEASUREMENTS

Some preliminary measurements have been made in order to verify basic points of the theoretical approach. The experimental setup including a peak detector and a real-time zero-crossing phase meter allows to measure in a chosen, one period time interval both amplitude and phase of tone-burst signals received in a constant distance from the source and to record them simultaneously versus the receiver angular position. All usual preservations concerning transients in the field formation as well as water tank reverberations being made[1], the measurements are equivalent to recording of the amplitude and the phase of steady-state directivity patterns.

Inhomogeneous media samples have been placed in a special perspex compartment, submerged in water. The measured sources emit from medium surface which coincides with water surface. The receiving transducer moves in water 270 mm underneath semi-cylindrical bottom of the compartment (R = 100 mm) made of thin 1.5 mm perspex plate. The receiver rotation axis coincides with the cylinder axis as well as with the source axis of symmetry. The correct behaviour of the measuring arrangement has been verified by comparison of directivity patterns measured in water in presence and in absence of the compartment bottom.

The measurements have been made in two kinds of inhomogeneous media : slightly inhomogeneous and highly diffusive ones.

Broadband (45 to 75 kHz) transducers have been used made of P 1-60 (Quartz et Silice, France) ceramics with two $\lambda/4$ glass and perspex layers. The receiving transducer of 7×7 mm^2 cross-section has been closed in hermetic, air-filled iron-sheet box that assures acoustic and electrical isolation of transducer from the water as well as electromagnetic blinding, receiving surface being coupled to water through thin mylar foil.

Directivity patterns of two sources have been measured at 56 kHz : the first of 7×50 mm^2 emitting surface is omnidirectional in the observation plane, the second, composed of 9 identical acoustically separated elements of 95×50 mm^2 effective emitting surface, constitutes a directive one.

It is worth remembering that directivity patterns of sources radiating from a pressure-released surface are governed by the Rayleigh-Sommerfeld formula and differ from predicted by the Rayleigh formula by the directional factor equal to the cosinus of the directional angle[1].

We have used starch as slightly inhomogeneous medium that could fill the assumptions made in our analysis. It has been found that hot, clotty starch is temperature inhomogeneous when cooling, thus having the desired random space fluctuations of acoustic wave propagation velocity.

The quasi-omnidirectional source has in water the directivity pattern that corresponds well to the theoretical one[1], fluctuations of its phase not exceeding $\pm$ 10° (Fig. 3). In starch both amplitude and phase fluctuate. The amplitude fluctuations are caused probably by refraction effects, that we had neglected in our actual theoretical model. The phase fluctuations introduced by inhomogeneities reach $\pm$ 25°.

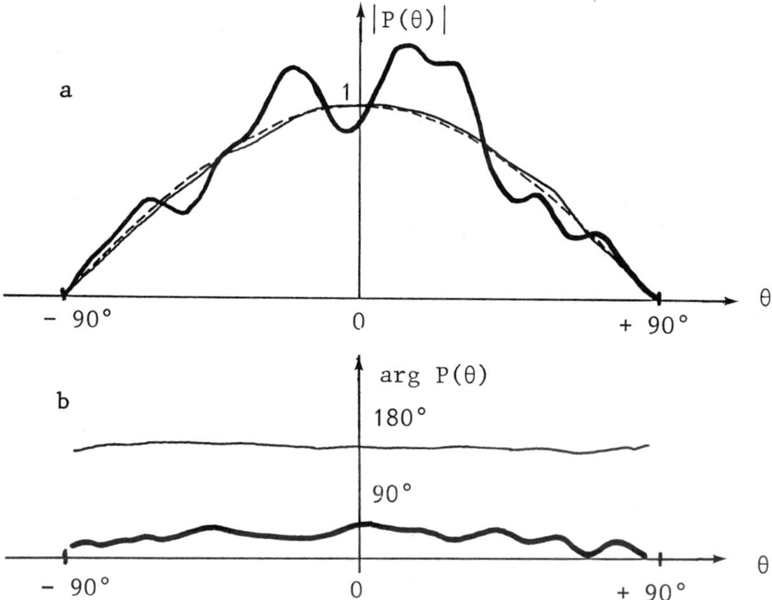

Fig. 3 - Amplitudes (a) and phases (b) of a narrow source directivity patterns measured in water (——) and in starch (▬▬), compared with theorical pattern in water (---).

The directive source pattern differing not much in water from the theoretical one, becomes strongly perturbed in starch (Fig. 4). The modification of the directivity pattern module is stronger in the paraxial zone than in tangent directions, the main lobe is attenuated, much weaker modifications being observed in the absolute level of the higher order lobes. On the other hand, the pattern phase fluctuations seem to be stronger for higher position angles.

In order to explain this phenomenon we have studied in a qualitive manner the influence of inhomogeneities localised in different zones of the field.

The inhomogeneous medium is presented in Fig. 5, each square corresponding to a different propagation velocity. Let us observe the propagation paths of various rays crossing the medium between the source points (x,0) and the field point $\vec{r}$. It can be assumed that the space correlation of the velocity fluctuations $f(\vec{r})$ decreases with the distance. In this case, it is easy to be seen that the relative phase fluctuations between the most distant rays contributing to the acoustic pressure constitution in the field, being stronger near the source (Zone A), become weaker when approaching the field points. When the distance between these rays becomes smal-

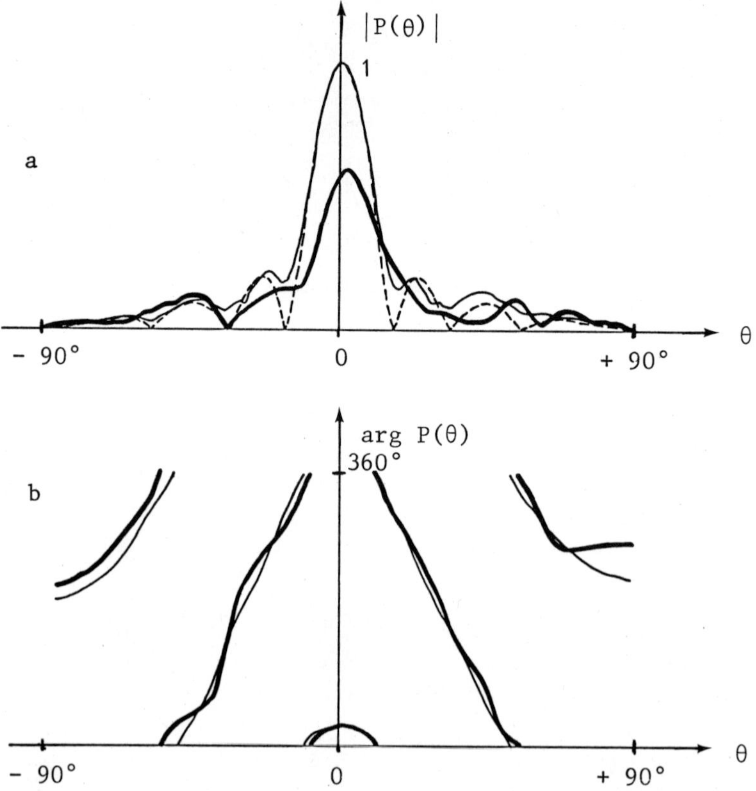

Fig. 4 - Amplitudes (a) and phases (b) of a 3.6 λ source directivity patterns measured in water (——) and in starch(▬▬), compared with the theoritical far-field pattern in water (---).

ler than the medium correlation length (Zone B), all of them sustain the same phase fluctuations.

It can be intuitively predicted that the fluctuating term $\tilde{s}(\vec{r}/x)$ of (11) depends on the distance and increases quicker in zone A which is much longer in the paraxial zone than at higher angles. This means that phases of contributions coming from different points of the source are the least correlated near the acoustic axis, the wavelets interference becoming less constructive in the main lobe, as well as less destructive at zero directions. At higher angles, where zone A is shorter, the directivity pattern amplitude is less perturbed, the mean term $\overline{s}(\vec{r})$ of (11) remaining responsible for the pattern phase fluctuations.

Since the above interpretation is an intuitive one it still requires an analytic confirmation.

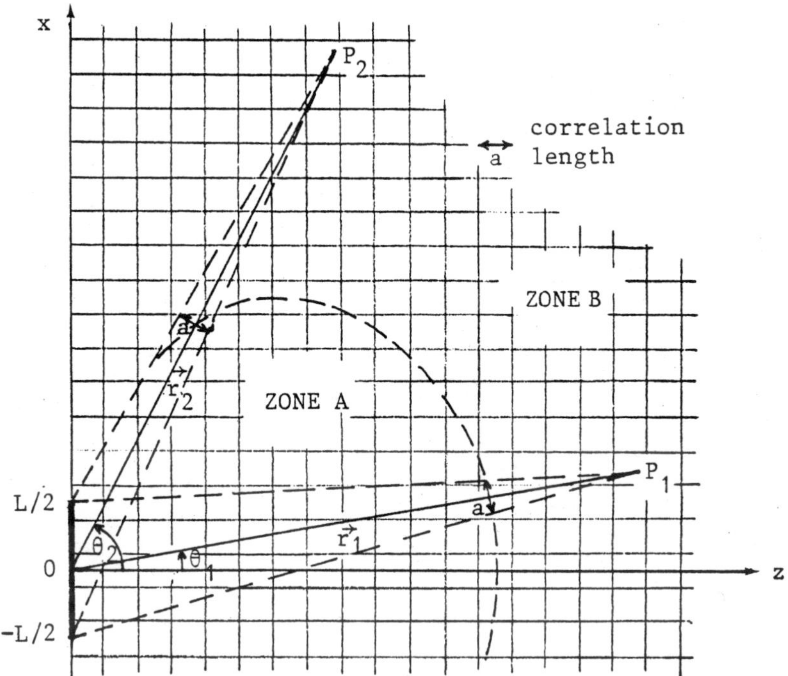

Fig. 5 - Zones of inhomogeneous medium

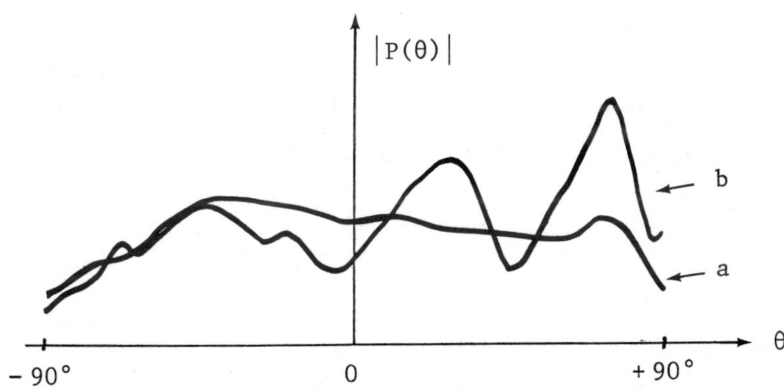

Fig. 6 - Directivity pattern amplitude measured in sand :
a) narrow 1 element source
b) broader 9 elements source

The same two sources have been used in the measurements of directivity patterns in a sample of wet sand, constituting in the interesting frequency band a diffusive, highly attenuating medium (Fig. 6). No essential difference can be observed between the patterns of narrow and broader sources. In diffusive media, multiple scattering effect causes high level phase fluctuations even between neighbouring rays that mean vanishing of any directivity in the source field.

CONCLUSION

The analitic approach to the problem of directivity patterns formation in heterogeneous media presented in the paper, together with the proposed criterion can constitute a basis for sistematic, detailed study of practical aspects of the problem.

The preliminary measurements verify some points of the theory. For example, it can be concluded that in such media, the main lobe is deteriorated quicker than the secondary ones. Moreover, inhomogeneous medium disturbs stronger higher frequency patterns, destructive influence growing with the distance. In highly diffusive media, where the phase correlation between neighbouring rays vanishes, no real directivity can be observed.

REFERENCES

1. B. Delannoy, H. Lasota, C. Bruneel, R. Torguet, E. Bridoux
The infinite planar baffles problem in acoustic radiation and its experimental verification, J. Appl. Phys. 50 : 5189-5195 (1979).

2. J.C. Bamber, R.J. Dickinson
Ultrasonic B-scanning : a computer simulation, Phys. Med. Biol. 25 : 463-479 (1980).

3. D.J. Vezzeti, S.O. Aks
Reconstruction from scattering data : analysis and improvements of the inverse Born approximation, Ultr. Im. 1 : 333-345 (1979).

4. M.C. Junger, D. Feit
Sound, Structures and their Interaction, MIT Press, Cambridge (1972).

5. U. Frisch
Wave propagation in random media, Probabilistic Methods in Applied Mathematics, Vol. 1, pp. 75-198, A.T. Bharucha-Reid, Ed., Academic Press, New York, (1968).

6. J.W. Goodman
Introduction to Fourier Optics, Mc Graw-Hill, New York, (1968).

TISSUE ULTRASONIC ATTENUATION WELL MODELIZED BY A MELLIN-CONVOLUTION

Michel Auphan, Jean-Marie Nicolas

Laboratoires d'Electronique et de Physique Appliquée
3, avenue Descartes
94450 Limeil-Brévannes (France)

ABSTRACT

It is shown that the incidence of attenuation in ultrasonic propagation within tissues can be considered as an integral transformation of transmitted acoustic signals, which transformation is known among mathematicians as a "Mellin-convolution".

Various methods are then proposed to solve the direct or the inverse problem relative to tissue propagation. Some of these methods make use of the Mellin-transformation.

As an illustration of the method an echographic computer simulation is described in a particular case where the inverse problem is solved i.e. the echographic signal corresponding to a non attenuating medium is restored.

INTRODUCTION

Our theoretical investigations about the ultrasonic propagation within biological tissues led us to an interesting and fruitful modelization of the tissue attenuation.

This modelization holds only for an attenuation proportional to frequency and homogeneous (i.e. identical for all the points of the propagation medium). In this case the radiative impulse response of a transducer can be obtained by an integral transformation of the radiative impulse response in non-attenuating medium. This integral transformation turns out to be a Mellin-convolution[1,2]. This tissue modelization is also valid for the roundtrip impulse response obtained by echoes on small point scatterers.

Let us suppose that we deal with a transducer of arbitrary shape but such that its transmitting operation corresponds to a velocity δ-function for the time $t = 0$ on the whole radiating area. This δ-function may also be assumed to be multiplied by any kind of weighing factor for each point of the radiative area. We know that in this case the velocity potential in a given point M within the propagation medium is obtained by an integral which has to be taken on the whole radiative area and which in addition to the weighing factors includes retarded Green functions of the type :

$$\frac{\delta(t - \frac{\rho}{c})}{\rho} \tag{1}$$

ρ being the distance between the point of the medium and a point of the radiative area.

In case of attenuation these Green functions have to be replaced by the function :

$$g\ (t - \frac{\rho}{c})$$

with :

$$g\ (t) = \frac{1}{\pi}\ \frac{\varepsilon\rho c}{\varepsilon^2\rho^2 + c^2t^2} \tag{2}$$

the Fourier transform of which is :

$$\hat{g}(\omega) = e^{-\varepsilon|\omega|\rho/c} \tag{3}$$

ω being the pulsation as it is well known. ε is the attenuation constant of the medium.

By using ρ as one of the two coordinates of the surface integral which has to be taken over the transducer radiative area one obtains the velocity potential in a point M which will be given by a simple integral over ρ of the type :

$$\psi_A(t) = \int_0^{+\infty} \frac{F(\rho)}{\pi}\ \frac{\varepsilon\rho\ d\rho}{\varepsilon^2\rho^2 + (ct-\rho)^2} \tag{4}$$

which in the case $\varepsilon \to 0$ implies :

$$F(\rho) = \psi_{NA}\ (\frac{\rho}{c})$$

$\psi_{NA}(t)$ being the velocity potential which would have been obtained in a non attenuating medium.

Hence :

$$\psi_A(t) = \int_0^{+\infty} \psi_{NA}\left(\frac{\rho}{c}\right) \frac{\varepsilon\rho}{\varepsilon^2\rho^2 + (ct-\rho)^2} \frac{d\rho}{\pi} \tag{5}$$

Putting : $\zeta = \frac{\rho}{c}$

$$\psi_A(t) = \frac{1}{\pi} \int_0^{+\infty} \psi_{NA}(\zeta) \frac{\varepsilon}{\varepsilon^2 + (\frac{t}{\zeta} - 1)^2} \frac{d\zeta}{\zeta} \tag{6}$$

or :

$$\psi_A(t) = \int_0^{+\infty} \psi_{NA}(\zeta)\, h\left(\frac{t}{\zeta}\right) \frac{d\zeta}{\zeta} \tag{7}$$

with :

$$h(x) = \frac{1}{\pi} \frac{\varepsilon}{\varepsilon^2 + (x-1)^2} \tag{8}$$

yielding the transformation which is called a "Mellin-convolution". Its definition is given by the following relation :

$f_3 = f_1 \underset{M}{*} f_2$ means

$$f_3(x) = \int_0^{+\infty} f_1(u)\, f_2\left(\frac{x}{u}\right) \frac{du}{u} = \int_0^{+\infty} f_1\left(\frac{x}{u}\right) f_2(u) \frac{du}{u} \tag{9}$$

Then we may write instead of (7) :

$$\psi_A(t) = \psi_{NA}(t) \underset{M}{*} h(t) \tag{10}$$

But in echographic problems the roundtrip impulse response $S_A(t)$ or $S_{NA}(t)$ is more useful than the radiative response $\psi(t)$. This roundtrip is generally defined by the echo due to a point scatterer and it can be shown that relation (10) holds for the roundtrip impulse responses and its validity domain covers all the targets made of a set (finite or infinite) of point scatterers without restriction for the multiple scattering.

DIRECT PROBLEM

In this case the problem of calculating the pattern of radiation or the roundtrip impulse response of a transducer is eased without loss of rigour. On the contrary the calculation of this impulse roundtrip response by means of a simple ordinary convolution by a Lorentzian function is only an approximation in which the distances between the scatterer and the various points of the transducer radiating area are assumed to be equal to a certain mean value taken as parameter for the Lorentzian.

Instead of calculating the integral (9) there are two other approaches of the direct problem.

The Mellin Transformation

The term of "Mellin-convolution" stems from its relation to the Mellin transformation. One demonstrates that (9) implies :

$$M\{f_3\} = M\{f_1\}\, M\{f_2\} \tag{11}$$

with the following definition of the Mellin transform :

$$M\{f(x)\}(s) = \int_0^{+\infty} f(x)\, x^{s-1}\, dx = F(s) \tag{12}$$

or :
$$M^{-1}\{F(s)\}(x) = \frac{1}{2\pi i}\int_{c'-i\infty}^{c'+i\infty} F(s)x^{-s}\, ds \tag{13}$$

Thus the roundtrip velocity impulse response $S_A(t)$ is given by :

$$S_A(t) = M^{-1}\left\{M\{S_{NA}(t)\}\, H(s)\right\} \tag{14}$$

$H(s)$ is the Mellin transform of $h(x)$ given by (8) and is obtained from tables of Mellin transforms [3] :

$$H(s) = \frac{(-1-i\varepsilon)^{s-1} - (-1+i\varepsilon)^{s-1}}{2i \sin \pi s}$$

The Use of a Logarithmic Time Scale

Let us put in (6) :

$$t = e^{v} \qquad S_A(t) = \tilde{S}(v)$$

$$\zeta = e^{i} \qquad S_{NA}(\zeta) = \tilde{S}_{NA}(u)$$

and we obtain :

$$\tilde{S}_A(v) = \frac{1}{\pi}\int_0^{+\infty} \tilde{S}_{NA}(u)\, \frac{\varepsilon du}{\varepsilon^2 + (e^{v-u}-1)^2} \tag{15}$$

which means a true convolution product :

$$\tilde{S}_A(u) = \tilde{h}(u) * \tilde{S}_{NA}(u) \tag{16}$$

with :

$$\tilde{h}(y) = h(e^y) = \frac{1}{\pi} \frac{\varepsilon}{\varepsilon^2+(e^y-1)^2} \tag{17}$$

So on this logarithmic scale the signal in an attenuating medium is obtained by a true convolution.

Of course these relations are only valid when one deals with velocity impulse responses of transducers and not electrical responses. But as the electrical roundtrip responses are generally calculated by means of the velocity impulse response which is then convolved by the acousto-electric response of the transducer, the only precaution to keep in mind is to apply this second true convolution after the Mellin-convolution i.e. on S_{NA}.

INVERSE PROBLEM

One of the main difficulties of the inverse problem (i.e. the problem of restoring the roundtrip signal $S_{NA}(t)$ which should have been obtained in a non attenuating medium starting from the knowledge of the roundtrip signal $S_A(t)$ corresponding to a known uniform attenuation) is that the electrical received signal has first to be deconvolved from the acousto-electric responses at both emission and reception. The inverse problem which will consist of a Mellin-deconvolution has obviously to be applied on the roundtrip velocity impulse response. And even when the transducer is known this first deconvolution is not always possible and adds much noise to the signal.

Let us now suppose that we have solved this first point and that we dispose of the roundtrip velocity impulse response $S_A(t)$. There are three ways of restoring $S_{NA}(t)$:

Mellin Transform

We write :

$$S_{NA}(t) = \mathcal{M}^{-1} \frac{\{S_A(t)\}}{H(s)} \tag{18}$$

and if we are able to calculate direct and inverse Mellin transforms without too much errors the problem is solved.

Use of Logarithmic Time Scale

Equation (16) has to be inverted. This implies a deconvolution by the function $\tilde{h}(y)$ which can be performed step by step provided that $\tilde{h}(y)$ is made causal in such a way that (15) becomes a Volterra equation of the first kind. For instance if :

$$\tilde{h}(y) = o \qquad \text{if } y < -a$$

we can write (16) with a finite integral :

$$S_A(v) = \frac{1}{\pi} \int_o^{a+v} \tilde{S}_{NA}(u)\ h(v-u)du \tag{19}$$

We know that this Volterra equation of the first kind is equivalent for numeric calculations to a Kramer system characterized by a triangular matrix easy to solve step by step.

Mellin Deconvolution

The same introduction of causality either by the addition of a small function or by just considering the function h(t) as negligible for $t < -a$ enable us to apply an upper limit to integral (7) which limits will include t. We find again a Volterra equation that we can try to solve step by step.

APPLICATION TO ECHOGRAPHY

We have made a computer simulation fitting the real examination as much as we could. We have then chosen a circular transducer with a typical acousto-electric response deduced from the Mason model but using the Cook-Redwood approximation. The simulated target of the roundtrip impulse response is made of two point scatterers one of which is off-axis as shown on figure 1.

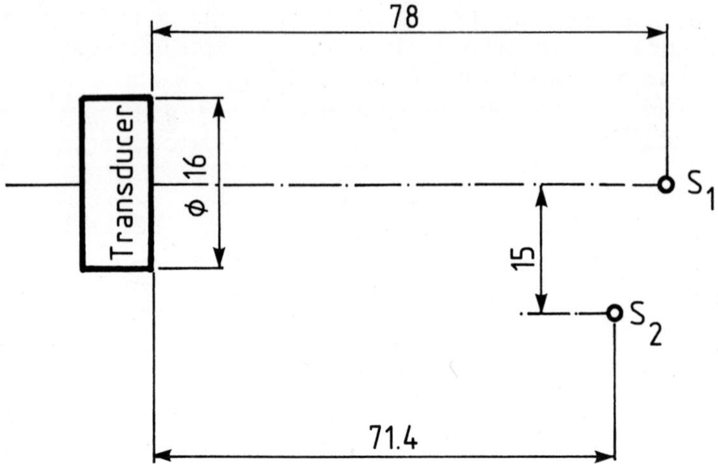

Fig. 1. Geometrical arrangement of the two scatterers S_1 and S_2. Lengths are in millimetres.

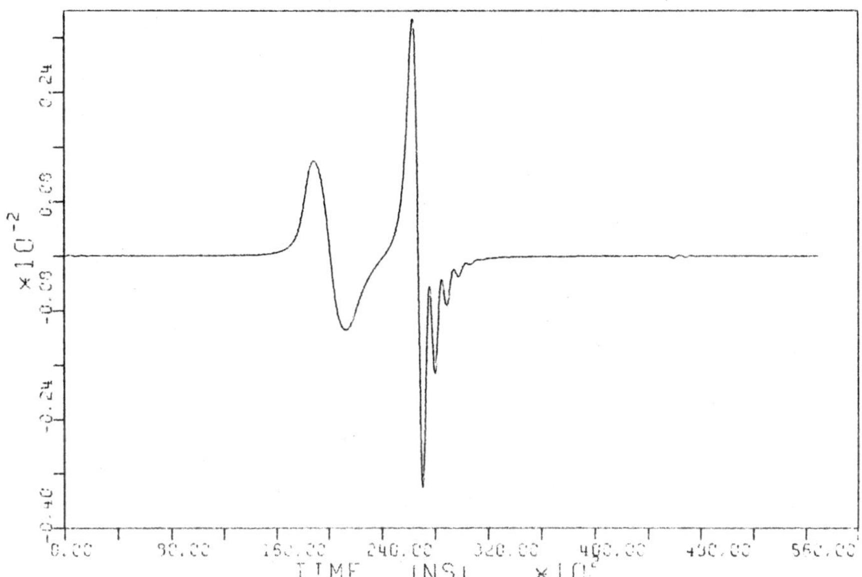

Fig. 2. Simulated roundtrip electrical response given by the circular transducer Ø 16 mm without matching layers but with a backing medium having an acoustic impedance of : $11\ 10^6$ Rayl.

The propagation medium has the same density and sound velocity as water but its attenuation factor is : 2 dB/cm/MHz

The ceramic is supposed to be a disc of lead metaniobate with the following specifications :

density : 6 200
coupling coefficient $K_t = 0.333$
dielectric constant $\varepsilon_{33}/\varepsilon_0 = 266.6$
thickness corresponding to 2.25 MHz

Figure 2 illustrates the echo electric signal issued from the transducer for a δ-shaped excitation[4]. Figure 3 shows the signal $S_A(t)$ obtained by deconvolving the electric signal of figure 2 by the acousto-electric transducer response. This deconvolution makes no problem since we deal with only simulated data. We have tested the two first methods previously mentioned for the solution of the inverse problem. As the second one turned out to be computer time consuming we have used the first one relying on direct and inverse Mellin transforms.

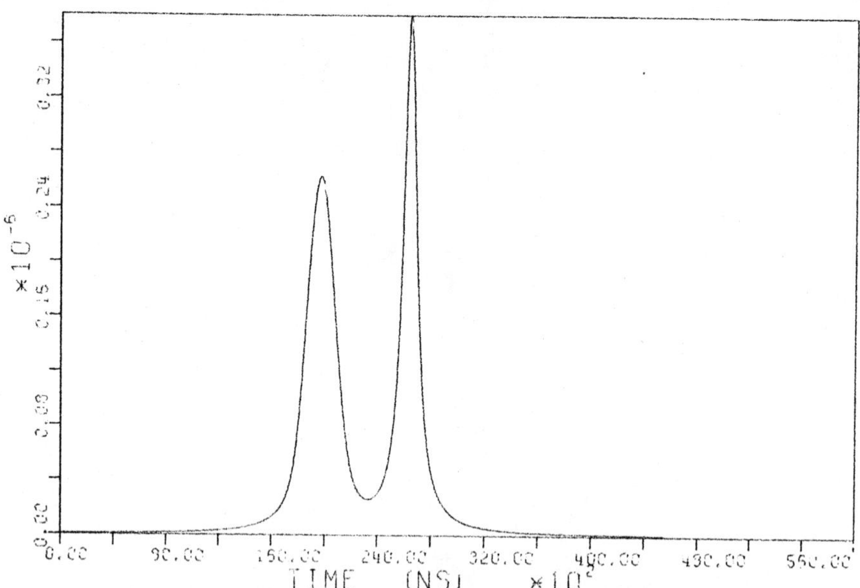

Fig. 3. Roundtrip impulse response $S_A(t)$ obtained from the signal of fig. 2 by an inverse convolution.

Actually direct and inverse Mellin-transforms are obtained by Fast Fourier Transforms in logarithmic scale according to known relations between Mellin and Fourier transforms.

$$\mathcal{M}\{f(x)\}\ (s) = \mathcal{F}\ \{f(e^x)\}\ (is)$$

$\mathcal{F}\ \{f(x)\}$ meaning the Fourier transform of f(x).

Figure 4 is the restored non attenuated signal S_{NA} obtained by application of relation (18). This figure is to compare with figure 5 which represents the roundtrip impulse response directly calculated for a non attenuating medium. The agreement is complete if one excepts the small oscillating artifacts due to the FFT method.

The curves corresponding to each scatterer are clearly separated. Obviously the wider and smaller peak correspond to the off-axis scatterer.

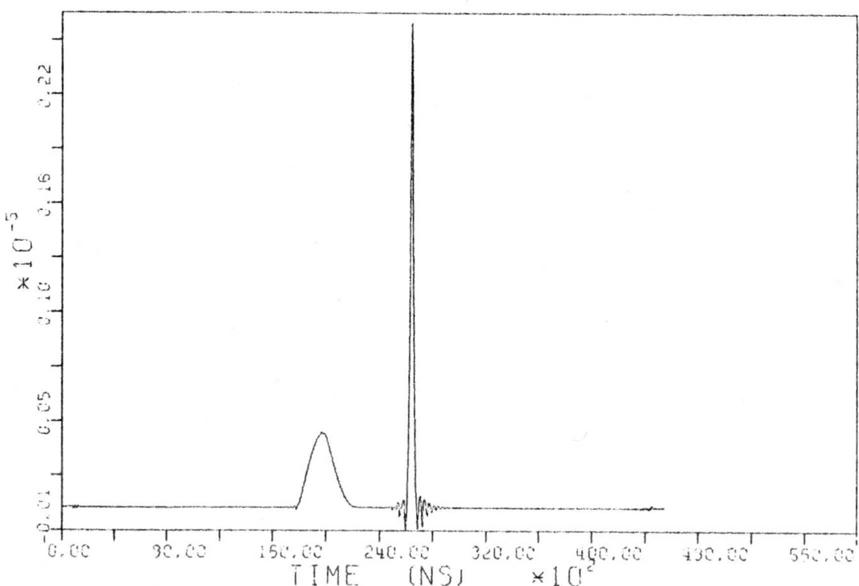

Fig. 4. Roundtrip impulse response $S_{NA}(t)$ restored for a non-attenuating medium by the Mellin transform process

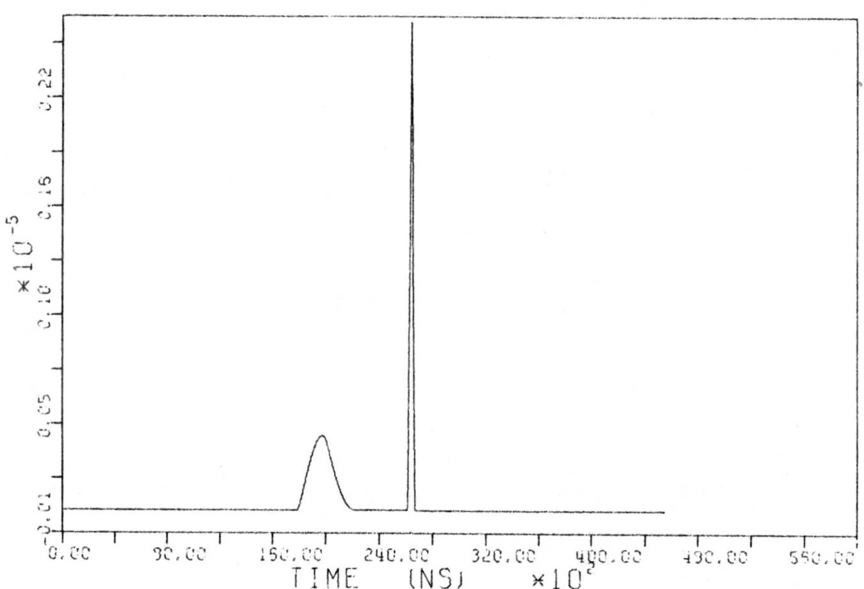

Fig. 5. Roundtrip impulse response directly calculated for a non attenuating medium

CONCLUSION

These methods will be for sure very fruitful for all the simulations of propagation within tissues since the direct transformations should be put in practice without much difficulties.

Furthermore we have shown that the inverse problem could be solved and this is encouraging for the study of echographic reconstructions which should take the attenuation into account.

REFERENCES

1. Zemanian, Generalized Integral Transformations, Wiley, 1968.
2. Colombo-Lavoine, Transformations de Laplace et de Mellin, Gauthier-Villars, 1972.
3. Bateman, Tables on Integral Transforms, vol. 1, McGraw Hill, 1954.
4. M. Auphan and H. Dormont, Pulsed Acoustic Radiation of Plane Transducers Ultrasonics 1977, 15, pp. 159-168.

THREE-DIMENSIONAL IMAGING OF SOFT TISSUE WITH DISPERSIVE ATTENUATION

J.M. Blackledge, M.A. Fiddy, S. Leeman* and L. Zapalowski

Physics Department, Queen Elizabeth College, University of London, Camden Hill Road, LONDON W8 7AH
*Medical Physics Department, Postgraduate Medical School, Hammersmith Hospital, Ducane Road, LONDON W12 OHS, U.K.

INTRODUCTION

The use of ultrasound is becoming an increasingly important and widely accepted method for non-invasive imaging of the human body and offers great potential for further developments in diagnostic medicine. Progress depends to a large extent on obtaining a more exact knowledge of the propagating ultrasonic fields and their interaction with tissue structures: but it is clear that the major factors describing these complex processes are the tissue sound velocity and specific acoustical impedance, the frequency dependent absorption of sound and its scattering by inhomogeneities in tissue density and elasticity.

There have recently been a number of publications that have considered the reconstruction of tissue parameters from weakly scattered ultrasonic fields. In particular, Mueller et al. (1979) have used diffraction methods based on a wave equation approach to the reconstruction of the velocity distribution. Norton and Linzer (1981), have considered the problem of reconstructing the reflectivity function of a three dimensional loss-less medium in which the scattering is sufficiently weak for the Born approximation to hold. In this presentation we examine the important effect of pulse deformation by absorption in soft tissue and show that even in this case, an exact three dimensional reconstruction procedure may be derived for the case of uniform dispersive absorption.

Absorption refers to acoustic wave energy loss as a result of a transformation into other energy forms. Two types of mechanisms are thought to be responsible for absorption: (1) Structural, thermal and other relaxation effects at the macromolecular level and (2) processes arising from the inhomogeneous nature of tissue, such as viscous relative motion (O'Donnell and Miller, 1979).

The majority of data on the ultrasonic properties of human and other mammalian tissues, show a roughly linear frequency dependence on attenuation. This refers strictly to energy loss as a result of both scattering and absorption, but in practice the measured values include contributions from other anomalous effects such as beam diffraction. If we suppose, as seems reasonable, that the scattering is weak, then a similar frequency variation must exist for absorption alone and the stress-strain loop for mammalian tissue must be regarded as being approximately independent of frequency. This requirement can be satisfied by considering a loss mechanism based on a hysteresis model. Such an approach deserves a more detailed description than is appropriate here, and will be considered elsewhere.

In this presentation, we consider viscous absorption only, which is erroneously regarded as being characterised by a frequency squared dependence on the linear absorption coefficient. However, such a model may be used as a basis for an exact reconstruction algorithm, since a judicious choice of parameters may produce a good fit to measured attenuation values for some soft tissues over the diagnostic frequency range 1-10 MHz and beyond. This is shown in Fig. 1 with attenuation values taken from Lang et al. (1978) , for the human breast. The treatment presented here, may be regarded as a purely phenomenological one, with no attempt being made to explain, on a physical basis, the actual values of the relaxation parameters which provide a fit to the experimental data.

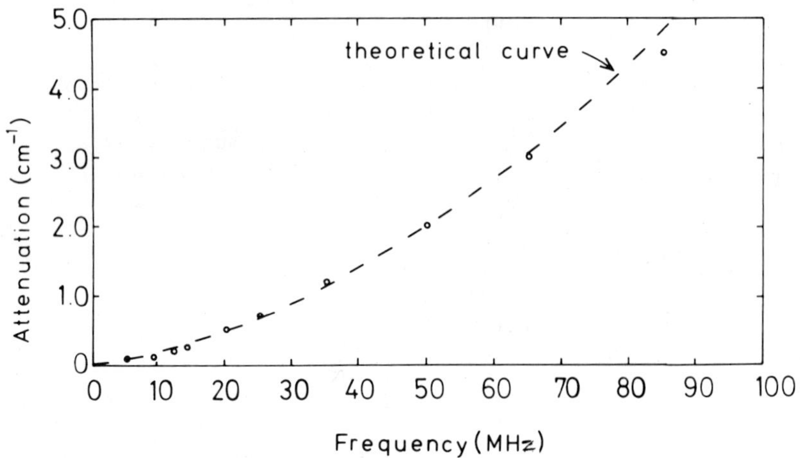

Fig. 1 Comparison of experimental data taken from Lang et al. (1978) of the attenuation against frequency for the human breast with values predicted by the proposed model for a relaxation time of 25ns and a sound velocity of 1414 ms^{-1}.

Our discussion will be restricted to high water content tissues (i.e. the soft tissues) which according to Fields and Dunn (1973), have density variations which are negligible compared to the compressibility fluctuations. The latter are thus the only scatter generators needed to be taken into account.

TISSUE MODEL

We consider a tissue model based on the following non-linear hydrodynamic equations

Conservation of mass: $$\frac{\partial \rho}{\partial t} + \nabla.(\rho \underline{v}) = 0 \quad (1)$$

Conservation of momentum: $$\rho(\hat{D}\underline{v}) - \nabla.\underline{\underline{T}} = 0 \quad (2)$$

Adiabatic equation of state: $$\hat{D}p - c^2 D\rho = 0 \quad (3)$$

where $\underline{\underline{T}}$ is the total material stress tensor, the convective derivative $\hat{D}$ is given by

$$\hat{D} = \frac{\partial}{\partial t} + \underline{v}.\nabla \quad (4)$$

and

$$c^2 = 1/(\rho\kappa) \quad (5)$$

We now assume that $\underline{\underline{T}}$ for soft tissue, can be represented by $-\underline{\underline{I}}p + \underline{\underline{\tau}}$ where $\underline{\underline{\tau}}$ is given by the Navier-Stokes velocity stress tensor $\underline{\underline{I}}\lambda\nabla.\underline{v} + \mu(\nabla\underline{v} + \underline{v}\nabla)$ in which λ and μ are the spatially variant Lamé parameters and $\underline{\underline{I}}$ is the unit tensor. Equations (1) - (3) can now be linearized by introducing a first order perturbation effect for small amplitude waves of the form

$$m = m_0 + m' \quad (6)$$

where m represents the physical parameters, m_0, the ambient values (for a homogeneous medium) and m' the perturbation where $m'/m_0 \ll 1$. Substituting the six perturbation equations represented by Eq.(6) into equations (1) - (3) and decoupling the resulting equations for the velocity vector field give

$$\rho \frac{\partial^2 \underline{v}}{\partial t^2} = \nabla\left(\frac{1}{\kappa}\nabla.\underline{v}\right) + \frac{\partial}{\partial t}\nabla.\left[\lambda\underline{\underline{I}}\nabla.\underline{v} + \mu(\nabla\underline{v}+\underline{v}\nabla)\right] \quad (7)$$

where for notational convenience, the primes on ρ, κ, λ, μ and $\underline{v}$ are surpressed. If we now consider the case where an inhomogeneous medium (soft tissue) defined by a domain Σ is embedded in a homogeneous medium (coupling medium) as in Fig. 2, the wave equation for the velocity vector field can be written as

$$\nabla\nabla.\underline{v} + \frac{\partial^2 \underline{v}}{\partial \overline{t}^2} + \frac{1}{\rho_0 \kappa_0}\frac{\partial}{\partial \overline{t}}\left[\lambda_0 \nabla\nabla.\underline{v} + \mu_0 \nabla.(\nabla\underline{v} + \underline{v}\nabla)\right]$$

$$= -\gamma_\rho \frac{\partial^2 \underline{v}}{\partial \overline{t}^2} - \nabla(\gamma_\kappa \nabla.\underline{v}) + \frac{1}{\rho_0 \kappa_0}\frac{\partial}{\partial \overline{t}}\left\{\lambda_0 \nabla(\gamma_\lambda \nabla.\underline{v}) + \mu_0 \nabla.\left[\gamma_\mu(\nabla\underline{v} + \underline{v}\nabla)\right]\right\} \quad (8)$$

where γ_ρ, γ_κ, γ_λ and γ_μ are defined by

$$\gamma_\rho(\underline{r}) = \begin{cases} [\rho(\underline{r}) - \rho_0]/\rho_0 & \underline{r} \in \Sigma \\ 0 & \underline{r} \notin \Sigma \end{cases} \qquad \gamma_\kappa(\underline{r}) = \begin{cases} [\kappa(\underline{r}) - \kappa_0]/\kappa(\underline{r}) & \underline{r} \in \Sigma \\ 0 & \underline{r} \notin \Sigma \end{cases}$$

$$\gamma_\lambda(\underline{r}) = \begin{cases} [\lambda(\underline{r}) - \lambda_0]/\lambda_0 & \underline{r} \in \Sigma \\ 0 & \underline{r} \notin \Sigma \end{cases} \qquad \gamma_\mu(\underline{r}) = \begin{cases} [\mu(\underline{r}) - \mu_0]/\mu_0 & \underline{r} \in \Sigma \\ 0 & \underline{r} \notin \Sigma \end{cases}$$

and $\overline{t} \equiv c_0 t$ where $c_0 = (\rho_0 \kappa_0)^{-\frac{1}{2}}$.

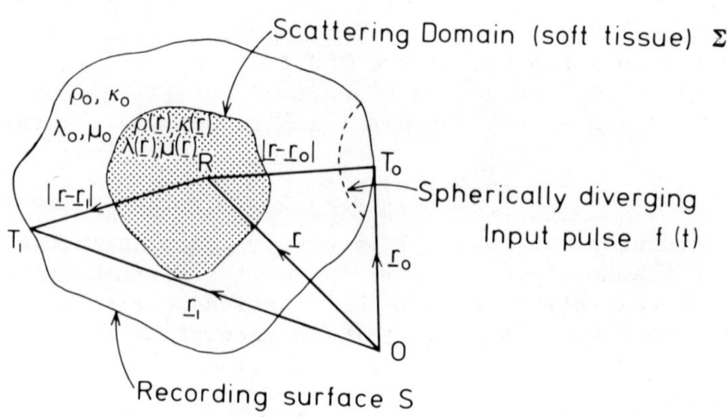

Fig. 2 The transducer (T_0) at $\underline{r}_0$ acts as a source of spherically diverging waves and as a receiver for the back-scattered radiation field and varies over a two dimensional surface S .

Using the vector and dyadic relationships

$$\nabla \wedge \nabla \wedge \underline{v} = -\nabla^2 \underline{v} + \nabla(\nabla . \underline{v})$$

$$\nabla \underline{v} + \underline{v} \nabla = 2 \nabla \underline{v} + \underline{\underline{I}} \wedge \nabla \wedge \underline{v}$$

and considering longitudinal waves only, as seems appropriate for human tissue (Frizzell and Carstensen, 1977) equation (8) can be reduced to the following

$$\nabla^2 \underline{\tilde{v}} + \frac{k^2}{1-i\ell k} \underline{\tilde{v}} = -\frac{\hat{F}\, \underline{\tilde{v}}}{1-i\ell k} \qquad (9)$$

where

$$\hat{F} = k^2 \gamma_\rho + \nabla \gamma_\kappa \nabla . + \frac{ik\lambda_0}{\rho_0 c_0} \nabla \gamma_\lambda \nabla . + \frac{2ik\mu_0}{\rho_0 c_0} \nabla . \gamma_\mu \nabla \qquad (10)$$

and $\underline{\tilde{v}}$ denotes the Fourier transform of $\underline{v}$ given by

$$\underline{v}(\underline{r}, t) = \int_{-\infty}^{+\infty} \underline{\tilde{v}}(\underline{r}, k)\, e^{-ikt}\, dt \qquad (11)$$

The relaxation length ℓ is defined in terms of λ_0 and μ_0 by $\ell = (\lambda_0 + 2\mu_0)/\rho_0 c_0$ where the numerator represents the compressional viscosity M_0 which can also be written in terms of the bulk, ζ_0, and shear, η_0, viscosities by $M_0 = \zeta_0 + \frac{4\eta_0}{3}$.

The operator $\hat{F}$ constitutes all the scatter generators in our model for tissue. Equation (9) is thus, the general wave equation for a longitudinal sound wave propagating through a medium with density, compressibility and viscosity variations. This equation is the basis for reconstruction techniques that would manifest information about these tissue parameters. The general case is complicated but we have made some considerable progress towards its exact solution. In the interest of a finite presentation we here concentrate our attention on a realistic tissue model, in which the absorption is uniform throughout and where $\gamma_\kappa \gg \gamma_\rho$ for soft tissue. In this case, we may construct a scalar wave equation for the velocity potential $\tilde{\phi}(\underline{r}, k)$ given by

$$\nabla^2 \tilde{\phi} + \frac{k^2}{1-i\ell k} \tilde{\phi} = -\frac{\gamma_\kappa(\underline{r})}{1-i\ell k} \nabla^2 \tilde{\phi} \qquad (12)$$

where

$$\tilde{\underline{v}} = -\nabla\tilde{\phi} \qquad (13)$$

FORMULATION OF THE INTEGRAL EQUATION FOR A SPHERICAL WAVE SOURCE

We now consider a transducer at a position $\underline{r}_0$ on a recording surface $\bar{S}$, to be a point source of radiation with time dependence $f(\bar{t})$. This demands, that we rewrite Eq. (12) in the form

$$\nabla^2\tilde{\phi} + \frac{k^2}{1-i\ell k}\tilde{\phi} = -\frac{\gamma_K(\underline{r})}{1-i\ell k}\nabla^2\tilde{\phi} - \tilde{f}(k)\,\delta^3(\underline{r}-\underline{r}_0) \qquad (14)$$

The outgoing spherical wave function when $\gamma_K(\underline{r}) = 0$ is then $\tilde{f}(k)\,g(\underline{r}|\underline{r}_0)$ where $g(\underline{r}|\underline{r}_0)$ is the outgoing Green's function given by

$$g(\underline{r}|\underline{r}_0) = \frac{1}{4\pi|\underline{r}-\underline{r}_0|}\exp\left[ik|\underline{r}-\underline{r}_0|/(1-i\ell k)^{\frac{1}{2}}\right] \qquad (15)$$

Substituting $\tilde{\phi} = \tilde{f}ge^u$ into Eq. (14) gives

$$\left[\nabla^2 u + 2\nabla u.\frac{\nabla g}{g} + \nabla u.\nabla u\right]\left[\frac{\gamma_K(\underline{r})}{1-i\ell k} + 1\right] = \frac{k^2\gamma_K(\underline{r})}{(1-i\ell k)^2}$$

$$+\frac{1}{g}\left[\left(\frac{\gamma_K(\underline{r})}{1-i\ell k} + 1\right) - e^{-u}\right]\delta^3(\underline{r}-\underline{r}_0) \qquad (16)$$

Assuming the validity of the Rytov approximation so that u varies sufficiently slowly for the term $\nabla u.\nabla u$ to be negligible relative to the other quantities and substituting $w/\tilde{f}g$ for u, Eq. (16) may be reduced to

$$\nabla^2 w + \frac{k^2}{1-i\ell k}w = \frac{k^2\tilde{f}}{(1-i\ell k)^2}\gamma_K(\underline{r})\,g + \tilde{f}\left[1 - \frac{w}{g} - e^{-\frac{w}{\tilde{f}g}}\right]\delta^3(\underline{r}-\underline{r}_0) \qquad (17)$$

when $\gamma_K \ll 1$. The solution to Eq. (17), by the usual Green's function method is then given by

$$w(\underline{r}_1|\underline{r}_0) = \frac{k^2\tilde{f}(k)}{(1-i\ell k)^2}\int_{\Sigma^3} g(\underline{r}|\underline{r}_1)\,\gamma_K(\underline{r})\,g(\underline{r}|\underline{r}_0)\,d^3\underline{r} \qquad (18)$$

where we have neglected boundary effects and considered the evaluation of the field at a point $\underline{r}_1$ on the surface S , as in Fig. 2. Thus, the total field $\phi(\underline{r}_1|\underline{r}_0, k)$ is given by the WKBJ type solution of the form

$$\tilde{\phi}(\underline{r}_1|\underline{r}_0, k) = \tilde{f}(k)\, g(\underline{r}_1|\underline{r}_0) \exp\left\{ \frac{k^2}{(1-i\ell k)^2 g(\underline{r}_1|\underline{r}_0)} \cdot \int_{\Sigma^3} g(\underline{r}|\underline{r}_1)\, \gamma_K(\underline{r})\, g(\underline{r}|\underline{r}_0)\, d^3\underline{r} \right\} \quad (19)$$

THE CASE OF BACK-SCATTERING

In order to evaluate the contribution to $\tilde{\phi}$ at $\underline{r}_0$ arising from the scattering within the domain Σ we write Eq. (19) in the form

$$\tilde{\phi}_{BS}(\underline{r}_0, k) = \lim_{r_1 \to r_0} \left\{ \tilde{f}(k)\, g(\underline{r}_1|\underline{r}_0) \sum_{n=0}^{\infty} \frac{k^{2n}}{(1-i\ell k)^{2n}} \frac{1}{n!} \cdot \frac{1}{g^n(\underline{r}_1|\underline{r}_0)} \left[\int_{\Sigma^3} g(\underline{r}|\underline{r}_1)\, \gamma_K(\underline{r})\, g(\underline{r}|\underline{r}_0)\, d^3\underline{r} \right]^n - \tilde{f}(k)\, g(\underline{r}_1|\underline{r}_0) \right\} \quad (20)$$

The expression for the back-scattered field is then given by

$$\tilde{\Phi}_{BS}(\underline{r}_0, k) = \int_{\Sigma^3} \gamma_K(\underline{r})\, g^2(\underline{r}|\underline{r}_0)\, d^3\underline{r} \quad (21)$$

where

$$\tilde{\Phi}_{BS}(\underline{r}_0, k) = \tilde{\phi}_{BS}(\underline{r}_0, k)\, (1-i\ell k)^2 / k^2 \tilde{f}(k) \quad (22)$$

and we note that

$$\lim_{r_0 \to r_1} \left\{ 1/g^n (\underline{r}_1 | \underline{r}_0) \right\} = 0 \qquad n = 1, 2, \ldots\ldots\infty \qquad (23)$$

The expression for $\widetilde{\Phi}_{BS}$ is identical to the expression which would have been obtained via the first Born approximation. Thus, for back-scattering, we might conclude that the Rytov approximation, if treated correctly as in Eq. (20) reduces to the Born approximation in the limit of back-scattering for weak scatter generators. Eq. (21) may now be reduced to a more convenient form by differentiating with respect to k and letting $k \longrightarrow k/2$ and $\ell \longrightarrow 2\ell$ which finally gives

$$\widetilde{\Theta}(\underline{r}_0, k) = \int_{\Sigma^3} \gamma_\kappa (\underline{r}) \, g (\underline{r} | \underline{r}_0) \, d^3\underline{r} \qquad (24)$$

where

$$\widetilde{\Theta}(\underline{r}_0, k) = 32\pi i \, \frac{(1-i\ell k)^{3/2}}{2 - i\ell k} \, \frac{d}{dk} \widetilde{\Phi}_{BS}(\underline{r}_0, k/2) \qquad (25)$$

Equation (24) is a linear integral equation which is the basis for our reconstruction algorithm and may be viewed as a linear mapping of the tissue compressibility function $\gamma_\kappa (\underline{r})$ to the observed back-scattered field $\widetilde{\phi}_{BS}(\underline{r}_0, k)$ where $\underline{r}_0$ is allowed to vary over some suitable two-dimensional surface. The measured data consists of values of the pressure $\widetilde{p}_{BS} (\underline{r}_0, k)$ detected by the transducer at $\underline{r}_0$ which is related to the velocity potential by

$$\widetilde{p}_{BS}(\underline{r}_0, k) = c_0 \rho_0 ik \widetilde{\phi}_{BS}(\underline{r}_0, k) + M_0 \nabla^2 \widetilde{\phi}_{BS}(\underline{r}_0, k) \qquad (26)$$

EXPLICIT INVERSION OF EQ.(24) FOR A SPHERICAL APERTURE

We require the exact solution to Eq. (24) for the tissue compressibility function $\gamma_\kappa (\underline{r})$ when the souce-receiver point $\underline{r}_0$ is allowed to vary over the surface of a sphere which encloses the scattering domain Σ. The geometry of this arrangement suggests that Eq. (23) be solved in a spherical coordinate system, via spherical harmonics, with the coordinates of $\underline{r}$ and $\underline{r}_0$ expressed as (r, θ, φ) and (r_0, θ_0, ϕ_0) respectively. The Green's function can then be expanded as a sum of spherical harmonic basis functions:

$$g(\underline{r}|\underline{r}_0) = \frac{i\xi}{4\pi} \sum_{m=0}^{\infty} (2m+1) J_m(\xi r) H_m^{(1)}(\xi r_0) P_m(\underline{n}.\underline{n}_0) \quad (27)$$

where

$$\xi = k/(1-i\ell k)^{\frac{1}{2}} \quad (28)$$

J_m, $H_m^{(1)}$ and P_m denote the Bessel functions, Hankel functions and Legendre polynomials respectively and the unit vectors $\underline{n}$ and $\underline{n}_0$ point in the directions of (θ, ϕ) and (θ_0, ϕ_0) respectively. Substituting Eq. (27) into Eq. (24), multiplying both sides by

$$(2n+1) P_n(\underline{n}'.\underline{n}) / H_n^{(1)}(\xi r_0) \quad (29)$$

and integrating over the surface of the sphere with respect to $\underline{n}_0$, Eq. (24) becomes

$$\int_{\Omega_0} d\Omega_0 \left\{ \widetilde{\Theta}(\underline{n}_0, k)(2n+1) P_n(\underline{n}'.\underline{n}_0) / H_n^{(1)}(\xi r_0) \right\}$$

$$= i\xi \int d^3\underline{r}\,(2n+1)\gamma_\kappa(\underline{r}) J_n(\xi r) P_n(\underline{n}.\underline{n}') \quad (30)$$

where we have used the identity

$$\int_{\Omega_0} d\Omega_0 P_m(\underline{n}.\underline{n}_0) P_n(\underline{n}'.\underline{n}_0) = \frac{4\pi}{2n+1} \delta_{mn} P_n(\underline{n}.\underline{n}') \quad (31)$$

and $d\Omega_0 = \sin\theta_0 d\theta_0 d\phi_0$ is an element of solid angle on the unit sphere and Ω_0 is the surface of the sphere. Multiplying both sides of Eq. (30) by $\xi J_n(\xi r')$, integrating with respect to ξ from $-\infty$ to $+\infty$ and summing n from 0 to ∞ gives

$$\gamma_\kappa(\underline{r}') = \int_{\Omega_0} d\Omega_0 \int_{-\infty}^{+\infty} d\xi \widetilde{\Theta}(\underline{n}_0, k) \frac{\xi}{i(2\pi)^2} \sum_{n=0}^{\infty} (2n+1) \frac{J_n(\xi r') P_n(\underline{n}'.\underline{n})}{H_n^{(1)}(\xi r_0)} \quad (32)$$

where we have exploited the identities

$$\int_{-\infty}^{+\infty} d\xi\ \xi^2 J_m(\xi r)\, J_m(\xi r') = \frac{\pi}{r^2}\ \delta^3(\underline{r} - \underline{r}') \qquad (33)$$

and

$$\sum_{n=0}^{\infty} (2n+1)\, P_n(\underline{n}.\underline{n}') = \left\{4\pi\,\delta(\phi - \phi')\,\delta(\theta - \theta')/\sin\theta\right\} \qquad (34)$$

If we now write $\tilde{\Theta}(\underline{n}_0, k) = \tilde{\Theta}(\underline{k})$ where $\underline{k} = k_0\underline{n}_0$ and suppress the primes, Eq. (32) can be written as

$$\gamma_K(\underline{r}) = \int d^3\underline{k}\,\tilde{\Theta}(\underline{k})\,\tilde{K}(\underline{k}, r) \qquad (35)$$

where

$$\tilde{K}(\underline{k}, r) = \frac{2-i\ell k}{2(1-i\ell k)}\ \frac{1}{ik(2\pi)^2}\sum_{n=0}^{\infty}\left\{(2n+1)\, J_n\left[kr/(1-i\ell k)^{\frac{1}{2}}\right].\right.$$

$$\left.P_n(\underline{n}.\underline{n}_0)/H_n^{(1)}\left[kr/(1-i\ell k)^{\frac{1}{2}}\right]\right\} \qquad (36)$$

and

$$d^3\underline{k} = k^2 d\Omega_0\, dk \qquad (37)$$

Equation (35) represents the exact solution to the three dimensional inverse scattering problem in the near-field. Any far-field approximation, although simpler, would be unrealistic for a medical imaging system, where the source-receiver point or array should be close to the scattering domain.

CONCLUSION

We have shown that the three dimensional scalar compressibility function for soft tissue (in particular, the human breast) with uniform viscous absorption characteristics, can be recovered from back-scattered waves measured as a function of time on a two dimensional surface. This has given some insight into the major obstacle of dealing with pulse deformation attributed to the absorption of a scattering medium. The problem to consider now, is the effect of spatially variant absorption and the incorporation of density fluctuations in tissue so that in the case of back-scattering, a measure of the tissue impedence can be obtained.

ACKNOWLEDGEMENTS

We would like to thank Dr S. Norton for communicating to us some of his thoughts on the use of the Rytov approximation in the case of back-scattering. We are also grateful to Professor R. Burge for extending to us the facilities of Queen Elizabeth College.

REFERENCES

Fields, S., Dunn, F., 1973, J. Acoust. Soc. Am., 54 (3) :809.

Frizzell, L.A., Carstensen, E.L., 1973, J. Acoust. Soc. Am., 60 (6) : 1409.

Lang, J., Zance, R., Gairard, B., Dale, G., Gros, Ch.M., 1978, Ultra. Med. Biol., 4:125.

Mueller, R., Kaveh, M., Wade, G., 1979, Proc. IEEE, 67 (4) :567.

Norton, S.J., Linzer, M., 1981, Proc. IEEE, BME-28:202.

Norton, S.J., private communication

O'Donnell, M., Miller, J.G., 1979, Ultrasonic Tissue Characterization II, M. Linzer, ed., NBS, Special Publication, 525.

ELASTIC WAVE IMAGING WITH THE AID OF AN INVERSION TRANSFORMATION

B.R. Tittmann

Rockwell International Science Center*
Thousand Oaks, CA 91360, U.S.A. and
G.P.S., University Paris VII, 75251 Paris Cedex05, France

F. Cohen-Tenoudji and G. Quentin

G.P.S., University Paris VII, 75251 Paris, France

INTRODUCTION

An inversion transformation has been developed for the extraction of geometrical parameters of a scatterer embedded in a solid. The transformation is shown to be useful in the $1 \leqslant ka \leqslant 4$ range for the construction of an image. Examples given are a 400 μm × 800 μm ellipsoidal void embedded in a Ti-alloy and a 800 μm spherical void with a small spherical perturbation also in Ti-alloy.

The inverse problem of extracting size and shape of an object from scattering data is important in many fields ranging from seismology, to underwater acoustics, ultrasonic NDE and radar. Although progress has been made towards solutions of the inverse problem for elastic waves, these efforts are still in the early stages with several approaches currently being evaluated.[1-4] In a recent letter[5] we proposed an approach based on a general inversion transformation $\mathscr{I}_A$ which when used on backscattering data produces a function A(t). For the special case of the Kirchhoff approximation, A(t) was interpreted as the cross-sectional area intercepted by a transverse plane moving in the same direction as that of an incident wave. The maximum value of A(t) was interpreted as the total cross section.

*Permanent address.

In this paper is presented the application of the technique to several sets of scattering data with a resulting reconstruction of an image.

THE INVERSION TRANSFORMATION - THEORY

The detailed discussion of the theoretical approach is described elsewhere[5,6] and is presented here only in the form of a summary. The inversion transformation $\mathscr{I}_A$ is defined by

$$A(t) \equiv \mathscr{I}_A\,[P(k)]$$

such that

$$A(t) = \int_{-\infty}^{+\infty} [P(k)/k^2]\,\exp(ikct)dk \quad .$$

Here P(k) is the pressure backscattered by a surface S when illuminated by an incident plane wave with wave number k and velocity $c = 2z/t$, where z is the position along the axis of propagation and t is time. In general, A(t) is difficult to derive explicitly but has been obtained[6] for the case of acoustic waves scattered by a fixed, rigid sphere. However, much progress has been made in computing values for P(k) for elastic waves scattered by obstacles of various shapes[7,8] and it is possible to carry out $\mathscr{I}_A$ on these theoretical data. In Fig. 1 is plotted the function A(t) as obtained from theoretical scattering amplitudes P(k) for an obstacle embedded in a solid, i.e., a void in the shape of an oblate spheroid with aspect ratio a:b = 2:1 with its semi-minor axis b = 200 μm along the direction of incidence. The origin of time is chosen at the centroid of the scatterer. The length of the semi-minor axis can be obtained from Fig. 1 by using the time difference Δt between the values when the function A(t) starts from zero and when t = 0 at the centroid. Also shown in Fig. 1 for comparison, is the graph of A(t) obtained from experimental scattering data.

EXPERIMENTAL RESULTS AND DISCUSSION

The experiments were performed on samples of Ti-alloy into which have embedded scatterers of a variety of shapes with the aid of the diffusion bonding process already described elsewhere.[9] The scattered waveforms were deconvolved of the transducer response with the use of waveforms reflected by a plane interface (metal-air) located at the same distance as the defect. In order to handle the typical band limitations of the experimental data, an iterative technique was developed to arrive at the experimental A(t). First, the phase of the scattered amplitude was forced to be zero. The results A(t) gave a rough estimate of the dimension of the scatterer. This estimate was used to calculate the slope of the phase at high frequencies, taking advantage of the knowledge from theory that the phase varies linearly for a spheroid at high frequencies, with the slope given by the distance from the front face to the centroid. The calculation of A(t) was then performed again, with the waveform signal shifted in time to have the correct slope for the phase at high frequencies but with a null still imposed in the unknown part at low frequencies. The function A(t) obtained was sufficiently stable with regard to this shift in time that the result in the second step was usually not affected by a slight error in the evaluation of the position of the centroid. The operation was now repeated several times to refine A(t) using the criterion that A(t) has to be positive for $t < 0$. In the process the position of the centroid was also refined.

The comparison shown in Fig. 1 shows that the iterative technique is able to produce a fairly accurate A(t) even for band-limited data in the range $1 \leq ka \leq 4$. Deviations for $t > 0$ could be due to the non-ideal shape of the void fabricated in the sample. It is to be noted that for negative values of time contributions

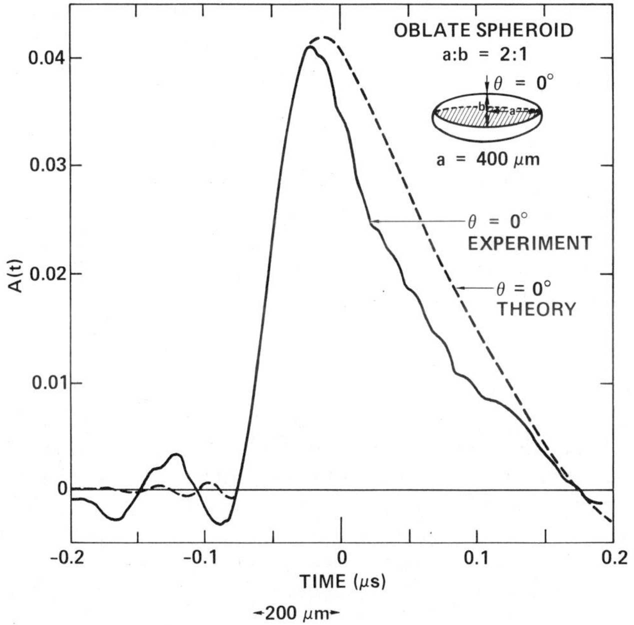

Fig. 1. A(t) for void in Ti-alloy in the shape of an oblate spheroid with longitudinal waves incident along $\theta = 0$.

from mode conversion from incident longitudinal to other wave types (transverse, creeping, etc.) are expected to be negligible but to play a key role for positive values of t. From a physical optics point of view, the part of the curve for $t < 0$ corresponds to the illuminated surface of the spheroid, whereas the part for $t > 0$ corresponds to the shadow regime.

When a wave is incident at an angle with respect to one of the axes of the spheroid, the process of calculating the dimensions becomes more complex. This is illustrated in Fig. 2a, where the line MP represents a wavefront just striking the spheroid at point M when incident at an angle θ, along the direction of the wave normal OP. It is obvious that when A(t) is obtained, the dimension derived is OP which may be related to the geometrical parameters by

$$OP = b\sqrt{(1 + \beta^2 \tan^2\theta)/(1 + \tan^2\theta)} \quad ; \quad \beta = a/b$$

The calculated points P for θ ranging in 10° steps from 0 to 70° are shown in Fig. 2a and the calculated lengths OP in Table 1.

Table 1. Comparison Between Theoretical and Experimental Determination of Geometrical Parameter OP

θ^*	0°	10°	20°	30°	40°	50°	60°	70°
OP_{exp} (25%-85%)†(μm)	198	194	205	261	294	355	370	359
AVG OP_{exp}§ (μm)	208	207	221	251	303	355	371	359
OP_{th}** (μm)	200	209	232,5	265,6	299	332	360,6	382

* angle between direction of incidence and minor axis of ellipse.
† obtained from data using 25%-85% criterion on A(t).
§ average obtained from some data using three different criteria.
** distance calculated for normal from centroid to wave front tangent to ellipse.

In the determination of OP from the experimental A(t) three criteria were used. One of these, illustrated in Fig. 1, is based on an empirical choice of 25% and 85% of the maximum value of A(t) for the start and end of the interval Δt and therefore OP. Another was the interval between the 25% point and the centroid ($t = 0$) obtained as a byproduct of the iterative technique. These two intervals correspond, respectively, to the time interval from the front face reflection to the onset of the shadow region on one hand and to the centroid on the other hand. These two intervals are the same only if the scatterer has inversion symmetry, as for example the oblate spheroid. For asymmetric scatterers the 25%-85% criterion provides additional information, since the location of the centroid may be far removed from the onset of the shadow region, as for example, for a cone with the tip pointed towards the incoming wave. The third criterion is based on the extraction of a feature in the scattered waveform usually identified as a creep wave signal. The delay of the creep wave which has traveled behind the scatterer with a known velocity gives the pathlength corresponding to one-half the perimeter. Table 1 compares the theoretical OP with the experimental OP based on the 25%-85% criterion alone and on the average of values obtained from all three criteria, labeled AVG OP_{exp}. These values have been used to obtain an image which is compared with the actual contour in Fig. 2b. With the values OP and the angles θ a family of tangent lines MP are constructed which correspond, as described above, to the wavefronts just striking the surface. The largest contour inscribed by all the tangent lines then forms the desired image which can be seen to be in good agreement with the actual contour. It is clear that additional experimental data for several different cuts through the oblate spheroid would allow the construction of a 3-D image.

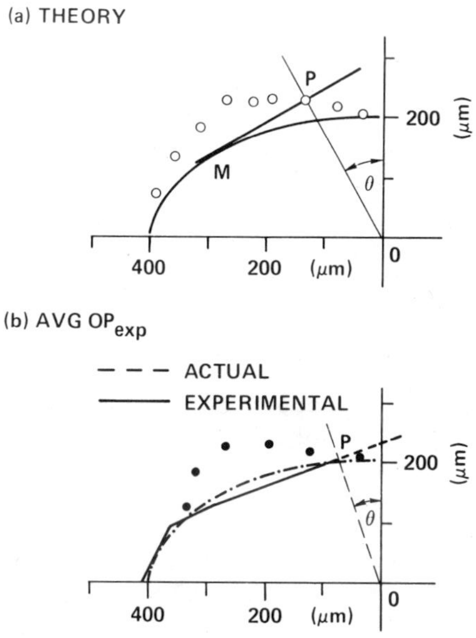

Fig. 2. (a) Theoretically derived contour of spheroid.
(b) Experimentally derived contour of spheroid.

In Fig. 3 is plotted A(t) for a more complex scatterer without inversion symmetry. It consists of two overlapping spherical voids identified as "lumpy" sphere. As shown in the schematic, the result is shown for backscattering at an angle of 15° from the common

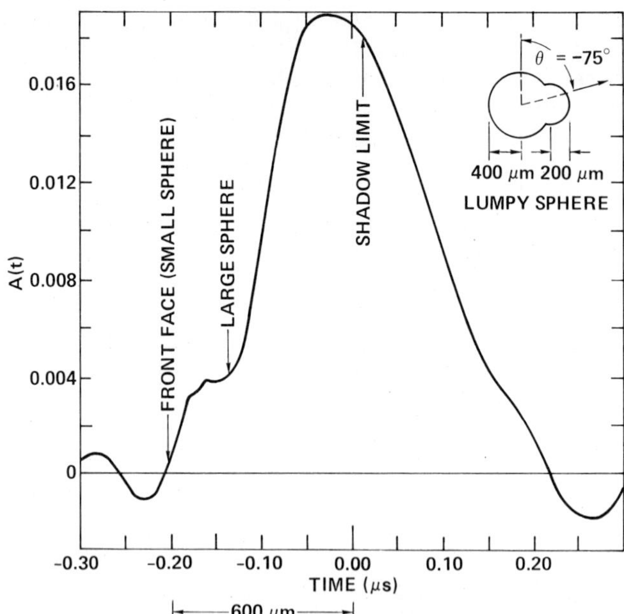

Fig. 3 A(t) experimentally obtained for spherical void with perturbation θ = 75°.

axis of the two spheres. For $t < 0$, A(t) clearly shows features identifiable with the presence of two overlapping voids. The features also allow estimates of their radii.

ACKNOWLEDGEMENT

This work was in part supported by the Rockwell International Science Center Independent Research and Development Program. The authors gratefully acknowledge Lloyd Ahlberg for his help in the experiments and the proofing of the final manuscript. One of the authors (B.R.T.) extends his appreciation to the Solid State Physics Group of Ecole Normale Superieure and to the Paris VII University for its support and its welcome as Invited Professor.

REFERENCES

1. J.M. Richardson, "The Inverse Problem in Elastic Wave Scattering at Long Wavelengths," _in_ "1978 Ultrasonics Symposium Proceedings," p. 759, IEEE, New York (1978).
2. J.H. Rose and J.A. Krumhansl, "Determination of Flaw Characteristics from Ultrasonic Scattering Data," _J. Appl. Phys._ 50:2951 (1979).

3. W.D. Cook, "Interrogation of Voids in Solids Utilizing Ramp Function Ultrasonic Pulses," in "Proceedings of the DARPA/AFML Review of Progress in Quantitative NDE," D.O. Thompson, Air Force Contract No. AFWAL-TR-80-4078, p. 454 (1980).
4. W.G. Neubauer, "A Summation Formula for Use in Determining the Reflection of Irregular Bodies," J. Acoust. Soc. Am., 35:279 (1963).
5. F. Cohen-Tenoudji, B.R. Tittmann, and G. Quentin, "A Technique for the Inversion of Backscattered Elastic Wave Data to Extract the Geometrical Parameters of Defects with Varying Shape," Appl. Phys. Letters, in press.
6. F. Cohen-Tenoudji, "These de Doctorat d'Etat es Sciences Physique," Elastic Wave Inversion Transformation, University of Paris VII, 1982.
7. J. Opsal, "Matrix Theory of Elastic Wave Scattering," in "Proceedings of the DARPA/AFML Review of Progress in Quantitative NDE," D.O. Thompson, Air Force Contract No. AFWAL-TR-80-4078, p. 328 (1980).
8. W.M. Visscher, "A New Way to Calculate Scattering of Acoustic and Elastic Waves. II. Application to Elastic Waves Scattered from Voids and Fixed Rigid Obstacles," J. Appl. Phys. 51(2):835 (1980).
9. B.R. Tittmann, H. Nadler, and N.E. Paton, "A Technique for Studies of Ductile Fracture in Metals containing Voids or Inclusions," Met. Trans. 7A:320 (1976).

NUMERICAL TECHNIQUES FOR THE INVERSE ACOUSTICAL SCATTERING PROBLEM IN LAYERED MEDIA

P.C. Pedersen, O.J. Tretiak, and Ping He

Biomedical Engineering and Science Institute
Drexel University
Philadelphia, Pa. 19104, U.S.A.

INTRODUCTION

Techniques for reconstruction of medium properties, mainly acoustic impedance or permittivity, based on the backscattered energy from an interrogating wave, have applicability in a wide range of scientific fields, such as in seismology, underwater acoustics, medical ultrasound, remote sensing and optics. Consequently, the classical inversion problem has in recent years been the subject of many investigations, especially in electromagnetics [1-3] and geophysics [4,5]. Theoretical treatment of the inverse acoustical scattering problem has been presented in papers and books [6-8].

The numerical techniques for reconstruction of the acoustic impedance profile of a layered medium may be divided into 2 categories: (a) reconstruction techniques which only take into account first order reflections and therefore are valid only for small acoustic impedance changes; (b) reconstruction techniques which consider both first order and higher order reflections and will yield accurate reconstruction also for large impedance changes.

In this paper, we will first briefly present a numerical technique for the direct acoustical scattering problem, i.e. the determination of the reflection impulse response for a layered medium from the given propagation velocity profile and specific acoustic impedance profile. The frequency dependent reflection coefficient is calculated by means of a transmission matrix method from which the reflection impulse response is obtained by the inverse Fourier transform. Next, we will discuss and give results for acoustic impedance profile reconstruction in non-attenuating media using two different numerical techniques, one from each of the aforementioned categories. The input function for both techniques is the reflection impulse response of the layered medium. One of the two reconstruction techniques is based on the well-known impediography equation [9-11] which ignores higher order reflections. The implementation, however, is simple, requiring only integration and exponentiation of the reflection impulse response. The other reconstruction technique which takes into account multiple reflections within the layered medium is based on a solution by Pierre Goupillaud [6,12] and will be termed Goupillaud's method. The algorithm is derived by dividing the inhomogeneous medium up into a large number of thin layers *with equal travel time*. The acoustic impedance profile is then reconstructed in small, discrete increments in which the magnitude of each increment is determined by operating on all the preceding increments.

All the derivations, presented in this paper, are based on the following assumptions:

Assumptions for the inhomogeneous medium:

(a) No attenuation losses
(b) Isotropic
(c) One dimensional impedance variation (in the x-direction)

Assumptions for the interrogating wave:

(a) Plane wave
(b) Longitudinal wave

NUMERICAL TECHNIQUE FOR THE DIRECT PROBLEM

The wave equation for an inhomogeneous medium interrogated by a plane wave of frequency ω may be written as [13]:

$$\frac{\partial \bar{P}(\omega,x)}{\partial x} = [\bar{A}(\omega,x]\bar{P}(\omega,x) \tag{1}$$

where $\bar{P}(\omega, x) = \begin{bmatrix} P^+(\omega,x) \\ P^-(\omega,x) \end{bmatrix}$, i.e. $\bar{P}$ is the sum of the forward component of the pressure wave, P^+, and backward component of the pressure wave, P^-.

$$\bar{A}(\omega,x) = \left[jk(\omega,x)\begin{bmatrix} -1 & 0 \\ 0 & 1 \end{bmatrix} + \alpha(x)\begin{bmatrix} 1 & -1 \\ -1 & 1 \end{bmatrix} \right] \tag{2}$$

$k(\omega,x) = \dfrac{\omega}{c(x)}$ is the wave number

$$\alpha(x) = \frac{1}{2z(x)} \frac{\partial}{\partial x} z(x) \tag{3}$$

where z(x) is the specific acoustic impedance in the inhomogeneous medium. Note that A is a constant only when $k(\omega,x)$ is a constant and $\alpha(x)$ is a constant; the latter being fulfilled for an exponential impedance profile.

$\overline{M}(x,x_N)$ is a 2 x 2 transmission matrix relating, for a layer of the inhomogeneous medium with thickness Δx, the forward and backward pressures at x with those at x_N, as shown in Fig. 1.

Hence, for constant frequency,

$$\overline{P}(x) = \overline{M}(x,x_N)\,\overline{P}(x_N) \tag{4}$$

From (1) and (4) we find

$$\frac{\partial \overline{M}(x,x_N)}{\partial x} = \overline{A}(x)\,\overline{M}(x,x_N) \tag{5}$$

When $\overline{A}(x)$ is constant, i.e. constant sound velocity, and exponential impedance profile, (5) can be solved [13], with the solution given in (6).

$$M(x,x_N) = e^{-\alpha\Delta x} \begin{bmatrix} \cos\psi + j\frac{k}{k^*}\sin\psi & \frac{\alpha}{k^*}\sin\psi \\ \frac{\alpha}{k^*}\sin\psi & \cos\psi - j\frac{k}{k^*}\sin\psi \end{bmatrix} \tag{6}$$

$$\psi = \Delta x\sqrt{k^2-\alpha^2}, \quad \Delta x = x_N - x, \quad k^* = \sqrt{k^2-\alpha^2}$$

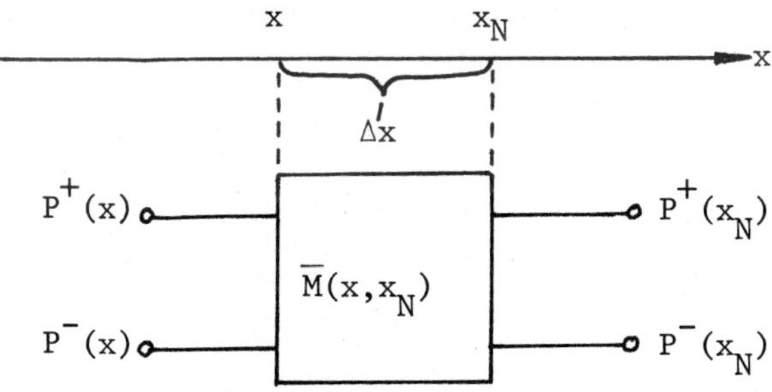

Figure 1. Relation of acoustic pressure wave at x with acoustic pressure wave at x_N through a transmission matrix $\overline{M}$.

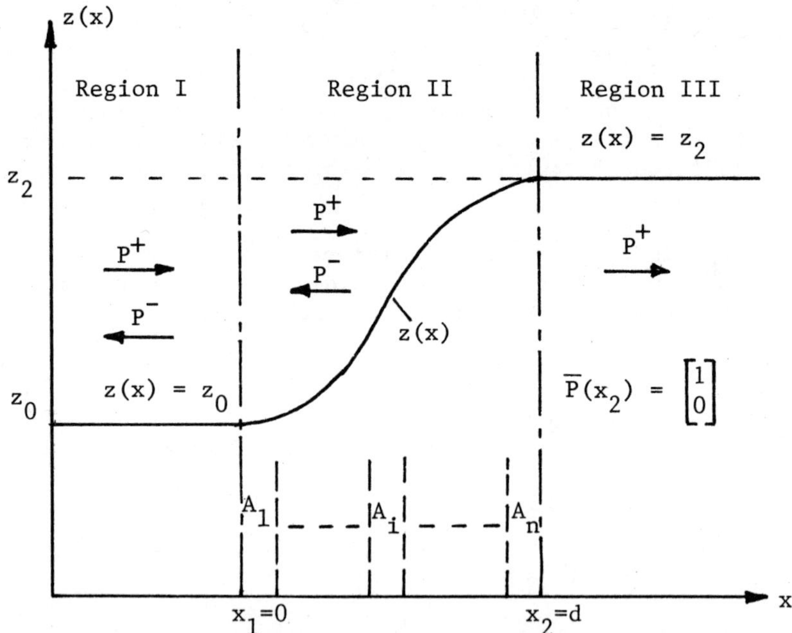

Figure 2. Acoustic impedance profile of a 3 layered medium in which the middle layer (Region II) is inhomogeneous, with a smoothly varying acoustic impedance in the x-direction. For calculation of the total transmission matrix, the inhomogeneous layer is divided into n parallel thin layers, A_1, A_2, ---, A_i, --, A_n.

We will analyze the reflection characteristics of an arbitrary inhomogeneous region (Region II in Figure 2) with finite width, d, bounded by plane, parallel regions (Regions I and III in Fig.3). The acoustic impedance in Region II is a smoothly varying function z(x), such that $z(0^+) = z_0$ and $z(d^-) = z_2$, i.e., there are no step changes in impedance at the boundaries.

The arbitrary impedance profile is divided into thin layers, and the actual impedance profile within each layer is approximated by an exponential impedance profile which produces the same total impedance change as the actual profile. If the sound velocity varies across the inhomogeneous region, the velocity within each thin layer must be approximated by a constant.

For an arbitrary thin layer, A_i, whose approximated impedance profile is stated in terms of α_i(see eq.(3)), the transmission matrix, $\overline{M}_i$, can be calculated by means of (6). Since Region III is homogeneous and semi-infinite, only a forward wave, P^+, exists in Region III. $P^+(x_2)$ is arbitrarily set equal to 1, and hence the boundary conditions at x_2=d are $\overline{P}(x_2) = \begin{bmatrix}1\\0\end{bmatrix}$. We may therefore write

$$\overline{P}(x_1) = \left[\prod_{i=1}^{n} \overline{M}_i\right] \overline{P}(x_2)$$

$$= \overline{M}(x_1,x_2)\ \overline{P}(x_2) \tag{7}$$

Inserting the above mentioned boundary condition, we obtain

$$\begin{bmatrix} P^+(x_1) \\ P^-(x_1) \end{bmatrix} = \begin{bmatrix} m_{11} & m_{12} \\ m_{21} & m_{22} \end{bmatrix} \begin{bmatrix} 1 \\ 0 \end{bmatrix} = \begin{bmatrix} m_{11} \\ m_{21} \end{bmatrix} \tag{8}$$

The frequency dependent reflection and transmission coefficients, $R(\omega)$ and $T(\omega)$, respectively, can be found as follows:

$$R(\omega) = \frac{P^-(x_1)}{P^+(x_1)} = \frac{m_{21}}{m_{11}} \tag{9}$$

$$T(\omega) = \frac{P^+(x_2)}{P^+(x_1)} = \frac{1}{m_{11}} \tag{10}$$

Next, the reflection impulse response, $r(t)$, can be calculated from (9) by evaluating $R(\omega)$ at a number of discrete frequencies, and then perform an inverse fast Fourier transform (IFFT):

$$r(t) = \text{IFFT}\ \{R(\omega)\} \tag{11}$$

The numerical technique, developed for the direct problem, will now be demonstrated by calculating the reflection coefficient, $R(\omega)$, and the reflection impulse response, $r(t)$, for two acoustic impedance profiles. These numerical examples are used in the direct problem only, and other, more complex impedance profiles will be considered for the inverse problem, described later in this paper.

The two impedance profiles have cubic and linear impedance variations, as shown and described in Figure 3. In both cases the inhomogeneous layer is 1 cm thick, with a constant sound velocity of 1500 m/sec and a total impedance change of a factor 3. Figure 4 shows the complex reflection coefficient and the impulse response calculated for these two acoustic impedance profiles.

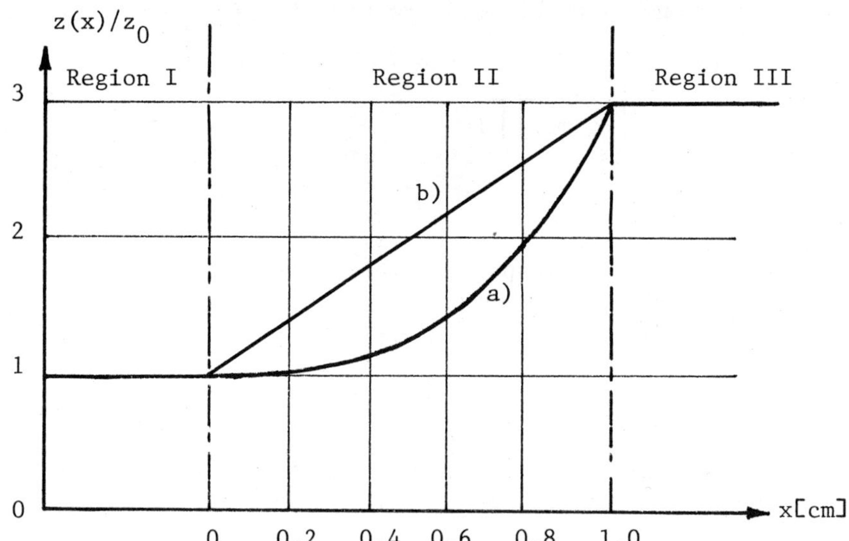

Figure 3. Acoustic impedance profiles used in numerical examples
a) Cubic impedance profile, $z(x) = z_0(1+2x^3)$
b) Linear impedance profile, $z(x) = z_0(1+2x)$

The choice of the number of layers into which the medium is divided is based on the desired smoothness of the resulting curves, and on computer time considerations. Thus, in generating curves for the reflection coefficient, the inhomogeneous region has been divided into 10 thin layers, with a frequency sampling interval of only 3 khz. When generating the input data for the inverse Fourier transform, yielding the reflection impulse response, r(t), we have used 20 thin layers and a frequency sampling interval of about 47 khz. As is seen in Figure 4, the effect of using a constant α within each thin layer is still apparent in form of the steplike shape of r(t).

Several interesting observations may be made from Figure 4. The reflection coefficient at dc, R(0), is equal to 0.5; in other words, the same as the reflection coefficient of an impedance step change of a factor of 3. As the frequency becomes very high, $R(\omega)$ approaches 0, i.e., the inhomogeneous layer here acts as a broadband impedance transformer.

NUMERICAL TECHNIQUES FOR THE INVERSE ACOUSTICAL SCATTERING PROBLEM

In this part of the paper, we will describe and compare two numerical techniques for reconstructing selected acoustic impedance profiles. Both reconstruction techniques use the reflection impulse response of the layered medium, the calculation of which was described in the previous part of this paper, as the input

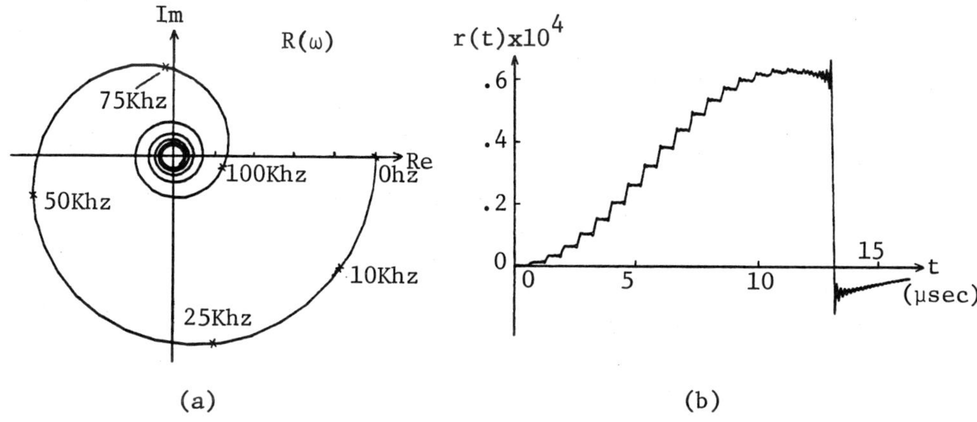

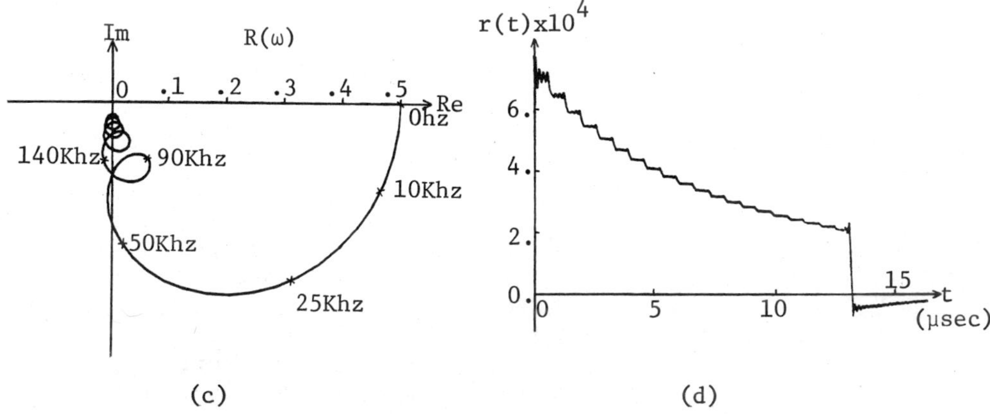

Figure 4. Reflection coefficient (a) and impulse response (b) of inhomogeneous medium with a cubic acoustic impedance profile. Reflection coefficient (c) and impulse response (d) of inhomogeneous medium with linear impedance profile. In both cases, the medium has a thickness of 1cm, and a sound velocity of 1500 m/s.

function. As stated in the introduction, one reconstruction method will be based on the impediography equation and the other on Goupillaud's method.

Review of the Reconstruction Techniques

The impediography equation has been extensively discussed in the literature [10,11]. It has been derived from the wave equation [14], and from a discrete reflection approximation [9], and may be stated as

$$\int_0^t r(t)dt = \frac{1}{2} \ell n \, [z(t)/z_0] \tag{12}$$

where r(t) is the impulse reflection coefficient, z_0 is the acoustic impedance of the incident side of the inhomogeneous region, and z(t) is the acoustic impedance at the point in the inhomogeneous region, reached in t/2 sec by the sound wave.

For the purpose of acoustic impedance profile reconstruction, eq.(12) may be written as

$$z(t) = z_0 \, \exp[2 \int_0^t r(\tau)d\tau] \tag{13}$$

The detailed derivation of Goupillaud's method for reconstruction of an acoustic impedance profile has been given by Ware and Aki [6] and Goupillaud [12], and only the key features will be outlined here.

Consider an inhomogeneous region (Region II in Figure 5), divided up into n *homogeneous* thin layers such that the travel time across each layer is constant and equal to Δt. The acoustic impedances of the halfspaces (Region I and III in Figure 5) are z_0 and z_{n+1} while the acoustic impedances of the n layers in the inhomogeneous region are z_1, z_2, ---, z_n. The reflection coefficients are r_0, r_1, ---, r_n, as shown in Figure 5. Because the inhomogeneous region is approximated by homogeneous layers of equal travel time, the reflected energy arrives in the form of a series of impulses with amplitudes R_0, R_1, R_2, ---, separated in time by $2\Delta t$. If we assume that the magnitude of the incident wave is unity, the amplitudes of the impulses may be calculated as:

$$\begin{aligned} R_0 &= r_0 \,, \quad @ \; t = 0 \\ R_1 &= r_1(1 - r_0^2), \quad @ \; t = 2\Delta t \\ R_2 &= r_2(1 - r_1^2)(1 - r_0^2) - r_0 r_1^2 (1 - r_0^2), \quad @ \; t = 4\Delta t \\ &\vdots \end{aligned} \tag{14}$$

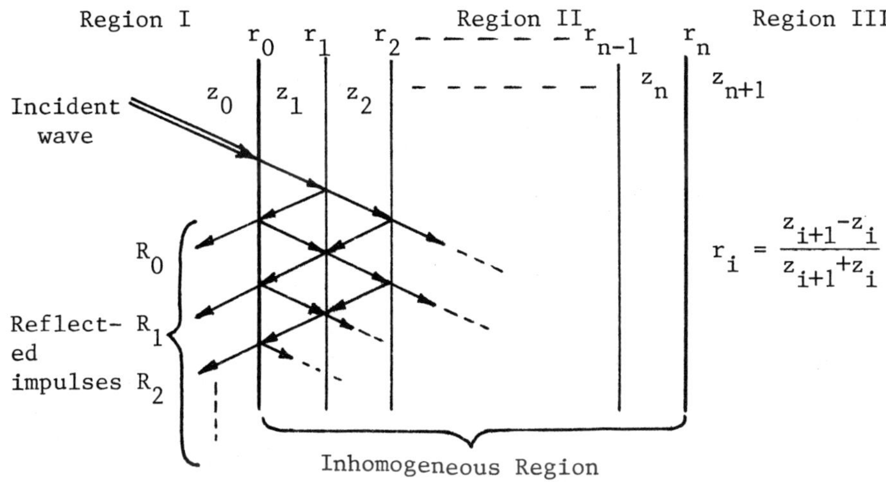

Figure 5. The division of an inhomogeneous region into n layers of equal travel time for the derivation of Goupillaud's method. r_0, r_1, ---, r_n are the reflection coefficients between the layers with impedances z_0 and z_1, z_1, and z_2, ---, z_n and z_{n+1}, respectively.

Solving Eq. (14) for r_0, r_1, --- yields

$$r_0 = R_0$$
$$r_1 = R_1(1 - r_0)^{-1} \quad (15)$$
$$r_2 = (R_2 + r_0 r_1 R_1)[(1-r_0^2)(1-r_1^2)]^{-1}$$

The key element in Goupillaud's method consists of a recursive algorithm which expresses an arbitrary reflection coefficient, r_k, in terms of r_0, r_1, ---, r_{k-1} and R_1, R_2,----, R_k; in other words, the determination of g_k in eq. (16).

$$r_k = g_k\,(r_0, r_1, ---, r_{k-1};\ R_1, R_2, ---, R_k) \quad (16)$$

Using eq. (16), the acoustic impedance profile is reconstructed in a discrete, recursive fashion as shown in eq. (17).

$$z_{k+1} = z_0 \prod_{i=0}^{k} \frac{1+r_i}{1-r_i} \quad (17)$$

Numerical Examples of Impedance Profile Reconstruction

We will now evaluate the performance of the two techniques,

based on acoustic impedance profile reconstruction of four selected impedance profiles.

Since only Goupillaud's method takes into account both first and higher order reflections, this reconstruction technique is expected *a priori* to be superior to the reconstruction technique, based on the impediography equation when the impedance variation is large. However, the numerical examples will illustrate over which range of impedance variation the reconstruction by means of the impediography equation will give accurate results, and will show the obtainable accuracy of Goupillaud's method.

Figures 6 to 9 present the four examples of acoustic impedance reconstruction. In all cases, the inhomogeneous layer is 1 cm thick, with a constant sound velocity of 1500 m/s. The reflection coefficients and the reflection impulse responses are calculated, as discussed in the previous part of this paper, but are not shown. The reconstructed impedance profiles are all obtained in terms of the round trip travel time, t. For the thickness and sound velocity, given above, t varies from 0 to 13.33 μsec.

Figure 6 shows the original and the two reconstructed acoustic impedance profiles, with the original impedance profile given as $z(t)/z_0 = \exp[230.26(tc/2)]$. With t varying from 0 to 13.33 μs, z(t) varies from z_0 to $10z_0$. The results of the reconstruction indicate that for an exponential impedance profile, the impediography equation performs well up to $2z_0$, i.e., up to 100% impedance variation. For larger impedance changes, the multiple reflections are significant, resulting in unacceptable reconstruction errors. The reconstruction by means of Goupillaud's method, on the other hand, gives a maximum reconstruction error of less than 5%.

Figure 7 illustrates the reconstruction of an impedance profile with sinusoidal impedance variation, given as $z(t)/z_0 = 2 + \sin[400\pi(t_c/2)-\pi/2]$. The results in Figure 6 are in agreement with Figure 7, that is, the impediography equation produces valid results when the total impedance variation is less than a factor 2. Goupillaud's method which includes all higher order reflections gives an accurate reconstruction over the full impedance profile.

Figures 8 and 9 present impedance profiles, composed of multiple exponential segments. In Figure 8, the impedance variation is only 10%, obtained by using $\alpha = \pm 5 \log_e(1.1)\text{cm}^{-1}$, while Figure 9 has a 100% impedance variation, obtained by using $\alpha = \pm 5 \log_e(2.0)\text{cm}^{-1}$. When the impedance variation is small, as in Figure 8, reconstruction with the impediography equation yields accurate results, even with the complex impedance profile, examined here. But for larger impedance variations, as in Figure 9, only reconstruction using Goupillaud's method is satisfactory, with maximum errors around 10%.

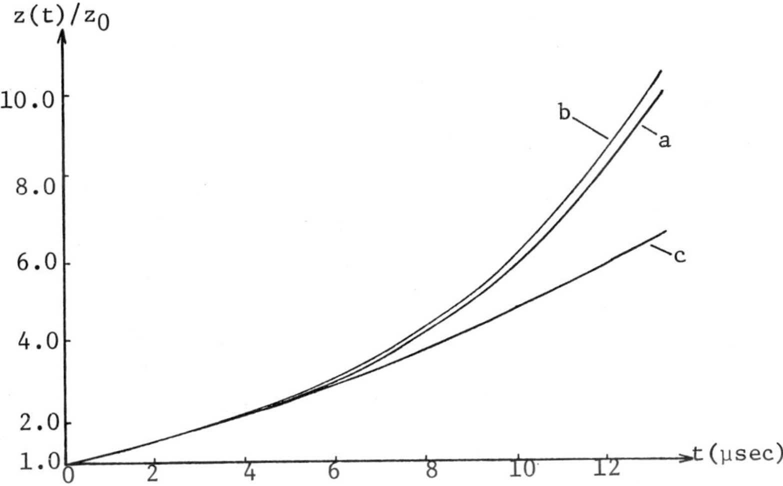

Figure 6. Original and reconstructed exponential acoustic impedance profiles as a function of the round trip travel time, for a 1cm thick layer. (a) Original impedance profile, $z(t)/z_0 = \exp[230.26(tc/2)]$, $0 \leq t \leq 13.33$ μsec, c=1500 m/s; (b) reconstructed impedance profile using Goupillaud's method; (c) reconstructed impedance profile using impediography equation.

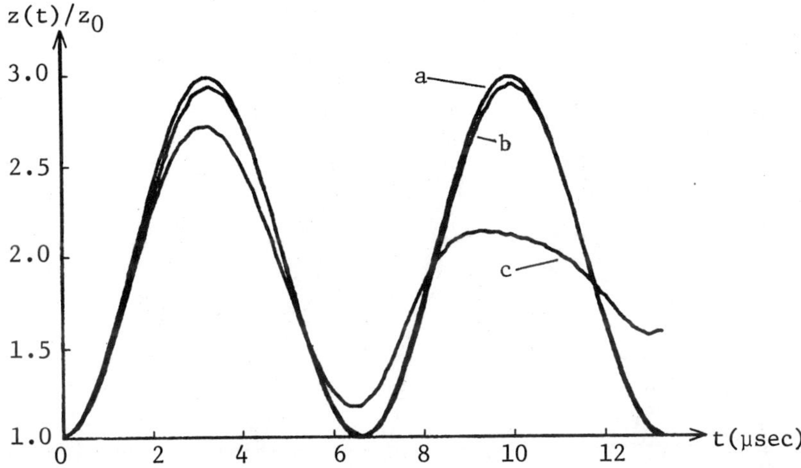

Figure 7. Original and reconstructed impedance profiles with sinusoidal impedance variation, as a function of the round trip travel time for a 1cm thick layer. (a) Original impedance profile, $z(t)/z_0 = 2 + \sin[400\pi(tc/2) - \pi/2]$, $0 \leq t \leq 13.33$ μsec, c=1500m/sec; (b) reconstructed impedance profile using Goupillaud's method; (c) reconstructed impedance profile using impediography equation.

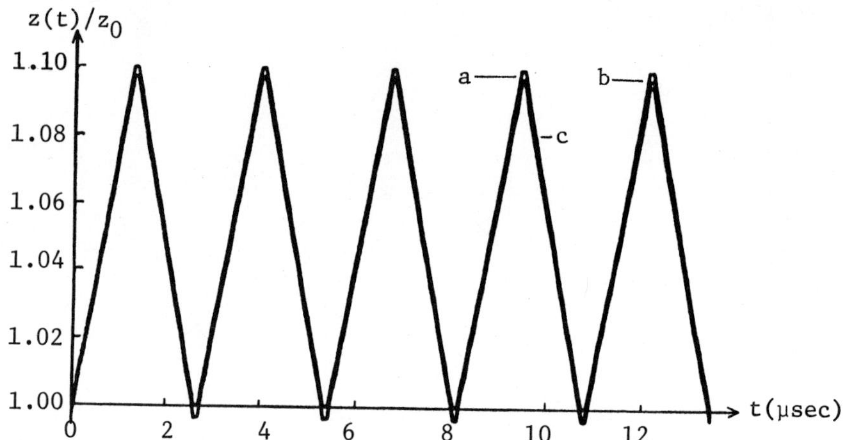

Figure 8. Original and reconstructed impedance profiles, consisting of multiple exponential segments with small variation, as a function of the round trip travel time for a 1cm thick layer. (a) Original impedance profile, $\alpha=\pm 5\ \log_e(1.1)\mathrm{cm}^{-1}$; (b) reconstructed impedance profile using Goupillaud's method; (c) reconstructed impedance profile using impediography equation.

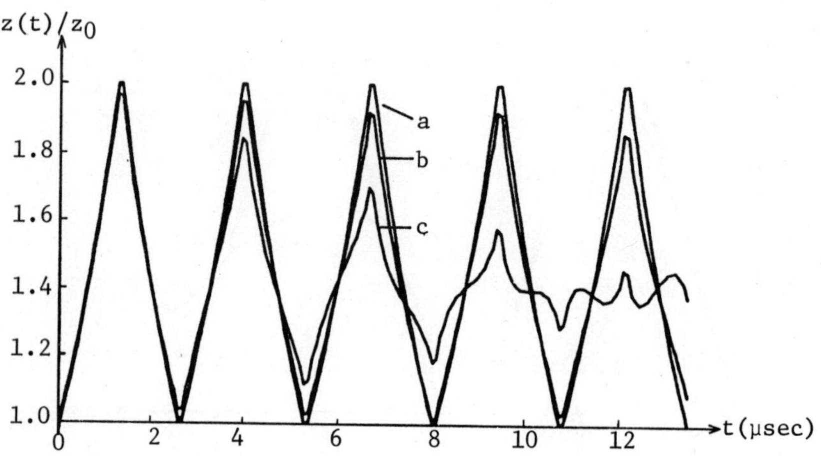

Figure 9. Original and reconstructed impedance profiles, consisting of multiple exponential segments with large variation, as a function of the round trip travel time for a 1cm thick layer. (a) Original impedance profile, $\alpha=\pm 5\ \log_e(2.0)\mathrm{cm}^{-1}$; (b) reconstructed impedance profile using Goupillaud's method; (c) reconstructed impedance profile using impediography equation.

The acoustic impedance profiles, given in Figures 6, 8 and 9, are constructed from exponential segments. This permits the calculation of the reflection coefficient, $R(\omega)$, to be carried out without the approximations by constants α's, as α in these impedance profiles is piecewise constant, thereby eliminating one of the error sources associated with generating the input functions for the reconstruction techniques. Only in Figure 7 which contains a sinusoidal impedance variation was the given impedance profile divided up into 30 thin layers for the calculation of the reflection coefficient.

Discussion of Error Sources

In this section we will give a brief discussion of potential error sources, including errors in generating the input function, i.e., the reflection impulse response, as well as in the reconstruction technique itself.

(a) A definite error source lies in the approximation of the actual impedance profile by an exponential impedance profile, when calculating the reflection coefficient, $R(\omega)$. This error source is not present, however, when the actual impedance is exponential, or composed of exponential segments, as in Figures 6, 8 and 9.

(b) In calculating the discrete values of $R(\omega)$ for the inverse FFT, the chosen frequency range must be wide enough to include all significant values of $R(\omega)$, as otherwise errors will be introduced. For the results, presented in this paper, $R(f)$ was evaluated in the range from 0 to 6 Mhz.

(c) The frequency interval between consecutive values of $R(f)$ must be small enough to prevent undersampling. Without considering multiple reflections, the duration of $r(t)$ is 13.33μs for the numerical examples presented here. Even when multiple reflections are significant, the value of $r(t)$ beyond 20μs is small. On this basis, the frequency sampling interval must be less than 50 khz. We have used 47khz which gives 256 data points from -6Mhz to +6Mhz. Still, reconstruction errors may be further reduced by using a smaller frequency sampling interval.

(d) Another potential error source, both for calculation of $r(t)$ and for the impedance profile reconstruction, is the computational error due to the finite word length of the computer which in our case is 32 bits. We have not evaluated the effect of the cumulative error.

Reconstruction of Impedance Profile as a Function of Spatial Distance

In the reconstruction of the acoustic impedance profile of an unknown medium, based on the reflection impulse response, r(t), from only one incident angle (typically normal incidence), we can only obtain the acoustic impedance (which is the product of medium density and sound velocity) as a function of the travel time of the sound wave. To be able to reconstruct the impedance as a function of spatial distance, the reflection impulse response must be determined for at least two different incident angles [14]. In this situation, it is possible to separate the sound velocity from the medium density by utilizing the refraction formula which determines the relationship between the direction of wave propagation and the sound velocity. As soon as the velocity can be determined separately, the reconstruction functions can be transformed from travel time to distance, thus allowing all the parameters, including acoustic impedance, to be reconstructed in the spatial domain.

SUMMARY

In this paper, we have evaluated two methods for reconstructing the acoustic impedance profile of an inhomogeneous medium, based on the reflection impulse response of the medium. The reflection impulse response is determined by an inverse FFT of the frequency-dependent reflection coefficient, $R(\omega)$, which is calculated by means of a transmission matrix technique. One reconstruction technique uses the impediography equation which is based on the assumption that higher order reflections are negligible. Hence, for smaller impedance variations the reconstruction by means of the impediography equation gives accurate results while larger impedance variations produce unacceptably large errors. The other reconstruction technique, Goupillaud's method, incorporates both first order and higher order reflections and therefore produces an accurate reconstruction also for large impedance variations.

ACKNOWLEDGEMENT

The work presented in this paper was supported by National Science Foundation Grant ECS-8025311.

REFERENCES

1. V.H. Weston, Electromagnetic inverse problem, in: "Electromagnetic Scattering", P.L.E. Uslenghi, ed., Academic Press, New York City, pp. 289-313 (1978).
2. Special issue on inverse methods in electromagnetics, IEEE Trans. Ant. and Prop., Vol. 29, No.2 (1981).

3. C.Q. Lee, Wave propagation and profile inversion in lossy inhomogeneous media, Proc. IEEE, 70:219 (1982).
4. J.F. Claerbout, "Fundamentals of geophysical data processing applications to petroleum prospecting, McGraw-Hill, New York City (1976).
5. J.G. Berryman and R.R. Greene, Discrete inverse methods for elastic waves in layered media, Geophysics, 45:213 (1980).
6. J.A. Ware and K. Aki, Continuous and discrete inverse-scattering problem in a stratified elastic medium. I. Plane waves at normal incidence, J. Acoust. Soc. of Am., 45:911 (1969).
7. L.M. Brekhowski, "Waves in layered media", 2nd ed., Academic Press, New York City (1980).
8. J.C. Hassab, Composition of propagated, reflected, and transmitted waves in arbitrary and continuously stratified environments, J. Sound and Vibration, 54:419 (1977).
9. H. Wright, Impulse response function corresponding to reflection from a region of continuous impedance change, J. Acoust. Soc. of Am., 53:1356 (1973).
10. J.P. Jones, Impediography: A new ultrasonic technique for diagnostic medicine, in: "Ultrasound in Medicine, Vol. 1", Plenum Press, New York pp. 489-497 (1975).
11. A.C. Kak, and F.J. Fry, Acoustic impedance profiling: An analytical and physical model study, in: "Ultrasonic Tissue Characterization, NBS Spec. Publ. 453" pp. 231-251. Washington D.C., US Govt. Printing Office (1976).
12. P.L. Goupillaud, An approach to inverse filtering of near-surface layer effects from seismic records, Geophysics, 26:754 (1961).
13. P.C. Pedersen, O. Tretiak, and P. He, Impedance-matching properties of an inhomogeneous layer with continuously changing acoustic impedance, J. Acoust. Soc. of Am., 72:327 (1982).
14. S.M. Candel, F. Defillipi, and A. Launay, Determination of the inhomogeneous structure of a medium from its plane wave reflection response, Part II: A numerical approximation, J. Sound and Vibration, 68:583 (1980).

PULSE AND IMPULSE RESPONSE IN HUMAN TISSUES

Lynda Hutchins and Sidney Leeman

Department of Medical Physics
Royal Postgraduate Medical School
Hammersmith Hospital, London W12 OHS

The interaction of ultrasound waves with human tissues is complex, and it is inevitable that simplifying physical and mathematical modelling must be invoked to aid our understanding of the processes involved. For pulse-echo methods, finite (bounded) transducers are used to generate three-dimensional (3-D) pulses which travel into the 3-D inhomogeneous medium that is tissue. For soft tissues, it is reasonable to disregard shear wave propagation, but attenuation of the ultrasound pulse is marked, and must be included. Moreover, since attenuation rises in a near-linear fashion with frequency, the shape of the pulse changes as it traverses tissue. Scattered waves are generated by tissue inhomogeneities (density, elasticity and absorption), and are coherently detected.

In practice, it is the back-scattered echoes which are most useful, and these can be shown (Leeman,1980) to arise mainly from fluctuations in the (characteristic)impedance However, the backscattering actually measured is embodied in the one-dimensional (1-D) voltage/time trace produced by the receiving transducers. Pulse-echo images are built up from many such processed 1-D lines, and "line-analysis" may lead also to a better understanding of the image formation process. The relationship of these "A-lines" to the true 3-D nature of tissues demands detailed theoretical arguments (Leeman,Hutchins,Jones,1982), but useful insights may be gained by modelling the structure of the 1-D echo sequence (output RF voltage waveform) itself. Some aspects of this are discussed here.

CONVOLUTIONAL MODELS

A common approach is to regard the 1-D A-scan trace, A(x), as originating from a tissue "impulse response", or "reflector sequence"

T(x), convolved with a 1-D pulse, P(x). Here, the entities are considered to be functions of a distance variable, x. The pulse shape is expected to change with distance in tissue, so that it is more accurately written as P(x;y) - that is, of shape $P(x;y_o)$ at depth y_o. Thus, providing linearity may be assumed

$$A(x) = \int T(y)\, P(x;y)\, dy \qquad (1)$$

Such integral expressions are cumbersome to manipulate, and further simplification is often sought. One approach is to neglect pulse distortion and to express the echo sequence as

$$A(x) = \int T(y)\, P(x-y)\, dy \qquad (2)$$

in order to arrive at the familiar convolutional integral. Frequency independent attenuation may be included by incorporating a simple exponential damping factor, or by assuming that it is conveniently allowed for by the time-gain-compensation facility provided in medical ultrasound scanners. More correctly, and this is the approach we follow here, (2) may be regarded as a good approximation to (1), provided that only a short segment of A(x) is considered. Over this segment, pulse distortions may be regarded as negligible. The size of the segment is clearly determined by the magnitude and frequency dependence of the attenuation, but this problem is not further explored here.

It should be emphasized that (2) represents a model for the (short segment) echo sequence, and that the convolutional components T(x) and P(x) are not, therefor , uniquely prescribed. Thus, it may be demanded that T(x) be relatively simple (e.g. a relatively sparse set of well-defined spikes, rather than a continuum), so that the complexity of the model fit to true data devolves to the pulse shape. P(x) may then bear little resemblance to any recognisable or acceptable pulse, and may well vary strongly from segment to segment.

Alternatively, it may be demanded that the pulse be of relatively simple shape, not too unlike the displayed 1-D echo from a plane reflector. P(x) would be required to vary smoothly from segment to segment. The complexity of the model fit to observed data now devolves to the tissue impulse response, T(x).

This last, intuitively appealing, approach is more in accord with theoretical models for 3-D pulse scattering from tissues (Leeman, Hutchins and Jones,1982), and is adopted here. P(x) is thus expected to have a relatively smooth, but peaked (at the carrier frequency), spectrum, while T(x) exhibits a complicated, and spiky spectrum. The pulse spectrum would be expected to be relatively narrow band, with respect to the extremely wide frequency range which the richly structured T(x) straddles. In this sense, bearing in mind the product nature of the convolution integral in Fourier space,the pulse spectrum may be regarded as a window through which (a portion of) the tissue spectrum may be viewed. The extended frequency range of T(x), relative to P(x), is a consistency requirement dictated by the experimental fact that highly structured echo sequences are obtained from tissue (attenuation allowing) when the central frequency of the probing pulse is moved over the entire diagnostic frequency range (and probably beyond).

Another consistency requirement is that the mean frequency of the pulse should drop with increasing penetration, i.e. as the short segment is shifted progressively deeper into tissue. This is dictated by the observed frequency-dependent behaviour of the attenuation. A final consistency requirement for the validity of the convolutional model is that it is possible to estimate either T(x) or P(x) directly from the short-segment echo sequence. Deconvolution then suffices to extract the other component.

The tissue impulse response may be estimated directly by spectral extrapolation methods (Papoulis and Chamzas,1979), and it is of importance as the input to quantitative imaging methods,such as impediography (Jones,1977), for which it provides an estimate of the effective impedance (Jones and Leeman,1982). Moreover, a knowledge of T(x), rather than A(x), is almost certainly the precursor to successful implementation of "structural" tissue characterisation methods (Jones and Leeman,1982).

Here, we concentrate on methods for pulse estimation. A knowledge of P(x) is important for tissue characterisation, since the decrease in its mean frequency with depth provides a possibly more robust estimate of tissue attenuation than short-segment Fourier analysis techniques devised by Jones (described in Jones and Leeman,1982). P(x) provides a convenient input to well-known deconvolution routines, such as Wiener filtering, whereby T(x) may be estimated. Such deconvolution procedures may also be regarded as (axial) resolution enhancement. We briefly describe below two methods for pulse estimation from short-segment echo sequences. The latter were obtained,"in vivo", from normal skeletal muscle, in the near field of a plane 1.5 cm diameter, 2.5 MHz transducer. Given the relatively complicated nature of the near field, we feel that a stringent test of the procedures is provided. The data analysed are from two .75cm adjacent echo segments, the proximal one commencing 1cm below the skin surface. Echoes were acquired with a standard, commercial NE4102 ultrasound scanner, digitised to 8 bits at 20 MHz, and analysed on a Perkin-Elmer 3220 computer. Data were acquired as part of a more extensive program of tissue characterisation of human skeletal muscle.

HOMOMORPHIC FILTERING

Homomorphic filtering is a subtle processing method particularly suited to analysis of convolution-type sequences. It basically relies on the logarithmic operation to transform the product spectrum of the pulse and reflector sequence into an additive one. A further Fourier transformation, into the so-called complex cepstral domain, allows a separation, in principle, between the smooth pulse, and spiky reflector sequence, spectra. The smooth pulse spectrum concentrates in the low-time region of the cepstral domain, while the spiky reflector spectrum tends to fill the higher cepstral range. Providing that the two cepstra are indeed separated, then the pulse cepstrum can be isolated by a suitable low-pass filter, and the pulse recovered by applying the inverse of the operations required to produce the cepstrum. The technique was devised by Oppenheim (see Oppenheim and Schafer,1975) and its application to pulse-echo sequences explained in simple fashion by Hutchins and Leeman (1981) and Jones and Leeman (1982).

Application of the method to muscle echoes allows satisfactory recovery of pulses (Fig. 1a). The power spectra of pulses recovered from adjacent echo segments also show "reasonable" behaviour (Fig.2). Note, however, that the downward shift in mean frequency is unexpectedly large. This is probably due to the recovered pulse spectrum from the proximal segment, which is centered somewhat higher than the expected 2.5 MHz.

In simulation studies, we have found (Hutchins and Leeman,1981) that homomorphic filtering is a useful method for pulse estimation, but rather less so for reflector sequence estimation. It is robust with respect to noise, and provides pulse estimates that are relatively uninfluenced by system control settings, such as time gain compensation. It may also be performed on quite short data segments. However, it is clear from the simulations that the method begins to fail when the separation of the first two reflectors in the sequence is somewhat less than a pulse-length. This "resolution" problem probably explains the unacceptable sensitivity of the method to the precise location of the initial point of the data segment being analysed. Successful pulse recovery cannot be guaranteed, "in vivo", in every case. The method is subtle, computationally demanding, and requires some on-line operator intervention and judgement. On balance, therefor, despite the sophistication of the method, we feel that a more direct approach to pulse estimation is required.

SPECTRAL SMOOTHING

It should be clear, from the consistency requirements demanded of the convolutional model, that the spiky nature of the observed echo segment spectrum arises from the tissue impulse response; the relatively smooth pulse spectrum acting, in some sense, as the enve-

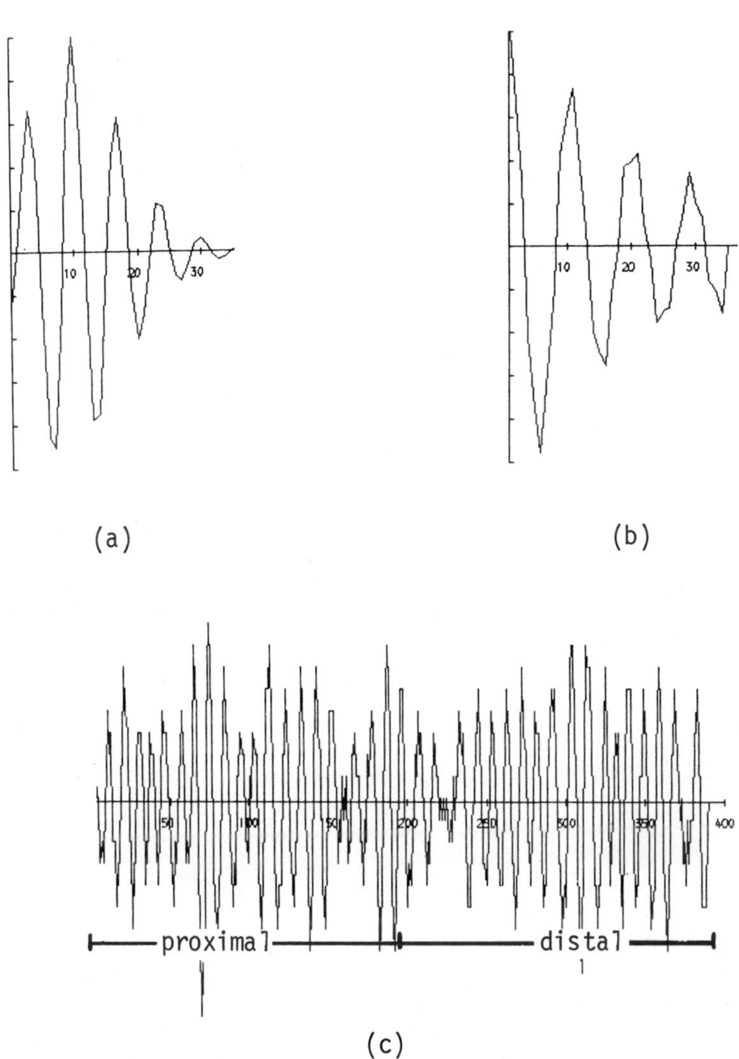

Fig. 1. Pulses recovered from echo sequence by (a) homomorphic filtering and (b) spectral smoothing. The (proximal) echo segment of "in vivo" data from which pulses are recovered is shown in (c).

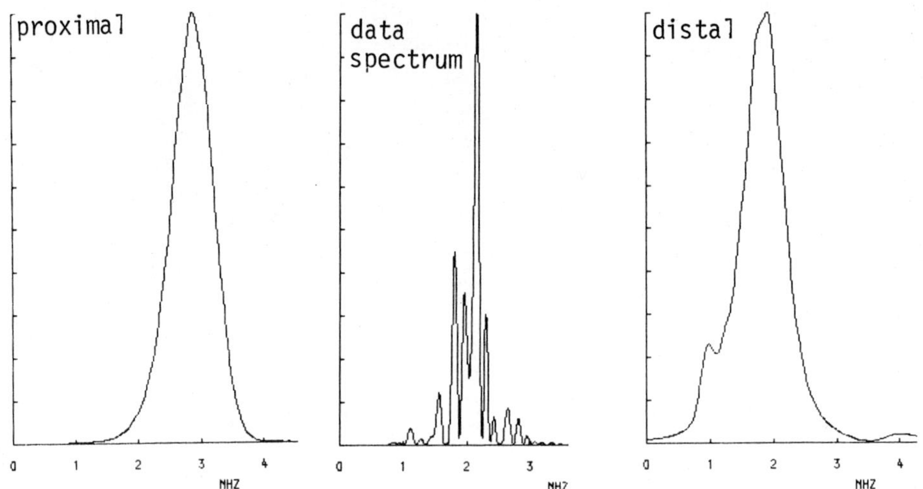

Fig. 2. Power spectra of pulses recovered by homomorphic filtering from the two adjacent (.75 cm) echo segments shown in Fig.1c. The power spectrum of the proximal data segment is also shown.

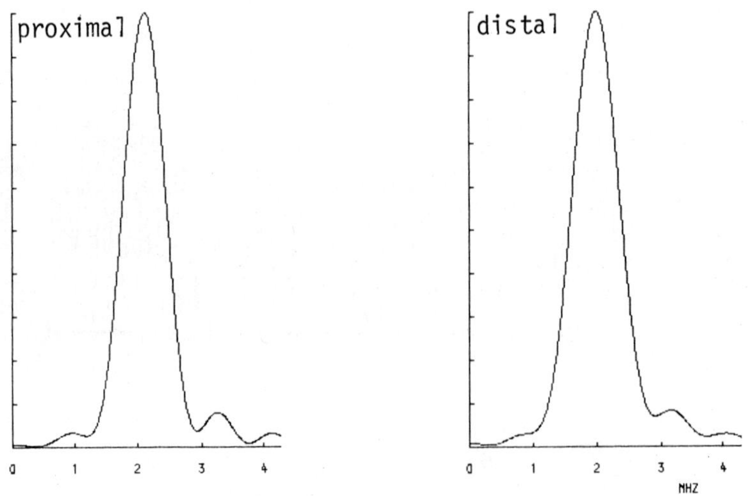

Fig. 3. Power spectra of pulses recovered by the spectral smoothing method, for the two echo segments shown in Fig. 1c.

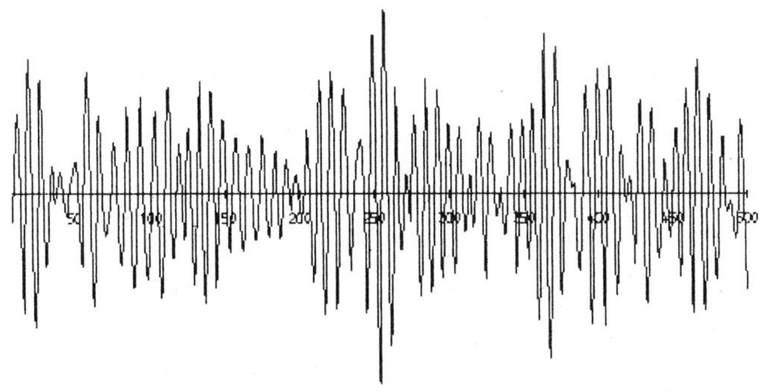

Fig. 4. "Echo sequence" generated by a 4th-order auto-regressive process.

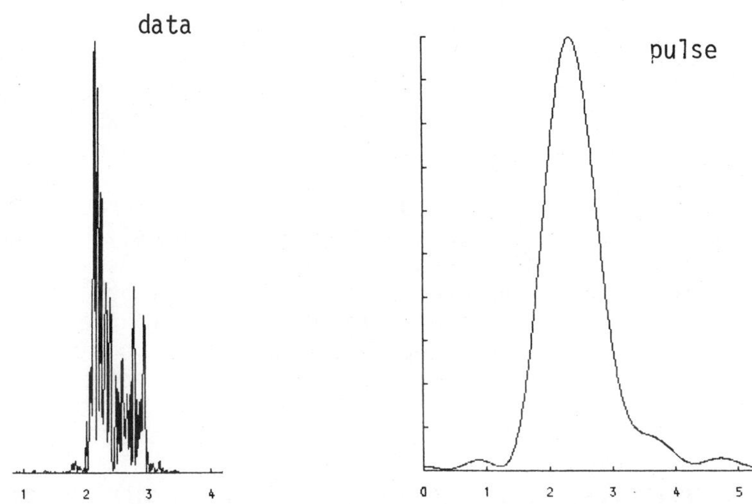

Fig. 5. Power spectrum of the AR sequence shown in Fig.4, with the "pulse spectrum" recovered via the spectral smoothing method.

lope of the observed echo spectrum. Thus, it may be expected that a good estimate of the pulse spectrum may be obtained by suitably smoothing the observed echo segment spectrum. We have devised an apparently successful procedure towards this end. The autocorrelation function of the echo segment is formed, and set to zero for lag numbers larger than the first major minimum in its envelope. This truncated autocorrelation function is then Fourier transformed, to yield an estimate of the pulse spectrum. The time domain pulse can be constructed from its amplitude spectrum by employing the digital Hilbert transform, as detailed in Oppenheim and Schafer (1975). The recovered pulse is constrained to be of minimum phase.

Estimated pulse spectra for the same two adjacent echo segments used to illustrate the homomorphic filtering technique are shown in Figure 3. Again, the spectra are quite "reasonable", and satisfy the consistency requirements outlined above. However, the proximal spectrum is centered near the expected frequency (2.5 MHz) and the downward shift in mean frequency between the two segments is much more acceptable than for the homomorphically filtered results. The downward shift measured by spectral smoothing, in <u>one</u> A-line, is very close to the .2 MHz/cm obtained by a short-time spectral analysis method (see the paper by Fink, in this volume). In our hands, the latter method demanded averaging over some 40 A-lines (1.5 cm long) in order to achieve convergent results. We are of the opinion that spectral smoothing has much to offer in reducing the total amount of data required for attenuation estimation by short-time spectral analysis. The (minimum phase) time domain pulse recovered from the amplitude spectrum of the proximal segment shown in Fig. 3, is demonstrated in Fig. 1b.

Spectral smoothing is rapid and easy to implement. It may be employed on band-limited data even when the underlying process is not convolutional, in the sense outlined above. As illustrated in Figs. 4 and 5, the technique may be successfully applied to data generated synthetically by an autoregressive (AR) process. Visual inspection of the data sequence shown in Fig. 4 does not reveal significant differences from "in vivo" muscle A-scans: indeed, AR and ARMA processes may well be employed to model observed echo sequences for tissue characterisation purposes. It seems to us that spectral smoothing is the preferred method for pulse spectral estimation, provided that the echo segment is sufficiently long for accurate <u>data</u> spectral estimates to be made. Its main disadvantage appears to be that the recovered time domain pulse is necessarily of minimum phase. However, this is clearly of no consequence when only spectral estimates are desired.

CONCLUSIONS

The representation of a measured 1-D echo sequence by a convolutional integral has been carefully analysed. Provided a number of consistency requirements are met, short data segments may be satisfactorily modelled as the convolution of a tissue impulse response and a 1-D pulse. Homomorphic filtering and spectral smoothing both provide methods for pulse estimation. Our results suggest that the latter technique is to be preferred for the analysis of "in vivo" muscle data.

REFERENCES

Hutchins,L., and Leeman,S., 1981, Pulse estimation from medical ultrasound signals, in: "Ultrasonics International 81", Z.Novak, ed., 427-433, IPC Press, Guildford.

Jones, J.P., 1977, Ultrasonic impediography and its application to tissue characterization, in: "Recent Advances in Biomedicine", D.N. White, ed., 131-156, Research Studies Press, Forest Grove.

Jones, J.P., and Leeman, S., 1982, Ultrasonic tissue characterization and quantitative ultrasound scatter imaging: methods and approaches, IEEE Trans. on Computers, in print.

Leeman, S., 1980, Impediography revisited, in: "Acoustical Imaging", vol.9, K. Wang, ed., 513-520, Plenum, N.Y.

Leeman, S., Hutchins, L., and Jones, J.P., 1982, Pulse scattering in dispersive media, in: "Acoustical Imaging", vol. 11, J.P. Powers, ed., Plenum, N.Y. (in press)

Oppenheim, A.V., and Schafer, R.W., 1975, "Digital Signal Processing", Prentice-Hall, Englewood Cliffs.

Papoulis, A., and Chamzas, C., 1979, Improvement of range resolution by spectral extrapolation, Ultrasonic Imaging, 1(2): 121, Academic Press, N.Y.

TWO-DIMENSIONAL DIFFRACTION SCANNING OF BOTH FRESH AND FIXED NORMAL AND CANCEROUS HUMAN HEPATIC TISSUE

D. Nicholas and A. W. Nicholas

Physics Department
Institute of Cancer Research/Royal Marsden Hospital
Downs Road, Sutton, Surrey, U.K.

INTRODUCTION

Previous clinical results have indicated that the technique of ultrasound diffraction is capable of *in vivo* tissue characterization, distinguishing both 'focal' and 'diffuse' disorders from normal tissue (Nicholas, 1979; Merton et al., 1982). This technique uses conventional pulse-echo ultrasound to investigate small regions of tissue by constraining the transducer to describe an arc centred on the volume of interest. The backscattered signals originating from the selected site are limited to one dimension, and as such are insufficient for complete appraisal of the acoustic scattering due to three-dimensional structures within the tissue volume.

In this paper an extended diffraction technique will be described which collects information over a two-dimensional surface, thereby improving the interpretation of the ultrasonic interaction with human soft tissues. This is achieved by utilising backscattered signals interrogated over a solid angle of transducer positions. At the same time the technique is further extended by scanning at a variety of frequencies. The results presented will be empirical but will emphasise the improved discrimination achieved by this extended technique in the investigation of cancerous, post-mortem liver tissue in both fresh and formalin fixed states.

APPARATUS AND TECHNIQUE

In order to evaluate more fully the backscattering from a complex three-dimensional tissue such as liver, it is necessary to construct a scanning head which permits movement of the transducer over two degrees of freedom. One could achieve the extra degree

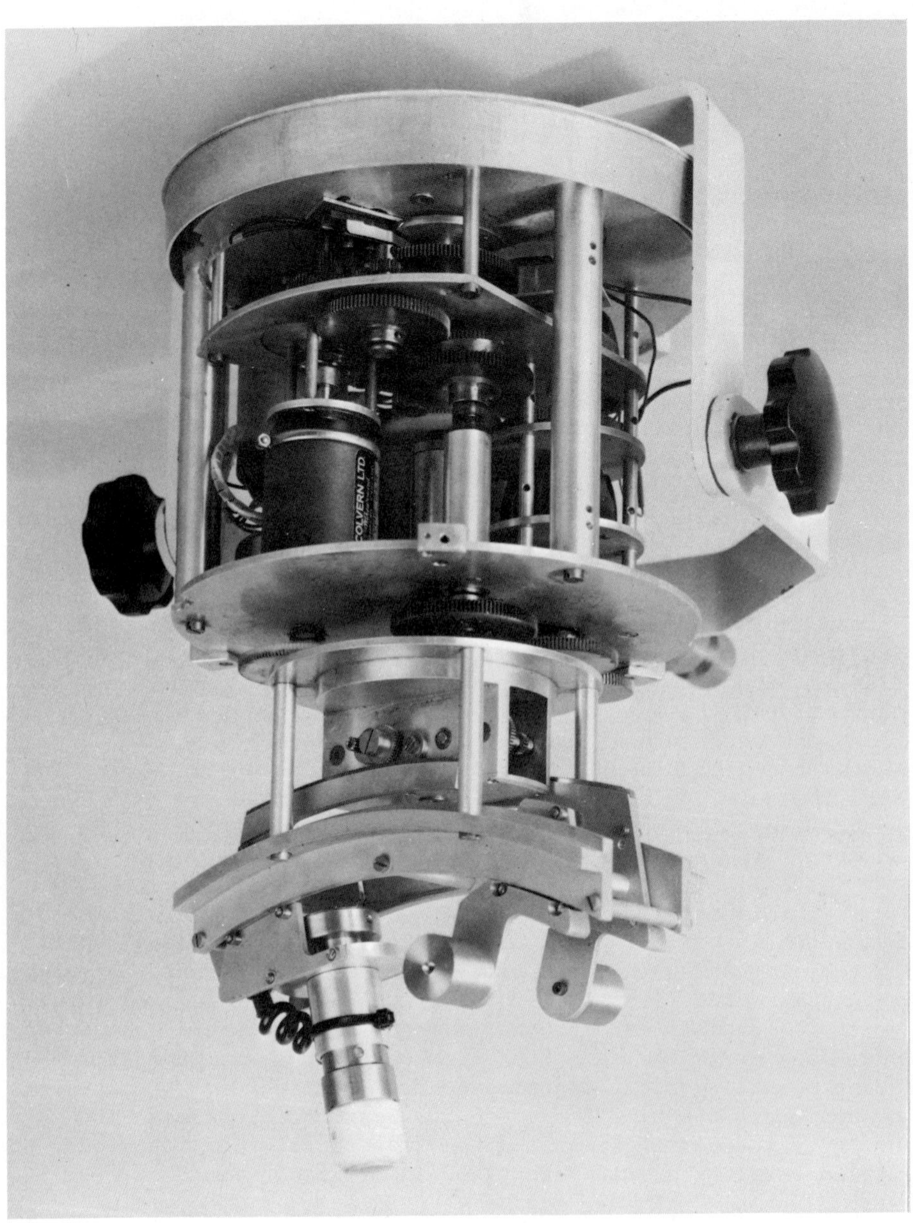

Fig. 1. Mechanics of scanning head facilitating two-dimensional isocentric scanning.

of freedom in a number of ways but we have chosen to adapt existing hardware. The apparatus illustrated in Figure 1 (modified from an isocentric C-scanner; McCready and Hill, 1971) permits the investigating transducer to move isocentrically over a spherical surface by spiralling in and out from a central position. The counterweight, shown in the photograph, moves in an opposing manner to balance the transducer. This mechanical arrangement extends our original isocentric sector scan to a two-dimensional isocentric scan with the point of pivot set within the focal zone of the transducer, 12 cm from its transmitting face.

The choice of isocentre is determined by a consideration of various factors and can only represent an optimal compromise. More detailed discussion of the apparatus characteristics has been reported elsewhere (Nicholas and Nicholas, in press) and will not be duplicated here.

Pulse Echo Diffraction

The purpose of the technique is to collect echoes originating from a specific tissue volume and to display their intensity as a function of interrogation aspect. To achieve this the received echo-trains (A-scans) are electronically time-gated to limit the signals to those originating from the small tissue volume (defined by the time-gate duration and beam-width) situated at the isocentre of the scanning motion. The gated signals are then frequency filtered to allow investigation of the scattering at a specific frequency. The electronic filter has a bandwidth of approximately 20 kHz and is tunable over the frequency range 1-5 MHz (Huggins and Phelps, 1977). Finally, the resulting signal is displayed as an intensity modulated point on a two-dimensional display at a position in the spiral raster corresponding to that of the scanning transducer. Several repeats of the scan at different frequencies (limited by the frequency range encompassed by our transducer operating in pulse-echo mode) provide three "dimensions" of acoustic information pertaining to the tissue volume under investigation, two in the angular backscattering domain and one in the frequency domain.

EXPERIMENTAL PROCEDURE

The tissue specimens, whole human livers, were obtained at post-mortem immediately after excision and transferred to the laboratory while immersed in water. The specimens were then either placed in physiological saline and left at a temperature of 7°C overnight, or in 10% formaldehyde until completely preserved. The latter tissues were examined after a minimum fixation period of two weeks.

In preparation for scanning, the organ is placed in a tank of degassed water at room temperature (20°C) and secured to a rubber mat, on the tank bottom, by steel pins. Vertical adjustment of the tissues within the water tank permits different tissue regions to be placed at the isocentre of the scanning motion.

As the objective of our measurements is to extract diffraction information from both normal and cancerous tissue, it is advantageous that the organ should first be scanned, by conventional ultrasound techniques, to determine the position of regions of potential interest. Since our system utilises a pulse-echo mode of operation we have incorporated additional electronics capable of providing conventional sector B-scan information. By locking the transducer in the central position and moving the whole scanning head horizontally, a rectilinear tomogram of a tissue cross-section can be produced. From such scans regions of interest are then selected for diffraction scanning. Furthermore, by carefully noting the scan plane chosen it is possible, at the conclusion of the experiment, to section the tissue and compare the sectional cut with the corresponding B-scan image. Not only does this permit the regions of the tissue corresponding to the sites from which the diffraction information was obtained to be excised for histological classification, but it also indicates the ability of conventional B-scanning to depict the differing anatomical structures.

RESULTS AND ANALYSIS

Figure 2 depicts typical two-dimensional diffraction patterns, termed 'interferograms' by us, from normal regions of hepatic tissue and from secondary liver neoplasms scanned at a frequency of 2.5 MHz. Simple visual comparison of these patterns indicates that the scans associated with the focal neoplasms (both freshly excised and formalin fixed) exhibit finer detail than the corresponding patterns associated with normal tissue. These findings complement our original one-dimensional results pertaining to the difference between cancerous and normal liver tissue (Nicholas and Hill, 1975).

To date we have attempted two simple forms of parameterization using polaroid records of the interferograms; the ratio of dark to light areas within a scan (termed 'contrast') and the number of discrete minima (dark regions) discernible within the image. Although the former has been accurately quantified by making graphical area measurements it will vary depending upon the intensity and contrast setting of the storage oscilloscope on which the data are recorded. In all these experiments these controls were set to produce a visually 'best' image, with good contrast and no 'blurring' due to too high an intensity setting, for scans performed at 2.5 MHz (the optimal frequency response of the transducer in use) and left unchanged for all measurements performed on a single organ. The sensitivity of the transducer at two other frequencies (2.0 and

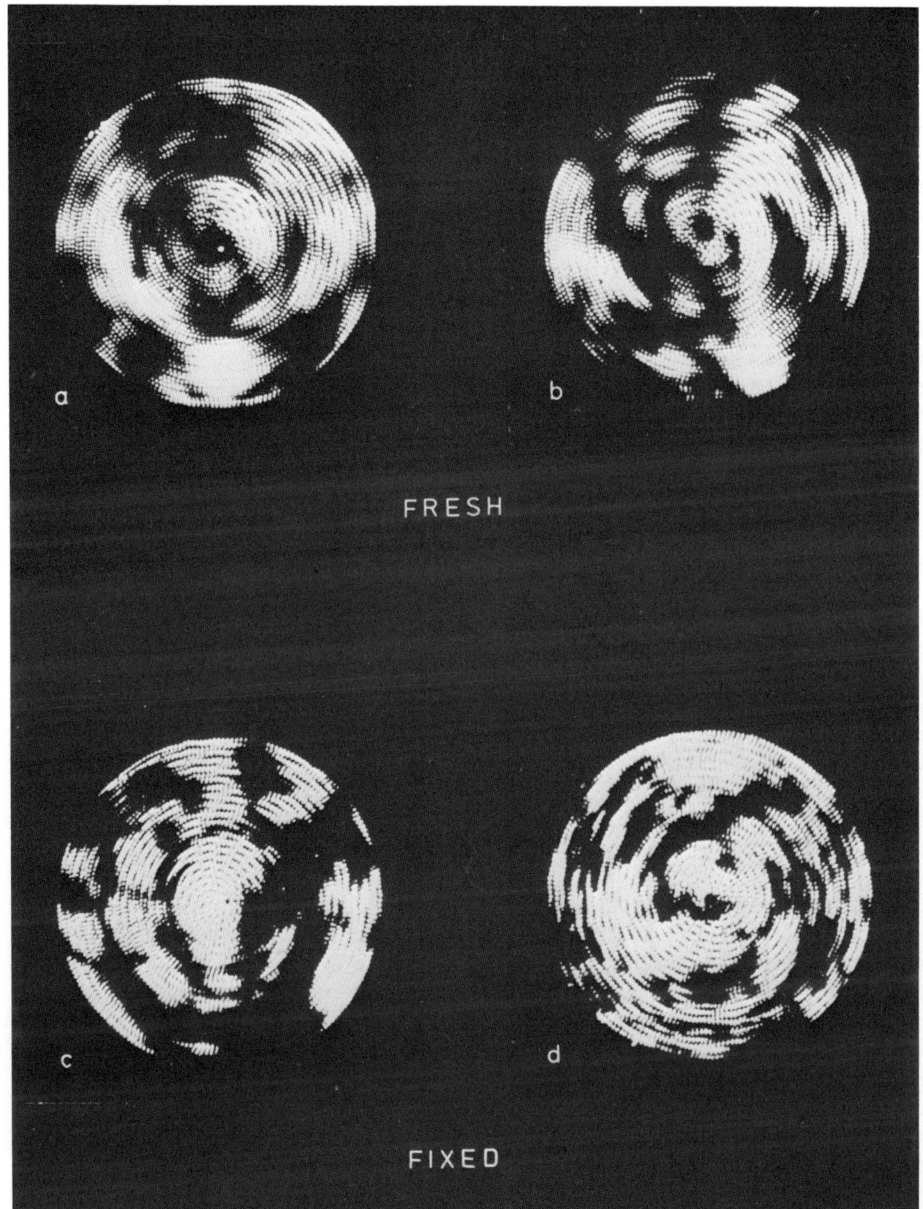

Fig. 2. Two-dimensional diffraction patterns, 'interferograms', obtained from 75 mm^3 volumes of post-mortem tissue at a frequency of 2.5 MHz.
(a) freshly excised normal liver tissue,
(b) freshly excised secondary liver neoplasm,
(c) formalin fixed normal liver tissue,
(d) formalin fixed secondary liver neoplasm.

3.0 MHz) was noted by examining the frequency spectrum of the pulse reflected from a plane Perspex* target oriented normal to the beam, 12 cm from the transducer face. The overall gain of the receiver electronics was then preset at each frequency to normalize for these sensitivity variations. Furthermore, the differing intensity of the backscattered echoes due to scanning at different depths within the tissue was corrected, to a first order approximation, by employing a conventional time-gain control to account for attenuation of the sound waves within the tissue.

The assessment of the number of discrete minima is somewhat subjective but, as all the assessments were performed as a 'blind' study, the relative differences between the results for different tissue types should be observer independent. This parameter is chosen because it relates to our previous studies and is fairly insensitive to small changes in the gain settings of the equipment.

Measurements were conducted on nine excised livers with a total of 52 different tissue sites scanned. Retrospective histological examination of these sites indicated that 20 were regions of cancerous involvement (secondary liver metastases of mixed origin; 13 fresh specimens and 7 fixed) whilst 32 were sites of normal liver parenchyma (26 fresh and 6 fixed). Measurements for each site were conducted at the three frequencies previously mentioned. Unfortunately our criteria for recording the diffraction patterns, outlined previously, occasionally led to unacceptable images at the frequencies of 2.0 and 3.0 MHz in that some scans were too 'dark', i.e. lacking in detail, to allow any realistic quantitation. These problems were specific to our limited data display facilities and do not imply any criticism of the technique.

Figures 3, 4 and 5 are scatter maps at 2.0, 2.5 and 3.0 MHz, respectively plotting the 'contrast' (ratio of 'dark' to 'white' area) versus the number of minima per steradian of scanning angle. One can immediately conclude that the more useful parameter for discriminating between the normal and neoplastic tissue is the number of minima per steradian, whilst the measure of 'contrast' is better for separating fresh and fixed tissues. Although Figure 4 seems to indicate that measurements at 2.5 MHz are best at discriminating between the four tissue types there is no reason to suppose that this frequency should be a more accurate discriminator than the other two, apart from its being the frequency at which the data acquisition was optimised in respect of signal to noise ratio.

Since all the features reported here indicate some degree of discrimination they have all been used, with equal weighting, in a stepwise discriminant analysis (Cooley and Lohnes, 1971). The resulting discriminant function has then been used to classify the

* Trade name for polymethylmethacrylate.

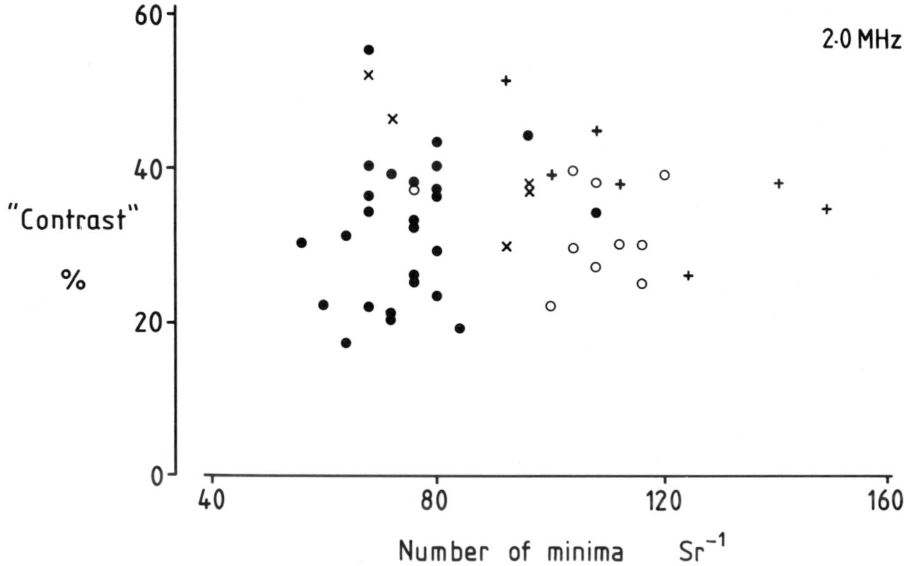

Fig. 3. Scatter map at 2.0 MHz of 'contrast' versus number of minima per steradian of scanning angle for normal liver tissue (fresh ● and fixed x) and liver metastases (fresh o and fixed +).

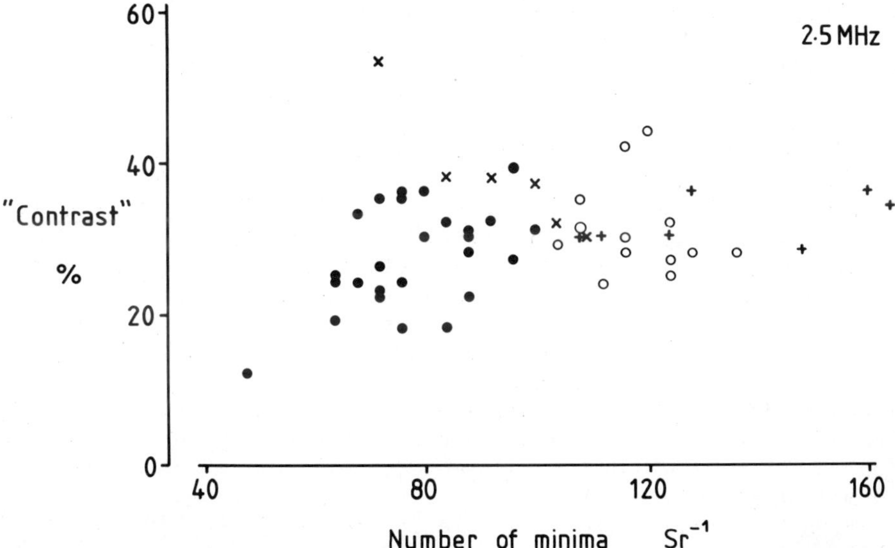

Fig. 4. Scatter map at 2.5 MHz of 'contrast' versus number of minima per steradian of scanning angle for normal liver tissue (fresh ● and fixed x) and liver metastases (fresh o and fixed +).

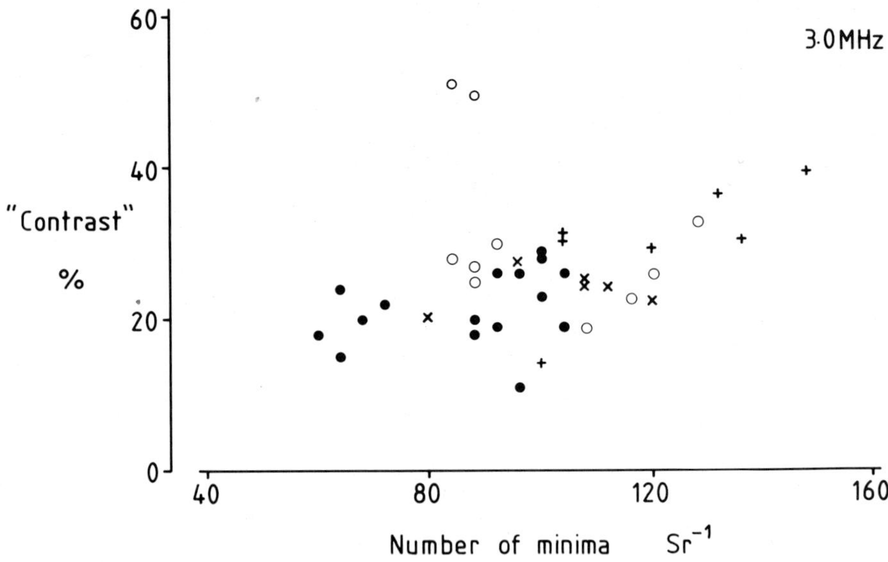

Fig. 5. Scatter map at 3.0 MHz of 'contrast' versus number of minima per steradian of scanning angle for normal liver tissue (fresh ● and fixed x) and liver metastases (fresh o and fixed +).

individual scans into one of the four categories. In our analysis the classifier uses a minimum distance criterion in which the six measured features are weighted according to the discriminant function and a discriminant score for each case computed. Figure 6 plots the discriminant scores for the four tissue types. These values are then compared with the average scores for each classification and the case assigned to the classification to which its score lies closest. In this trial the discriminant function is derived from the 35 cases where all six variables were recorded. The final computer classification uses all 52 cases and, as shown in Table I, achieves an overall success rate of 83% with only one focal neoplasm being incorrectly classified as normal. The separation of fresh and fixed tissue is less well achieved with eight misclassifications. However, the fact that an 85% success in separating fresh and fixed tissues is achieved indicates that fixation does have a significant effect on this form of data, whereas the measurements of Bamber et al. (1979) suggest that formalin fixation has little effect on some of the bulk acoustic properties of tissue.

In this context 'success' is defined in relationship to ability of the diffraction data, when subjected to discriminant analysis, to separate (in multi-variate space) the four sets of tissue types ('normal' and 'cancerous'; fresh and fixed) which constitute this

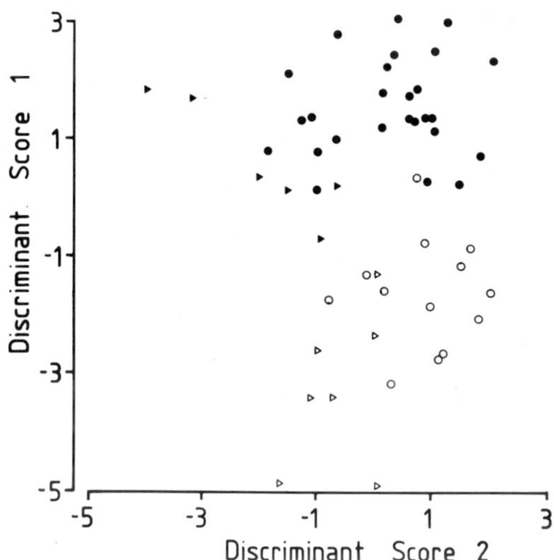

Fig. 6. Plot of discriminant scores for the 52 individual tissues. Normal liver tissue (fresh ● and fixed ▶). Liver metastases (fresh O and fixed ▷).

Table 1. Tissue Classification Ability of 2-D Diffraction

Histology	Interferograms	Normal Liver Fresh	Normal Liver Fixed	Tumour Fresh	Tumour Fixed	Total
Normal Liver	Fresh	21	5	0	0	26
	Fixed	0	6	0	0	6
Tumour	Fresh	1	0	10	2	13
	Fixed	0	0	1	6	7

83% correctly classified

training set of 52 tissue specimens. A similar ability to separate these histologies in subsequent blind trials is strongly implied but not, of course, demonstrated.

CONCLUSION

In this preliminary in vitro study we have described a two-dimensional diffraction scanning technique which yields information pertaining to the structural organisation of the small tissue volume examined. To date we have only reported on the empirical nature of our diffraction scans, termed 'interferograms', and limited ourselves to classifying freshly excised and formalin fixed focal cancers and normal liver tissue. The 83 percent overall accuracy of our results is especially notable in the light of the rather crude quantification of the patterns. The results not only confirm our previous one-dimensional findings (Nicholas and Hill, 1975; Nicholas, 1979) but provide an improved specificity and interpretation.

The existing display system is currently being supplemented by incorporating an analogue scan conversion memory into the system to provide a full 'grey-scale' presentation of the images. Furthermore digital acquisition of the scans will be provided to enable more sophisticated analysis and discrimination.

The full potential of this technique has still to be realised yet its present successes and relative ease of use warrant its consideration as a powerful tool for improving our understanding of acoustic interactions with human soft tissues.

REFERENCES

Bamber, J. C., Hill, C. R., King, J. A. and Dunn, F., 1979, Ultrasonic propagation through fixed and unfixed tissues, Ultrasound Med. Biol. 5:159.

Cooley, W. W. and Lohnes, P. R., 1971,"Multivariate Data Analysis", Wiley, New York.

Huggins, R. W. and Phelps, J. V., 1977, Bragg diffraction scanner for ultrasonic tissue characterization in vivo, Ultrasound Med. Biol., 2:271.

McCready, V. R. and Hill, C. R., 1971, A constant depth ultrasonic scanner, Brit. J. Radiol., 44:747.

Merton, J., Nicholas, D., Hill, C. R., Grover, S., Queenan, M. and Cosgrove, D. O., 1982, Ultrasonic diffraction scanning of the thyroid, Ultrasound Med. Biol., 8:145.

Nicholas, D., 1979, Ultrasonic diffraction analysis in the investigation of liver disease, Brit. J. Radiol., 52:949.

Nicholas, D. and Hill, C. R., 1975, Tissue characterization by an acoustic Bragg scattering process, in: "Ultrasonics International 1975", IPC Science and Technology Press, London, pp.269-272.

Nicholas, D. and Nicholas, A. W., in press, Two-dimensional diffraction scanning of normal and cancerous human hepatic tissue in vitro. Ultrasound Med. Biol.

AN APPROACH TO TISSUE CHARACTERIZATION BASED ON WAVE THEORY USING A NEW VELOCITY ANALYSIS TECHNIQUE

P.R. Mesdag, D. de Vries, A.J. Berkhout

Delft University of Technology
Dept. of Applied Physics, Group of Acoustics
P.O.B. 5046, 2600 GA Delft, The Netherlands

INTRODUCTION

In this paper we will discuss an approach to tissue characterization which was developed within our group.
First our approach will be outlined.One of the basic steps we use here is a high resolution Synthetic Focusing procedure (Berkhout et al.,1982). For this procedure, and also for tissue characterization itself, correct knowledge of the sound propagation velocity is essential. In the second part of this paper a method will be put forward to determine values of the sound propagation velocity with little prior knowledge of the medium.This method is based on the minimization of the entropy of the synthetically focused image.
Now an optimally focused image can be produced,which may be analysed in detail.In the final part of this paper we will discuss a method of analysis,based on two dimensional Fourier transformation of the r.f.data of part of the image.

OUR APPROACH

Our approach to tissue characterization consists of four steps.
Step 1: Measurement and Storage of the r.f.data.
By means of a linear array backscatter technique the r.f.data of a two dimensional image are gathered and stored into the memory of a computer system (Ridder et al.,1981).
Step 2: Synthetic Focusing.
As a second step to our approach <u>all</u> data is treated by a synthetic focusing procedure (Berkhout et al.,1982).This step is performed on an interactive basis with the sound velocity analysis technique which we discuss in this paper.Synthetic focusing of correctly measured

data produces an image with maximal detail and minimal distortion.

Step 3: Selection of an area of interest.

After synthetic focusing we select a sample volume from within our image.Due to the high resolution and the high dynamic range of the image the information coming from the selected area is effected only minimally by information coming from the rest of the image.

Step 4: Performance of any analysis procedure.

In the final step of our approach the selected sample volume may be analysed.

In our view there are two main groups of acoustic parameters which may be used for tissue characterization.

-Firstly there are bulk parameters such as the spatially averaged sound propagation velocity and absorption.For correct estimation of these bulk parameters a high lateral resolution is _not_ essential. In transmission one only need measure the amplitude and the arrival time of a pulse.In reflection however,the duality between depth and time of flight must be overcome before one can estimate these bulk parameters (Kossoff,1976).With our velocity analysis procedure this duality is overcome,only assuming lateral homogeneity for the sound velocity within the region of interest.

-Secondly there are the local reflectivity parameters.The higher the resolution in axial and in lateral sense,the better these local parameters can be studied.Consequently our approach is pre-eminently suited to study these local parameters,which in our view carry more information to characterize tissue than the above mentioned bulk parameters.In the final part of this paper two dimensional Fourier analysis will be put forward as one way to study these local parameters.

VELOCITY ANALYSIS

Why and How

A very important bulk parameter in the above tissue characterization approach is the _propagation velocity_,because it plays a double role.Firstly,accurate knowledge of the velocity is necessary in the synthetic focusing procedure (step 2) to get correct depth localization and optimal lateral resolution (De Vries and Berkhout, 1982).Secondly,the propagation velocity is a specific material constant that can be used to characterize tissues.

From geophysical literature,several velocity determination methods are known,usually applying coherency criteria (e.g. Hubral, 1976; Wapenaar et al.,1982).These methods,however,are only valid for layered media.For scattering media as tissue structures,velocity analysis techniques are still to be developed.In the following we introduce such a technique,based on the _Minimum Entropy_ criterion.

Information Measures based on Minimum Entropy

The concept of Minimum Entropy was introduced in geophysical literature by Wiggins (1978) as a means for deconvolution of seismic registrations.Minimum entropy has to be interpreted as "minimum chaos" or "maximum ordering" or "maximum information".In this conception,a N-point data set has maximum information when it consists of one spike and N-1 zeros.An increasing number of spikes,especially when they have equal amplitude,makes the information decrease (or: the entropy increase).In other words: the more sparse and spiky a data set,the higher the information and the lower the entropy.

Deeming(1981) gives a survey of several information measures defined in seismic literature,based on the Minimum Entropy idea. In general,such a measure V_t,applied to one seismic trace containing N discretized values y_i of a time function $y(\tau)$,has the form

$$V_t = \frac{1}{N}\sum_{i=1}^{N} z_i F(z_i); \quad z_i = \frac{y_i^2}{\overline{y^2}}; \quad \overline{y^2} = \frac{1}{N}\sum_{i=1}^{N} y_i^2, \tag{1}$$

$F(z_i)$ being some function monotonically increasing with z_i.It is easily seen that high values of z_i,i.e. a few high peaks in y_i,yield a high value of V_t.If,as always in practice,the trace contains the response of some temporal wavelet,the value of V_t is also influenced by the bandwidth and the phase of this wavelet: increasing bandwidth and decreasing phase yield a higher information measure.

Usually,an information measure is calculated over a multi-trace registration.Then the contributions V_t of the individual traces are summed,often after weighting with an appropriate weighting function g_t:

$$V = \sum_{t=1}^{M} g_t V_t. \tag{2}$$

Minimum Entropy and Velocity Analysis

Wave propagation from a reflectivity distribution on subsurface S_1 (depth ζ_1) to surface S_0 (depth ζ_0) can be described,in the space-frequency domain,as a spatial convolution along the lateral directions ξ and η (Berkhout,1980):

$$p(S_0) = w(S_0/S_1) * p(S_1), \tag{3}$$

where p denotes pressure and $w(S_0/S_1)$ is a forward propagation operator describing the propagation from S_1 to S_0.The propagation velocity between S_1 and S_0 is one of the parameters determining $w(S_0/S_1)$.

Inverse extrapolation, as applied in synthetic focusing, can be interpreted as a lateral deconvolution of recording $p(S_0)$ with a spatial filter $f(S_1/S_0)$ that compensates for the propagation effects described by $w(S_0/S_1)$, thus yielding an estimation of the pressure distribution at S_1:

$$\langle p(S_1)\rangle = f(S_1/S_0) * p(S_0) = f(S_1/S_0) * w(S_0/S_1) * p(S_1). \quad (4)$$

For optimal inversion, the resulting <u>spatial wavelet</u> $f(S_1/S_0)*w(S_0/S_1)$ should approximate the spatial delta pulse. In practical situations, band-limited inversion must occur, leading to a band-limited zero-phase spatial wavelet. If, however, the propagation velocity distribution between S_0 and S_1 is not accurately known, $f(S_1/S_0)$ will be chosen non-optimal and the phase spectrum of the spatial wavelet will not be zero. This gives rise to lateral broadening of this wavelet and hence to a decrease of the lateral resolution.

In the space-time domain, the increasing phase of the spatial wavelet with increasing velocity error appears as a lateral extension of the focused result. This can be seen in Fig. 1, where the acoustic image of a <u>point diffractor</u> in a homogeneous medium is shown, after synthetic focusing with an operator $f(S_1/S_0)$ containing a velocity value deviating 0,2,4,6,8 and 10% from the true value. Note that the length of the significant information in the individual traces is not much affected by velocity errors.

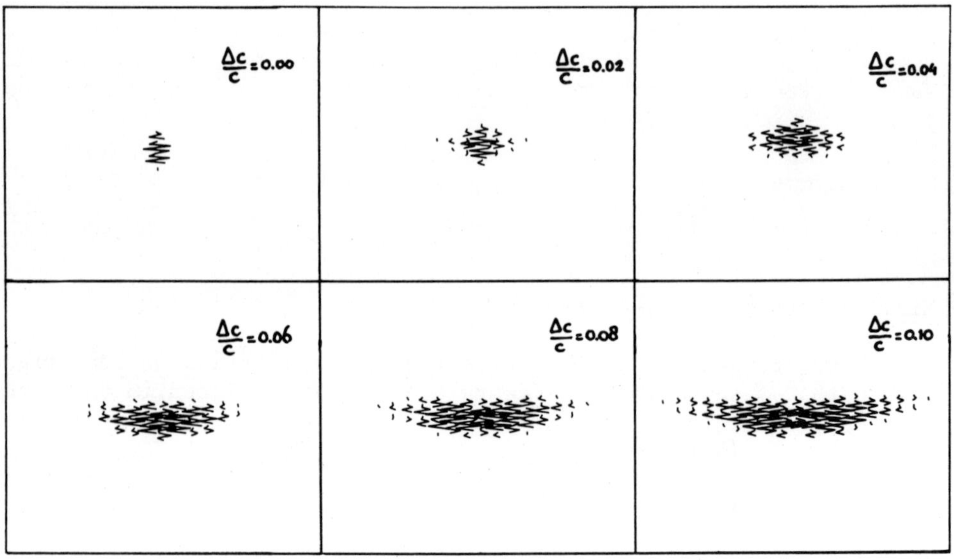

<u>Fig.1.</u>: Acoustic image of a point diffractor after synthetic focusing with different velocity errors.

Now, bearing in mind the features of information measures based on the Minimum Entropy (ME) criterion discussed above, one might expect that such measures, if applied laterally to the focused patterns of Fig. 1, i.e. to the transposed space-time data matrices, will show a maximum value if the correct propagation velocity value has been inserted into the inverse operator $f(S_1/S_0)$. This expectation is confirmed in Fig. 2a, where a multi-trace ME information measure V with appropriate choice of $F(z_i)$ and appropriate trace weighting has been determined for the transposed data of a focused point-diffractor as a function of the relative velocity error. Fig. 2b gives the same information measure, but now applied to the time traces without transposition. Note that for this choice V is not dependent on velocity errors.

Results on Simulated and Measured Data

Above we considered the use of ME information measures after synthetic focusing of the zero-offset response of one point diffractor. Since this response is identical to the common midpoint response of a plane reflector, the conclusions about the applicability of ME information measures for velocity analysis may be extended to layered structures as well.

In the context of our tissue characterization research, we calculated ME information measures, after focusing with different velocity values, for several simulated clusters of point diffractors embedded in a homogeneous medium. In short, our conclusions are:

- correct velocity value substitution in the synthetic focusing operator corresponds with a significant peak of ME information measures

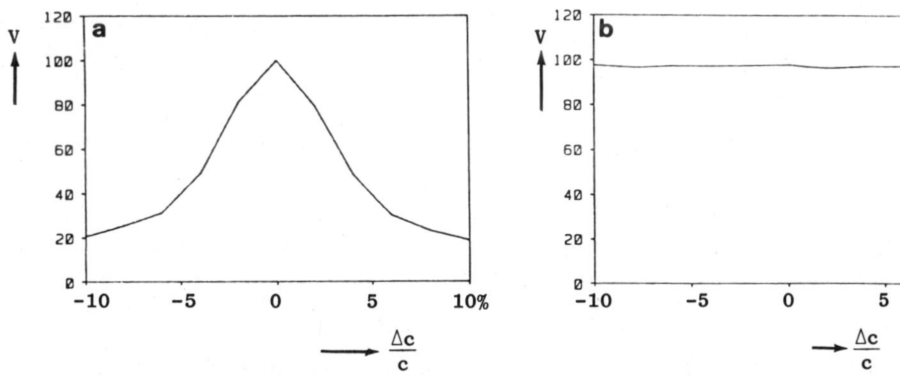

Fig.2.: Information measure for the focused point diffractor of Fig.1, as a function of relative velocity error.
a: application to transposed space-time data matrix;
b: application to non-transposed data matrix.

calculated for the transposed focused data set, under certain conditions concerning average diffractor interval, aperture width and aperture angle;

- for larger numbers of point diffractors, also ME information measures calculated for the non-transposed focused data set yield velocity information;
- ME velocity analysis results are not much affected by the presence of noise.

Our ME velocity analysis technique was also applied, after synthetic focusing, to the measured zero offset response of a tissue phantom, consisting of nylon wires and anechoic "cysts" embedded in a graphite doped gel. We found a bulk velocity of 1488 m/s, which is in excellent agreement with the results of other experimental determinations.

TWO DIMENSIONAL FOURIER ANALYSIS OF THE FOCUSED IMAGE

After completion of the synthetic focusing procedure with an optimal sound velocity, the image may be analysed as a whole, or in part. In this section we will analyse images with increasing complexity. First the responses of all models were simulated by computer and synthetically focused. Secondly the same models were built up of steel wires (d = 100 μm), scanned in a watertank and synthetically focused. Finally the results of the simulated and of the measured data were compared with each other.

One single point scatterer

First we will consider one single point scatterer in a homogeneous medium. Mathematically this point scatterer may be represented by a delta pulse in two dimensional space with coordinates x_0 and z_0:

$$r(x,z) = \delta(x - x_0)\ \delta(z - z_0), \qquad (5)$$

where x is the axis along the transducer surface and z is the axis into the medium.

If our measurement system had no limitations we would be able to reconstruct this delta pulse accurately. Instead, it is well known that any measurement system smears the acoustic energy from one point scatterer over a large portion of the image. This results in what is generally referred to as the point spread function (PSF) of a point scatterer. As our measurement system stores and processes the r.f. data in a large digital memory we have the r.f. PSF at our disposal. Berkhout et al. (1982) compared the lateral resolution of simulated an measured PSFs as a function of the distance from the transducer surface. Here it was found that at distances greater than 40 mm there was no significant difference between measurements and simulations.

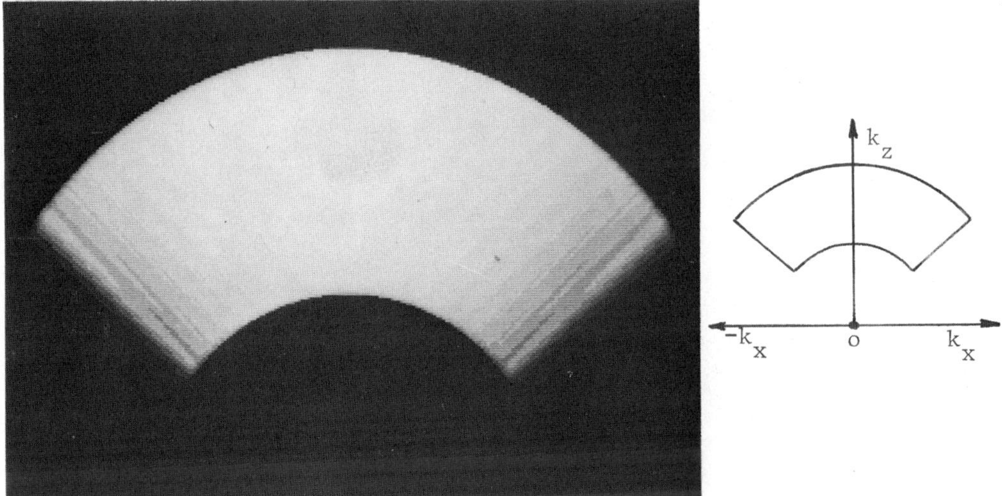

Fig.3.: Two dimensional Fourier spectrum of one point scatterer,as measured with a transducer with infinitely small elements, which emit a white spectrum (simulated).

For further analysis the two dimensional Fourier transform of the r.f. PSF was computed.Ideally we would find the Fourier transform of equation (5),which would be a white spectrum

$$|\tilde{r}(k_x,k_z)| = 1.0 \quad \text{for all } k_x \text{ and } k_z \ , \tag{6}$$

where k_x is the spatial wave number in the x-direction and k_z is the spatial wave number in the z-direction.Due to the system limitations we are not able to reconstruct a white spectrum.In Fig. 3 we see the two dimensional spectrum of a PSF of a point scatterer situated at 40 mm from a transducer with an aperture of 100 mm.The origin of the k_x - k_z domain of figure 3 is situated in the middle at the bottom of the picture.The k_x axis runs along the horizontal and the k_z axis along the vertical.The measurement was simulated by using transducer elements with an omnidirectional sensitivity,emitting a pulse with a white spectrum between 2 and 4.5 MHz.

Two kinds of limitations can be seen in Fig. 3.The circular limitations are due to the limited frequency band.As

$$k^2 = k_x{}^2 + k_z{}^2 \ , \tag{7}$$

where $k = 2\pi f/c$,the lower frequency of 2 MHz determines the inner circle and the higher frequency of 4.5 MHz determines the outer circle.The linear limitations result from the limited aperture.In the case of the model of Fig. 3 no signal could be received at greater angles than approximately 50° ($\alpha = tg^{-1}(0.5 \times 100 \text{ mm} / 40 \text{ mm})$).

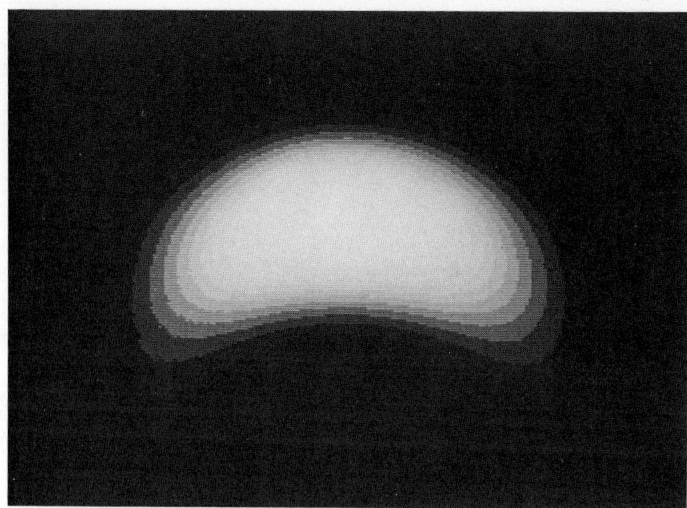

Fig.4.: Two dimensional Fourier spectrum of one point scatterer,as measured with a transducer with elements of finite size, which emit a pulse with a realistic spectrum (simulated).

For this measurement situation we have now defined our two dimensional acoustic window under ideal circumstances.Outside this window no information can be retrieved after data acquisition.In real situations the information content within the acoustic window can only be reduced even more.
For a finite element size (0.3 mm) and a realistic spectral content of the emitted pulse,the two dimensional spectrum is shown in Fig. 4. The main difference between the ideal and the real situation is a more rapid fall off of the signal towards the sides of the acoustic window.The realistic spectrum of the pulse results in a radial fall off and the beam forming of the finite element gives an angular fall off.
Exactly the same situation was scanned in a watertank and processed in the manner described above.The result is shown in Fig. 5.

Two point scatterers

In the previous paragraph we considered the response of one infinitely small discontinuity.Here we will consider two of these discontinuities,which may be represented by

$$r(x,z) = \delta(x - x_0)\ \delta(z - z_0) + \delta(x - x_1)\ \delta(z - z_1)\ , \tag{8}$$

where we have chosen $|x_1 - x_0|$ and $|z_1 - z_0|$ both to be 1.0 mm.
If again there were no system limitations we would be able to reconstruct these deltapulses accurately.We then would be able to measure

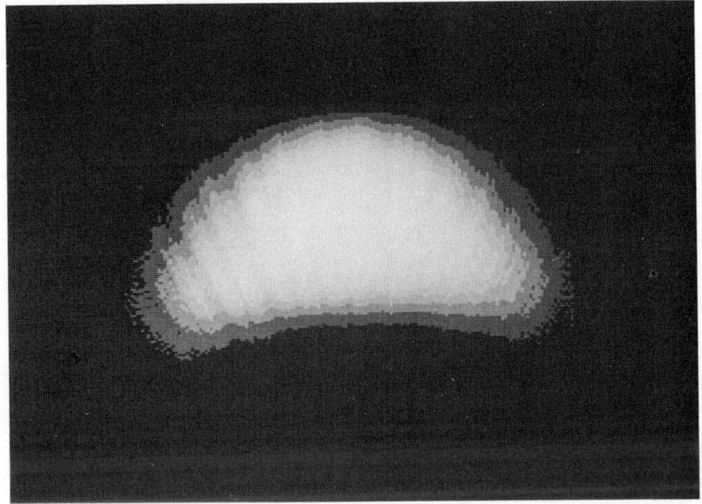

Fig.5.: As Fig. 4, but measured in a watertank.

a two dimensional Fourier spectrum as shown in Fig. 6.It can easily be shown that this pattern is defined by

$$|\tilde{r}(k_x,k_z)| = 2 \cos \tfrac{1}{2}\{k_x(x_1 - x_0) + k_z(z_1 - z_0)\} \ . \tag{9}$$

Due to system limitations the delta pulses of equation 8 will be convolved by the PSF of one single point scatterer.In the Fourier domain such a convolution is represented by a multiplication.Consequently the diffraction pattern of equation 9 is multiplied by our acoustic window.

The response of two point scatterers at a distance of 80 mm from a transducer with an aperture of 50 mm was simulated by computer,synthetically focused and transformed into the Fourier domain. The resulting two dimensional spectrum is shown in Fig. 7.
The aperture limitations of the acoustic window are larger here than in the previous model.Here the maximum angle under which we still may expect to receive a signal is approximately 17° ($\alpha = tg^{-1}$(0.5 x aperture / depth of scatterer)).
Here too the exact situation was measured in a watertank and the measurements were processed as mentioned above.Fig. 8 shows the result.

An array of scatterers

As a final example an array of scatterers was scanned.The two dimensional Fourier spectrum of the focused result is shown in Fig.9. Even though the closest distances between the wires are the same as

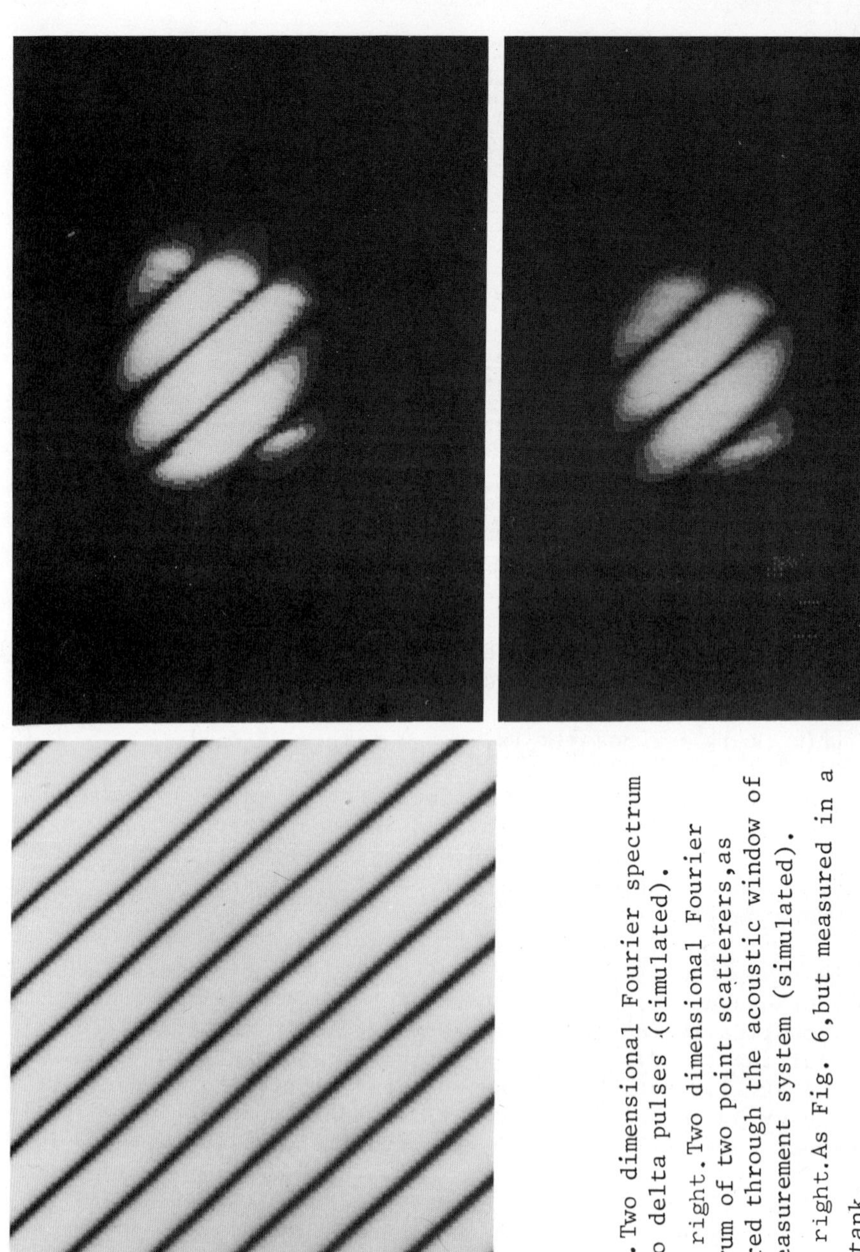

Fig.6.: Above. Two dimensional Fourier spectrum of two delta pulses (simulated).

Fig.7.: Above right. Two dimensional Fourier spectrum of two point scatterers, as measured through the acoustic window of our measurement system (simulated).

Fig.8.: Below right. As Fig. 6, but measured in a watertank.

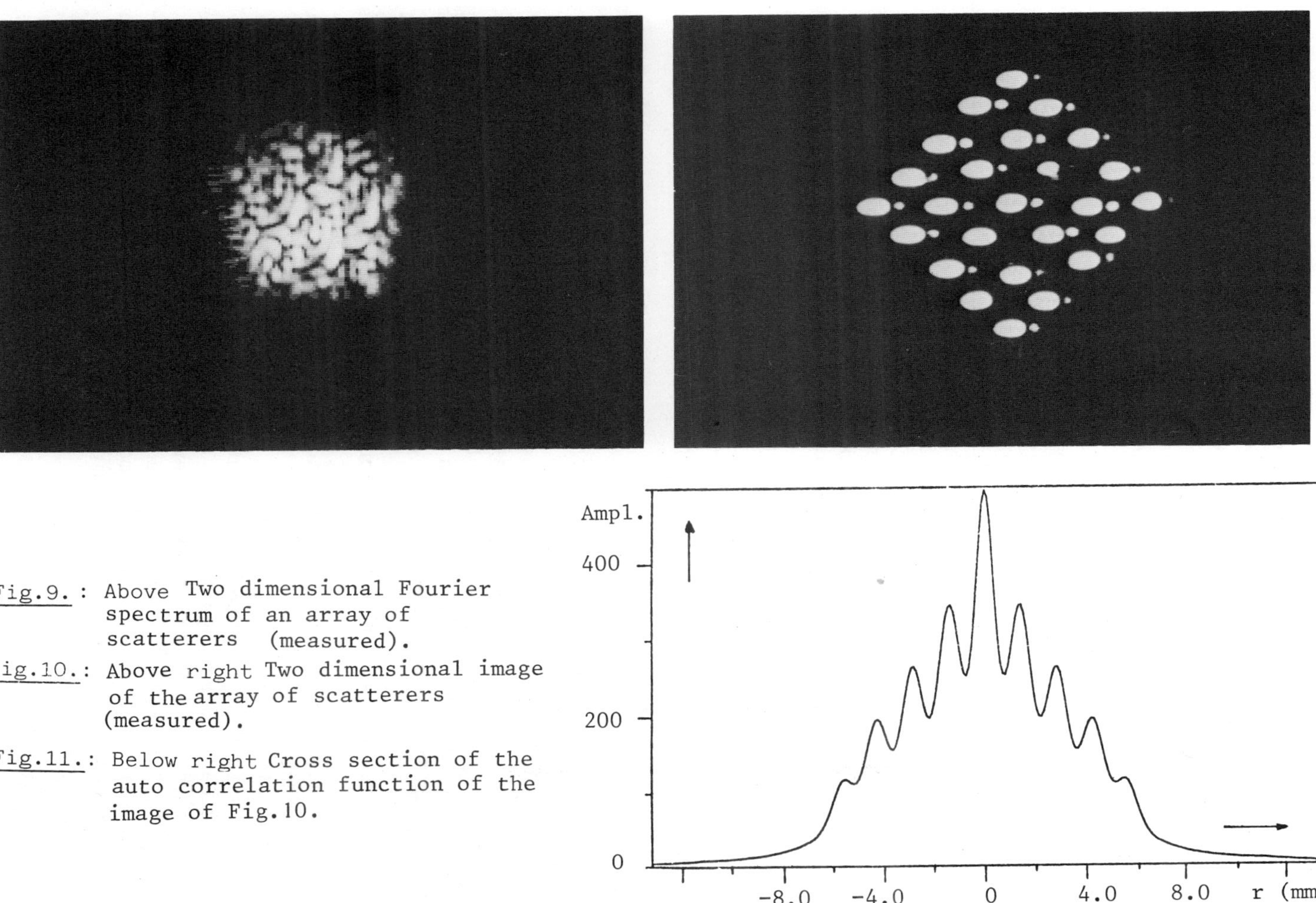

Fig.9. : Above Two dimensional Fourier spectrum of an array of scatterers (measured).

Fig.10.: Above right Two dimensional image of the array of scatterers (measured).

Fig.11.: Below right Cross section of the auto correlation function of the image of Fig.10.

in the model of two scatterers (1.4 mm),the diffraction pattern of Fig. 9 shows more rapid and incoherent fluctuations than those in Figs.6,7 and 8.The reason for this must be sought in the small variations in the positioning of the wires in relation to an ideal lattice.
In other words: when the positioning of the scatterers is not exactly according to an ideal lattice,the conditions for Bragg diffraction are not fulfilled.

If we wish to analyse an image which contains a fair amount of scatterers,like this one,it does not suffice to look at the amplitude spectrum alone.Here we must incorporate phase information too.
The original focused image contains all phase information.As can be seen from Fig. 10 the individual wires are still easily resolved.
An alternate way to analyse the image is by studying the auto correlation function of the original image (or: the inverse Fourier transform of the squared spectrum).Hereby we are able to estimate the characteristic distances within the image.
The two dimensional auto correlation function of the envelope of the original image of Fig. 10 was computed.Then a cross section through the origin,which made an angle of 45° with the x and the z axes was plotted in Fig. 11.From Fig. 11 we see that this cross section of the image has characteristic distances of approximately 1.4,2.8,4.2 and 5.6 mm,corresponding to all possible distances within the image.

CONCLUSIONS

We developed a new approach to tissue characterization,based on wave theory,consisting of four steps.To determine propagation velocity distributions,which play an important role in our tissue characterization procedure,we introduced a new technique based on the Minimum Entropy criterion,which can be applied in scattering as well as in layered structures.

From our two dimensional Fourier analysis procedure it follows that

- the measurements are in very good agreement with the computer simulations.
- analysis of single realizations of backscattered diffraction patterns (either measured,or computed by Fourier analysis)
  1) clearly describes the basic limitations of a measurement system,
  2) gives full information on very simple models and
  3) fails completely with more complex media.
- analysis must be done of the image including correct phase information,e.g. by calculating the auto correlation function.

REFERENCES

Berkhout,A.J.,1980,"Seismic Migration",Elsevier,Amsterdam-Oxford-New York.

Berkhout,A.J.,Ridder,J.,Graaff,M.P.de,1982,New Possibilities in Data Measurement,Signal Processing and Information Extraction; Philosophy and Results,1982 Acoustical Imaging,Vol 12.

Deeming,T.J.,1981,Deconvolution and Reflection Coefficient Estimation using a Generalized Minimum Entropy Principle, unpublished.

De Vries,D.,Berkhout,A.J.,1982,Influence of Velocity Errors on Computerized Acoustic Imaging, unpublished.

Hubral,P.,1976,Internal Velocities from Surface Measurements in the Three-Dimensional Plane Layer Case,Geophysics,41(2):233.

Kossoff,G.,1976,Reflection Techniques for Measurement of Attenuation and Velocity,Ultrasonic Tissue Characterization,N.B.S. Spec. Publ.435,ed.M.Linzer:135-139.

Ridder,J.,Wal,L.F.v.d.,Berkhout,A.J.,1981,Synthetic Focussing by means of Wave-Field Extrapolation a New Imaging Technique for Medical Ultrasound,1981 Ultrasonics Symposium Proceedings,Vol 2,ed.B.R.McAvoy:627-631. I.E.E.E.

Wapenaar,C.P.A.,Berkhout,A.J.,1982,Velocity Determination in Layered Systems with Arbitrary Curved Interfaces by means of Wave-Field Extrapolation of CMP-data, unpublished.

Wiggins,R.A.,1978,Minimum Entropy Deconvolution,Geophysics 16:21.

SHORT TIME FOURIER ANALYSIS AND DIFFRACTION EFFECT IN BIOLOGICAL TISSUE CHARACTERIZATION

Mathias Fink, François Hottier

Laboratoires d'Electronique et de Physique Appliquée
3, avenue Descartes
94450 Limeil-Brévannes (France)

ABSTRACT

This study presents a new method based on short time Fourier analysis giving a local unbiased estimation of the attenuation in biological tissue. Short time Fourier analysis concepts are well adapted to process time echographic signals which are non stationary. The interest of Short Time Fourier Analysis is to give an estimation of the signal spectral composition as a function of time. It will be shown that the time dependence of the spectral centre of gravity allows to deduce easily the frequency dependant attenuation. The near field low pass filtering effect due to diffraction is emphasized. A numerical calibration technique developed for each transducer which corrects these effects is described. Experimental results obtained from tissue like phantom and from in vivo liver tissue are presented.

INTRODUCTION

The attenuation of ultrasound in soft tissue,which is strongly frequency dependant, appears to be one of the most important parameter in tissue characterization. The work of Kuc[1,2] has shown that normal and pathological liver can be differentiated on the basis of their frequency dependant attenuation. It is also known that normal, infarcted and ischemic heart tissue can equally be differentiated by the frequency dependant attenuation. Thus the development of accurate techniques for in vivo determination of attenuation is an important objective of tissue characterization.

The method developed by Kuc compared averaged power spectra at different depths for a serie of A echographic lines. It requires quite a large amount of data to obtain a reliable value of attenuation and it is only adapted to thick organs as the liver. Another method has been proposed by Ophir which compares the echographic power spectrum coming from two C planes[3].

This paper presents a new method based on the short time Fourier analysis of echographic data, providing a local unbiased estimation of the attenuation slope. The short time Fourier analysis has the major interest of providing an estimation of the signal spectral composition as a function of time. In echographic mode, it permits to follow the depth dependence of the local tissue roundtrip transfer function. It will be shown that the depth dependence of the spectral centre of mass allows easily the determination of the frequency dependant attenuation. Moreover the non invariant transfer function due to diffraction process which increases the low frequency components of the spectrum in the transducer near field has also been assessed and can be corrected in some cases.

The algorithms have been at first tested on simulated echograms derived from a 1D tissue model , taking into account the exact roundtrip impulse response of a discrete distribution of scatterers embedded in an attenuating medium. Experimental results obtained from a tissue like phantom are presented ; they show up the influence of diffraction and attenuation. Preliminary estimation of in vivo liver attenuation has been performed by this technique and are consistent with the values found by Kuc and various authors.

DETERMINATION OF THE LOCAL TISSUE ROUNDTRIP TRANSFER FUNCTION

The echographic signal observed as a voltage function $E(t)$ depends on the voltage function $e(t)$ applied to the electrical terminals of the transducer, through a set of transfer functions which depend on the transducer acoustoelectric transfer, on its geometry and on the propagating medium (velocity, frequency dependent attenuation, spatial distribution and frequency response of the scatterers). These various transfer functions may be classified in two sections.

a) The transfer functions which are independent of the scatterers position. These are the acoustoelectric transfer functions in transmission $I^T(\nu)$, in reception $I^R(\nu)$ and the frequency dependent scattering amplitude $u(\nu)$. For the subsequent development, it will be convenienceto write the product of these different transfers as $G(\nu) = I^T(\nu)\ I^R(\nu)\ u\ (\nu)$. In the time domain, a Fourier transformation of $G(\nu)$ gives $g(t)$ which is the impulse response of a point scatterer.

b) The transfer functions associated with the ultrasonic propagation in the medium (diffraction and attenuation) which depend strongly on the scatterers location relatively to the transducer. It is well known that the directivity pattern of a transducer depends on frequency and thus, for a broadband excitation, a scatterer will experience a frequency dependent acoustical force. A diffraction transfer $H_o(\vec{r}_o,\nu)$ may be defined in transmit mode as the velocity potential observed at any point $\vec{r}_0$, for a unity sine-shaped velocity excitation of the transducer at frequency ν.

For a planar piston transducer embedded in an infinitely rigid baffle, the aperture function of which is $O(x,y)$, the Rayleigh Sommerfield choice of the green function gives a transfer function as :

$$H_0(\vec{r}_0,\nu) = \iint O(x,y) \frac{\exp(ikR)}{2\pi R} ds \tag{1}$$

where k is frequency dependent ($k = 2\pi c\nu$) and $R = |\vec{r}-\vec{r}_0|$

In biological tissue, the frequency dependant attenuation $\alpha(\nu)$ modifies the directivity pattern differently at each frequency and it will be observed a radiative transfer in such medium $H(\vec{r}_0,\nu)$ different from $H_0(\vec{r}_0,\nu)$. It is the modification of this radiative transfer linked to the attenuation which has the major interest. Neglecting the velocity dispersion linked to the attenuation[4] we may replace in (1) the usual monochromatic spherical wave by an attenuated wave and thus the new radiative transfer is :

$$H(\vec{r}_0,\nu) = \iint O(x,y) \frac{\exp\left((i2\pi c\nu - \alpha(\nu))R\right)}{2\pi R} ds \tag{2}$$

An important simplification of this formula may be obtained if all points on the radiating aperture are approximately equidistant from the scattering point. Thus by defining $R(x,y,y) = \bar{R}$ we may separate completely the effect of diffraction and of attenuation in the radiative transfer :

$$H(\vec{r}_0,\nu) = \exp(-\alpha(\nu)\bar{R}).H_0(\vec{r}_0,\nu) \tag{3}$$

In the time domain, a radiative impulse response $h(\vec{r}_0,t)$ may be defined[5] as the time Fourier transform of $H(\vec{r}_0,\nu)$. In echographic mode the two-way roundtrip impulse response is then given by the autoconvolution product of $h(\vec{r}_0,t)$.

In order to generalize these concepts to the determination of the echographic response of any tissue, we must model a tissue, as a continuous distribution of scatterers, where each differential volume dV has a spherical wave scattering amplitude $U(\vec{r}_0,\nu)dV$. However we consider that the frequency dependence of the scattering coefficient is independent of the spatial coordinates, that is to say that dispersion is uniform ; we then write :

$$U(\vec{r}_0,\nu) = s(\vec{r}_0)\, u(\nu) \tag{4}$$

The echo coming from this medium is then obtained through the coherent summation of individual echoes.

Integrating throughout the scattering volume we obtain :

$$E(t) = e(t) \otimes g(t) \otimes \int (h(\vec{r}_0,t) \otimes h(\vec{r}_0,t))\; s(\vec{r}_0)dV \tag{5}$$

in which $\otimes$ defines a convolution.

The tissue radiative transfer function is then obtained by Fourier transforming this expression and may be written as :

$$\int (H(\vec{r}_o,\nu))^2 \, s(\vec{r}_o) \, dV \tag{6}$$

In order to evaluate the frequency dependent attenuation of the tissue, we need to define a local roundtrip transfer function which only take into account the scatterers located in a thin tissue slice. For a constant velocity medium, it is then possible to associate with a given roundtrip time τ_i, a scatterer slice located on an isochrone surface from the transducer aperture. By example, for a plane disc transducer, these slices are, in the near field, disc plane surfaces located in the geometrical shadow of the transducer. In the far field, out of shadow scatterers contribute significantly to the echographic response, and the isochrone surfaces become more complicated.

Estimation of the frequency dependent attenuation may be obtained in the case of an ideal medium, made of two well separated scatterer slices whose echoes are non overlapping. Selection of two non overlapping slices, by the use of two windows correctly positioned along the A line will give by Fourier transformation two spectra $\tilde{E}(\nu,\tau_i)$ which differ only by the radiative transfer associated with the various roundtrip times τ_i. Separating as in (3) diffraction and attenuation effects in the radiative transfer gives

$$\tilde{E}(\nu,\tau_i) = \tilde{e}(\nu)\; G(\nu)\; \exp(-\alpha(\nu)c\tau_i)\; \mathcal{H}_o(\nu,\tau_i) \tag{7}$$

where the symbol ~ means Fourier transform and where the diffraction part of the transfer may be written as :

$$\mathcal{H}_o(\nu,\tau_i) = \int (H_o(\vec{r}_o,\nu))^2 \, s(\vec{r}_o) \, \delta(R - c\tau_i/2) \, dV \tag{8}$$

Restriction of the transfer to a thin isochrone slice is obtained through the distribution $\delta(R - c\tau_i/2)$. Exact estimation of the attenuation needs the knowledge of this local diffraction transfer at least for two slices of the medium. However the exact estimation of the diffraction integral $\mathcal{H}_o(\nu,\tau_i)$ needs an a priori knowledge of $s(\vec{r}_o)$ which is one of the unknown data of the problem. An important simplification can be done if the distribution of scatterers is relatively uniform in the transducer field. Then we can approximate this transfer function by an average transfer function which corresponds to an uniform scattering medium ($s(\vec{r}_o)=1$) in each slice.

$$\bar{\mathcal{H}}_o(\nu,\tau_i) = \int (H_o(\vec{r}_o,\nu))^2 \, \delta(R - c\tau_i/2) \, dV \tag{9}$$

Computation of this transfer function for different depths z_i shows that for a 10 mm diameter plane transducer, diffraction effects increase the spectrum low frequency components as the distance transducer-slice decreases (fig. 1). Then diffraction modifies the spectral composition of the signal in the transducer near field and some spectral correction will be needed to get an unbiased value of the attenuation.

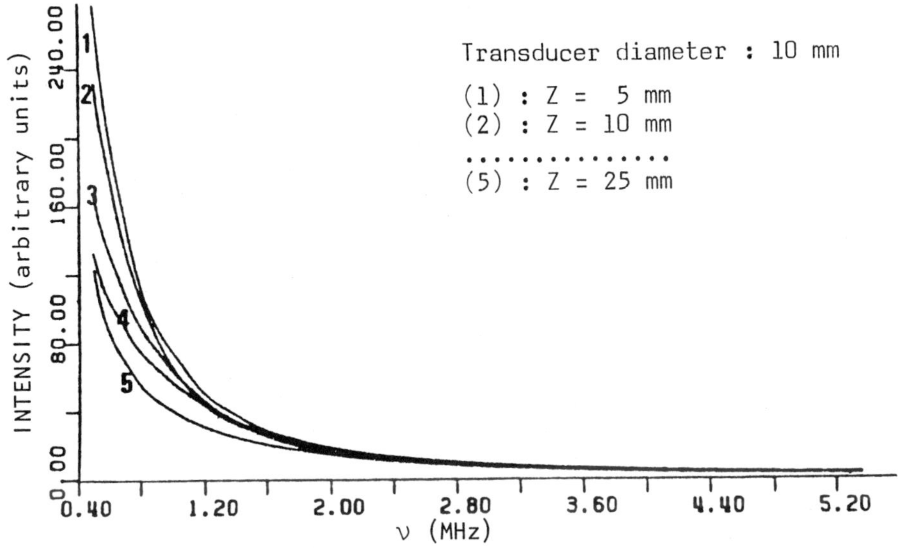

Fig. 1 Diffraction transfer functions

If we assume that diffraction transfer is known and has been corrected on the experimental spectra $\tilde{E}(\nu,\tau_i)$, we may find easily the frequency dependant attenuation. For biological medium where the attenuation is approximatively frequency linear ($\alpha(\nu) = \beta|\nu|$), different choices for the product $\tilde{e}(\nu)\ G(\nu)$ can be done. Gaussian shape filters are of practical use.

$$\tilde{e}(\nu)\ G(\nu) \propto \exp - \frac{(\nu-\nu_o)^2}{2\sigma^2} \tag{10}$$

and gives for the diffraction corrected value of $E(\nu,\tau_i)$:

$$\tilde{E}_c(\nu,\tau_i) \propto \exp - \frac{(\nu - (\nu_o+\Delta\nu_i))^2}{2\sigma^2} \tag{11a}$$

with :
$$\Delta\nu_i = \beta\ c\ \sigma^2\ \tau_i \tag{11b}$$

that is to say that the corrected echographic spectrum from non overlapping slices are also of Gaussian shape, but with a mean frequency translated towards the low frequency. Comparison of the two mean frequencies may be done and will give the exact value of β.

THE SHORT TIME FOURIER ANALYSIS

In the general case of a 3D scattering medium, the artificial selection of two well separated slices cannot be done because of the time duration of the transducer acoustoelectric response. In the echographic signal there is a strong overlapping between different slices. However the local transfer function varies slowly with depth, and a Short Time Fourier Analysis (STFA) can give a good estimation of the local tissue roundtrip transfer function versus the roundtrip time.

STFA consists of sampling the signal by a sliding window. For each position of the window, the sampled signal is Fourier transformed and its spectrum is computed. The results are finally displayed in a three dimensional plot where the successive spectrum modules are drawn as a function of time. The STFA of the echographic signal is then defined as :

$$\varepsilon(\nu,\tau) = \int E(t)\, w(\tau-t)\, e^{i2\pi\nu t}\, dt \qquad (12)$$

where $E(t)$ is the echographic signal, w is the window and τ the time at which the short time Fourier transform is performed. This two dimensional function is a good approximation of $\tilde{E}(\nu,\tau)$ defined in (7) for completely separated tissue slices.

STFA has been, at first, applied to simulated echographic data from a simple 1D tissue model, taking into account frequency dependant attenuation, acoustoelectric transfer of some transducers, and different statistics of scattering centres. Diffraction effects were not included in this first simulation. The first simulated datum corresponds to a tissue of depth 3.5 cm long consisting in a random distribution of Rayleigh scatterers embedded in a .8dB/cm/MHz attenuation medium. A 2 μs long rectangular window has been used. A direct observation of these snapshots, similar to the ones of fig. 2, shows a shift of the local spectrum content towards low frequency as a function of time. This effect is all the more important as the medium attenuation is high. At this stage it is necessary to define an indicator in order to evaluate the time evolution of the running spectrum. Taking into account the formula (11) which gives the mean frequency shift for gaussian pulse, it appears that the centre of gravity of the successive spectra is a standard parameter which could be very well correlated to the attenuation of the medium.

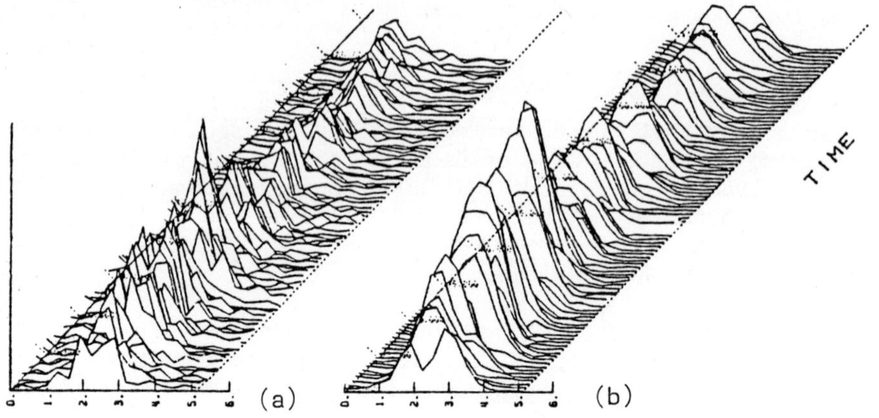

Fig. 2 STFA of echographic data observed from an "echobloc" phantom (2.5 MHz transducer) - (a) rectangular window ; (b) Hamming window

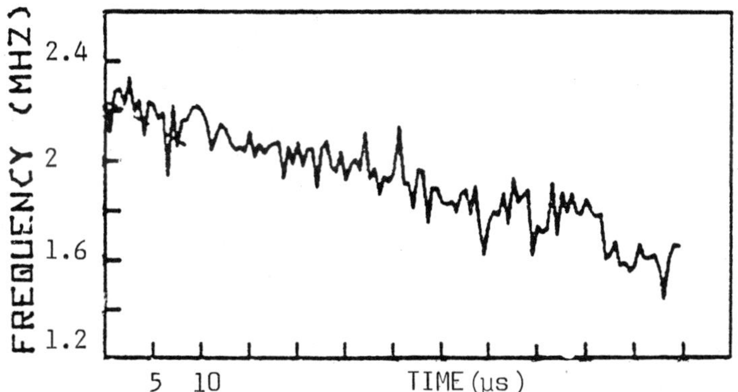

Fig. 3 Spectral centre of gravity position versus time from simulated data (.8 dB/cm/MHz)

This indicator is defined as :

$$\nu_G(\tau) = \frac{\int_{\nu_1}^{\nu_2} \nu \;\; \varepsilon(\nu,\tau)\, d\nu}{\int_{\nu_2}^{\nu_1} \varepsilon(\nu,\tau)\, d\nu} \qquad (13)$$

We have plotted on figure 3 the position of the centre of gravity of the running spectrum corresponding to the simulated data (.8dB/cm/MHz). The expected shift with time towards low frequency is well observed. High frequency oscillations are superimposed to this overall shift, which depend on the type and on the length of the sampling window. However, by taking into account the exact value of the transducer bandwidth (σ), formula 11b gives for β a value of .8dB/cm/MHz which shows the interest of the method.

EXPERIMENTAL RESULTS

This method has been used to process experimental digitized data. An "echobloc" tissue like phantom, simulating a liver tissue, was chosen for this purpose. A home mode acquisition system was made, with a A/D converter operating at a sampling frequency of 20 MHz and coding on 10 bits. All the recorded data have been obtained with 10 mm diameter unfocused transducers equipped with two $\lambda/4$ matching layers. The power spectrum can be approximated by a Gaussian curve.

Spectrograms from tissue like phantom (echobloc)

Different echographic A lines from the echobloc have been processed by STFA. The variations of the signal spectral composition as a function of time are shown in figures 2a,b in a time-frequency representation. The signal has been sampled by a 2 μs long rectangular and Hamming window respectively. The damping of the high frequency component is shown on these figures and the use of a Hamming window leads to more regular shape for the running spectra. Moreover, the occurence of "double peaked" spectrum is much more frequent in the case of a rectangular window. That means that the 2 μs long rectangular window generally includes more than one "isolated" echo and then interference effects caused by the periodicity of the medium appear and modulate the spectrum. The corresponding Hamming window which has a better time resolution is less sensitive to this phenomenon.

We have studied the time evolution of the running spectral centre of gravity. One of this evolution is illustrated in the dotted line in figure 4. It must be noticed that two different behaviours are observed. In the near field of the transducer, diffraction filtering is the most important, and as explained before, it increases the low frequency component of the signal. A positive slope is observed in this region. After some depth, the attenuation transfer becomes the most important and a negative slope is then obtained, which is well correlated to the medium attenuation.

Correction of the running spectrum may be done by the time variant inverse filter $1/\bar{\mathcal{H}}_0(\nu,\tau)$ defined in (9). Such a correction may be quite effective for some A lines. A constant negative slope is then obtained for the whole field and the value of β deduced from this slope is equal to .49 dB/cm/MHz which is in very good agreement with the expected attenuation. We must however notice that for some A line the average transfer $\bar{\mathcal{H}}_0$ is not efficient. This is due to the fact that the exact radiative roundtrip transfer is non deterministic(9);it depends on the statistical behaviour of the scatterers which is included in the function $s(\vec{r}_0)$. In some cases, the defined average transfer may be very different from the exact transfer. To be really efficient the inverse transfer function for diffraction correction has to be applied on averaged curves.

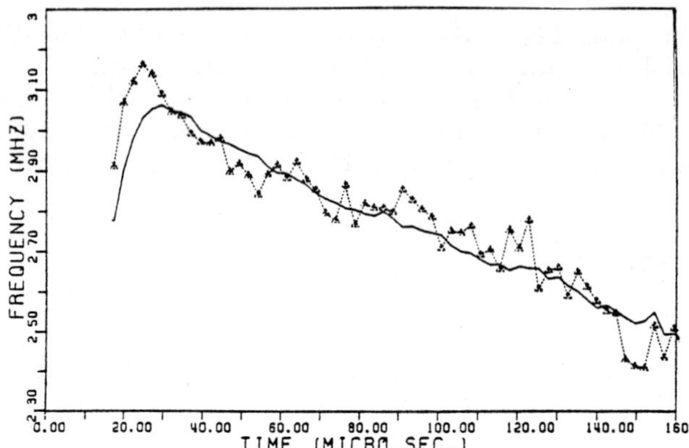

Fig. 4 Spectral centre of gravity position from "echobloc" data (3.5 MHz transducer). The dotted line corresponds to one A line, the solid line to an average on 32 A lines.

Neglecting these diffraction effects, we have studied more precisely one part of the echographic A line, located in the far field. Its duration corresponds to a tissue length of 5.6 cm. The length of the window controls the thickness of the tissue slice considered in the determination of local transfer. If the echoes issued from individual scatterers were not overlapping, it would be possible to use a window corresponding to the duration of the echo returning from an individual scatterer (2 μs). That would give a band limited signal representing the complete transfer of the analysed tissue slice. Generally, the tissue slice contains a random array of scatterers which leads to overlapping echoes on the RF echographic A line. Drastic variations of the centre of gravity position will then be observed depending on the distribution of the scatterers. Echoes overlapping inside the sliding window modify strongly the slope of the spectrum. A longer window,including more scatterers,would reduce this random noise. The sliding window will then experience more stationary signals as it moves along the tissue. A new scatterer taking into account by the moving window will introduce less modification for longer window. The role of the window appears clearly by inspecting the following centre of gravity versus depth curves with the same A line.

Rectangular and Hamming windows with different durations have been used. The effect of the window length is shown on table I. The slope of the centre of gravity versus depth curve is calculated by using standard least square analysis techniques. The spreading of the experimental centre of gravity position with respect to the regressive line is estimated by the expression :

$$\delta^2 = \sum_{i=1}^{N} (y_i - Ax_i - B)^2$$

(x_i, y_i) being the frequency-depth coordinates of the successive centre of gravity positions and (A,B) the coefficients of the regression line.

In order to get stable estimations for these two quantities (slope and δ^2) an average on a certain number of data is needed. 32 different A lines have been considered. The mean of the slopes and of the δ^2 indicators are given in table I. The attenuation deduced from (11b) with σ equal to .57 MHz is also given. The solid line on figure 4 shows the effect of averaging the centre of gravity curves on 32 successive A lines.

As shown in the table I the values found for the resulting attenuation are not very dependant of the particular choice of the window. On the contrary the δ^2 quantity is about four times more important for a 2.5 μs than for a 10 μs window. For this latter case a stationary value of the attenuation is obtained after an average on five different A lines, for a 5.6 cm length of tissue.

Table I

window length (μs)	type of window	slope cm/MHz	δ^2 MHz x MHz	attenuation dB/cm/MHz
10 μs	rectangular	.018	.23 E-02	.49
10 μs	Hamming	.019	.38 E-02	.52
5 μs	rectangular	.018	.52 E-02	.49
2.5 μs	rectangular	.017	1.01 E-02	.46

Experimental results on in vivo liver

The data are collected by using a B scan system to locate a suitable part of the organ. A sample corresponding to a thickness of 6 cm is selected on a healthy worker. A Gaussian slope transducer is used with a central frequency of 3.5 MHz and a σ equal to .67 MHz. A time-frequency snapshot of a part of the A line is shown in fig. 5. It shows a simultaneous decrease of the overall signal energy as a function of time and a selective frequency shift. An average of 5 centre of gravity versus time curves, obtained with a 10 μs long Hamming window is presented on fig. 6. The slope deduced from this latter curve gives an attenuation of .45 dB/cm/MHz taking into account the σ of the transducer.

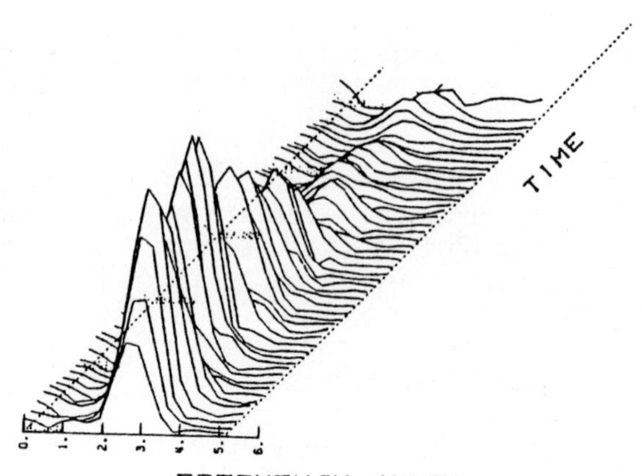

Fig. 5 STFA of echographic data observed in vivo from liver (3.5 MHz transducer)

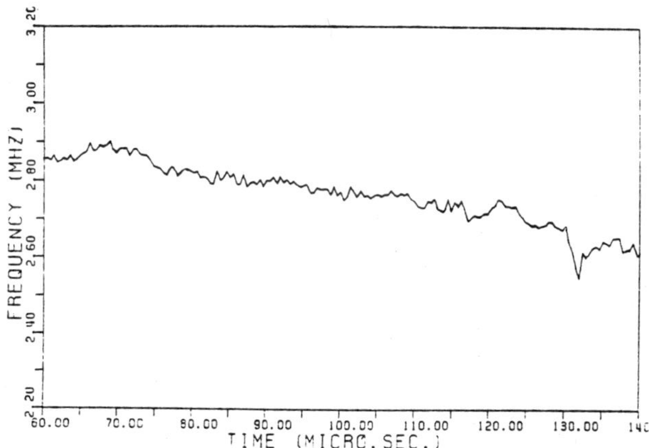

Fig. 6 Average of 5 spectral centre of gravity curves obtained from in vivo liver

CONCLUSION

These results are consistent with the values found by Kuc. However this method is able to work with a smaller amount of data. In this case it will be informative on smaller piece of tissue (~ 2 cm) which corresponds to the need of the clinical field.

On the other hand these concepts may be extend to a 2D map of the spectral centre of gravity introducing a new imaging technique, adding quantitative information to the classical B mode imaging.

REFERENCES

1. R.B. Kuc, Statistical Estimation of the Acoustical Attenuation Slope for Liver Tissue, Ph. D Thesis, Columbia University, 78-2331 (1977).
2. R. Kuc and M. Schwartz, Estimating the Acoustic Attenuation Slope for Liver from Reflected Ultrasound Signals, IEEE Trans. Sonics Ultrasonics vol.SU-26, 353-362, Sept.1979
3. J. Ophir and N.F. Maklad, A New Stochastic C-Scan Technique for Attenuation Coefficient Measurements in Tissue Equivalent Material, presented at the 23rd Annual Meeting of the American Institute of Ultrasound in Medicine, San Diego, Calif, Oct. 1978.
4. M.O. Donnel, E.T. Jaynes, J.G. Miller, Mechanisms : Relationship between Ultrasonic Attenuation and Dispersion Tissue Characterization Meeting, Gaithersburg 1978.
5. M. Fink, Theoretical Study of Pulsed Echographic Focusing Procedures, Acoustical Imaging vol. 10, 437-453 (1982), Plenum.

TISSUE CHARACTERISATION USING ACOUSTIC MICROSCOPY

D.A. Sinclair and I.R. Smith

Department of Electronic and Electrical Engineering
University College London
Torrington Place, London WC1E 7JE

ABSTRACT

It has been recognised for some time in clinical ultrasound that the speckle-like appearance of regions of relatively homogeneous tissue in B scans can provide significant diagnostic information - particularly in the case of diffuse disease. Improved understanding of the scattering and diffractive processes experienced by an acoustic wave propagating through such tissue regions can be expected to enhance the clinical assessment of the corresponding B scans.

The scanning acoustic microscope provides a suitable method for imaging variations in the acoustic properties of *in vitro* samples with a resolution comparable to the acoustic wavelength. We discuss here techniques for the derivation of quantitative values for velocity and impedance from acoustic microscope image data. In particular we demonstrate the validity of a paraxial theory of microscope operation and show that enhanced sensitivity can be achieved by careful selection of the microscope coupling liquid. Maps of acoustic velocity and impedance are derived for human liver.

INTRODUCTION

This paper investigates an approximate technique[1,2,3] for deducing the elastic constants of a liquid-like object from its image in a scanning acoustic microscope (SAM)[4,5]. The motivation behind the measurement of these parameters is twofold. First, it has been shown that determination of the impedance and velocity of tissue can form the basis of a method of characterising pathology[6,7,8]. Second, it has been demonstrated[8] that diffraction plays a major role in the formation of a B scan image of regions of soft tissue. In particular, recent results that exploit the differences in diffraction patterns from healthy and diseased regions of soft tissue[9,10] show that a knowledge of the acoustic structure of such tissue on a scale somewhat finer than the wavelengths used clinically (1.5mm to 150μm) will bear directly on clinical diagnostic methodology.

THEORY

In the acoustic microscope the image data is a function of both the beamshape produced by the lenses and of the acoustic properties of the object. One method of developing a paraxial theory for the microscope, (due to Bennett[1]), described the beamshape of the lens with a zero order gaussian mode. The object properties were assumed to be those of a uniform liquid slab, similar in properties to the fluid coupling between the lenses. The object transmissivity was represented by a Taylor expansion truncated in the spirit of paraxial approximation above the 2nd order terms. It was found that the expressions for the image data were then identical to those for the case of a single, normally incident, plane wave propagating through the sample.

To examine the consequences of this theory let the impedance and longitudinal velocity of the object be z and v. These are normalised to the corresponding values for the lens to object coupling fluid. The microscope output has amplitude, A, and phase, ϕ, and the thickness of the object is d. The wavenumber in the lens to object coupling fluid is $k = 2\pi/\lambda$. The microscope output, under the conditions of the paraxial theory described above, is given by

$$A^2 = \frac{(2z/(z^2+1))^2}{(2z(z^2+1))^2\cos^2(kd/v) + \sin^2(kd/v)} \qquad (1)$$

$$\tan\phi = \frac{(z^2+1)}{2z}\tan(kd/v) \qquad (2)$$

In the above expressions A and ϕ are normalised to conditions when no object is present.

While the above expressions are only strictly true for normal incidence upon a uniform object, the paraxial theory has been found to be valid for soft tissue objects and focussed beams with half angles up to 30°. This implies the acoustic beam will have a diffraction limited waist of about one wavelength at the object plane. Thus the theory will remain valid for an inhomogeneous sample provided the properties of that sample do not change significantly on a scale comparable to a wavelength.

The main advantage of the paraxial theory is the relative ease with which equations 1-2 can be inverted to obtain the object properties as a function of the image data:

$$z = \frac{\sin\phi \pm \sqrt{1 - A^2}}{\sqrt{(A^2 - \cos^2\phi)}} \qquad (3)$$

$$v = \frac{kd}{\pm\cos^{-1}((\cos\phi)/A) \pm 2\pi m} \qquad (4)$$

where m is an arbitrary integer.

These expressions show that, for a particular amplitude and phase of the microscope output, there are many solutions for the object impedance or velocity. This is a consequence of adopting what is essentially a lossless transmission line model for propagation in the object. The ± signs in the expression for the velocity are due to the fact that an extra half wavelength section can be inserted in the model without changing the amplitude and phase at the microscope output. The delay through the object is increased by the extra section, but since only the modulus 2π value of the phase can be recorded in the microscope, this means that there is a series of possible values of v for an object of given thickness. It can also be seen that, by substituting $Z_o = 1/Z$ in equations 1-2 that there are two values of normalised impedance that can, for a given object velocity and thickness, give rise to a particular object amplitude and phase of transmission. This is why either sign of the square root can be taken in equation 3.

A further difficulty occurs when the object thickness approaches an integral number of half wavelengths since the amplitude and phase of the microscope output tend to 1 and 0 respectively. Substitution of A=1 and ϕ=0 in equation 3 gives an indeterminate division. This means that when the object thickness is close to an integral number of half wavelengths the microscope output provides no information on the impedance of the sample. In fact, it is possible to remove these ambiguities by making measurements at several different frequencies[1]. It is then possible to determine the impedance and velocity of the sample uniquely - but at the cost of substantially increased experimental complexity.

From the above considerations it can be seen that it is necessary to have target values for the object velocity and impedance when processing the image data, and hence some *a priori* knowledge is required about the object properties. The velocity and impedance computed from the experimental data that lie nearest to the target values are then assumed to be the correct values.

Since macroscopic measurements[7] in the literature suggest that, at least on a large scale, soft tissue acoustic properties do not vary by more than a few tens of percent it is important that the method has good sensitivity and accuracy. The differential forms of equations 3-4 relate the fractional error in computed velocity $\delta v/v$ and the impedance $\delta z/z$ to the experimental errors as

$$\frac{\delta z}{z} = \frac{(z^2+1)^3 A\delta\phi}{4z^2(1-z^2)\sin^2\phi} + \frac{(z^2+1)^3(1-A^2)\cos\phi\delta A}{4z^2(1-z^2)\sin^3\phi} \quad (5)$$

$$\frac{\delta v}{v} = \frac{-v.\sin\phi\delta\phi}{kd\sqrt{(A^2-\cos^2\phi)}} - \frac{Av.\cos\phi\delta\phi}{A\sqrt{(A^2-\cos^2\phi)}} + \frac{\delta d}{d} \quad (6)$$

These equations show that there are certain conditions under which the errors in reconstructed velocity and impedance rise dramatically. This occurs when either $\phi \to m\pi$, $A \to \cos\phi$ or when $z \to 1$. The first two of these conditions apply to both the velocity and impedance error. They occur when $(kd/v) \to m\pi$ and in this case the object is an integral number of half wavelengths thick. Examination of these equations suggests that careful selection of the impedance and velocity of the lens to object coupling fluid might help to minimise the experimental error bars. We chose here only to optimise the impedance since the velocity is restricted to be close to that of tissue by the paraxiality constraints. The minimal fractional error occurs when the sample is a quarter wavelength thick and then equations 5-6 conveniently become functions of impedance only:

$$\frac{\delta z}{z} = \frac{(1+z^2)^2 \delta A}{2z(1-z^2)} \quad (7)$$

$$\frac{\delta v}{v} = \frac{-2(z^2+1)\delta\phi}{2\pi z} \quad (8)$$

These expressions are plotted in Figure 1. Macroscopic measurements[7] suggest that soft liver tissue should have a velocity of around 1575 m/s and an impedance of approximately 1.65 MRayls. Water is a convenient and commonly used coupling fluid and for the liver parameters above would correspond to $z = 1.1$ and $v = 1.05$. However, it can be seen from the graph that this does not result in the best sensitivity. In fact a liquid such that $z = 2.25$ would be significantly better and on the whole it would be better to err to the higher rather than lower impedance side. The velocity should remain close to $v = 1$ to retain the paraxial approximations. This implies a low density liquid with a fairly slow acoustic velocity. One possible candidate is iso-pentane, $(CH_3)_2CH.CH_2CH_3$ (density 620 kg/m^3, velocity 1016 m/s) - this gives $z = 2.6$ and $v = 1.5$.

It is an interesting point that our criterion for minimum error is also, of course, a criterion for maximum useful contrast which we can exploit in normal microscopy.

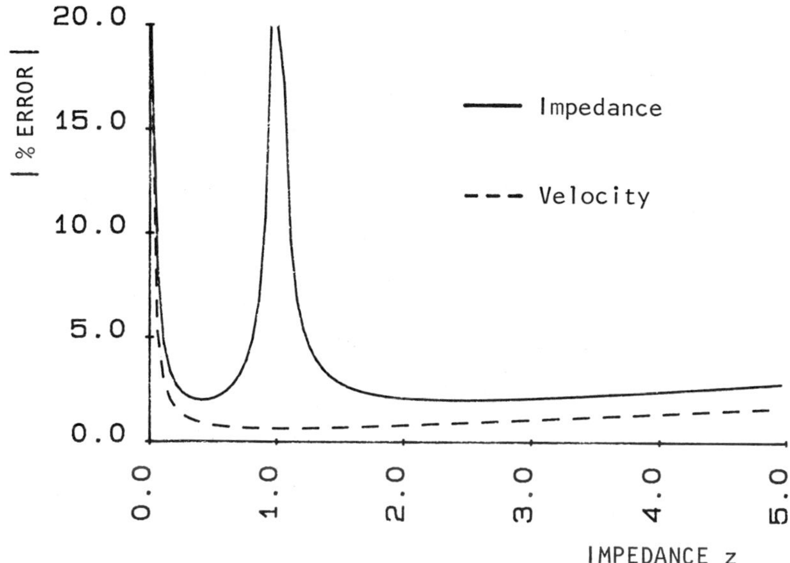

Fig.1. Variation in computed impedance error as a function of coupling liquid impedance.

An estimate of the expected error bars in the impedance and velocity values computed from the image data can be obtained in the following way. A certain object velocity, impedance and thickness is selected and the expected microscope A and ϕ are computed. A small perturbation is added to these values which are then input to the algorithm which computes the objects impedance and velocity, and the maximum difference between the computed and original impedance and velocity values is recorded. This procedure has been carried out for water and isopentane and the soft liver parameters above over a range of thicknesses, and the results are shown in Figure 2. The amplitude and phase errors were 1 in 100 and this is of comparable order to the experimental error in the images shown below. It can be seen that isopentane gives a significantly reduced spread of error in the reconstructed impedance and velocity.

EXPERIMENTAL RESULTS

We have examined a thin slice (420μm thick) of formalin fixed metastatic human liver in a conventional transmission SAM[3]. Coherent detection[11] was employed to record amplitude and phase images simultaneously at an acoustic frequency of 11.5 MHz. The scanning mechanism permitted removal of the object from the beam at regular intervals to allow a phase reference reading to be taken. This was used to compensate for velocity changes in the microscope due to temperature variations. This procedure also enabled the measured amplitude and phase, A_m and ϕ_m, to be normalised to the values obtained in the absence of the object, A_o and ϕ_o. The

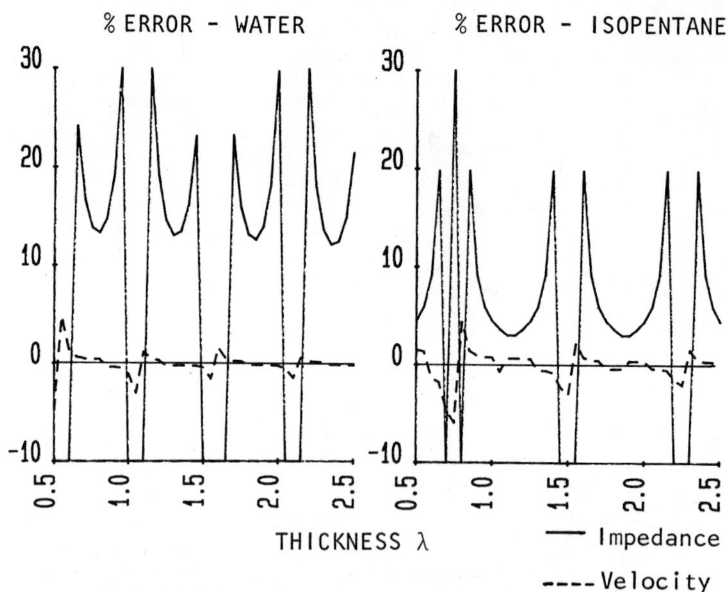

Fig.2. Maximum error in computed impedance and velocity as a function of sample thickness (scaled in coupling liquid wavelengths) for a 1% error in amplitude and phase.

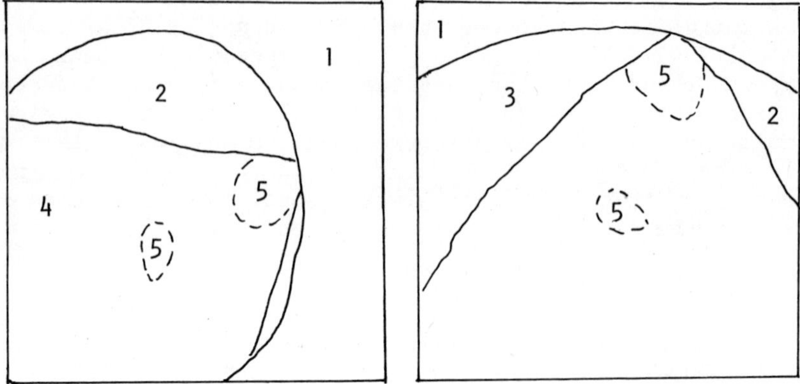

Fig.3. Schematic of the major components in the images in figures 4 to 7: (1) Aluminium mounting frame, (2) water between mylar films, (3) air bubble, (4) tissue section, (5) regions of abnormal pathology.

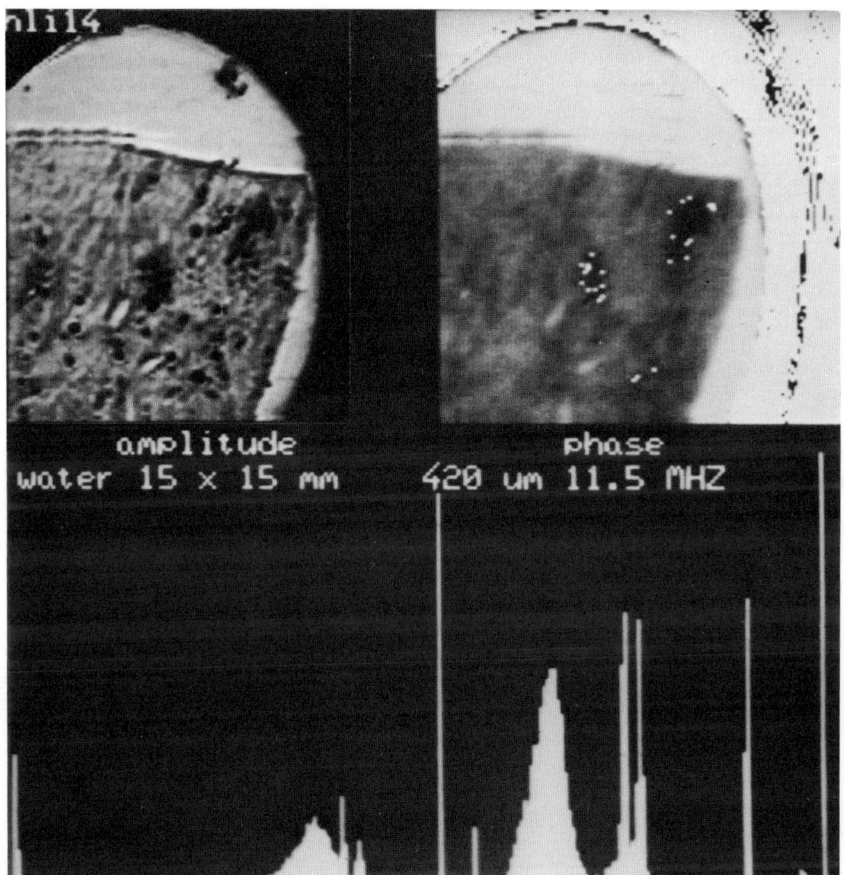

Fig.4. Amplitude (left) and phase images of a 420 μm slice of fixed metastatic liver with water coupling between the lenses and object. The field of view is 15 x 15 mm. A histogram of pixcell values is shown below each image.

normalised values are then given by

$$A = A_m/A_o \tag{9}$$

$$\phi = \phi_m - \phi_o + kd \tag{10}$$

Amplitude and phase images of the sample for water coupling between the lenses are shown in figure 4. Figure 5 shows the corresponding impedance and velocity map as determined from equations 3-4. In computing these images it has been assumed that z is always greater than unity. This seems a reasonable assumption since although soft tissue has a high water content generally, the solid material in the tissue would be expected to increase the velocity above that of pure

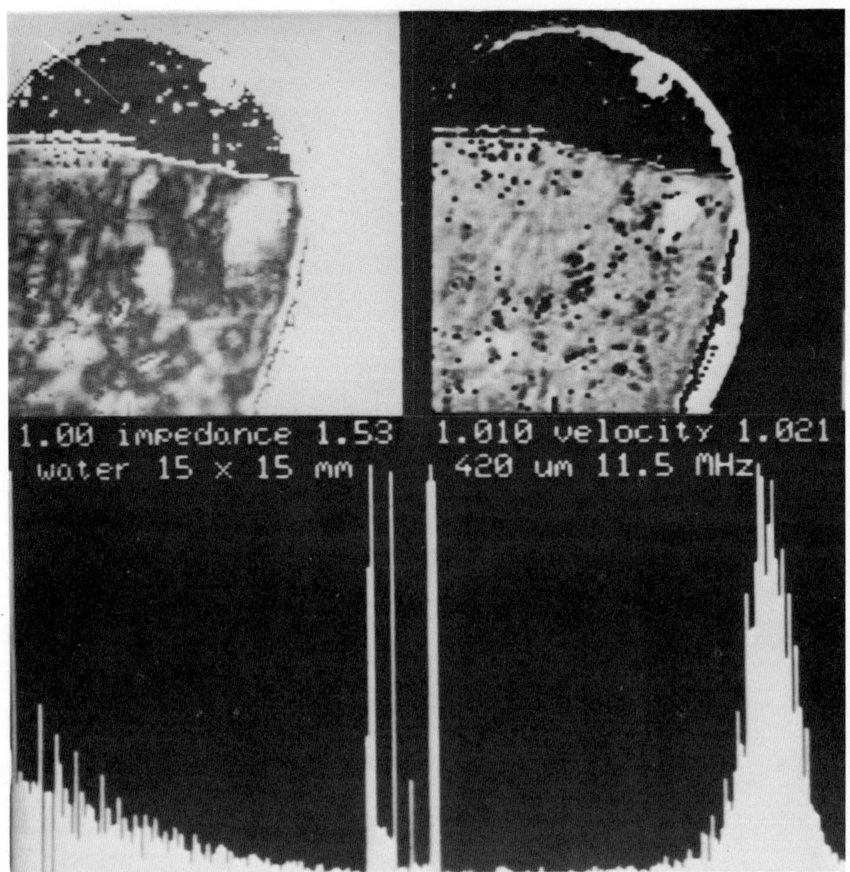

Fig.5. Impedance (left) and velocity maps computed from the images in figure 4. A histogram of pixcell values is shown below each image.

water. Regions of abnormal tissue ((5) in figure 3) show a high impedance and high velocity of approximately 2.25 MRayls and 1500 m/s. There are some locations in the images for which impedance and velocity values cannot be computed. For instance this can occur if $\cos\phi > A$. In these cases the z and v have been set to 1 in the impedance and velocity pictures and this is the explanation of the single pixcells dotted about the image that are drastically different from their neighbours. In addition there are some small holes in the sample and in a number of these an air bubble has become trapped - these can be identified by looking for low transmissivity in that region in the amplitude image.

Figure 6 shows amplitude and phase images of the sample for isopentane coupling between the lenses. Note that the sample is mounted between two thin (2μm) mylar films and the region between these films contains water. These films are held by a stainless

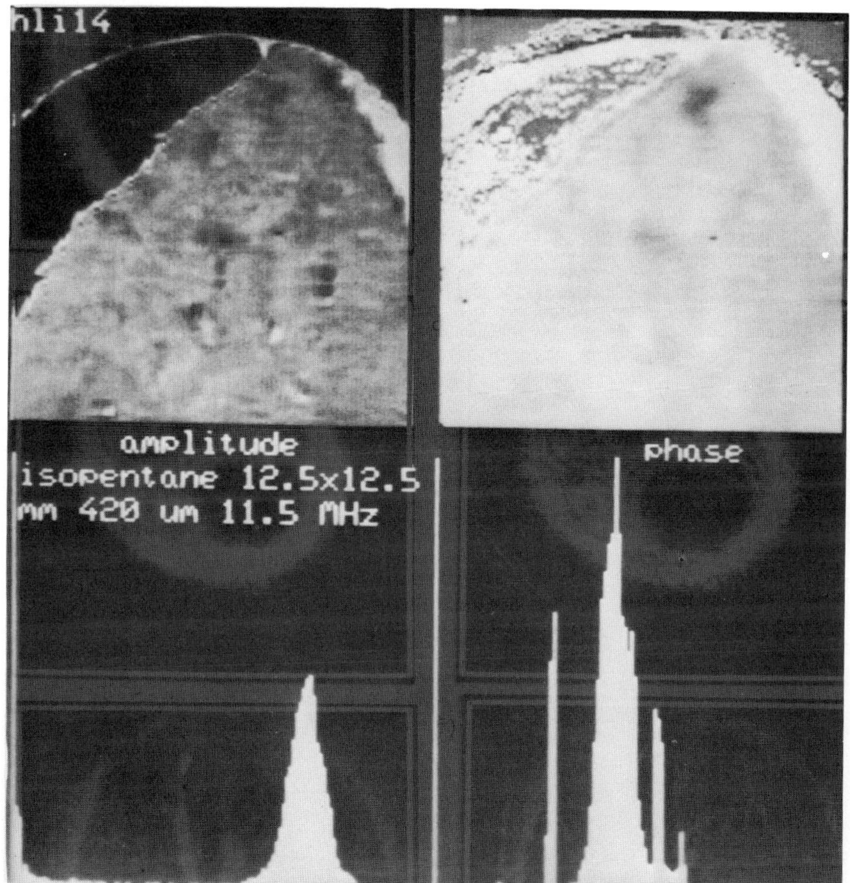

Fig.6. Amplitude (left) and phase images of a 420 μm slice of fixed metastatic liver with isopentane coupling between the lenses and object. The field of view is 12.5 by 12.5 mm. A histogram of pixcell values is shown below each image.

steel jig in which is provided a circular viewing window - the arc of which can be seen in the top region of each image. While this arrangement allows intimate acoustic coupling between the isopentane and the tissue, it prevents direct physical contact between the two. Isopentane will certainly dissolve the fats in the tissue and may also denature some of the protein, although this latter effect is not particularly important here since the sample is already fixed. Figure 7 shows the corresponding impedance and velocity map. The abnormal areas show up again as areas of high impedance (around 2.4 MRayls). The velocity in the tissue is again around 1500 m/s. However, in contrast to the water coupled results above, the abnormal regions show a lower velocity than the normal regions. The source

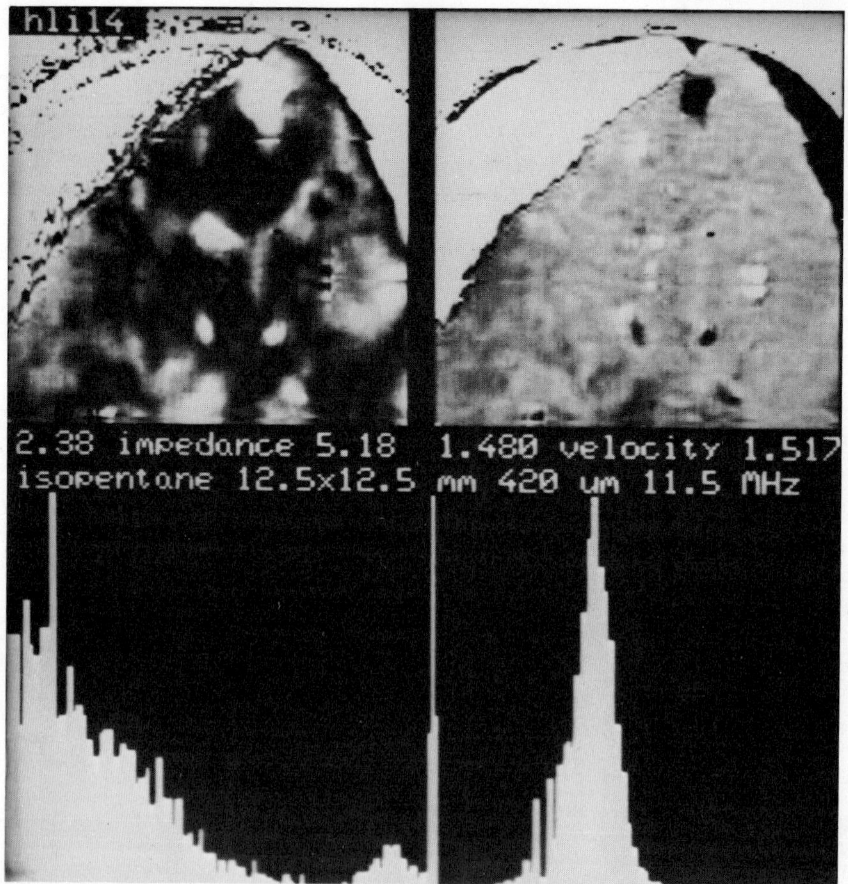

Fig.7. Impedance (left) and velocity maps computed from the images in figure 6. A histogram of pixcell values is shown below each image.

of this discrepancy is yet to be discovered - one possible cause is a selection of the incorrect value of m in equation 4 - this may possibly be avoided by a careful selection of the object thickness.

The error analysis of figure 2 suggests that the accuracy of the impedance and velocity measurements is of the order of 15% and 3% respectively in the case of the water coupling, and 4% and 2% in the case of isopentane.

DISCUSSION

The above images show large variations in impedance that are significantly greater than those reported in macroscopic experiments in the literature. Some of these variations may be accounted for by the omission of scattering and loss from the theory. The spread of velocity is very much less and this is a consequence of the fact that the value of m in equation 3 may be chosen at will to select the nearest possible computed value of velocity to the target value. This restricts the range over which the computed velocity may vary. If the velocity is to be correctly deduced, then a more sophisticated algorithm will have to be developed and this will probably involve taking into account information from adjacent pixcells. A reduction in the value of d will also help reduce the ambiguity in m. However in the single sample examined here the regions of abnormal pathology did consistently show a significantly higher impedance than the surrounding tissue.

ACKNOWLEDGEMENTS

We would like to thank E.A. Ash, H.K. Wickramasinghe and S.D. Bennett for many useful suggestions. We are particularly grateful to C.R. Hill and D. Nicholas for their helpful discussions. This work was supported by the Medical Research Council. D.A. Sinclair would like to thank the Rank Prize Funds for the award of a Research Fellowship.

REFERENCES

1. S. D. Bennett, "Coherent Techniques in Acoustic Microscopy", PhD Thesis, University of London (1980).
2. S. D. Bennett, "Approximate Materials Characterisation using Coherent Acoustic Microscopy", (1981) unpublished.
3. D. A. Sinclair and I. R. Smith, "Tissue Characterisation using Scanning Acoustic Microscopy", Proc. 6th Int. Symposium on Ultrasonic Imaging and Tissue Characterisation", Gaithersburg, Maryland, USA (1981), published in Ultrasonic Imaging 2:4 p44-46 (1981).
4. D. A. Sinclair, I. R. Smith and H. K. Wickramasinghe, "Recent Developments in Scanning Acoustic Microscopy", The Radio and Electronic Engineer, 52:10 (1982).
5. C. F. Quate, A. Atalar and H. K. Wickramasinghe, "Acoustic Microscopy with Mechanical Scanning - A Review", Proc. IEEE SU 67:8, p 1092-1114 (1979).
6. E. C. Gregg and G. L. Palagallo, "Acoustic Impedance of Tissue", Invest. Radiol. 4, p 357-363 (1969).
7. J. P. Jones, "Impediography: a new ultrasound technique for Diagnostic Medicine", in "Ultrasound in Medicine", ed. D.N. White, 1, 489-97, Plenum Press (1975).
8. S. A. Goss, R. L. Johnston and F. Dunn, "Comprehensive compilation of Empirical Ultrasonic Properties of Mammalian Tissues", JASA 64:2, p 423-457 (1978).
9. C. R. Hill, R. W. Huggins and D. Nicholas, "Scattering of Ultrasound by Human Tissues", in "Ultrasound: its Applications in Medicine and Biology", ed. F.J. Fry, ch 9, Elsevier (1978).

10. F. Lizzi and D. J. Coleman, "Ultrasonic Spectral Analysis in Opthalmology", in "Recent Advances in Ultrasound and Biomedicine", ed. D.N. White, 1, Research Studies Press (1977).
11. S. D. Bennett and E. A. Ash, "Differential Imaging with the Acoustic Microscope", IEEE Trans. on Sonics and Ultrasonics, SU 28:2 (1981).

ULTRASONIC DOPPLER VESSEL IMAGING IN THE DIAGNOSIS OF ARTERIAL DISEASE

S.J. Calil, J.C. Graham, V.C. Roberts

Biomedical Engineering Department
King's College Hospital Medical School, LONDON SE5

INTRODUCTION

The need for a reliable method of investigating patients at risk from vascular disease, particularly that which affects the arteries which supply the head, has resulted in the development of many instruments based on the use of Doppler shifted ultrasound. These devices range from the simplest non-directional CW Doppler velocimeters to the complex multigated pulsed Doppler imaging systems (Fish, 1975).

In the application of the various techniques to the investigation of the cerebrovasculature, most of the methods have proved relatively insensitive to minor degrees of stenosis which are nevertheless likely sites of thrombus initiation. Small degrees of intimal thickening lead to flow disturbance and vibrations in and near the vessel wall. Early work by Lee et al (1970) on the detection of these vibrations lead others to investigate the possibilities in more detail (Foreman et al, 1970; Miller et al, 1980). Techniques developed so far, have been based on phonoangiography. Although useful, they can fail to detect minor degrees of stenosis because of tissue interference with the acoustic transmission of the vibrations (Fredberg et al, 1974). The majority of investigators have only been able to detect the effects of flow disturbances downstream from stenoses of greater than 50% area reduction (Kirkeeide et al, 1977; Young et al, 1973; Duncan et al, 1975). A few in vitro studies have investigated the possibility of assessing lower degrees of stenosis (Khalipha et al, 1981; Cassanova et al, 1978).

The purpose of this present investigation has been to evaluate the potential of using a range-gated pulsed Doppler velocimeter to study the flow disturbances caused by stenoses which produce effective area reductions of as little as 20%. The application of Doppler ultrasonic velocimetry to the qualitative and quantitative assessment of disturbances of blood flow in the arteries of the human body is, however, limited by many factors. These include the signal processing system used, the direction of the ultrasonic beam with respect to the blood vessel, the ratio of ultrasonic beam to vessel width and the method and corrections used for the final evaluation of velocity, flow or vessel calibre. These present investigations were therefore in two parts:

(1) An assessment of the ability and accuracy of a pulsed Doppler vessel imaging system to measure vessel calibre, volume flow and velocity under steady and pulsatile flow.
(2) An assessment of the flow disturbances caused by stenoses both in vitro and in vivo.

EVALUATION OF THE SYSTEM

Ultrasonic imaging velocimeter:

The pulsed Doppler system used (MAVIS, Picker International) has an insonation frequency of 4.8 MHz with a focal length of 3.0 cm. The live beam is divided into 30 adjacent sample gates each of approximately 0.64 mm and the pulse repetition rate of the instrument is 5 kHz. Its depth can be step varied each 10.24 mm. up to a maximum of 71.68 mm.

A vessel can be imaged by moving the probe over the skin surface and when the detected Doppler signal exceeds a pre-set amplitude and frequency an image is produced on a colour TV monitor screen. The position of the beam on the screen is related to the position of the artery by a probe position computer. To distinguish between flows toward and away from the probe, two different colours are used. A further colour is and to show areas where flow is occurring in both directions.

For flow calculation a set of operational procedures is followed to obtain the beam/vessel angle. For the final result the instrument takes an average of up to 32 cardiac cycles. After calculations and corrections performed by an internal microcomputer, the mean flow is displayed on a second screen, together with the mean velocity, vessel width, beam vessel angle and the pulsatility index. It also displays an averaged and corrected graphic of the flow velocity during an entire cycle and the cross sectional velocity profile. This latter can be displayed each 25 msec of the cycle.

Hydraulic test rig:

The flow rig consisted of a recirculating system with liquid being pumped from the downstream reservoir to an upper reservoir. An in-line pneumatic flow inducer provided pulsatile flow at up to 82 beats per minute. A by-pass system was used to provide steady flow with the flow inducer turned off. This bypass was also used to control the pulse wave shape in the test section, and the flow wave could thus be continuously varied from steady to pulsatile, the latter with forward and reverse components of variable magnitude. A Penrose silicone rubber tube 25 cm long, 6mm ID (wall thickness 0.02mm) and suspended in a water tank was used for the first stage calibrations.

For all the tests reported here, a Newtonian fluid was used composed of 5% of silcolapse 5000 and 95% of distilled water. The Reynolds number ranged, for steady flow, from 424 to 1516 and for pulsatile flow from 471 to 1323. Due to the tendency of the silicone to settle or to agglutinate, the emulsion was constantly filtered to prevent particles bigger than 40 μm entering the test section. The ultrasonic probe was held absolutely steady by means of a mechanical device which allowed fine probe position adjustments in three orthogonal planes. All mean indicated flows were compared against a timed collection.

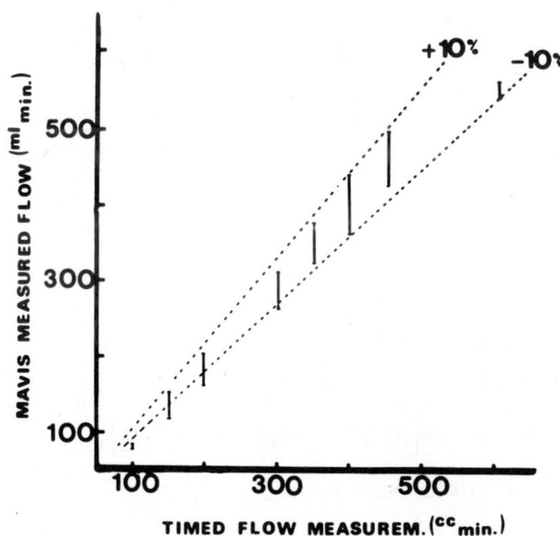

Fig 1 Calibration under steady flow conditions.

Experimental results:

As can be seen from Figure 1, over the range studied, the majority of the measurements are within ±10%. During the course of this investigation we also noticed that if care is not taken with the angle of ultrasonic incidence when measuring high velocities, aliasing can cause very distorted velocity and width evaluation. However, these errors appear not to be reflected in the volume flowrate estimates, a degree of self cancellation being apparent.

For pulsatile flow with no reverse component 33% of the points were outside the ±10% range (r = 0.91). For pulsatile flow with a reverse component of about 30% of the cardiac pulse, the results became distorted, 61% of the points lying outside the ±10% tolerance range as shown in Figure 2. The correlation coefficient of this calibration dropped to 0.88.

To investigate whether vessel wall movement was the main cause for this increased variability we used a nylon tube having substantially the same ID but with a 1mm wall thickness. The results were worse than those indicated in Figure 2 under pulsatile flow conditions, the correlation coefficient being only 0.76. However, under steady flow conditions the correlation between actual and indicated flow fell within the ±10% range indicated by Figure 1.

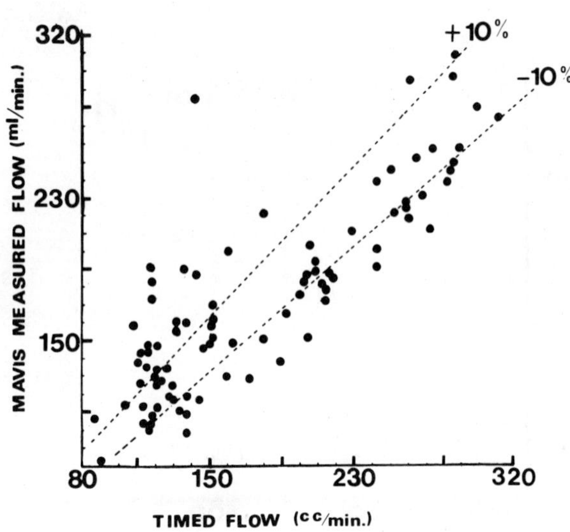

Fig 2 Calibration with pulsatile flow (with reverse component).

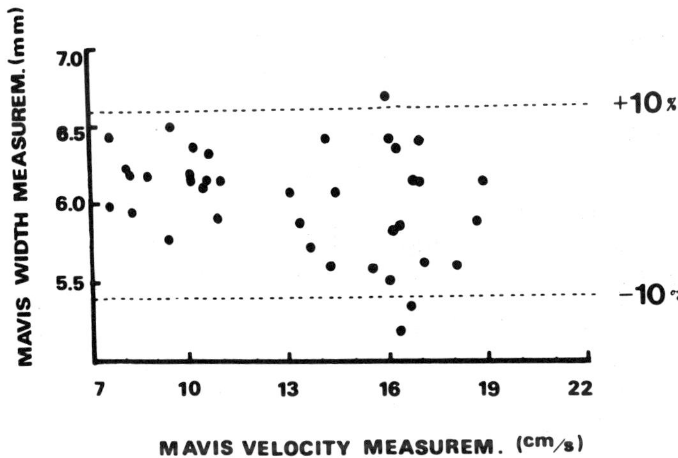

Fig 3 Variation of width estimation with flow velocity.

Further investigations were made of the sensitivity of the indicated width measurement to changes in mean velocity. The results obtained under pulsatile flow conditions are shown in Figure 3 and indicate that the measurement of width, like volume flow rate, lies within ±10%. However, it was also found that this accuracy held for pulsatile flow with a negative component.

The final stage of the system calibration was to repeat the tests using an umbilical vein graft of 5.5mm calibre in the test section. The results were better than all the previous ones under steady and pulsatile flow conditions. For the latter the correlation coefficient rose to 0.98 and only 16% of the points were outside the ±10% limits.

The results obtained during this first stage of the investigation indicate that the accuracy of MAVIS is within an acceptable range for it to be of clinical use in situations where no negative flow components are to be found. Furthermore, investigations have indicated that under conditions of disturbed flow, the computed values of velocity and volume flowrate may be seriously in error, the magnitude of the error depending on the degree of flow disturbance.

EXPERIMENTAL APPARATUS AND PROCEDURE FOR WALL MOVEMENT INVESTIGATION

The apparatus used for the detection of flow disturbance and vessel wall movement is as described above. Stenotic models were constructed in three different forms, each of the three shapes being constructed with a different degree of area reduction varying from 15% to 75%:

(a) Axisymmetric smooth contoured stenoses of rigid acrylic
(b) Axisymmetric sharp edged stenoses of rigid acrylic
(c) Nonsymmetric smooth contoured stenoses of soft vinyl

The axisymmetric stenoses were constructed by machining a solid acrylic rod to the required shape and area reduction and then inserting and fixing the stenosis into the length of Penrose tubing at its centre point. The nonsymmetric models were made by casting a vinyl mould compound (Vinamold) around a brass form and inserting the stenosis as before and glueing it in place.

Experimental procedure:

Having the facility of 3-D imaging, a lateral view of about 5cm of vessel was built up around the stenotic area to find out if any flow disturbance could be detected. The Doppler probe was then positioned over the flow test section and traversed in a plane perpendicular to the flow axis in order to build an image of the cross section. Having established this the probe was then moved so that the ultrasound beam intersected the centre of the cross section. By scanning through the 30 channels of the pulsed Doppler and searching for those which are located at the boundaries of the vessel, it was possible to obtain just the signals from sites near or at the vessel wall.

The output Doppler shift signals normally available from MAVIS are high pass filtered below about 200 Hz. However, a significant proportion of the turbulent spectrum is likely to occur at low frequencies. Accordingly, an external amplification and band pass filtering system had to be used to obtain frequencies ranging from 20-800Hz and the signals obtained prior to the normal filter networks. The output of this system was then fed to a DEC 11/40 computer through an analog to digital interface. The signal coming out of each MAVIS sample gate close to the wall of the test section was sampled at 2.56 kHz and the data train broken into 12 complexes of 256 points. Each complex was subjected to Fourier analysis and the 12 resulting data blocks used to produce the ensemble average. The data is presented as log/log plots of energy density against frequency.

Experimental results:

We present in this study a representative selection of the results obtained from the stenotic models under steady flow conditions. The models used were axisymmetric and nonsymmetric with smooth contours. The Reynolds number for all of the experiments was kept around 1900, which is just below the turbulence transition point for the whole flow stream.

In Figure 4 are shown the power spectra obtained from the non-symmetric stenoses. In this figure, as in those which follow, position 1 is 1cm upstream of the stenosis position, 2 at the downstream edge of the stenosis and position 3 is 1cm further downstream from position 2. Figure 4a shows the power spectra obtained from the 50% stenosis. As can be seen there is an increase in power of over 14db at the low frequency end of the spectrum produced immediately downstream of the stenosis. That the turbulent energy is propagated downstream of the stenosis is shown by the fact that the low frequency power is still almost 3db greater than the pre-stenotic level.

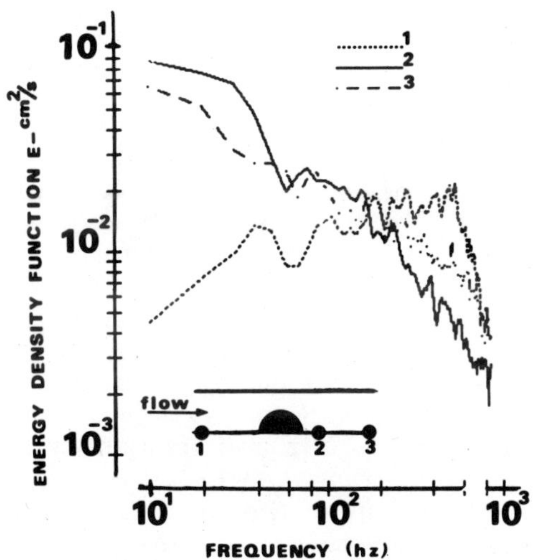

Fig 4a Power spectra obtained from a 50% non-symmetrical stenosis.

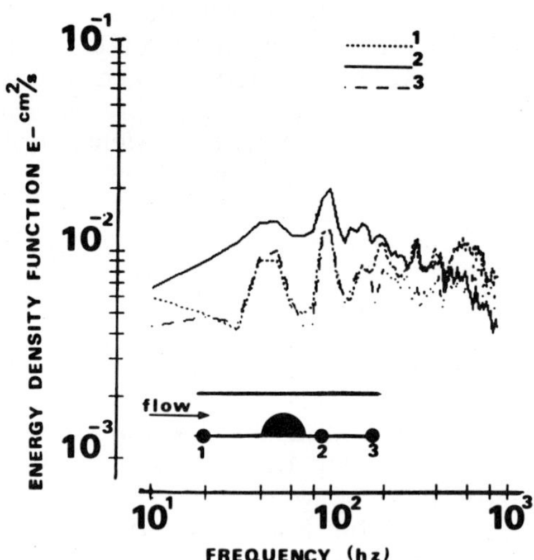

Fig 4b Power spectra obtained from a 20% non-symmetric stenosis.

Figure 4b shows the corresponding spectra for the 20% stenosis. Although the signal power is smaller at the post-stenotic position 2, it is still some 3db greater than the pre-stenotic level, with the same overall pattern of a fall off at the higher frequencies. In this case, however, the power levels up and downstream of the stenosis are substantially the same, indicated that the turbulent fluctations have disappeared within just over 1 tube diameter from the obstruction. It is also possible to note the presence of system noise in these signals (substantial components at 50Hz and its harmonics).

Figures 5a and 5b show the results obtained from 50% and 30% rigid axisymmetic stenoses. In these figures we can see that the signal power immediately downstream of the stenosis is elevated at low frequencies and that the increase in low frequency energy is greatest for the higher degree of stenosis. In the case of the more severe stenosis the turbulent power at the highest frequencies is greatest downstream from the stenosis.

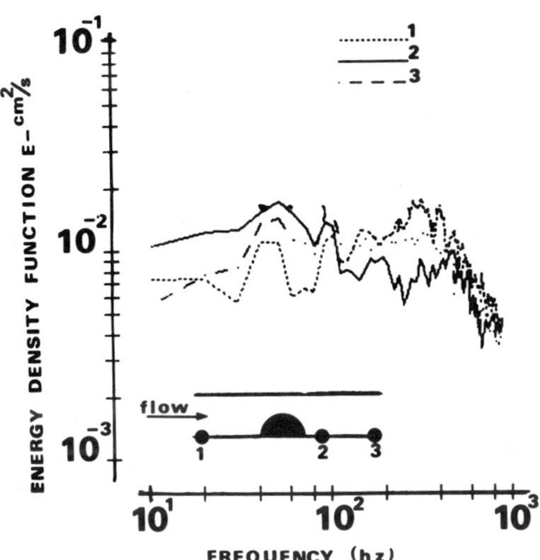

Fig 5a Power spectra obtained from a 50% rigid axisymmetric stenosis.

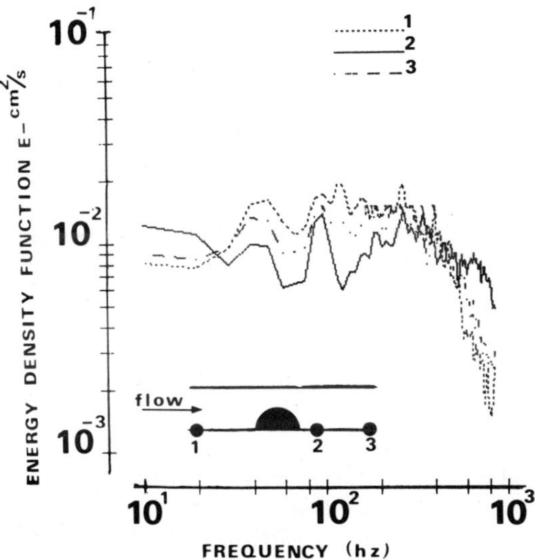

Fig 5b Power spectra obtained from a 30% rigid asymmetric stenosis.

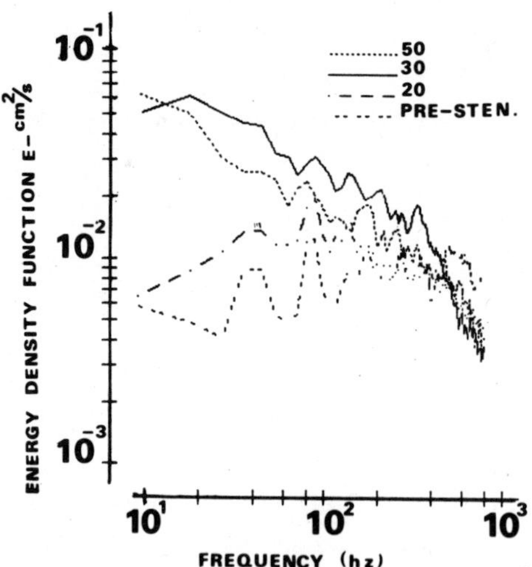

Fig 6 Composite power spectra for 20%, 30% and 50% non-symmetric stenoses.

In Figure 6 we see a composite showing the energy spectra obtained for the non-symmetric stenosis for 50%, 30% and 20% stenosis from the position immediately downstream of the stenosis. Also superimposed is the pre-stenotic spectrum for the 20% stenosis (approximately equivalent to the unstenosed spectrum). This Figure illustrates very clearly the effects of increasing stenosis and shows the sensitivity of the analysis methods in detecting small degrees of stenosis - particularly by inspection of the low frequency components of the spectrum.

CLINICAL APPLICATION

Figure 7 shows the spectra obtained close to a small defect in the vessel wall of a section of common iliac artery which was tested under steady flow conditions in the hydraulic rig. The lower of the two curves is the intrinsic noise in the Doppler processing system (note the components at 50Hz and its harmonics). The solid line is the spectrum obtained at the wall of the vessel close to the lesion. The amplitude of the signals are again particularly elevated at the low frequency end of the spectrum.

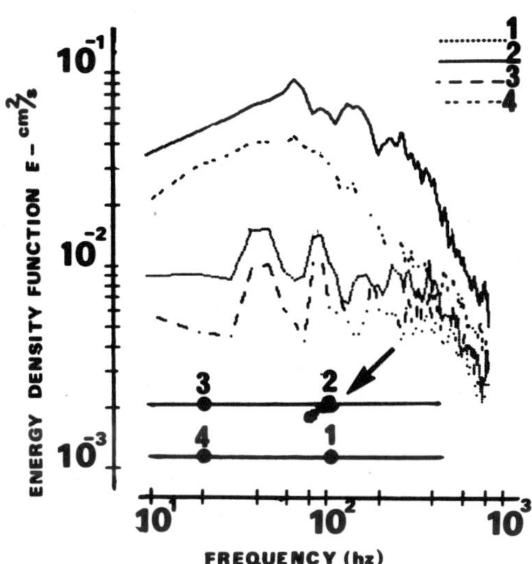

Fig 7 Spectra obtained just downstream of a defect in the wall of a section of external iliac artery.

CONCLUSION

We have evaluated the ability of an ultrasonic vessel imaging system to measure velocity, volume flowrate and vessel calibre under conditions of steady and pulsatile flow. Our data indicates that under flow conditions where there is no reverse flow component, the assessment of vessel calibre and volume flowrate has a likely accuracy of ±10%. However, under conditions of reverse flow, which is found in much of the human vasculature and certainly in the presence of disease (causing local turbulence), the estimates are likely to be more seriously in error. This error is a function of the data manipulation carried out within MAVIS and indicates the need for corrective reprogramming of the system.

The imaging capability of MAVIS is, however, unaffected by the presence of reverse flow and turbulent phenomenon can clearly be demonstrated. However, our work indicates that the unambiguous image resolution of flow disturbances caused by small degrees of stenosis is not yet possible. In this situation the frequency analysis of the low level signals close to the vessel wall provides a powerful means of detection. The presence of a higher signal level for points more distal from stenosis is evidence of the effects generated by a confined jet, i.e. a spectral peak which increases in intensity as the flow progresses downstream (Cassanova et al, 1978). Indeed, our results suggest that the frequency analysis technique is electronically sensitive to stenoses of less than 30% - the very area where all other methods fail. Further work is now needed fully to evaluate the potential of this method in vivo.

REFERENCES

Cassanova, R.A., Giddens, D.P. Disorder distal to modelled stenoses in steady and pulsatile flow. J. Biomech. 11: 441-453, (1978).

Duncan, G.W., James, O.G., Dewey, C.F., Myers, G.S., Lees, R.S. Evaluation of carotid stenosis by phonoangiography. New Eng. J. Med., 13: 1124-1128, (1975).

Fish, P.J. Multichannel direction resolving Doppler angiography. Proc.2nd Europ.Congress Ultrasound in Med. Exerpta Medica, Amsterdam, 153-159, (1975).

Foreman, J.E.K., Hutchison, K.J. Arterial wall vibration distal to stenosis in isolated arteries of dog and man. Circ. Res. 26: 583-589, (1970).

Fredberg, J.J. Pseudo sound generation at atherosclerotic constriction in arteries. Bull. Math. Biol. 36: 143-155, (1974).

Kalipha, A.M.A., Giddens, D.P. Characterisation and evolution of post stenotic flow disturbances. J. Biomech. 14: 279-296, (1981).

Kirkeeide, R.S., Young, D.F. Wall vibrations induced by flow through simulated stenoses in models and arteries. J. Biomech. 10: 431-441, (1977).

Lee, R.S., Dewey, C.F. Phonoangiography: a new non-invasive diagnostic method for studying arterial disease. Proc. Nat. Acad. Sci., 67: 935-942, (1970).

Miller, A., Lees, R.S., Kistler, J.P., Abbott, W.M. Spectral analysis of arterial bruits (phonoangiography): experimental validation. Circulation, 61: 515, 520 (1980).

Young, D.F., Tsai, F.Y. Flow characteristics in models of arterial stenoses - I, steady flow. J. Biomech., 6: 395-410 (1973).

ACKNOWLEDGEMENTS

The authors are indebted to the Brazilian National Council for the Development of Science and Technology (CNPq) and the Wates Foundation for their support with these investigations.

REAL-TIME TWO-DIMENSIONAL BLOOD FLOW IMAGING USING A DOPPLER ULTRASOUND ARRAY

James W. Arenson, Richard S.C. Cobbold, K. Wayne Johnston

Institute of Biomedical Engineering
University of Toronto,
Toronto, Ontario, M5S 1A4, Canada

INTRODUCTION

It is generally agreed that a substantial percentage of strokes result from arterial stenosis in the region of the carotid bifurcation. Current ultrasound methods to detect and assess the severity of carotid stenoses are based either on imaging, Doppler signal analysis, or a combination of the two. Of the imaging methods, three modalities can be identified: B-mode, pulsed Doppler, and continuous wave (CW) Doppler. The limitations of each method may be partially overcome either by using combinations of the imaging modalities or by combining one modality with Doppler signal analysis.

Imaging by B-mode provides a structural view of the vessel wall and surrounding tissue with intensities that depend on the scattering that occurs at the various interfaces. A normal vessel lumen remains echo free due to the relatively small amount of scattering by blood. If the artery is completely occluded by thrombosed blood, the image may be substantially unchanged. Moreover, while calcified plaque in the vessel wall may be clearly imaged, a soft fatty plaque, whose acoustic properties are similar to the arterial wall may not be distinguishable. To overcome these problems, most commercial B-mode carotid scanners incorporate a Doppler subsystem to enable flow velocity estimates to be made at selected sites in the image. Specifically, Doppler recordings detect the increased frequencies which occur at the site of a stenosis and/or the disturbed flow which is present beyond the narrowing.

Barber et al. [1] described a duplex system, in which a pulsed Doppler transducer attached to the B-mode scan head enabled flow velocities to be measured in manually selected regions of the sector scan image. Significant further improvements [2,3] enable color coded blood velocity information to be superimposed on the B-scan image. A somewhat similar system but using a linear B-scan and an electro-mechanically positioned Doppler probe was described by Green et al. [4]. A scheme that uses a compound B-scan head in a fluid filled cavity with a pliable membrane has also been described [5].

Although duplex scanning can potentially provide both structural and flow information, the system is complex and expensive. Moreover, the imaging procedure may be fairly difficult due to a bulky transducer and registration problems between the B-scan and the Doppler beam. To alleviate these problems, a number of groups have demonstrated imaging systems using Doppler exclusively, in which the flow velocity is represented by either the intensity or the color on a CRT display. In the system described by Hokanson et al. [6] and improved by Miles et al. [7], a single pulsed Doppler probe is mounted on an articulated position sensing arm to map out either lateral or multiple cross-sectional views of the flow velocity. Fish [8] described a multigate pulsed Doppler system which uses parallel processing from a large number of adjacent range positions. Also using an articulated arm, but employing CW Doppler, Reid et al. [9] produced lateral views of projected blood flow. Curry and White [10] improved on this by using three colors to display three ranges of peak flow velocities. Using a real-time frequency analyzer Coghlan and Taylor [11] extended this to sixteen velocity-coded grey scale levels.

In lateral imaging, the pulsatile nature of the blood flow generally makes it necessary for, each point in CW systems, or each line for multichannel pulsed Doppler displays, to be measured over an entire cardiac cycle. Since there is a finite rate at which a Doppler probe can be manually moved without introducing large Doppler artifacts from stationary targets, all the Doppler imaging systems described above require long scanning times. Typically, a bilateral carotid study can take from 20 to 60 minutes, during which time the patient must remain still.

Hottinger and Meindl [12] suggested a rather complex multi-element array approach to overcome the problem of scanning time. Their array orientation would have resulted in perpendicular insonnation of the carotid arteries, thereby imaging wall motion rather than blood flow.

GENERAL APPROACH

This paper presents a new approach to Doppler imaging using a linear stepped array that insonnates the carotid arteries at an optimal angle and which should achieve a major reduction in scan time. The basic principles of the array [13] can be summarized, with the help of Fig. 1, as follows. The crystals on the array are all canted at approximately 30^0 from the face of a normal linear array, producing a stepped linear array with a step height equal to an integral number of wavelengths. This arrangment allows multiple elements to be driven in parallel to form a well collimated beam, and yet steers the beam through the required 30^0 while using only a small number of relatively large crystals with simple

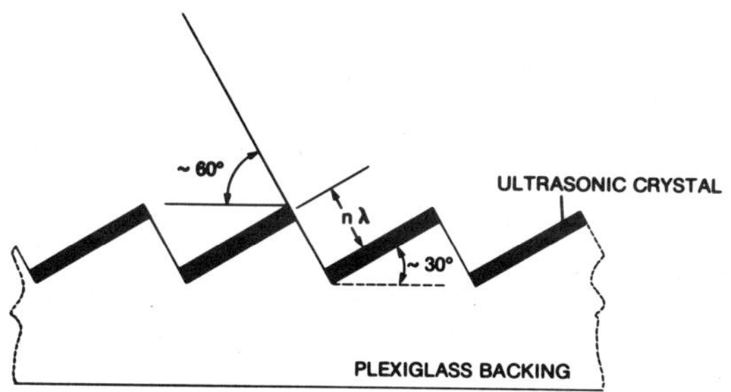

Figure 1. Crossectional view illustrating the basic principles of the stepped array construction.

electronic phasing. Further, this linear arrangement of crystals allows for a simple electronic implementation for driving and receiving. An identical crystal combination can be incrementaly addressed through the array, providing a complete line of information. Deflecting the beam with a long rotating mirror located in front of the array allows a two-dimensional area to be scanned.

The decision to use a long pulse CW mode of operation, rather than a short pulse range gated mode, was based on several considerations. It was felt that although additional depth information could be obtained with a pulsed system, the increased system complexity, longer scanning times, the more difficult clinical procedure, and the problem of interpretation, would not justify its use. The pulse length chosen (~0.8 ms) was sufficiently long so that the significant Doppler frequencies could be determined, and short enough to achieve real-time imaging.

Fig. 2 illustrates the manner in which an image is obtained. Each successive frame represents the display at a later instant of time. A line of information is first imaged by sequentially scanning down the array, and then the mirror is rotated through a small angle to allow an adjacent line to be imaged.

TRANSDUCER DESIGN

The size of the imaging window (~2x6 cm) was chosen to be large enough to view an entire carotid bifurcation, and small enough to be sensed by a small portable hand-held transducer.

It was decided that the prototype beamwidth should be less than 5 mm, consistent with that measured on a commercial linear B-scan machine capable of imaging carotid arteries, and with the imaging systems described in the introduction. Using a computer aided design program [14], it was concluded that the six crystal grouping shown in Fig. 3 represents the minimum number of crystals that can realize the required beamwidth and beam orientation while using CW Doppler. The two middle crystals

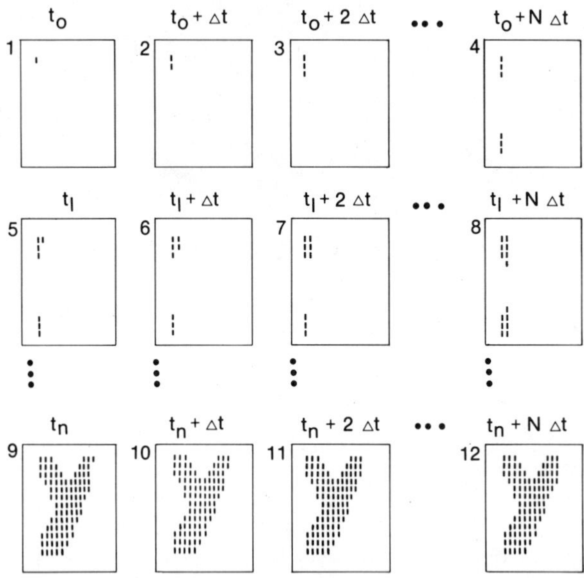

Figure 2. Sketch showing the manner in which a complete frame is generated to produce an image of a carotid bifurcation: N= number of crystal groups, n= number of rotations of the mirror, Δt= time per step down the array.

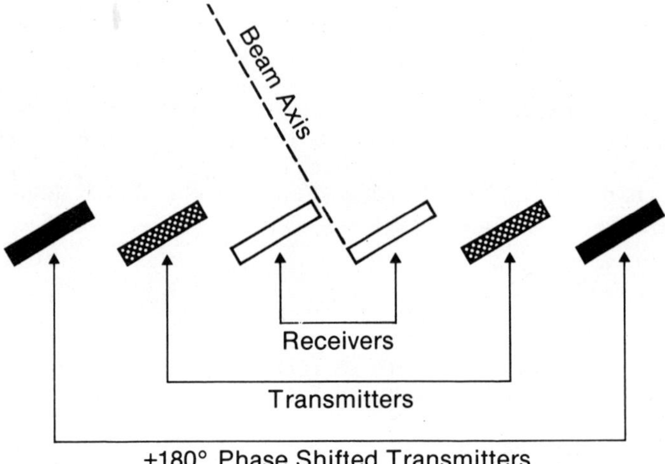

Figure 3. Crystal grouping arrangement chosen for the initial array design. Receiver phasing is not used, but the transmitter pairs are phase shifted by 180 degrees to produce some degree of focussing.

are receivers and the four outer crystals are transmitters. The two outermost transmitters are phased by $+180^0$ relative to the inner transmitters in order to steer the ultrasonic beam towards the beam axis thereby improving the focussing.

The array consists of thirty-two PZT-5 5 MHz crystals, each 2x8 mm in size. They are individually mounted on a machined plexiglass backing with a four wavelength step height. The array is in turn mounted within a small water-filled cavity, with a stainless steel mirror located in front. A window machined in the watertight casing is covered with a thin molded Spandex membrane to couple the transducer to the often difficult contours of the neck.

A small stepping motor rotates the mirror through ten 1^0 steps. The motor and mirror are connected by a three-bar linkage specially designed to linearly transform the motor's 7.5^0 steps into 1^0 mirror rotations.

SYSTEM DESIGN

Fig. 4 shows a system block diagram for the two-dimensional Doppler imager [15]. The thirty-two crystals are wired to three identical 32 channel multiplexers in parallel. These three multiplexers connect the 0^0 and 180^0 drivers and the receiver to the proper group of 6 crystals. The timing and switching controller incrementally steps this switching scheme down the array, and at the same time generates the Y-axis sweep for the CRT display. At the end of each sweep, the controller steps the motor and increments the X-axis on the display. After 10 lines have been imaged the motor direction and X sweep reverses.

Under ideal circumstances, the output of the receiver amplifier should connect directly to the Doppler demodulator to detect the Doppler shifted signals. In the case of simple Doppler systems using a dedicated receiver and transmitter crystal pair, blood flow signals are typically -74dB to -80dB relative to signals reflected from stationary targets. This small signal-to-noise (SNR) ratio is just within the limited dynamic range of standard Doppler demodulators. Unfortunately, with an array of crystals, this very small SNR is further reduced by additional acoustic and electronic feedthrough, and noise. Furthermore, the physical differences in the crystal groupings relative to each other as well as to insonnated stationary targets give rise to variations in the RF signal amplitude and phase. If this signal were fed directly to a standard Doppler demodulator, even assuming it had sufficient dynamic range, the output would saturate due to these large signal variations between crystal groups. To avoid having to use 27 separate demodulators that are tailored to each crystal grouping, an RF preprocessor was designed. It conditions the received signal so that a single Doppler demodulator sequentially detects the flow velocity signals from each crystal group.

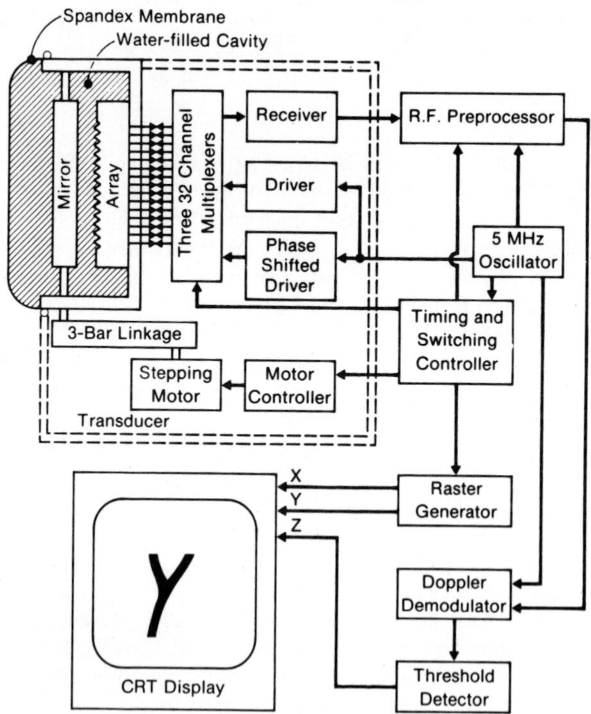

Figure 4. Block diagram of the system used for the initial prototype.

The preprocessor shown in Fig. 5 attempts to establish a constant amplitude and phase for the RF signal detected from each receiver pair in the array. Any variations from a reference amplitude level and phase shift are corrected by a voltage controlled amplifier and voltage controlled phase shifter respectively. This is done by monitoring and correcting the amplitude and phase of the incoming signal for a time corresponding to that required for receiving a reflection from just inside the tissue. From this time on the correction voltages controlling the amplifier gain and phase shifter are held constant, allowing the small Doppler signals to pass through with no correction. The large 5 MHz carrier signal is removed from this amplitude and phase corrected signal by simple subtraction. This signal is then demodulated with a simple mixer. Any errors in the automatic gain and phase controllers become apparent as dc offsets after demodulation. These small offsets are removed by a zero-voltage restoring circuit, allowing the train of Doppler signals from the array to be further amplified.

A level detector establishes when to brighten a display pixel. An increase in either the amplitude or frequency of the Doppler signal causes the level detector to be activated for a longer time. Thus, the

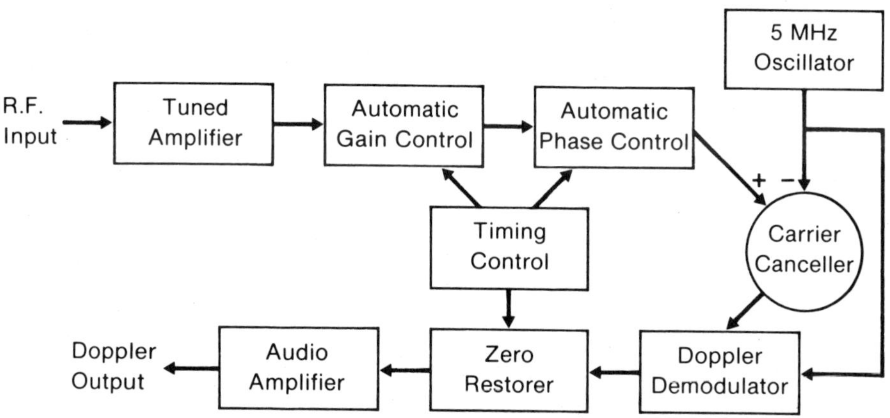

Figure 5. Block diagram of the RF preprocessor used to automatically control the amplitude and phase of the received signal. A large part of the carrier is subtracted before Doppler demodulation.

integrated intensity of a pixel is greatest with a high amplitude, high frequency signal.

RESULTS

The preliminary system images in-vitro flow over a projected region of approximately 2x6 cm at a frame rate of 5/sec. Each frame consists of 10x27 pixels. Fig. 6 shows the measured field response for all crystal groupings on the array at a distance of 5 cm from the array. This is the expected distance to the carotid artery when account is taken of the water path length. For the sake of clarity, only the main lobe pattern for each crystal grouping in the array is shown. As is clear from the figure, several crystal groups are not as well behaved as others. This likely results from discrepancies in mounting the crystals during the array assembly. The non-uniformity of multiplexer characteristics may also contribute to the measured variations.

Fig. 7 shows photographs (left) and the corresponding flow images (right) for a number of in-vitro flow systems. At the top is a bifurcation fabricated from tygon tubing (ID=0.10"). In the centre and bottom are excised feline arterial and venous sections respectively. The bifurcations used were the abdominal aorta to left and right common iliac arteries and the inferior vena cava to left and right common iliac veins. A 1% (by weight) solution of Sephadex in water was pumped through the systems by a peristaltic pump at a rate of approximately 600 ml/minute. The flow images were photographed directly from the CRT screen, using a 5 s exposure to average out the pulsatile nature of the flow.

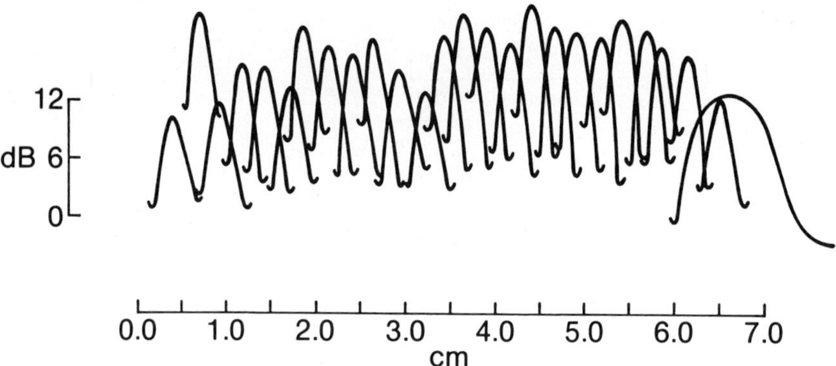

Figure 6. Superimposed main lobe patterns for all crystal groupings in the array. The measurements were made with a small vibrating target in a water tank.

DISCUSSION AND CONCLUSIONS

We have presented a new approach to real-time two-dimensional blood flow imaging. The prototype system, based on a stepped linear array of Doppler crystals, has been used to image in-vitro flow from a 2x6 cm projected region at the rate of 5 frames/sec: each frame consisting of 10x27 pixels. As can be seen in Fig. 7, the resolution is good enough to clearly image major blood vessels. The primary use of this image will be to quickly locate the carotid vessels and to detect major stenoses, and then, with a joystick control, to carry out detailed spectral analysis of selected regions using CW excitation.

Ongoing work is directed towards further reducing system noise, improving the array design and signal processing, in the expectation that these will enable in-vivo blood flow imaging to be realized.

ACKNOWLEDGEMENTS

This work is supported by NSERC, and the Ontario Heart Foundation.

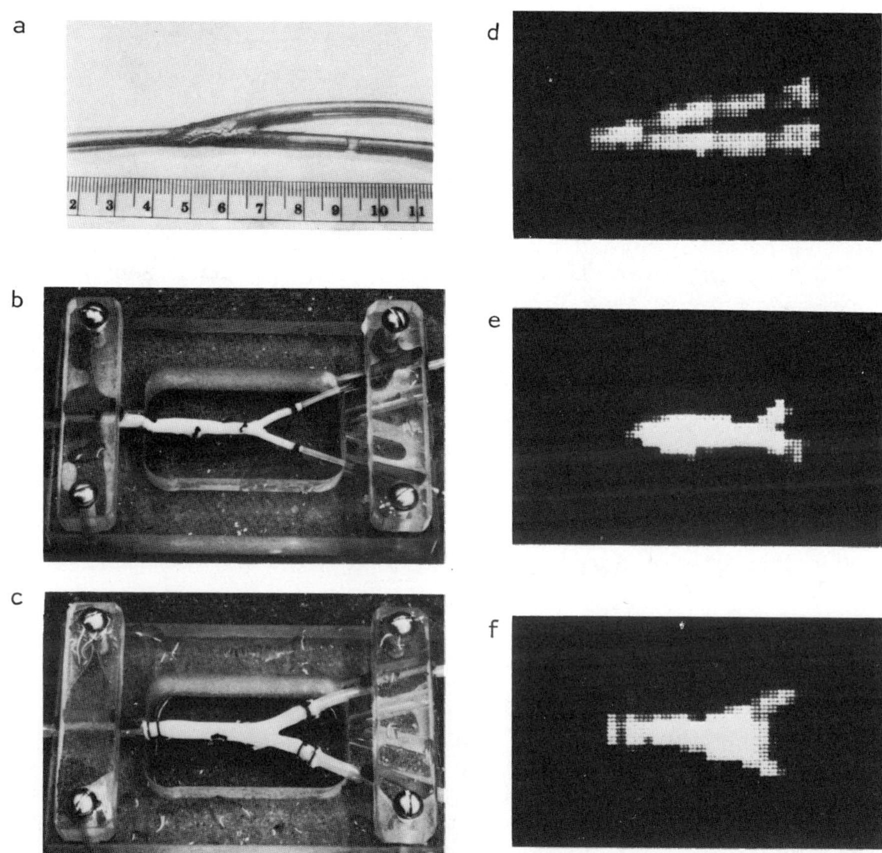

Figure 7. In-vivo blood flow systems (left), and Doppler imaged flow (right): a,d) tygon tubing; b,e) feline abdominal aorta to left and right common iliac arteries; c,f) feline inferior vena cava to left and right common iliac veins. Scale shown is in centimetres.

REFERENCES

1. Barber, F.E., Baker, D.W., Nation, A.W.C., Strandness, Jr., D.E., and Reid, J.M., "Ultrasonic Duplex Echo-Doppler Scanner", IEEE Trans Biomed Eng, BME-21:109-113, 1974.
2. Phillips, D.J., Powers, J.E., Eyer, M.K., Blackshear, Jr., W.M., Bodily, K.C., Strandness, Jr., D.E., and Baker, D.W., "Detection of Peripheral Vascular Disease Using Duplex Scanner III", Ultrasound Med Biol, 6:205-218, 1980.
3. Eyer, M.K., Brandestini, M.A., Phillips, D.J., and Baker, D.W., "Color Digital Echo/Doppler Image Presentation", Ultrasound Med Biol, 7:21-31, 1981.

4. Green, P.S., Schaefer, L.F., Taenzer, J.C., Holzmer, J.F., Ramsey, Jr., S.D., and Suarez, J.R., "Real-time Ultrasounic B-scan Imaging and Doppler Profile Display System and Method", US patent No. 4,141,347, February, 1979.
5. Waxman, A.S., and Havlice, J.F., "Ultrasonic Imaging Apparatus", US patent No. 4,231,373, November, 1980.
6. Hokanson, D.E., Mozersky, D., Sumner, D.S., and Strandness, Jr., D.E., "Ultrasonic Arteriography: A New Approach to Arterial Visualization", Biomed Eng, 6:420, 1971.
7. Miles, R.D.,, Russel, J.B., and Sumner, D.S., "Computerized Ultrasonic Arteriography: A New Technique for Imaging the Carotid Bifurcation", IEEE Trans Biomed Eng, BME-29:378-381, 1982.
8. Fish, P.J., "Multichannel Direction-resolving Doppler Angiography", Proc 2nd European Cong Ultrasonics Med., Munich, Exerpta Medica, Amsterdam, pp 153-159, 1976.
9. Reid, J.M., Spencer, M.P., and Davis, D.L., "Ultrasonic Doppler Imaging System", Ultrasound in Med. (ed. D. White and R. Brown), Vol. 3B, pp 1227-1235, Plenum Press, New York, 1975.
10. Curry, G.R., and White, D.N., "Color Coded Ultrasonic Differential Velocity Arterial Scanner (Echoflow)", Ultrasound Med Biol, 4:27-35, 1978.
11. Coghlan, B.A., and Taylor, M.E., "A Carotid Imaging System Utilizing Continuous Wave Doppler-Shift Ultrasound and Real-time Spectrum Analysis", Med Biol Eng Comput, 6:739-744, 1978.
12. Hottinger, C.F., and Meindl, J.D., "An Ultrasonic Scanning System for Arterial Imaging", IEEE Ultrasonics Symp., pp 86-87, 1973.
13. Arenson, J.W., Cobbold, R.S.C., and Johnston, K.W., "A Linear Stepped Doppler Ultrasound Array for Real-time Two-dimensional Blood Flow Imaging", IEEE Ultrasonics Symp., pp 775-779, 1980.
14. Arenson, J.W., Cobbold, R.S.C., and Johnston, K.W., "Computer Aided Design of Ultrasonic Imaging Arrays", Digest 8th CMBES Conf., pp 162-163, 1980.
15. Arenson, J.W., Cobbold, R.S.C., and Johnston, K.W., "Ultrasound Imaging Apparatus", US Patent Application 06/203875, Nov. 1980.

MEASUREMENT OF BLOOD FLOW USING ULTRASOUND

J M Evans, R Skidmore, J P Woodcock and P N Burns

Department of Medical Physics
Bristol General Hospital
Bristol, UK

INTRODUCTION

At present the only non-invasive method for calculating absolute volume flow in a specific blood vessel, transcutaneously, is based on ultrasonic techniques.

DOPPLER ULTRASOUND

If an ultrasound beam is directed through the skin at a blood vessel, the incident ultrasound is scattered in all directions by the blood cells. Since the blood cells are moving, there will be a difference in frequency between the transmitted ultrasound and that detected at a receiving transducer. This difference in frequency is the Doppler shift frequency (Δf) which is given by

$$\Delta f = \frac{2fv \cos \theta}{c} \tag{1}$$

f = transmitted frequency
v = velocity of blood ($v << C$)
θ = angle between ultrasound beam and flow direction
c = velocity of sound in tissue

However, because there is a velocity profile across the blood vessel, the Doppler signal is a spectrum rather than a single frequency. The Doppler shift spectrum, therefore, depends on the velocity distribution of the blood across the vessel cross-section, and the angle θ. To calculate absolute volume flow, it is necessary to know the diameter of the blood vessel, the angle of inclination of the ultrasound beam to the direction of blood flow, and the spatial and temporal mean velocity of the blood.

As volume flow, Q, is the product of mean velocity and lumen area using Equation (1) it can be expressed as

$$Q = \frac{\overline{\Delta} f c \pi d^2}{8f \cos \theta}$$

where

$\overline{\Delta} f$ = mean Doppler shift frequency
d = lumen diameter
Q = volume flow m^3/sec

Two techniques have been established which calculate volume flow. The first is the velocity profile method, requiring the measurement of the instantaneous average velocity, simultaneously, at a number of points across the lumen. This can only be achieved by using a multigated pulsed Doppler flowmeter (Fish, 1975). The second technique, and the one described in this paper, is the uniform sensitivity method requiring the spatial mean velocity to be calculated directly (Gill, 1979; Eik-Nes et al, 1980). The measurement of the instantaneous average velocity at a given position in a blood vessel, over the cardiac cycle, depends upon the assumption of uniform insonation of the whole of the blood vessel cross-section. If the sensitivity of the ultrasonic field intersecting the blood vessel is a constant at all points in the field then each streamline will contribute equally to the Doppler signal. Uniform insonation is approximated to by using a variable gate pulsed Doppler flowmeter, and the gate width is adjusted to insonate the whole cross-section of the blood vessel.

In order to locate the sample volume over the vessel of interest, the Doppler flowmeter is linked to a real-time pulse echo ultrasonic imaging system. The image is obtained from a rotating transducer scanhead housing 3 transducers and is displayed as a 90° sector image. When the transducers are stationary the Doppler ultrasound beam is steered by means of a joystick control on the scanhead. A bright dot on the frozen pulse-echo image indicates the position of the sample volume along the axis of the beam. The angle θ and the lumen diameter are both measured directly from this image. The Doppler shift signals are processed through a digital signal processing system to determine average velocity and hence volume flow. An analogue mean frequency processor has also been developed and initial results are comparable to the digital technique.

DESIGN OF A VARIABLE GATE PULSED DOPPLER FLOWMETER

A simplified diagram of the pulsed Doppler electronics is shown in Fig. 1. A crystal oscillator and digital counters are used to provide high frequency stability for the demodulator reference signals and pulse repetition frequency (prf). Counters are also used to generate accurate range delay and pulse width time intervals. Optical encoders provide a way of changing the sample depth position and gate size while ensuring jitter free operation. A digital display of gate position and width calibrated in mm is provided. Care was taken in the design and construction of the receiver and demodulators to obtain a low noise high gain system. Phase quadrature demodulation is used to obtain directional information from the received signals. Quadrature signals, ϕ_1 and ϕ_2, are digitally generated from the master oscillator to ensure accurate phase separation. Information regarding the direction of flow is contained in the relative phase of the two Doppler signals.

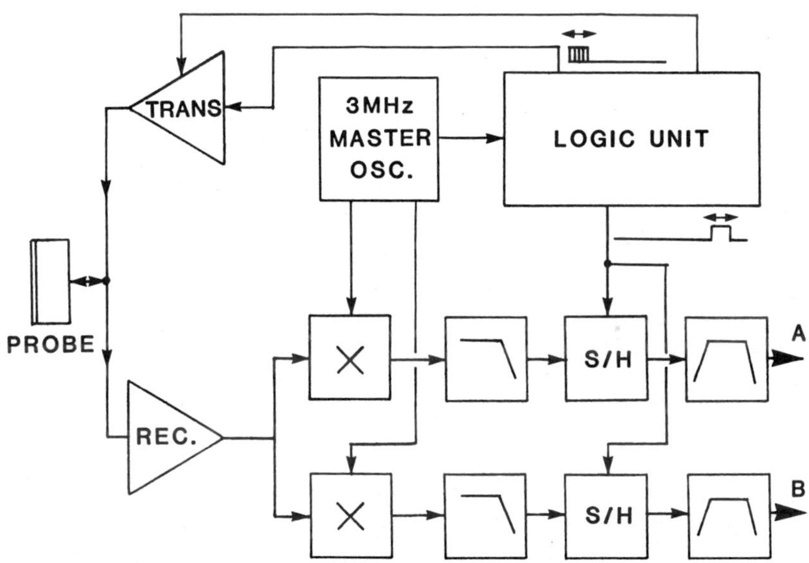

Fig. 1. Block diagram of pulsed Doppler flowmeter.

Integrating type sample and holds are used to average the Doppler information over the sample gate width. This has the advantage of improving the signal to noise ratio of the total system. Eight-pole low-pass filters are used to smooth the reconstructed Doppler shift signals and eliminate any prf breakthrough.

Switchable high-pass filters have been included to reject low frequency wall movements.

For optimal signal to noise ratio the duration of the sample gate should match the duration of emission (Peronneau et al, 1974). This is achieved digitally, but, since the transmission width will vary, the average acoustic output power will also change. To compensate, the transmitter amplifier supply voltage is driven from a $1/\sqrt{x}$ function where x is the transmission width. Hence average power is kept constant regardless of gate width and prf.

The Equivalent Input Noise (EIN) expressed as

$$\text{EIN} = \frac{\text{Noise (rms)}}{\text{Gain}}$$

was measured as 0.3 μV rms with a wide gate and 1 μV with a very narrow gate. The theoretical minimum noise from a 50 Ω source over the same bandwidth is 0.07 μV, but this does not take into account the noise contribution from the wide band rf stages. Phase error in the quadrature signals was found to be less than 1° over the maximum audio bandwidth of 10 kHz.

MEAN FREQUENCY ESTIMATION

To calculate the mean frequency of the Doppler shift spectrum a digital signal processing system is employed (Fig. 2). It consists of a hard-wired Fast Fourier Transform (FFT) analyser connected to a PDP-11/23 computer. The FFT analyser calculates a 128 point 12-bit spectrum every 8 mS. This information is transferred in digital form to a 5M byte disk from which the data can be processed 'off-line'. Such tasks as noise thresholding and brick-wall filtering can be achieved easily with the data stored in this form. The mean frequency of each spectrum is then calculated and displayed graphically. By entering the angle and diameter measurements previously obtained from the ultrasound image, volume flow is computed over a selected number of cardiac cycles. However, as the FFT analyser cannot accept directional information in the form of quadrature channels A and B, pre-processing of the Doppler shift signals is required. This is achieved by frequency domain processing (Coghlan and Taylor, 1976). The two channels A and B are multiplied by an audio quadrature carrier frequency and the outputs summed. This single output then represents directional flow as frequencies above or below the carrier frequency, and is connected to the FFT analyser.

The mean frequency estimation of the Doppler signals has also been investigated using an inexpensive analogue processing technique (Arts and Roevros, 1972). As this is a directional processor,

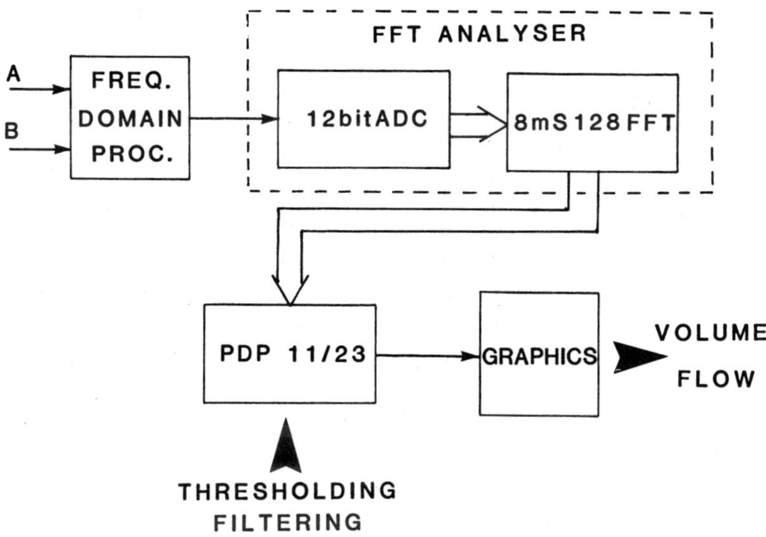

Fig. 2. Digital signal processing for the calculation of mean frequency and volume flow.

pre-processing in the frequency domain is not required. Its operating principle is described by the equation

$$d/dt(A \sin \omega t) = A\omega \cos \omega t$$

Thus when a Doppler signal is differentiated, the effect is to weight the amplitude of each component by its frequency. It can be shown (Atkinson and Woodcock, 1982) that if two Doppler shifted components of amplitude A and B are in quadrature so that

$$V_A = A \cos(\omega_A t + \phi_A) + B \cos(\omega_B t + \phi_B)$$

$$V_B = -A \sin(\omega_A t + \phi_A) - B \sin(\omega_B t + \phi_B)$$

where ω_A and ω_B are the Doppler shifts and ϕ_A and ϕ_B are the phases of the two components, the output, V_o, of the mean frequency processor after filtering is

$$V_o = \frac{A^2\omega_A + B^2\omega_B}{A^2 + B^2}$$

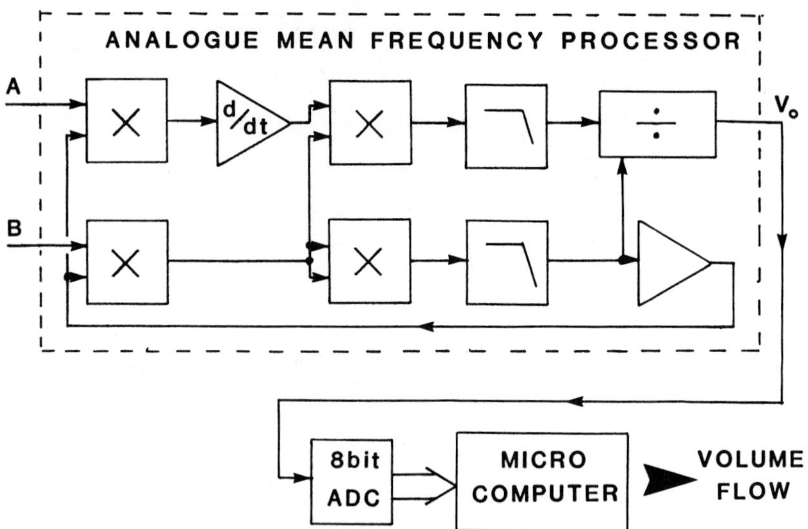

Fig. 3. Analogue mean frequency processor and microcomputer for volume flow calculation.

Now V_O is the normalized first moment of the Doppler power spectrum $(P(\omega))$. This generalises to more complex spectra in which

$$\bar{\omega} = \frac{\int_0^\infty \omega P(\omega)\, d\omega}{\int_0^\infty P(\omega)\, d\omega}$$

A practical implementation of this technique has been developed and tested. Filtered white noise was used to test the processor over a 30 dB dynamic range and the result compared to the digitally calculated mean frequency of the noise. The two methods agreed to within 4 per cent.

The output of the analogue mean frequency processor is then connected to a microcomputer via an 8 bit analogue to digital converter (Fig.3). The waveforms are displayed graphically and after selection, volume flow is computed. Results obtained to date have been within 6 per cent of the digital technique.

LIMITATIONS AND ACCURACY

Several factors impose limits on the application of this system. The minimum measurable Doppler frequency is determined by the high pass filters in the flowmeter, while the highest frequency is determined by the Nyquist limit. In practice, the highest Doppler shift obtainable, without aliasing, is less than half the pulse repetition frequency.

The minimum size of vessel which can be dealt with is determined by the signal to noise ratio or sensitivity of the flowmeter and the accuracy of the lumen diameter measurement. With the present imaging system, lumen diameter can be measured to within ± 0.5 mm. The size and uniformity of the sample volume restricts the maximum vessel size.

Several factors can cause errors in the volume flow calculation The two most important are measurement of the lumen area and angle θ. The vessel may not be of circular cross-section and, if an artery, this section will change in dimensions during each heart cycle. A 10 per cent error in diameter measurement causes a 20 per cent error in the flow estimate. The angle θ should be kept as small as possible. For example an error of 5° causes a 15 per cent error in flow with an angle θ of 60°. In order to achieve even insonation of the vessel a wide gate Doppler system must be used and the transducer operated in the far field. However, a radiation pressure plot of the transducer in water revealed that the -3 dB lateral beam width at its focus of 7 cm is only 5 mm. Mean frequency estimation is further effected by the wall filter characteristics of the flowmeter which will cause an overestimate of the flow.

CONCLUSIONS

The design and the use of a variable gate pulsed Doppler flowmeter for the measurement of blood flow has been described and the sources of errors discussed. Preliminary in vivo measurements have shown that this system is capable of obtaining satisfactory results with known errors, and holds promise for blood flow measurement in the adult abdomen as well as the pregnant uterus. Present work is concentrating on in vitro calibration using flow rigs. A mean frequency processor has been developed, compared with an extensive digital processing technique and found to operate correctly except in very noisy signal conditions. Grey scale M-mode will reduce the errors in the measurement of lumen diameters. Due to the limitation of the scanhead geometry an angle θ of greater than 50° is usually used. Therefore by attaching a separate Doppler transducer to the scanhead these angle errors could be significantly reduced.

ACKNOWLEDGEMENTS

Thanks are due to Elizabeth Pitcher for her help and patience in obtaining radiation pressure plots of the transducer and to Dave Ford for the construction of the transducer tank.

REFERENCES

Arts, M.G.J. and Roevros, J.M.G.J., 1972, On the instantaneous measurement of blood flow by ultrasonic means, Med. Biol. Engng, 10: 23-24.

Atkinson, P. and Woodcock, J.P., 1982, in: Doppler Ultrasound and Its Use in Clinical Measurement, Academic Press, London.

Coghlan, B.A. and Taylor, M.G., 1976, Directional Doppler techniques for detection of blood velocities, Ultrasound Med. Biol., 2: 181-8.

Eik-Nes, S.H., Brubahk, A.O. and Ulstein, M.K., 1980, Measurement of fetal blood flow, Br. Med. J., 280: 283-4.

Fish, P.J., 1975, Second European Congress on Ultrasonics in Medicine, Munich, p. 153. Excerpta Medica, Amsterdam.

Gill, R.W., 1979, Pulsed Doppler with B-mode imaging for quantitative blood flow measurement, Ultrasound Med. Biol., 5: 223-35.

Peronneau, P.A., Bournat, J.P., Bugnon, A., Barbet, A. and Xhaard, M., 1974, Theoretical and practical aspects of pulsed Doppler flowmeter: real-time application to the measure of instantaneous velocity profiles in vitro and in vivo, in: Cardio-Vascular Applications of Ultrasound, R.S. Reneman, ed., North Holland/American Elsevier.

LARGE AREA DOPPLER ARRAY FOR THE RAPID INVESTIGATION OF THE BREAST

M. Halliwell*, P.N. Burns*, P.N.T. Wells*
and A.J. Webb†

* Bristol General Hospital, U.K.
† Bristol Royal Infirmary, U.K.

INTRODUCTION

Breast cancer is an unfortunately common affliction especially in the western world where it represents the major cause of death in women between 40 and 45 years of age. The annual death rate is 27 per 100,000 of population and, during their lifetimes, one in 13 of all women will be affected by the disease (Silverberg, 1980). Despite improvements in surgery, radiotherapy and chemotherapy an apparent slight increase in incidence has resulted in the overall mortality remaining constant over the last 50 years. About half of all the individuals presenting with a malignant lump in their breast will die within five years. However, if consideration is given only to those individuals with small lumps (about 1 cm in diameter as opposed to the average presentation size of 3 cm), then treatment can give an over 90 per cent cure rate (Haagensen, 1971). The difference arises because of the occurrence of distant metastases. The likelihood of metastases increases with size so that any technique which results in the earlier detection of breast lumps will lead to an improved overall survival and reduce the personal tragedy that is often caused by this disease.

Several methods are available for the detection of breast lumps. Palpation is capable of detecting centimetre-sized lumps under average conditions but in large, nodular or dysplastic breasts this becomes more difficult. Mammography is an excellent technique but runs into difficulties with very large or small breasts and dense or fibrous ones. In addition, the slight hazard associated with ionising radiation means that mammography is not regularly performed on individuals under the age of 35. Thermography, which relies on the surface temperature alterations caused by vascular changes due to the lesion, is not a particularly sensitive investigation and diaphanography, transillumination of the breast, is currently under evaluation. Pulse-echo ultrasound has the capacity to detect 2 mm fluid lesions but solid lesions have to be more than a centimetre in diameter to be detectable with current equipment; it too is under evaluation.

The Doppler method described in this paper offers a very rapid investigation technique, suitable for those individuals for whom palpation and mammography are least appropriate, those with dense, lumpy, very large or small breasts, and the young woman of child-bearing age.

Malignant lesions in the breast cause considerable changes in the local vasculature. Many new small arterial branches (arterioles) develop to feed the growing rind which exists around the necrotic and avascular core of the mass. In addition several arterio-venous shunts also appear near the mass. Using Doppler ultrasound backscattered from the blood close to the lesion to investigate the speed direction and quantity of flow has led to the result that it is possible to differentiate malignant lesions from normal breast tissue. The three main differences between normal breast arterial flow and that around malignant lesions are: the average speed of blood flow is greater; the total backscattered energy is greater; the flow signals seem to come from a randomly oriented set of small arterial vessels which give rise to a "characteristic" medium frequency component to the audible signal (Burns et al, 1982). Current work on these findings has been performed with a small Doppler transducer of about ½ sq cm area. This is ideal for obtaining the signals but not for investigating the total volume of a breast in a clinical situation. Development of this method as a diagnostic tool has led to the construction of a Doppler transducer array of area 9 sq cm to overcome the problem of a lengthy investigation time. A further reduction in the duration of the examination will be effected with the 40 sq cm array now under construction.

METHOD

The 9 sq cm array was constructed by mounting nine pencil transducers in a 3 x 3 format using a Perspex mount so that they were as closely spaced as possible. Many modes of operation, with regard to transmission and reception of the signals, are possible. The first actually used was the simple one of exciting each transducer in turn. This was really as part of a feasibility study to ensure that characteristic signals could still be obtained with a transducer no longer capable of large changes in angulation. Each transducer was connected in turn via a simple switching box to a Doppler flowmeter and each Doppler signal was evaluated by

Figure 1. The face of the array showing the nine pencil Doppler transducers.

the operator using headphones. Each transmitting element had a different resonant frequency but the quality factor was so low that the same oscillator frequency could be used for each. Using the same receiver for each receiving element too resulted in an overall system with a sensitivity variation of about 20 dB.

The other mode of operation used after the initial feasibility study involved a common transmitter, activating every transmitting element simultaneously, and nine separate receiving channels, one for each receiving element. This avoids the problems of timing

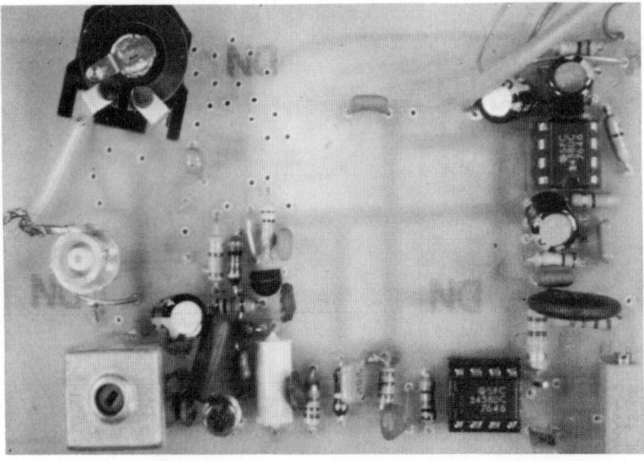

Figure 2. The components of one complete Doppler receiver.

inherent in any switched or multiplexed system looking at a slowly time-varying signal. The sensitivity variation is still a problem but can be minimised in the receiver chains, at some expense to the overall sensitivity.

The output can be displayed in a variety of ways; summed as an audible output representing the total Doppler signal, sampled a channel at a time as an audible output or displayed as a crude image. The imaging potential of a real-time Doppler array is obviously considerable as well as its capacity to speed apprecia-tion of the total Doppler information. Because only nine pixels were available to form the image an array of nine high brightness light emitting diodes was constructed. This kind of display has the advantage over a cathode ray tube display that it can be

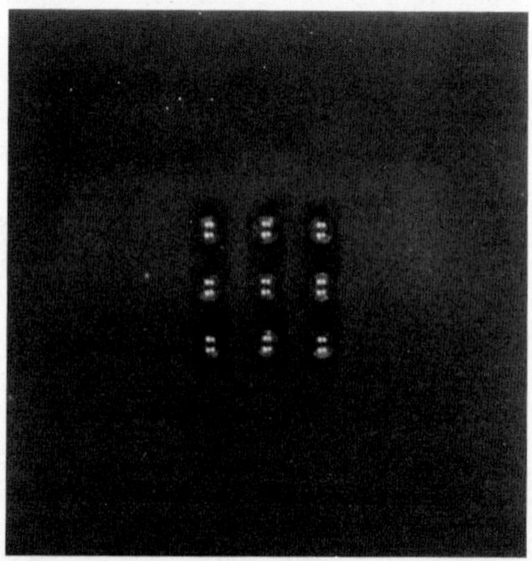

Figure 3. The LED array used to display the signal output nine parallel Doppler channels.

mounted in close proximity to the transducers so that the bright-ness modulation can be linked, by the operators hand-eye coordina-tion, with small movements of the arrays.

An operator variable threshold was included to determine the signal level above which the light emitting diodes registered. This allowed for the inter-individual variation in normal arterial signal strength. With the array over the normal side the threshold was increased until the arterial signals were just displayed. Then the contralateral, suspect, site was examined; any abnormal signals would be expected to show up as brighter illumination of the dis-play. This is a direct development of the single transducer tech-nique where it has been shown that it is essential to compare signals at contralateral sites on the same individual, the inter-individual variation is very large.

RESULTS

The performance of this array has been qualitatively assessed in terms of the behaviour of the single transducer technique as a detector of Doppler signals from both normal and abnormal breast vasculature. Using the simple switching box it has been shown that characteristic signals can be obtained from malignant lesions with the restricted angulation capacity of the array structure. The large variation of sensitivities, however, precluded any estimate of the loss of sensitivity introduced by this physical constraint.

The nine parallel receiving channels have recently been completed and initial results show that this kind of processing is capable of producing a real-time Doppler image. However, because of the coarse separation of the picture elements in this prototype device static images are unimpressive whereas the live operator is able to generate a mind's-eye view of the vascular situation.

CONCLUSION

A real-time continuous wave Doppler array has been used to generate images of the vasculature surrounding a malignant breast lesion. The image is coarse but in practice this kind of array would not be required to generate a high definition vascular map; it would be used solely as an indicator of the presence somewhere in the breast of an abnormal vasculature. From this point of view the physical overlapping of the Doppler beams, which occurs to some extent with this array but which will be of considerable importance in the larger, more closely packed array, is of little practical significance. In effect, as far as a cancer detection system is concerned, flooding the breast with ultrasound and using one transducer to produce a summed Doppler signal would be a perfectly satisfactory device (if the characteristic elements could be satisfactorily extracted from the total signal). In other words, an image is not really necessary.

At present the display is modulated by total signal strength, this means that it is sensitive to skin movement noise as the array position is altered slightly. Other parameters which will be used for display will include maximum frequency, mean frequency and flow direction.

ACKNOWLEDGEMENTS

The authors are indebted to Mr D Follett, Mr G Garland, Mr D Ford, Mr R Jordan and Mr A Rankin for much help and advice. Financial assistance came from a Medical Research Council grant held by Dr P N T Wells.

REFERENCES

Burns, P.N., Halliwell, M., Wells, P.N.T. and Webb, A.J., 1982, Ultrasonic Doppler studies of the breast, Ultrasound Med. Biol. 8: 127-43.

Haagensen, C.D., 1971, Diseases of the Breast, 2nd edn., W.B. Saunders, Philadelphia, London, Toronto.

Silverberg, E., 1980, Cancer Statistics, 1980, Ca-30, 23-38.

ULTRASOUND COMPUTERIZED TOMOGRAPHY USING TRANSMISSION AND REFLECTION MODE: APPLICATION TO MEDICAL DIAGNOSIS

Dietmar Hiller and Helmut Ermert

Department of Electrical Engineering
University of Erlangen-Nuremberg
Cauerstr. 9, D-8520 Erlangen, FRG

INTRODUCTION

The concept of computerized tomography (CT) is not only applicable to X-rays, but also to other kinds of rays, for example to ultrasound /1-8/ and to microwaves /10/. These non-ionizing waves are known to be safe for medical diagnostic purposes. Ultrasound computerized tomography (USCT) is possible in different modes: in the transmission-mode acoustic absorption as well as acoustic velocity can be used to reconstruct as a cross-sectional distribution. Furthermore, parameters which characterize the dependence on frequency of these quantities can be displayed. In the reflection-mode the distribution of reflectivity can be reconstructed using the CT-concept /4-6/. Additionally, the measurement of diffracted signals allows tomographic imaging of inhomogeneous media /8/.

Different modes of USCT have different properties and different ranges of application. In some cases, several modes are applicable to one organ. The resulting cross-sectional images present different information about this object, the combination of the results may lead to an improvement of diagnosis. Additionally, the result of one USCT mode can be used to reduce the errors in the cross-sectional image obtained by another USCT mode. Therefore, using an iterative technique, an improvement of images can be achieved.

ULTRASOUND COMPUTERIZED TOMOGRAPHY (USCT)

Transmission-mode

The transmission-mode of ultrasound computerized tomography is equal to the well-known X-ray-CT. The principle is illustrated in Fig. 1. Ultrasonic multielement-arrays can be used as transmitting and as receiving apertures in a rotatable configuration. In X-ray-CT a single projection can be interpreted as an one-dimensional "shadow" of the cross-section. In transmission-mode USCT acoustic attenuation as well as the velocity of acoustic propagation can be reconstructed as a cross-sectional distribution from attenuation measurements and "time-of-flight"-measurements, respectively. Additionally, multifrequency or broadband techniques allow the measurement of the dispersion of the medium. The ultrasonic signals are assumed to propagate approximately along straight lines as thin rays without any lateral extension. A further assumption is the neglection of all discontinuities. Reflections at discontinuous boundaries cause attenuation and simulate absorption even in the case of lossless material.

In the computerized reconstruction each value of a single projection is backprojected along the straight path of the ray, which produced this special value. No information about the distribution of parameters along the propagation axis is available from a single projection. Only the backprojection of projections obtained from different angles leads to the original cross-sectional distribution. A simple summation of backprojected projection results in a blurred reconstructed image.

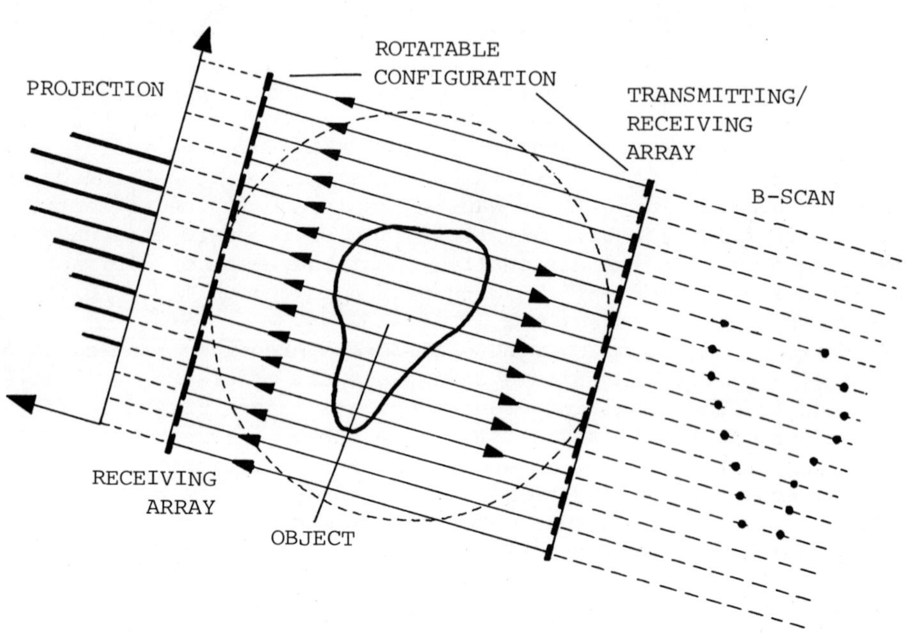

Fig. 1 Principle of USCT (transmission mode and reflection mode).

Fig. 2 illustrates, that the image of a single point in the original distribution is not a point but a function, which has to be corrected by a suitable filter function in order to achieve an optimized point-spread-function. These techniques have been reported in /1,2/.

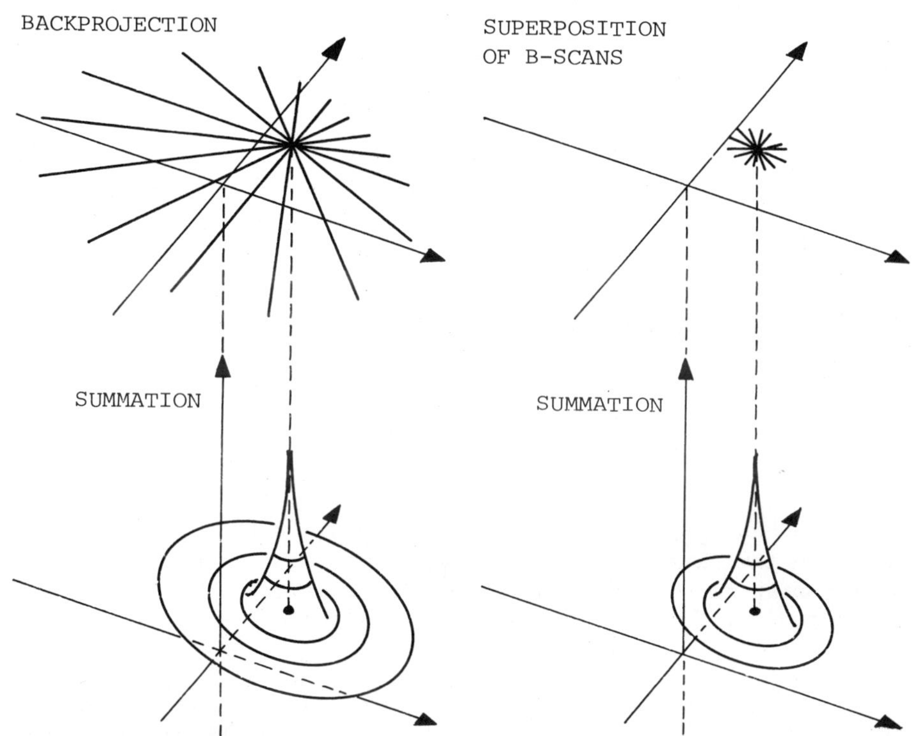

Fig. 2. Point-spread-function of transmission-mode

Fig. 3. Point-spread-function of reflection-mode

Reflection mode

The simplest way to explain this kind of CT-mode is to consider again the arrangement in Fig. 1. Only one rotating ultrasonic array is now working and picks up a discrete number of conventional B-scans from different angles. Each B-scan represents an image of the same cross-section with different resolution in axial and in lateral direction corresponding to the actual orientation of the array. Computerized superposition of all single B-scans eliminates the poor lateral resolution and makes use of the good axial resolution for all directions. Fig. 3 illustrates, that a simple summation of B-scans leads to a blurred reconstructed image. The superposition of single B-scan point-spread-functions causes the same effect as in the case of transmission-mode CT the superposition of backprojected projections. A filtered backprojection of B-scans is therefore necessary to obtain high quality images from reflection-mode CT /4,5/.

Mixed mode

Reflection-mode USCT does work exactly if the velocity of ultrasonic signals is constant in the whole imaged cross-sectional area. Additionally, this area is assumed to contain single "isolated" point scatterers without any mutual interaction. If the velocity is not constant and if the cross-section consists of several areas with different velocities, echos from a single point-scatterer penetrate these different areas and are not superimposible to a single point image. In such a case the distribution of sound velocity which has been obtained by transmission-mode CT can be used for the computerized reconstruction of a reflection-mode CT image with better accuracy /9/. Also the distribution of acoustic absorption obtained by transmission-mode CT is conceivable to be used for correction of amplitudes in a reflection-mode CT image.

PROPERTIES OF USCT-MODES

Generally, the resolution of conventional imaging systems can be calculated, it depends on the frequency and the bandwidth of signals, on the aperture size and the distance between object and aperture. In the case of optical or quasi-optical systems the objects are assumed to be surrounded by an idealized transparent, homogeneous medium. In the case of ultrasonic imaging for medical diagnosis the biological tissue is medium as well as object. For example, inhomogeneities in the tissue are objects and have to be imaged while other inhomogeneities in the surroundings affect the quality of the image. Since ultrasound computerized tomography is based on highly idealized conditions for correct operation, resolution not only depends on the parameters mentioned above but also on the degree of homogeneity and discontinuity of the tissue. Quantitative evaluation of the resolution of some USCT modes has been presented elsewhere /5/. In this chapter only some qualitative properties shall be discussed.

Transmission-mode USCT has a relative poor resolution. If the straight path propagation model is valid, the resolution, which is not dependent on the position and direction, is approximately equal to the spatial resolution of the projections. Inhomogeneity of the tissue leads to several effects which have an influence on the resolution and image quality: discontinuities produce reflection as well as diffraction. Reflection causes attenuation of the transmitted signal which cannot be distinguished from true absorption in the reconstructive procedure. Diffraction and curvilinear propagation in soft inhomogeneities also simulate absorption because the signals may slightly miss the correct receiving element. Furthermore curvi-

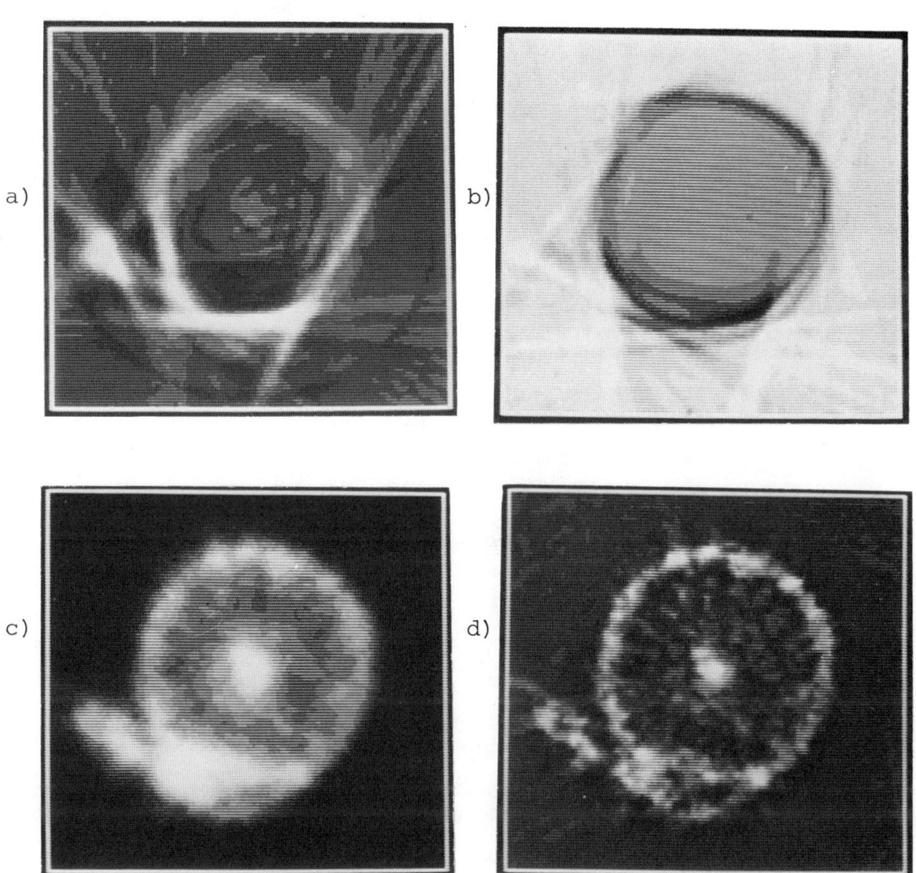

Fig. 4. Canine testicle, in vitro, 3.5 MHz
a) transmission-mode (absorption)
b) transmission-mode (velocity)
c) reflection-mode (summation)
d) reflection-mode (filtered summation)

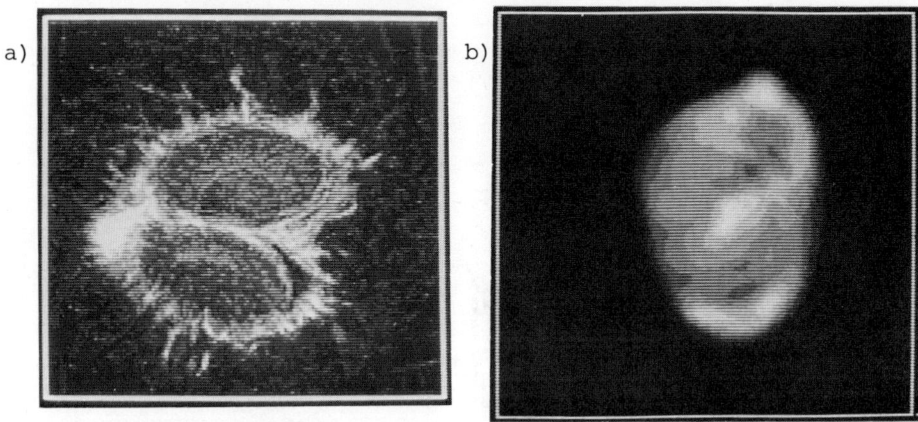

Fig. 5. Human testicle, 3.5 MHz
a) normal, in vivo, reflection-mode (filtered summation)
b) carcinoma, in vitro, transmission-mode (velocity)

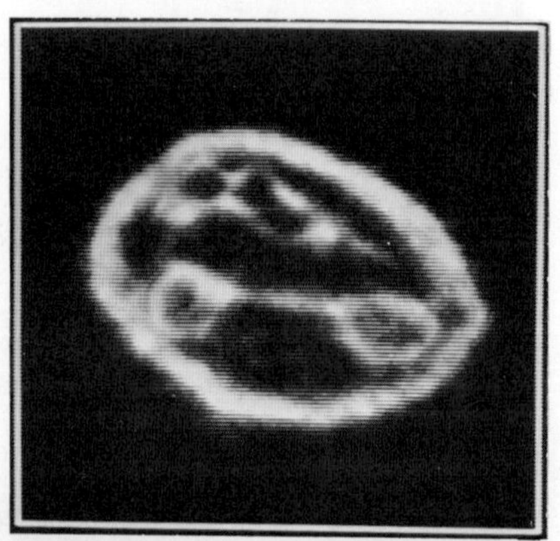

Fig. 6. Human forearm, in vivo, reflection-mode (summation), 2.25 MHz

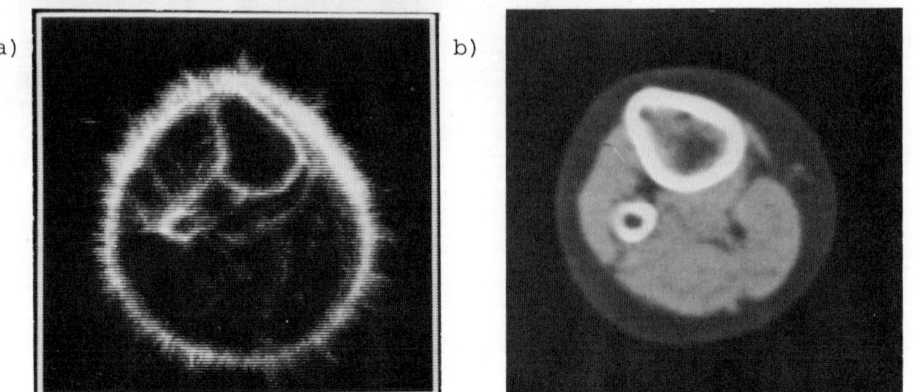

Fig. 7. Human shank, in vivo
a) reflection-mode (summation), 2.25 MHz
b) x-ray-CT

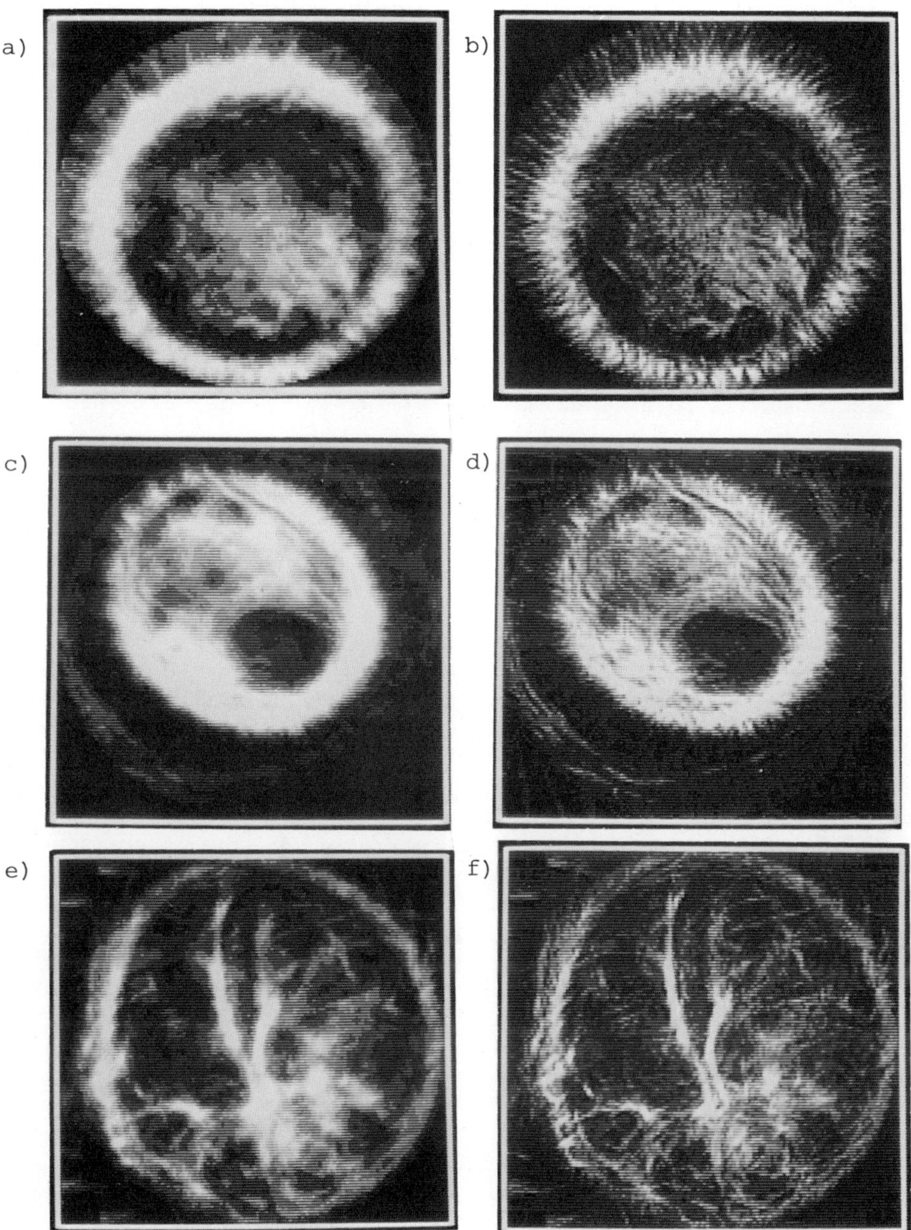

Fig. 8. Female breast, in vivo, reflection-mode, 3.5 MHz
a) normal, summation
b) normal, filtered summation
c) fibroadenoma, summation
d) fibroadenoma, filtered summation
e) carcinoma, summation
f) carcinoma, filtered summation

linear propagation leads to wrong velocity-values in time-of-flight measurements. The advantage of transmission-mode CT is the quantitative measurement of distribution of acoustic parameters. These quantitative results are useful for tissue characterization. Transmission-mode USCT is applicable to low contrast media with smooth inhomogeneities. Ray-tracing techniques are possible for image improvement /7/ in order to take curvilinear propagation into account.

Reflection-mode USCT produces high quality images from echo measurements with an isotropic resolution. The resolution is in the order of magnitude of the axial resolution of B-scan systems. This mode workes exactly if the medium is assumed to be homogeneous containing ideal scatterers. Experimental work confirmed the feasibility of reflection-mode USCT to the cross-sectional imaging of soft tissues. Beyond that, contours inside the cross-section which produce specular reflection can be imaged; every point of the circumference contributes to this image because of the diversity of angles, from which the object is insonified. Moreover, non transparent areas do not lead to a significant image degradation. It is possible to image cross-sections containing "acoustic obstacles" as long as other interesting parts of the object can be viewed by 180°.

EXPERIMENTS

An experimental setup has been devised, that allows for simultaneous acquisition of transmission and reflection data. The setup is built around a modified conventional real-time B-Scanner (SIEMENS MULTISON 400). Two identical multielement transducer arrays are mounted on a turntable, facing each other with approximately 15 cm distance. One is acting as a transmitter/receiver, the other is used in receive-only mode for the transmission measurement. Two different turntables with 2.25 MHz and 3.5 MHz arrays are available, depending on the application. The turntables can be rotated by means of a stepper motor under computer control. The arrays are operated in a water bath, with the water serving as the coupling medium. The digitized transmission and reflection data is fed to the computer. A special data processor allows for the reconstruction of the collected data while the turntable is proceeding to its next position. When the mechanical scanning is done, the image is also ready for output.

With the transmission data, tomograms of the distribution of attenuation, acoustic velocity or the frequency dependence of attenuation can be reconstructed. The reflection data can be presented as a conventional real-time image, scans can be superimposed to give a conventional compound image or the scans can be reconstructed with a CT algorithm to give a reflection mode tomogram.

MEDICAL APPLICATION

Organs that are suited for USCT imaging fall into 2 classes: Those, which can be investigated by reflection USCT only, e.g. limbs or the thyroid gland, and those, where both modes can be applied, like testicles or female breast. Three of these potential applications for USCT techniques were experimentally investigated: cross-sectional imaging of limbs, testicles and breast.

Orthopaedy (Fig.6; Fig.9)

"In-vivo" experiments performed on human forearms and legs gave very good results. This indicates the possibility for an exact quantitative measurement of the cross-section of bones using reflection USCT. This may be useful for the distinction between physiological and pathological torsions of bones, which today is usually done by X-ray CT methods. The images obtained by USCT are comparable in quality and resolution to those obtained by X-ray CT. With an iterative technique it is even possible to determine the wall-thickness of bones.

Urology (Fig.4; Fig.5)

Other objects under investigation were excised testicles from dogs "in-vitro" as well as excised tumourous human testicles "in-vitro". The intact testicle tissue showed a very homogeneous distribution of acoustic velocity and a low attenuation, while in the tumourous specimen a strongly inhomogeneous acoustic velocity was found. The reconstruction of attenuation yielded images of poor resolution. Reflection mode USCT resulted in images with fine details. An "in-vivo"-reflection mode image of a human testicle has a good resolution, a change to higher frequencies might further improve the results.

Gynaecology (Fig.8)

Cross-sectional imaging of the female breast is probably the most promising application for USCT methods. Experimental investigations were performed on healthy and tumourous breasts (fibroadenoma and cancer) "in-vivo". Because the breast is a highly structured organ, it was difficult to detect lesions in low-resolution transmission images. Again however, reflection mode tomograms are able to detect the tumourous areas, which differed significantly from normal tissue.

CONCLUSIONS

From the present knowledge it seems, that Ultrasound Computerized Tomography is not useful for cross-sectional imaging of the whole human body, but it is suitable for imaging "small" organs, using the safety of ultrasonic rays as an additional advantage. Reflection mode USCT applied to medical objects leads to an image quality and resolution better than obtained with conventional ultrasonic imaging systems. Transmission mode USCT yields images of parameters, which contain totally different information. Although the resolution in this mode is lower, the information may serve as a supplement for reflection mode USCT and it may be useful in tissue characterization.

ACKNOWLEDGEMENT

These investigations were supported by the Deutsche Forschungsgemeinschaft (DFG), Bonn-Bad Godesberg (Research grant ER 94/2-2). The medical investigations were carried out in cooperation with clinical institutions at the medical faculty of the University of Erlangen-Nuremberg.

REFERENCES

1/ J. F. Greenleaf, S.A. Johnson, R. C. Bahn, B. Rajagopalan, S. Kenue, Introduction to Computed Ultrasound Tomography, in: Computer aided tomography and ultrasonics in medicine, ed. J. Raviv, North Holland Publ. Comp., Amsterdam, New York, Oxford (1979) 125 - 136

2/ R. K. Mueller, M. Kaveh, G. Wade, Reconstructive tomography and applications to ultrasonics, Proc. IEEE, vol. 64 (1979) 567 - 587

3/ D. Hiller, H. Ermert, The application of transducer arrays in ultrasound computerized tomography, Ultrasonics Intern. 1979, Conference Proceedings, IPC Science and Technology Press, Guildford, 540 - 544

4/ D. Hiller, H. Ermert, Tomographic reconstruction of B-scan images, in:" Acoustical Imaging",vol. 10, ed. P. Alais, A.F. Metherell, New York (1982) 347 - 364

5/ D. Hiller, H. Ermert, Ultrasound Computerized Tomography using Transmission and Reflection Mode: An Experimental Comparison, Ultrasonics Intern. 81, IPC Science and Technology Press, Guildford, 235 - 240

6/ S. J. Norton, M. Linzer, Ultrasonic reflectivity tomography: reconstruction with circular transducer arrays, Ultrasonic Imaging, vol. 1 (1979) 154 - 184

7/ H. Schomberg, An Improved Approach to Reconstructive Ultrasound Tomography, J. Phys. D: Appl. Phys. 11 (1978), 181 - 185

8/ R. K. Mueller, M. Kaveh, R. D. Iverson, A new approach to acoustic tomography using diffraction techniques, in:"Acoustical Imaging", vol. 8, ed. A. F. Metherell, New York (1980) 615 - 628

9/ S. A. Johnson, J. F. Greenleaf, B. Rajagopalan, R. C. Bahn, Ultrasound Images corrected for Refraction and Attenuation:

A Comparison of new high Resolution Methods, in: Computer aided tomography and ultrasonics in medicine, ed. J. Raviv, North Holland Publ. Comp., Amsterdam, New York, Oxford (1979) 55 - 72

10/ H. Ermert, G. Fülle, D. Hiller, Microwave Computerized Tomography, 11th European Microwave Conference, Amsterdam, 1981, Conference Proceedings, 421 - 426

SYNTHETIC APERTURE TOMOGRAPHIC IMAGING FOR ULTRASONIC DIAGNOSTICS

A.P. Anderson and M.F. Adams

University of Sheffield
Department of Electronic & Electrical Engineering
Mappin Street, Sheffield S1 3JD

INTRODUCTION

The tomographic mode of imaging is a powerful aid to non-invasive diagnostics. If the probing radiation penetrates sufficiently through the object, then image slices showing internal structure may be obtained.

The medical applications of tomography are most widely represented at the present time, particularly X-ray CAT scanning in which the ability of X-rays to penetrate and travel in straight lines through an object allows the application of projection techniques for image reconstruction. If long wavelength diagnostic imaging, i.e. using acoustic waves and microwaves, could also acquire a tomographic capability, many other non-destructive testing applications would emerge.

Although the pulse-echo or B-scan mode of acoustic imaging is naturally tomographic, it is not possible, in principle, to obtain the diffraction limit of resolution. Moreover, the images obtained are dependent on operator ability.

Projection tomographic techniques have been successfully applied to acoustical imaging in which it is assumed that either attenuation or velocity can be measured as a projection along the straight line connecting transmitter and receiver[1]. However, these techniques, which neglect diffraction effects, are also incapable of diffraction limited resolution.

Attempts to produce more rigorous and accurate solutions to the structure of an object from its scattered radiation may be grouped under the comprehensive title 'inverse scattering'. In general, inverse scattering techniques require either multi-frequency (including pulse-echo) methods[2,3,4] or multiview/single frequency methods[5], and are sometimes known as 'diffraction tomography'. In this paper we will concentrate on a multiview/single frequency approach which is an image space solution analogous to the convolution-back projection approach to projection tomography, already investigated by the authors in the microwave regime[6]. This approach provides an interesting comparison to the frequency domain approach[5]. Therefore we shall be discussing transmission tomography rather than backscatter tomography[7] which has a useful although limited tomographic role.

IMAGE SPACE SOLUTION FOR DIFFRACTION TOMOGRAPHY

The acoustic pressure, u_t, in an inhomogeneous fluid of constant density, ρ_0, may be written[8] :

$$u_t(\underline{r}) = u_i(\underline{r}) + \int_{V_1} f(\underline{r}_1)\, u_t(\underline{r}_1)\, g\,(\underline{r}, \underline{r}_1)\, dv_1 \tag{1}$$

where, as illustrated in Fig.1, $u_i(\underline{r})$ is the incident acoustic pressure, g is the ideal isotropic medium Green's function

$$g(\underline{r},\underline{r}_1) = \frac{\exp(jk_o|\underline{r} - \underline{r}_1|)}{4\pi|\underline{r} - \underline{r}_1|} \tag{2}$$

and the inhomogenities contained within the volume v_1 are described by :

$$f(\underline{r}_1) = k_o^{\,2} \left(\frac{\chi(\underline{r}_1) - \chi_o}{\chi_o}\right) \tag{3}$$

$\chi(\underline{r}_1)$ is the compressibility within v_1 and χ_o is the value for the surrounding homogeneous fluid. In a loss free fluid the acoustic velocity, $c(\underline{r}_1)$ is given by :

$$c^2(\underline{r}_1) = \frac{\chi(\underline{r}_1)}{\rho_o} . \tag{4}$$

$k_o = 2\pi/\lambda_o$ is the propagation constant in the homogeneous fluid.

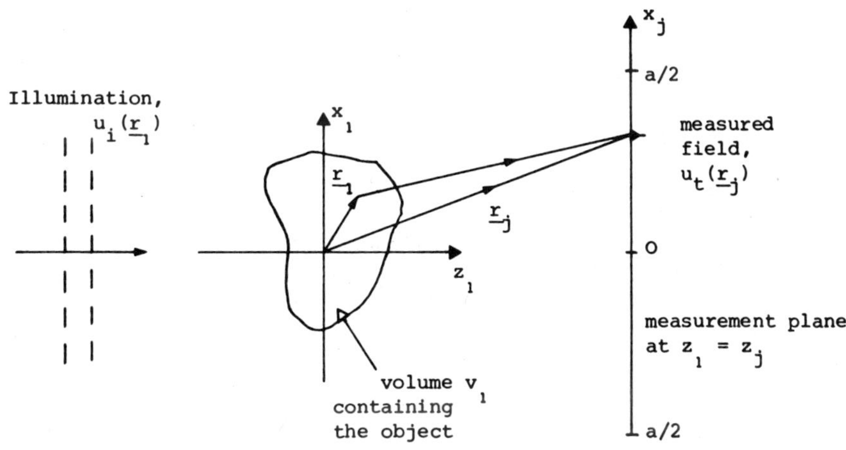

Fig.1 Data acquisition scheme for one view of the object.

The field u_t is measured over the plane at $z_1 = z_j$ and it will be assumed that the measurement aperture is infinite (although the effects of a finite aperture may be included[9]). The field due to scattering by the object, $u(\underline{r})$, is simply the integral part of eqn.(1), i.e. :

$$u(\underline{r}) = u_t(\underline{r}) - u_i(\underline{r}), \tag{5}$$

and on the measurement plane at $z_1 = z_j$ the field is defined as $u_j(x_j, y_j)$ where

$$u_j(x_j, y_j) = u(\underline{r}_j) = u(\underline{r})\Big|_{z=z_j} \tag{6}$$

The inverse scattering is to solve eqn.(1) for the scattering properties of the object, $f(\underline{r}_1)$. However, no closed form solution for eqn.(1) exists so that it is necessary to adopt an approximation. The first-order approximation, often known as the Born approximation, is to assume that the object is a weak scatterer so that the field inside v_1 is given by the incident field u_i. A fictive object may now be modelled by an ensemble of point scatterers, each of which re-radiates the incident wave field modified by the scattering strength at that point, $f(\underline{r}_1)$, as a diverging spherical wavefront. With this approximation the scattered field on the measurement plane becomes :

$$u_j(x_j,y_j) \simeq \int_{V_1} f(\underline{r}_1)\, u_i(\underline{r}_1)\, g\,(\underline{r}_j,\underline{r}_1)\, dv_1 \tag{7}$$

Our approach to the solution of eqn.(7) is a multiview technique analogous to the convolution-back projection process which is a common method in the solution of an object from its projections along straight lines. The measured data or 'view', u_j, may be used to generate a 3D image by a focusing or inverse diffraction technique. This stage corresponds to the back projection stage in projection tomography.

As shown in Fig.1, the illumination is considered to be a plane wave so that :

$$u_i(\underline{r}_1) = \exp(jk_o z_1). \tag{8}$$

A focused image is generated by first performing a 2D Fourier transform upon the measured data and recognising that eqn.(7) is in the form of a convolution integral :

$$\begin{aligned} U_j(s_x,s_y) &= \mathcal{F}_{2D}\{u_j(x_j,y_j)\} \\ &= \int F(s_x,s_y,z_1).\exp(jk_o z_1).G(s_x,s_y,z_j-z_1)\, dz_1 \end{aligned} \tag{9}$$

The 2D Fourier transform of the spherical wave function g is given by[10] :

$$\begin{aligned} G(s_x,s_y,z_j-z_1) &= \mathcal{F}_{2D}\{g(x_j,y_j,z_j-z_1)\} \\ &= \frac{j\lambda_o}{4\pi}\ \frac{\exp\{jk_o m(z_j-z_1)\}}{m} \end{aligned} \tag{10}$$

$$\text{where } m = \sqrt{1 - (\lambda_o s_x)^2 - (\lambda_o s_y)^2} \tag{11}$$

The 2D Fourier transform of the object is given by :

$$F(s_x,s_y,z_1) = \iint f(x_1,y_1,z_1)\, \exp\{-j2\pi(s_x x_1 + s_y y_1)\}\, dx_1 dy_1 \tag{12}$$

Now $U_j(s_x,s_y)$ represents the plane wave expansion of the measured scattered field distribution which, since the propagation process may be regarded as a linear filter acting on this expansion[11], may be back propagated to any image plane by application of an inverse diffraction filter, B_j (acting on the radiating components only). The focused image plane, U, is then given by :

$$U(s_x,s_y,z) = U_j(s_x,s_y).B_j(s_x,s_y,z) \tag{13}$$

where $B_j(s_x,s_y,z) = \frac{4\pi m}{j\lambda_o} \cdot \exp\{-jk_o m(z_j-z)\}.\exp(-jk_o z)$

when $s_x^2 + s_y^2 \leqslant 1/\lambda_o^2$

$$= 0 \text{ when } s_x^2 + s_y^2 > 1/\lambda_o^2 \tag{14}$$

The plane of the image at depth z is then given by an inverse 2D Fourier transform upon U so that :

$$h(x,y,z)\big|_z = \mathcal{F}_{2D}^{-1}\{U(s_x,s_y,z)\} \tag{15}$$

Combining eqns. (9) - (14) the focused image plane in the frequency domain is given by :-

$$U(s_x,s_y,z) = \int F(s_x,s_y,z_1)\exp\{-jk_o(z-z_1)\}\exp\{jk_o m(z-z_1)\}dz_1 \tag{16}$$

This expression is a convolution integral in z and a multiplication in s_x,s_y which, on inverse 2D Fourier transformation, as in eqn.(15), produces the image plane in image space :

$$h(x,y,z) = \iiint f(x_1,y_1,z_1).p(x-x_1,y-y_1,z-z_1)dx_1dy_1dz_1 \tag{17}$$

$$\text{where } p(x,y,z) = \iint \exp(-jk_o z)\exp(jk_o mz)\exp\{j2\pi(s_x x + s_y y)\}ds_x ds_y \tag{18}$$

Thus it is seen that the 3D image volume generated from one 'view' of the measured data may be considered as a convolution of the object with a defocusing function, p ;

$$h(x,y,z) = f(x,y,z) \circledast p(x,y,z) \tag{19}$$

The form of p, obtained by computer simulation, is shown in Fig.2a in 2D. This distribution is an aperture-limited image of a point scatterer from a linescan of the scattered field at $z_j = 16\lambda_o$ with aperture, $a = 16\lambda_o$. The function p is the single view imaging system point spread function (PSF).

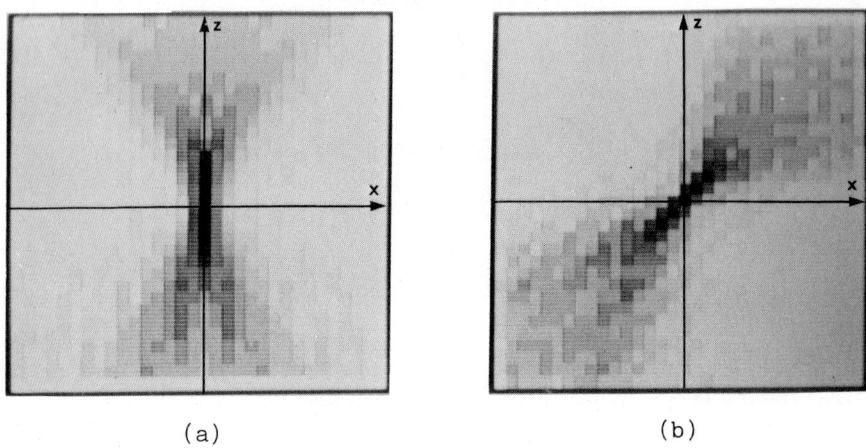

Fig.2. The defocusing function, p, for focused images at views of (a) 0° and (b) 45°

It can be seen from Fig.2a that the imaging system PSF exhibits high lateral resolution but the resolution in depth is considerably worse. To overcome this a multiview approach is adopted in which the scan and illumination are considered to rotate with respect to the fixed object and measurement of the scattered field repeated for this new 'view'. Equivalently, and more practically, the object may be rotated with respect to the fixed scan/illumination. The new view, at angle α say, may then be focused to yield a 3D image as before and the image volume, h_{α}, can be represented by the convolution :

$$h_{\alpha}(x,y,z) = f(x,y,z) \circledast p_{\alpha}(x,y,z), \tag{20}$$

where p_{α} is the defocusing function p rotated to the view angle. Fig.2b illustrates p_{α} in 2D for a view at $\alpha = 45^{\circ}$ rotation.

The next stage is the generation of an intermediate 3D image volume given by the sum of all constituent images generated from each view :

$$h_t(x,y,z) = \sum_{\text{all } \alpha} h_\alpha(x,y,z)$$

$$= f(x,y,z) \circledast \sum_{\text{all } \alpha} p_\alpha(x,y,z) \quad (21)$$

The final stage is an attempt to remove the effect of the overall convolving function given in eqn.(21). This is done by further pursuing the analogy with the convolution-back projection technique which was introduced earlier. It is known that in this case and in 2D that the overall convolving function tends to a function $^1/_r$ in the limit of an infinite number of views. The effect of this function can be removed by applying a 'ramp' filter in the frequency domain[13]. Similarly in diffraction tomography it is noted that the convolving function for one view (e.g. see Fig.2) may, in the region of the scatterer, be approximated by a line extending in the view direction. So that the overall function given by eqn.(21) is a summation of rotated lines which in the limit of a large number of views and in 2D will tend to the function $^1/r$. Therefore the ramp filter will suffice as an attempt to deconvolve eqn.(21). Firstly a 2D Fourier transform is performed on the summed image :

$$H_t(s_x,s_z) = \mathcal{F}_{2D}\{h_t(x,z)\}$$

$$= F(s_x,s_z) . \{\mathcal{F}_{2D}\Big[\sum_{\text{All } \alpha} p_\alpha(x,z)\Big]\}. \quad (22)$$

H_t is then ramp filtered so that :

$$H_f(s_x,s_z) = H_t(s_x,s_z) . A(s_x,s_z) , \quad (23)$$

where $A(s_x,s_z) = |\rho|$ and $\rho = \sqrt{s_x^2 + s_z^2}$. (24)

The final image is generated by an inverse 2D Fourier transform :

$$h_f(x,z) = \mathcal{F}_{2D}^{-1} \{H_f(s_x,s_z)\}. \quad (25)$$

In practice it has been found necessary to apply a gaussian weighting to the ramp filter to avoid the over-emphasis of high spatial frequencies so that :

$$A(s_x,s_z) = |\rho| . \exp(-2\pi\sigma^2\rho^2), \quad (26)$$

where the parameter σ determines the resolution of the filter.

FREQUENCY DOMAIN INTERPRETATION OF DIFFRACTION TOMOGRAPHY

A further analogy with traditional projection tomography may be found in the frequency domain approach to diffraction tomography[5,12]. The frequency domain interpretation of projection tomography is based on the Projection Theory of Fourier transforms. The equivalent process for diffraction tomography may be obtained by performing a 3D Fourier transform on the focused image from one view as given by eqn.(19) :

$$H(s_x,s_y,s_z) = F(s_x,s_y,s_z) \cdot P(s_x,s_y,s_z) \qquad (27)$$

where $P(s_x,s_y,s_z) = \mathcal{F}_{3D}\{p(x,y,z)\}$

$$= \delta(s_z - \frac{(m-1)}{\lambda_o}) \qquad (28)$$

i.e. the 3D Fourier transform of the convolving function from one view is a hemispherical delta function distribution of radius $^1/\lambda_o$ and centre $0,0,^{-1}/\lambda_o$.

This result means that the measured scattered field maps onto a hemispherical surface in the frequency domain of the object. As the object rotates with respect to the scan/illumination, the hemispherical surface rotates with respect to the object and hence completely maps the frequency domain within the bandlimit $\rho \leq \sqrt{2}/\lambda_o$. Fig.3 illustrates this interesting correspondence between the image space and frequency domain approaches by display of the 2D Fourier transform of the focusing functions shown in Fig.2. They are semi-circular delta function distributions rotated to the appropriate view angle. Fig.3 also serves to illustrate that the effect of a finite aperture is to attenuate the high frequency components of the measured data[9].

The frequency domain process gives confidence in the use of a ramp filter (as indicated by eqn.(23)). The Fourier transform of the overall convolving function within eqn.(22) is given by the sum of the rotated semi-circles (two of which are shown in Fig.3). If a sufficiently large number of views through a total angular range of 360° are used, this sum tends to the function $^1/\rho$.

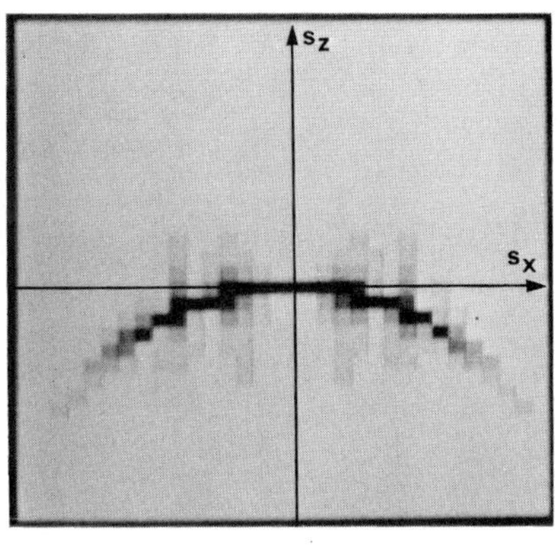

(a) View angle 0°

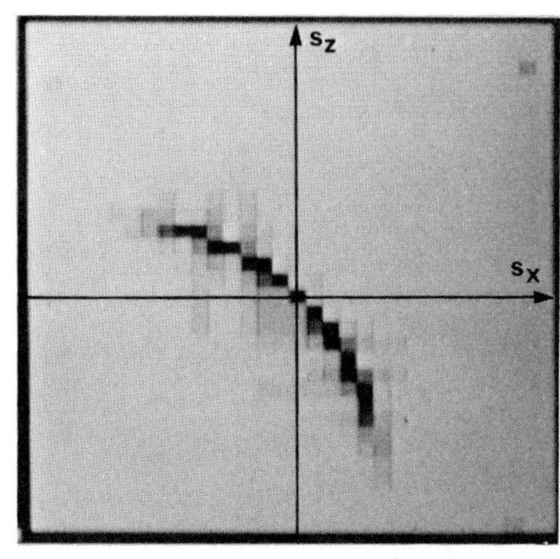

(b) View angle 45°

Fig.3. 2D Fourier transform of the single view defocusing functions shown in Fig.2.

EXPERIMENTAL RESULTS

To demonstrate the technique of section (2), ultrasonic experiments were performed at frequencies of 1MHz and 5MHz in a water tank. The objects were chosen to be uniform in the y direction so that 1-D scans would be sufficient (though not ideal). Since this technique is a synthetic aperture process, the receive transducer was lensed so that it approximated to an isotropic receiving probe. Initially measurement of the illumination function at a frequency of 5MHz (λ = 0.3mm) was performed with the object removed. The resulting amplitude and phase along the aperture is shown in Fig.4.

The object was then replaced and rotated with respect to the fixed scan/illumination to provide 25 views at successive increments of 7.2° through 180°. Each view contained 64 samples in an aperture of 77mm and at a distance, z_j = 83mm. If the measured field at a particular view is u_t, then the field due to scattering by the object, u, is found by modifying eqn.(5) so that :

$$u(x_j) = \{u_t(x_j) - u_i(x_j)\}/u_i(x_j) \tag{29}$$

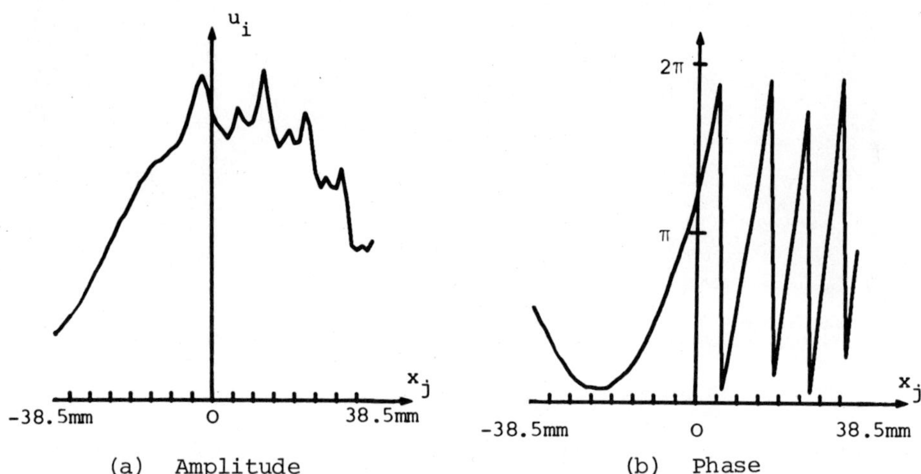

Fig.4. Amplitude and phase of the illumination function on the recording aperture at 5MHz.

This modification is necessary because, as may be seen from Fig.4, the illumination is not a plane wave over the object region. The normalization of eqn.(29) was performed on each of the 25 views and the resulting data were used to generate 2D focused images of the object, which is shown in Fig.5a. It is an ensemble of parallel 1mm diameter metal wires. The focused image at a particular view, 52°, is shown in Fig.5b. Fig.5c illustrates the summation of all 25 focused images and Fig.5d is the result of the ramp filtering process described by eqns.(22) to (26). The resolution in Fig.5d is approximately 2mm (7λ) and is primarily determined by the receive transducer being too directive[9].

In order to investigate the imaging process on a more general target, a block of reticulated polyurethane foam of the type suggested as material for the construction of tissue-like phantoms[14], was used. The block of foam had an internal structure which, as shown in Fig.6 and schematically in Fig.7, simulated both scattering and velocity modulation effects in tissue. The frequency was lowered to 1MHz to reduce attenuation.

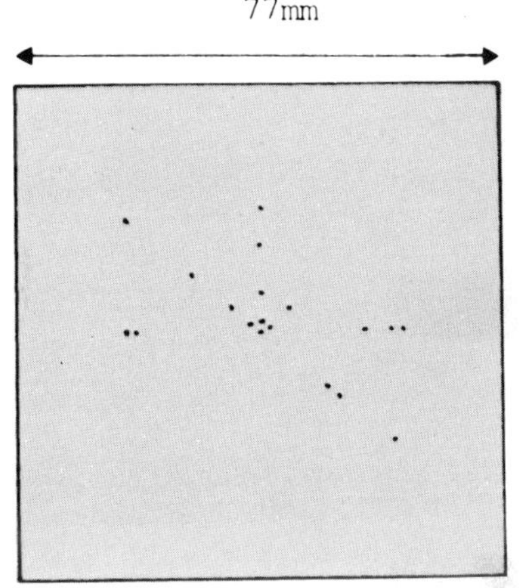

(a) Position of wires within the image area.

(b) Focused image from one view at 52°.

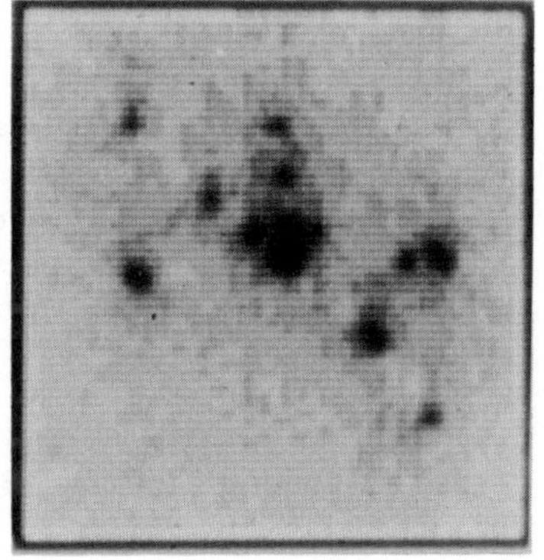

(c) Addition of 25 focused images.

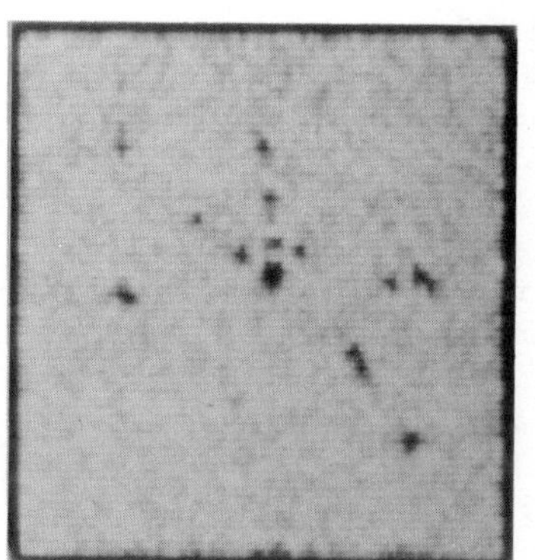

(d) Ramp filtered image.

Fig.5. Focused image technique for reconstruction of an ensemble of metal wires.

Fig.6. Foam phantom used at a frequency of 1MHz (λ_0=1.5mm)

25 views of the object through a total angular range of 180° were recorded. Each view contained 64 samples in an aperture of 102mm at a scan distance, z_j, of 115mm. Each view was subjected to the normalization procedure of eqn.(29) and then used to generate a focused image. The result of the ramp filter (using the gaussian taper of eqn.(26) where $\sigma = .75\lambda_0$) on the summation of all 25 focused images is shown in Fig.8. The amplitude of the image is displayed in Fig.8a where all 4 wires are well resolved and the hole is visible though somewhat misshapen. However, it is interesting to note that the phase of the image, as displayed in Fig.8b, shows very clearly the internal structure of the object.

The foam material is known to have an attenuation of approximately 0.5dB/cm at 1MHz and to present a 1% velocity perturbation (with respect to water)[14]. There are several possible explanations for the appearance of the images of Fig.8, none of which have, as yet, been fully explored. The object distribution, as given by eqn.(3), is a real function which will not allow a phase variation in its image (other than 0 or π). The small phase variation in the image may be due to attenuation in the object. If attenuation may be introduced by the existence of a complex compressibility (i.e. inclusion of a loss factor comparable to that for a complex permittivity in the microwave case) then the object, $f(\underline{r}_1)$ will be complex and hence the image distributions of Fig.8 may be justified accordingly.

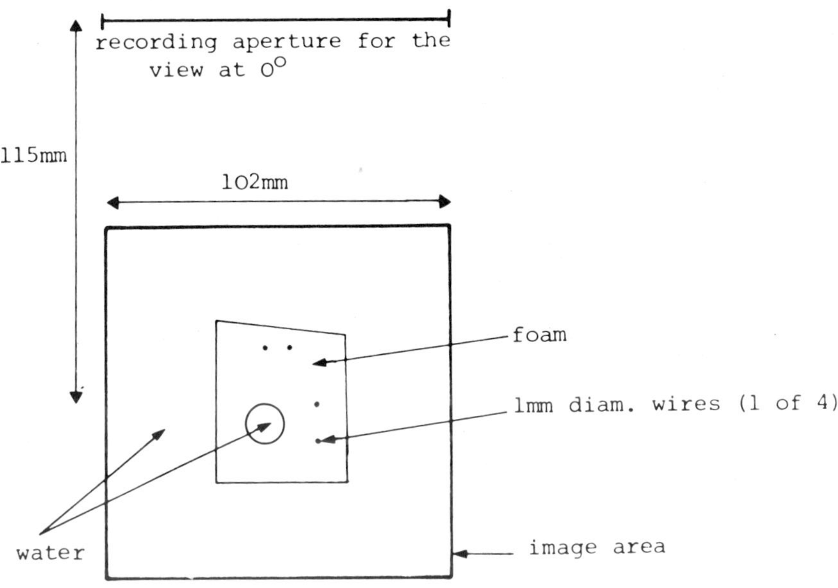

Fig.7. Schematic diagram of the foam phantom and its relationship to the recording aperture for one view.

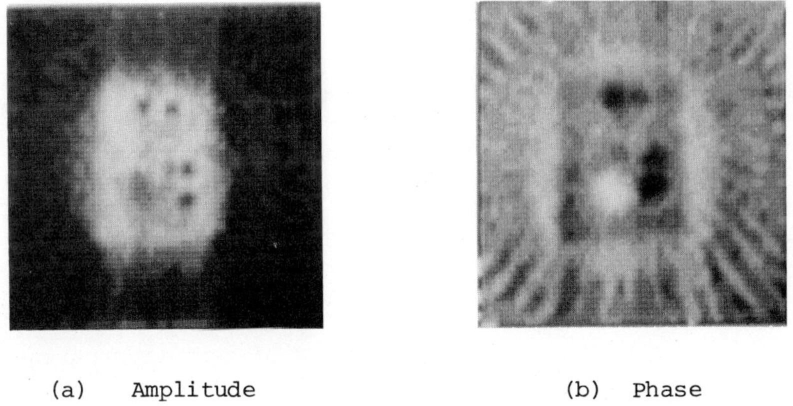

(a) Amplitude (b) Phase

Fig.8. Reconstruction of foam object by focused technique.

ACKNOWLEDGEMENTS

The authors would like to thank Mr. I. Price of the Northern General Hospital, Sheffield, for supply of the foam material and also to acknowledge the financial support of the Science and Engineering Research Council.

REFERENCES

1. J.F. Greenleaf, R.C. Bahn, Clinical imaging with transmissive ultrasonic computerised tomography, IEEE Trans. Biomed. Eng. 28:177 (1981)
2. S.A. Johnson, J.F. Greenleaf, M. Tanaka, B. Rajagopalan, R.C. Bahn, Algebraic and analytic inversion of acoustic data from partially or fully enclosing apertures, in : "Acoustical Imaging Vol.8", A.F. Metherell ed., Plenum Press (1978)
3. S.J. Norton, M. Linzer, Ultrasonic reflectivity imaging in three dimensions : Exact inverse scattering solutions for plane, cylindrical and spherical apertures, IEEE Trans. Biomed. Eng. 28:202 (1981)
4. W.J. Berkhout, J. Ridder, L.F. V.D. Wal, Acoustical imaging by wave field extrapolation. Pt.1 - Theoretical considerations, in : "Acoustical Imaging Vol.10", P. Alais, A.F. Metherell, eds., Plenum Press (1982)
5. R.K. Mueller, M. Kaveh, G. Wade, Reconstructive tomography and applications to ultrasonics, Proc. IEEE. 67:567 (1979)
6. M.F. Adams, A.P. Anderson, Synthetic aperture tomographic (SAT) imaging for microwave diagnostics, Proc. IEE, pt.H, 129:83 (1982)
7. M.F. Adams, A.P. Anderson, Tomography from ultrasonic diffraction data : Comparison with image reconstruction from projections, in : "Acoustical Imaging Vol.10", P. Alais, A.F. Metherell, eds., Plenum Press (1980)
8. P.M. Morse, K.U. Ingard, "Theoretical Acoustics", McGraw-Hill (1968)
9. M.F. Adams, "3D image reconstruction techniques for long wavelength diagnostics", Ph.D Thesis, University of Sheffield, U.K. (1982)
10. J.R. Shewell, Wolf, E, Inverse diffraction and a new reciprocity theorem, J. Opt. Soc. Am. 58:1596 (1968)
11. J.W. Goodman, "Introduction to Fourier Optics", McGraw-Hill, (1968)
12. E. Wolf, 3D structure determination of semi-transparent objects from holographic data, Optics Commun. 1:153 (1969)
13. H.J. Scudder, Introduction to computer aided tomography, Proc. IEEE, 66:628 (1978)
14. R.A. Lerski, T.C. Duggan, J. Christie, A simple tissue-like ultrasound phantom materials, British J. Radiol. 55:156 (1982)

A CLINICAL PROTOTYPE ULTRASONIC TRANSMISSION TOMOGRAPHIC SCANNER

J. F. Greenleaf*, J. J. Gisvold+, and R. C. Bahn†

*Department of Physiology and Biophysics,
+Department of Diagnostic Radiology, and
†Department of Anatomic Pathology, Mayo Clinic,
Rochester, MN 55905

INTRODUCTION

Ultrasound is a non-invasive and non-ionizing method of imaging soft tissue which may have better patient acceptance than x-ray mammography and, in the B-scan mode, may be useful in scanning patients with radiographically dense breasts.[1] Several investigators have been developing ultrasound B-scan technology for imaging the breast.[2,3]

Among others,[4] we have been studying ultrasonic computer assisted tomography for the past several years[5] with the rationale that quantitative images of speed and attenuation obtained with tomography should provide more information than qualitative images of backscatter obtained with B-scans or qualitative images of x-ray attenuation obtained with x-ray mammography.

The purpose of this paper is to report a prototype transmission ultrasonic computerized tomography scanner designed to obtain data required for computing images of acoustic speed and attenuation within coronal planes through the breast. The scanner was designed to obtain data quickly and with no discomfort to the patient. To conserve design effort, the scanner utilizes mechanical scanning and rotation with individual transducers and is of relatively simple design. To date, more than 160 patients have been scanned with good reliability and without adverse consequences to the patient.

METHODS

The scanner is a "first generation" translate-rotate mechanism in which four transducers (Figure 1) are mounted above one another vertically. The patient lies prone with a breast suspended in the water tank and the transducers are scanned in a coronal plane through the breast.

Fig. 1. Four transducer pairs are placed vertically, separated by 14 mm and are scanned in a translate-rotate fashion obtaining 201 samples on each of 60 profiles. The scanner then lowers by 7 mm and repeats the scan resulting in data for eight sets of speed and attenuation images of coronal planes separated by 7 mm through the breast. Some patients having large breasts require the scanner to be lowered by 49 mm and the procedure is repeated to obtain a total of 15 independently imaged planes.

Usually the top transducer pair are 3.5 MHz transducers since the breast is largest near the chest wall. The remaining three pairs of transducers have center frequencies near 5 MHz. The receiver transducers have lenses focused at about 10 centimeters into the 20 cm gap between the transducers. The transducers scan in four parallel planes separated by 14 mm.

The four transmitters are excited sequentially with 300 volt pulses as they are scanned across the breast. The pulses are triggered from an optical position encoder which determines when the transducers have travelled a predetermined distance from the last pulse (Figure 2).

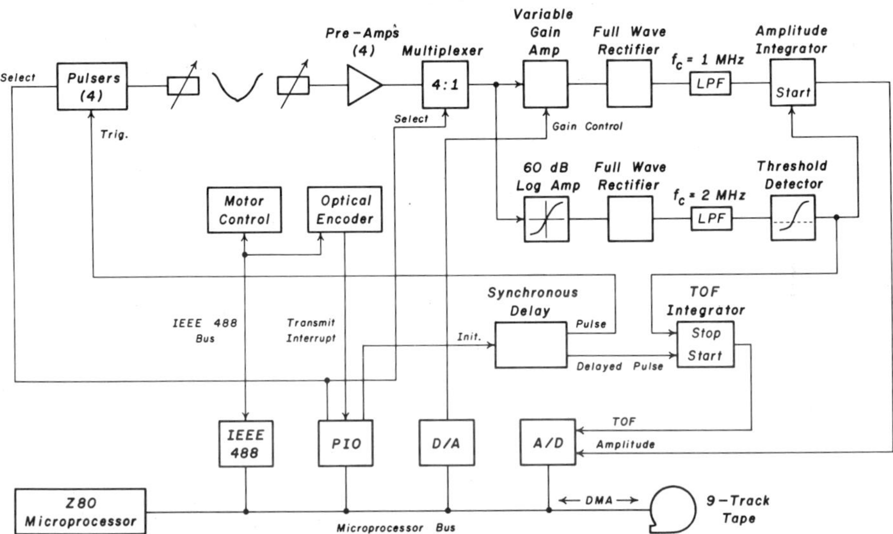

Fig. 2. Schematic of the signal chain and microprocessor control of the scanner. Each of the four pairs of transducers has an attenuator-preamplifier, variable gain amplifier, full-wave rectifier, and low-pass filter. The variable gain amplifier is controlled by the microprocessor depending on the measured amplitude of the previous signal and maintains the signal level within the linear range of the integrator. The position of the transducer is measured with an optical encoder which is monitored by the microprocessor so that transmit triggers are sent when the transducers have moved an appropriate distance. To minimize jitter, the transmit triggers are aligned with the 100 MHz clock of the synchronous delay. Data are stored on the 9-track tape with a home-made DMA interface.

Each receiver chain consists of a computer-controlled attenuator, with a 20 dB amplifier, and a 4:1 multiplexer and then an amplifier with computer-controlled gain, a full-wave detector, a low pass filter, and then either an integrator for measuring the amplitude (area of the signal in the first two microseconds of the received signal) or a threshold detector for the time-of-flight detection.

The time-of-flight detector consists of an integrator which is started with a trigger from a digital delay (which is synchronized with the transmitter) and which is turned off by the threshold detector. The resulting voltage is held and digitized to 14 bits. The amplitude signal is also digitized to 14 bits.

Previous to scanning each patient, the arrival time "walk", due to change in signal amplitude, and the receivers are calibrated by switching the attenuator over a range of 80 dB in 5 dB steps for each of eight gain levels in the variable gain amplifier while transmitting through water. The resulting calibration tables are used by the microprocessor to control the gain of the amplifier as the scanner traverses the breast thus maintaining a good dynamic range.

For each traverse of the transducers, the system obtains 201 samples over a distance of 20 cm. In approximately two and a half minutes, the scanner takes 60 views, separated by 3°, and then lowers by 7 mm and obtains another set of data for four additional sets of images of speed and attenuation, the positions of which are interlaced between the four planes of the first scan. Thus, in the average sized breast, we obtain images of speed and attenuation in eight planes separated by 7 mm in about 5 minutes.

The data are reconstructed off-line using a filtered back projection algorithm which assumes straight line propagation of the energy from the transmitter to the receiver. Each image requires about 20 seconds of computing time on a 7/32 Perkin Elmer computer using an AP120B Floating Point Systems array processor.

The resulting reconstructions are displayed with a digital display device (Grinnell) and viewed on a TV monitor from which the speed and relative attenuation can be measured in small regions of the tissue using a numerical "biopsy" cursor and associated program. The reconstructions are 118 pixels on a side and are displyed with density resolution of six bits. All reconstructions for each patient are stored on floppy discs for ease of access.

RESULTS

We have tested the scanner using a phantom made by Professor Earnie Madsen at the University of Wisconsin. Figure 3 illustrates an x-ray reconstruction of the phantom using a 1 cm thick slice. The phantom has several tumor and cyst like inclusions to simulate the speed and attenuation of breast tissues.

The ultrasound reconstructions of speed and attenuation in the plane shown in Figure 3 are shown in Figure 4. Because ultrasonic energy does not travel either in a plane or in a straight line, the

(E. Madsen)

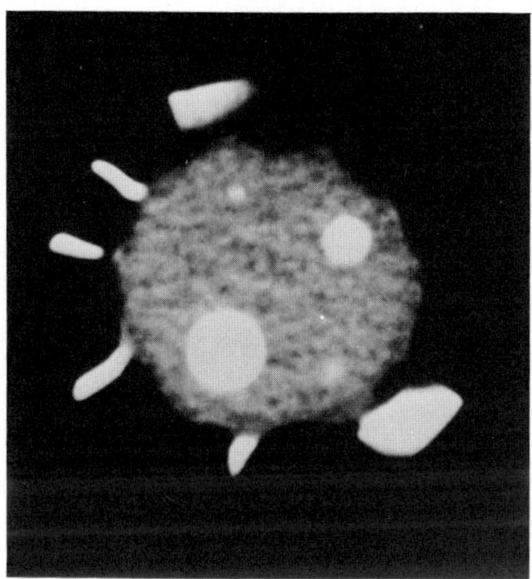

Fig. 3. Reconstruction of x-ray densities within coronal plane through breast phantom. Cyst-like and tumor-like lesions are visible. Phantom is described elswhere in detail.[5] (Reproduced with permission from J. F. Greenleaf and R. C. Bahn, 1980 Ultrasonics Symposium Proceedings, pp 966-972 (November) 1980.)

ultrasound reconstructions have aberrations causing blurring and distortion. However, the images are fairly accurate geometrically if somewhat in error quantitatively. These images can be compared to earlier reconstructions of the same phantom which were reported elsewhere.[6] A great improvement in the fidelity of the images can be appreciated.

Reconstructions of speed and attenuation within coronal planes through a breast at the level of a cancer are shown in Figure 5. The x-ray mammogram indicated a 3 cm indeterminant mass at the upper inner quadrant in the right breast. At surgery was found a grade IV adenocarcinoma. The lesion in the ultrasound image is apparently the bright region at 2:00 in the image of speed in the right breast.

The ultrasound images shown in Figure 6 are from a patient in which the x-ray mammogram indicated a 2.5 cm cancer at the lower inner quadrant of the left breast. Surgery found a grade IV adenocarcinoma, infiltrating ductal type, 1.6 cm in diameter. The lesion is apparently at 7:00 in the images of the left breast. The region of high speed and attenuation at 3:00 in the same breast was found to have apparently been due to a rib upon subsequent evaluation of the excised breast and associated images.

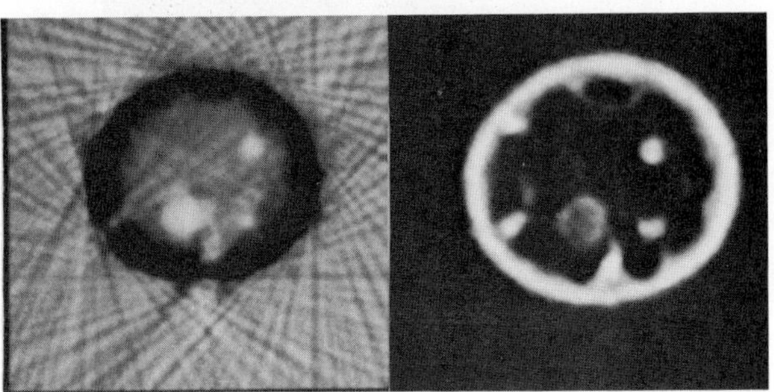

Fig. 4. Ultrasound reconstruction of speed (left) and attenuation (right) through coronal plane imaged by x-ray in Figure 3. Resolution is much higher than obtained previously.[5]

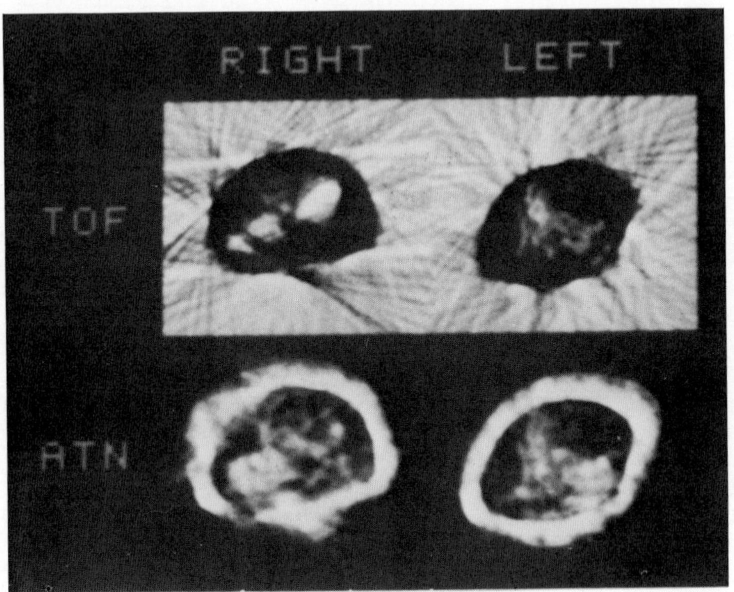

1/27

Fig. 5. Ultrasound reconstruction of speed (upper panels) and attenuation (lower panels) in breasts of patient having grade IV invasive adenocarcinoma (3 cm) in the upper inner quadrant of the right breast.

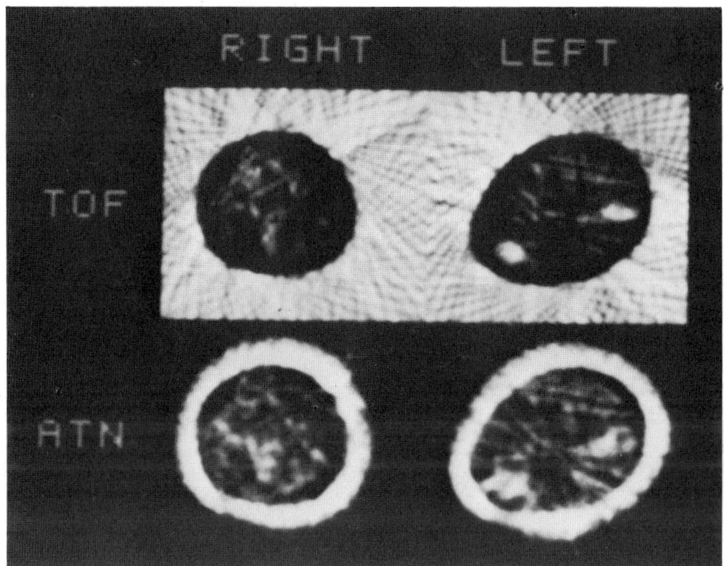

2/3

Fig. 6. Ultrasound reconstruction of speed and attenuation in breasts of a woman having grade IV adenocarcinoma, 1.6 cm in diameter, in the lower inner quadrant of the left breast.

DISCUSSION

Transmission tomographic ultrasound can be used to obtain quantitative images of speed, and qualitative images of attenuation, in the breasts of women using a simple mechanical scanner and electronics and using reconstruction methods that assume the energy travels in a straight line.

However, acoustic energy propagates according to a wave equation rather than as particles as in the case of x-ray, therefore, the waves can be deflected by a variety of mechanisms such as reflection, refraction, and diffraction. The assumption that the energy travels in a straight line greatly simplifies the mathematics, but may over simplify the model for wave propagation.

The process of solving the wave equation for the distribution of material properties that cause the deflection of energy, given measurements of the scattered wave, is called "inverse scattering" since it is the inverse of solving for the scattered wave given the material characteristics of the scatterers. Several investigators are studying inverse scattering methods as they pertain to ultrasonic energy.[7-9] Several reviews of ultrasonic tomography methods are available.[10,11]

Most methods of inverse scattering require special geometries (e.g., plane wave insonification) or often, these methods require making difficult measurements (e.g., absolute phase) in order to be useful. One method of backward propagating the received waves back to the center of the object previous to calculating arrival time, has been demonstrated by us with promising results.[12]

Whether better approximations to the equations governing wave propagation will ultimately result in images with better fidelity is yet to be seen, although some preliminary results are available from simulations[8] and from some simple experiments.[12]

Clinical results from the first 160 patients scanned with the scanner reported here will be reported elsewhere.

It is as yet unclear whether inverse scattering methods will increase the fidelity of these images and indeed, it is unclear whether such increased fidelity will increase the diagnostic utility of the ultrasound methods over the x-ray mammography methods.

REFERENECES

1. T. G. Frazier, C. Cole-Beuglet, A. Kurtz, B. Goldberg, and S. Ryan, Further evaluation by ultrasound of mamographically determined breast dysplasia, J. Surg. Oncol. 19:69 (1982).

2. T. Tellings, T. S. Reeve, G. Kassoff, B. Barraclaugh, and T. Croll, Ultrasonic assessment of symptomatic patients with breast disease, in: "Proceedings of the 25th Annual Meeting of AIVM," 1980, New Orleans, LA.

3. Proceedings of the 2nd International Congress on the Ultrasonic Examination of the Breast, London, June 22-23, 1981, Institute of Cancer Research, Clifton Avenue, Sutton, Surrey SM2 5PX, United Kingdom. Ultrasound Med.Biol. 8, no.4. 1982.

4. P. L. Carson, C. R. Meyer, A. L. Scherzinger, and T. V. Oughton, Breast imaging in coronal planes with simultaneous pulse echo and transmission ultrasound, Science 214:1141 (1981).

5. J. F. Greenleaf and R. C. Bahn, Clinical imaging with transmissive ultrasonic computerized tomography, IEEE Trans. Biomed. Eng. BME-28(2):177 (1981).

6. E. L. Madsen, J. A. Zagzebski, G. R. Frank, J. F. Greenleaf, and P. L. Carson, Anthropomorphic breast phantoms for assessing ultrasonic imaging system performance and for training ultrasonographers: Part II, J. Clin. Ultrasound 10:91 (1982).

7. R. K. Mueller, Diffraction tomography I: The wave equation, Ultrasonic Imaging 2:213 (1980).

8. R. K. Mueller, M. Kaveh, and R. D. Iverson, A new approach to acoustic tomography using diffraction techniques, in: "Acoustical Imaging," A. F. Metherell, ed., Plenum Press, New York (1980)

9. A. J. Devaney, A filtered backprojection algorithm for diffraction tomography, (unpublished).

10. R. K. Mueller, M. Kaveh, and G. Wade, Acoustical reconstructive tomography and applications to ultrasonics, Proc. IEEE 67:567 (1979).

11. J. F. Greenleaf, Computerized transmission tomography, in: "Methods of Experimental Physics - Ultrasound," P. D. Edmonds, ed., Academic Press, New York (1981).

12. J. F. Greenleaf, P. J. Thomas, and B. Rajagopalan, Effects of diffraction on ultrasonic computer-assisted tomography, in: "Acoustical Imaging," J. Powers, Plenum Press, New York, Vol. 11 (In Press).

ULTRASONIC TOMOGRAPHY FOR DIFFERENTIAL THERMOGRAPHY

M. J. Haney and W. D. O'Brien, Jr.

Bioacoustics Research Laboratory
Department of Electrical Engineering
University of Illinois
1406 W. Green Street
Urbana, Illinois 61801 USA

INTRODUCTION

This paper describes work in progress in the study of ultrasound computer aided tomography (UCAT) and its application to differential thermography. There are many situations in which it is desirable to determine the amount of induced heating generated by applied hyperthermia (microwave or ultrasound). However, it is not always possible or safe to insert a temperature sensitive probe into the subject. The application of assessing tissue temperature from the temperature dependence of ultrasonic speed has been suggested by others (Bowen et al., 1979; Nasoni et al., 1979; Rajagopalan et al., 1979). But it may be possible to refine the assessment of temperature change from the simultaneous determination of the acoustic speed and the ultrasonic attenuation coefficient. A method is outlined for producing maps of temperature change after heating.

Pulses of ultrasound are transmitted through the subject. Time of flight and frequency content measurements are made to analyze the speed of propagation and attenuation coefficient of the regions of the subject. Algebraic reconstruction, based on an interpolated model of the ray paths, is used to form transit time (inverse speed) and attenuation coefficient images. (An error

---

This research is funded in part by a grant from the National Institutes of Health (GM 24994), and by an unrestricted gift from Ultrasonic Research, Inc.

estimation is performed to be used in later analyses.) Differential images (before and after heating) are used to estimate temperature change.

EQUIPMENT

Our apparatus permits equiangle divergent beam data to be collected. Signals from a Perkin Elmer 7/32 computer are sent to a SYM-1 microcomputer, which in turn controls the positioning of two vertical posts in a water filled tank (see Figure 1). These posts trace out horizontal fan shaped sectors at arbitrary viewing angles.

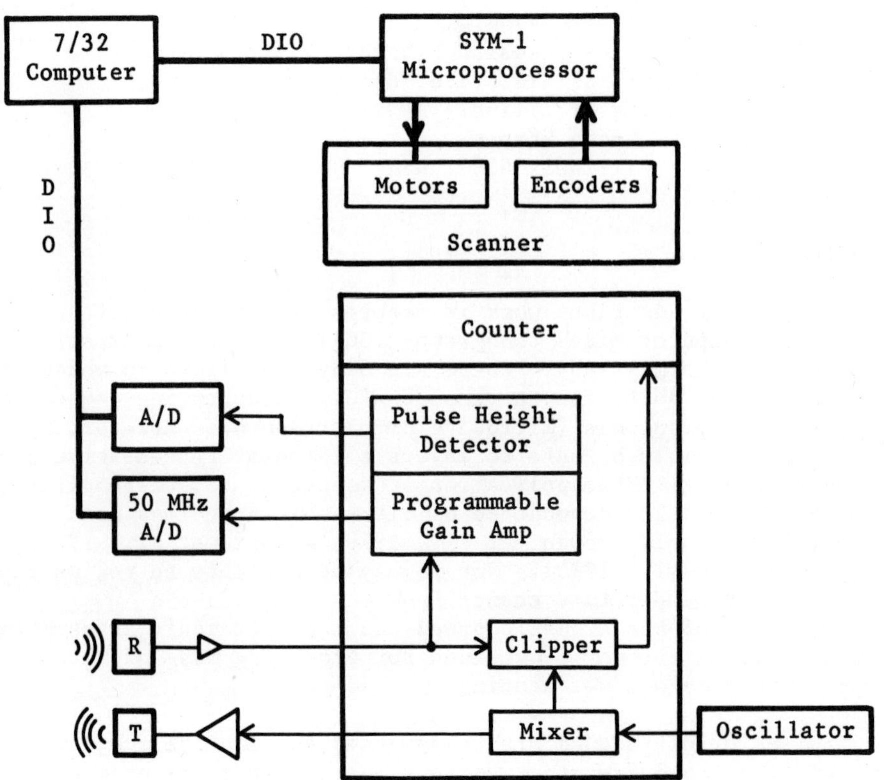

Figure 1: Equipment Block Diagram

A Hewlett-Packard 8660B Frequency Synthesizer supplies a reference signal (1 to 10 MHz) to a wave packet forming mixer (1/2 to 128 cycles). Pulses from this unit are amplified and fed to a Panametrics ultrasonic transducer mounted on one of the posts in the tank. The other post holds a receiving transducer connected to a preamplifier and filters, then to a pulse height detector and time of flight counter (Hewlett-Packard 5328A Universal Counter). Two analog to digital convertors are used to digitize the pulse height and the pulse itself. Samples are placed in the tank between to two transducers in a 20 cm diameter sample region.

METHODS

There are many alternatives in collecting ultrasound tomography data. Using threshold level detection, coarse time of flight measurements can be made. The received ultrasound pulse is digitized and recorded (8 bits @ 50 MHz) for fine time of flight and frequency content analyses.

The time of flight measurement represents the line integral of the transit times for the pulse through the intervening tissue. Tomographic reconstruction of this measurement yields the time per spatial resolution unit (inverse speed) of the tissue regions. Although one can normalize the time of flight with respect to the transit time through water, and thus reconstruct indices of refraction, the normalization would only have to be reversed in later stages of calculation. Note: the digitization of the received pulse allows for more accurate time of flight measurements by cross correlation methods (more accurate than threshold level detection).

Three methods are readily available for studying absorptive attenuation. The least accurate but most simple is to use narrow band pulses and measure the signal height within the received pulse. This method measures both absorption and scattering. Two more absorption sensitive (scattering insensitive) approaches are suggested by Kak (1979). By obtaining the frequency spectrum from the FFT of the received pulse, the attenuation can be determined by the ratio of spectral energies, or from the shift of the median frequency of the received pulse compared to a reference pulse transmitted through water. In all of the methods, the natural log of the attenuation gives the integral of the attenuation coefficient through the intervening tissue.

Reconstruction is based on an algebraic model of the physical system. Consider the time of flight per unit spatial resolution, or the attenuation coefficient, as a function f(x,y). Then the measured data $g(\chi,\phi)$ for a divergent beam tomography system are given by

$$g(\chi,\phi)=\int f(x,y)ds \qquad [1]$$

Represented as a discrete sum,

$$g(\chi,\phi)=\sum\sum a(\chi,\phi;i,j)f(i,j) \quad [2]$$

where $a(\chi,\phi;i,j)$ is the weight that $f(i,j)$ contributes to the (χ,ϕ) projection (see Figure 2).

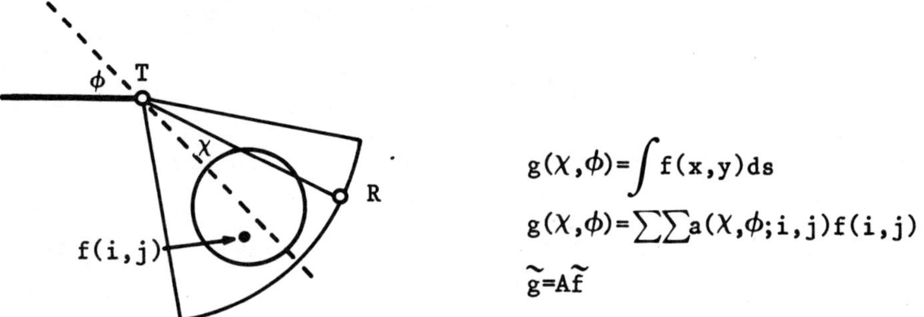

Figure 2: Fan Beam Geometry

Most models assume $a(\chi,\phi;i,j)$ is equal to 0 or 1, or the length of the portion of the ray that passes through a box (pixel) around point (i,j) of the object (see Figure 3). These assumptions contribute a nontrivial amount of aliasing error in the name of computational convenience. In analogy to one dimensional signal processing, these assumptions are equivalent to representing a function by a similar valued stairstep (piecewise constant) function.

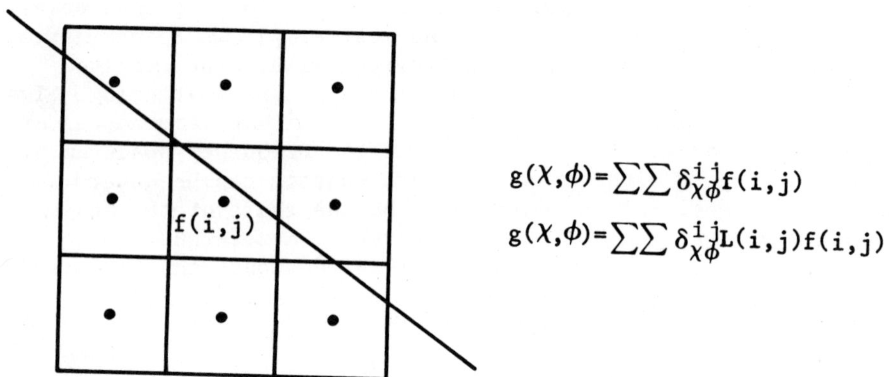

Figure 3: Conventional Models

Ideally, the contribution of each pixel should be calculated from the line integral of the function in that pixel. However, this would require an interpolation calculation involving every pixel in the image. Computationally, this would be very expensive.

A simple compromise is available through linear interpolation. For the region between any 4 sample points (e.g. (x,y) bounded by (i,j), (i+1,j), (i,j+1), and (i+1,j+1)):

$$f(x,y)=(i+1-x)(j+1-y)f(i,j) + (x-i)(j+1-y)f(i+1,j) + (i+1-x)(y-j)f(i,j+1) + (x-i)(y-j)f(i+1,j+1) \quad [3]$$

The line integral of the projection through this region can be evaluated explicitly. Assume the ray passes through the points (x0,y0) and (x0+Δx,y0+Δy), both on the edges of the region (see Figure 4). Then,

$$\begin{aligned}\int f(x,y)ds = {} & f(i,j)[\ L(i+1-x0)(j+1-y0) - L(i+1-x0)\Delta y/2 - L(j+1-y0)\Delta x/2 + L\Delta x\Delta y/3] \\ & + f(i+1,j)[\ L(x0-i)(j+1-y0) - L(x0-i)\Delta y/2 + L(j+1-y0)\Delta x/2 - L\Delta x\Delta y/3] \\ & + f(i,j+1)[\ L(i+1-x0)(yo-j) + L(i+1-x0)\Delta y/2 - L(y0-j)\Delta x/2 - L\Delta x\Delta y/3] \\ & + f(i+1,j+1)[\ L(x0-i)(y0-j) + L(x0-i)\Delta y/2 + L(y0-j)\Delta x/2 + L\Delta x\Delta y/3]\end{aligned} \quad [4]$$

where L is the euclidean length of the portion of the ray passing through the region.

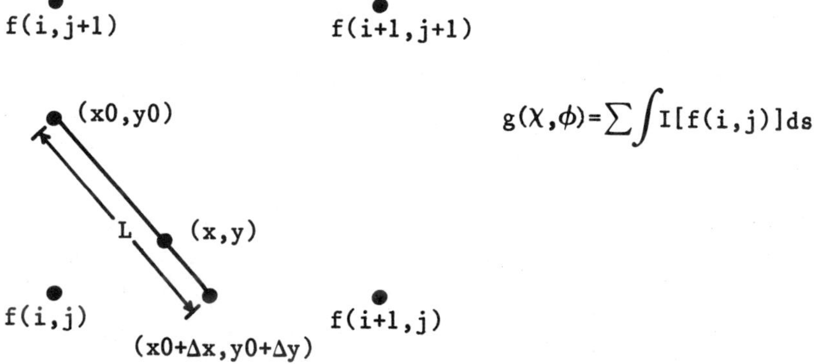

Figure 4: Proposed Linear Interpolation Model

Three notes are in order. First, although more involved, the calculation of $a(\chi,\phi;i,j)$ for interpolation can be performed once and saved, to be used for many reconstructions. Thus the computational expense is relatively unimportant. Second, in the previous assumptions, only a small proportion of the $a(\chi,\phi;i,j)$ are nonzero. In this integrated interpolation, only approximately twice as many values are nonzero. Thus sparse array data storage and processing methods can still be used. Finally, to compare to one dimensional signal processing, this interpolation is equivalent to approximating a function by a similarly valued sawtooth (piecewise linear) function, with correspondingly less aliasing.

Using vector notation,

$$\tilde{g}=A\tilde{f}+\tilde{n} \tag{5}$$

where $\tilde{n}$ represents the noise present in the collected data $\tilde{g}$. Reconstruction is performed using the projection iterative method (Huang, 1977). Let g(m) be the m′th value of the vector $\tilde{g}$, and let $\tilde{a}(m)$ be a vector formed by the m′th row of A. Then the k+1 iteration is given by:

$$\tilde{f}^{k+1}=\tilde{f}^{k}+\frac{[g(m)-\tilde{f}^{k}\tilde{a}(m)]}{\tilde{a}(m)\tilde{a}(m)}\tilde{a}(m) \tag{6}$$

where m = k modulo M (the number of values in $\tilde{g}$). After each M iterations, let

$$\tilde{n}=\tilde{g}-A\tilde{f}^{k} \tag{7}$$

Termination of the iteration is controlled jointly by $\tilde{n}$ satisfying the statistics of the noise (measured independantly) and $\tilde{f}$ reaching a cyclic limit. Nonnegative value constraints and matrix order shuffling are used to improve the rate of convergence.

Algebraic reconstruction has been chosen for several reasons. First is flexibility. Missing data (data lost due to drop outs or rejected because of saturation) can be handled by simply eliminating rows of the A matrix. Second, uniformity of data collection is not required. Horn′s theoretical approach (1978) requires reasonably continuous data. Davidson and Grunbaum′s approach (1979) works for parallel beam data if all of the angles are known ahead of time. Thus run time difficulties (such as drop outs) result in large amounts of additional computation. Algebraic reconstruction works with what is available. Third, the majority of calculations involved are vector additions and dot products. The single instruction multiple data (SIMD) super computers of tomorrow will readily and efficiently handle such calculations. Finally, algebraic methods of reconstruction allow a better approach to estimating error.

Two error estimates are readily available by first computing the pseudo inverse matrix V for the matrix A (Hall, 1979). By rearranging the original vector equation,

$$\tilde{g}=A(\tilde{f}+\tilde{e}) \rightarrow V\tilde{g}-\tilde{f}=\tilde{e} \tag{8}$$

From this, the error vector is given by $\tilde{e}=V\tilde{n}$. The second approach is a sensitivity analysis:

$$\sum \frac{f(i)}{g(j)} * \sigma^2 = V\tilde{\sigma}^2 = \tilde{\sigma}_f^2 \qquad [9]$$

where σ^2 is the varience of the through water sound measurements. A third method, suggested by Katz (1978), can be used, but it only estimates the total error over the entire image. This can be used to estimate the overall error in the temperature calculation, but a point by point error estimate is desired for later phases of the thermographic computations.

Temperature change estimation depends on the observation that the speed of propagation and attenuation coefficient are both temperature sensitive. For example, with water, the speed of propagation, measured in meters per second, can be approximated by c = 1403 + 5T + higher order terms (Kinsler and Frey, 1962), where T is measured in degrees celsius. Attenuation coefficients exhibit similar behavior (Fry and Dunn, 1962). Thus if one were to reconstruct a speed of propagation image c1(x,y) before heating, and another c2(x,y) after heating,

$$c1(x,y)-c2(x,y)=kc(x,y)\Delta T(x,y) \qquad [10]$$

where kc(x,y) is the thermal coefficient for the tissue at the point (x,y), and ΔT is the change in temperature. A similar calculation can be performed for attenuation images.

As a first approximation, setting kc(x,y) and ka(x,y) equal to the thermal coefficients for water will yield an estimate of differential temperature. A better estimate can be obtained by using those values of kc and ka available through the literature (Erikson et al., 1974, or Goss et al., 1978). Data available on the temperature dependence of the ultrasonic attenuation coefficient and the speed of sound in vivo are quite sparse. However, what is available, mostly in vitro, suggests that the speed of sound increases slowly with temperature in the range of 0.5 to 2 m/s per degree C around 40 degrees for non-fatty soft tissue (Bamber and Hill, 1979; Bowen et al., 1979; Nasoni et al., 1979; Rajagopalan et al., 1979) and that the ultrasonic attenuation coefficient decreases slowly with temperature for some non-fatty tissues and increases for others with this dependence strongly dependent upon frequency and ambient temperature (Dunn and Brady, 1973 and 1974; Bamber and Hill, 1979). Thus, before the use of ultrasonic propagation property measurement can be used to assess tissue temperature change and related parameters, a much more detailed database must be obtained.

Given a table of thermal coefficients for common tissues, it is necessary to classify each point (x,y) according to its tissue type (liver, kidney, etc.), forming a map of the tissues conforming to the image. The easiest approach to this is to have a human operator outline and identify the tissue regions from the reconstructed images. These outlines, with the identified tissues, form a classification map which directs the selection of kc and ka from the table of literature values. It is not necessary to classify every point of the image. Only the regions of interest need be identified.

ADDITIONAL STUDY

The final phase of this study will involve automated tissue region classification (instead of or in addition to operator specified tissue region classification) for assigning thermal coefficients. Parametric classification, based on the transit time and attenuation coefficient calculations, can be used to identify points of maximal confidence. Nonparametric (Bayesian) classification using an expected value - expected position model appropriate to the region being studied can also be used to identify points of high confidence. The error estimate plays an important role here, since it directly affects the confidence of the classification. Once these candidate points have been classified, constrained region growing can be used to classify neighboring points. Operator intervention in any of these steps would improve the results. Once a classification map has been generated, thermal coefficients can be assigned as described in the previous section.

REFERENCES

Bamber, J. C. and Hill, C. R., 1979, Ultrasonic attenuation and propagation speed in mammalian tissues as a function of temperature, Ultrasound in Medicine and Biology, vol. 5, no. 2, pp 149-157.

Bowen, T., Conner, W. G., Nasoni, R. L., Pifer, A. E., and Sholes, R. R., 1979, Measurement of the temperature dependence of the velocity of ultrasound in soft tissue, in: "Ultrasonic Tissue Characterization II," M. Linzer, ed., U.S. Government Printing Office, Washinton, D.C., NBS Special Publication 525, pp 57-61.

Davidson, M. E. and Grunbaum, F. A., 1979, Convolution algorithms for arbitrary projection angles, IEEE Trans. Nuclear Science, vol. NS-26, No. 2, pp 2670-2673.

Dunn, F. and Brady, J. K., 1973, Pogloshchenie ul′trazvyeka v biologicheskikh sredakh, Biofizika 18, 1063. Translation, 1974, Ultrasonic absorption in biological materials, Biophysics 18, 1128.

Dunn, F. and Brady, J. K., 1974, Temperature and frequency dependence of ultrasonic absorption in tissue, _Proceedings 8th International Congress on Acoustics_, vol. I, p 366c.

Erikson, K. R., Fry, F. J., and Jones, J. P., 1974, Ultrasound in medicine - a review, _IEEE Trans. Sonics and Ultrasonics_, vol. SU-21, no. 3, pp 144-169.

Fry, W. J. and Dunn, F., 1962, Ultrasound: analysis and experimental methods in biological research, _in_: "Physical Techniques in Biological Research," Academic Press, New York.

Goss, S. A., Johnston, R. L., and Dunn, F., 1978, Comprehensive compilation of empirical ultrasonic properties of mammalian tissue, _J. Acoust. Soc. Am._, 64, pp 423-457.

Hall, E. L., 1979, "Computer Image Processing and Recognition," Academic Press, New York.

Horn, B. K. P., 1978, Density reconstruction using arbitrary ray sampling schemes, _Proc. IEEE_, vol. 66, no. 5, pp 551-562.

Huang, T. S., 1977, Algebraic methods of image restoration, _in_: "Digital Image Processing and Analysis," J. C. Simon, and A. Rosenfeld, ed., Noordhoff, Leyden.

Kak, A. V., 1979, Computerized tomography with x-rays, emission, and ultrasound sources, _Proc. IEEE_, vol. 67, no. 9, pp 1245-1272.

Katz, M. B., 1978, Questions of uniqueness and resolution in reconstruction from projections, _in_: "Lecture Notes on Biomathematics," Springer-Verlag, New York.

Kinsler, L. E. and Frey, A. R., 1962, "Fundamentals of Acoustics," Wiley and Sons, Inc., New York.

Nasoni, R. L., Bowen, T., Conner, W. G., and Sholes, R. R., 1979, In vivo temperature dependence of ultrasound speed in tissue and its applications to noninvasive temperature monitoring, _Ultrasonic Imaging_, vol. 1, no. 1, pp 34-43.

Rajagopalan, B., Greenleaf, J. F., Thomas, P. J., Johnson, J. A., and Bahn, R. C., 1979, Variation of acoustic speed with temperature in various excised human tissues studied by ultrasound computerized tomography, _in_: "Ultrasound Tissue Charaterization," M. Linzer, ed., U.S. Government Printing Office, Washington, D.C., NBS Special Publication 525, pp 227-233.

FURTHER RESULTS ON DIFFRACTION TOMOGRAPHY USING RYTOV'S APPROXIMATION

M. Kaveh*, M. Soumekh*, Z. Q. Lu*†, R. K. Mueller*

and J. F. Greenleaf††

* Department of Electrical Engineering
University of Minnesota
Minneapolis, Minnesota 55455

† On leave from Nankai University
People's Republic of China

†† Mayo Graduate School of Medicine
Rochester, Minnesota 55901

ABSTRACT

The inhomogeneous Helmholtz equation is expressed in terms of a general transformation of the propagating wave. Rytov's transformation is then seen as a natural one in the inversion of this equation. An updated inversion algorithm is presented and the problem of phase determination is discussed. Finally, simulation and experimental results are presented to show some of the problems with the Rytov approach in acoustic tomography.

INTRODUCTION

Diffraction tomography has been formulated as an approximate inversion of the inhomogeneous Helmholtz equation. The most basic assumption in this approach, of course, is that this type of wave equation is a reasonable representation of the mechanisms that generate the sensed signals. Furthermore, practical inversion algorithms have been devised based on the further assumption that the parameters of the object to be imaged deviate slightly from similar parameters of the homogeneous medium surrounding the object. Two methods have received special attention in diffraction tomography: methods based on Born and Rytov approximations (see for example references 1 and 2). The appealing feature of these methods is the fact that the inversion algorithms are Fourier transform based and computationally efficient. Computational efficiency and algorithmic simplicity, however, is accompanied by limitations on the ranges of validity for these methods.

In a recent paper[3], a comparison of tomograms based on the Born and Rytov approximations showed the rapid breakdown of Born's method for very simple objects of interest. Rytov's method, however, showed relatively good reconstructions of the parameter distribution within these objects[3]. This approximation, nonetheless, has peculiarities of its own (although less serious than Born's approximation) and requires special signal processing considerations in practice.

The aim of this paper is to investigate diffraction tomography based on Rytov's approximation more closely, both theoretically and experimentally. This is done to pinpoint more accurately the sources of error in the images in the hope of obtaining possible remedies for them. The presentation is as follows. A more general transformation of the sensed signal is first presented, that includes, as special cases, the Born and Rytov formulations. An updated inversion algorithm is then presented that considerably reduces interpolation errors compared to our previous reconstruction algorithm. Finally, several examples of Rytov tomograms are given and some of the sources of error in the reconstructions are delineated.

THE WAVE EQUATION AND A GENERALIZED TRANSFORMATION

As has been done previously[1-3], we consider a soft, isotropic scattering medium, single frequency plane wave insonification and a two-dimensional wave equation given by

$$\nabla^2\psi + k^2\psi = 0 \tag{1}$$

where ∇^2 is the Laplacian operator, $k(x,y)$ is a complex wave number including velocity of propagation changes and possible attenuation in the medium under study and $\psi(x,y)$ is the propagating wave. We model $k^2(x,y)$ as[4]:

$$\begin{aligned} k^2(x,y) &= k_0^2[1 + f(x,y)] \\ k_0 &= \frac{\omega_0}{C_0} + i\varepsilon \;,\quad \varepsilon \ll 1, \; |f(x,y)| \ll 1 \end{aligned} \tag{2}$$

k_0 is the complex (small attenuation) wave number of the homogeneous medium surrounding the object and $f(x,y)$ is the complex function describing the object that is to be imaged. The imaging procedure is to estimate $f(x,y)$ from measurements of $\psi(x_0,y)$ along a linear array at x_0, external to the object, for different angular orientations of the object[1-3]. This can be accomplished, approximately, by a preliminary transformation on $\psi(x,y)$.

The aim of expressing equation (1) in terms of a function of ψ is to approximate the resulting equation as a linear constant coefficient system relating the resulting function of the measurement to $f(x,y)$, this in turn making computationally efficient reconstruction possible. Thus, let $\psi_0 = A_0 e^{ik_0 x}$ be the incident wave. Following our experience with Born and Rytov approximations we define χ as a differentiable scalar function of $\alpha = \frac{\psi}{\psi_0}$. Thus let

$$\chi = g(\alpha) \text{ and } \chi_1 = \psi_0 g(\alpha) \tag{3}$$

Expressing equation (1) in terms of χ_1, α and $g(\cdot)$ one obtains[5]:

$$\nabla^2 \chi_1 + k_0^2 \chi_1 = -\psi_0 \frac{dg(\alpha)}{d\alpha} \alpha k_0^2 f + \psi_0 \frac{d^2 g(\alpha)}{d\alpha^2} (\nabla\alpha)^2 \tag{4}$$

The operator on left hand side of equation (4) is a familiar one. This is the system used in the inversion based on Born and Rytov approximations. Indeed appropriate choices of the transformation $g(\cdot)$ do lead to these two formulations. The general form of equation (4), however, gives one a chance of investigating other as well. This, without computational concerns about the actual inversion algorithm beyond the initial transformation.

An interesting feature of equation (4) is the manner in which the function of interest, $f(x,y)$, and other extraneous terms appear. Following the approach in references[1-4], the ideal equation that results in a distortionless reconstruction of $f(x,y)$ is of the form

$$\nabla^2 \chi_1 + k_0^2 \chi_1 = -b\psi_0 f \tag{5}$$

where b is a constant. Therefore, for a given $g(\cdot)$, the terms appearing on the right hand side of (4) that are extra to that in (5) will be considered as distortion terms. A good transformation would minimize these extra terms.

We will now reexamine the Rytov and Born approximations as results of special forms of $g(\cdot)$ and consider a class of transformations which includes the former two as limiting cases[5].

i) Rytov transformation. One obtains this form by noting from equations (4) and (5) that for b = 1

$$\frac{dg(\alpha)}{d\alpha}\alpha = 1 \quad \text{and } g(\alpha) = \ell n\alpha \tag{6}$$

It is important to note the very desirable property that this transformation removes the modulation of f(x,y) by a spatially dependent function that depends on the scattered wave. The distortion term in this approximation results from $D_r = -(\nabla\alpha)^2/\alpha^2\psi_0$.

ii) The Born transformation. In this case $g(\alpha) = \alpha - 1 = \frac{\psi-\psi_0}{\psi_0}$ and the distortion term is due to $D_b = -(\psi-\psi_0)k_0^2 f$ resulting in the very limited range of validity of this approximation[3].

iii) An intermediate transformation[5]. Here we consider an asymptotic expansion for $\ell n\alpha$ given by:

$$g(\alpha) = n(\alpha^{\frac{1}{n}} - 1) \tag{7}$$

This transformation is interesting in that for n = 1 it gives the Born approximation and for $n \to \infty$ it approaches the Rytov transformation. The resulting wave equation is given by

$$\nabla^2\chi_1 + k_0^2\chi_1 = -\psi_0\alpha^{\frac{1}{n}}k_0^2 f - \psi_o(1 - \frac{1}{n})\alpha^{\frac{1}{n}}(\frac{\nabla\alpha}{\alpha})^2 \tag{8}$$

But

$$\alpha^{\frac{1}{n}} = 1 + \frac{\ell n\alpha}{n} + \frac{1}{2!}(\frac{\ell n\alpha}{n})^2 + \dots \tag{9}$$

Thus, for n such that $\left|\frac{\ell n\alpha}{n}\right| << 1$, the distortion generating term is:

$$D_i \simeq -\psi_0\frac{\ell n\alpha}{n}k_0^2 f - \psi_0(1 - \frac{1}{n})\alpha^{\frac{1}{n}}(\frac{\nabla\alpha}{\alpha})^2 \tag{10}$$

It should be noted from (10) that the phase, ϕ, of α plays an important role in the nature of the distortion. If $\frac{\ell n|\alpha|}{n} << 1$ but $\frac{\phi}{n}$ is not very small f-dependent distortion similar to that in Born's approximation results.

INVERSION

In this section we briefly outline the interpolation scheme used in the basic diffraction tomography algorithm[1,2]. This technique is superior to that previously reported[1,2] and resembles the familiar convolution-back propagation method used in straight path tomography. Since the left hand side of the equation for the methods discussed are identical the algorithm is general and can be used on all the three transformations given earlier.

Assume a lossy plane-wave, $\psi_0 = e^{ik_o x}$, incident on the object. Equation (5) can be transformed into the two dimensional S-domain by taking the Laplace transform of both sides. Let $F(\mu,\lambda)$ and $L(s_1,s_2)$ be two-dimensional Fourier and Laplace transforms of $f(x,y)$ and $S(\nu,\theta)$ the Fourier transform of $\chi_1(x_0,y)$ for the object at rotation angle θ. Using appropriate contour integration one can show that [1,2],

$$L(s_1,s_2) = AS(\nu,\theta)\sqrt{k_0^2-\nu^2}\, e^{-s_0 x_0} \tag{11}$$

and

$$\begin{aligned} s_0 &= i(k_0-\sqrt{k_0^2-\nu^2}) \\ s_1 &= \gamma_1 + i\mu_1 = i[(k_0 - \sqrt{k_0^2-\nu^2})\cos\theta + \nu\sin\theta] \\ s_2 &= \gamma_2+i\lambda_1 = i[(k_0 - \sqrt{k_0^2-\nu^2})\sin\theta + \nu\cos\theta] \end{aligned} \tag{12}$$

The Jacobian of transformation from (s_1,s_2) to (ν,θ) is given by:

$$J(\nu,\theta) = \frac{k_0|\nu|}{\sqrt{k_0^2-\nu^2}} \tag{13}$$

We now use the fact that $ds_1,ds_2 = J(\nu)d\nu d\theta$ and substitute (13) into (11) to obtain:

$$f(x,y) =$$

$$A\int_0^{2\pi}\int_{-\infty}^{\infty}|\nu|\,S(\nu,\theta)e^{-s_0x_0}e^{(\gamma_1+i\mu_1)x+(\gamma_2+i\lambda_1)y}\,d\nu d\theta \tag{14}$$

Taking the Fourier transform of equation (14) leads to

$$F(\mu,\lambda) =$$

$$A\int_0^{2\pi}\int_{-\infty}^{\infty}|\nu|\,S(\nu,\theta)e^{-s_0x_0}H_1(\mu_1,\mu)H_2(\lambda_1,\lambda)d\nu d\theta$$

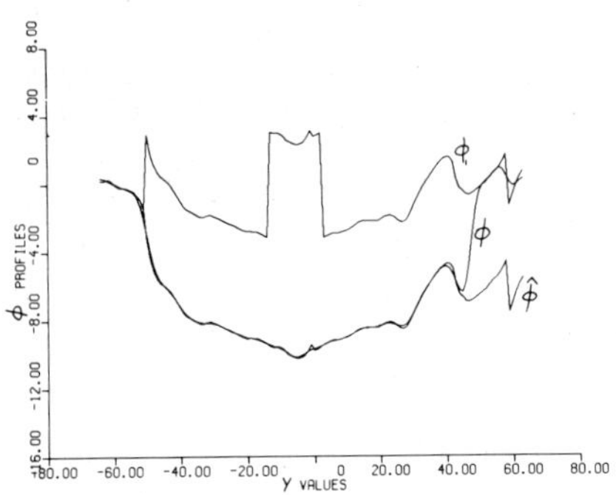

Figure 1. Phase profile of a gelatine cylinder. ϕ_1: Principal value; ϕ adjacent profile corrected unwrapped phase; ϕ incorrect unwrapped phase.

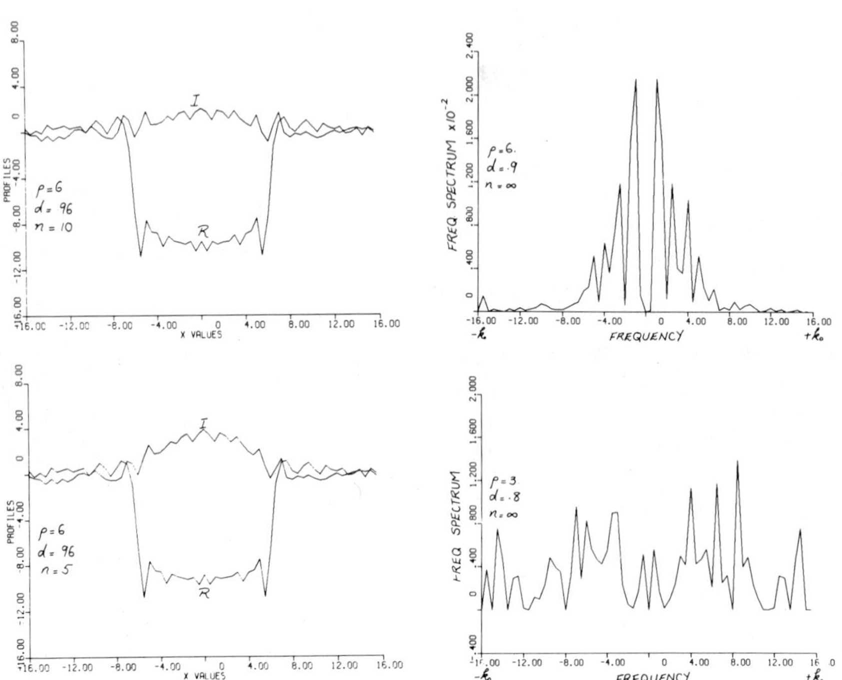

Figure 2. Reconstructed profiles of a cylindrical object using the intermediate transformation.

Figure 3. Frequency spectra of the error in the reconstructed slices of cylindrical objects.

where

$$H_1(\mu_1,\mu) = \int_{-\infty}^{\infty} e^{(\gamma_1+i\mu_1-i\mu)x} dx \qquad \text{and}$$

$$H_2(\lambda_1,\lambda) = \int_{-\infty}^{\infty} e^{(\gamma_2+i\lambda_1-i\lambda)y} dy$$

For $\varepsilon << 1$, and digital reconstruction on a regular grid of Fourier pixels, (λ,μ), the Kernels H_1 and H_2 may be approximated by rectangles of the size of the grid width $\Delta\nu$ at (λ,μ). $F(\mu,\lambda)$ then may be approximated by

$$F(\mu,\lambda) = A \sum_{(\mu_1,\lambda_1)\ \varepsilon\ \mathrm{grid}(\mu,\lambda)} |\nu|\, S(\nu,\theta) e^{-s_0 x_0} \tag{15}$$

where s_1 is given by (13) and $f(x,y) = \mathrm{IDFT}\ [F(\mu,\lambda)]$.

It is worth noting that equation (14) can be used to obtain a filter back propagation formulation in spatial domain[7]. However, due to its algorithmic efficiency we use the frequency domain interpolation discussed above.

Phase Determination

A major problem in the practical use of the Rytov and the intermediate transformation (7) is the determination of the absolute phase, $\phi(y_i,\theta_j)$, of α for each profile θ_j, where y_i is the discrete position of the ith sensor. Since many measurement techniques result in the principle value of the phase $\phi_1(y_i,\theta_j)$, one has to use an appropriate "phase-unwrapping" algorithm to obtain ϕ from ϕ_1. This, of course, requires some reasonable physical constraints on and/or apriori information about the phase. We use the following criteria to accomplish the determination of $\phi(y_i,\theta_j)$

i) $\phi(y_i,\theta_j) \simeq 0$ for y_i removed from the shadow of the object.

ii) $\phi(y_i,\theta_j)$ must be continuous for each θ_j in the y direction.

iii) $|\phi(y_i,\theta_j) - \phi(y_i,\theta_{j\pm1})|$ should be small.

We use property i) to commence the unwrapping process, resorting to property ii) to add or subtract 2π radians where a phase wrap is detected[3]. If $\phi(y_i,\theta_j)$ does not return to zero, property iii) is invoked starting with a properly unwrapped phase profile to correct the calculation of the profile adjacent to it. Figure 1 shows an example of this process on a phase profile from measurements on an 80 wavelength diameter gelatine cylinder with a 6 wavelength water-filled hole. Investigation of other methods to

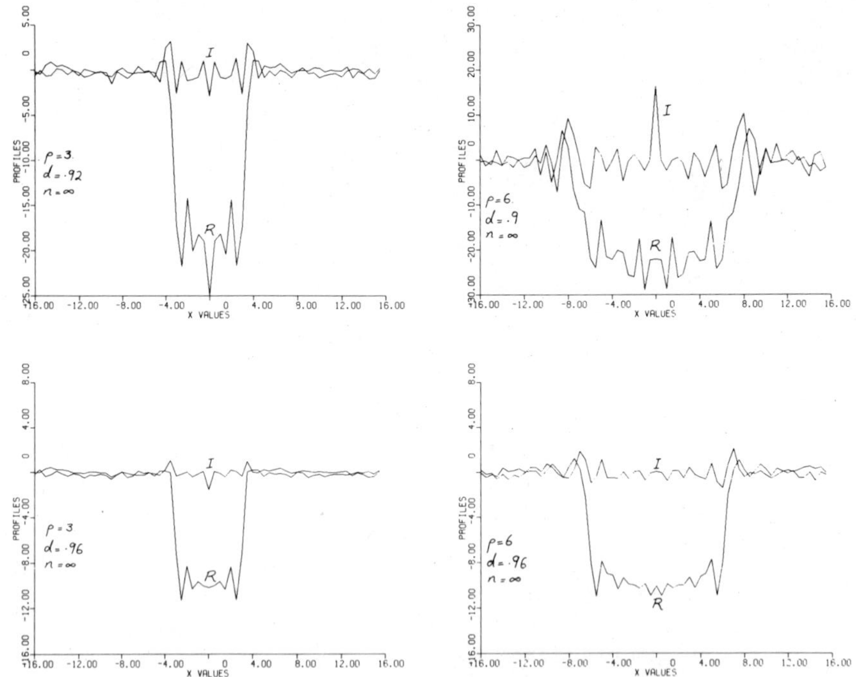

Figure 4. Unfiltered reconstructed slices of cylindrical objects using the Rytov transformation.

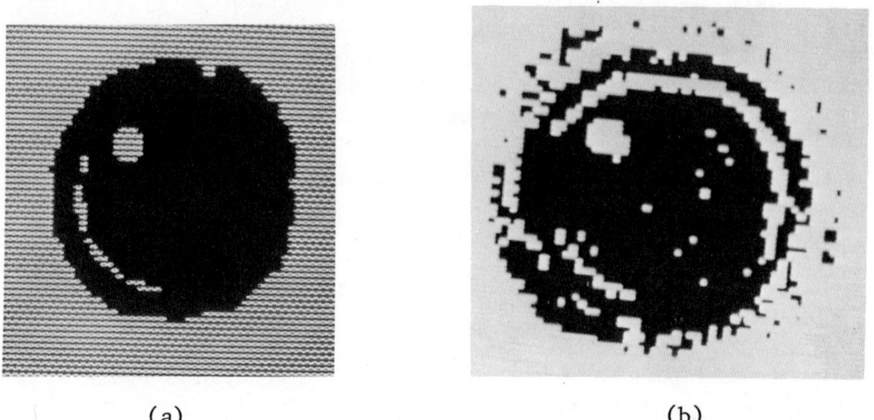

(a) (b)

Figure 5. Rytov tomograms of an 80 wavelength gelatine cylinder with a 6 wavelength hole. Images thresholded for binary representation. (a) correct phase unwrapping, (b) incorrect phase unwrapping for 12 profiles.

aid the phase determination algorithm is underway including the use of the function in formula (7) as a side information for the Rytov transformation.

EXAMPLES AND DISCUSSION

The examples shown in this section are mainly based on simulated data obtained by a solution of the Helmholtz equation for uniform cylindrical soft scatterers. These examples are presented to show the effect of the error terms in the reconstructed Rytov images. These examples also indicate the type of problems expected when imaging an object with a uniform disturbance superimposed on a more random structure. Rytov reconstruction based on actual experiments are presented as well.

The first set of examples show radial "slices" of tomograms of objects with a uniform cylindracal k parameter given by

$$k(x,y) = \begin{cases} dk_0, & x^2 + y^2 \leqslant \rho^2 \\ k_0 & \text{elsewhere} \end{cases}$$

where k_0 is the (real) wavenumber of the homogeneous medium surrounding the object and d is a constant. According to the earlier definition of the disturbance parameter to be imaged, $f(x,y) = 1-d^2$. Since the main objective of these examples is to show the inherent reconstruction errors no additional filtering is done to smoothe the images. In practice we use an appropriate low-pass window (usually Hamming or Kaiser) to reduce the high frequency interpolation and approximation errors. This results in considerably improved image quality.

Figure 2 is from a tomogram of a cylinder with d = .95 using the intermediate transformation (iii) and n=5 and 10. Note that for n=5, the average of the imaginary part of the reconstructed f is non-zero. This is a result between Born and Rytov techniques. The low frequency error is the contribution of Born's method and the high frequency error is mainly due to the Rytov transformation. For n=10 the result is nearly identical to Rytov's ($n=\infty$).

Figure 4 illustrates the model approximation error that is generated in a tomogram using the Rytov transformation. Although it is very difficult, if not impossible, to infer knowledge of the error D_R for a general object an approximate analysis[6] has shown an increase in the bandwidth and energy of this term with an increase in $|f(x,y)|$. An examination of Figures 4(a)-(d) indicates the wideband error term. This term is obviously concentrated within the object. Much of the disturbance outside the object area is due to interpolation errors. Note that $|f(x,y)|$ plays a dominant role in generating D_R, whereas the size, ρ, of the object has a relatively minor effect on it. Figure 3 is a plot of the magnitude square of the Fourier transform of a slice of D_r versus spatial frequency. The bandwidth expansion of D_r with increased $|f(x,y)|$ is obvious.

The next set of examples were obtained through experimental measurements. Figure 5 shows tomograms of an 80 wavelength diameter gelatine cylinder with a 6 wavelength water-filled hole. The images have been enhanced into a binary format to highlight their differences. Figure 5(a) is obtained from unwrapped phases without using adjacent angular profile correction. This results in 12 divergent profiles from a total of 64. Figure 5(b) is the tomogram based on corrected phase profiles. It should be mentioned that these images were not low pass filtered to show the different errors in the two schemes.

CONCLUSIONS

A more general view of diffraction tomography was presented as consequence of a general transformation of the normalized measured wave. Born and Rytov techniques were presented as special cases of this transformation. An updated reconstruction algorithm was then given that accounts for possible attenuation in the medium surrounding the object. In the zero attenuation instant this reduces to a technique similar to the familiar convolution back propagation method in straight-path tomography. Finally, examples of tomograms based on the Rytov transformation were presented. These examples gave an indication of someof the dominant errors one can expect in these tomograms.

ACKNOWLEDGEMENT

This work was supported by the National Science Foundation under Grants ENG76-84521 and ECS-7926008.

REFERENCES

1. M. Kaveh, R. K. Mueller and R. D. Iverson, Ultrasonic tomography based on perturbation solutions of the wave eqution, Computer Graphics and Image Processing, 9:105 (1979).
2. R. K. Mueller, M. Kaveh and G. Wade, Reconstructive tomography and applications to ultrasonics, IEEE Proceedings, 67:567 (1979).
3. M. Kaveh, M. Soumekh and R. K. Mueller, A comparison of Born and Rytov approximations in acoustic tomography, in: "Acoustical Imaging, Vol. 10," J. Powers, ed., Plenum, New York (1981).
4. M. Kaveh, M. Soumekh and R. K. Mueller, ICASSP '82 Proceedings, Paris (1982).
5. Z. Q. Lu, Internal Memo, Department of Electrical Engineering, University of Minnesota (1982).
6. M. Soumekh, Ph.D. Dissertation, University of Minnesota (1983).
7. A. Devaney, A filtered back propagation algorithm for diffraction tomography, (1982). (unpublished).

A BACKWARD PROJECTION ALGORITHM WHICH CORRECTS FOR OBJECT MOTION IN A SCANNING ACOUSTICAL IMAGING SYSTEM

Hua Lee
Glen Wade

Department of Electrical and Computer Engineering
University of California, Santa Barbara
Santa Barbara, California 93106

ABSTRACT

Motion correction in acoustical imaging is commonly performed in the spatial frequency domain by a filtering process to remove the degradation due to the motion. Construction of the motion-correction filter is difficult and often unrealizable for complicated nonlinear object motion.

The backward projection algorithm in acoustical imaging permits independent data processing for all detected data samples. Because of this, it can provide for motion correction by means of a technique equivalent to spatially displacing the receiving positions. In this paper we present a method for removing motion degradation due to both linear and nonlinear motions. The resolution limit after the motion has been corrected can be determined from an equivalent aperture derived from the motion.

The algorithm which corrects for motion is a modified version of the conventional backward projection algorithm for image reconstruction. By properly grouping the received sample positions, the required modification can be substantially simplified.

INTRODUCTION

Motion blur is a common effect contributing to image degradation. Motion correction is therefore an important subject in image reconstruction. Motion-correction algorithms have been developed for various types of imaging systems and for different types of object motion [1,2]. Removal of the effect of motion can be accomplished by means of spatial domain filtering or space domain-phase correction.

For imaging systems with scanning detection arrays, data samples are detected at various instants of time. The object motion introduces no distortion to the resultant wavefield and the motion effect therefore can be completely removed.

Backward projection performs image reconstruction by applying a pseudo-inverse operator to each data sample independently. Motion correction can be accomplished by shifting the aperture to the equivalent receiving position. This paper reviews the backward projection algorithm and gives a modification of it for eliminating motion effects. The analysis can be applied to both linear and nonlinear motion. Examples are given to demonstrate the removal of image degradation due to various kinds of object motion.

In order to reduce the complexity in image reconstruction, grouping is introduced. The aperture is divided into small zones in which the sample spacing is uniform and the normalization factor, approximately constant. This approach improves computational efficiency both in the normalization process and in the matched filtering associated with the backward projection algorithm.

BACKWARD PROJECTION METHOD

The backward-projection method can be used to reconstruct images in most linear imaging systems. The derivation, analysis, and processing can be formulated in both analog and discrete forms [3]. Consider a linear data acquisition system in which the detected wavefield $g(\bar{x})$ is related to the unknown object distribution $f(\bar{x})$ by a three-dimensional convolution integral

$$\begin{aligned} g(\bar{x}) &= f(\bar{x}) \circledast h(\bar{x}) \\ &= \int_R f(\bar{\alpha})\, h(\bar{x}-\bar{\alpha})\, d\bar{\alpha} \end{aligned} \qquad (1)$$

where $h(\bar{x})$ is the impulse response which is also known as the Green's function in free-space wave propagation, R denotes the object region, and $\bar{x}$ is the space coordinate vector defined as $\bar{x} = [\underline{x},\underline{y},\underline{z}]$ [4]. For a single data sample detected at the position $\bar{x}_o$, we evaluate equation (1) at $\bar{x}_o$ and have

$$g(\bar{x}_o) = \int_R f(\bar{\alpha})\, h(\bar{x}_o - \bar{\alpha})\, d\bar{\alpha} \qquad (2)$$

The complex data sample $g(\bar{x}_o)$ is called a projection of the object distribution with the weighting $h(\bar{x}_o - \bar{\alpha})$.

We wish to solve equation (2) for $f(\bar{x})$, the object distribution, but the solution is not unique. The optimal estimate is obtained by applying the pseudo-inverse operator [5] as follows:

$$\tilde{f}(\bar{x},\bar{x}_o) = n(\bar{x}_o)\ g(\bar{x}_o)\ h^*(\bar{x}_o - \bar{x}) \tag{3}$$

where h^* is the conjugate of the impulse response and $n(\bar{x}_o)$ denotes the normalization factor defined as

$$n(\bar{x}_o) \triangleq \left[\int_R |h(\bar{x}_o - \bar{x})|^2\ d\bar{x} \right]^{-1} \tag{4}$$

The factor $n(\bar{x}_o)$ is a real and positive scalar analogous to the Gram matrix for a single detected sample. It is a function of the receiving position with respect to the object region. Equation (3) is the minimum norm solution to equation (2) with $n(\bar{x}_o)$ compensating for the divergence due to the wave propagation and $h^*(\bar{x}_o - \bar{x})$ performing the focusing. The optimal estimate $\tilde{f}(\bar{x},\bar{x}_o)$ is commonly called a backward projection based upon the projection $g(\bar{x}_o)$.

The resultant image of the object distribution can be formed by integrating the backward projections from all receiving positions over the aperture

$$\hat{f}(\bar{x}) = \int_A \tilde{f}(\bar{x},\bar{x}_o)\ d\bar{x}_o \tag{5}$$

$$= \int_A n(\bar{x}_o)\ g(\bar{x}_o)\ h^*(\bar{x}_o - \bar{x})\ d\bar{x}_o$$

where A denotes the receiving aperture. For systems employing discrete data acquisition, the integral in equation (5) is replaced by a summation.

Alternatively, we may write equation (5) as a convolution between $n(\bar{x})\ g(x)$ and $h^*(-\bar{x})$.

$$\hat{f}(\bar{x}) = [n(\bar{x})\ g(\bar{x})] \circledast h^*(-\bar{x}) \tag{6}$$

Equation (6) shows that backward projection can be interpreted as matched-filtering of the normalized received wavefield [6]. By Fourier transforming equation (6), we obtain a simple multiplication in the spatial-frequency doman.

Because the normalization factor is real and positive and the impulse response is even, equation (6) can also be written as

$$\hat{f}^*(\bar{x}) = [n(\bar{x})\ g^*(\bar{x})] \circledast h^*(\bar{x}) \tag{7}$$

Since $\hat{f}$ and $\hat{f}^*$ have the same amplitude distribution, the image reconstruction process can be regarded as using $[n(\bar{x})\ g^*(\bar{x})]$ as a secondary source distribution and propagating its wavefield back to the object region where $\hat{f}^*(\bar{x})$ is then formed.

MOTION EFFECT

Suppose the object is moving with a known velocity $\bar{v}(t)$ which can be linear or nonlinear. The displacement due to the motion over the time interval (t_o,t) is

$$\bar{d}(t) = \int_{t_o}^{t} \bar{v}(t)\ dt \tag{8}$$

Under this condition the object distribution now is not only a function of space variables, but also of time. We use the object distribution at t_o, $f(\bar{x},t_o)$, as a reference. The object function at time t can be wrttcn as

$$f(\bar{x},t) = f(\bar{x} - \bar{d}(t),t_o) \tag{9}$$

Accordingly, equation (1) becomes

$$g(\bar{x},t) = f(\bar{x},t) \circledast h(\bar{x}) \tag{10}$$

$$= f(\bar{x} - \bar{d}(t),\ t_o) \circledast h(\bar{x})$$

$$= f(\bar{x},t_o) \circledast h(\bar{x} - \bar{d}(t))$$

From equation (10), it is seen that the motion effect can be transfered from the object functon to the impulse response. Thus the backward projection algorithm can be generalized for moving objects by modifying the associated impulse response.

For a data sample detected at position $\bar{x}_o$ and time t, the corresponding normalization factor for a moving object is

$$n(\bar{x}_o,t) = [\int_R |h(\bar{x}_o - (\bar{x} - \bar{d}(t)))|^2 d\bar{x}]^{-1} \quad (11)$$

$$= [\int_R |h((\bar{x}_o + \bar{d}(t)) - \bar{x})|^2 d\bar{x}]^{-1}$$

$$= n(\bar{x}_o + \bar{d}(t), t_o)$$

The backward projection resulting from a single detected sample $g(\bar{x}_o,t)$ can be formulated by modifying equation (3).

$$\tilde{f}(\bar{x}_o,\bar{x}) = n(\bar{x}_o,t)\, g(\bar{x}_o,t)\, h^*(\bar{x}_o - (\bar{x} - \bar{d}(t))) \quad (12)$$

$$= g(\bar{x}_o,t)\, n(\bar{x}_o + \bar{d}(t), t_o)\, h^*((\bar{x}_o + \bar{d}(t)) - \bar{x})$$

For a displacement $\bar{d}(t)$ due to object motion, equation (12) shows that backward projection from the detected sample at $\bar{x}_o$ is equivalent to that from $\bar{x}_o + \bar{d}(t)$ for the object if it were stationary. Therefore, the backward projection algorithm can be modified for a moving object by shifting the coordinate system from $\bar{x}_o$ to $\bar{x}_o + \bar{d}(t)$. The equivalent receiving position will be used to compute $n(\bar{x}_o + d(t), t_o)$ for the normalization process and $h^*(\bar{x}_o + d(t) - \bar{x})$ for focusing. Hence, motion correction for a scanning data acquisition system becomes a mathematical rearrangement of the receiving positions. Corresponding to the object motion, an equivalent synthetic aperture is formed by the shifted receiving positions. The resolution of the resultant images is determined by the size of the equivalent aperture.

Because holographic detection systems with scanning receivers collect data samples at various receiving positions and time instants, the motion effect then can be removed by applying the backward projection algorithm assuming a rearrangement of the receiving positions in accordance with the displacement due to the object motion.

We will now perform motion correction for three different cases to demonstrate the effectiveness. For simplicity, we limit both the receiving aperture and the object region to a single dimension. In following examples we consider an imaging system with a receiver which scans over an aperture of 100 wavelengths long. The distance from the aperture to the object region is also 100λ where λ is the wavelength.

Case 1: Cross-Range Nonlinear Motion.

A point object is oscillating in the cross-range direction with the oscillation amplitude 5λ. The frequency of the oscillation is two cycles for the total scanning period. Fig. (1-a) shows the image reconstruction without motion correction and Fig. (1-b) shows it with motion correction.

Case 2: Nonlinear Motion in the Range Direction.

A point object is oscillating in the range direction with the amplitude and frequency the same as in Case 1. Fig. (2-a) shows the image reconstruction without motion correction and Fig. (2-b) shows it with motion correction.

Case 3: Nonlinear Motion in Both Directions.

The motion of the point object in this case consists of two oscillation components having different directions, different frequencies, and different amplitudes. In the cross-range direction, the oscillation has a frequency of three cycles for the total scanning period and oscillation amplitude of λ. In the range direction, the oscillation amplitude is 2λ and the frequency is four cycles for the total scanning period. Fig. (3-a) shows the image reconstruction without the motion correction and Fig. (3-b) shows it with motion correction.

DIGITAL RECONSTRUCTION AND GROUPING

The backward-projection algorithm consists of two major parts, the normalization process and the matched filtering. The normalization process involves multiplying the detected wavefield in the space domain with a precomputed positional factor. The matched filtering is usually performed in the spatial-frequency domain by Fourier transforming equation (6). When conventional holographic reconstruction is replaced by digital computation, the discrete version of the algorithm is employed. As far as image reconstruction is concerned, the major difference between the analog and the digital approaches lies in the process of the matched filtering.

In the case of stationary objects, the size of the receiving aperture and the object region govern the complexity of the computation. All the data sequences in the matched filtering process must have finite length. If the upper and lower bounds of the aperture are $\bar{x}_{Amax}$ and $\bar{x}_{Amin}$, and that of the object region are $\bar{x}_{Rmax}$ and $\bar{x}_{Rmin}$ respectively, then the section of the impulse response involved in matched filtering is bounded between ($\bar{x}_{Rmax}$ -

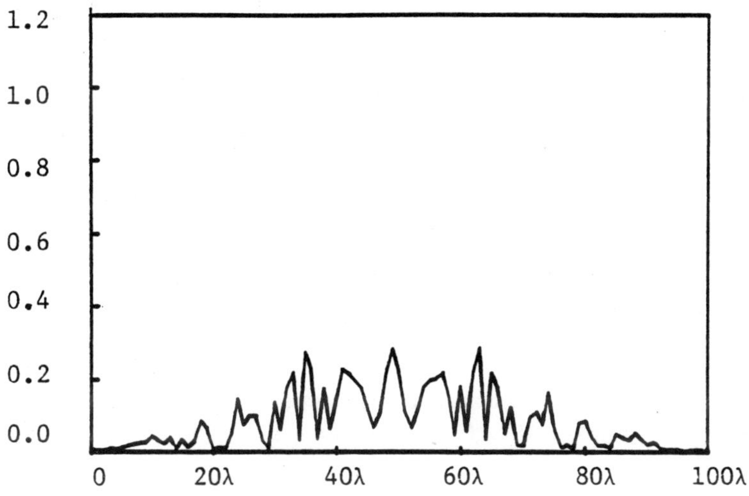

Fig.(1 a) Backward-projected image of a point object moving in the cross-range direction without motion correction.

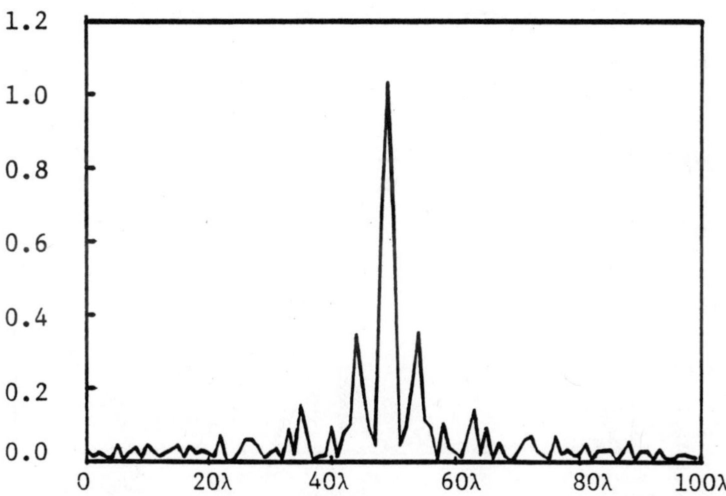

Fig.(1 b) Backward-projected image of the above point object with motion correction.

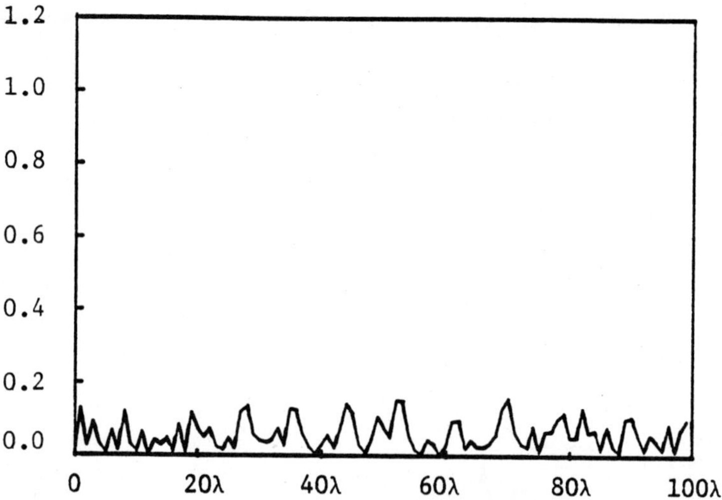

Fig.(2 a) Backward-projected image of a point object moving in the range direction without motion correction.

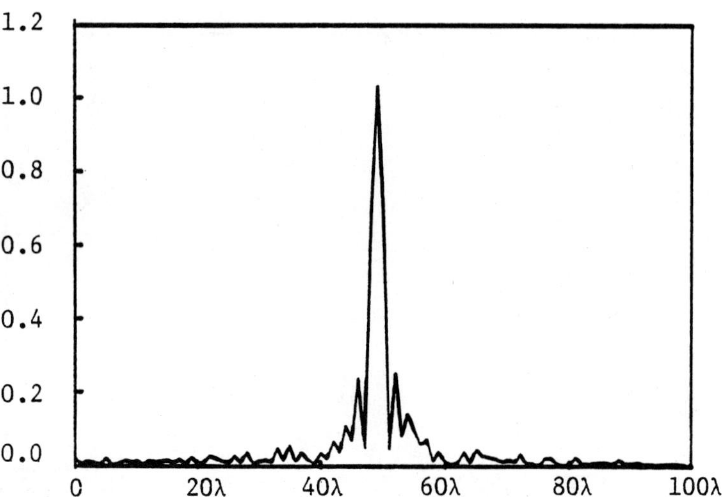

Fig.(2 b) Backward-projected image of the above point object with motion correction.

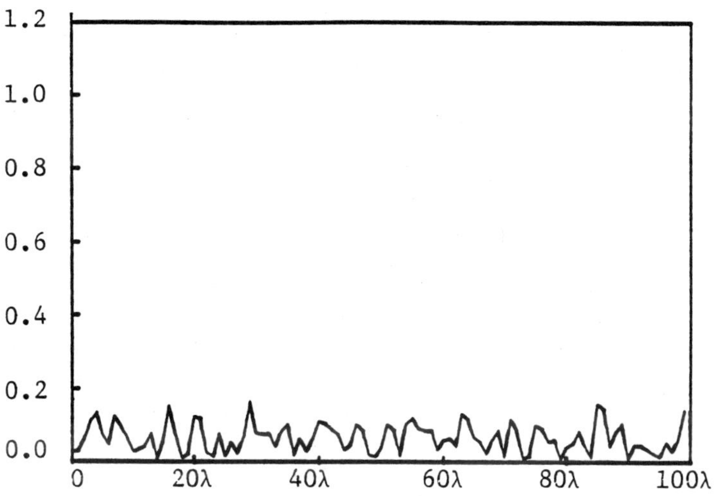

Fig.(3 a) Backward-projected image of a point object moving in both the range and cross-range directions without motion correction.

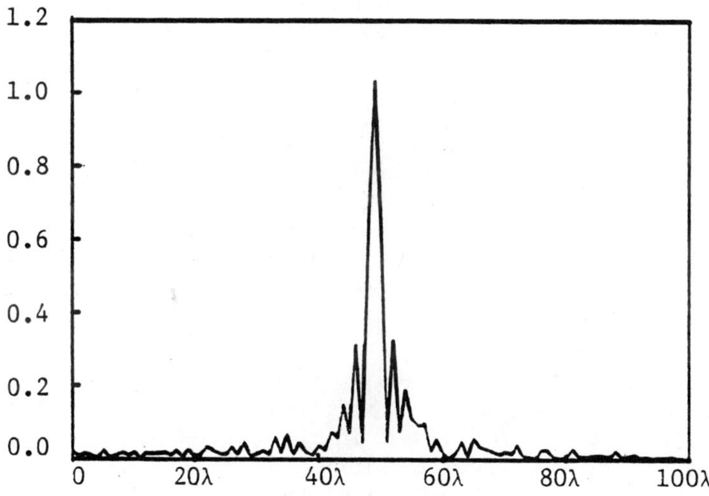

Fig.(3 b) Backward-projected image of the above point object with motion correction.

$\bar{x}_{Amin}$) and ($\bar{x}_{Rmin}$ - $\bar{x}_{Amax}$). It can be seen that if the size of the aperture and the object region are L_A and L_S respectively, the size of the impulse response sequence is $L_A + L_S$. In order to avoid undersired overlapping, the FFT associated with the convolution must have its length greater than $2L_A + L_S$. This requirement is applicable to all directions. With the size of the object region fixed, the computational complexity depends on the aperture size.

To image stationary objects, the scanning receivers usually detect the wavefield in a compact format in which the sample density is uniform. For moving objects, the receiving positions are shifted to form the equivalent synthetic aperture due to the object motion. Within the equivalent synthetic aperture, the sample density is not generally uniform. To perform the digital computation, the detected samples must be padded with zeros to form an array with uniform sample spacing. The padding will considerably increase the size of the sequences, as a result, increase the complexity of the image reconstruction.

This problem can be resolved by grouping the equivalent receiving positions. Grouping can be defined as a three-dimensional partitioning |7|. It divides the equivalent receiving aperture into several small zones. The zones contain all data samples. The size of the zones may vary and the sample distribution in them may be different. However, the sample density in each zone should eb approximately uniform.

The backward projection algorithm can be applied to all the zones independently to reconstruct the image. If the zones are all small compared to the object region, the normalization factor for the receiving positions within a single zone will have little variation. Only one normalization factor is then needed for each zone. Each zone can have an independent coordinate system so that zero padding is minimized.

When grouping is properly performed, the normalization factors are calculated for the zones instead of for the individual receiving positions. The sizes of the sequences involved in the matched filtering is reduced since the number of zeros in the padding is minimized. As a result, the computation for the image reconstruction is significantly simplified.

CONCLUSION

This paper presents an extended version of the backward projection algorithm to include motion correction. It is applicable to both linear and nonlinear object motion. We have shown it to be effective by means of various examples. Data grouping can be introduced to reduce computational complexity.

For imaging systems with scanning detectors, the analysis shows that the motion correction can be performed by mathematically rearranging the receiving positions to correspond to the object motion. The resolution of the resultant images is governed by the equivalent synthetic aperture after the rearrangement. The analysis is useful in providing a clear viewpoint for specific cases as, for example, in determining the resolving ability of an imaging system in which the object is caused to vibrate with a designated motion.

REFERENCES

[1] N.B. Tse, L. Schlussler, J. Fontana, and G. Wade, "Computer-Corrected Reconstruction of Acoustic Holograms of Nonuniformly Moving Objects," _1977 Ultrasonics Symposium Proceedings_, October 1977, pp. 193-197. (IEEE Group on Sonics and Ultrasonics).

[2] N.B. Tse, L. Schlussler, J. Fontana, and G. Wade, "Image Reconstruction of Acoustic Holograms for Moving Objects," _Proceedings of the International Optical Computing Conference 1977_, SPIE Vol. 118, August 1977, pp. 35-44.

[3] H. Lee, C. Schueler, G. Wade, and J. Fontana, "Digital Reconstruction of Acoustical Holograms in the Space Domain with a Vector Space Approximation," _Acoustical Imaging_, Vol. 9, K. Wang Ed., Plenum Press, New York, 1980, pp. 631-641.

[4] J.W. Goodman, _Introduction to Fourier Optics_, McGraw-Hill, New York, 1968, p. 44.

[5] D.G. Luenberger, _Optimization by Vector Space Methods_, John Wiley and Sons, New York, 1968, pp. 163-164.

[6] A.D. Whalen, _Detection of Signals in Noise_, Academic Press, New York, 1971, pp. 167-175.

[7] A. Gersho, "Asymptotically Optimal Block Quantization," _IEEE Transactions on Information Theory_, Vol. 25, No. 4, July 1979, pp. 373-380.

3-D ACTIVE INCOHERENT ULTRASONIC IMAGING

Takayoshi Yokota and Takuso Sato

The Graduate School at Nagatsuta
Tokyo Institute of Technology
4259 Nagatsuta, Midori-ku
Yokohama-shi, 227 Japan

ABSTRACT

Aiming at high resolution and speckle noise reduced high quality ultrasonic imaging, the concept of active incoherent imaging with proper nonlinear processings[3-5] was proposed in the previous paper.[1] The treatment there, however, has been restricted to one dimensional cases.

In this paper, the concept is extended to general three dimensional cases and the corresponding practical imaging algorithm is proposed. First, it is shown that if continuous ultrasonic waves with relatively random phases are transmitted from the array elements arranged densely on a hemisphere, which contains the object under observation at its center,then the resulting wave field in the three dimensional object region is almost incoherent, with effective coherence length of the order of the used wavelength. The wave field reflected by the object are detected by using the transducers on the same hemisphere and the spatial coherence function of them is derived. Then the desired 3-D image of the object for incoherent illumination is reconstracted by means of 3-D inverse Fourier transform of the observed coherence function.

As the method is essentially incoherent imaging, we can expect speckle noise reduced images as in the previous cases. A new image reconstruction algorithm which combines parametric and non-parametric maximum entropy methods is proposed to get high resolution image. The effectiveness of the proposed method is demonstrated by computer simulations and experiments.

1. INTRODUCTION

Reflection coefficients of objects for ultrasonic waves are generally random complex values. Hence the conventional coherent imaging systems have suffered to some extent from speckle noice. But if we can generate at each point on the object sufficient large number of sample wave fields with the same magnitude and random phases and the image of the average intensity of the field can be derived, then a desired imaging which is free from speckle noises must be realised. This is the basis idea of our method.

The samples of reflected wave fields with random phases can be generated by illuminating the object by plane waves with relatively random phases from many different directions. That is, incoherent wave fields can be generated actively on the object by these processes. Then the image of the average intensity distribution on the object can be reconstructed by back propagating the coherence function detected at some area distant from the object. It corresponds essentially to the spectral estimation from the observed coherence function. Hence new algorithm for 2-D and 3-D nonlinear spectral estimation can be applied at this stage to get desired spectrum even from coherence function observed only in a restricted area.

These processes are discussed in this paper. The details of the principle, formulation are shown in the following and the usefulness of this method is demonstrated by numerical analyses and experiments.

2. PRINCIPLE

2.1 Active Generation of 3-D Quasi-Incoherent Wave Field on the Object

If we transmit continuous ultrasonic waves of amplitude $a(x_T,y_T)$ from a point $P_T(x_T,y_T,z_T)$ on a hemisphere with radius R_O, the incident wave field $f_{in}(\xi,\eta,\nu)_{x_T,y_T}$ on a point $Q(\xi,\eta,\nu)$ around the center of the sphere is given by

$$f_{in}(\xi,\eta,\nu)_{x_T,y_T}= K\cdot a(x_T,y_T)\cdot \exp\{-jk\sqrt{(x_T-\xi)^2+(y_T-\eta)^2+(z_T-\nu)^2}\} \tag{1}$$

For the transmissions of waves from all points on the hemisphere of maximum radius R_T the equivalent coherence function $\Gamma_{f_{in}}(\xi,\xi',\eta,\eta',\nu,\nu')$ between wave fields at two points $Q(\xi,\eta,\nu)$ and $Q'(\xi',\eta',\nu')$ on the object is given by

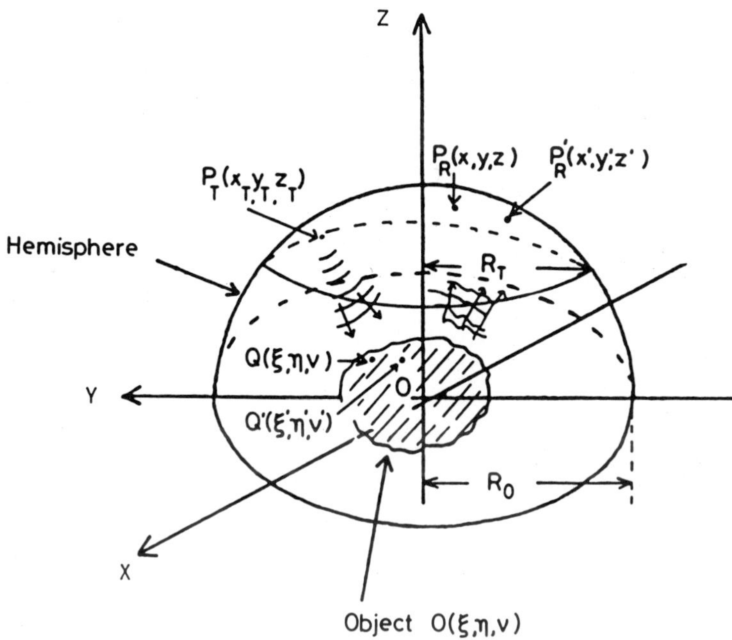

Fig. 1 Geometry of active generation of 3-D quasi-incoherent wave field on the object.

$$\Gamma_{f_{in}}(\xi,\xi',\eta,\eta',\nu,\nu') = \iint_{x_T^2+y_T^2 \le R_T^2} dx_T dy_T \; f_{in}(\xi,\eta,\nu)_{x_T,y_T} \cdot f^*_{in}(\xi',\eta',\nu')_{x_T,y_T} \tag{2}$$

In Fresnel region, Eq. (2) is reduced to

$$\begin{aligned}\Gamma_{f_{in}}(\xi,\xi',\eta,\eta',\nu,\nu') &= 2\pi\cdot K^2\cdot \exp\{-jk(\xi^2-\xi'^2+\eta^2-\eta'^2+\nu^2-\nu'^2)/2R_0\} \\ &\times \int [\, r\cdot a^2(r)\cdot \exp\{jk(\nu-\nu')\sqrt{R_0^2-r^2}\,/R_0\} \\ &\times J_0\{k(r/R_0)\sqrt{(\xi-\xi')^2+(\eta-\eta')^2}\}]\, dr \end{aligned} \tag{3}$$

If we choose $a^2(r)=1/\{2\pi K^2\cdot R_0\sqrt{R_0^2-r^2}\}$ and $R_T=R_0$, the transverse and longitudinal components of $\Gamma_{f_{in}}$ are given by

<u>Transverse Component</u>

$$\Gamma_{f_{in}}(\xi-\xi',\eta-\eta',\nu=\nu') = \mathrm{sinc}\{k\sqrt{(\xi-\xi')^2+(\eta-\eta')^2}\} \tag{4}$$

Longitudinal Component

$$\Gamma_{f_{in}}(\xi=\xi', \eta=\eta', \nu-\nu') = \exp\{jk(\nu-\nu')/2\} \operatorname{sinc}\{k(\nu-\nu')/2\} \quad (5)$$

These relations show that the effective correlation volume of $\Gamma_{f_{in}}$ can be made of the order of the wavelength of the used ultrasonic waves. Thus, in this way an incoherent wave field can be generated actively on the object.

2.2 Image Reconstruction

The reflected wave fields s(x,y,z) for the transmission from $P_T(x_T, y_T, z_T)$ and detection at $P_R(x,y,z)$ on the same hemisphere is given by

$$s(x,y,z)_{x_T,y_T} = \iiint d\xi d\eta d\nu\, O(\xi,\eta,\nu)\cdot f_{in}(\xi,\eta,\nu)_{x_T,y_T} \times \exp\{-jk\sqrt{(x-\xi)^2+(y-\eta)^2+(z-\nu)^2}\} \quad (6)$$

Hence, the equivalent coherence function $\Gamma_s(x,x',y,y',z,z')$ between detected wave fields is given by

$$\Gamma_s(x,x',y,y',z,z') = \iint_{x_T^2+y_T^2 \leq R_T^2} dx_T dy_T\, s(x,y,z)\cdot s^*(x',y',z') \quad (7)$$

In Fresnel region, Eq. (7) is reduced to

$$\begin{aligned}\Gamma_s(x,x',y,y',z,z') = &\iiiint\!\!\iint d\xi d\xi' d\eta d\eta' d\nu d\nu' \\ &\times O(\xi,\eta,\nu)\cdot O^*(\xi',\eta',\nu')\cdot \Gamma_s(\xi,\xi',\eta,\eta',\nu,\nu') \\ &\times \exp\{jk(x\xi-x'\xi'+y\eta-y'\eta'+z\nu-z'\nu')/R_0\} \\ &\times \exp\{-jk(\xi^2-\xi'^2+\eta^2-\eta'^2+\nu^2-\nu'^2)/2R_0\} \quad (8)\end{aligned}$$

If the wave field on the object is incoherent, then, $\Gamma_s \simeq \delta(\xi-\xi')\cdot\delta(\eta-\eta')\cdot\delta(\nu-\nu')$, and Eq. (8) is reduced to

$$\Gamma_s(x,x',y,y',z,z') \simeq \iiint d\xi d\eta d\nu\, |O(\xi,\eta,\nu)|^2 \times \exp[jk\{(x-x')\xi+(y-y')\eta+(z-z')\nu\}/R_0] \quad (9)$$

This relation shows that the detected coherence function Γ_s is the 3-D Fourier spectrum of the intensity distribution of the waves reflected on the object. Hence, the image can be reconstructed as the 3-D inverse Fourier transform of the detected coherence function.

2.3 A New Superresolution Image Reconstruction Algorithm

In practical uses of this method, however, the following conditions must be considered; i) data acquisition only over a small area on the hemisphere is allowed in most practical cases, and ii) reduction of the number of repetitions of transmission and reception is desired.

To satisfy these requirements, we considered the following procedure. The area on the hemisphere for transmission and detection of ultrasonic waves is restricted to a small one. Then the resolution in the range direction can be realized by using pulsed waves. Hence, a set of C-mode images gives 3-D structure of the object. The lateral resolution reduced due to the restricted area is improved through proper nonlinear spectral estimation algorithm.

The details of this new algorithm is as follows.

Let us assume that the intensity distribution under consideration consists of the following two components.

i) $I_b(\xi,\eta)$: smooth background which can be represented within the Rayleigh limit of the observed coherence range.

ii) $I_p(\xi,\eta)$: peak-like components which can not always be represented by the Rayleigh limits.

Then, the model of the intensity distribution is given as follows.

$$I(\xi,\eta) = I_b(\xi,\eta)+I_p(\xi,\eta) \tag{10}$$

$$I_b(\xi,\eta) = \sum_{l=-N_2}^{N_1} \sum_{m=-N_2}^{N_2} R_b(l,m) \exp[-j2\pi\{(l\xi+m\eta)/W\}] \tag{11}$$

$$I_p(\xi,\eta) = \sum_{l=-\infty}^{\infty} \sum_{m=-\infty}^{\infty} R_p(l,m) \exp[-j2\pi\{(l\xi+m\eta)/W\}] \tag{12}$$

$$(N_1,N_2) \leq (K_1,K_2) \text{ : observation region} \tag{13}$$

$$R_b(l,m)=0 \quad \text{for } |l|\geq N_1,\ |m|\geq N_2 \tag{14}$$

R_b and R_p are the equi-sampled coherence functions which correspond to I_b and I_p, respectively. W is an area size to be imaged.

We estimate I_p by maximizing its entropy with respect to several different consistency conditions over the observed coherence function.

Background Order Estimation

The background orders are determined by using the following measure.

$$FBO(\hat{N}_1,\hat{N}_2) = \sum_{i=1}^{N} \sum_{k=1}^{N} \hat{I}_{pn}(\xi_i,\eta_k,\hat{N}_1,\hat{N}_2) \cdot \log \hat{I}_{pn}(\xi_i,\eta_k,\hat{N}_1,\hat{N}_2) + \log N^2 \quad (15)$$

where $$\hat{I}_{pn}(\xi_i,\eta_k,\hat{N}_1,\hat{N}_2) = \hat{I}_p(\xi_i,\eta_k,\hat{N}_1,\hat{N}_2) / \sum_{i'=1}^{N} \sum_{k'=1}^{N} \hat{I}_p(\xi_{i'},\eta_{k'},\hat{N}_1,\hat{N}_2) \quad (16)$$

This measure evaluates the decrease of the -plog(p) type entropy of the image $\hat{I}_p$ which is obtained by the entropy maximization under the consistent constraint over the region as illustrated in Fig. 2. The background order is estimated so that the value of FBO of Eq. (15) is maximized, since in this case the information about the peak-like components may be maximized.

Fig. 2 Consistency region

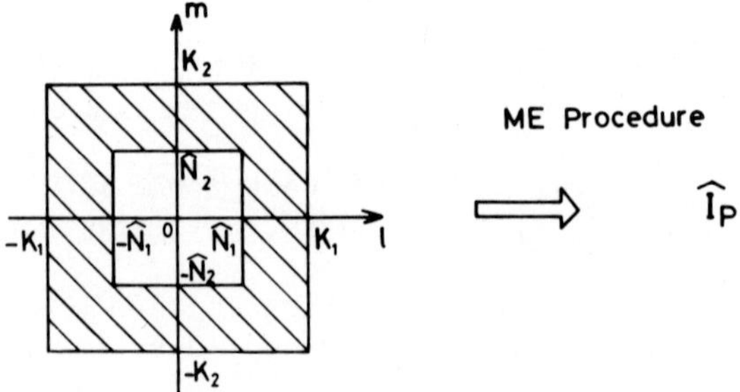

Entropy Maximization Procedure

As the entropy maximization procedure for the image, we used Werneck et al's method.[2] That is, we estimate Ip so that the following object function J is maximized.

$$J = \sum_{i=1}^{N} \sum_{k=1}^{N} \log \hat{I}_p(\xi_i,\eta_k,\hat{N}_1,\hat{N}_2) - \lambda \sum_{(l,m) \subseteq D} |R(l,m) - \hat{R}_p(l,m)|^2 \quad (17)$$

where $D=(K_1,K_2)-(\hat{N}_1,\hat{N}_2)$: consistency region,

λ is a positive constant

And we used gradient method for the maximization of J.

Estimation of Background Energy

Once the optimum background orders $\hat{N}_{1opt}$, $\hat{N}_{2opt}$ are estimated, our next task is the estimation of the background coherence function $R_b(l,m)$ as defined in Eq. (11). First, we estimate the background energy $R_b(0,0)$ by using the following measure.

$$FBE(\alpha) = -\{1-E_n(\alpha)\}\cdot H_n(\alpha) \tag{18}$$

where
$$\alpha = \{R(0,0)-\hat{R}_b(0,0)\}/R(0,0) = \hat{R}_p(0,0)/R(0,0) \tag{19}$$

$$H_n(\alpha) = \sum_{i=1}^{N}\sum_{k=1}^{N} \log[\hat{I}_p\{\xi_i,\eta_k,\hat{N}_{1opt},\hat{N}_{2opt},\alpha R(0,0)\}] - \sum_{i=1}^{N}\sum_{k=1}^{N} \log[\frac{1}{N^2}\sum_{i'=1}^{N}\sum_{k'=1}^{N} \hat{I}_p\{\xi_{i'},\eta_{k'},\hat{N}_{1opt},\hat{N}_{2opt},\alpha R(0,0)\}] \tag{20}$$

$$E_n(\alpha) = \sqrt{\frac{|\alpha R(0,0)-\hat{R}_p(0,0)|^2 + \sum\sum_{(l,m)\subseteq D} |R(l,m)-\hat{R}_p(l,m)|^2}{|\alpha R(0,0)|^2 + \sum\sum_{(l,m)\subseteq D} |R(l,m)|^2}} \tag{21}$$

$H_n(\alpha)$: normalized log(p) type entropy of the estimated intensity distribution $\hat{I}_p$.
$E_n(\alpha)$: normalized root mean square consistency error.
$\hat{I}_p\{\xi_i,\eta_k,\hat{N}_{1opt},\hat{N}_{2opt},\alpha R(0,0)\}$: maximum entropy estimate of I_p under the consistent condition over the region $(l,m)\subseteq D$ and a new value $\alpha R(0,0)$ as the total power.

The optimum value α_{opt} is determined so that $FBE(\alpha)$ is maximized. $\hat{R}_b(0,0)$ and $\hat{R}_p(0,0)$ are given as follows.

$$\hat{R}_b(0,0) = (1-\alpha_{opt})\cdot R(0,0) \tag{22}$$

$$\hat{R}_p(0,0) = \alpha_{opt}\cdot R(0,0) \tag{23}$$

The other parameters of the background are estimated by interpolating the coherence function $\hat{R}_p(l,m)$.

The final estimate which includes both of $\hat{I}_b$ and $\hat{I}_p$ is obtained by

$$\hat{I}(\xi_i,\eta_k) = \hat{I}_b(\xi_i,\eta_k) + \hat{I}_p(\xi_i,\eta_k)$$

where $$\hat{I}_b(\xi_i,\eta_k) = \hat{R}_b(0,0) + \sum_{(l,m) \subseteq (\hat{N}_{1opt},\hat{N}_{2opt})-(0,0)} \sum \{R(l,m)-\hat{R}_p(l,m)\} \times \exp[-j\frac{2\pi}{N}\{(i-\frac{N}{2})l+(k-\frac{N}{2})m\}] \quad (24)$$

The algorithm is summarized in Fig. 3.

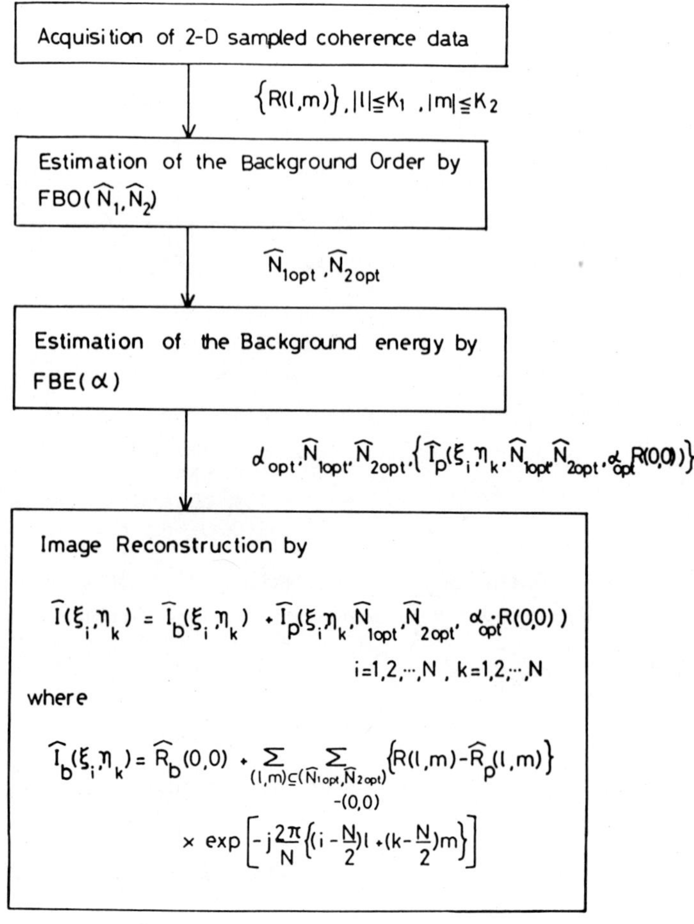

Fig. 3 The algorithm of our new 2-D superresolutional image reconstruction.

3. NUMERICAL ANALYSES

In order to show the effectiveness of our active incoherent ultrasonic imaging method together with the new 2-D super-resolution image reconstruction algorithm, we performed several computer simulations of C-mode image reconstruction.

The geometry used for the simulation is illustrated in Fig. 4. We used a scanning transmitter and 2-D equi-spaced rectangular detecting array. The objects to be imaged are shown in Fig. 5. Upper parts of the objects have constant or linear phase with constant amplitude, hence these parts correspond to specular objects. Lower parts of the objects have phase distributions which follow a Gauss-Markovian random process, hence these parts correspond to diffusive objects.

Fig. 6 shows the result for object I. In this case the transmissions are carried out from each mesh point of interval 15 mm on rectangle $R_x \times R_y$. Image is reconstructed by averaging the intensity images obtained in each transmission. This process is exactly the same with the inverse Fourier transform of the detected coherence function.

The results show the effectiveness of this method for the reduction of speckle noises in the diffusive parts and proper reconstruction can also be observed even for the parts with specular reflectivity.

Fig. 7 is the result for object II. In this case the transmissions are carried out from points on a circle of radius R_T at each 9 degrees.

This result shows also the effectiveness of the active incoherent wave field generation. Fig. 7 (d) shows the result when our new 2-D superresolution image reconstruction algorithm is applied to the image (c). A sharp image which is close to the original is obtained.

Fig. 8 is the demonstration of the effect of our new 2-D superresolution image reconstruction algorithm. In this case, the object was assumed completely incoherent one, and the corresponding coherence function was derived in the restricted area. Neither the conventional direct inverse Fourier transform method nor the conventional 2-D maximum entropy method give high contrast desired images. The image obtained by our new algorithm, however, is very close to the original object.

7.5 mm
Scanning Transmitter
(Circular or rectangular mesh)
x
y
R_T
Receiving Transducer Array of 8x8 elements
Z = 200 mm
Wave Length λ =1.5 mm
ξ
η
Object Plane

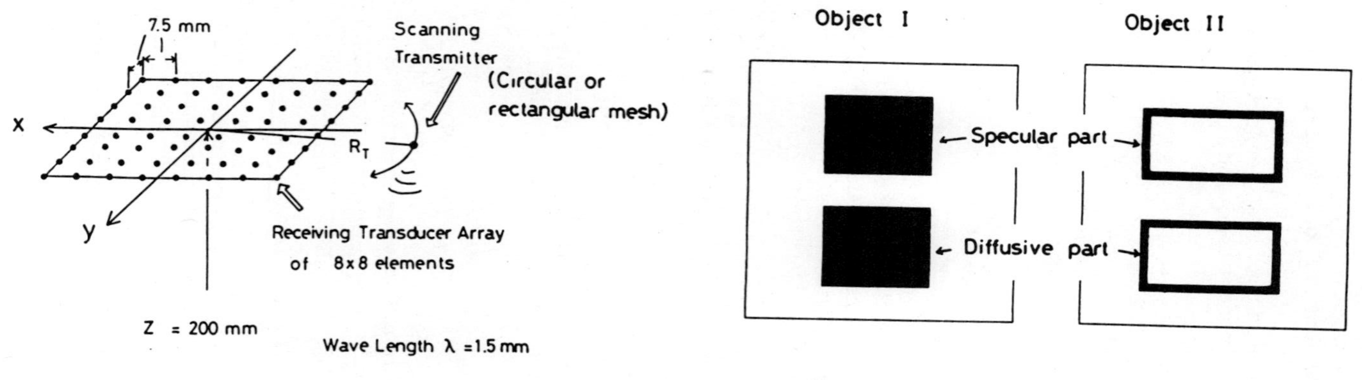

Fig. 4 Imaging geometry used for the computer simulation.

Object I
Object II
Specular part
Diffusive part
Model
Specular part : Uniform Amplitude
: Constant Phase (Object I)
Linear Phase (Object II)
Diffusive part : Uniform Amplitude
: Gauss-Markovian Random Phase
$\phi(r_1 - r_2) = \pi^2 \cdot \exp(-\alpha|r_1 - r_2|)$

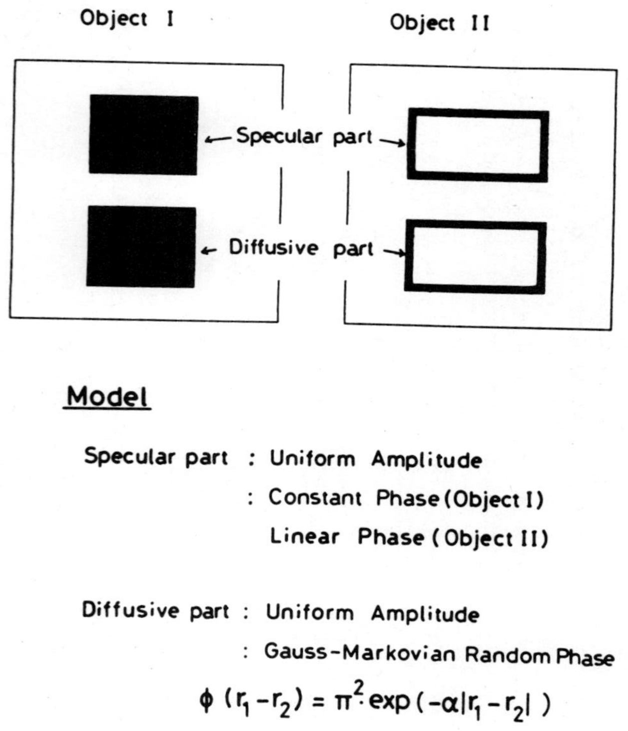

Fig. 5 Objects used for the computer simulation.

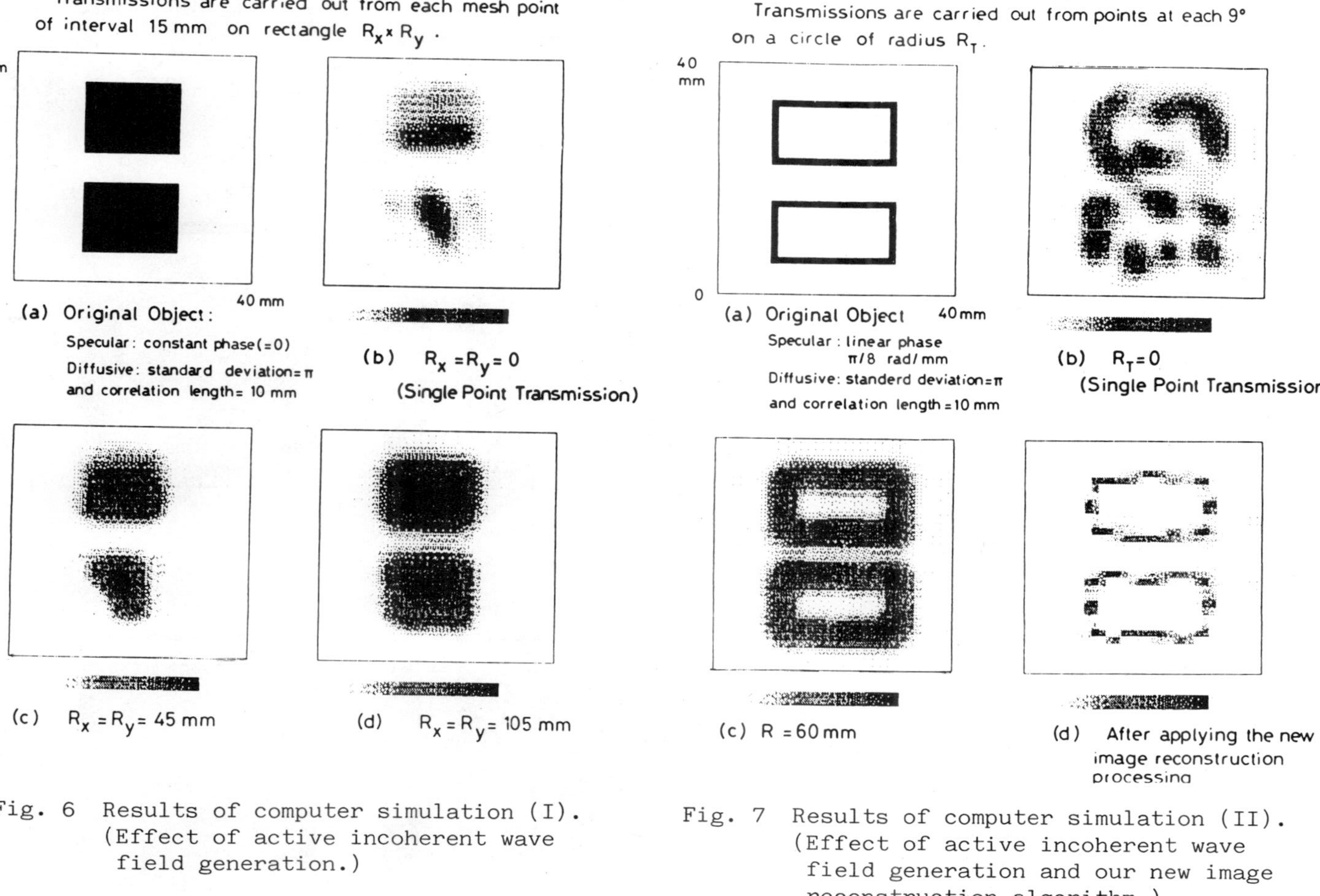

Fig. 6 Results of computer simulation (I). (Effect of active incoherent wave field generation.)

Fig. 7 Results of computer simulation (II). (Effect of active incoherent wave field generation and our new image reconstruction algorithm.)

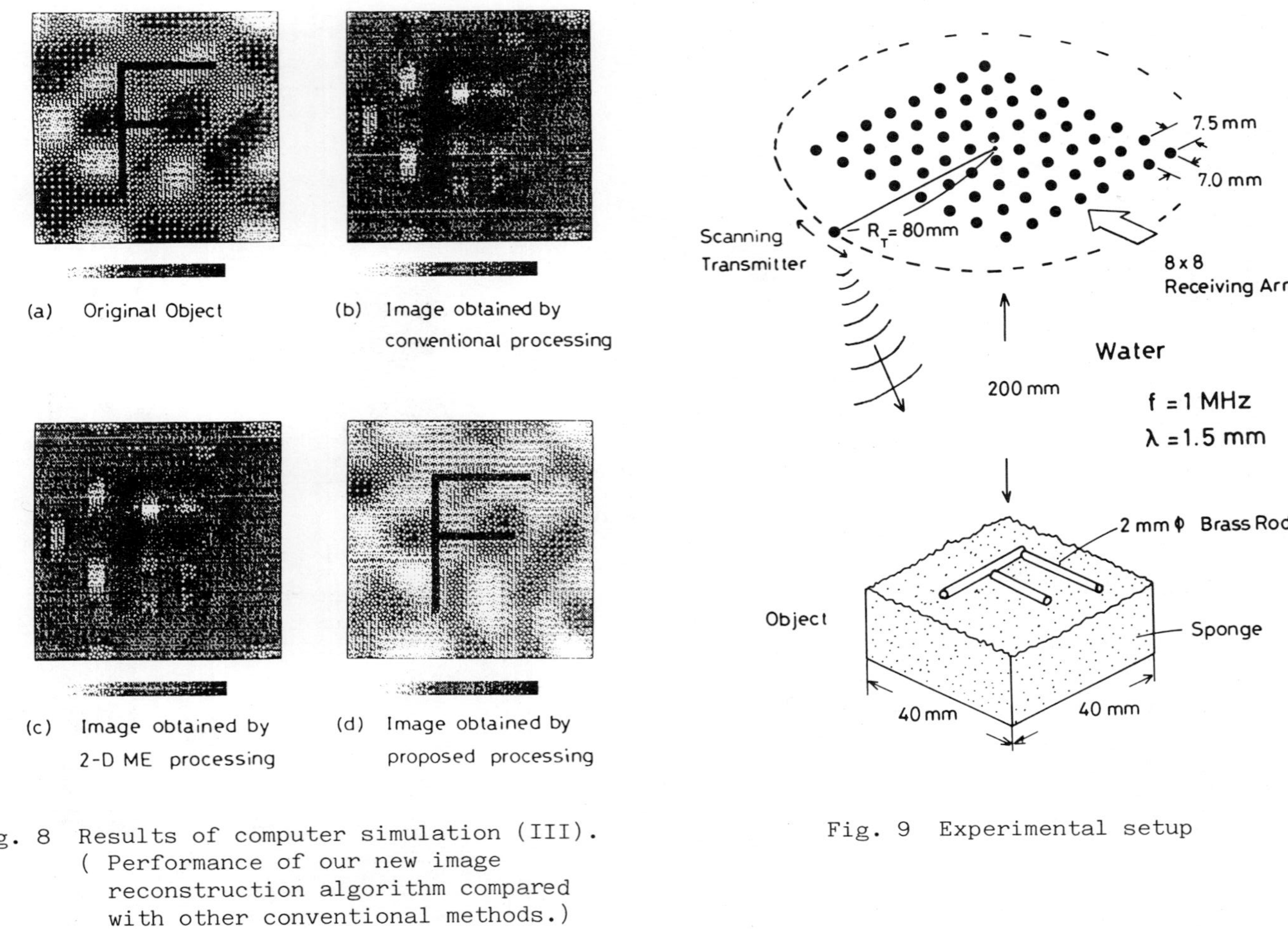

Fig. 8 Results of computer simulation (III). (Performance of our new image reconstruction algorithm compared with other conventional methods.)

Fig. 9 Experimental setup

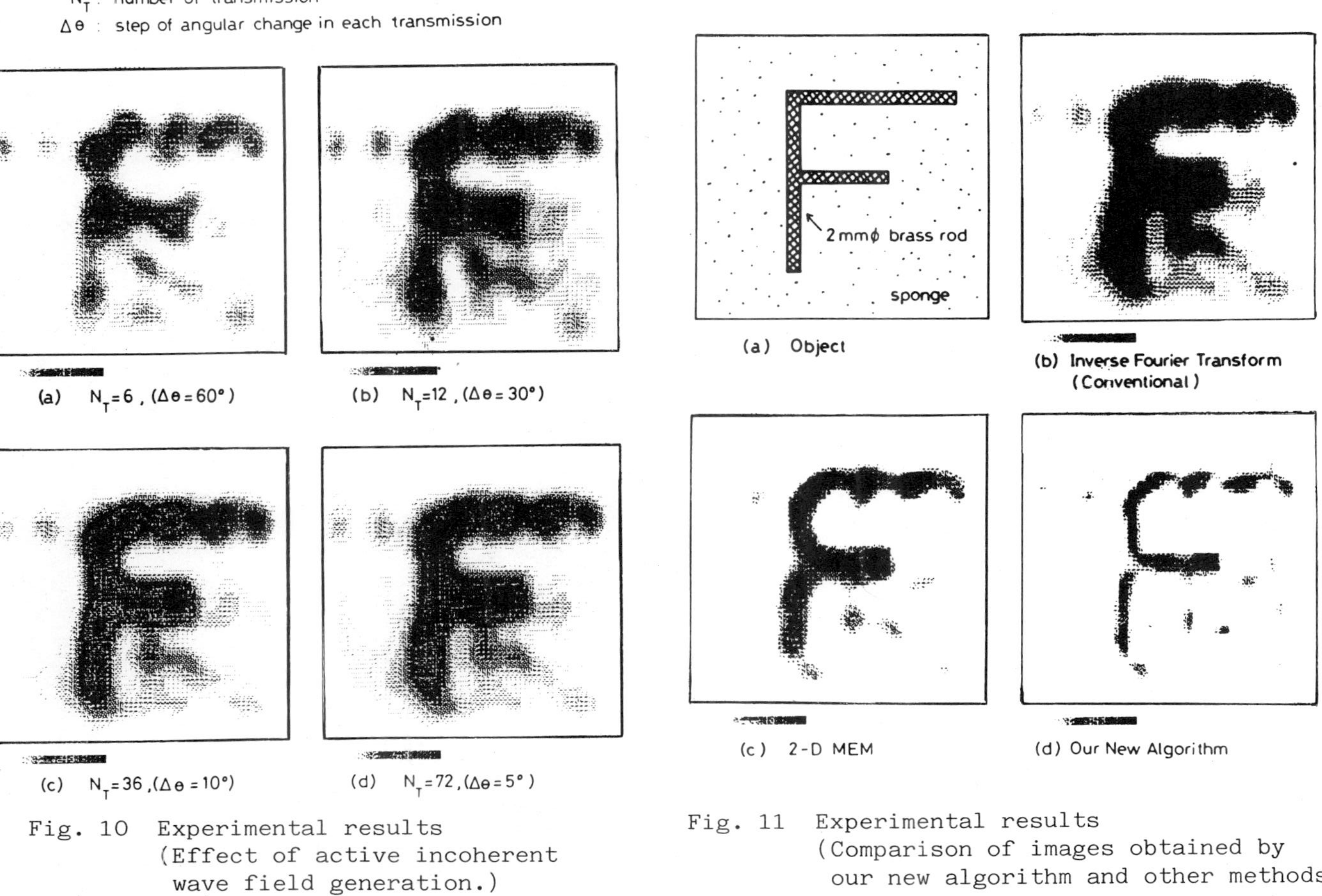

Fig. 10 Experimental results (Effect of active incoherent wave field generation.)

Fig. 11 Experimental results (Comparison of images obtained by our new algorithm and other methods.)

4. EXPERIMENTAL RESULTS

Fig. 9 shows the experimental setup used for C-mode active incoherent ultrasonic image reconstruction.

The object (letter F) made of 2 mm ϕ brass rod is mounted on a sponge with random surface. The transmission of the ultrasonic pulsed waves is from a transmitter scanned on a circle of radius 80 mm. The receiving rectangular array has 64 elements. Its Rayleigh resolutions are about 5 mm.

Fig. 10 shows the obtained results. In this case the image reconstruction was carried out by averaging the intensity images obtained in each transmission. This is equivalent with the direct inverse Fourier transform of the coherence function as mentioned before. With the increase of the number of transmissions the speckle noises are reduced and more stable images are obtained.

Fig. 11 shows the effect of the new image reconstruction algorithm. 2-D maximum entropy method increases the resolution a little but the improvement of the resolution is limited due to the existence of the background. The image obtained after applying our new algorithm gives sharp image with high contrast.

5. CONCLUSION

In this paper, we showed the principle of 3-D active incoherent ultrasonic imaging system which uses transmission/reception on a restricted area on a hemisphere and new nonlinear image reconstruction algorithm. The results of numerical analyses and experiments showed the usefulness of this method both for the speckle noise reduction of diffusive objects and faithful image reconstruction of specular objects.

REFERENCES

1. Takuso Sato, Takayoshi Yokota and Osamu Ikeda, An Optimum Ultrasonic Imaging System Using ARMA Processing, Acoustical Imaging, vol. 11, J.P. Powers ed., Plenum Press 1982 (in press).
2. S. J. Werneck and L. R. D'Addario, Maximum Entropy Image Reconstruction, IEEE Trans. Comput., C-26:351 (1977).
3. J. Capon, High-Resolution Frequency-Wavenumber Spectrum Analyses, Proc. IEEE, 57:1408 (1969).
4. G. L. Duckworth, Adaptive Array Processing For Acoustic Imaging, Acoustical Imaging, vol. 9, Prenum Press, New York, (1980).
5. "Nonlinear Methods of Spectral Analyses", S. Haykin, ed., Springer-Verlag, New York, (1979).

ACOUSTICAL IMAGING USING THE PHASE OF ECHO WAVEFORMS

L. Ferrari, J. Jones, V. Gonzalez and M. Behrens

Department of Radiological Sciences
University of California Irvine
Irvine, California 92717 U.S.A.

INTRODUCTION

All conventional ultrasound imaging equipment, including static B-scanners and real time units, apply some form of envelope detection to the received radio-frequency (rF) signal in order to produce an image. Unfortunately, the envelope detection process essentially destroys all phase and spectral information which is readily available on each rF A-line waveform. Since such additional data offer a new dimension of information which has proven to be useful if not necessary, for quantitative tissue characterization, it is worthwhile to consider techniques for incorporating both phase and spectral information into an image. Here we propose one method for utilizing phase information in a fundamental way which produces a new class of ultrasound images. In brief, envelope detection is replaced by a frequency demodulation process to produce a signal with phase rather than amplitude information. This signal is fed to a standard digital scan converter, with appropriate storage algorithms, to produce an image. These phase or FM images have a totally different appearance from conventional amplitude or AM images. In-vivo scans of human subjects and various gray scale test objects show recognizable anatomical features (which overlay precisely with amplitude images) but significantly different texture and structural patterns. A clear measure of attenuation is also noted. In this paper we will provide some physical justification for the origin and utility of phase information, describe our method (which we have implemented in hardware) for producing phase or FM images, show several FM images of human subjects and test objects, compare AM and FM images of the same target, and discuss various hardware considerations related to this method.

MOTIVATION FOR FM IMAGING

It is clear that an ultrasonic echo waveform contains amplitude information. In fact a B-mode ultrasonogram is merely a mapping of echo intensities. To produce this mapping or ultrasonic image, the rF ultrasound waveform is treated as a standard AM signal. That is, it is detected (or envelope detected) as a necessary first step in the production of an image.

However, an ultrasound waveform can be viewed as both an AM and an FM signal. This is equivalent to saying that the signal contains both amplitude and phase information. The FM aspects of an ultrasound waveform are clearly demonstrated if we measure the mean frequency (in a short time window) as a function of depth for a single A-line waveform. Such a demonstration is shown in Figure 1. Here a single rF A-line of normal human liver, recorded in vivo, has been subjected to short-time Fourier Analysis and the mean-frequency plotted as a function of acoustic travel time (or depth). The non-uniform variation in the mean-frequency with depth are a clear indication of the presence of FM in the signal. Thus, an rF ultrasound waveform can be viewed as both an AM and an FM signal.

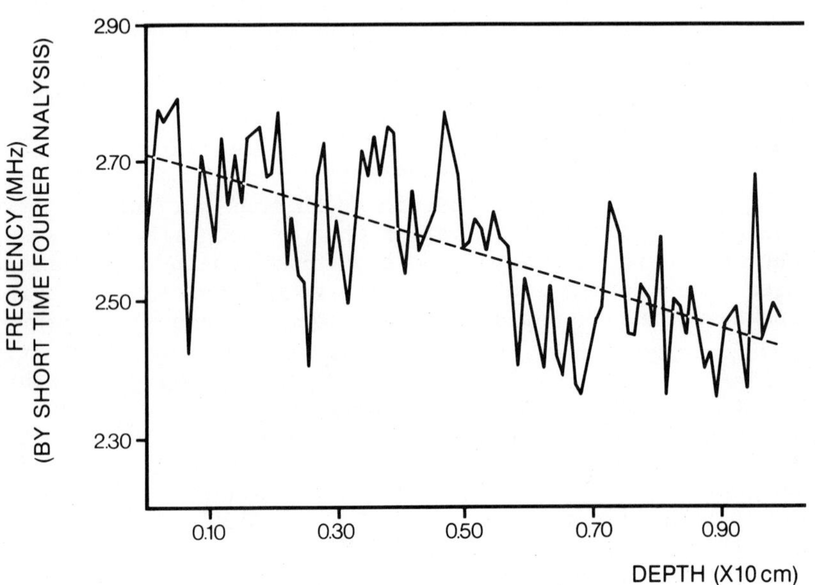

Figure 1. The mean-frequency (solid curve) as a function of depth for a single A-line from normal human liver recorded in vivo. Frequency modulation is evident as well as the overall effect of frequency dependent attenuation (dotted curve).

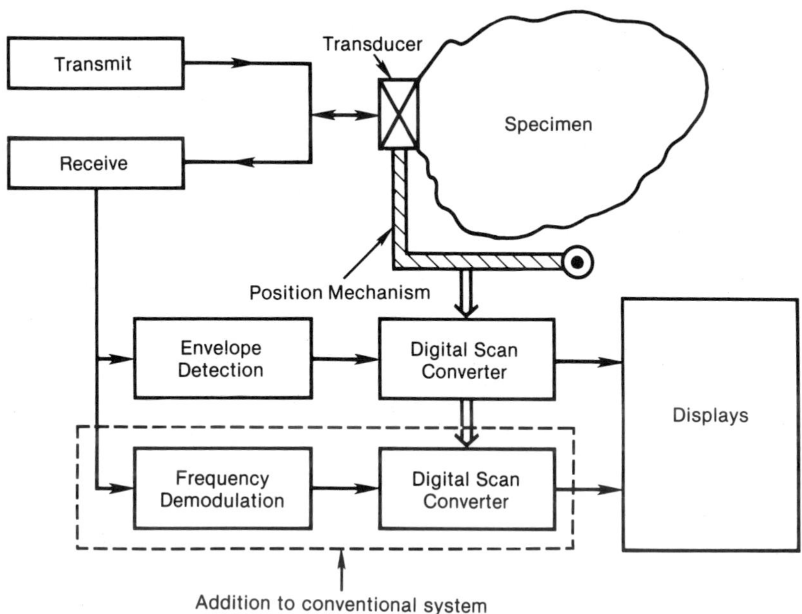

Figure 2. Block diagram of ultrasonic imaging system using frequency demodulation. An FM demodulator was added to a conventional ultrasound scanner (Philips 5580).

If standard AM techniques are used to produce a conventional ultrasound image than it would seem that standard FM techniques could be used to produce a new class of ultrasound images, utilizing phase rather than amplitude information. It is precisely this concept of FM imaging which forms the basis of the present work.

Treating the rF ultrasound waveform as an FM signal (as well as an AM signal) led us to replace envelope detection in a conventional scanner (Philips 5580 B-scanner) with frequency demodulation. Thus, the conventional AM processing of envelope detection is replaced by the conventional FM processing of frequency demodulation. A block diagram of the instrumentation is shown in Figure 2. For simplicity the demodulator used was a commercial unit (JVC) designed for video recording applications. Although this unit covered the bandwidth of the ultrasound signals it was clearly not designed for this particular application and represents a sub-optimal design.

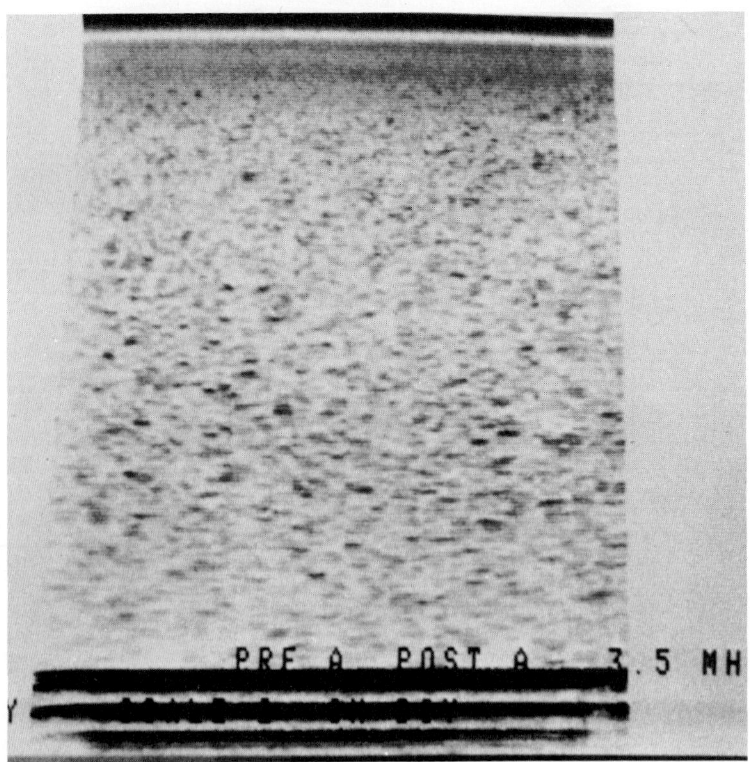

Figure 3. Conventional B-mode ultrasonogram (AM image) of a gray-scale test object designed to simulate soft tissue, especially liver.

EXPERIMENTAL RESULTS

The equipment described in Figure 2 was used to obtain conventional (or AM) ultrasound images as well as FM images from various test objects and several human subjects. Figures 3-6 show a representative example of the results. The accompanying figure captions describe the results in some detail.

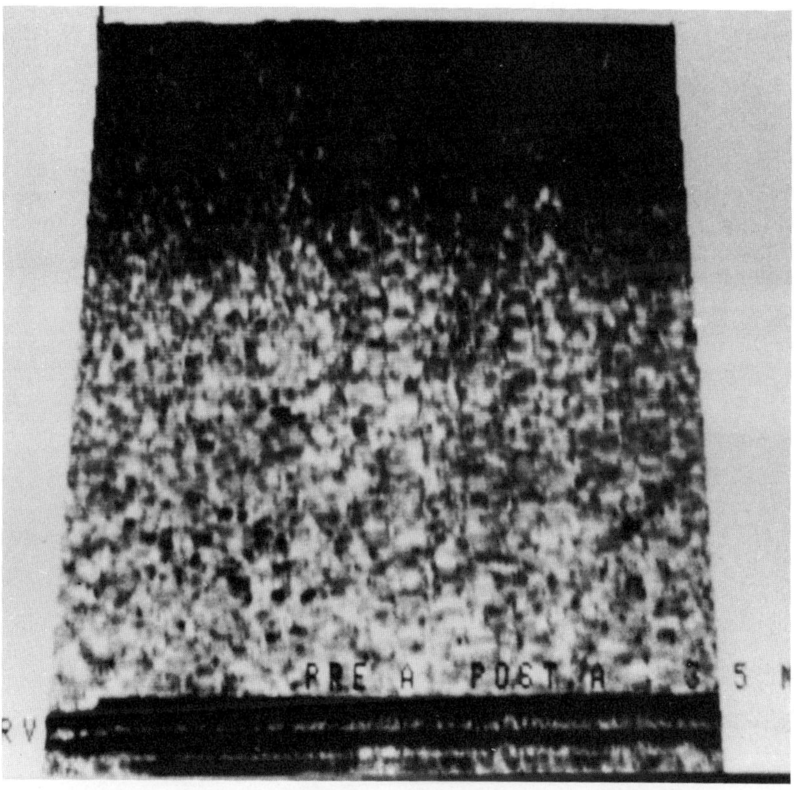

Figure 4. FM image of the same test object as Figure 3. Note the significant difference in texture between the two images. The black band at the top of this figure is a photographic artifact. In the original image one sees a gradual lightening with depth which is indicative of frequency dependent attenuation.

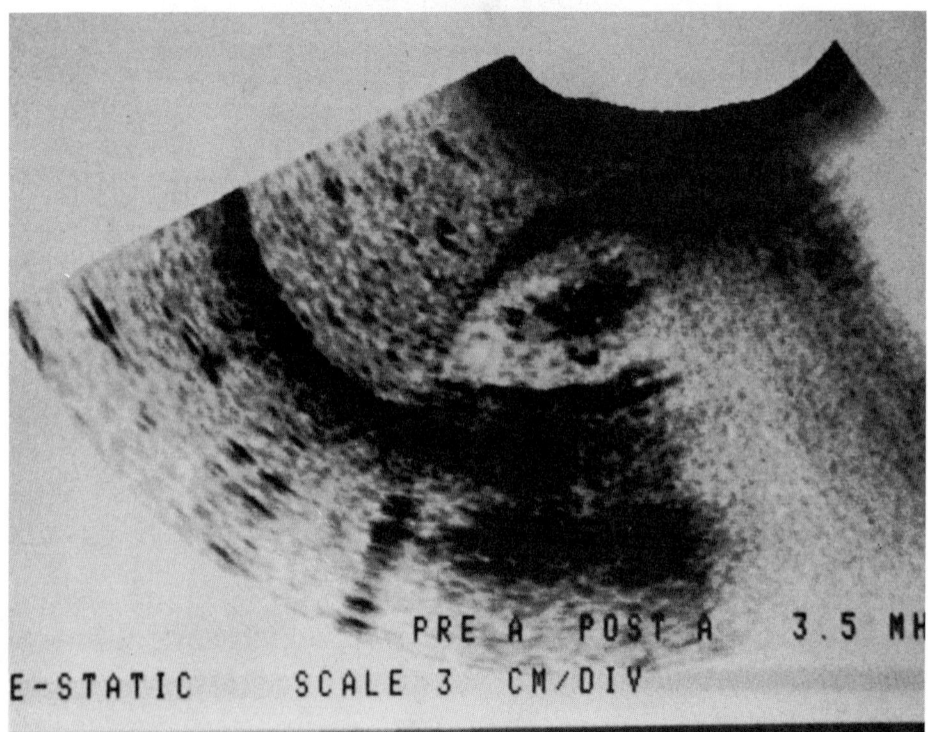

Figure 5. Conventional B-mode ultrasonogram (AM image) of a normal human subject. The liver is at the upper left, the kidney in the upper center, and the diaphragm the dark line in the center extending from the left (under the liver) to the right center of the image.

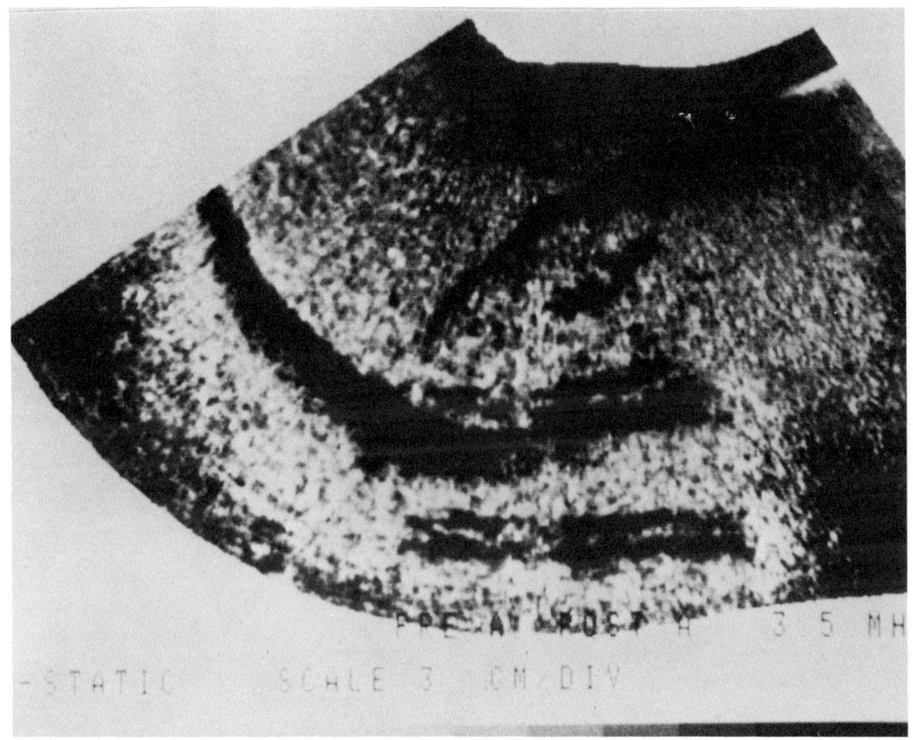

Figure 6. FM image of the same structure as Figure 5. Note that the anatomical features are identical but that the texture patterns are quite different.

CONCLUSIONS

We have proposed and implemented a simple technique for incorporating phase information into an acoustical image. The technique is based on our observation that a backscattered ultrasound waveform (A-line waveform) contains frequency modulated (FM) information as well as amplitude modulated (AM) information, utilized in current imaging systems. Our present implementation of an FM imaging system utilizes a sub-optimal frequency demodulator. Work now in progress is designing instrumentation more appropriate to this application. Extensive clinical trials and a critical evaluation of the utility of FM imaging will follow.

NDE IMAGING WITH MULTIELEMENT ARRAYS

R.C. Addison, K.A. Marsh, J.M. Richardson, and C.C. Ruokangas

Rockwell International Science Center

Thousand Oaks, CA 91360

An ultrasonic phased array system will be described. This is a digitally based system consisting of two 2.5 MHz, 32 element piezoelectric transducer arrays connected to a multiplexer. The multiplexer can select contiguous groups of 16 elements within either array as transmitters and/or receivers. The sixteen receiving elements are each connected to a separate A/D converter and buffer memory. The beam forming for the 16 received signals is done by an array processor which is capable of implementing a wide range of signal processing techniques associated with the synthesis of the beam. The output of the array processor can be stored on the disk memory of a mini-computer and post processing can be performed on the signals prior to displaying them on a color graphics display.

The system is intended to provide a flexible technique for exploring the use of multielement arrays in NDE applications. The digital post-processing capabilities of the system permit many of the practical difficulties encountered with available multielement transducer arrays to be overcome. This signal processing capability plus the availability of two transducer arrays also allows a variety of imaging techniques to be investigated.

Results have been obtained with several flaws within metal parts using sector scanning and a differential imaging technique. Current work is concerned with the use of both array transducers to synthesize larger apertures and thus improve the detectability and resolution of flaws within metal parts.

INTRODUCTION

The phased array system described here has been specifically developed to explore the feasibility of using multi-element arrays for NDE applications. It needs to be emphasized that even though the use of phased ultrasonic array systems is widespread in the medical community,[1-5] there is currently little use of them for NDE applications. This situation arises because the requirements placed on an array system for NDE applications are quite different than those needed for medical applications. The applications of arrays in NDE are limited by the current technology relating to array transducer fabrication. Limitations are also imposed by the materials that are to be inspected. A thorough discussion of this aspect of the problem has been given in the paper by Addison.[6]

The intent of the design of this system is that it be useful in a number of different configurations so that many different applications can be investigated. Generally the usefulness of an array for NDE applications stems from the capability of the array to rapidly scan an ultrasonic beam and to change its focus to accomodate the curved surfaces of different part. This latter characteristic will be referred to as beam agility. The array is also useful as a data acquisition system for synthetic aperture and image reconstruction techniques.

We will give a description of the architecture of the array system and some of the hardware features. The system is controlled by a minicomputer and utilizes an array processor for beam forming. The flexibility that is associated with these components will be described.

The system has been used in a sector scan mode to image voids in metal parts. We will describe a differential imaging technique that is useful in overcoming difficulties that are associated with this mode of operation. In addition a probabilistic image reconstruction technique is described. This technique allows the combination of data acquired over a diversity of angles. The quality of the image reconstructed in this way is improved over that obtained with a sector scan.

SYSTEM CONCEPT

The rapid beam scanning capability of an array can be exploited in several different ways in an NDE application. If the array transducer is sufficiently long, it can be used to scan over a part or segment of a part by electronically switching the active group of transducers within the array. Generally in this type of application the part is a long metal plate that passes beneath the array transducer in a direction that is orthogonal to the scan direction of the array. In a similar way it is possible to inspect flat circular disks by orienting the long dimensions of the array (the dimension parallel to the scan direction) along a radius of the disk and offsetting the transducer to one side of the center. If the plate has a two dimensional contour or if the disk has a circularly symmetric contour, simple electronic switching of the transducer elements is no longer applicable. Now the array system must be capable of changing the angle of the ultrasonic beam and preferably also its focus in order for the beam to enter the part perpendicular to the surface.

An example of this mode of operation is shown in Fig. 1. Here the beam is always entering the part normal to the surface. The beam agility feature can be used to compensate for focusing effects caused by curvature of the metal surface. The tilting and translation of the beam is done electronically and avoids problems with mechanical inaccuracies and backlash that are frequently encountered in mechanical drives. The technological capability for making a linear array transducer that would be long enough to scan over the entire radius of a turbine disk has not been demonstrated at this time. Instead to assure the performance of the array, we have found that it is necessary to use array transducers of length 1/2 to 1 inch that are suitable for phased array applications. To preserve some of the capabilities of the longer array transducer, two of the shorter transducers each containing 32 elements are used with a mechanically variable spacing between them as shown in Fig. 2. Although this severely limits the use of the array system for contour following applications, it can be used to demonstrate the feasibility of the concept.

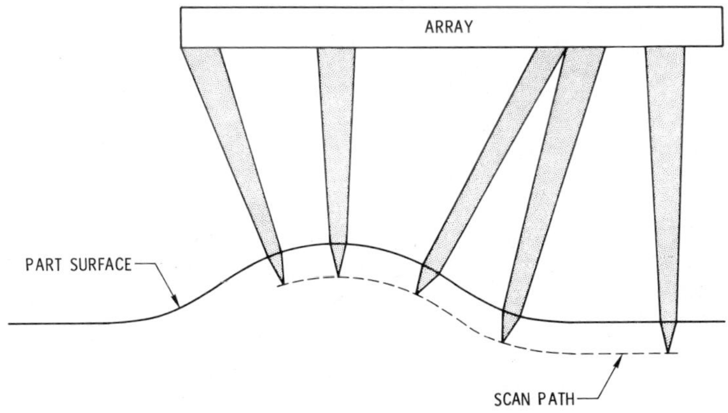

Fig. 1 Contour scanning mode.

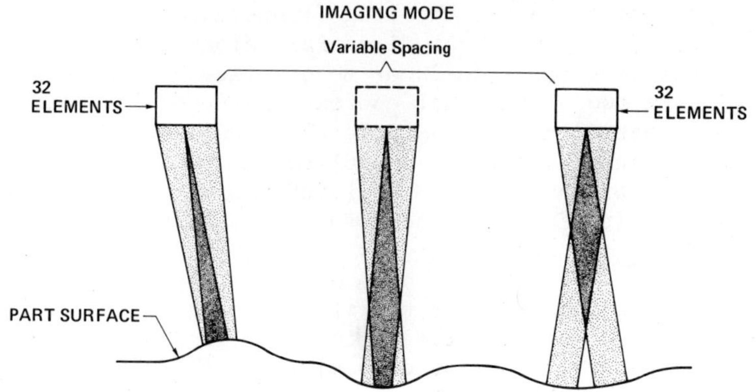

Fig. 2 Contour scanning - two arrays.

The array system can also be used to interrogate an isolated flaw indication. A single group of array elements can be used to generate a sector scan to produce a B-scan image of the flaw; data can be acquired using two arrays to reconstruct an image of the flaw; or the array can be used to obtain scattering data from the flaw that can be used with various long wavelength inversion techniques. This scattering mode, shown in Fig. 3, can acquire data using either a pulse-echo or a pitch-catch technique.

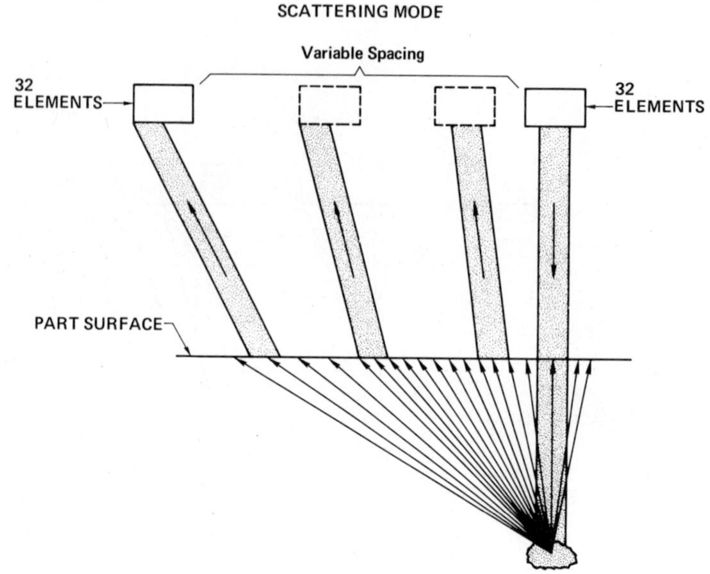

Fig. 3 Scattering data acquisition mode - two arrays.

DESCRIPTION OF SYSTEM

The electronics for driving this array have a somewhat different objective than some of the array based systems that are currently available, which acquire and display an image in as little time as possible. This system must be sufficiently flexible to be used in a wide variety of situations and be capable of operating in many different modes. In essence it is to serve as a test vehicle for exploring the feasibility of different NDE imaging concepts. To accomplish this the acquired waveforms must contain a minimum of spurious signals caused by crosstalk or non-linearities. It is desirable to display a single image and be able to recognize the important details of the object under study. In designing and constructing the system we have weighed the features providing flexibility much more than those providing speed or compactness. This was done in the belief that the essential features for a particular application could always be abstracted to produce a fast or a compact system. The two important questions are: is the system reaching its potential as a data acquisition device? And, how does its performance compare with alternative techniques? The details of the system have been previously described.[7,8] Here we will provide an overview of the salient features and the principles of operation.

The block diagram in Fig. 4 delineates the major subassemblies of the array system and their functions. The minicomputer sends out a set of codes that define the state of the system prior to triggering a transmit-receive cycle. These codes are stored in the control memory. The timing and control block interprets the codes and carries out the actions they specify. Five codes are required to specify the state of the system. The particular set of 16 contiguous transmit elements and independent set of 16 contiguous receive elements must be selected from among the 64 elements that are available. The set of time delays that define the transmit beam direction must be selected. The gain of each of the 16 receive channels must be selected. Finally the delay after the trigger before turning on the 16 A/D converters must be specified. The timing and control block and the transmit trigger block provide the delays for the 16 selected transmitter elements. A feature of this circuitry is the use of digitally controlled delay lines to decrease the quantization level of the specified delay to 4 nsec. This fine delay increment is useful in reducing the sidelobes in the transmit beam that can be caused by coarse quantization of the delays.

The trigger signals activate the selected pulsers. Upon receipt of a trigger, a pulse which can be varied in amplitude from 5 to 400 volts is applied to each of the selected elements. The returning ultrasonic signal from the object under investigation is received by each of the 16 selected receive elements. The received

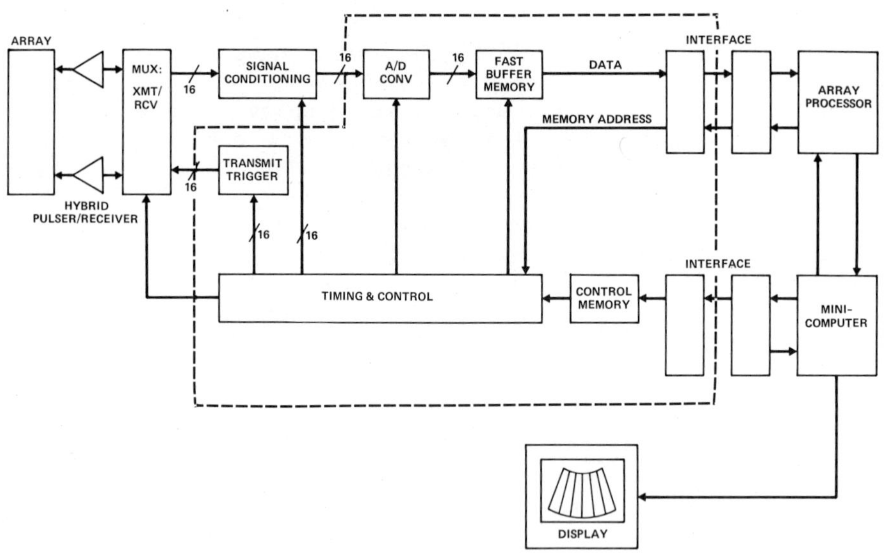

Fig. 4 Ultrasonic phased array system block diagram.

signals are each amplified 26 dB by a low noise, low distortion hybrid preamplifier. Several design limitations dictated that the pulser/receiver be on the array side of the multiplexer. This means that a pulser/receiver is required for each array element or a total of 64 for the entire system. This could only be achieved by designing and building the pulser/receivers as hybrids. Two complete hybrid pulser/receivers were mounted on a single substrate. The package shown in Fig. 5 is a 3.4 cm × 2 cm × 0.5 cm hermetically sealed can with 24 pins.

After leaving the receiver, the signals are passed through the receiver multiplexers to one of sixteen receiving channels. The elements of one channel are shown in Fig. 4. The first block contains a signal conditioning circuit. The signal conditioning circuit serves as an anti-aliasing filter and adjusts the peak-to-peak amplitude of the signal as well as its dc level to make it compatible with the A/D converter. The digitally controlled attenuator is located here. It provides up to 48 dB of attenuation in 3 dB steps.

The next block is the 8 bit, 18 MHz A/D converter which digitizes the signal and loads it into an 8 × 1K-bit fast random access buffer memory. All 16 of the A/D converters are started synchronously after a delay corresponding to an ultrasonic round trip time that is selected by the operator. The contents of each of the sixteen memories can be sequentially clocked through the interface to

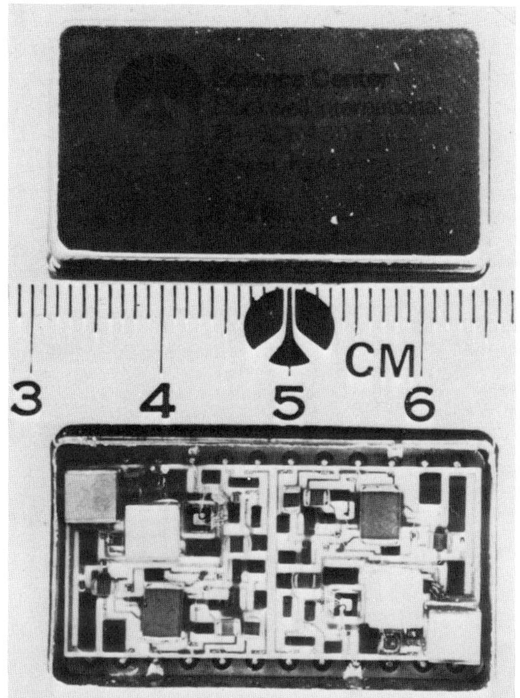

Fig. 5 Photograph of hybrid pulser/receiver.

the array processor at a rate compatible with the operation of the array processor and minicomputer (approximately 0.5 MHz). The individual waveforms from each of the receive channels can either be passed through the array processor to the memory of the minicomputer, or combined with the other waveforms in the array processor. Both capabilities are useful. For image reconstruction algorithms that are experimental in nature, it is preferable to have the raw data present in the minicomputer to facilitate the optimization of the algorithm. For beam forming operations used for making sector scans the waveforms are combined in the array processor.

After the beam-forming or image reconstruction operation is completed, the processed signals are stored on the disk memory of the minicomputer. The signals are recalled and displayed on a color display unit using a program that is specific to the application. Generally the images take the form of B-scan displays.

It is essential that the hybrid pulser/receivers and the multiplexer circuitry be near the array transducers. Consequently they are located in a waterproof box with dimensions of 36 cm × 36 cm

× 41 cm. This box and the array transducers are held in place by a frame mounted on the scanning bridge of an existing bridge and carriage assembly as shown in Fig. 6. This assembly provides rectilinear motion in the horizontal plane under computer control. The box can be manually raised and lowered using a rack and pinion drive. The spacing between the two transducer assemblies can be changed manually through a threaded drive assembly. The spacing can be varied from zero inches (i.e., the array housings are physically touching) to a maximum of 15 cm.

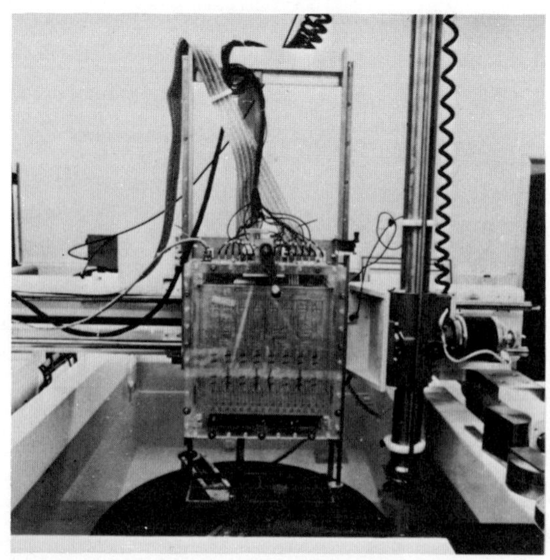

Fig. 6 Multiplexer, pulser/receiver and transducer assembly mounted on test bed scanning bridge.

SOFTWARE UTILIZED FOR CONTROL OF ARRAY AND AP400

The interactive minicomputer resident software package, ARRAY, allows a user to control and coordinate the activities of both the array digitizer electronics (ADE) and the Analogic AP400 array processor (AP). The package supplies the user with the ability to easily command the acquisition of a full set of waveforms with a minimum of user input, as well as to single step, i.e., command by single instruction, the initiation of the ADE and acquisition of data from the ADE via the AP400. The existing version of the software does not take advantage of all of the capabilities of the ADE hardware. It provides access to a subset of the hardware that will permit the acquisition of single waveforms and sector scans. ARRAY consists of both FORTRAN 5 and assembly language modules. It was written for a Data General ECLIPSE S/200.

For the standard acquisition of data from a range of angles, the user is queried for sector scan definition which includes the total number of waveforms, the number of degrees in the sector, and the starting angle. In addition, the output disk file name for the storage of waveforms, the initialization parameters for the ADE, and the number of channels to utilize must be specified. The transfer of these input parameters to the ADE, the subsequent acquisition of data, and the activities of the array processor are transparent to the user. ARRAY notifies the user of the completion of signal processing for the entire sector.

The array processor manufactured by Analogic and designated the AP400 is used for the high speed acquisition of data from the ADE and serves as the "beam former" when standard sector scan data is acquired. Although the array processor can be used in an interactive program which allows the user to specify, in order, each function that is to be executed, the unit is used with ARRAY in a mode that executes a set of commands. This module acquires the data from each of the 16 channels of the ADE and combines them in a specific way to synthesize a beam returning at a specified angle. This synthesis is accomplished by shifting each of the waveforms by a time delay that will permit the waves to sum coherently in the direction specified. For a sector scan, this direction corresponds to that of the transmitted beam. As each waveform is shifted, it is added to a running sum until all 16 waveforms have been combined. There are several ways available to shift the waveforms. Perhaps the most straightforward technique is to shift the waveform in the time domain by an integral number of sample intervals. The sample interval for the ADE is 55 nsec. When a beam is formed with this coarse quantization of the delay, the side lobe level becomes considerably higher than desired.

As the quantization of the time delay is decreased, the side lobe amplitudes are reduced. The best way to decrease the quantization, which implies some sort of interpolation between samples, is to perform the time shifting in the frequency domain. The signals are Fourier transformed and then multiplied by the complex phase factor $\exp[i\omega\tau]$ where τ is the desired time shift. The signal is then inverse Fourier transformed and summed to synthesize the beam in the desired direction.

DISPLAY OF DATA

The display processor unit contains a 480 × 512 pixel memory which is mapped onto the 3:4 aspect ratio CRT of a television monitor. The memory is filled with data derived from data in the disk memory of the minicomputer. The method used to store the data in the memory of the display processor, i.e., the angles and positions of the vectors corresponding to the ultrasonic waveforms is controlled by a software routine that is specific to the application. Each of the pixels can be characterized by an 8 bit word to provide a level of gray or a color according to a code that is chosen by the operator.

In the case of a sector scan a pattern like that shown in Fig. 7 is displayed. The scaling of the sector segment is automatically deduced from the maximum and minimum angles of the sector, the number of lines within the sector, the number of degrees in the sector, the starting angle, and the delay time before the waveform digitization begins. With metal specimens there is a significant refraction of the acoustic beam as it enters the metal. This has been discussed in reference 6. To prevent distortion in the image, the software correctly accounts for the refraction angles and the increase in acoustic velocity within the metal.

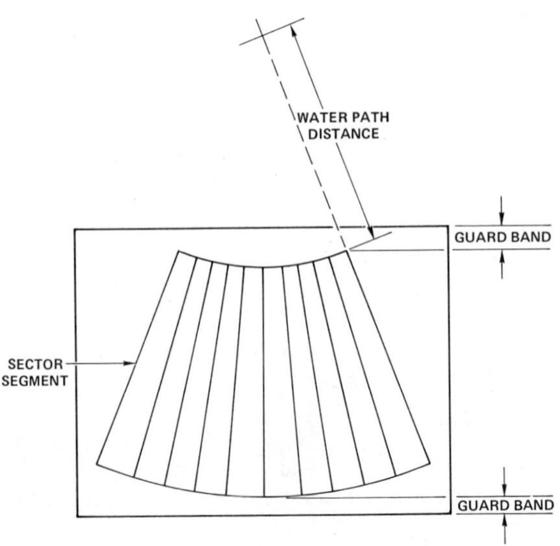

Fig. 7 Sector scanning pattern.

OPERATION OF PHASED ARRAY SYSTEM

For the imaging of flaws in metals the loss of longitudinal resolution due to the increase in velocity imposes severe constraints on the use of relatively low frequency arrays. Particularly troublesome is the low level of the flaw signal relative to the front face echo. In the case of a 1200 micrometer diameter spherical void, the front face signal must decay to a level that is 40 dB below its peak value before the flaw signal is detectable without some sort of post processing. Transient response measurements of the array transducer only extended to a level -30 dB below the peak value, but even at this level the elapsed time was 3.75 μsec. This corresponds to a flaw depth of one-half inch in a metal like titanium. This severely restricts the use of real time arrays for detecting flaws in metals except in cases where the flaw signals are very strong or the flaws are located one inch or more below the surface.

One possibility for minimizing the front surface echo is to propagate through it at an oblique angle. The extent to which this is useful depends on the magnitude of the angle, whether the part can be tilted relative to the array, and the strength of the target. It has been observed that the reflection from a flat part remains reasonably strong over the entire range of ±10 degrees that has typically been used for our sector scans. This results in an imaging artifact that appears as an arc of a circle that is tangent to the surface of the part at the point where the angle of incidence of the beam is zero degrees. Since the amplitude of the backscatter from the oblique intersection of the beam with the flat surface is very small relative to a specular reflection, it is usually not visible in the image. Thus a typical image obtained from a part appears like the one shown in Fig. 8.

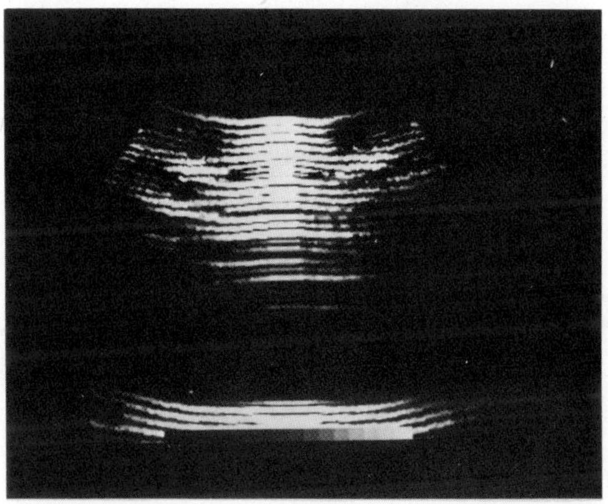

Fig. 8 Unprocessed image obtained from a titanium part with a flat front surface.

As it stands this image is not very useful. The "ring down" from the front surface echo is obscuring the flaw that is present and the artifact created by the reflection from the front surface is creating the illusion that the surface is curved. Most of these problems can be overcome by using a differential imaging technique.

This means that a reference image is obtained of a region of the part that contains no flaw but has essentially the same surface reflection characteristics as the region containing the flaw. The reference image is then subtracted from the flaw image to eliminate the problems that are created by the front surface echo. When this technique is used with the image in Fig. 8, the differential image shown in Fig. 9 is obtained. The subtraction has suppressed the front surface echos by about 25 dB and the image of the 1200 μm spherical void stands out clearly.

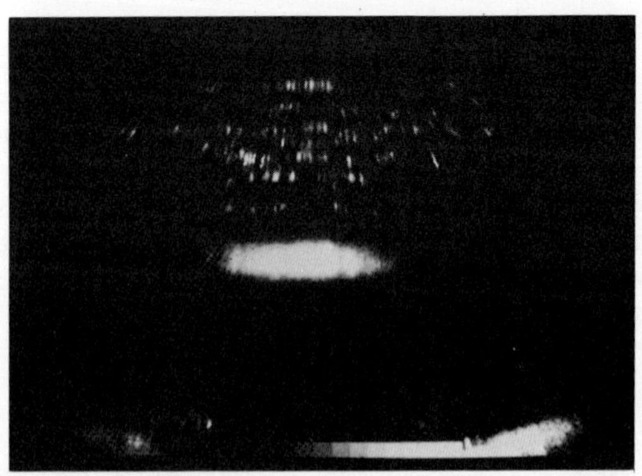

Fig. 9 Differential image of a 1200 μm spherical void in a titanium disk.

The lateral extent of the flaw reflects the width of the ultrasonic beam that results from a 7.2 mm wide transducer when it propagates through a 50 mm water path and a 12.5 mm titanium path. (Some idea of the scale in Fig. 9 can be obtained from the fact that the distance between the front and back surface of the titanium sample is 25 mm.) This does not result from any characteristic of the array other than its aperture. Figure 10 shows a B-scan image of the same flaw that was obtained using a differential imaging technique with a single element transducer whose aperture was 7.2 mm. The streaks which demonstrate the difficulty in obtaining a differential image when the transducer has to be mechanically moved, are caused by slight irregularities in the motion of the transducer. Note that the width of the flaw is about the same as in Fig. 9.

PROBABILISTIC IMAGING

The performance of the array system is satisfactory when using it for simple sector scanning. However, the resolution is not very good because of the small aperture. This can be improved by using

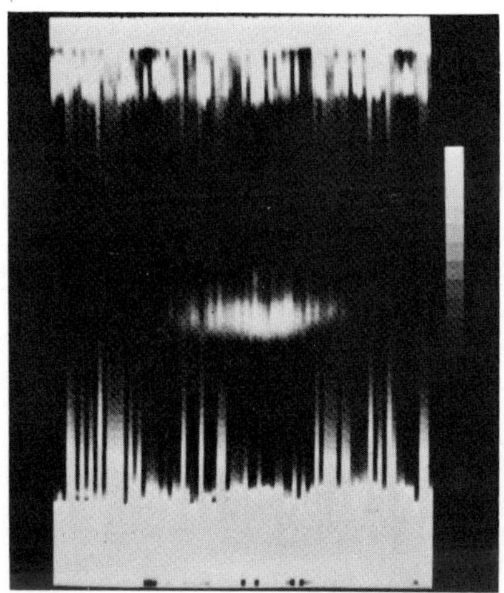

Fig. 10 B-scan of a 1200 μm spherical void obtained with a single element transducer having the same aperture as the array.

both array transducers spaced some distance apart to subtend a larger numerical aperture at the target. The mechanical stability of the arrays, as evidenced by comparing Figs. 9 and 10, makes them good choices for data acquisition devices.

Theory

To utilize the data acquired in this way, we have chosen to develop a new image reconstruction technique that applies estimation theory to imaging.[9] The basic concept is easy to state: we start with a probabilistic measurement model (representing an *a priori* statistical ensemble of possible inhomogeneities and measurement errors) and then find the most probable inhomogeneity given the actual measurements of the scattered wave.

The stochastic measurement model for the case of longitudinal-to-longitudinal backscatter is given in the time-domain by

$$f_n(t) = p(t) * R(t,\vec{e}^{\,i}_n) + \nu_n(t) \quad , \tag{1}$$

where the symbols are defined as follows:

$f_n(t)$ = possible waveform for n^{th} measurement at the time t,

$p(t)$ = reference waveform (transducer response),

$\nu_n(t)$ = error in n^{th} measurement (noise),

$R(t,\vec{e}^{\,i}_n)$ = impulse response function for nth measurement with incident direction $\vec{e}^{\,i}_n$.

The time t is confined to the interval [0,T] where T is the observation time. The measurement error $\nu_n(t)$ is assumed to include both grain scattering and electronic noise. Here we make the simple assumption that the $\nu_n(t)$ are Gaussian random processes with the properties

$$E\ \nu_n(t) = 0$$

$$E\ \nu_n(t)\ \nu_{n'}(t') = \delta_{nn'}\ C_\nu(t - t') \quad . \tag{2}$$

Here we make the somewhat artifical assumption that $p(t)$, $R(t,\vec{e}^{\,i}_n)$, and $C_\nu(t)$ are all periodic with period T.

The impulse response function is more conveniently discussed in the temporal frequency domain in which case we consider the scattering amplitude defined by

$$A(\omega,\vec{e}^{\,i}_n) = \int_0^T dt\ R(t,\vec{e}^{\,i}_n)\ \exp\ (i\omega t) \tag{3}$$

For L → L backscatter in the Born approximation we can write

$$A(\omega,\vec{e}^{\,i}_n) = \alpha\omega^2\bar{\zeta}(\vec{q}_n(\omega)) \quad , \tag{4}$$

where $\overline{(\cdot)}$ denotes the spatial Fourier transform of $(\cdot)$ and $\bar{\zeta}(\vec{k})$ is given by

$$\bar{\zeta}(\vec{k}) = \int_{D_s} d^3\vec{r}\ \zeta(\vec{r})\ \exp\ (-i\vec{k}\cdot\vec{r}) \quad , \tag{5}$$

for an arbitrary spatial frequency $\vec{k}$ and where α is a constant dependent only upon the properties of the host medium. The function $\zeta(\vec{r})$ is the deviation of the longitudinal acoustic impedance, i.e., $\zeta(\vec{r}) = \delta(\rho c_L)$. We assume that it vanishes outside of the rectangular localization domain, D_s, defined by the inequalities

$$-1/2\ L_x \leq x < 1/2\ L_x$$

$$-1/2\ L_y \leq y < 1/2\ L_y$$

$$-1/2\ L_z \leq z < 1/2\ L_z \tag{6}$$

The spatial frequency $\vec{q}_n(\omega)$, associated with the nth measurement, is given by the expression

$$\vec{q}_n(\omega) = -\frac{2\omega}{c_L}\vec{e}_n^{\,i} \tag{7}$$

where c_L is the propagation velocity of longitudinal elastic waves.

Thus we have a distribution of impedance deviation completely enclosed in a box of volume $V_s = L_x L_y L_z$, insonified by ultrasonic waves of frequency ω as shown in Fig. 11. Note that the spatial Fourier transform of the distribution of impedance deviation $\zeta(\vec{r})$ can be represented in terms of the basis functions of the box as

$$\zeta(\vec{r}) = \frac{1}{V_s}\sum_q \overline{\zeta}(\vec{q})\, e^{i\vec{q}\cdot\vec{r}} \quad , \tag{8}$$

where the $\vec{q}$ values are on a regular grid such as that shown in Fig. 12a and are given by the expression

$$\vec{q} = 2\pi\left(\frac{P_x}{L_x}\vec{e}_x + \frac{P_y}{L_y}\vec{e}_y + \frac{P_z}{L_z}\vec{e}_z\right) \quad , \tag{9}$$

with P_x, P_x, P_z representing integers and $\vec{e}_x$, $\vec{e}_y$, $\vec{e}_z$ representing unit vectors. These $\vec{q}$ values will be referred to as the "gridded" spatial frequencies. In contrast the spatial frequencies associated with the nth measurement lie on a line corresponding to the nth measurement direction as shown in Fig. 12b. The spacing between these values depends on the sampling rate and is in general different than the grid spacing. The "gridded" spatial frequencies can be related to the "measurement" spatial frequencies by noting that

$$\overline{\zeta}(\vec{q}_n(\omega)) = \int_{D_s} d^3\vec{r}\, \zeta(\vec{r}) \exp(-i\vec{q}_n \cdot \vec{r}) \tag{10}$$

and substituting Eq. (8) for $\zeta(\vec{r})$ to obtain

$$\overline{\zeta}(\vec{q}_n) = \sum_{\vec{q}} \Lambda\,(\vec{q}_n(\omega) - \vec{q})\overline{\zeta}(\vec{q}) \quad . \tag{11}$$

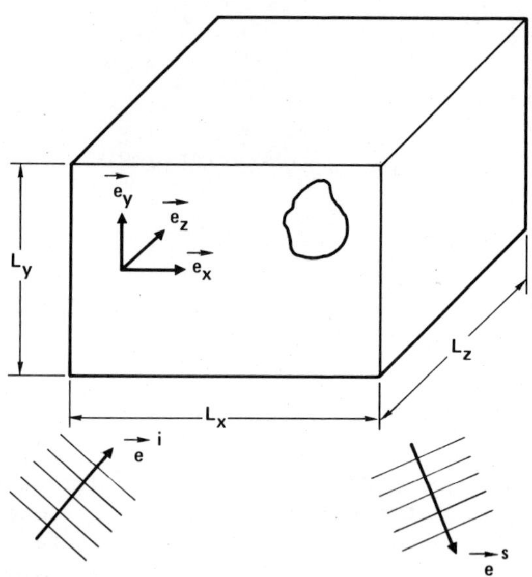

Fig. 11 Localization domain and geometry for probabilistic image reconstruction.

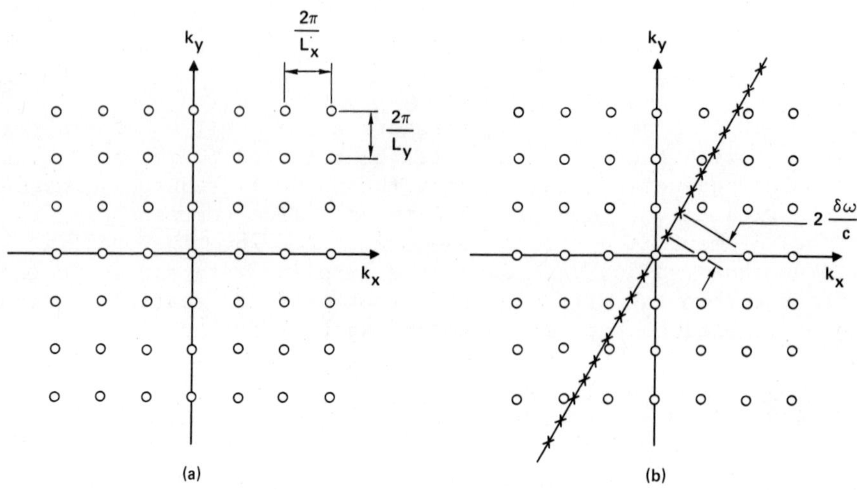

Fig. 12 (a) Location pattern of "gridded" spatial frequencies.
(b) Location of "measured" spatial frequencies along direction of nth measurement.

The quantity $\Lambda(\vec{q}_n - \vec{q})$ is given by

$$\Lambda(\vec{q}_n(\omega) - \vec{q}) = \text{sinc}\left[\frac{q_{nx} - q_x}{2} L_x\right] \text{sinc}\left[\frac{q_{ny} - q_y}{2} L_y\right] \text{sinc}\left[\frac{q_{nz} - q_z}{2} L_z\right] . \tag{12}$$

Equation (11) is the interpolation formula needed to go from the "gridded" spatial frequencies to the measurement spatial frequencies. Note that this formula is not arbitrary but follows as a natural consequence of the assumptions that have been made.

We assume finally that the a priori statistical properties of $\zeta(\vec{r})$, the impedance derivation are given by the assumption that it is a Gaussian random process in D_s with a modified stationarity property consistent with cyclic boundary conditions; namely we assume the properties

$$E\,\zeta(\vec{r}) = 0$$

$$E\,\zeta(\vec{r})\zeta(\vec{r}') = C_\zeta(\vec{r} - \vec{r}') \tag{13}$$

where $C_\zeta(\vec{r})$ is assumed to have the periodicity of the localization domain D_s.

We wish to find the optimal estimate for the impedance deviation $\zeta(\vec{q})$ evaluated at each of the "gridded" spatial frequencies. The derivation for this expression is well-known in linear estimation theory.[10] We find that the best estimate $\hat{\zeta}(\vec{q})$, i.e., the most probable function $\zeta(\vec{q})$ given the scattering measurements represented by the waveforms $f_n(t)$, is given symbolically by the expression (in matrix notation):

$$\hat{\bar{\zeta}} = C_{\bar{\zeta}} M^\dagger (M C_{\bar{\zeta}} M^\dagger + C_{\breve{\nu}})^{-1} f \tag{14}$$

in which $\hat{\bar{\zeta}}$ represents the estimate and f the set of n scattering measurements, $C_{\bar{\zeta}}$ and $C_{\breve{\nu}}$ represent the covariance matrices of $\zeta(\vec{q})$ and $\breve{\nu}(\omega)$ respectively, where $\breve{\nu}(\omega)$ is the temporal Fourier transform of $\nu(t)$. The symbol $M^\dagger$ denotes the Hermitian transpose of M. The symbol M represents a matrix whose terms are obtained from the expression

$$M(\omega,n;\vec{q}) = \breve{p}(\omega)\alpha\omega^2 \; \Lambda \; (\vec{q}_n(\omega) - \vec{q}) \quad . \tag{15}$$

The reader is reminded that the ω are evaluated along the nth measurement direction at the sampled frequencies. The $\vec{q}$ are evaluated at the "gridded" spatial frequencies. The quantity $p(\omega)$ is the temporal Fourier transform of $p(t)$, the transducer response. Once the values of $\zeta(\vec{q})$ are found, the final step is to perform an inverse Fourier transform on the $\zeta(\vec{q})$ to obtain the image $\zeta(\vec{r})$.

Experimental Results

To test the probabilistic imaging technique we have used synthetic data to represent the scattering waveforms from a spherical void. There were 16 × 16 "gridded" spatial frequencies. The covariance of the measurements, C_ν, was assumed to be a constant, i.e., the noise was assumed to be white. The covariance of the image, C_ζ, was assumed to be constant in the spatial frequency domain. The bandwidth of the transducer extended from 1.0 MHz to 3.0 MHz. A set of four measurements equally spaced in angle over a 90 degree sector were used as input data and an image of a void of radius 1200 μm was constructed within a field-of-view of 4800 μm. The result is shown in Fig. 13, which shows that the flaw was reasonably well resolved. Certainly the resolution is superior to that obtained experimentally in Fig. 9.

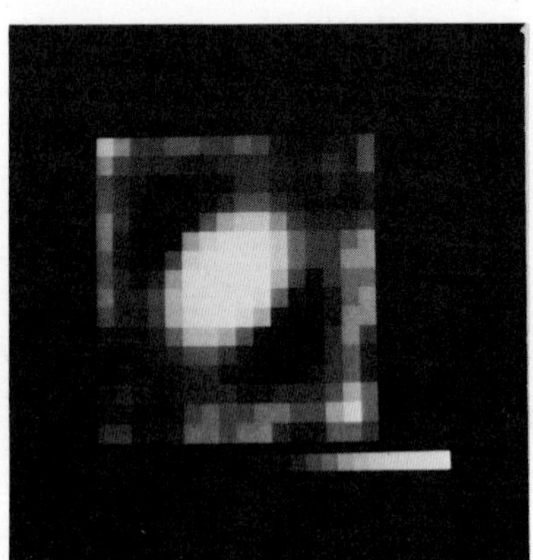

Fig. 13 Probabilistic image reconstruction using synthetic data.

We have also tested the imaging technique using real data obtained from a 635 μm diameter wire in water. The two array transducers were located symmetrically about the wire as shown in Fig. 14. The 64 array elements were divided into 4 groups of

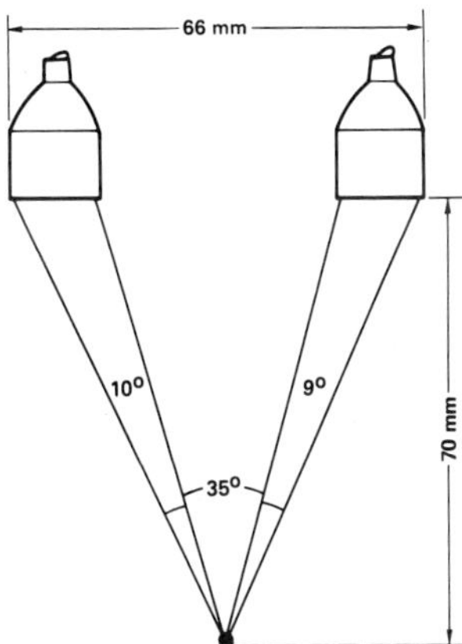

Fig. 14 Experimental arrangement for data acquisition for probabilistic image reconstruction.

four. Each mini-array was used in the pulse echo mode to interrogate the wire target in the directions indicated. Thus a total of 4 waveforms were obtained over a sector of 54 degrees. The reconstructed image of the wire is shown in Fig. 15, whose field of view is 2560 μm. Positive values only are shown. The orientation is the same as Fig. 14, the wire being located at image center. In addition to the wire image, some spurious sidelobes are evident, which are partly due to the limited range of pulse-echo directions. Although the wire was unresolved by a conventional sector-scan image, it was resolved by a factor of 2 in the reconstructed image, thus the reconstruction technique has improved the resolution by a factor of 6.

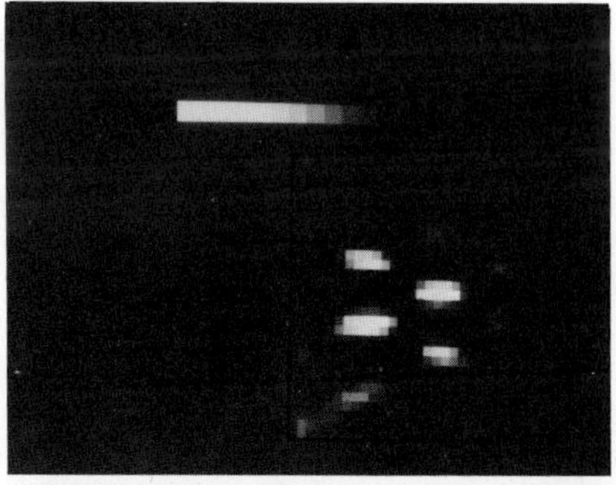

Fig. 15 Probabilistic image reconstruction using experimental data.

SUMMARY

A digitally-based ultrasonic phased array system has been built. It utilizes two, 32 element transducer arrays operating at 2.25 MHz. It features parallel processing of the 16 receiving channels, an array processor for beam forming, and minicomputer control of the system. A custom hybrid pulser/receiver was designed and fabricated specifically for the system.

The system has been used for imaging known flaws in metal parts using a sector scan mode of operation and a differential image processing technique. To improve the spatial resolution of the images, a probabilistic image reconstruction technique is being developed. Preliminary results of this technique were presented.

ACKNOWLEDGMENT

The author acknowledges the technical contribution of R.B. Houston toward the successful design and fabrication of the electronic circuitry for the array system. He was assisted in this task by W.E. Peterson, J. Liska, and V. Nance. The apparatus for holding the array transducers and the multiplexer, pulser/receiver assembly was fabricated by H.E. Feathers. We also thank R.K. Elsley for helpful discussions and advice regarding the operation of the array processor and suggestions regarding the software for the display.

The development of the array system was supported by the Defense Advanced Research Projects Agency. The image reconstruction work was supported by Rockwell International internal research and development funding.

REFERENCES

1. J.C. Somer, "Electronic Sector Scanning for Ultrasonic Diagnosis," Ultrasonics, July 1968, pp. 153-159.
2. F.L. Thurstone and O.T. Von Ramm, "Electronic Beam Steering for Ultrasonic Imaging," 2nd World Congress on Ultrasonics in Medicine, June 4-8, 1973, Rotterdam, Netherlands, Excerpta Medica, Amsterdam, Netherlands, 1974, pp. 43-48.
3. W.A. Anderson et al., "A New Real Time Phased Array Sector Scanner for Imaging the Entire Adult Human Heart," Ultrasound in Medicine, 3B, 1977, pp. 1547-1558.
4. C.B. Burckhardt et al., "A Simplified Ultrasound Phased Array Sector Scanner," Echocardiogrphy, Third Symposium, Martinus Nijhoff Publishers, The Hauge, pp. 385-393.
5. H.E. Karrer, J.F. Davis, J.D. Larson, R.D. Pering, "A Phased Array Acoustic Imaging System for Medical Use," 1980 Ultrasonic Symposium Proceedings, IEEE Cat. No. 80CH1602-2, pp. 757-762, 1980.

6. R.C. Addison, Jr., "Multi-element Arrays for NDE Applications," to be published in V. 11 of Acoustical Imaging, Ed. J. Powers, Plenum Press.
7. R.C. Addison, Jr., R.B. Houston and C.C. Ruokangas, "Test Bed for Quantitative NDE - Imaging Results," Review of Progress in Quantitative Nondestructive Evaluation, (Plenum Press, New York), V. 1, pp. 801-810, 1982.
8. R.C. Addison, Jr., "Ultrasonic Test Bed for Quantitative NDE," Final Report for Contract No. F33615-78-C-5164 (to be published as an Air Force Technical Report).
9. J.M. Richardson and Jack C. Gysbers, "Application of Estimation Theory to Image Improvement," 1977 Ultrasonic Symposium Proceedings, IEEE Cat. No. 77CH1264-1SU, pp. 212-218, 1977.
10. M.H. Buonocore, W.R. Brody and A. Macovsky, "A Natural Pixel Decomposition for Two-Dimensional Image Reconstruction," IEEE Transactions on Biomedical Engineering, V. BME-28, No. 2, pp. 69-78, 1981.

ACOUSTICAL HOLOGRAPHIC SCANNERS

E. G. LeDet and C. S. Ih*

The Johns Hopkins University
Applied Physics Laboratory
Laurel, Md. 20707 USA

*University of Delaware
Department of Electrical Engineering
Newark, De. 19711 USA

ABSTRACT

Optical holographic scanners have been actively researched because of their high resolution, fast scanning speeds and mechanical simplicity. This paper describes the application of holography to acoustical scanning. The theory of acoustical holographic scanners is developed and optimal design procedures are discussed. Construction of a two-dimensional scanner operating at 10.5 and 10.23 MHz, with a total resolution of 2500 pixels evenly distributed on a 75 by 75 mm scan plane, is detailed, and its performance measured. Experimental results indicate that holographic scanning can be successfully applied to acoustical radiation.

DESCRIPTION OF SCANNER

The scanner consists of a rotating disc with a series of holograms around its periphery and a fixed, concave auxiliary reflector (AR), oriented so that the scanner has spherical symmetry about the axis of rotation of the disc. Each hologram directs acoustical radiation, from a normally incident reconstruction beam, to the AR, which reflects it to a unique scan plane spot. Disc rotation causes the spot to follow a curved locus. A complete revolution brings successive holograms into the field-of-view of the beam, resulting in a series of scan lines. Thus, a raster scan is generated by a simple, continuous disc rotation. For a fixed disc size, the number of scan lines can be increased by using multiple rows of holograms (serial multiplexing) and/or by imaging multiple scan lines per hologram (field multiplexing). In addition, the scanner can be designed to operate at infinite conjugation, in which case the holograms reconstruct collimated beams.

The scanner is based upon research on holographic laser beam scanners with auxiliary reflectors. Initially, convex AR's were used,[1] which required projection lenses for high resolution operation.[2] Comparable performance without lenses has been demonstrated for concave AR scanners.[3,4] The concave AR functions as a positive lens, producing a real, magnified image of the hologram. This image is the system exit pupil; its size determines the scanner's resolution.

ANALYSIS AND DESIGN

Front and side views of the scanner are shown in Figure 1. Assume N holograms per row distributed over n_1 rows on the disc. Lower case letters denote linear parameters normalized with respect to R, the mean disc radius. The tangential and radial hologram dimensions are defined as

$$h_t = 2\sin(\tfrac{1}{2}\alpha_h) , \quad h_r = h_t/\cos\psi , \tag{1}$$

respectively, where $\alpha_h = 2\pi/N$ is the horizontal scan angle and ψ is the principal angle.

The exit pupil is the hologram image, formed by the AR. First-order geometric optics gives

$$p_b = p_a r_a/(2p_a - r_a) , \quad p_e = (p_b/p_a)h_t = \delta h_t \tag{2}$$

for its location and diameter, respectively, where r_a is the AR radius of curvature and δ is the magnification factor. The following relationships exist for the quantities shown in Figure 1:

$$r' = \tfrac{1}{2}n_1 h_r + 1 \qquad \phi = \tfrac{1}{4}\pi + \tfrac{1}{2}\psi$$

$$r_m = r_a \sin\phi \qquad d_m = p_a \cos\psi$$

$$p_a = (r_m - 1)/\sin\psi \qquad d_b = d_m - r_a\cos\phi$$

$$r_e = p_b - r_m .$$

The linear parameters can be expressed as functions of the magnification factor through[5]

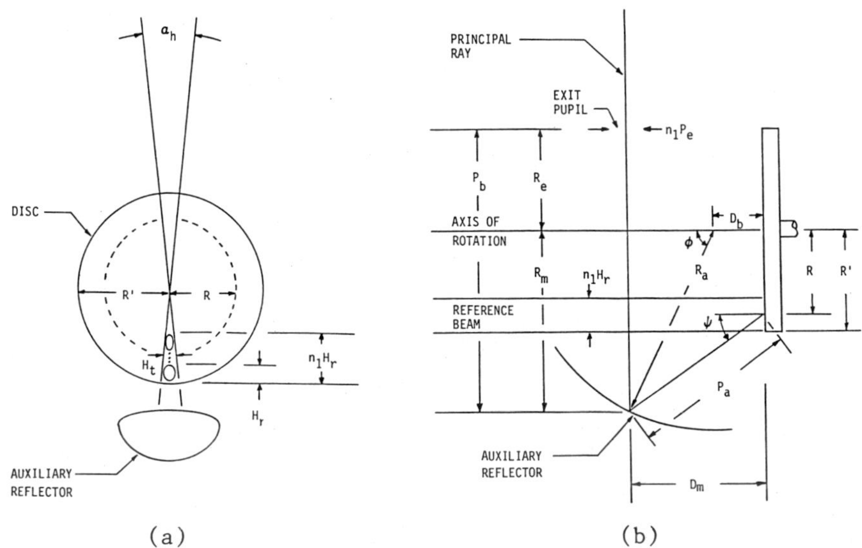

Figure 1. Front (a) and side (b) views of the holographic scanner.

$$r_a = \frac{2\delta}{(2\sin\phi-\sin\psi)\delta-\sin\psi} . \tag{3}$$

Finite Conjugation

For finite conjugation scanners, the scan plane is located at a finite radius, r_s, from the axis of rotation. The reconstructed spots have diameters

$$o_s = 1.22\lambda(r_s-r_e)/(\delta h_t) \tag{4}$$

determined by diffraction, where $\lambda=\Lambda/R$ is the normalized source wavelength. With a field multiplicity of n_2, the vertical (continuous) and horizontal (discrete) scan plane dimensions are, respectively,

$$l_v = Nn_1n_2o_s , \quad l_h = \alpha_h r_s . \tag{5}$$

By defining the scan plane aspect ratio as $\mu=l_h/l_v$, the total resolution can be expressed as

$$T = \mu(Nn_1n_2)^2 . \tag{6}$$

The total resolution is limited by the scanner geometry, illustrated in Figure 2. The pupil diameter shown in the figure represents an entire column of holograms. The AR axial size must be greater than this diameter. It is given by

$$km_a = 2k[d_b-d_m+(r_a^2-r'^2)^{\frac{1}{2}}] \ . \tag{7}$$

k is a constant used to ensure that the AR does not intersect the reference beam. The left-most and right-most spots must lie within lines A-E and B-Y; rays emanating from spots outside these bounds, passing through the pupil, do not all intersect the AR. Moreover, the scan radius must be large enough that rays originating at the various spot positions are within the bounds.

The restrictions are expressed mathematically by considering similar triangles F-G-H and F-Y-I for the minimum scan radius and Y-K-J and M-K-L for the vertical length, with the following results:

$$r_{s_{min}} = \frac{km_a r_e + n_1 p_e r_b}{km_a - n_1 p_e} \tag{8}$$

$$l_{vg} = \frac{km_a(r_s - r_e) + (n_1-2)p_e(r_s+r_b)}{r_e+r_b} \ , \tag{9}$$

where

$$r_b = [r_a^2-(d_b-d_m+\tfrac{1}{2}km_a)^2]^{\frac{1}{2}} \ . \tag{10}$$

The aspect ratio for l_{vg}, refered to as the geometric aspect ratio, is

$$\mu_g = 2\pi\rho r_{s_{min}}/(Nl_{vg}) \ , \tag{11}$$

where ρ is the scan radius factor given by

$$\rho = r_s/r_{s_{min}} \ . \tag{12}$$

The scan plane aspect ratio must be greater than the geometric aspect ratio.

It is convenient to remove the wavelength dependence from total resolution; the normalized total resolution, using Equations 4, 5, 6 and the definition of the aspect ratio, is

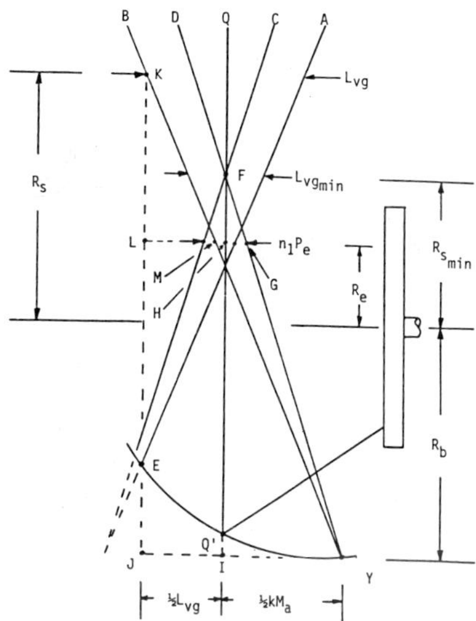

Figure 2. Geometric limitations on the finite conjugation scanner.

$$\tau = \frac{1.22\lambda}{4\pi^2} T = \frac{n_1 n_2 \delta}{N} [1-(r_e/\rho r_{s_{min}})]^{-1} , \tag{13}$$

which is subject to the restrictions

$$km_a > n_1 p_e , \quad \rho \geq 1 , \quad \mu \geq \mu_g . \tag{14}$$

The normalized total resolution is plotted as a function of magnification and scan radius in Figure 3.

Infinite Conjugation

For infinite conjugation, the holograms reconstruct collimated beams that irradiate an angular sector of the scan area above the exit pupil. The diffraction-limited angular resolution is

$$\alpha_s = 1.22\lambda/p_e , \tag{15}$$

and the scan area is bounded horizontally by α_h and vertically by

$$\alpha_v = N n_1 n_2 \alpha_s . \tag{16}$$

An analysis of the angular restrictions, similar to that in the pre-

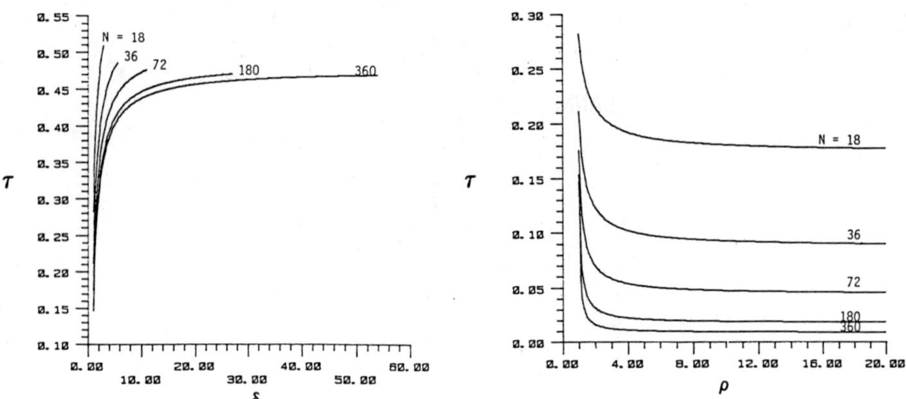

Figure 3. Normalized total resolution as a function of magnification and scan radius factors.

ceeding section, leads to a normalized total resolution of

$$\tau = n_1 n_2 \delta / N \,, \tag{17}$$

subject to

$$km_a > n_1 p_e \,, \quad \mu \geq \mu_g \,, \tag{18}$$

where $\mu = \alpha_h / \alpha_v$ and

$$\mu_g = 2\tan^{-1}\left[\frac{km_a - 2(n_1 - 1)p_e}{2(r_b + r_e)}\right] . \tag{19}$$

Design

For scanners, the total resolution, wavelength and aspect ratio are usually specified as design parameters. Then the disc radius necessary to achieve the required resolution can be computed from Equation 13, giving

$$R = \frac{1.22\Lambda\mu N^3 n_1 n_2}{4\pi^2 \delta}[1-(r_e/\rho r_{s_{min}})] \tag{20}$$

for the finite conjugation case. By letting $\rho \to \infty$, the same expression can be used for infinite conjugation scanners. Equation 20 is the design equation, which can be used to design an optimal scanner - one having the minimum disc radius for a given total resolution. This is accomplished by choosing the largest magnification factor and, for the finite case, the smallest scan radius factor that satisfy the scan plane aspect ratio and radius restrictions, a process which can be automated.[5]

For all but the largest scan radii, some correction to the shapes of the holograms is necessary in order to reconstruct circular spots. The first-order theory presented above does not account for aberrations in images formed by the AR. The correction is applied by numerically tracing rays from the spot position, through a circular pupil, to their termination at the disc. In all cases, this results in elongated holograms that require larger discs than those computed from the first-order theory.

EXPERIMENTAL SCANNER

A 10.5 MHz scanner was designed using the procedure discussed above, for a total resolution of 2500 pixels, an aspect ratio of 1, and a 300 mm scan radius. It has a 153 mm diameter disc containing 25 holograms with 8.5 by 32.5 mm dimensions. The diffraction-limited spot diameter is 1.5 mm, so the scan plane is 75 by 75 mm. Both the radius of curvature and diameter of the AR are 150 mm.

A novel second order field multiplexing technique is used in this scanner. The holograms reconstruct lines separated by a spot diameter. Then the source frequency is shifted, on alternate revolutions, between 10.5 and 10.23 MHz, to interlace the scan lines. The latter frequency is computed from the grating equation using the principal angle of a spot displaced one diameter off the principal ray axis.

The holograms were calculated using previously developed numerical techniques,[6,7] and drawn with a computer-controlled plotter. Briefly, the procedure involves locating a point source at the required scan plane position and identifying its corresponding hologram area on the disc. Then the phase delays of rays from the source, which reflect off the AR onto the disc within the hologram area, are computed. Fringes are formed by connecting points of constant delay that are spaced by integral wave numbers. Since the reference beam phase delay is constant, its contribution is ignored. The hologram fringe patterns were drawn with the aid of cubic spline interpolation, which, coupled with a high resolution plotter, produced significantly smoother fringes than those reported in References 6 and 7.

Photographic transparencies of the plots, reduced to the hologram dimensions, were mounted in a mask that shaped them properly for circular spot reconstruction. The assembly was contact printed into high resolution photoresist spin coated on a circular printed circuit board made of phenolic backed copper. After development, the disc was etched, which left acoustical holograms on its surface. 10 MHz ultrasound is almost totally reflected by the copper, but propagates through the phenolic. An 8% diffraction efficiency was measured for these holograms. A photograph of the disc is shown in Figure 4.

The auxiliary reflector is a conventional optical mirror. Its reflection at 10 MHz is excellent, because the scanner geometry is such that the angles of incidence of most rays intersecting the AR from the holograms are greater than the critical angle for total internal reflection.

EXPERIMENTS

The disc and AR were arranged in a water tank, with the reconstruction beam supplied by an air backed PZT transducer. The ultrasound was transmitted through the rear of the disc.

Reconstruction spot sizes and positions were measured, for each of the holograms at the two operating frequencies, with a detector consisting of a 10 MHz receiving transducer and a 1 mm acoustical pinhole. The longitudinal resolution of the center spot was 4.7 mm and the measured scan radius was 290 mm. The detector was maintained at this radius so that the measurements would reflect performance at a flat scan plane. The center spots for the 10.5 and 10.23 MHz frequencies had diameters of 2.2 and 2.4 mm, respectively. With the exception of the two left-most spots, maximum spot diameters were 3.9 mm for the higher frequency and 6 mm for the lower one, and the spots were within 10% of their expected locations. The two left-most spots could not be imaged properly because a significant portion of the energy diffracted by their holograms was not incident on the AR.[5]

Two copper objects were imaged with the scanner to qualitatively assess its performance in imaging systems. To avoid scanning the receiver, it was fixed along the principal ray axis and the exit pupil was imaged to its aperture by a pair of acoustical lenses with the test objects sandwiched between them. Thus, all energy emerging from the exit pupil was incident at the receiver. The disc was rotated by a synchronous motor and the amplified output of the detector was sampled with an A/D converter and stored in a microprocessor. Drawings of the test objects and the images are shown in Figure 5. The images are plotted assuming the diffraction-limited diameter for each pixel, and were generated by thresholding the sampled values.

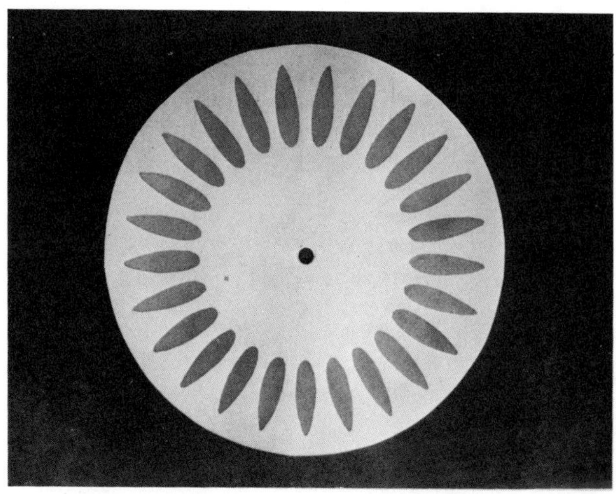

Figure 4. Photograph of the experimental scanner disc.

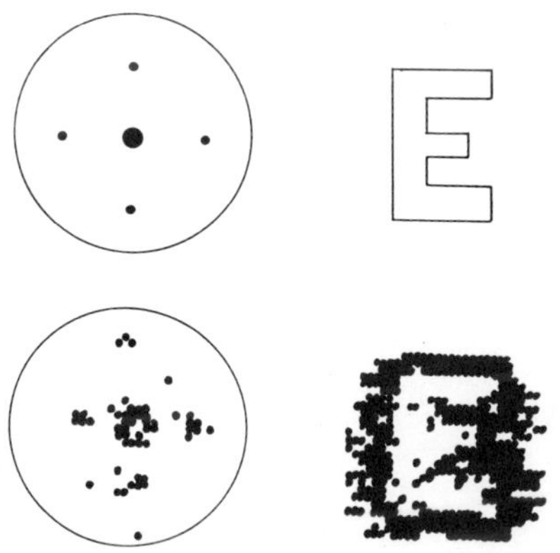

Figure 5. Test objects and acoustical images.

CONCLUSION

The acoustical holographic scanners discussed in this paper can be used in many two-dimensional scanning applications. They can be configured for either finite or infinite conjugation scanning and, in either case, operated actively or passively. Because their raster scan patterns are produced by a stationary source and a continuous disc rotation, they are capable of high resolution performance at rapid scanning speeds.

ACKNOWLEDGEMENT

We are grateful to the American Cancer Society for their partial support through an Institutional Research Grant to the University of Delaware.

REFERENCES

1. C. S. Ih, Holographic laser beam scanners utilizing an auxiliary reflector, Appl. Opt. 16:2137 (1977).
2. C. S. Ih, Design considerations of 2-D holographic scanners, Appl. Opt. 17:1582 (1978).
3. C. S. Ih, E. G. LeDet and N. S. Kopeika, Characteristics of holographic scanners utilizing a concave auxiliary reflector, Appl. Opt. 20:1656 (1981).
4. C. S. Ih, N. S. Kopeika and E. G. LeDet, Characteristics of active and passive 2-D holographic scanner imaging systems for the middle infrared, Appl. Opt. 19:2041 (1980).
5. E. G. LeDet, An Acoustical Holographic Scanner, PhD dissertation, University of Delaware (1981).
6. K. Yen, Computer-generated holograms and mirror-imaging system analysis using ray tracing, Master's thesis, University of Delaware (1980).
7. C. S. Ih, N. Kong and T. Giriappa, Computer-produced holograms for scanners utilizing an auxiliary reflector, Appl. Opt. 16:1582 (1978).

SECOND TIME AROUND ECHO IMMUNITY FROM PSEUDO STEREOSCOPIC HOLOGRAPHIC IMAGING

D.I. Shaw, J.C. Bennett, A.P. Anderson

Department of Electronic & Electrical Engineering

University of Sheffield, England

The idea of using two widely-spaced co-linear arrays as a means of obtaining high resolution imagery of the sea-bed has recently been proposed and explored [1,2]. The pseudostereoscopic holographic imaging system used is typically configured as shown in Fig.1 and provides a sideways-looking mode of underwater mapping. Although there are apparent similarities between this arrangement and the side scan sonar, the new approach is not reliant on forward motion of the transducer arrays. Consequently it can also be used to image its environment from a static location, thus exhibiting a characteristic common with other forms of ultrasonic imaging systems [3,4].

The pseudostereoscopic system in its most basic form utilises a single frequency CW waveform. However, in order to reduce the period required for data acquisition and provide a freeze-frame or quasi-real-time capability, pulsed transmissions are used. The exploitation of such pulse-echo techniques is encountered in many radar and sonar systems and generally introduces the problem of multiple-time-around returns [5] which can lead to ambiguity in the interpretation of results, although various schemes such as modulation of the prf have been used for distinguishing the desired signal. In the case of the pseudostereoscopic imaging system, however, it is not necessary to resort to these sophistications, and it is the intention in this contribution to demonstrate the inherent immunity to second-time-around returns.

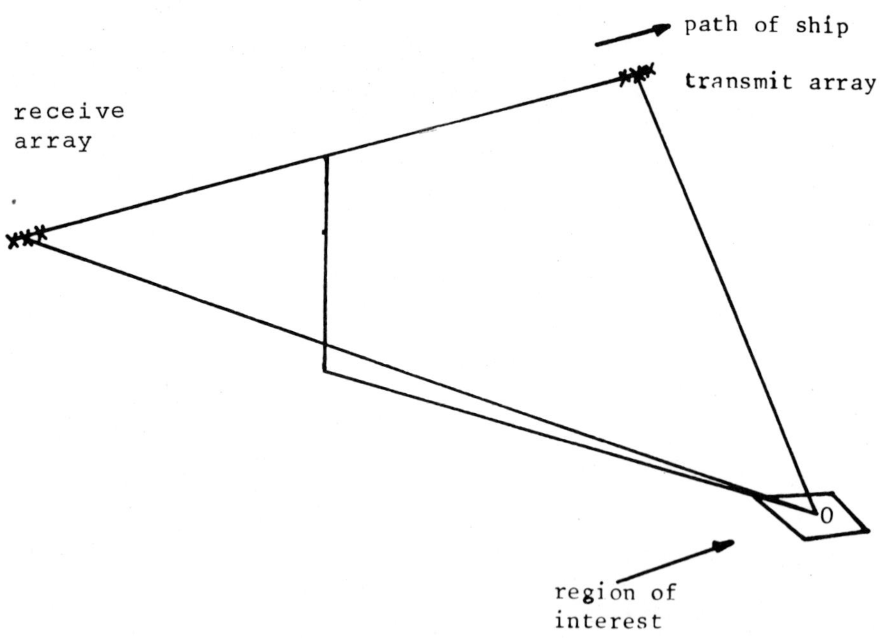

FIG.1

The analysis of second-time-around echo immunity is based on a previous theoretical treatment of the pseudostereoscopic imaging system [1]. Fig.2 shows the essential system geometry looking at a point P in a field of view centre O. The array lengths ℓ_1 and ℓ_2 are assumed to be small compared to the ranges R_1 and R_2. For any particular transmit element the received signal is a primary return combined with the second-time-around echo originating from the previous transmit element. This process can be conveniently represented by primary and displaced secondary transmit apertures contributing primary and second-time-around returns respectively. The second-time-around echo aperture is shown widely separated from the primary transmit aperture for diagrammatic clarity. In practice, however, the two apertures would be displaced by one element with the second element of the second-time-around echo aperture coincident with the first element of the transmit aperture when the outermost element is the first to transmit.

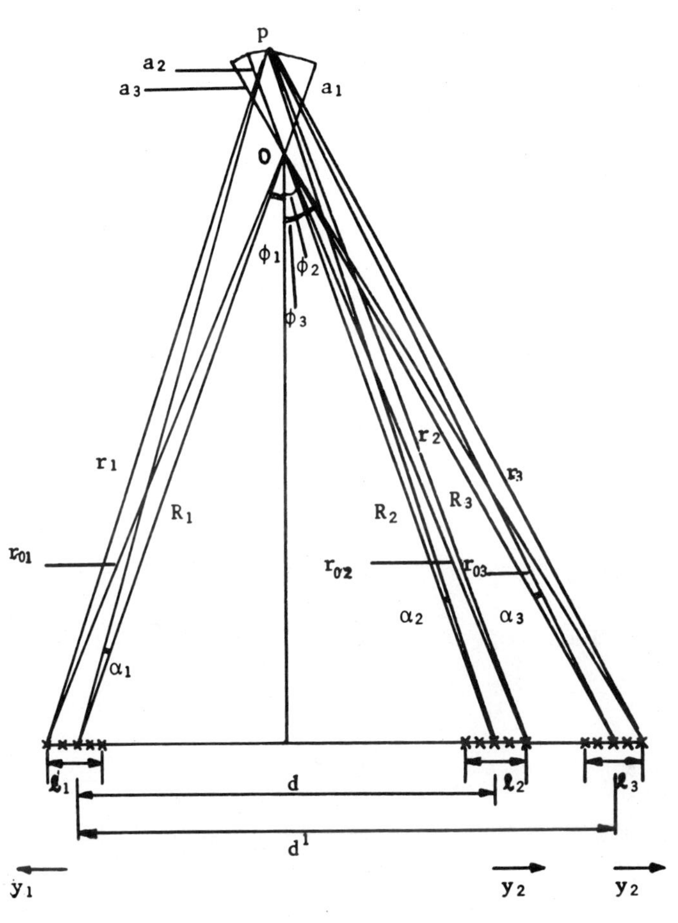

receive aperture

transmit aperture

second time around aperture

FIG.2

Consider the phase of a signal received at element y_1 reflected from P after illumination from y_2 in the transmit aperture.

The total path length is $(r_1 + r_2)$ where

$$r_1^2 = \{(R_1 + a_1)\cos\alpha_1 + y_1\sin\phi_1\}^2$$

$$+ \{y_1\cos\phi_1 - (R_1 + a_1)\sin\alpha_1\}^2$$

Invoking the paraxial approximation for each array aperture :

$$r_1 = R_1 + a_1 + \frac{y_1^2}{2(R_1 + a_1)} + y_1\sin(\phi_1 - \alpha_1) \quad (1)$$

and since

$$\frac{y_1^2}{2(R_1 + a_1)} \simeq \frac{y_1^2}{2R_1}$$

then $$r_1 = (R_1 + a_1) + y_1\sin(\phi_1 - \alpha_1) + \frac{y_1^2}{2R_1} \quad (2a)$$

Similary $$r_2 = (R_2 + a_2) + y_2\sin(\phi_2 - \alpha_2) + \frac{y_2^2}{2R_2} \quad (2b)$$

$$r_3 = (R_3 + a_3) + y_2\sin(\phi_3 - \alpha_3) + \frac{y_2^2}{2R_3} \quad (2c)$$

The phase of the received signal is :

$$f(y_1,y_2) = \exp j\frac{2\pi}{\lambda}(r_1 + r_2) \quad (3)$$

Correcting the phase to steer and focus the arrays on centre O leads to :

$$f'(y_1,y_2) = f(y_1,y_2)\exp -j\frac{2\pi}{\lambda}(r_{o_1} + r_{o_2}) \quad (4)$$

where $$r_{o_1} = R_1 + y_1\sin\phi_1 + \frac{y_1^2}{2R_1}$$

and $$r_{o_2} = R_2 + y_2\sin\phi_2 + \frac{y_2^2}{2R_2}$$

If however the received signal is that of a second-time-around echo and is corrected as in equation 4 the signal will be consistent with that having originated from an aperture shifted one element spacing to the right and thus can be considered as a second-time-around echo aperture.

The phase of the received second-time-around echo will be :

$$p(y_1,y_2) = \exp j\frac{2\pi}{\lambda}(r_1 + r_3) \quad (5)$$

correcting in accordance with equation 4 gives :

$$p'(y_1,y_2) = p(y_1,y_2)\exp j\frac{2\pi}{\lambda}(r_{o_1} + r_{o_2}) \quad (6)$$

Hence

$$f'(y_1,y_2) = K_1\exp j\frac{2\pi}{\lambda}\{y_1(\sin(\phi_1 - \alpha_1) - \sin\phi_1) + y_2(\sin(\phi_2 - \alpha_2) - \sin\phi_2)\} \quad (7)$$

and

$$p'(y_1,y_2) = K_2\exp j\frac{2\pi}{\lambda}\{y_1(\sin(\phi_1 - \alpha_1) - \sin\phi_1) + y_2(\sin(\phi_3 - \alpha_3) - \sin\phi_2)\} \quad (8)$$

where $K_1 = \exp j\frac{2\pi}{\lambda}(a_1 + a_2)$ and $K_2 = \exp j\frac{2\pi}{\lambda}(R_3 - R_2 + a_1 + a_3)$

are constant phase factors.

Suppose the target reflectivity is composed of independent scatterers with complex distribution $\underset{\sim}{D}\{u_1,u_2\}$ where

$$u_1 = \{\sin(\phi_1 - \alpha_1) - \sin\phi_1\} \quad (9)$$

$$u_2 = \{\sin(\phi_2 - \alpha_2) - \sin\phi_2\} \quad (10)$$

The received signal neglecting space attenuation is given by :

$$g(y_1,y_2) = K_T\iint \underset{\sim}{D}(u_1,u_2)\exp j\frac{2\pi}{\lambda}(r_1 + r_2)du_1du_2 \quad (11)$$

where K_T is the net phase factor for the target distribution.

Correcting the phase as in the step from equations 4 to 7 :

$$g'(y_1,y_2) = K_T\iint \underset{\sim}{D}(u_1,u_2)\exp j\frac{2\pi}{\lambda}(y_1u_1 + y_2u_2)du_1du_2 \quad (12)$$

Since $g'(y_1,y_2)$ is the '2D' Fourier transform of $D(u_1,u_2)$ the image distribution for aperture lengths ℓ_1,ℓ_2 is $\tilde{g}$iven by :

$$G(u_1,u_2) = \underset{\sim}{D}(u_1,u_2)\ \{\ell_1\ell_2 \text{sinc}(\frac{\ell_1 u_1}{\lambda})\,\text{sinc}(\frac{\ell_2 u_2}{\lambda})\} \tag{13}$$

for uniformly weighted array elements.

Repeating steps from equations 11 to 13 for the second-time-around echo aperture leads to :

$$G'(u_1,u_2) = \underset{\sim}{D}(u_1,u_3)\ \{\ell_1\ell_2 \text{sinc}(\frac{\ell_1 u_1}{\lambda})\,\text{sinc}(\frac{\ell_2 u_3}{\lambda})\} \tag{14}$$

where

$$u_3 = \{\sin(\phi_3 - \alpha_3) - \sin\phi_2\} \tag{15}$$

Although expressed in the classical form the imaging equation (eqn.13) will generate a distorted field of view unless the area is properly mapped in terms of $\phi_1,\phi_2,\alpha_1,\alpha_2$.

Mapping the image derived from second-time-around echoes in terms of ϕ_1,ϕ_2,α_1 and α_2 will produce an image with a positional error. However, since the shift between the transmit aperture and the second-time-around echo aperture is only one element spacing then $\phi_2 \simeq \phi_3$ and $\alpha_2 \simeq \alpha_3$ which then leaves the image only slightly displaced from its true position. Furthermore since this image appears close to its correct position it can be identified as being twice as far away from the area being viewed. With this knowledge the image can be mapped using $\phi_1,\phi_3,\alpha_1,\alpha_3$ and be positioned correctly.

The above treatment can also be applied to the possible error in measuring the separation of the two apertures. If, referring to Fig.2, the measured distance between the two apertures is d and the true separation is d' then ϕ_2 and α_2 will be used instead of ϕ_3 and α_3 for mapping the positions of the targets. This type of error is non recoverable and will result in the image being shifted away from its true position. The magnitude of this positional error will be directly related to the measurement error. The properties of this imaging system are such that this type of error will not defocus the image processed from the data.

Validation of the imaging principle illustrated by equation 13 has been carried out [2] using mechanically synthesised apertures at an ultrasonic frequency of 1MHz in a small swimming pool as shown in Fig.3. The apertures were 1.25m apart and 96mm in length. Two planar aluminium targets, each 50mm square, were located at a range of 3m from the apertures. Fig.4a shows the target distribution and Fig.4b shows the data received by transmitting sequentially across the transmit aperture whilst recording the data received across the whole receive aperture for each transmit position. The two-dimensional array of data was phase

FIG.3

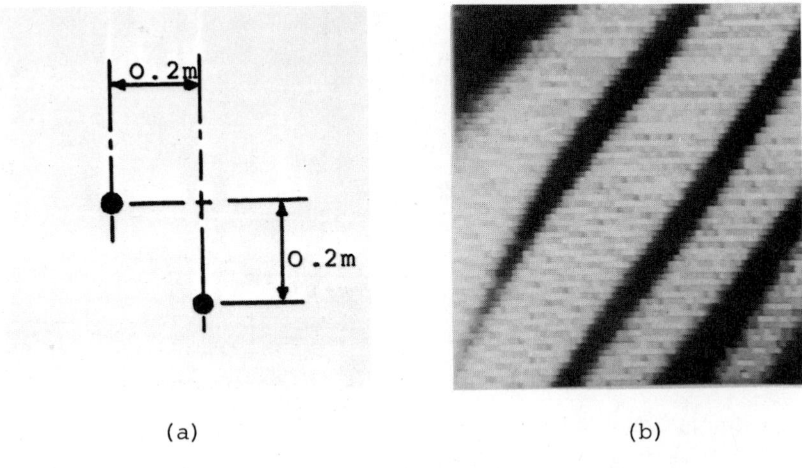

(a) (b)

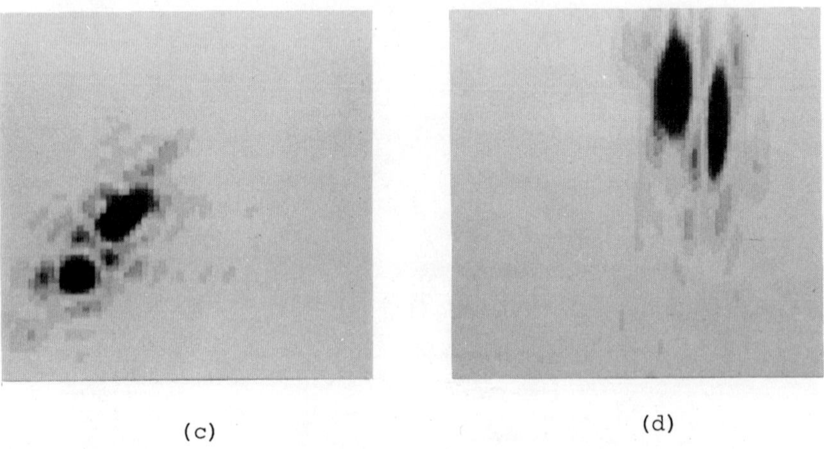

(c) (d)

FIG.4

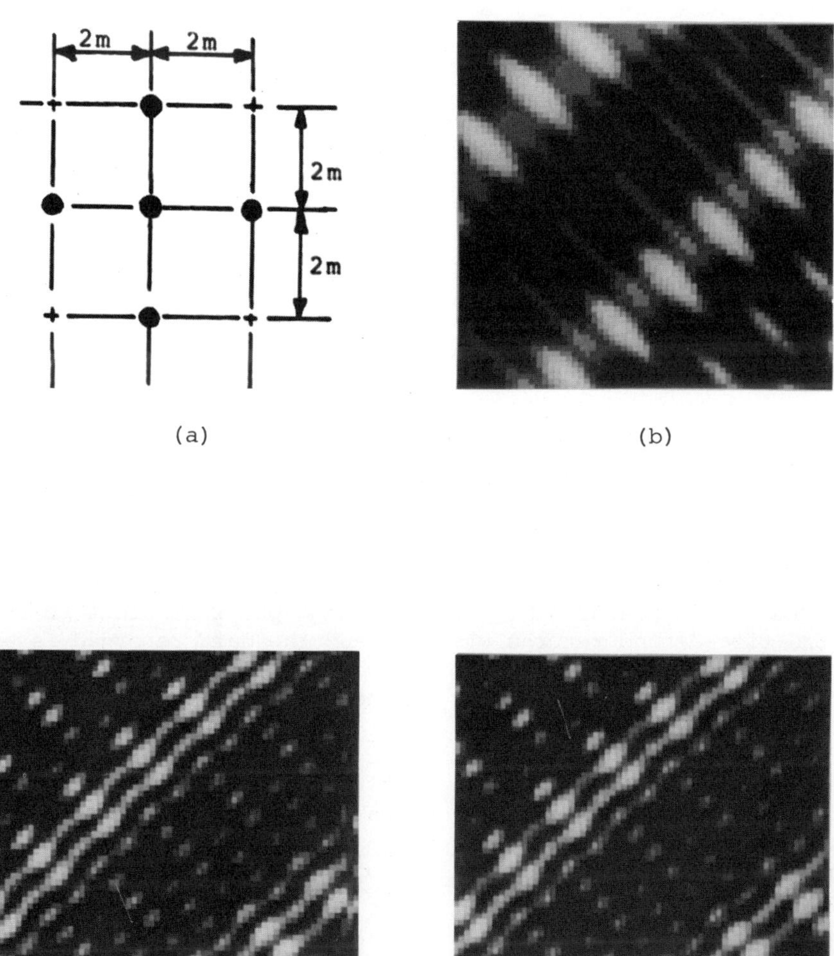

(a) (b)

(c) (d)

FIG.5

corrected as indicated by equation 4 and the resultant image obtained following a subsequent '2-D' Fourier transformation is shown in Fig.4c. Mapping the image in Fig.4c onto a Cartesian coordinate system results in the target distribution shown in Fig.4d. Here the targets are seen to be clearly resolved with axial and transverse definitions governed by the aspect geometry of the data acquisition layout.

Although essentially a CW imaging system, the requirement of a reasonable data acquisition time necessitates the use of bursts of sound sequentially from each of the transmit transducers, giving rise to the creation of second-time-around echoes. Unfortunately since the data acquisition system presently in use is composed of mechanically scanned apertures and not transducer arrays an experimental validation cannot yet be presented. However, to demonstrate the validity of the approach, computer simulations of the system operation under conditions where second-time-around echoes exist are now provided.

Simulations were carried out for a system having apertures of 0.32mm separated by 20m at a wavelength of 5mm. The five target centred shown in Fig.5a was centres 30m distant from the apertures and on axis. Fig.5b shows the received data for this target when the prf used does not give rise to second-time-around echoes. A further single on-axis target was then additionally considered and located at a range of 60m. Once again a prf sufficient to avoid second-time-around returns was used and the data recorded is shown in Fig.5c. By changing the prf such that returns from the 60m target contributed second-time-around echoes the data of Fig.5d was obtained.

Figs.6a, b and c show the resultant images of 5 target clusters derived from the data shown in Figs. 5b, c and d respectively and show no discernable differences between them. Figs. 6d and e show the images of both target groups together when derived from the data shown in Figs. 5c and 5d, and once again no perceptible differences are present.

This property of pseudostereoscopic holographic imaging allows a reduction in data acquisition time by launching a CW burst from sequential transmit transducers in rapid succession. The received returns will be in the form of direct and multiple time around echoes but can be processed to produce images of areas larger than those produced from the primary returns alone. This also reduces the severe problem of correcting for platform motion during the time the sound propagates to and from the target region and leaves only the period of transmission and reception of the sound waves. This period can be reduced by shortening the pulse length radiated by each transmit transducer and processing

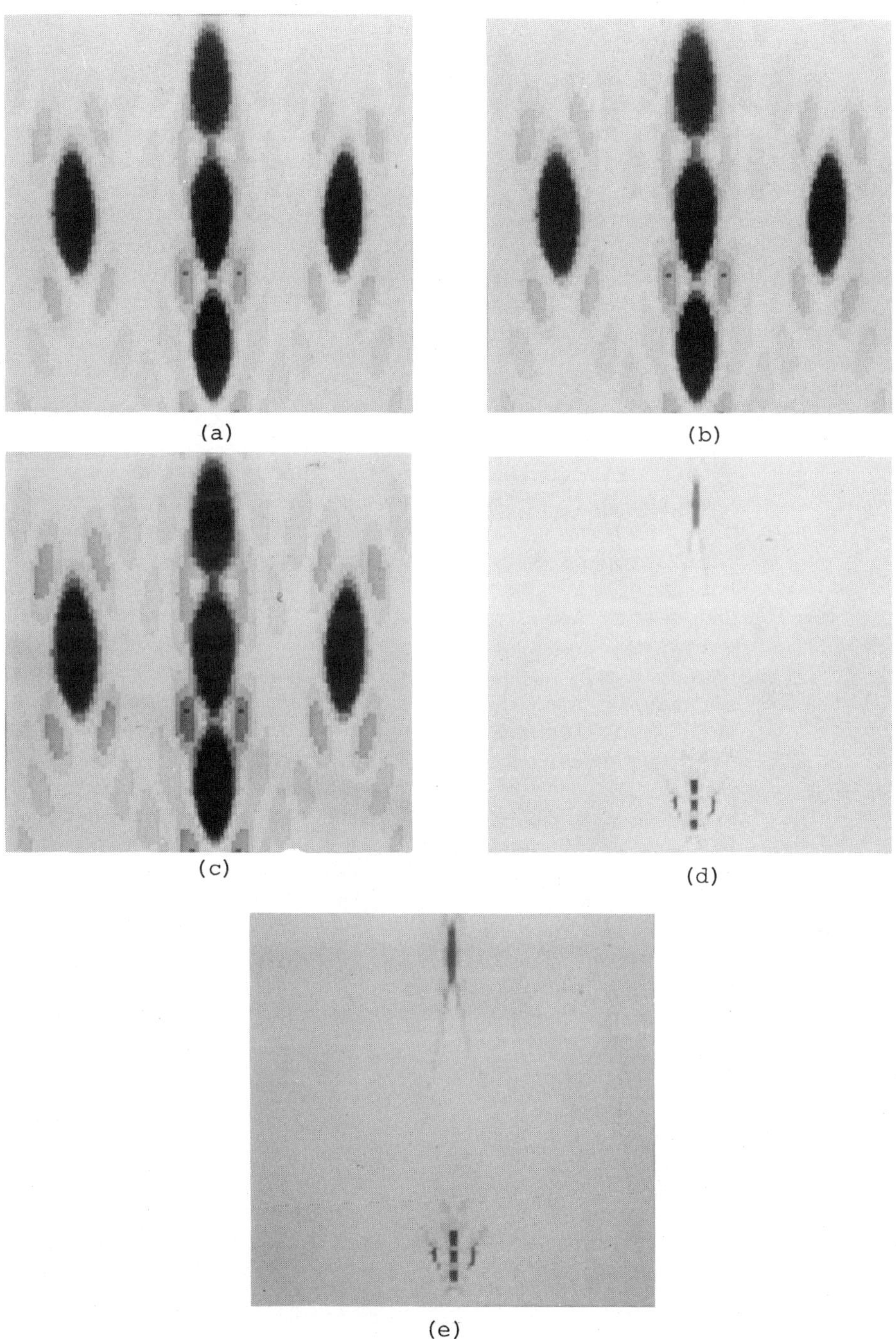

(a) (b)

(c) (d)

(e)

FIG. 6

the received data for higher orders of multiple time around echoes.

In conclusion, it has been demonstrated before that the pseudostereoscopic imaging system can be operated in a CW mode and that the image characteristics obtained in practice conform with those expected from the theory. It has now been shown theoretically and by a computational model that the technique is tolerant to the second-time-around returns experienced in pulse-echo systems. This inherent immunity will enable the proposed system to provide an attractive real-time feature which can be combined with its high resolution area-mapping capability.

REFERENCES

1. Shaw, D.I., Anderson, A.P., Bennett, J.C. : "Pseudostereoscopic holographic imaging system : a possible approach to high resolution underwater mapping", Electronics Letters, 29th October 1981, Vol.17, No.22, pp.841-842
2. Shaw, D.I., Anderson, A.P., Bennett, J.C. : "Experimental validation of a pseudostereoscopic holographic system for underwater mapping", Proc. Institute of Acoustics, Underwater Acoustics Group, Portland, December 1981, pp.4/1-4/7
3. Takuso Sato, Kimo Sasaki,Masato Horiuchi : "Superresolution ultrasonic imaging system using focused beam illumination and algebraic reconstruction", J. Acoust. Soc. Am. 68(4), October 1980, pp.1149-1153
4. Hildebrand, B.P. : "Advances in acoustical holography", Proc. Soc. Photo-Optical Instrumentation Engineers, 1980, Vol.215, pp.116-128
5. Skolnik, M.I. : "Introduction to radar systems", McGraw-Hill, 1962, pp.130-131

UHB IMAGING

E. Alasaarela[1], K. Tervola[1], J. Ylitalo[1] and J. Koivukangas[2]

[1]Dept. of Electrical Engineering, University of Oulu, 90570 Oulu 57, Finland

[2]Dept. of Neurosurgery, University of Oulu, 90570 Oulu 57, Finland

ABSTRACT

A new ultrasonic holographic imaging method, UHB imaging has been developed. This method is based on the combination of the holographic principle and conventional B-scan technique. This combining has been achieved using computerized numerical image reconstruction based on wavefront backward propagation. An important feature of this reconstruction process is a wavefront curvature compensation which leads to optimal lateral and longitudinal resolution. The theory of UHB imaging has been verified firstly by simulated UHB imaging in a computer and secondly by constructing an UHB apparatus and producing experimental UHB images of a resolution test object and human brain specimens.

INTRODUCTION

A new ultrasonic imaging method, UHB imaging, is under development at the University of Oulu. This method is a high resolution two-dimensional imaging method which is appropriate for a wide range of applications, especially for medical imaging and non-destructive testing of metal structures.

The starting point of the UHB research has been an attempt to improve the performance of conventional medical B-scan imaging. The factors restricting the image quality in B-scanning are:

- poor lateral resolution out of focus distance
- it requires interpretation due to artefacts
- compensation of propagation errors of ultrasound is impossible
- image processing is available only for the resulting image, not for the ultrasound wavefronts

The development work has been based on the idea of numerical wavefront propagation in the image space. The basic theory of this image reconstruction method has been presented by Boyer et al (1). UHB imaging is a combination of conventional B-scan technique and one-dimensional ultrasonic holography based on numerical image reconstruction by the wavefront backward propagation method. The name 'UHB' has been devised from the initials of Ultrasonic Holographic B-scan.

HOLOGRAMS AS WAVEFRONT RECORDINGS

Holography is an imaging method which is distinguished from other methods by the fact that both the amplitude and the phase of the wavefront reflected from the object is recorded. The insonifying wavefront must be monochromatic and coherent because of the measurement of phase.

In ultrasonic holography, when using numerical reconstruction methods, it is explicit to examine the holographic principle using a complex presentation form of wavefronts. In this sense the generation of holograms in a hologram plane involves merely the measurement of the wavefront scattered by the object.

The wavefront is measured as a complex wavefield, i.e. both the amplitude and the phase of the wave are measured on each measurement point. Alternatively, the real and imaginary parts can be measured.

When the numerical reconstruction method is used, the hologram must be stored in digital form in a computer memory. The stored digital data includes the complex information on the object wave, and therefore also the three-dimensional pictoral information on the object itself. It can thus be called a "hologram" even though it seems to bear no visual resemblance to conventional optical holograms which are usually complicated fringe patterns on photographic films.

IMAGE RECONSTRUCTION FROM HOLOGRAMS BY BACKWARD PROPAGATION

In the numerical reconstruction process the image is calculated numerically from the hologram data in a computer. Reconstruction by backward propagation is based on a Fourier transform and a spectrum shift theorem (2). The spatial frequency spectrum of the measured wavefront is calculated using the Fourier transform. This spectrum is shifted into the plane of the original object by multiplying it by a phase factor, and finally the wavefront in the object plane is calculated from the shifted spectrum using an inverse Fourier transform. Figure 1 illustrates the process.

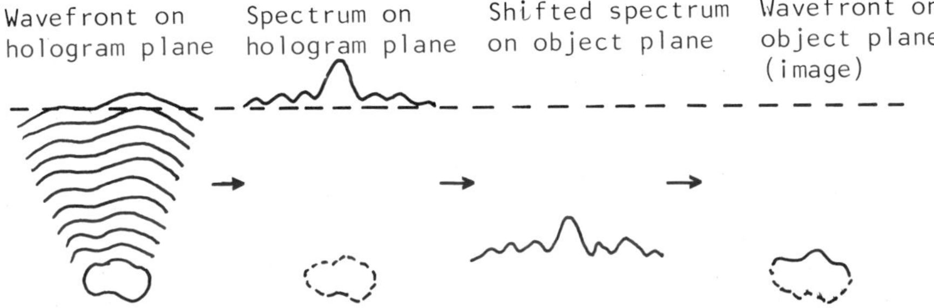

Fig.1 Simplified principle of numerical reconstruction by backward propagation of a wavefront

PRINCIPLE OF UHB IMAGING

Figure 2 illustrates the principle of the combination of holographic and B-scan imagings used in the UHB. In ultrasonic holography, while using numerical image reconstruction, the recording of a hologram involves merely the measurement and storing of the wavefront in a hologram plane (see fig. 2a). Of course, the object has first been insonified with monochromatic coherent ultrasound. The wavefront is recorded as a complex wavefield, as was mentioned earlier. In practice this can be realized by scanning the hologram plane with a point like transducer, sending a wide angle ultrasound burst at each point, and recording the amplitude and phase values of the reflected echoes. A three-dimensional image is processed from these data in a computer.

B-scan imaging is a two-dimensional imaging technique where narrow angle ultrasound pulses are sent into the

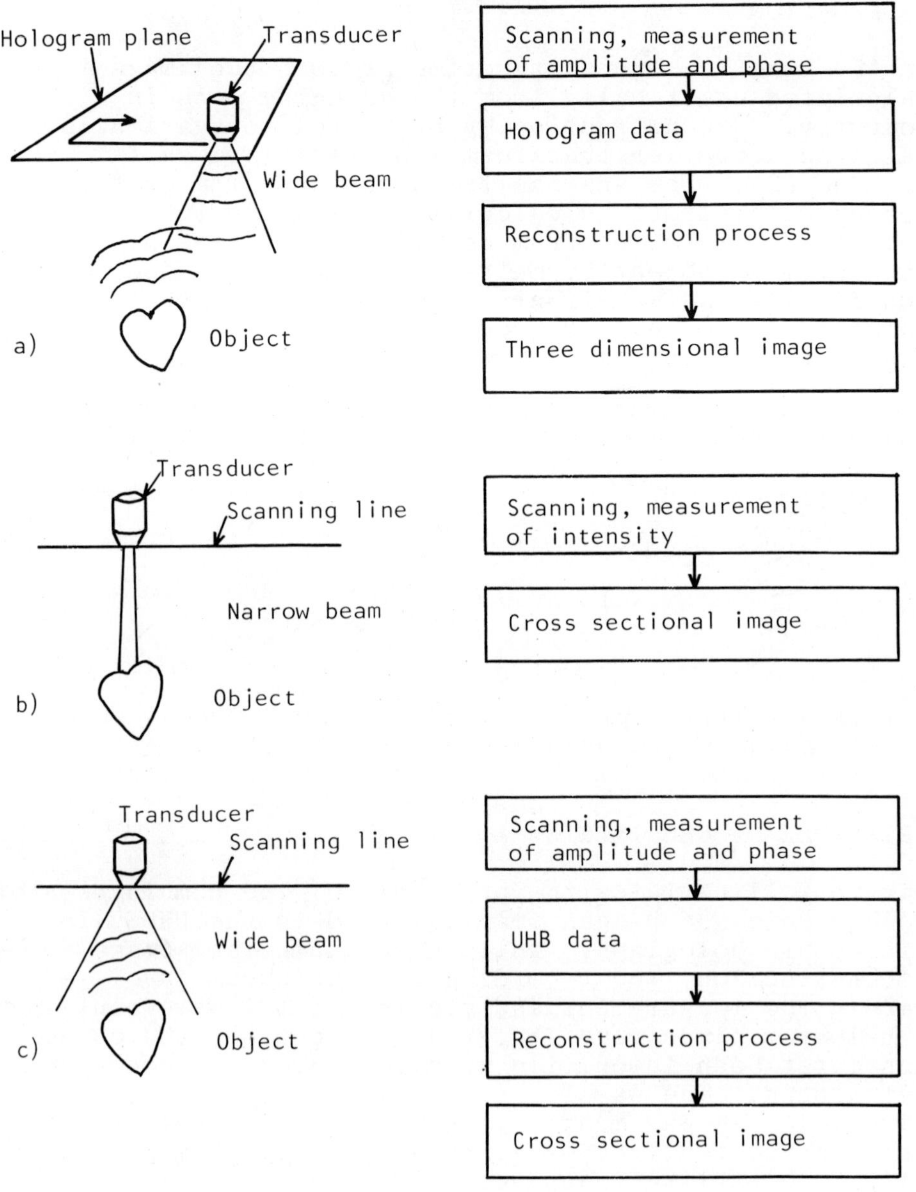

Figure 2 (a) Holographic imaging process
(b) B-scan imaging process
(c) UHB imaging process

object and the intensities of the reflected echoes are recorded as a function of propagation time. The transducer scans along a horizontal line above the object (see Fig. 2b). The display shows an image whose abscissa gives the position of the transducer on the scanning line and ordinate gives the depth with aid of the propagation time. The brightness of the display points are proporational to the intensity of the echoes.

In the UHB (see Fig.2c) a wide angle ultrasound burst is sent into the object as in holographic imaging and the relected echoes are measured by the function of the propagation time as in B-scan imaging, but now both the amplitude and phase of the reflected echoes are recorded. This is repeated at each measurement point of the scanning line. All the holographic data (called UHB data) are stored in a computer for a reconstruction process. In the reconstruction process the UHB data are first rearranged so that the measurement results corresponding to each depth are collected together to form one-dimensional wavefields which represents the wavefronts reflected from each depth. An image is reconstructed by propagating these wavefields backward to their own places in image space. This is done in the computer using a special program. An important part of this program is the part of curvature compensation which affects that the echoes arriving at different angles to the transducer are ascribed to the correct depth of the imaging area.

The whole UHB-image, whose axis correspond to the axis of the B-image, is obtained by calculating the intensity lines from these wavefields and by scanning these lines on the television monitor one after the other.

The main advantage of the UHB is good longitudinal and lateral resolutions. In fact, in the UHB each image point has been mathematically focussed. Thus by using a high quality transducer it is possible to achieve a resolution of the order of ultrasound wavelength all over the image.

The second advantage of the UHB is the possibility of using a wide range of image filtering and error compensation processes since the complex wavefields and their spectra are available in the reconstruction process.

UHB APPARATUS

An apparatus was constructed for verifying the theory of UHB imaging and for experimentally studying its imaging properties. Figure 3 shows the overall construction of the UHB apparatus.

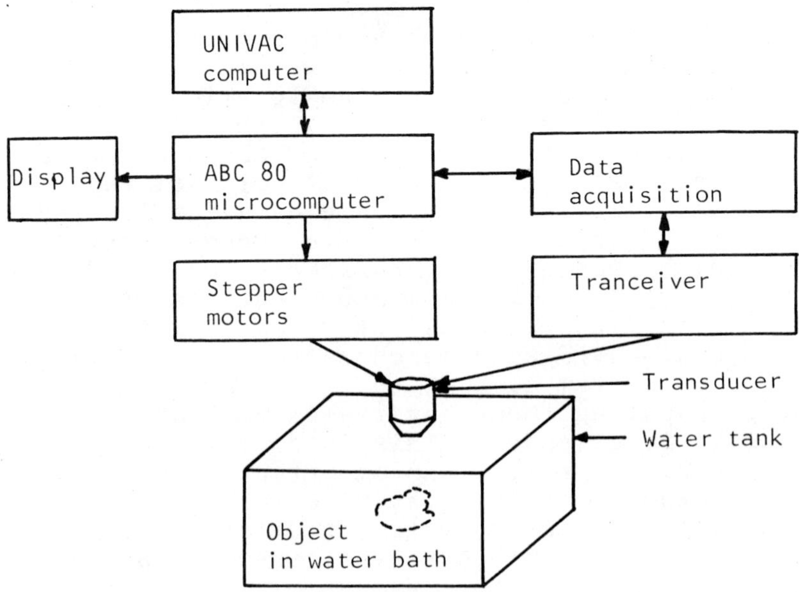

Figure 3 Overall construction of the UHB apparatus

The apparatus is controlled by a microcomputer system, ABC 80. It scans the transducer by means of stepper-motors. At each scanning point it gives the start command to the tranceiving system (tranceiver/transducer/data acquisition) and reads the measurement results from the memory of the data acquisition unit. After filling the memory of the ABC 80 it automatically opens the communication path to the main computer of the University (UNIVAC) and sends the results into its memory. After the recording of all UHB data the reconstruction process is started in the UNIVAC. Finally the image is transferred to the display unit.

EXPERIMENTS AND A COMPUTER SIMULATION

Figure 4 shows a resolution test object which can be used to evaluate both lateral and longitudinal resolution.

The wires of three test groups have been spaced at intervals of 5,4,3,2 and 1 mm.

Figure 5 shows the simulated UHB image of the resolution test object. The image has been formed in the computer the UHB data has been calculated using the derived UHB mathematics and the image has been reconstructed using the same reconstruction as in real measurements. The ultrasound frequency was assumed to be 4MHz.

Figure 6 is an experimental UHB image of the resolution test object. The ultrasound frequency was 4MHz. The resolution is not yet near the theoretical limit (0.7mm): there is noise in our system and the beam angle of our transducer is too narrow (about 15°).

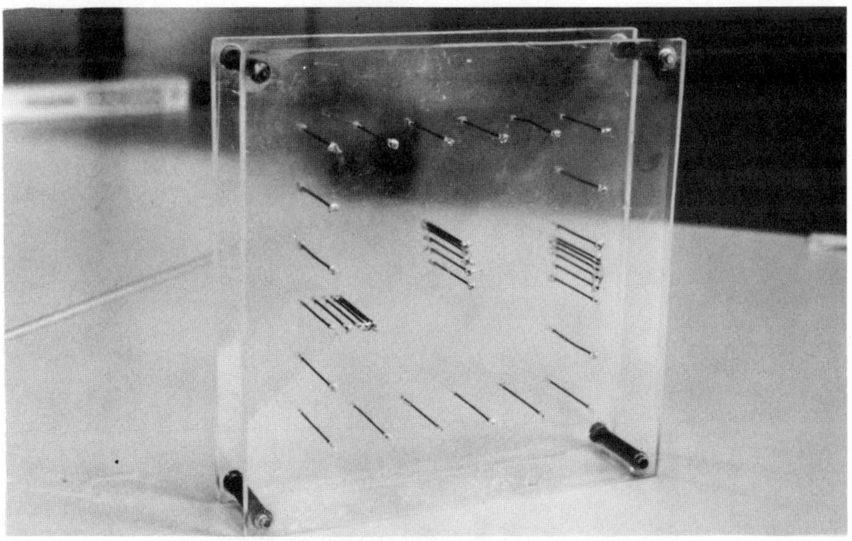

Figure 4 A resolution test object

Figure 7 shows one of the human brain specimens imaged with the UHB system. The specimen was a 10mm thick slice when the UHB image was measured. It was sectioned along the scanning line for the photograph. The cancer can be easily distinguished. A more rigorous analysis of this and many other UHB experiments will be presented later (3).

SUMMARY

Poor resolution is a problem in medical ultrasonic imaging especially in the lateral direction. A special

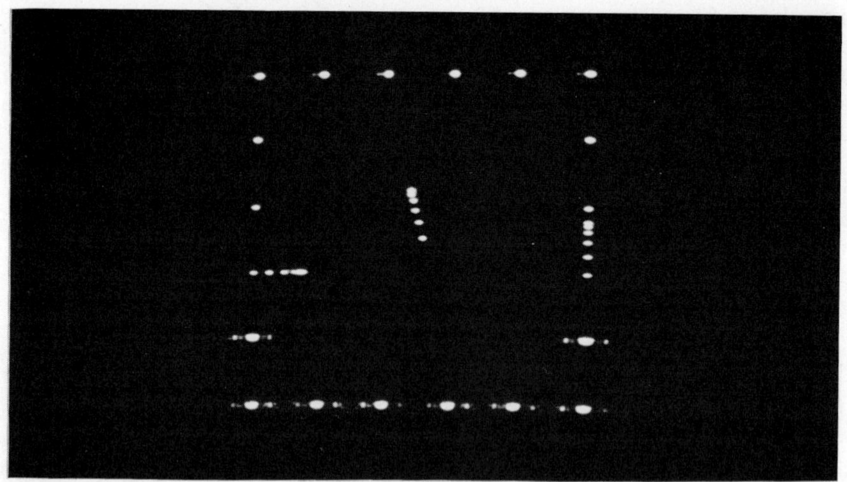

Figure 5 Simulated UHB image of the resolution test object of Figure 4

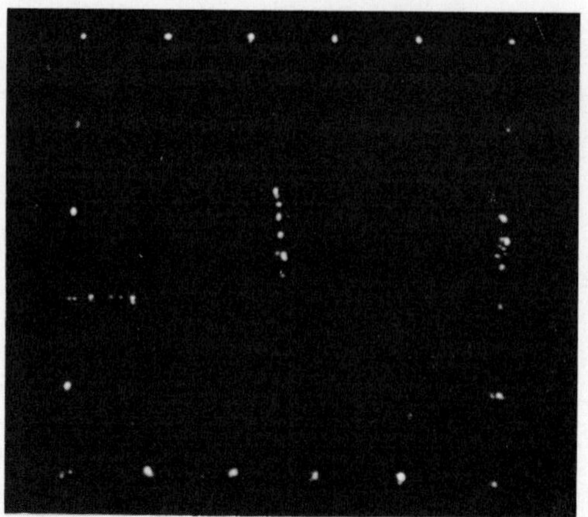

Figure 6 Experimental UHB image of the resolution test object of Figure 4

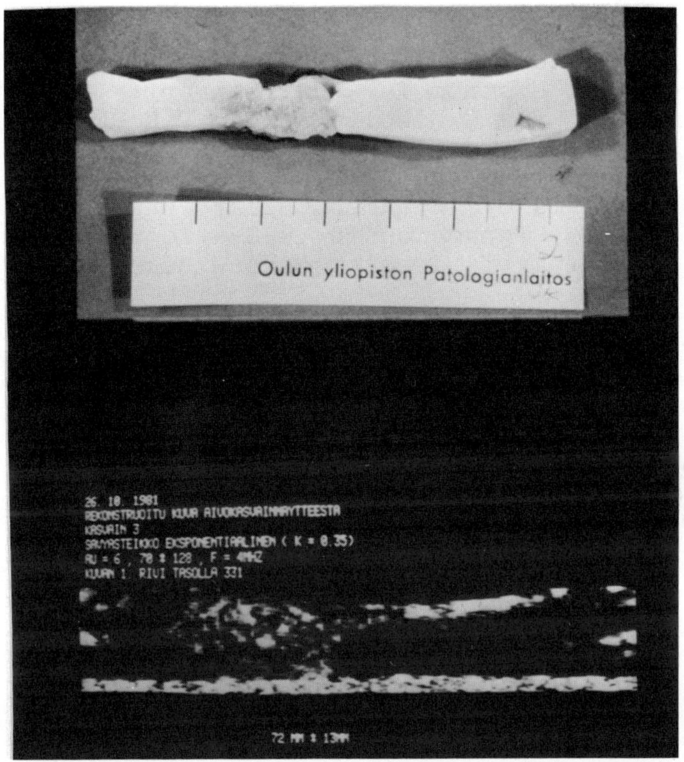

Figure 7 Brain specimen with cancer above (sectioned along the scanning line) and the UHB image of it

feature of holography is its good lateral resolution. By combining the conventional B-scan technique with holographic imaging, based on the numerical wavefront backward propagation method, it is possible to acheive good resolution in any direction all over the image. UHB imaging has another advantage. While each image point is numerically focused and processed, it is possible to use a wide range of image filtering and error compensation methods.

The main problems of the UHB are the phase incoherence of ultrasound at greater depths, which may force to use a lower ultrasound frequency; high speed required for recording the UHB data, and the sophisticated computer system required for real time UHB imaging.

ACKNOWLEDGEMENTS

The authors would like to thank Professor A Tauriainen and Professor S Nystrom for their expert scientific criticism, Dipl Eng M Rikola and Eng P Houkanen for their most valuable design and construction work. The financial support of the Finnish Academy, OMP Ylityma Oy, ORION Ylityma Oy, Finnish Culture Fund, Eemil Aaltosen Saatio, Tanno Tonningin Saatio and Oulun Yliopiston Tukisaatio is also acknowledged.

REFERENCES

1. A L Boyer et al., Computer reconstruction of images from ultrasonic holograms. Acoustical Holography, Vol 2 Ed A F Metherell et al., Plenum Press, New York (1970).
2. J W Goodman, Introduction to Fourier optics. McGraw - Hill Book Company, San Francisco (1968).
3. J Koivukangas, Doctoral thesis (to be published in 1983)

EXPANSION OF ACOUSTIC HOLOGRAM APERTURES USING ARMA MODELLING TECHNIQUES

R E Abdel-Aal, C J Macleod and T S Durrani

Department of Electronic and Electrical Engineering
University of Strathclyde
Glasgow G1 1XW, Scotland

ABSTRACT

As distinct from optical holography, the wavelengths in acoustic holography are large and therefore the numerical aperture is small for a given physical aperture. Further, the number of points at which the hologram can be sampled is limited due to time considerations when mechanical scanning is used and the high cost of building large arrays to sample the hologram. In order to achieve adequate resolution large apertures are necessary, therefore, a requirement exists for investigating signal processing techniques for increasing aperture size from available data.

In this paper a method is proposed for enlarging a limited hologram aperture in the space domain by extrapolating, computationally, additional points outside the available aperture using estimation techniques. The paper is concerned with two particular aspects of hologram extrapolation:

(i) Moving Average (MA)/Autoregressive (AR) processing techniques based on the available data.

(ii) The effect of disturbing noise in the measured hologram signals on the estimation accuracy.

Results obtained by computer simulation show the effectiveness of this method for enlarging the aperture with corresponding improvement in resolution with and without disturbing noise.

INTRODUCTION

Signal processing aimed at increasing lateral resolution is of particular importance in acoustic imaging because numerical apertures (i.e. aperture values expressed in terms of multiples of the wavelength) are normally much smaller than those available in optical imaging. Improving the resolution in this way would achieve considerable savings in time and cost by reducing the area over which the data is collected. Moreover, situations arise in the field of acoustic imaging where the available aperture is physically restricted, for example, when imaging the inside of the chest through narrow spaces between the ribs. A considerable amount of such signal processing can be conveniently carried out using a digital computer, which most holographic imaging systems use nowadays for reconstructing images.

The possibility of resolution beyond the diffraction limit has been recognised in optics for a number of years (1), and several techniques have been developed for this purpose (2). These include a process of analytic continuation on the spatial frequency spectrum of the image (3), and an object restoration technique based on expanding the aperture function in terms of prolate spheroidal wave functions (4). Other methods which are more robust against noisy environments include the error energy reduction method proposed by Gerchberg (5) and the maximum entropy technique by Frienden (6).

A number of techniques to improve lateral resolution in acoustic holography have been reported in the literature. A sequential estimation method based on the minimization of a mean square criterion function in the frequency domain have been used by Takuso Sato et al (7) to expand the hologram aperture. Other methods attempt to improve resolution without expanding the available aperture, such as a more exact reconstruction method proposed by Williams et al (8), (9). Another method is that suggested by Ikeda et al (10) where the resolution is improved by artificially increasing the frequency of the hologram signals before reconstruction.

This paper describes a new method for enlarging a limited hologram aperture in the space domain using estimation techniques. The method is based on modelling the hologram signal over the available aperture and using the model so constructed to predict new points outside that aperture. The models used are in general of the ARMA type, containing both space variant and space invariant terms. The paper also describes a technique for correcting the predicted signals in order to improve the prediction accuracy. The effect of noise in the hologram signals is discussed and numerical examples are given to illustrate the effectiveness of the method as a tool for improving resolution when imaging single or multiple-point objects with and without disturbing noise.

NATURE OF THE PROBLEM

Hologram signal and simulation of noise

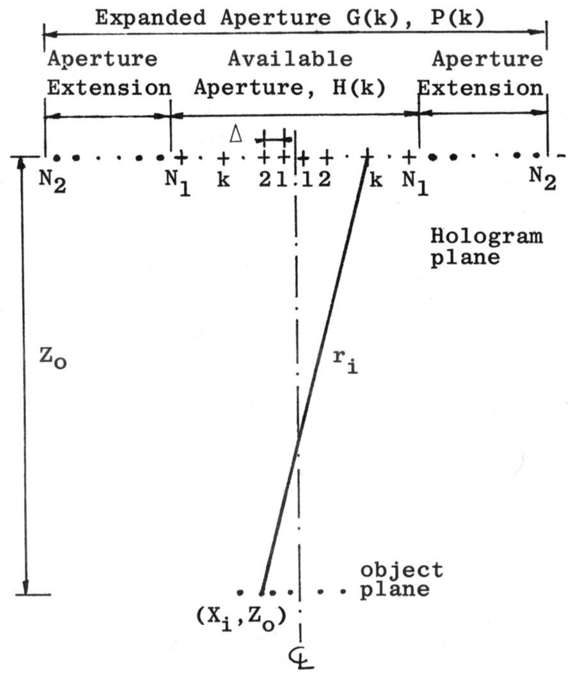

Fig 1. Geometry of the imaging system

Consider the holographic imaging system shown schematically in Fig 1. For simplicity the object is assumed to consist of a finite number of point sources which radiate acoustic energy coherently. The hologram is sampled at 2 N_1 uniformly distributed points in the available aperture, spaced at distance Δ wavelengths. The object lies in a plane parallel to the hologram at distance Z_o from it, Z_o is assumed to be known beforehand.

The complex hologram signal at sampling point k in the avavilable aperture is given by:

$$H(k) = \sum_{i=1}^{\ell} \frac{A_i}{r_i} \exp K r_i = \alpha_k + j \beta_k$$

where ℓ is the number of points in the object, K is the wavenumber, $K = 2\pi/\lambda$, λ is the wavelength, A_i is an amplitude factor, and r_i is the length of the vector joining point i on the object to point k on the aperture.

$$\alpha_k = \sum_{i=1}^{\ell} \frac{A_i}{r_i} \cos (K r_i), \quad \beta_k = \sum_{i=1}^{\ell} \frac{A_i}{r_i} \sin (K r_i)$$

We assume in practice that the hologram is obtained by measuring the quantities α_k and β_k at the sampling points. The measured quantities, α'_k and β'_k, will contain random noise which can be simulated by relating α'_k and β'_k to α_k and β_k as follows:

$$\alpha'_k = \alpha_k + \eta_k$$

$$\beta'_k = \beta_k + \xi_k$$

where η_k and ξ_k are white noise processes with zero mean and variance prescribed by the input signal to noise ratio. The noise level is expressed throughout as the relative peak amplitude of η_k wrt the hologram signal α_k. This is taken to be equal to that of ξ_k wrt β_k.

Prediction and the definition of errors

This paper is concerned with expanding the aperture in the space domain by predicting new points. The prediction of signals to the right of the available aperture is based on a model derived from the signals at the RH half of the available aperture. Similarly the LH half is used to predict points to the left. The following definitions are made, (see Fig 1):

- H(k), $k \leqslant N_1$ is the true hologram signal over the small available aperture
- G(k), $k \leqslant N_2$ is the predicted hologram signal over the expanded aperture,

$$G(k) = \begin{cases} H(k) & k \leqslant N_1 \\ \text{Predicted signal at point } k & N_1 < k \leqslant N_2 \end{cases}$$

- P(k), $k \leqslant N_2$ is the true hologram signal over the expanded aperture.

Let $G(k) = g_1(k) + j\, g_2(k)$

$P(k) = p_1(k) + j\, p_2(k)$

The percentage error at point $k(k=N_1+1, N_1+2, \ldots N_2)$ is defined as:

$$E(k) = 100\ \{(g_1(k)-p_1(k))^2+(g_2(k)-p_2(k))^2\}^{\frac{1}{2}} / \{(p_1(k))^2+(p_2(k))^2\}^{\frac{1}{2}} \qquad (1)$$

$$k = N_1+1,\ N_1+2,\ \ldots\ N_2$$

A better expression for the error, which takes into account its effect on the image reconstruction, is obtained by weighting the error in eqn (1) by the ratio between the modulus of the true hologram signal at point k and the modulus of the maximum signal over the whole of the true aperture, P_{max}, where:

$$P_{max} \triangleq \max\ \{|P(k)|\} \quad k = 1, 2, \ldots N_2$$

This leads to:

$$E(k) = 100\ \{(g_1(k)-p_1(k))^2+(g_2(k)-p_2(k))^2\}^{\frac{1}{2}} / P_{max} \qquad (2)$$

THE MOVING AVERAGE (MA) MODEL

Referring to Fig 1, consider a single-point object and assume a total available aperture containing $2N_1$ points and an expanded aperture of $2N_2$ points. Considering the RH side of the aperture, the Moving Average (MA) or space variant model for the hologram signal in the available aperture is obtained by relating the signal at point k to that at point (k-1) by:

$$H(k) = \psi(k)\, H(k-1), \qquad k = 2, 3, \ldots N_1 \tag{3}$$

where $\psi(k)$ is a function of the position of point k relative to the centre of the aperture. If $\psi(k)$ is assumed to be a polynomial function in the distance d_k from point k to the centre expressed in wavelengths (see Fig 1), then:

$$\begin{aligned}\psi(k) &= a_1 + a_2(d_k)^1 + a_3(d_k)^2 + \ldots + a_N(d_k)^{N-1} \\ &= \sum_{i=1}^{N} a_i (d_k)^{i-1}\end{aligned} \tag{4}$$

where a_i is a complex coefficient, and N is the number of terms in the polynomial.

Writing eqn (3) for all values of k and substituting for $\psi(k)$ from eqn (4) we have the following set of equations which are solved for the model coefficients $a_1, a_2, \ldots a_N$:

$$\begin{bmatrix} H(2) \\ H(3) \\ \cdot \\ H(N_1) \end{bmatrix} = \begin{bmatrix} H(1) & H(1)d_1 & H(1)d_1^2 & \cdot & H(1)d_1^{N-1} \\ H(2) & H(2)d_2 & H(2)d_2^2 & \cdot & H(2)d_2^{N-1} \\ \cdot & \cdot & \cdot & \cdot & \cdot \\ H(N_1-1) & H(N_1-1)d_{N_1-1} & H(N_1-1)d_{N_1-1}^2 & \cdot & H(N_1-1)d_{N_1-1}^{N-1} \end{bmatrix} \begin{bmatrix} a_1 \\ a_2 \\ \cdot \\ a_N \end{bmatrix}$$

For an exactly determined system, the number of terms in the polynomial is assumed to be equal to the number of equations, i.e. $N = N_1-1$.

To use the model for prediction we assume that the relationship in eqn (3) holds outside the available aperture, with $\psi(k)$ now defined by the model coefficients $a_1, a_2, \ldots a_N$. Thus the signal at the first predicted point $G(N_1+1)$ is obtained from the signal at the last point in the available aperture $H(N_1)$ by:

$$G(N_1+1) = \psi(N_1+1)\, H(N_1)$$

$$= \{ \sum_{i=1}^{N} a_i (d_{N_1+1})^{i-1} \} H(N_1)$$

Similarly $G(N_1+2)$ is obtained from $G(N_1+1)$, ... and so on. The process of constructing and using the model is repeated for the left half of the aperture with prediction from right to left.

The model has been used to expand a total aperture of 16 points (N_1=8), spaced at 1λ, four times (N_2=32) to image a 1-point object on the centre at a range of 100λ. Fig 2a shows the percentage prediction error, as defined by eqn (2), over the aperture. The images reconstructed from the small available aperture, the expanded predicted aperture, and the expanded true apertures are shown in Fig 2b normalised to the same peak value. Backward wave propagation has been used throughout to obtain the reconstructed images. The images

(a) Percentage prediction error over the aperture

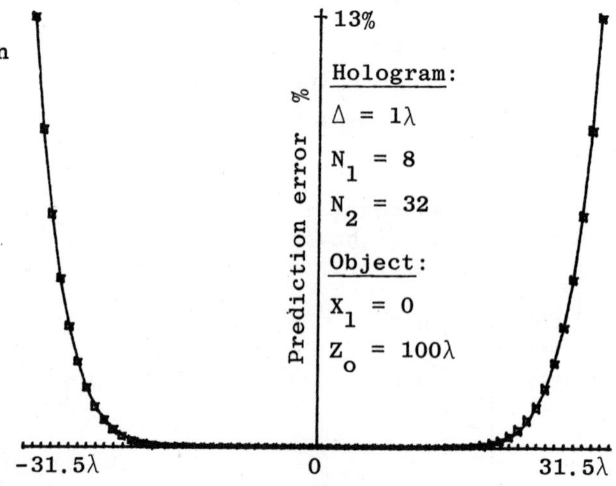

(b) Intensity of reconstructed images normalised to the same peak value

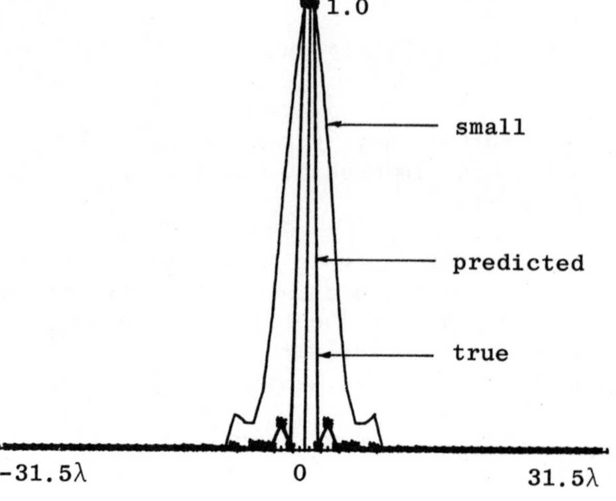

Fig 2. Expanding an aperture four times using an MA predictive model to image a 1-point object

obtained from both the true and predicted holograms are in good agreement with a corresponding improvement in the width of the point image over the case of the small aperture.

For the case of 1-point object at the centre in the Fresnel zone and beyond, it was found that the prediction error is inversely proportional to the object range. It is interesting to note that for this case the spatial frequency over the aperture also varies inversely with range. Obviously, the higher the spatial frequency the greater the errors that would be expected in prediction due to the larger variations in the signal from one point on the aperture to the next.

Several types of polynomials have been tried for the model function ψ. Orthogonal Chebyshev, Legendre and Hermite polynomials gave almost identical results to those obtained with ordinary polynomial expressions. In the case of the Fourier series polynomial, it was found that the predication accuracy depends on the choice of the spatial fundamental frequency of the series, which should be optimized for maximum accuracy.

THE MOVING AVERAGE/AUTOREGRESSIVE (ARMA) MODEL

When attempting to use the MA model for predicting the hologram of a 2-point object the errors were too large to allow for useful expansion ratios. The concept of the ARMA model was then introduced. In this model, the signal at one point is related not only to the signal at the previous point by a space variant function but is also linearly related to the signals at a number of the preceding points. Assuming an ordinary polynomial expansion for the space variant function we have:

$$H(k) = \{\sum_{i=1}^{p} a_i (d_k)^{i-1}\} H(k-1) + \sum_{j=1}^{q} c_j H(k-q+j-1) \qquad (7)$$

with

$$k = 2, 3, \ldots N_1$$

$$H(k-q+j-1) = 0 \quad \text{for} \quad j \leqslant q-k+1$$

where:

p is the number of MA (space variant) terms, q is the number of AR (space invariant) terms, $a_i (i=1, \ldots, p)$ are the MA coefficients, $c_j (j=1, \ldots, q)$ are the AR coefficients.

As both the MA and the AR parts of the model have to share the number of degrees freedom, and assuming an exactly determined system, then:

$$p+q = N = N_1-1$$

Writing eqn (7) for all values of k over the available aperture we have N equations which can be solved for the model coefficients a_1, a_2, ... a_p, c_1, c_2, ... c_q. The model is then used to predict points outside the available aperture. First, $G(N_1+1)$ is obtained from $H(N_1)$ together with signals at the preceding q-1 points. In general, $G(N_1+\ell)$ is obtained as follows:

$$G(N_1+\ell) = \{ \sum_{i=1}^{p} a_1 (d_{N_1+\ell})^{i-1} \} G(N_1+\ell-1) + \sum_{j=1}^{q} c_j \; G(N_1+\ell-q+j-1)$$

It was found that for the case of 1-point objects the optimum model is an MA model (no AR terms). For 2-point objects, however, an ARMA model gives better results. Fig 3 shows the reconstructed images obtained when doubling the size of a 16 point total aperture, spacing 1λ, to image 2 points separated by a distance of 6λ, range

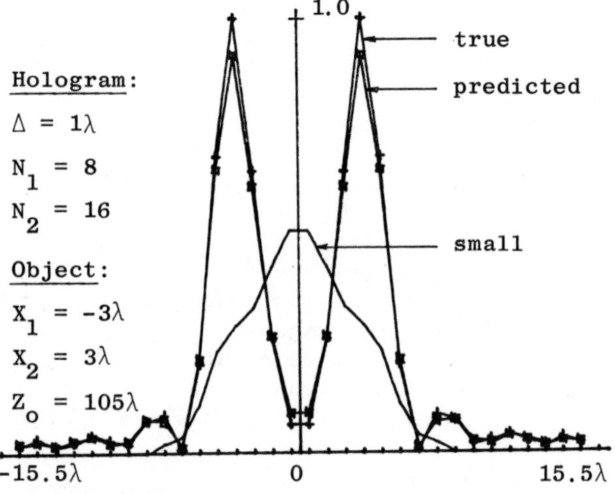

Fig 3. Intensity of reconstructed images obtained when expanding an aperture twice using a 5-2 ARMA predictive model to image a 2-point object

105λ. The prediction is performed using the optimum ARMA model with 5 MA terms and 2 AR terms. The figure shows that while the small available aperture fails to resolve the two points, they are clearly resolved with the expanded predicted aperture which gives an image similar to that obtained from the expanded true aperture.

THE CORRECTIVE MODEL

An attempt has been made to improve the prediction accuracy by correcting the predicted signals using a corrective model based on comparing the predicted data with corresponding data which are known to be true. To construct such a model the known data over one half of the available aperture is related to the data at the same points obtained by prediction using a predictive model based on the other half. The corrective model is then used to correct predicted data outside the available aperture. The performance of the corrective model depends upon the structure of both the predictive and corrective models employed and a search for the optimum model is often required.

THE EFFECT OF DISTURBING NOISE

When random perturbations were used to simulate noise in measured hologram signals, the prediction errors were too large and no satisfactory images were obtained even for relative noise amplitudes as low as 1%. The increased errors are due to the sensitivity of the matrix solution to perturbations in the data vector and it was observed that this sensitivity increases as the matrix size increases. It was also noticed that the larger the number of AR terms in an ARMA model the less the sensitivity of the matrix colution to noise. Referring to eqn 7, for a totally AR model (p=0, q=N), the system matrix reduces to a triangular form. It was found that the solution of this form of matrix exhibits better stability with noise than in the case of a square matrix.

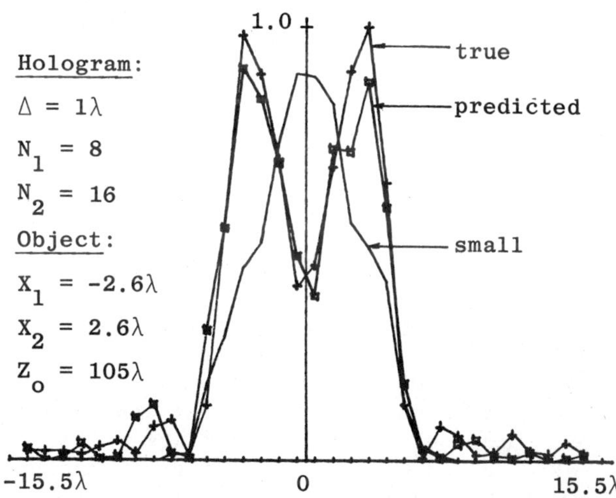

Fig 4. Intensity of reconstructed images obtained when expanding an aperture twice to image a 2-point object. Relative noise amplitude: 10%. Predictive model: AR.

Fig 4 shows the image reconstructions obtained when using an AR predictive model to double the size of a 16 point aperture, spacing 1λ, in the presence of noise with relative amplitude of 10%. The object consists of two points spaced at 5.2λ, range 105λ. Although the AR model is not the optimum model without noise, it gives better results in the presence of noise and the two points in the object are clearly resolved.

To image 1-point objects in the presence of noise the MA model had to be modified such that the system matrix takes a triangular form. Although this achieves better stability with noise, it also reduces the prediction accuracy of the model and a corrective model of a triangular matrix form had to be employed. Fig 5 shows the image reconstructions obtained when doubling the size of a 16 point apertures, spacing 1λ, to image a 1-point object at the centre, range = 100λ with the hologram signals perturbed by noise having 10% relative amplitude. It is clear from the figure that in this case

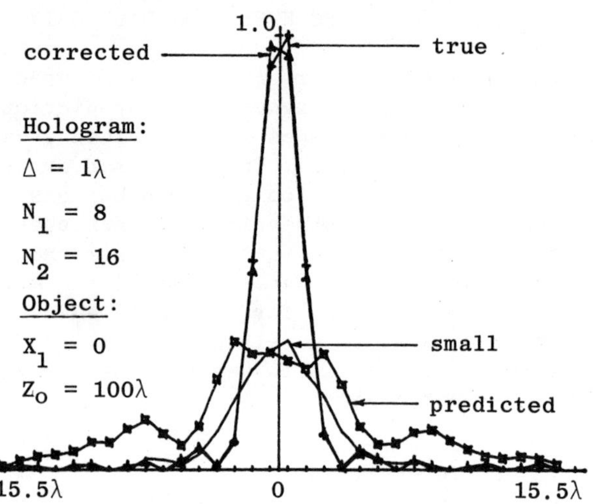

Fig 5. Intensity of reconstructed images obtained when expanding an aperture twice to image a 1-point object. Relative noise amplitude: 10%. Predictive and corrective model: MA type, triangular matrix.

the use of a corrective model is essential for achieving the improvement expected from expanding the small aperture.

COMMENTS

In this paper a method for increasing the lateral resolution in acoustic holography is described. The approach is essentially based on modelling the hologram signal at the available aperture and this aperture is then expanded in the space domain by using the model to predict the hologram signal at new points outside the aperture. An MA model gives optimum prediction in the case of 1-point objects while for multiple-point objects an ARMA model gives better results in general. The accuracy of determining the estimated signals can be increased by employing a corrective model based on the knowledge of both predicted and true data in part of the available aperture.

The effect of noise in the hologram signals is also discussed and it was found that a triangular model matrix has better stability with noise than a square matrix. This, however, is at the expense of some deterioration in the prediction accuracy. Typical results obtained by computer simulation show the effectiveness of the technique described for improving the resolution with and without noise for both single and multiple-point objects.

ACKNOWLEDGEMENT

The work described is supported by a research grant, Grant No. GR/B/88686, from the UK SERC. This support is gratefully acknowledged.

REFERENCES

1. J.W. Goodman, Introduction to Fourier optics, McGraw Hill, New York, 1968.
2. P.N. Keating, T. Sawatari, and G. Zilinskas, "Signal processing in acoustic imaging". Proc. IEEE, Vol 67, 1979, pp 496-510.
3. J.L. Harris, "Diffraction and resolving power". J. Opt. Soc. Am., Vol 54, 1964, pp 931-936.
4. C.W. Barnes, "Object restoration in a diffraction-limited imaging system". J. Opt. Soc. Am., Vol 56, 1966, pp 575-578.
5. R.W. Gerchberg, "Super resolution through error energy reduction". Opt. Acta., Vol 21, 1974, pp 709-720.
6. B.R. Frieden, "Restoring with maximum likelihood and maximum entropy". J. Opt. Soc. Am., Vol 62, 1972, pp 511-518.
7. T. Sato, K. Sasaki, and K. Uemura, "Super-resolution imaging using a sequential estimation of a hologram in an extended area from data obtained in a limited area". J. Acoust. Soc. Am., Vol 65, 1979, pp 976-984.
8. E.G. Williams, J.D. Maynard, and E. Skudrzyk, "Sound source reconstructions using a microphone array". J. Acoust. Soc. Am., Vol 68, 1980, pp 340-344.
9. E.G. Williams and J.D. Maynard, "Holographic imaging without the wavelength resolution limit". Physical Review Letters, Vol 45, 1980, p 554-557.
10. O. Ikeda and T. Sato, "On method of superresolution of a limited number of point targets using ultrasonic scanning". J. Acoust. Soc. Am., Vol 57, 1975, pp 334-337.

ARRAY SYSTEMS FOR UNDERWATER VIEWING BY ACOUSTICAL HOLOGRAPHY

J P N Wei, D J Zhang, Z Q Sun and Z Q Zhou

Wuhan Institute of Physics, Academia Sinica
People's Republic of China

Experiments on underwater viewing by acoustical holography at Wuhan Institute of Physics, Wuhan, China are described. Early experiments were conducted in an anechoic water tank with scanning linear arrays. The first satisfactory reconstructed image was obtained by a scanning linear array of 100 elements at a frequency of 1 MHz from simple objects at a distance of half a meter by the end of 1972. Other experiments were conducted at objective distances ranging from several meters to several tens of meters in anechoic water tanks and in a swimming pool, under different conditions and different modes of scanning. Results found to be in accord with theory. A scanning linear array acoustical holographic imaging system could also give sharp reconstructed images to moving objects under certain conditions. Such systems are useful for many applications.

This paper also describes the design, characteristics and preliminary experiments of our square-array acoustical holographic imaging systems, one of which is a 64 x 64 square-array system working at a frequency of 312.5 kHz with an angular resolution of 21' and an angular field of view of 23° 4'. The designed working distance is 3 to 50 meters. The receiving array of high sensitivity and uniformity, and a transmitting transducer with smooth main lobe is also described.

The prospects of these systems in underwater viewing is also discussed.

INTRODUCTION

It is widely believed that the acoustic method is the only practical method in underwater viewing, especially in turbid water. Among the various acoustical methods maybe acoustical holography is the most promising method in underwater viewing.

The idea of using the holographic method in the field of acoustic imaging was first introduced by Greguss[1], Thurstone[2] and Mueller[3]. It was Wade[4] who first suggested the use of a scanning linear array in a system for practical underwater search, and it was Aoki[5] who first published a paper giving a reconstructed image of a simple object using a scanning linear array acoustical holographic device.

In order to develop a practical system for underwater viewing, experimental work on acoustical holography at Wuhan Institute of Physics was switched to array systems using scanning linear arrays at the beginning of 1972, and a year later, square array systems have also been designed and built. The following is a brief account of the experimental work done in our Institute,

EARLY WORK WITH A SCANNING LINEAR ARRAY

The earliest work with a scanning linear array was done in a small anechoic water tank. A scanning linear array of 100 receiving transducers was used. The acoustic frequency was 1 MHz and the distance from the transmitting transducer to the scanning array was one meter with the object at about midway between them. The object was a piece of sheet metal with the letter E or two Chinese characters carved out from it. Manual scanning was used and the hologram was photographed line by line and at the end, a hologram of 100 x 100 pixels was obtained. It was reconstructed under laser light. Fig. 2 gives the object, hologram and reconstructed image thus obtained at the end of 1972. The letter E is 100 x 100 mm^2 with the stroke 20 mm in width. The two Chinese characters combined is 100 mm in height and 80 mm in width, and the stroke is 15 mm in width.

Measurements on the dimensions of the object and the reconstructed images, together with the data on the configuration of the setup, gives the resolution of the system which agrees with the theory quite well.

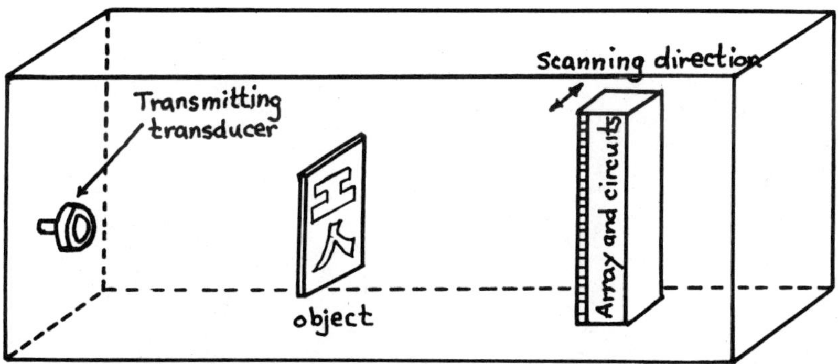

Fig. 1. Schematic of the scanning linear array acoustical holography system.

FURTHER EXPERIMENTS WITH SCANNING LINEAR ARRAYS

A Different modes of scanning

Early experiments were carried out exclusively with the scanning receiving array mode, and the scanning was by steps so that the hologram took the form of a two-dimensional array. As the objective distance became larger, the array must be made larger to get satisfactory resolution. Thus, it was much easier to scan the transmitting transducer which was much lighter. As it is well known, scanning the transmitter gives the same hologram as scanning the receiver, later on we always used the scanning transmitter mode instead of scanning the receiver. Furthermore, a grid-like hologram is easier to make and gives the high order images in one dimension only, this mode of scanning was adopted throughout our experiments.

B Different Objective Distances

A series of imaging experiments were performed for different objective distances up to 53 meters. They all gave reasonably recognizable reconstructions. Figs. 3 to 5 are the objects, holograms and reconstructed images for object to hologram-plane distances of 3m, 18.3m and

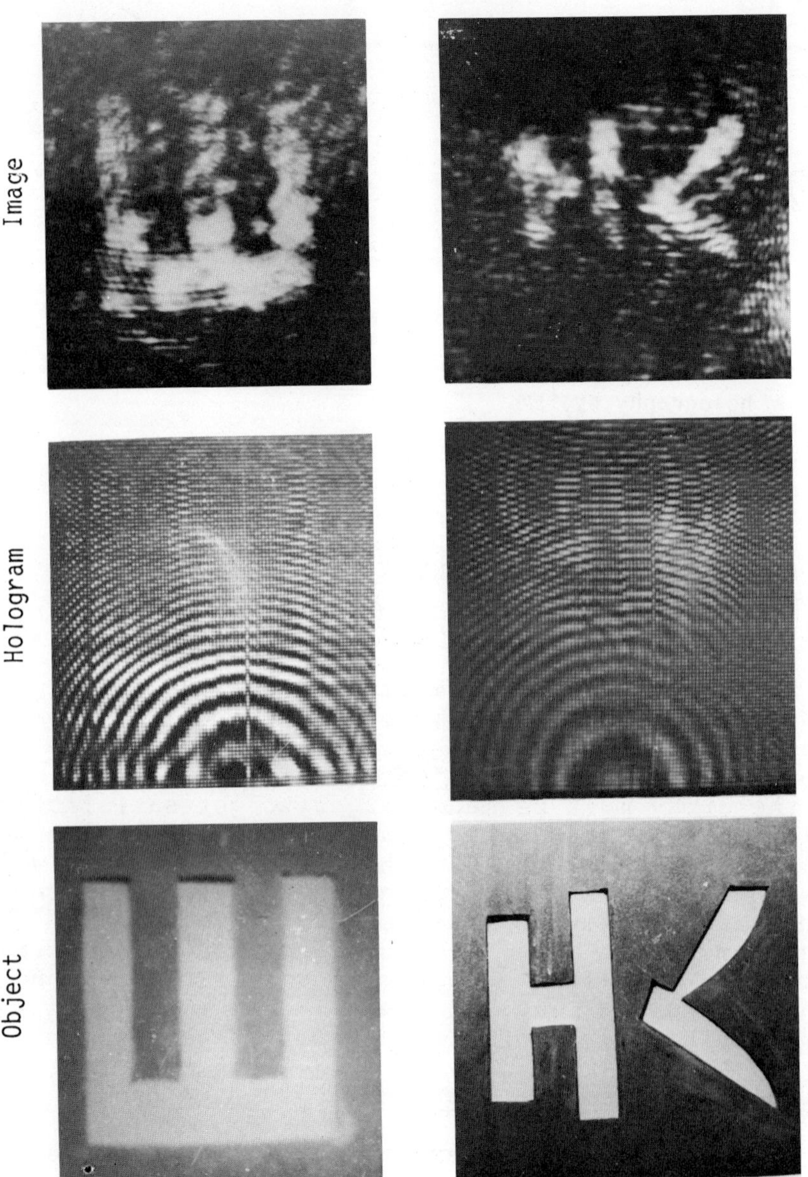

Fig. 2. Objects, hologrames and reconstructed images of E and simple Chinese character.

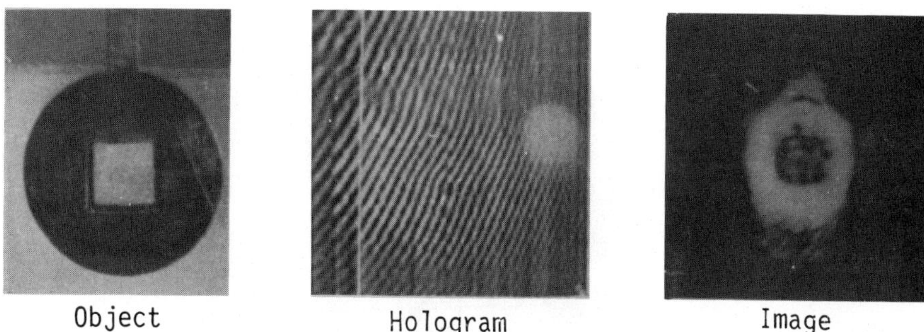

Fig. 3. Object, hologram and reconstructed image of a round piece of sheet metal with a square hole. Diameter of circle is 12.8cm, side of square is 4.8cm.

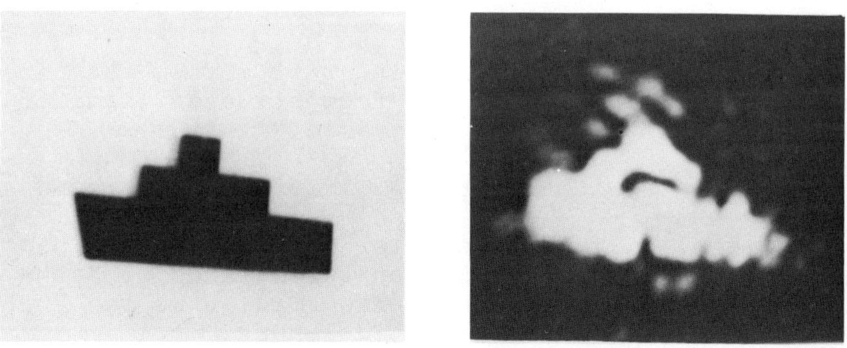

Fig. 4. Object and reconstructed image of an object at a distance of 18.3 m from the hologram plane.

53 m. Fig. 3 gives the object, hologram and reconstructed image of a round piece of sheet metal with a square hole at the center at a distance of 3 m from the array. Fig. 4 gives the object and reconstructed image of a flat piece of metal at a distance of 18.3 m from the scanning array. This experiment was also conducted in an anechoic water tank.

In the above experiments, CW sound was used. With pulsed sound and time gating, experiments need not be conducted in anechoic water tank. Fig. 5 gives the

Fig. 5 Imaging of a cylinder in a swimming pool. Distance from object to array is 53 m

result of the imaging of a cylinder 53 m from the linear array in a swimming pool. The reconstructed image of the cylinder is clearly recognizable.

C Experiments with Object on Different Background

A metal cylinder one meter long and with a diameter of 12 cm was put horizontally on three kinds of background, pebbles, sand or mud. From Fig. 6 we see that they all give clear reconstructed images.

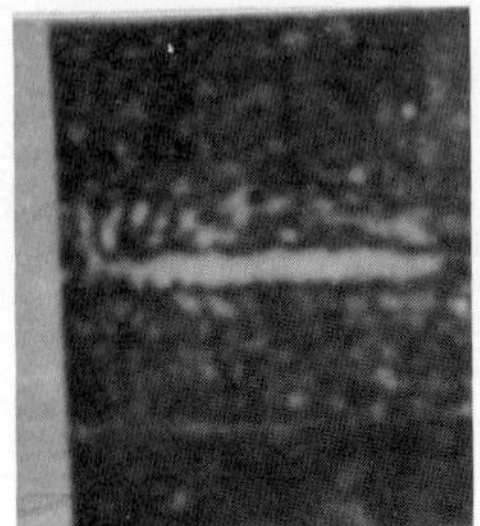 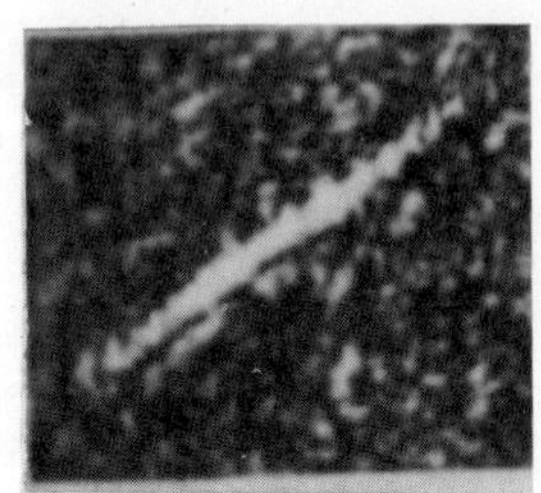 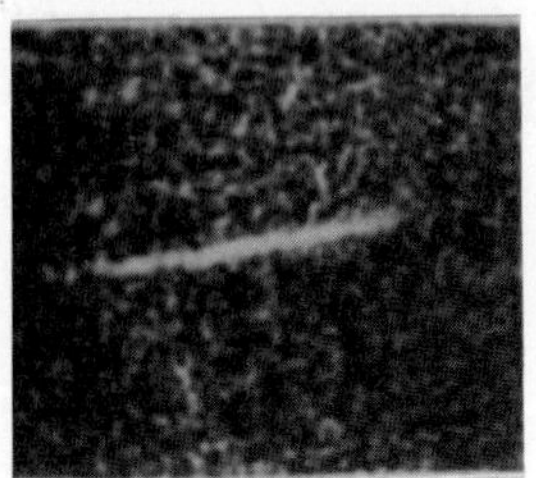

Fig. 6 Reconstructed images of a metal cylinder on different background.

D Imaging of Moving Object with a Scanning Linear Array

It is often assumed that in order to get a satisfactory reconstruction from an acoustical holographic system using a scanning receiver, the relative motion of the object and a reference point in the hologram plane should be exceedingly small during the whole scanning period. In fact, it is not necessary for a scanning linear array system because the linear array itself is electronically scanned at a very high speed. The motion of the object gives only a phase shift to each line of the hologram so that it only causes a distortion of the reconstructed image similar to the distortion of a moving object taken with a camera with a focal plane shutter. Zhou et al have analyzed the problem mathematically and got the result that

1) if the motion of the object is parallel to the direction of the scanning, the reconstructed image is either elongated or compressed;

2) if the motion of the object is perpendicular to the direction of the scanning but parallel to the hologram plane, the reconstructed image will experience a rotation;

3) if the motion of the object is perpendicular to the hologram plane, the reconstructed image will have a displacement.

Fig. 7 gives the result when the motion of the object is along the direction of the scanning, we see that the circle is compressed as predicted. Fig. 8 gives the result when the motion of the object is perpendicular to the motion of the scanning array but parallel to the hologram plane. We see that the reconstructed image gives a rotation according to theory. Fig. 9 gives the result when the object moves perpendicular to the hologram plane. A displacement of the position of the image is obtained.

SQUARE-ARRAY ACOUSTICAL HOLOGRAPHIC SYSTEMS

In order to obtain experience in designing a relatively big experimental square-array acoustical holographic system for underwater viewing, a rather small square-array system was built in 1973. It was a 16 x 16 square-array system. It was small enough to be built easily but still big enough to be experimented on. We used that system to test the performance of the constituent parts of the system as well as its imaging capability. According to the data obtained, we

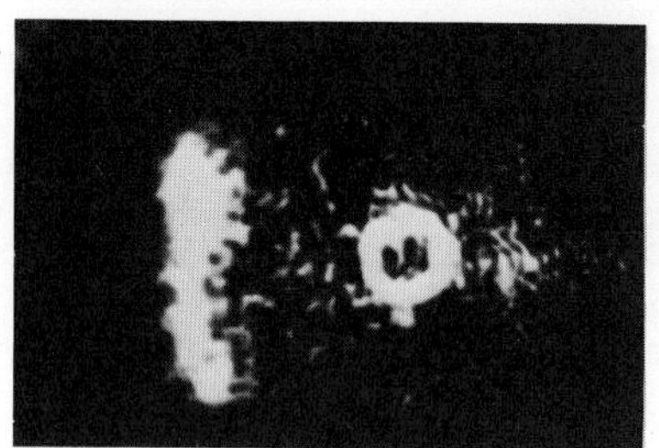
stationary

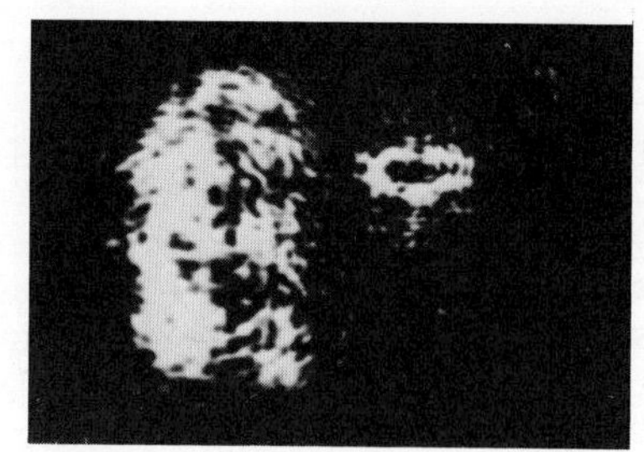
moving

Fig. 7. Result of the motion of object opposite to the scanning.

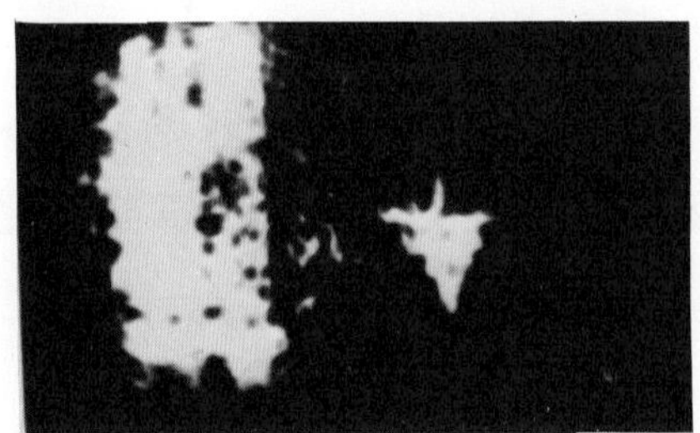
stationary

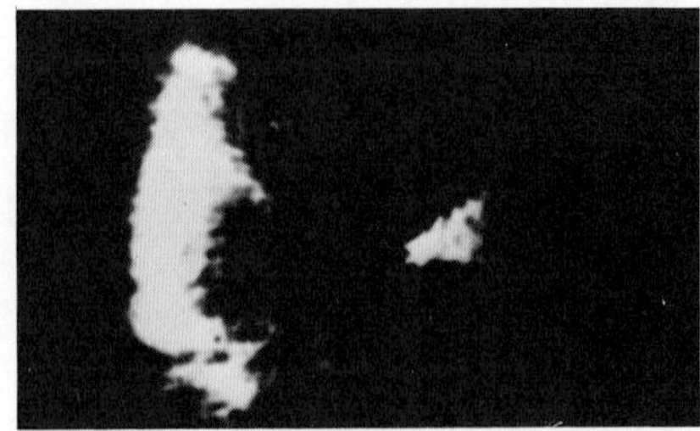
moving

Fig. 8. Result of the motion of object perpendicular to the scanning direction but parallel to the hologram plane.

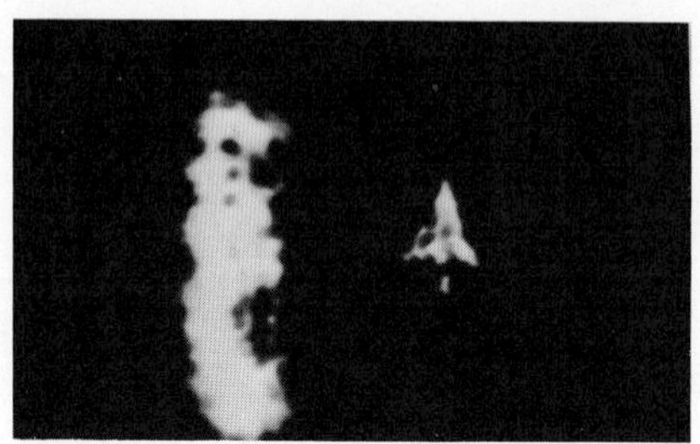
stationary

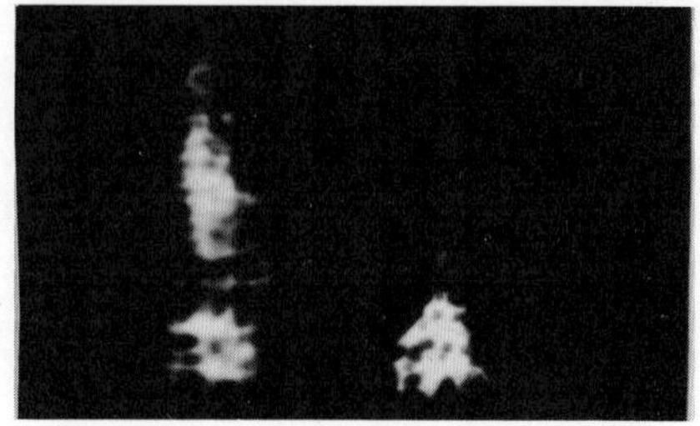
moving

Fig. 9. Result of the motion of object perpendicular to the hologram plan.

designed and built a 64 x 64 square array system which we shall describe below.

A Transmitting Transducer

A good transmitting transducer should give a wide enough and smooth main lobe which satisfies the need of the field of view of the holographic system. Our transmitting transducer was designed according to the method developed by Zhang and Xia in our laboratory. According to their calculations, the directivity of our transducer is given by:

$$D(\theta) = \frac{\int_0^b q(R_1) R_1 \exp\left(-jk\frac{R_1^2}{2A}\cos\theta\right) J_0\left(kR_1\sqrt{1-\frac{R_1^2}{4A^2}\sin\theta}\right) dR_1}{\int_0^b q(R_1) R_1 \exp\left(-jk\frac{R_1^2}{2A}\right) dR_1}$$

$$q(R_1) = \exp\left(-\frac{R_1^2}{b^2}\right)$$

where A = 100mm, a = 35 mm, h = 6 mm, k = 1.31, b = $a^2 + h^2$ and R1 is the distance from any point to the center of the transducer. Below is the directivity of the transducer.

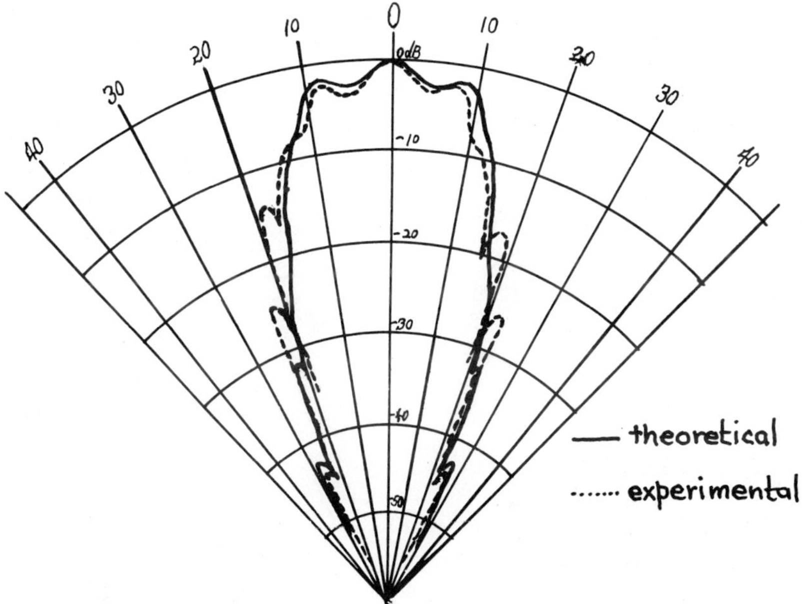

Fig.10. Directivity of the transmitting transducer.

B The Receiving Array

The receiving array is an evenly spaced square array of 64 x 64 elements. Each element was a transducer so carefully selected that its resonance frequency was 339 kHz, thus it would ensure that the characteristics of the transducer could be the same at its operating frequency. After that, the transducers were put into place and the space between transducers was filled with sound absorbing material so that the acoustic coupling between adjacent transducers was reduced to a minimum. Then the whole array was carefully measured for its sensitivity and phase for each element. Any element whose sensitivity and phase deviated too much from the mean was replaced. Table 1 gives the main parameters of the receiving array thus made.

C Preliminary Experiments

It is quite easy to measure the angular resolution and the angular field of view of an imaging system simply by the imaging of two rods put at an angle with each other. Measuring the place where the two rods are just separable will give the necessary information for calculating the angular resolution, and measuring the separation of two adjacent orders of the recon-

Table 1 Main Parameters of the Array

Number of elements	4096
Aperture	760.5mm
Average sensitivity	-105.5 db (V/μbar)
Deviation of sensitivity	± 0.45 db
Deviation of phase	1° 29'
Acoustic coupling	-36 db
Electrical coupling	-60 db

structed images will give information for calculating the angular field of view. In our case, the angular resolution thus calculated is 21' and the angular field of view 23°4'. These values agree quite well with those calculated from the aperture and the separation of individual receiving transducers. Figs. 11 and 12 give two of the holograms and reconstructed images obtained by this system.

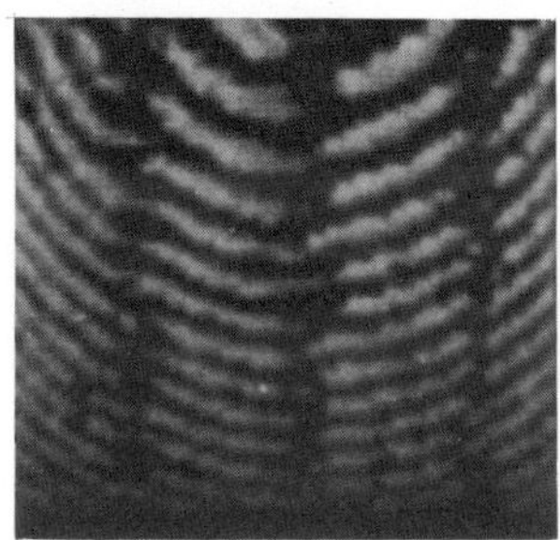

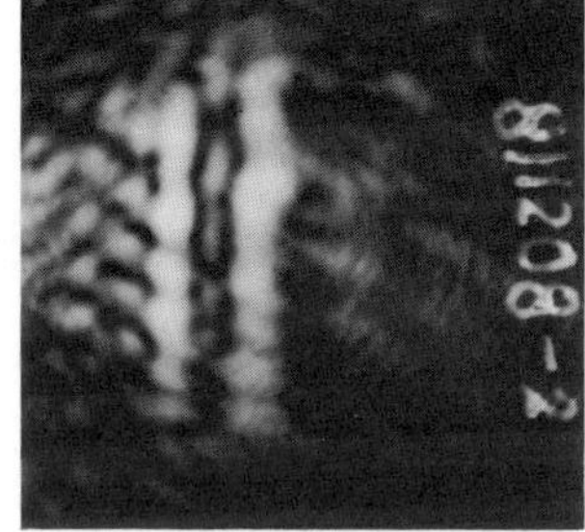

Fig.11. The hologram and its reconstructed image of two parallel cylinders. Distance from the object to the array is 3.7m.

Fig.12. The hologram and its reconstructed image of two cylinders at an angle.

CONCLUSIONS AND ACKNOWLEDGEMENTS

Preliminary tests on our array systems indicate that they have good imaging capability up to 50 meters. The square-array system, when completed, should be able to be used for identification of underwater objects up to a distance of several tens of meters. The scanning linear array systems should have many applications in NDE even for slowly moving objects.

The authors wish to thank W.G. Du, G.L. Lü, K.K. Xü and Z.Q. Wang for their participation in various phases of the experiments and L. Li, X.M. Ho and X. Su for the designing and building of the electronic circuits.

REFERENCES

1. P. Greguss, Ultraschall-hologramme, Research Film, 5:330 (1965)
2. F.L. Thurstone, Ultrasound holography and visual reconstruction, Proc. Symp. Biomed. Eng, 1:12 (1966)
3. R.K. Mueller and N.K. Sheridon, Sound holograms and optical reconstruction, App. Phys. L. 9:328 (1966)
4. G. Wade, M. Wollman and R. Smith, An acoustical holographic system for underwater search, PIEEE 57:2051 (1969)
5. T. Iwasaki, Y. Aoki and M. Suzuki, Ultrasonic holography by one dimensional receiver array, Oyo Buturi 42:679 (1973)
6. Linear-array Group, Linear-array underwater acoustical holographic imaging, Wuhan Inst. Phys. Tech. Report, 1973 No. 2, p.1
7. A.F. Metherell and S. Spinak, Acoustical holography of nonexistant wavefronts detected at a single point in space, App.Phys. Letters 13:22 (1968)
8. Z.Q. Zhou, Z.Q. Wang, S.W. Tung and Z. Xie, The scanning holographic imaging of moving objects using a linear array, Acta Acustica, 1982 in press.
9. D.J. Zhang and L.J. Xia, Theoretical calculation and experimental research of directivity characteristics for convex sperical face transducers, Wuhan Inst. Phys. Tech. Report. 1980 No. 2.

ACOUSTIC DETERMINATION OF SUB-BOTTOM DENSITY PROFILES USING A PARAMETRIC SOUND SOURCE

L.F. van der Wal[1], D.Ph. Schmidt[1] and A.J. Berkhout[2]

[1]Institute of Applied Physics, Delft, The Netherlands

[2]Department of Applied Physics, University of Technology, Delft, The Netherlands

ABSTRACT

Pulse-echo techniques may be well suited to acquire continuous recordings of sub-bottom density profiles. To obtain an accurate estimate of the sub-bottom reflectivity function, an acoustic system should have a broad frequency spectrum for adequate depth-resolution (a), sufficient energy at low frequencies for a pre-specified depth-penetration (b) and a narrow beamwidth to minimize the influence of scattering noise (c).
In this paper the use of a non-linear (parametric) sound source to meet these requirements is discussed. The system covers a frequency band of 2,5 - 22,5 kHz and has a beamwidth < 6°. To allow for calibrated recordings the system is used in combination with spot measurements, using a nuclear density gauge.
Main emphasis will be given to the processing of the raw acoustic data. Averaging, inverse filtering and correlation with nuclear spot measurements will be discussed on the basis of examples, using in-situ recordings.

1. INTRODUCTION

To keep the Dutch harbours of Rotterdam and Europoort accessible to very large tankers, the approach channels should have a guaranteed depth of 22,5 m. Since this depth is far below the natural equilibrium depth, a rapid sediment accumulation occurs. For this reason the Netherlands Rijkswaterstaat (RWS) is obliged to carry out constant maintenance dredging at very high annual costs. Since 1973 several research projects have been started by RWS and the municipality of Rotterdam in order to minimize dredging costs.

In one of these projects, it was shown that, within certain limits, ships can safely sail through layers of unconsolidated sediment. As a result of this research project the depth of the sediment layer with a density of 1200 kg/m^3 is defined as the "nautical depth". During weekly hydrographic surveys nuclear density gauges (1) are used to measure density profiles, from which the nautical depth of the approach channels and the harbours is derived.
A typical result of a nuclear gauge measurement is displayed in Figure 1, showing sediment density as function of depth as well as the gauges inclination.

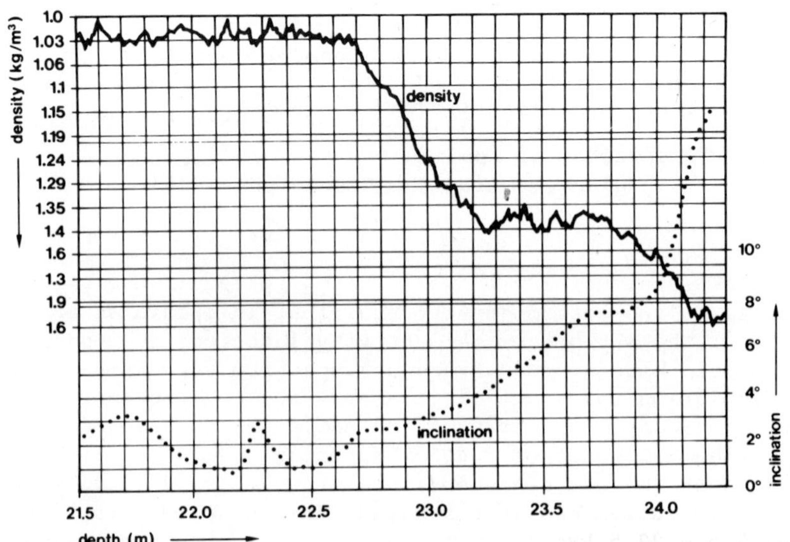

Figure 1: Nuclear gauge density measurement, showing density and gauge inclination as function of depth.

These measurements are used to produce combined echo-depth/density charts.
Thusfar intermediate density values are obtained by linear interpolation (2).
As was pointed out by Kirby and Parker in 1974 (3), acoustic echo-sounding techniques may be used in combination with nuclear density measurements to obtain continuous, calibrated recordings of the sub-bottom structure. An accurate acoustic system may need only few nuclear spot measurements for calibration, thus minimizing time and costs of the hydrographic surveys.
In this paper the outline of such a combined nuclear/acoustic measuring system, which can provide a continuous recording of the bottom density profile, is discussed.

2. ACOUSTIC DENSITY MEASUREMENTS

2.1 Requirements

In 1979 a study was started to investigate the feasibility of the application of pulse-echo techniques in density measurements and to determine the requirements an acoustic density measurement system would have to meet (4,5).
Starting-point in this study was the assumption that the bottom density profile can be calculated from its acoustic reflectivity function. Results of the study confirmed this assumption and showed that - to achieve the best possible estimate of this reflectivity function - the acoustic measurement system should meet the following requirements:

a. A broad frequency spectrum, for adequate depth-resolution.
b. Sufficient energy at low frequencies for pre-specified depth-penetration.
c. A narrow beamwidth to minimize the influence of scattering noise.

Within a single conventional transducer assembly the above mentioned requirements are difficult to meet: a broad spectrum in the low frequency range e.g. conflicts with a narrow beamwidth, when using small transducer sizes.
Parametric sound sources however, which make use of non-linear effects in acoustic wave propagation, may be well suited to fulfil these conditions.

2.2 Parametric Sound Source

The first practical application of non-linear effects in acoustic wave propagation were described in 1957 by Westervelt (6,7).
He suggested that the non-linear interaction of two collimated high-frequency sound waves, will generate a difference frequency sound wave, originating from the common volume of the two primary waves, as if an end-fire array of secondary sources was created within that volume. Theoretical considerations and practical aspects have been reported since by many authors in the literature (8,9,10).
As stated by Muir and Willette (11), the non-linear or so-called "parametric" effect "can be used to advantage in the design of acoustical systems having a requirement for high directivity, in combination with either a low operating frequency or a small transducer size".
Thus a parametric sound source could be well suited to meet the requirements stated in the previous section.
Following successful tests with a prototype system in 1977, it was decided that the core of the acoustic density measurement system should be formed by a parametric sub-bottom profiler, initially developed at the Birmingham University and further developed and manufactured by Ulvertech Limited (Ulverston, Cumbria, U.K.).

The single primary frequency of the system is about 180 kHz, and, using a switchable on/off modulation, the difference frequency band can be adjusted to three different ranges: 2,5 - 7,5 kHz, 7,5 kHz - 12,5 kHz and 17,5 - 22,5 kHz.
The lateral beamwidth in the difference frequency range is < 6°, while no significant side-lobes occur at levels exceeding -30 dB. The complete transducer assembly is mounted on a stabilized platform to minimize the influence of ship movement.

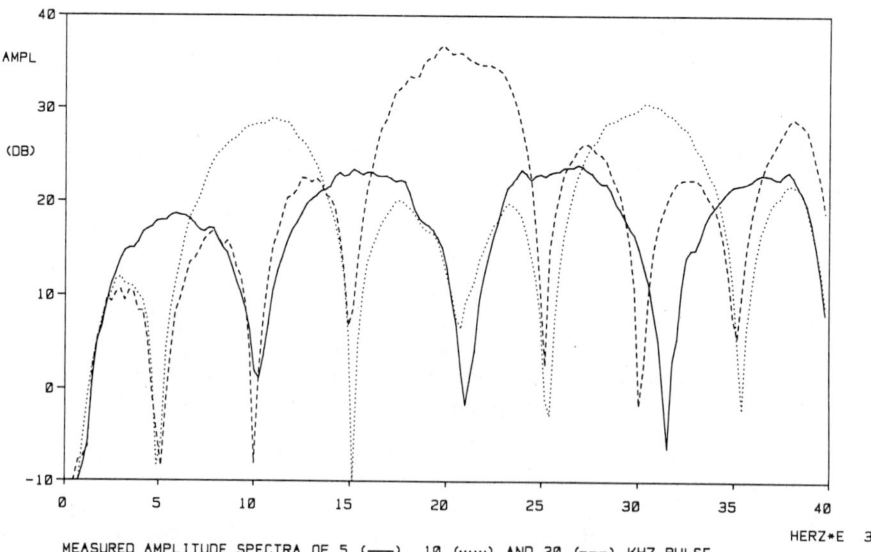

Figure 2: Measured amplitude spectra of selectable difference frequencies (5, 10 and 20 kHz), at 10.0 m from the transducer.

Measured amplitude spectra of the selectable difference frequencies are shown in Figure 2. The acoustic signals were recorded in a waterbasin (12.0 x 6.0 x 5.0 m) at 10.0 m from the transducer, using a B&K hydrophone. Due to the special type of modulation of the primary frequency, uneven harmonics of the difference frequencies are generated as well. The measured spectra are in good agreement with theory, as can be seen from Figure 3. Figure 3 shows the measured amplitude spectrum of the 10 kHz difference frequency (note the 30 kHz band is present as well), together with a one-dimensional computer simulation based on the quasi-linear approach of Westervelt (12). Both spectra show very good agreement.

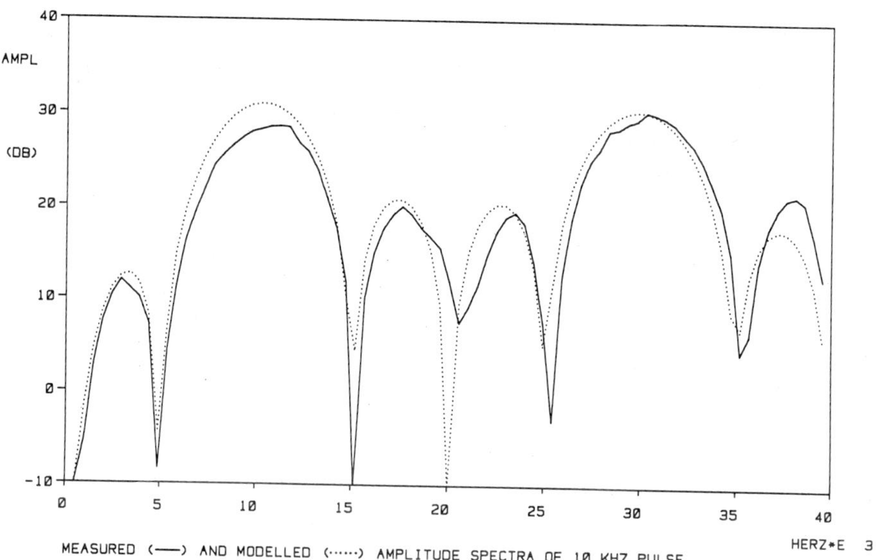

Figure 3: Measured and modelled amplitude spectra of 10 kHz difference frequency, at 10.0 m from the transducer.

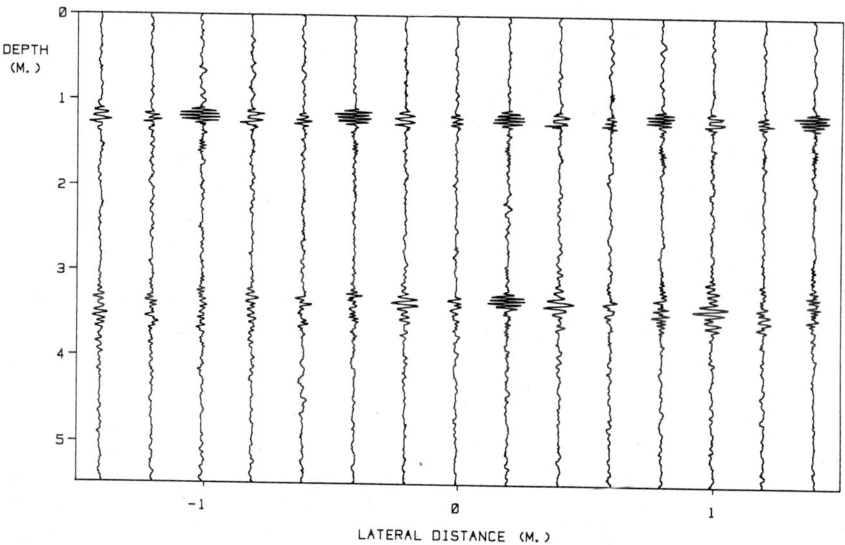

Figure 4: Part of a recorded data set, showing 5, 10 and 20 kHz difference frequency registrations, after digitization.

2.3 Data Acquisition and Processing

At this stage of the research project pulse-echo signals are recorded on an analog recorder and processed off-line on a mini-computer.
The different frequency bands (5, 10 and 20 kHz) are transmitted sequentially and the first processing step consists of an A/D-conversion.
Figure 4 shows part of a data set after digitization. Data were recorded sailing over a distinct silt layer with a speed of ca. 9,0 km/h. At a pulse repetition rate of 12,5 Hz (50,0 m depth range), the lateral distance between successive recordings equals ca. 0,2 m.
After digitization the recorded echo signals are filtered, using a band-pass filter with a bandwidth of 5,0 kHz. Next the echo signals are corrected for "heave"-movements of the ship, using the high frequency (180 kHz) bottom trigger. Figure 5 shows the data set of Figure 4, after filtering and heave correction.

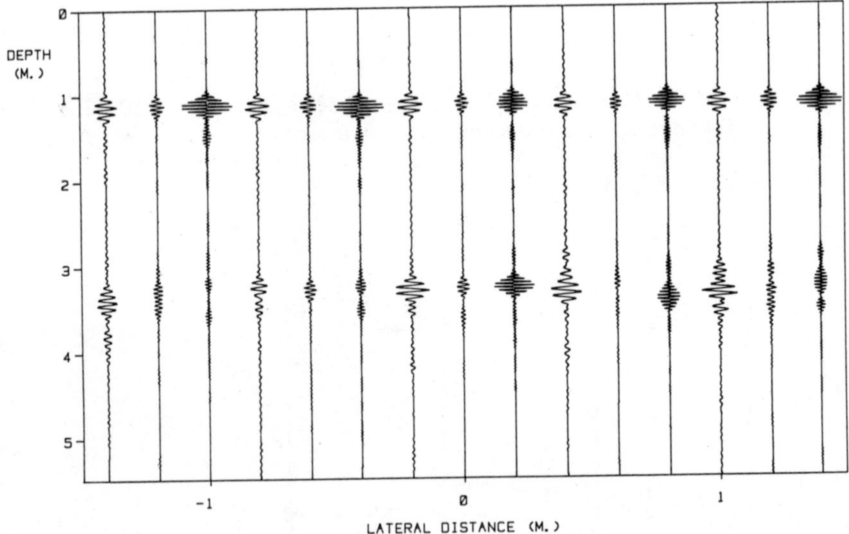

Figure 5: Data set of Figure 4, shown after digital filtering and "heave"-correction.

Figure 5 shows 10, 15 and 20 kHz signals, instead of the expected 5, 10 and 20 kHz signals. The 5 kHz difference frequency showed a very low signal to noise ratio under sailing conditions. Although there is no clear explanation for this loss in signal strength, it was decided to use the first uneven harmonic frequency band, i.e. the 15 kHz band, instead (see also Figure 2).

The third step in the data processing consists of a so-called "spectral equalization" of the data of interest. The process of spectral equalization is explained from Figure 6.

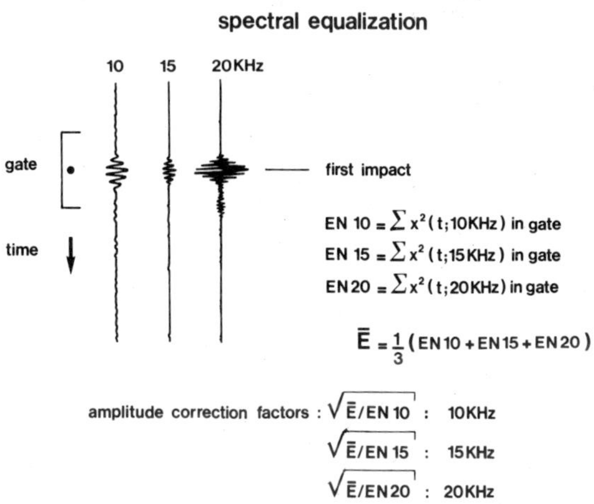

Figure 6: Schematic illustration showing the basic principle of spectral equalization.

Starting at the first silt impact a time window (gate) is applied and the energy within is calculated for all three frequency bands, i.e. EN 10, EN 15 and EN 20.
With E = 1/3 (EN 10 + EN 15 + EN 20), the following apodization factors are applied to the data samples in the successive recordings:

$$10 \text{ kHz} : \sqrt{E/EN\ 10}\ ,$$

$$15 \text{ kHz} : \sqrt{E/EN\ 15}\ ,$$

$$20 \text{ kHz} : \sqrt{E/EN\ 20}\ .$$

With respect to the first silt impact, these apodization factors may be regarded as a correction of differences in transmitted acoustic energy and differential losses (absorption, spherical spreading) within the water situated above the silt layer. To all following data samples - the gate is shifted from sample to sample in the time-direction - the apodization expresses a compensation for relative differences in absorption.

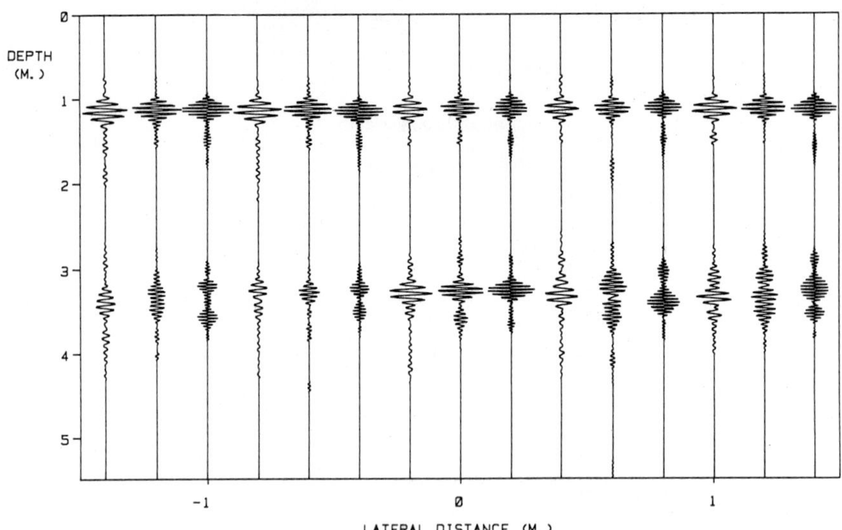

Figure 7: Data set of Figure 5, shown after spectral equalization.

Figure 7 shows the same data set as shown in Figure 5, but after spectral equalization.

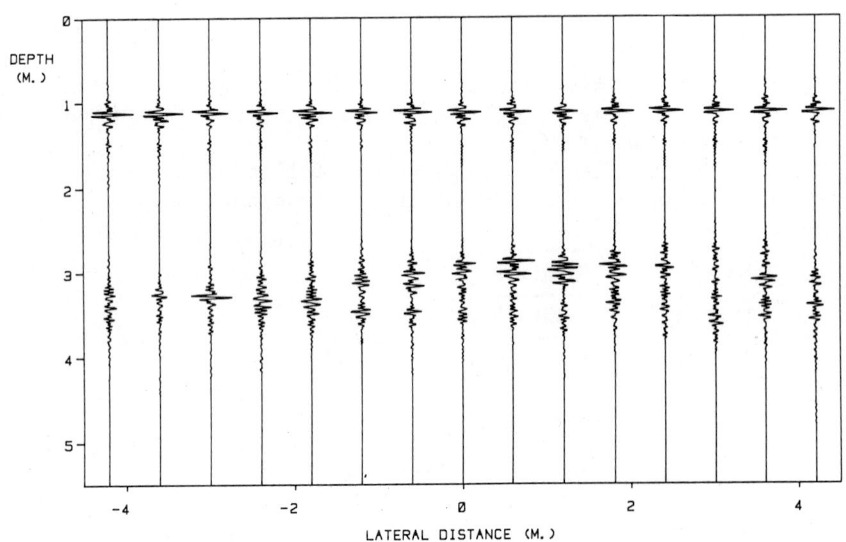

Figure 8: Data set showing r.f.-registrations, after combination of three different frequency bands (10, 15 and 20 kHz).

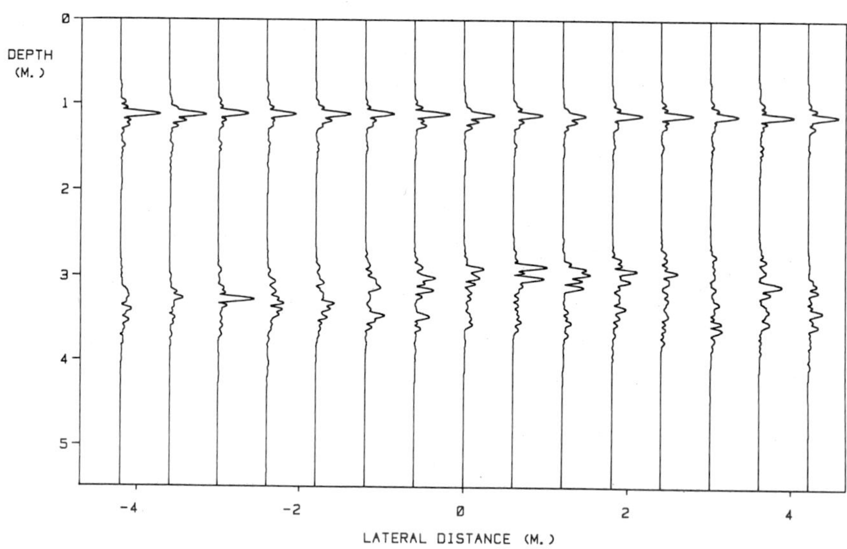

Figure 9: Data set of Figure 8, shown after envelope detection.

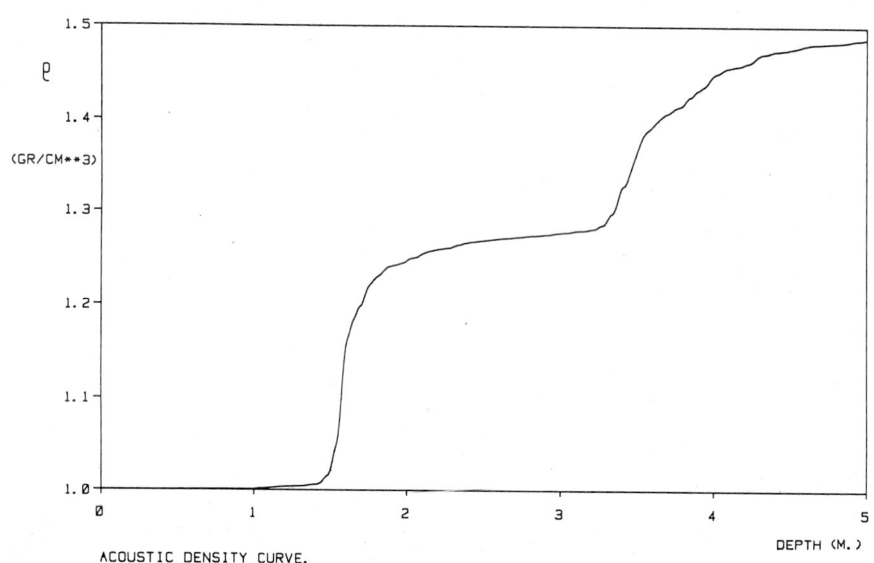

Figure 10: Acoustic density curve, obtained by integration of envelope detected registrations shown in Figure 9.

In the final stages of the processing line the different frequency bands are summed, as shown in Figure 8, and envelope detected, as shown in Figure 9.
Data shown in Figure 8 represent a r.f.-estimate, of the reflectivity function of the silt layer. Integrating the envelope detected data of Figure 9 results in an estimate of the density variations occuring within the silt layer (4), the so-called "acoustic density curves". A typical acoustic density curve is shown in Figure 10.
To be able to allot calibrated density values to the acoustic curves, they should, at given positions, be correlated with spot measurements from the nuclear density gauge, e.g. using a least squares criterion. Figure 11 shows a nuclear gauge measurement, carried out at the same position where the acoustic density curve shown in Figure 10 was obtained. In spite of the large amount of "jitter" to be seen on the nuclear curve, the similarity with the acoustic curve is evident.

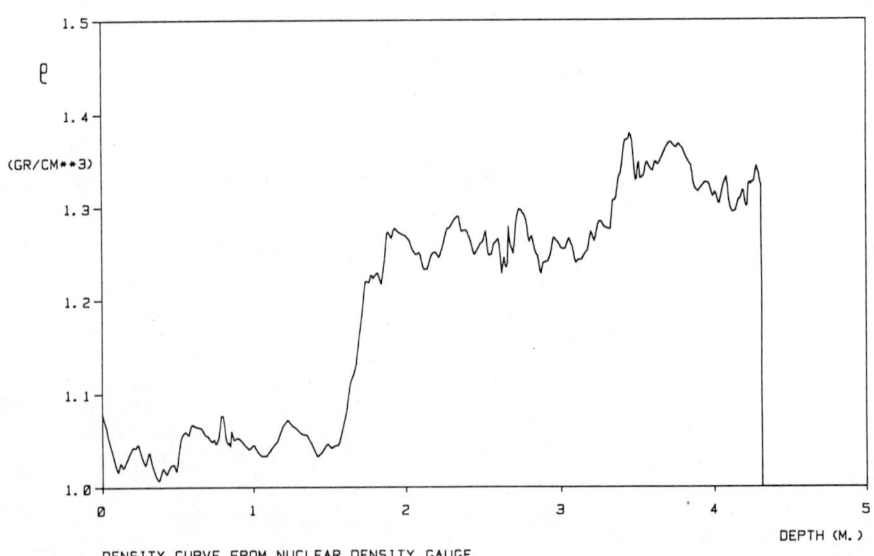

Figure 11: Nuclear gauge density curve. Measurement was carried out at exactly the same position as the acoustic measurement shown in Figure 10.

3. CONCLUSIONS

In this paper an acoustic density measurement system is described, to be used under sailing conditions in the Europoort area. The system is based on a parametric sub-bottom profiler (Ulvertech Ltd., U.K.), which is used for data acquisition.

The acquired acoustic recordings show very good depth-resolution, sufficient depth-penetration in the sediment and a narrow beamwidth. The acoustic density curves are in good agreement with nuclear gauge measurements, although the amount of data processing is quite large.
Thusfar data processing is done off-line in the laboratory on a mini-computer. Work is in progress to develop an on-line processing unit, which enables the generation of acoustic density curves aboard ship in real time. Following data reduction and storage, the acoustic density curves will be correlated with nuclear spot measurements in a processing centre ashore, resulting in continuous, calibrated lateral recordings of the sub-bottom density profile.

ACKNOWLEDGEMENTS

The authors would like to thank their colleagues of the Netherlands Rijkswaterstaat for their indispensable assistance during the measurements in the Europoort area.
This work is supported by the Hydrographic Survey Division, Lower Rhine Directorate of the Netherlands Rijkswaterstaat.

REFERENCES

1. Parker, W.R., Sills, G.C. and Paske, R.E.A. "In situ Nuclear Density Measurements in Dredging Practice and Control", Symposium on Dredging Technology, 17-19 Sept. 1975, published by BHRA Fluid Engineering, Cranfield, Bedford, England.

2. Bakker, D.J. "Density Measurements in conjunction with Echo-Sounding", The Hydrographic Journal, no. 14, pp. 21-27, April 1979.

3. Kirby, R. and Parker, W.R. "Seabed Density Measurements related to Echo-Sounder Records", The Dock and Harbour Authority, pp. 423-424, March 1974.

4. Schmidt, D.Ph. and Janssen H.C. "Acoustic Silt-Density Measurements, a Feasibility Study", Institute of Applied Physics Tech.rep.no. 910-281, July 1980 (in Dutch).

5. Schmidt, D.Ph., Janssen, H.C. and Berkhout, A.J.B. "Prediction of Sediment Density Profiles by means of Echo-Sounding", ULTRASONICS INTERNATIONAL 81, pp. 165-170, published by IPC Science and Technology Press Limited, Guildford,Surrey, England, 1981.

6. Westervelt, P.J. "Scattering of Sound by Sound", J.Acoust.Soc. Am., Vol. 29, pp. 199-203, Febr. 1957.

7. Westervelt, P.J. "Scattering of Sound by Sound", J.Acoust.Soc. Am., Vol. 29, pp. 934-935, August 1957.

8. Bellin, J.L.S. and Berger, R.T. "Experimental Investigation of an End-Fire Array", J.Acoust.Soc.Am., Vol. 34, pp. 1051-1054, August 1962.

9. Berktay, H.O. "Possible Exploitation of Non-Linear Acoustics in Underwater Transmitting Applications", J. Sound Vib., Vol. 2, no. 4, pp. 435-461, Oct. 1965.

10. Smith, B.V. "An Experimental Study of a Parametric End-Fire Array", J. Sound Vib., Vol. 14, No. 1, pp. 7-21, Jan. 1971.

11. Muir, T.G. and Willette, J.G. "Parametric Acoustic Transmitting Arrays". J.Acoust.Soc.Am., Vol. 52, pp. 1481-1486, 1972.

12. Westervelt, P.J. "Parametric Acoustic Array". J.Acoust.Soc.Am. Vol. 35, pp. 535-537, April 1963.

ABNORMAL BACKSCATTERING OFF LOW ROUGHNESS SURFACE OF METALLIC OBJECT IMMERSED IN WATER *

J.F. Gelly - C. Maerfeld

Thomson-CSF DASM/DTAS
BP 53
06801 Cagnes/Mer - France

P. Maguer

GESMA
DCAN
29240 Brest - France

ABSTRACT

A C-Scanned Acoustic Camera has been made and was described previously [1] . Using it, we were able to visualise almost polished metallic object where diffusion echoes (coming from backscattering) were more than 60 dB below the specular echo. However, this small signals appear to be higher than predicted.

An experimental analysis is described here leading to the demonstration that fringing wave and specially Rayleigh wave reflections are responsible for this phenomenon.

Acoustical imaging has been proposed for the observation of metallic immersed objects in turbid water such as exist in a busy harbour, where optical cameras become ineffective. Those objects, man made, have geometrical shapes with a surface corrugation amplitude far below one millimeter and thus small, compared with the wavelength of the highest frequency which can be used (a few MHz).

Therefore, a strong echo can be observed for specular object orientation giving mirror-like images, where you cannot recognise the shape of the object. Furthermore, backscattering echoes which allow for good images are very low.

In order to evaluate the possibility of making C Images (like optical) we have built an experimental camera as shown in Fig. 1 and 2 [1] . It is made of a plexiglass lens and a 64 element linear array, electronically scanned in the X direction.

*Work supported by DCAN/GESMA France.

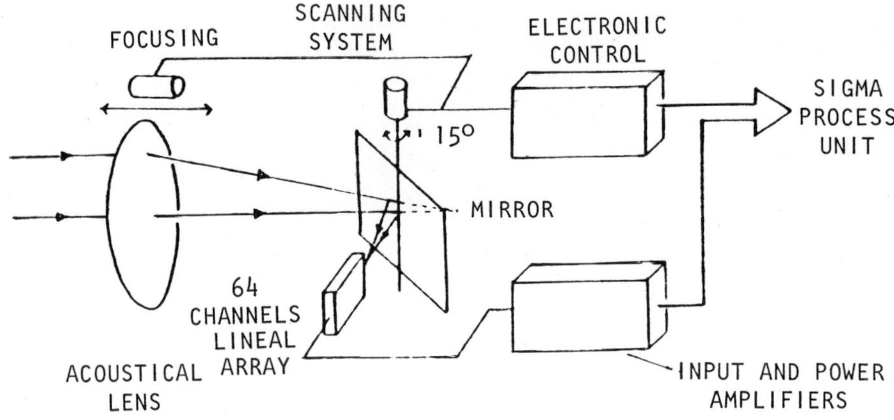

Fig. 1 Schematic of the immersed parts of the acoustical camera

Fig. 2 View of the acoustical camera

Scanning in the Y direction is performed by a rotating mirror. The depth of field is limited by time gating. Acoustic waves are launched either by the array or by external projectors. In both cases, the wave reflected by the object has the same direction as the incident wave. It is the pure surface backscattering of the object which is visualised. The working frequency is 2 MHz. The angular resolution is 0.2° giving a resolution of 1.7 cm at a distance of 5 m.

We describe here the observation made while imaging almost polished objects. We found that the backscattering energy in given directions is higher than predicted by the more widely used theories [2, 3].

Following Beckman theory [3], the expression of the intensity of the echo of an incident plane wave backscattered in the same direction by a corrugated plate is :

$$\langle I_s \rangle = e^{-g} \left\{ f^2(\theta) + \frac{2\pi L^2 g}{S \cos\theta} \, \mathrm{exp} \, \frac{-k^2 L^2}{2} \right\} \quad (1)$$

where $g = (2kh \cos\theta)^2$

θ is the incidence angle, k the wave number.
h is the RMS height of the defects and L the correlation length.
S is the insonified surface.
Intensity of the backscattered wave is normalised to the maximum of the specular reflected wave for a plane surface.
$f(\theta)$ represents the specular echo and is negligible as soon as $\theta \neq 0$.
The second terms give the incoherent diffused field backscattered by the defects of the surface.

OBSERVATION OF SPHERES OF DIFFERENT ROUGHNESS

Different aluminium spherical targets were built - the surface roughness being between 0.1 μ (polished surface) and 20 μ.

The acoustic image of the 20 μ rugosity sphere of diameter 200 mm is given in fig. 3 when excitation was provided by the array. Central specular reflection is evident. But the whole sphere is clearly visible with an unexpectedly high level of backscattering specially at the edge.

Fig. 4 represents the video central line in the same case. Peaks (very reproducible) are visible and not predicted by formula (1). Those peaks are mainly responsible for the good image quality obtained.

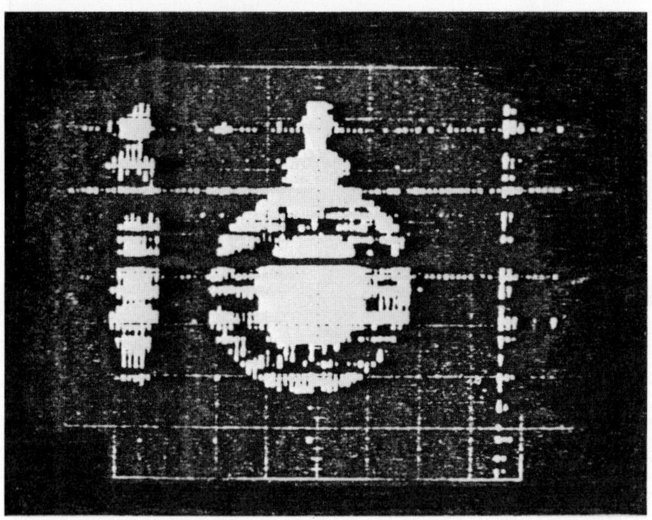

Fig. 3 Acoustical image of a 200 mm diameter sphere at a range of 2.5. m (mean roughness of the surface is 20 μ)

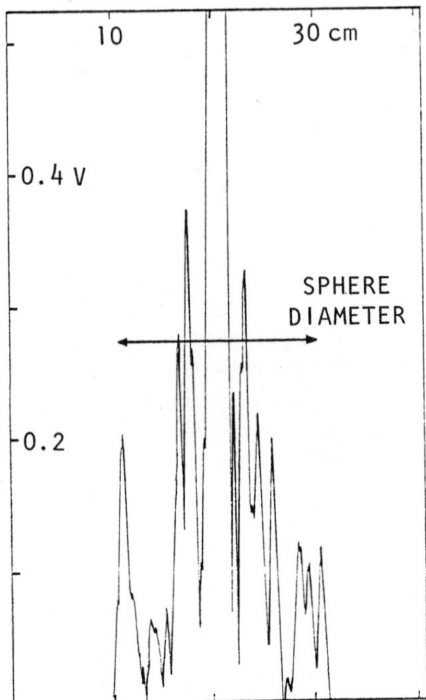

Fig. 4 Record of the field amplitude in the image plane of the camera along the diameter of the sphere imaged on fig. 3.

Fig. 5 shows the same central line but for a 0.1 μ polished sphere. It is now impossible to recognize the object. The edges have vanished. But lateral peaks at the same place as for the first experiment are still visible. We have to point out that those peaks are neither multiple reflection inside the sphere, nor creeping waves turning around the object since they are not delayed in time.

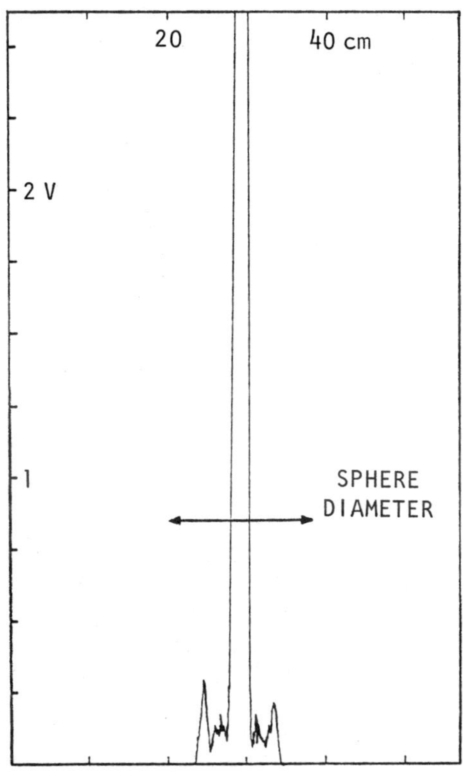

Fig. 5 Same as fig. 4 for a 200 mm polished sphere at a range of 4 m.

PLATE OBSERVATION

For more accurate experiments, plane plates 20 mm thick, 200 x 200 mm wide, were investigated.

Plate n° 1 is polished (roughness 0.5 μ crest to crest) and taken as reference.
Plates n° 2,3,4 were corrugated by sand projection according to :

- plate	roughness c.c.
2	12 - 15 μ
3	15 - 20 μ
4	10 - 40 μ

The camera was used differently in a "sonar mode" i.e. only one element was used for transmission and reception of the 2 MHz pulse 8 μs length. The plate is 4 m. distant from the lens and can be rotated by an angle α (for $\alpha = 0$, the plate is parallel to the lens). At a given α, the plate surface is scanned by rotating the camera mirror. The sensitivity is such that diffused reflexions as low as 90 dB below the specular echo can be observed.

a) finite plate

Fig. 6 shows the results for different angle α. For $\alpha = 0$, a strong echo is observed, small echoes appear coming from the edge of the plate. At $\alpha = 5°$, only edge echoes are observed. The edge closer to the camera giving a stronger signal. At 10°, the situation is reversed : the stronger signal comes from the further edge. Further more, signals appear in the middle of the plate. These signals are less important for $\alpha = 15°$ and reappear for $\alpha = 25°$. At 30°, the far edge gives a very strong echo. At $\alpha = 35°$, the signals vanish and only small reflexion from the plate edges are seen.

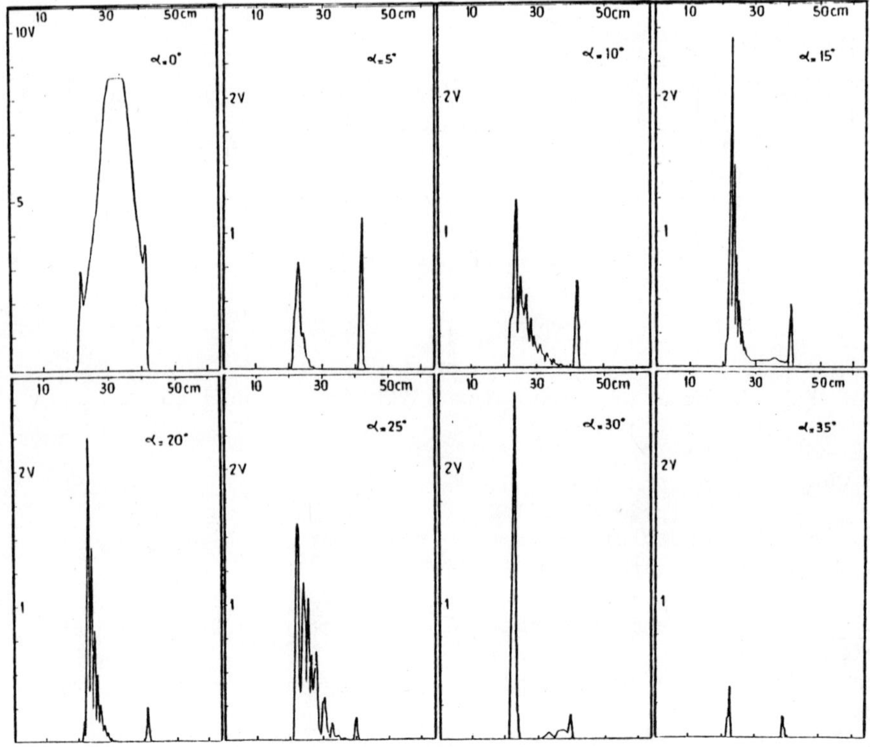

Fig. 6 Field backscattered by the plate 1 at 2.8 m for different values of incidence α (record of the detected signal versus observation point on the object plane).

The interpretation is quite simple. The abnormal backscattering observed at 10° and 30° occurs for incidence angle where fringing waves can be emitted. i.e. for the critical angles for longitudinal and shear waves obeying the equation :

$$k \sin\alpha = k_f$$

$k = \frac{\omega}{c}$ incident wave number, c=sound velocity in water,

k_f : fringing wave number ; $k_f = \frac{\omega}{v_l}$, $\frac{\omega}{v_s}$.

v_l being the longitudinal wave velocity, v_s being the shear or Rayleigh wave one.

In aluminium, those critical angles are respectively 12°, 29° or 31°. As shown in fig. 7, a part of the incident energy is converted into fringing waves which propagate to the edge and then is reflected. But the propagation is leaky. The wave re-radiates bulk waves in water and thus a part of the energy reaching the observation point is reemitted toward the camera. The increase of the amplitude as the observation point gets closer to the far edge is a measure of the fringing wave decay. The longitudinal wave decay is quite low while the shear or Rayleigh wave decay is strong and thus, for the right incident angle, the edge echo is very strong.

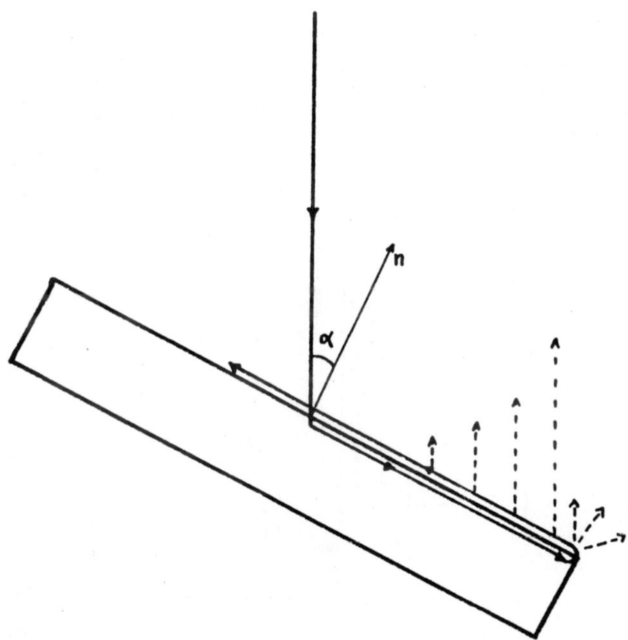

Fig. 7 Mechanism of excitation, reflexion and re-radiation of a leaky surface wave (fringing wave).

b) infinite plate

To see that would be the backscattering of an infinite plate, we gated the observation at a time corresponding to the reflexion on the analysed surface. The resulting signal, far lower, as previously observed, appears in fig. 8 at $\alpha = 30°$. Those echoes are 60 dB below the specular echo. Its variation, versus the incidence angle α is shown on fig. 9, and is compared with the polished plate. In all cases, a strong maximum appears for $\alpha = 30°$ a smaller one exists around 15°/20°.

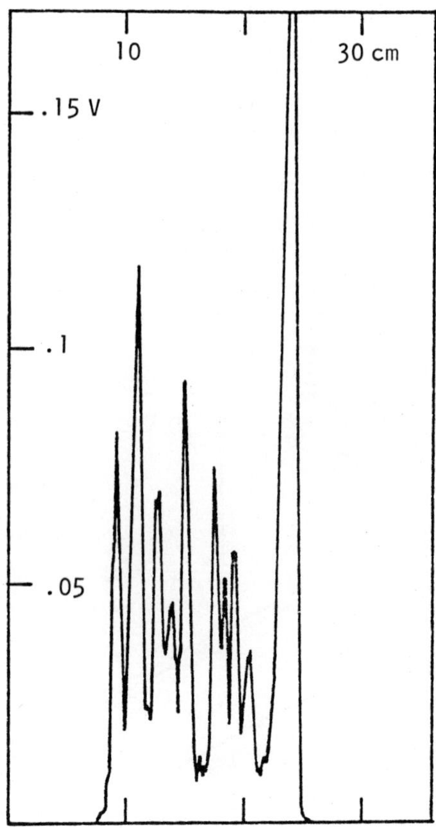

Fig. 8 Same as fig. 6 for $\alpha = 30°$ on plate 2 showing the low level backscattering.

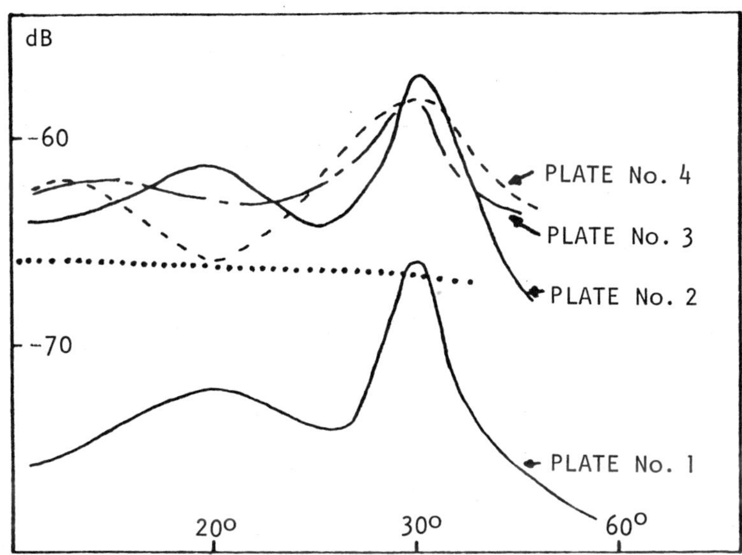

Fig. 9 Comparison of diffuse backscattering level v s. incidence angle for plate 1,2,3,4 (0 dB reference is the specular echo), dotted line: Beckman's theory for h = L = 15 μ.

This behavior is not explained by Beckmann theory. As the maximum occurs for the critical angle 30°, we deduced that the phenomenon is similar to the previous case of finite plate. i.e. it is the reflected Rayleigh wave which radiates a wave in the same direction as the incident wave (fig. 10).

The reflexion of Rayleigh wave on corrugated surfaces is studied else where. This effect explains the peaks observed in the case of the sphere.

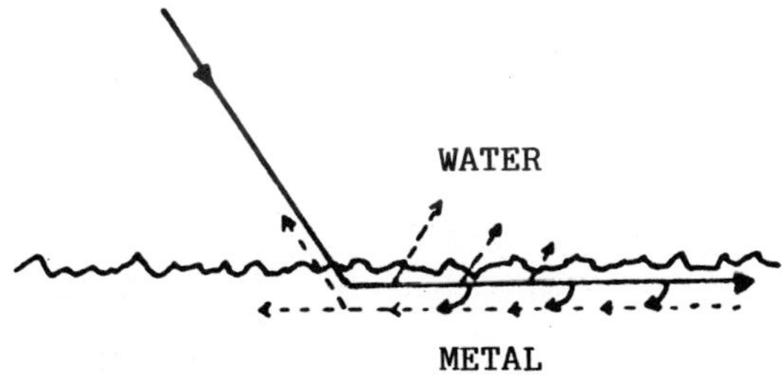

Fig. 10 Schematic representation of the origin of the abnormal backscattering by the rough surface observed near a critical angle.

CONCLUSION

The backscattering of acoustic waves on low roughness metallic surface is higher than predicted by the Eckart theory.
Fringing waves excited and diffused by surface corrugations seem to be responsible of these observations.

As a fortunate consequence, we have been able to visualize rather polished metallic object immerged in water with a C scanned camera.

ACKNOWLEDGMENT

We would like to thank Mr L. WOLNERMAN for his technical assistance.

REFERENCES

1. "An Underwater Focused Acoustic Imaging System" - P. Maguer, J.F. Gelly, C. Maerfeld and G. Grall - Acoustic Imaging - Vol. 10 (1982) p 607
2. "The Scattering of Sound from the Sea Surface" - C. Eckart - J.A.S.A. May 1953 p 566
3. "The Scattering of Electromagnetic Waves from Rough Surfaces" P. Beckmann - Electromagnetic Waves - vol 4 - Pergamon Press - (1963)

LIST OF PARTICIPANTS

R. Abdel-Aal
University of Strathclyde
Dept. of Electronic Science
Royal College Building
George Street
Glasgow G1 1XW, UK

M.F. Adams
University of Sheffield
Dept. of Electrocnic &
Electrical Engineering
Mappin Street
Sheffield S1 3JD, UK

R.C. Addison
Rockwell International Science
Center
1049 Camino Dos Rios
PO Box 1085
Thousand Oaks
California 91360, USA

P.M. Alais
Laboratoire de Mecanique
Physique
2 place de la Gare de Ceinture
78210 Saint-Cyr-l'Ecole
France

E. Alasaarela
University of Oulo
Dept. of Electrical Engineering
90570 Oulo 57
Finland

A.P. Anderson
University of Sheffield
Dept. of Electronic & Electrical
Engineering
Mappin Street
Sheffield S1 3JD, UK

J.A. Archer-Hall
University of Aston
Birmingham

V. Arat
University of Houston
Electrical Engineering Dept.
Houston TX 77004, USA

E.A. Ash
University College London
Dept. of Electrical and
Electronic Engineering
Torrington Place
London WC1E 7JE, UK

M. Auphan
Laboratoires d'Electronique
et de Physique Appliquee
3 avenue Descartes
94450 Limeil-Brevannes, France

A. Ayoola
King's College London
Electronic & Electrical
Engineering Dept.
Strand, London WC2R 2LS, UK

S. Bailey
Ultrasonics
PO Box 63
Westbury House
Bury Street, Guildford
Surrey GU2 5BH, UK

J.C. Bamber
Institute of Cancer Research
Physics Division
Clifton Avenue
Sutton, Surrey SM2 5PX, UK

R.H.T. Bates
University of Canterbury
Electrical Engineering Dept.
Christchurch, New Zealand

M Bele
CGR Ultrasonics
52 Bd Gallieni
92133 Issy-les-Moulineaux
France

M. Berson
Lab. Biophysique Medicale Tours
2 Bis Boulevard Tonnelle
37032 Tours, Cedex, France

J.M. Blackledge
Queen Elizabeth College
Dept. of Physics
Campden Hill Road
London W8 7AH, UK

L. Bond
University College London
Dept. of Electronic & Electrical Engineering
Torrington Place
London WC1E 7JE, UK

F.H. Breimesser
Siemens AG
ZFE TPH 42, Postfach 3240
D8520-Erlangen,
Federal Republic of Germany

H. Brettel
GSF
Hornstr. 8
D-8000 Munchen 40,
Federal Republic of Germany

G.A.D. Briggs
University of Oxford
Department of Metallurgy
Parks Road, Oxford OX1 3PH, UK

Y. Brun
Enertec Schlumberger
26 Rue de la Cavee
92 Clamart, France

C. Bruneel
University de Valenciennes
59326 Valenciennes, Cedex, France

S.F. Burch
UK Atomic Energy Authority
AERE, Harwell, Oxfordshire
OX11 ORA, UK

N. Burton
University College London
Torrington Place
London WC1E 7JE, UK

N.L. Bush
Institute of Cancer Research
Clifton Avenue,
Sutton, Surrey, UK

D.S. Cairns
Diagnostic Sonar Ltd.
Baird Road, Kirkton Campus
Livingston EH54 7BX,
Scotland

S.J. Calil
King's College Hospital Medical School,
Dept. of Biomedical Engineering
Dulwich Hospital
East Dulwich Grove,
London SE22, UK

D. Carpenter
Ultrasonics Institute
5 Hickston Road
Millers Point, Sydney 2000
Australia

P.L. Carson
University of Michigan
Dept. of Radiology (Box 13)
University Hospital
Ann Arbor MI 48109, USA

A Carini
IROE-CNR
Via Panciatichi 64
50127 Florence, Italy

P Cervenka
Laboratoire de Mecanique Physique
2 place de la Gare de Ceinture
78210 Saint-Cyr-l'Ecole
France

J.A. Champion
National Physical Laboratory
Div. of Materials Appls.
Teddington, Middx TW11 OLW, UK

S.E. Childs
CERL
Kelvin Avenue
Leatherhead, Surrey KT22 7SE, UK

S Christie
University of Exeter
Physics Dept.
Stocker Road,
Exeter, Devon, EX4 4QL, UK

N. Chubachi
Tohoku University
Dept. of Electrical Engineering
Faculty of Engineering
Sendai 980, Japan

R.S.C. Cobbold
University of Toronto
Institute of Biomedical Engineering
Toronto M5S 1A4, Ontario, Canada

M. Colline
RATP, C/o Mme Roger
Service des Relations
Exterieures-33 Ter, Quai
Des Grands Augustins
75006 Paris, France

T.E. Constandinou
BICC R & EL
38 Ariel Way, London W12 7DM, UK

A.I. Courtman
Institute of Cancer Research
22 Broom Close
Broom Road, Teddington, Middx, UK

D.I. Crecraft
The Open University
Faculty of Technology
Milton Keynes MK7 6AA, UK

D.L. Cunningham
Sigma Research Inc.
2950 Gw. Way, Richland WA 99352
USA

C Curry
Advanced Technology Laboratories
Bellevue, Washington, USA

K. Dalland
Norwegian Underwater Technology Centre
Gravdalsveien 255, N 3034
YTRE Laksevag/Bergen, Norway

G. Dasani
Picker International
East Lane, Wembley, Middx. UK

G.J.S. Davies
Open University, Seasons
Gardenfield Lane,
Berkhamsted, Herts HP4 2NN, UK

M.P. de Graff
Delft University of Technology
Dept. of Applied Physics
Group of Acoustics
Delft, the Netherlands

N. de Jong
Erasmus Universiteit
Rotterdam, Postbus 1738
the Netherlands.

P.N. Dennigh
2 Grange Hill Road
Kings Norton
Birmingham B38 8RG , UK

D de Vries
Delft University of Technology
Dept. of Applied Physics
Group of Acoustics,
Delft, the Netherlands

R.J. Dickinson
GEC Hirst Research Centre
c/o MRC Cyclotron Unit
Ducane Road, London W12 OHS, UK

E. D'Ottari
Isbibuto Acusbaca C.N.R.
Via Cassia 1216, Rome, Italy

P. Durouchoux
DRET, 26 bd. Victor
75996 Paris, Armee, France

T.S. Durrani
University of Strathclyde
Dept. of Electronic Science
Royal College Building
George Street, Glasgow G1 1XW
Scotland

H. Ermert
Universitat Erlangen-Nurnberg
Institut fur Hochfrequenztechnik
Cauerstr 9, D 8520 Erlangen
Federal Republic of Germany

J.M. Evans
Bristol General Hospital
Dept. of Medical Physics
Ultrasonics Unit,
Guinea Street, Bristol BS1 6SY
UK

G. Faber
TH-Delft
Lab. Voor Techn. Natuurkunde
PO Box 5046, Delft
the Netherlands

F. Faridian
University College London
Electronic Engineering
Torrington Place
London WC1E 7JE, UK

G.W. Farnell
McGill University
Dept. of Electrical Engineering
817 Sherbrooke Street West
Montreal, Canada H3A 2K6

M. Fink
Laboratoires d'Electronique
et de Physique Appliquee
3 avenue Descartes - BP 15,
94450 Limeil-Brevannes, France

G.L. Fitzpatrick
Spectron Development Labs. Inc.
1010 Industry Drive,
Seattle, WA 98188, USA

S. Flax
c/o Gillain
International General Electric
Co. of New York Ltd.
(Medical Systems)
Jubilee House,
120 Blyth Road, Hayes, Middx. UK

D.R. Fox
Thorn EMI
Central Research Laboratories
Sonics Dept.
Schoenberg House,
Trevor Road, Hayes UB3 1HH, UK

L. Frizzell
University of Illinois
Bioacoustics Research Lab.
1406 West Green Street
Urbana, Illinois 61801, USA

B.Froelich
Enertec Schlumberger
26 Rue de la Cavee
92 Clamart, France

R.A. Giblim
University College London
Dept. of Electronic &
Electrical Engineering
Torrington Place
London WC1, UK

R.P. Glover
Picker International
East Lane, Wembley
Middx. UK

B Granz
Forschungslaboratorien der
Siemens AG
P-Gossenstrasse 100
D-8520 Erlangen
Federal Republic of Germany

D.T. Green
Ministry of Defence
Perme Westcott
Aylesbury, Bucks MP18 ONZ, UK

J.F. Greenleaf
Mayo Foundation
Dept. of Physiology & Biophysics
Rochester, MN 55905, USA

M.J. Greenwood
Ferranti Instrumentation Ltd.
St. Mary's Road
Moston, Manchester, UK

G.E. Haines
British Telecom
R.I.S. Research Centre
Martlesham Heath
Ipswich IP5 7RE, UK

T.J. Hall
Institute of Cancer Research
F Block, Clifton Avenue
Sutton, Surrey, UK

M.Halliwell
Bristol General Hospital
Dept.of Medical Physics
Guinea Street
Bristol BS1 6SY, UK

W.V. Harrison
ADR Ultrasound
2626 G. Roosevelt,
Tempe, AZ 85282, USA

S.O. Harrold
Portsmouth Polytechnic
Dept. of Electrical &
Electronic Engineering,
Anglesea Building,
Anglesea Road, Portsmouth PO1 3DJ
UK

D. Dauden
LPMO-CNRS
32 Ave de l'Observatoire
25000 Besancon, France

G. Hayward
University of Strathclyde
Dept. of Electronic Science
& Telecommuniccations,
Royal College, George Street
Glasgow GI, UK

C.R. Hill
Institute of Cancer Research
Physics Division
Clifton Avenue, Sutton, Surrey
SM2 5PX, UK

D. Hillar
Universitat Erlangen-Nurnberg
Institut fur Hochfrequenztechnik
Cauerstr 9, D-8520 Erlangen
Federal Republic of Germany

H. Hoffmann
F Hoffmann-La Roche & Co. AG
Grenzacherstr. 124
Dept. ZFE, CH 4002, Basle
Switzerland

K.R. Holdstock
University of Bath
Dept. of Electrical Engineering
Claverton Down, Bath, Avon, UK

R. Hunik,
Kema NV
Utrechteweg 310
Arnhem, the Netherlands

J.W. Hunt
Ontario Cancer Institute
Dept. of Medical Biophysics
University of Toronto
5000 Sherbourne Street
Toronto, Canada M4X 1K9

L Hutchins
Royal Postgraduate Med. School
Dept. of Medical Physics
Hammersmith Hospital
Ducane Road, London W12 OHS, UK

C. Ilett
University of Oxford
Dept. of Metallurgy & Science
of Materials
Parks Road, Oxford OX1 3PH, UK

K.A. Ingebrigtsen
ELAB, 7034 Trondheim-NTH
Norway

K. Ito
Jichi Medical School
Dept. of Clinical Pathology
3311-I Yakushi-Cho
Kawachi-Gun
Tochigi-Pref. Japan

A. Iwata
c/o Prof. J. Dudek
Institut fur Medizinsche
Statistik und Dokumentation
Heinrich-Buff-Ring 44,
D-6300 Giessen
Federal Republic of Germany

E. Jacobsen
NUTEC
Gravolalsv 255
N 5034 Ytre Laksevag, Norway

S. Jasim
King's College, London
Dept. of Electrical Engineering
Strand, London WC2, UK

N. Johnson,
University College London
Dept. of Electrical Engineering
Torrington Place,
London WC1E 7JE, UK

S.A. Johnson
University of Utah
Dept. of Bioengineering
Salt Lake City, Utah 84112, USA

J.P. Jones
University of California, Irvine
Dept. of Radiological Sciences,
Irvine CA 92717, USA

C. Joseph
Thomson-CSF
Route due Conquet
29283 Brest Cedex
France

S..G. Joshi
Marquette University
Dept. of Electrical Engineering
Milwaukee, WI 53233, USA

C. Kasai
Aloka Co. Ltd.
6-22-1 Mure, Mitaka-shi,
Tokyo 181, Japan

M Kaveh
Univeristy of Minnesota
Dept. of Electrical Engineering
123 Church Street SE
Minneapolis MN 55455, USA

T. Konishi
Aloka Co. Ltd.
6-22-1, Mure, Mitaka-City,
Tokyo, Japan

J. Kushibiki
Tohoku University
Dept. of Electrical Engineering
Faculty of Engineering
Sendai 980, Japan

O. Lannuzel
Thomson CSF
Route du Conquet
BP 128, 29283 Brest Cedex, France

D.A. Linkens
University of Sheffield
Dept. of Control Engineering
Mappin Street, Sheffield S13 8P,
UK

P. Lloyd
Ferranti Instrumentation Ltd.
St. Mary's Road, Moston,
Manchester, UK

H. Lasota,
Institut Industriel du Nord
59650 Villeneuve D'Ascq.
Cedex, France

E.G. Le Det
Johns Hopkins University
Applied Physics Laboratory
Johns Hopkins Road, Laurel
MD 20707, USA

D.T.A. Lines
Thorn EMI PLC
Central Research Labs.
Sonics Dept., Trevor Road
Hayes, Middx. UB3 1HH, UK

A. Lovik
U&niversity of Trondheim
ELAB, N-7034 T-NTH, Norway

C.J. Macleod
University of Strathclyde
Dept. of Electronic Science
Royal College Building
George Street, Glasgow G1 1XW

C. Maerfeld
Thomson-CSF
Division Activites Sous-Marines
Chemin des Travails
BP 53
06801 Cagnes-sur-Mer Cedex 01
France

A.E. Marble
Technical University of Nova Scotia
PO Box 1000, Halifax
Nova Scotia, Canada

R. Martinez-Ona
Tecnatom SA
KM 19 CNI
San Sebastian de los Reyes
Madrid, Spain

J.F. McDonald
Rensselaer Polytechnic Institute
JEC 6026, ECSE Dept.
Troy, NY 12181, USA

F.A. McDonald
Southern Methodist University
Physics Dept.
Dallas, Texas 75275, USA

J.A. McKnight
Rm. RD1/116
Risley Nuclear Power
Development Laboratories,
UKAEA, Risley, Warrington
WA3 GAT, UK

P.R. Mesdag
Delft University of Technology
Dept. of Applied Physics
Group of Acoustics
PO Box 5046, 2600 GA Delft
the Netherlands

A. Metherell
South Bay Hospital
816 Emerald Bay, Laguna Beach
California 92651, USA

A.J. Miller
GEC Hirst Research Centre
East Lane, Wembley,
Middlesex HA9 7PP, UK

P. Nauth
Gesellschaft zue Forderung
der Forscungs an de DKD
Aukammallee 33, D-62 Wiesbaden
Federal Republic of Germany

D.K. Nassiri
Institute of Cancer Research
The Royal Marsden Hospital
Downs Road, Sutton, Surrey
SM2 5PT, UK

D. Nicholas
Royal Marsden Hospital
Physics Dept., Downs Road
Sutton, Surrey, UK

M. Nikoonahad
University College London
dept. of Electrical &
Electronic Engineering,
Torrington Place, London
WC1E 7JE, UK

B. Nongaillard
Universite de Valenciennes
Laboratoire OAE
59326 Valenciennes, Cedex, France

C.P. Oates
Newcastle General Hospital
Dept. of Medical Physics
Westgate Road,
Newcastle upon Tyne, UK

D. O'Brien
University of Illinois
Bioacoustics Research Lab.
Dept. of Electrical Engineering
Urbana, IL 61801, USA

F. Patat
Lab. Biophysique Medicale Tours
2 Bis Boulevard Tonnelle
37032 Tours, Cedex, France

M.S. Patterson
Ontario Cancer Institute
Dept. of Medical Biophysics
University of Toronto
500 Sherbourne Street
Toronto, Canada M4X 1K9

P.C. Pedersen
Drexel University
Biomedical Engineering &
Science Dept.
Philadephia PA 19104, USA

P. Pesque
LEP
3 Avenue Descartes
94450 Limeil-Brevannes, France

D. Phipps
Racal-Decca Ltd.
New Business & Advanced
Development Division,
9 Davis Road, Chessington
surrey KT9 1TB, UK

F. Pino
University College London
Electrical Engineering Dept.
Torrington Place, London WC1E 7JE
UK

E.J. Pisa
Philips Ultrasound
2722 S. Fairview
Santa Ana, CA 92704, USA

S.M. Posso
University of Newcastle upon Tyne
Dept. of Electrical Engineering
NE1 7RU, UK

J.P. Powers
Naval Postgraduate School
EE Dept. Code 62PO
Monterey CA 93940, USA

G. Quentin
University of Paris VII
France

J.M. Reeves
Portsmouth Polytechnic
Dept. of Electrical & Electronic Engineering
Anglesea Building,
Anglesea Road, Portsmouth
PO1 3DJ, UK

J.M. Reid
Drexel University
Dept. of Biomed Engineering,
Philadelphia PA 19104, USA

J. Ridder
Delft University of Technology
Dept. of Applied Physics
Group of Acoustics
PO Box 5046, 2600 GA Delft
the Netherlands

E. Roberts
The Electricity Council
Research Centre
Capenhurst, Chester, UK

V.C. Roberts
King's College Hospital Medical School
Dept. of Biomedical Engineering
Dulwich Hospital
East Dulwich Grove
London SE22, UK

D. Rugar
Stanford University
Ginzton Lab. Stanford CA 94305
USA

M. Salahi
King's College London
Electronic & Electrical Engineering Dept.
Strand, London WC2R 2LS, UK

M.G. Sambrook
Institute of Cancer Research
Dept. of Physics, F. Block
Royal Cancer Hospital
Clifton Avenue, Sutton,
Surrey, UK

S. Sepehr
Portsmouth Polytechnic
Dept. of Electrical & Electronic Engineering
Anglesea Building
Anglesea Road, Portsmouth
PO1 3DJ, UK

D.I. Shaw
University of Sheffield
Dept. of Electrical Engineering
Mappin Street
Sheffield S1 3JD, UK

D.A. Sinclair
University College Londong
Department of Electrical & Electronic Engineering,
Torrington Place
London WC1E 7JE

P. Sirotti
University of Trieste
Istituto di Elettrotecnica ed Elettronica
Via A Valerio 10, 34127 Trieste
Italy

I.R. Smith
University College London
Dept. of Electrical & Electronic Engineering
Torrington Place
London WC1E 7JE, UK

M. Soumekh
University of Minnesota
Dept. of Electrical Engineering
123 Church Street SE
Minneapolis MN 55455, USA

A. Stamberger
Elektroniker
Switzerland Press
Stoneswood Cottage
Limpsfield Common
Surrey RH8 0QY, UK

R.W.B. Stephens
Chelsea College
49 West Hill Road
Wandsworth, London SW18 1LE, UK

K. Tervola
University of Oulu
Dept. of Electrical Engineering
90570 Oulo 57, Finland

B.R. Tittmann
Rochwell International Science Center
1049 Camino dos Rios
PO Box 1085, Thousand Oaks
California 91360, USA

E.A. Trautenberg
Siemens Research Labs.
Siemens AG
ABT ZT ZFE TPH 42
Guenther-Scharowsky-Strasse
D 8520 Erlangen
Federal Republic of Germany

M. Tristam
10 Hanmer Walk, London N7, UK

C.W. Turner
King's College London
Eleronic & Electrical Engineering Dept.
Strand, London WC2R 2LS, UK

M.U. Ueda
Topkyo Institute of Technology
Research Laboratory of PME
Nagasuta, Midori-ku
Yokohama 227, Japan

L.F. van der Wal
Institute of Applied Physics
Stielfjesweg 1, 2628 CK Delft
the Netherlands

W.A. Verhoef
K U Nymegen
Biofysisch Laboratorium
Oogheelkunde
Philips van Leydenlaan 15
Nymegen, the Netherlands

K. Verhulst
Institute of Applied Physics
PO Box 155, 2600 AD Delft
the Netherlands

W. Vollmann
Philips GmbH Forschungslaboratorium
Vogt-Koelln-Strasse 30
Postfach 54 08 04
D-2000 Hamburg 54
Federal Republic of Germany

R.C. Waag
The University of Rochester
Medical Center
Rochester, New York 14642, USA

G. Wade
University of California
Dept. of Electrical & Computer Engineering
Santa Barbara, CA 93106, USA

R.D. Weglein
6317 Drexel Ave. CA 90098, USA

J.P.N. Wei
Wuhan Institute of Physics
The Chinese Academy of Sciences
PO Box 241
Wuhan, Hubei
People's Republic of China

P.N.T. Wells
Bristol General Hospital
Dept. of Medical Physics
Guinea Street, Bristol BS1 6SY, UK

G.C. Wetsel
Southern Methodist University
Physics Dept. Dallas
Texas 75275, USA

J.F. Whiting
Medical Physics, QIT
PO Box 2434
Brisbane, Queensland,
Australia

H.K. Wickramasinghe
University College London
Dept. of Electrical &
Electronic Engineering,
Torrington Place,
London WC1E 7JE, UK

V. Wilke
Carl Zeiss
PO Bos 1369
7082 Oberkochen
Federal Republic of Germany

A.F.G. Wyatt
University of Exeter
Dept. of Physics
Exeter, Devon, UK

T. Yamamoto
Hokkaido University
Dept. of Electrical Engineering
Faculty of Engineering
N 13 W 8, Sapporo 060, Japan

J. Ylitalo
University of Oulo
Dept. of Electrical Engineering
90570 Oulu 57, Finland

T. Kokota
Tokyo Institute of Technology
The Graduate School at Nagatsuta
4259 Nagatsuta
Midori-ku, Yokohama-shi, 227 Japan

G. Yue
University College London
20 Grove Road
Mile End, London E3 5AX, UK

J. Zieniuk
IPPT-PAN
00.049 Swietokrzyska 21
Warsaw, Poland

L. Zapalowski
Queen Elizabeth College
Dept. of Physics
Campden Hill Road
London W8 7AH, UK

INDEX